ORIGIN OF THE MOON

Library of Congress Cataloging-in-Publication Data

Origin of the moon.

Papers presented at the Conference on the Origin of the Moon, held in Kona, Hawaii, October 1984.
Includes index.
1. Moon—Origin—Congresses. I. Hartmann, William K. II. Phillips, R. J. (Roger J.), 1940– . III. Taylor, G. Jeffrey, 1944– . IV. Conference on the Origin of the Moon (1984: Kailua Kona, Hawaii)
QB580.075 1986   523.3   86–10629
ISBN 0-942862-03-1

ORIGIN OF THE MOON

edited by

W. K. HARTMANN   R. J. PHILLIPS   G. J. TAYLOR

LUNAR & PLANETARY INSTITUTE
HOUSTON

The Final Stages of Lunar Formation

> "Somewhere near 4.5×10^9y ago, we had to arrive at a
> volatile-and iron-depleted moon near ten Earth-radii, in prograde
> orbit with an orbital inclination to Earth's equator near 10°. Almost
> everyone agrees such a stage occurred; there is less agreement on
> how it came about."
> —*Stan Peale,*
> *in a talk summarizing the Conference*
> *on the Origin of the Moon, Kona, 1984.*

Rather than illustrate a specific theory of lunar origin, this view attempts to show the earliest stage on which most researchers agree, in the spirit of Peale's remark. Perhaps it is a slightly earlier stage. The Moon, heavily cratered, is shown somewhat inside the ten Earth radii to which it can be back-traced dynamically. Its shadow is prominent on the Earth. It is in the final stages of accreting a disk of circumterrestrial debris. The inner part of the disk forms a relatively stable circumterrestrial ring system within which little accretion occurs. The Moon is still sweeping up a more diffuse outer ring-swarm, which itself has partially accreted into independent small bodies. The view occurs during the hour-long unfolding of a basin-forming impact on the Moon; smaller impacts are visible on Earth. In keeping with the "nuclear winter" theory of impact effects, shrouds of dark haze hang over cratered Earth as a result of the very high impact rate of heliocentric debris associated with the terminal stages of accretion. Two bright comets are visible in the background sky, the result of the high flux of icy planetesimals scattered from the outer solar system into the inner solar system. Further in the background are nearby stars and nebulosity, part of the short-lived molecular cloud and open star cluster complex that spawned the sun; they have just witnessed the subject of this book, the origin of the Moon.

- PAINTING BY WILLIAM K. HARTMANN

The Original Moon

Four and a half aeons ago
  a dark, dusty cloud deformed.
Sun became star; Earth became large,
  and Moon, a new world, was born.

This Earth/Moon pair, once linked so close,
  would later be forced apart.
Images of young intimate ties
  we only perceive in part.

Both Earth and Moon were strongly stripped
  of their mantle siderophiles.
But Moon alone was doomed to thirst
  from depletion of volatiles.

Moon holds secrets of ages past
  when planets dueled for space.
As primordial crust evolved
  raw violence reworked Moon's face.

After the first half billion years
  huge permanent scars appeared;
ancient feldspathic crust survived
  with a mafic mantle mirror.

But then there grew from half-lived depths
  a new warmth set free inside.
Rivers and floods of partial melt
  resurfaced the low 'frontside'.

Thus evolved the Original Moon
  in those turbulent times.
Now we paint from fragments of clues
  the reasons and the rhymes:

    Sister planet;
    Modified clone;
    Captured migrant;
    Big Splash disowned?

The Truth in some or all of these
  will tickle, delight
  temper, and tease.

—CARLE PIETERS

TAYLOR'S AXIOM:

The best models for lunar origin
are the testable ones.

TAYLOR'S COROLLARY:

The testable models for lunar origin
are wrong.

—S. Ross Taylor,
paraphrased by Sean Solomon,
at the Conference on the Origin of the Moon,
Kona, 1984

Preface

*W*hen the decision was made to explore the Moon and bring back lunar samples, scientists hoped that the resulting information would clarify the origin of the body that poets and lovers have pondered for 10,000 years. Indeed, solving the mystery of the Moon's origin was billed as a major goal of lunar exploration. As it turned out, neither the Apollo astronauts, the Luna vehicles, nor all the king's horses and all the king's men could assemble enough data to explain circumstances of the Moon's birth.

It had been hoped that structural properties of the Moon and, more importantly, the chemistry of primordial lunar rocks, would clarify the formative processes. Astronauts were trained to look for these 4.5-b.y.-old "genesis rocks." These rocks would be the vaunted "rosetta stones" of planetary science, giving in several languages—chemical, isotopic, petrologic—the circumstances of planet and satellite births. Unfortunately, while a few chips from this era have been found and while isotopic techniques have affirmed that the Moon formed essentially during the brief planet-forming era 4.5 b.y. ago, "genesis rocks" were scarce or absent on the lunar surface. The Moon had hidden her secrets behind a veil of dust. A layer of powdery regolith blankets virtually the whole Moon, including the nine landing sites that have been sampled so far. The extraordinarily intense bombardment rate associated with the sweep-up of planetary debris, about 4.5–4.0 b.y. ago, pulverized most of the primordial material, and with it most of our clues.

Yet, as Carlé Pieters' poem attests, the Moon continues to tease us. Where did the blasted thing come from? Prior to Apollo and Luna, as reviewed in the historical papers in Part I by Stephen Brush and John Wood, the paradigm view was that theories of lunar origin could be grouped in three classes: capture, fission, and coaccretion. These are described in more detail throughout this volume. The concensus after the lunar voyages was that none of these theories worked, at least in their classical forms.

Immediately after the lunar voyages, a flurry of papers pointed out some basic similarities and differences between the Moon and the material of the Earth's mantle. Basic facts included the Moon's depletion in iron, depletion in volatiles, and possible augmentation in refractory elements. A few authors tried to fit these facts into the older theories or new ones. For a while, it was popular to say that the differences, such as volatile deficiencies, ruled out an origin from terrestrial material. Perhaps the most sophisticated new theorizing was that of A. E. Ringwood, who postulated that very hot silicate vapor that was spun off or somehow thrown out of the Earth could condense and accrete into a body resembling the Moon. In 1975 W. K. Hartmann and D. R. Davis and in 1976 A. G. W. Cameron and W. R. Ward proposed that Earth mantle material might have been blown out of the Earth, heated, pulverized, and/or vaporized, and might have thus provided the basic material for forming the Moon. A few papers and a major book by Ringwood followed up some of these ideas, but little further work was done in the decade following the last manned lunar landing in 1972.

*T*he idea for holding this conference had its roots at a meeting of the Lunar and Planetary Sample Team (LAPST) held in November, 1982, at the Lunar and Planetary Institute. A LAPST subcommittee was brainstorming about the logical subject for their next topical workshop, to follow in their series of meetings, which started in 1979: the Conference on the Lunar Highlands Crust and the Apollo 16 Workshop (November, 1980), Magmatic Processes of Early Planetary Crusts (August, 1981), Lunar Breccias and Their Meteoritic Analogs (November, 1981), and Pristine Lunar Highland Rocks and the Early History of the Moon (October, 1982); the central theme of these workshops was the origin and early evolution of the lunar crust. "Why not expand the focus to the origin of the entire Moon?" one of the members asked. The idea was instantly appealing.

It seemed to us that the decade since the last manned landing had provided time to reflect further on the origin of the Moon and indeed had provided some revisions and refinements to the earliest conclusions drawn after the lunar landings. One substantial advance, for example, was the analysis of meteoritic and lunar oxygen isotope data, showing that only the Moon and the Earth shared identical oxygen isotope ratios, while materials from other parts of the solar system had different ratios. This seemed an important boundary condition on the origin of the lunar material. It was thus agreed that LAPST should at some point organize a conference on lunar origin.

G. Jeffrey Taylor, who was chairman of the LAPST subcommittee in 1982, was nominated to be the LAPST organizer. Such an interdisciplinary topic needed experts in other areas of planetary science, such as geophysics and dynamics. Geophysicist Roger Phillips, then the LPI Director, was appointed as a second conference convener. LAPST also wanted to tie the conference to a meeting of the Division of Planetary Sciences (DPS) of the American Astronomical Society, in order to promote dialog between the geochemical/geophysical lunar sample community and the planetary astronomy community. DPS agreed to cosponsor the conference and William Hartmann joined the organizers as the DPS representative. This trio met in November, 1983, and mapped out plans for the Conference on the Origin of the Moon, to be held the following October.

In addition to the three coconveners, committee members planning the conference included William Boynton, Alan Harris, Lonnie Hood, Pamela Jones, Günter Lugmair, and Graham Ryder. It appeared to us that by announcing a major topical conference specifically devoted to the problem of lunar origin, we might encourage new thinking from leading data analysts as well as planetary theorists. In this, we believe the effort was eminently successful. The conference was held in Kona, Hawaii in October, 1984, immediately following the meeting there of the Division of Planetary Sciences of the American Astronomical Society. The two-and-a-half day conference attracted 58 papers by 62 authors from 7 countries and California. We found that the papers naturally fell into several categories. We started with invited reviews on basic topics (helping workers in different disciplines learn to talk to each other—no small problem at a gathering of dynamicists and geochemists). These were followed by sessions on dynamical, geochemical, and geophysical constraints. The conference ended with two sessions entitled "My Model of Lunar Origin," intended to encourage authors to present formally, but also to discuss informally, their ideas about processes of origin and evolution.

We have been able to follow a similar organization in this book. We begin in Part I with two major historical reviews covering the very early period (Brush) and the more recent period of lunar exploration (Wood). These are followed by Parts II, III, and IV on dynamical, geochemical, and geophysical constraints; each of these parts leads off with a review paper that we recommend as a general overview. In Parts III and IV the reviews are followed by a group of contributed papers that might be thought of as basic data papers constraining the Moon's origin. In the final Parts, V, VI, VII, and VIII, are contributed papers dealing with various theoretical models or processes of lunar origin, grouped loosely in categories including capture and fission processes, large planetesimals during the Earth-forming period, impact-trigger processes, and evolution of a circumterrestrial swarm.

We have arrived at these categories to direct the reader to the basic thrusts of the meeting, but the categories are not perfect. There is some overlap, since some of what we classified as "geochemical constraint" papers, such as those by S. R. Taylor and J. Larimer, go on to discuss models of origin, and some "model" papers, such as those by H. Wänke and G. Dreibus and by Ringwood, give a careful review of geochemical data. We hope the authors will not object to our rough categorization, which is intended only as an aid to the reader.

W hat are the results? Are we any closer to understanding the origin of the Moon than we were when the first Apollo expedition took off from Earth or when the last Apollo expedition returned from the Moon? Certainly there has been a shift of paradigm and perhaps even a convergence from the pre-Apollo views. The conference was widely reported in the press (e.g., Kerr, 1984, *Science*), and these accounts (as well as those of our own three summarizers at the conference—S. J. Peale, S. C. Solomon, and D. J. Stevenson) suggested that two major ideas now dominated the field. First was an emphasis on the giant impact model as solving many of the earlier problems such as producing a Moon of mantle-like composition with Earth-like oxygen isotope ratios. Indeed, we were surprised by this unexpected convergence as we read the abstracts that were prepared independently by various authors and submitted to us in mid-1984. While the conference and this book included papers discussing variants of the three classical scenarios of capture, fission, and coaccretion, aspects of the giant impact model were treated in at least seven papers at the conference by workers such as Ringwood, Wänke and Dreibus, Hartmann, Cameron, W. Kaula and A. Beachey, Stevenson, and G. W. Wetherill, and subsequently in this book by H. J. Melosh and C. P. Sonett in addition. Recently, groups at Los Alamos (W. Benz, W. L. Slattery, and Cameron) and Sandia (Melosh) have been able to generate "films" of giant impacts, and we have been able to include, in short notes, sequences of frames from these films; they vividly stimulate our imaginations.

A second idea that gained new popularity was a coaccretional "composition filter" model developed primarily by a consortium in Tucson (R. Greenberg, C. Chapman, Davis, M. J. Drake, Hartmann, F. Herbert, J. Jones, and S. J. Weidenschilling), and also independently advanced in somewhat different form by J. Wasson and P. H. Warren. In these scenarios, a disk of debris around the Earth evolves over time with its composition being biased toward silicates by the preferential capture of fine silicate dust from heliocentric orbits. This alternative might be viewed not so much as an independent, complete, and competing theory of the origin of the Moon, since the origin of the circumterrestrial swarm is not fully specified, but rather as a potentially critical evolution process in a disk of whatever origin. The authors themselves allow that this disk might have been produced by large impact processes on or near the Earth, but it might also have originated by other means. In any case, the evolution of a swarm of debris around Earth appears to be a critical part of the modern problem.

Aside from the direct issue of lunar origin, this book reflects a paradigm expansion in planet formation theory, beyond problems of early small-body accretion alone, toward problems of the late planet-forming stage and the roles of large planetesimals, circumplanetary disk systems, etc.

In spite of growing interest on the part of a wide variety of authors in an impact-triggered swarm of debris, we cannot yet claim that the origin of the Moon is understood. As summarized neatly by Taylor's sardonic axiom and corollary quoted above, we want testable models, and the models offered so far do not offer easy tests without a return to the Moon, nor are they wholly convincing as they stand now.

We are not yet satisfied! We hope that this book points the way toward future tests and future explorations. We have observed that our conference and the preparations for this book have spurred a number of continuing studies. Perhaps in a year or two we can reconvene to consider what has been accomplished, either at a topical workshop or in a special session at some meeting. We hope that an assessment of this book will lead to plans for future tests: on the Earth, in lunar orbit, and on the lunar surface. What is the total iron content of the Moon? What are the best values of the Fe/Mg ratio? Can oxygen isotope data be further refined? What does lunar paleomagnetism tell us? If a giant impact really occurred, would any material be truly left in orbit to form the Moon? What temperature distributions are produced in such material? Can we conduct element-by-element inventories and show that the Moon could have been formed by either modified Earth-mantle material or captured meteoritic material? If not, could a combination of altered Earth-mantle material and captured meteoritic material do the job? Could a small captured "embryo" satellite, such as an Earth equivalent of Phobos, have played a role in sweeping up material injected into a circumterrestrial swarm by some process? If a giant impact occurred, were circumterrestrial debris produced as direct crater ejecta, or more by a spin-up of the Earth, yielding an event resembling fission? Or, alternatively, can the old ideas of pure fission or pure capture be revived? Or have we missed some alternative—will a wholly new model suddenly emerge from the morass of today's partial models? Truly, in Carlé Pieters' words, "the truth in some or all of these will tickle, delight, temper, and tease."

William K. Hartmann
Tucson, Arizona

ACKNOWLEDGMENTS

The editors wish to thank the staff of the Lunar and Planetary Institute for producing this book, especially Renee Dotson, Stephanie Tindell, Karen Hrametz, Lisa Bowman, Bethany Lee, Pamela Thompson, Carl Grossman, Rhonda Diem, Donna Chady, and Sharon Adlis. We are also grateful to the following people for thoroughly and enthusiastically reviewing the papers in this book:

T. Ahrens
S. Banerjee
B. G. Bills
A. B. Binder
D. Bogard
A. Boss
W. V. Boynton
C. Chapman
D. R. Davis
J. Delano
M. Drake
R. Durisen
B. Goldstein
C. Goodrich
W. A. Gose
R. Greenberg
L. Grossman
A. Harris
L. Haskin
K. Holsapple

L. L. Hood
K. Housen
O. James
J. Jones
W. Kaula
K. Keil
S. Kieffer
R. Korotev
J. Larimer
D. Leich
J. Longhi
H. J. Melosh
C. Meyer
W. McKinnon
J. Morgan
S. Mueller
H. Newsom
P. Nicholson
L. Nyquist
H. Palme

D. Papanastassiou
S. J. Peale
G. Ransford
A. E. Ringwood
C. T. Russell
G. Ryder
P. Salpas
B. Sharpton
J. Shervais
S. C. Solomon
S. R. Taylor
D. Turcotte
P. Warren
S. J. Weidenschilling
G. Wetherill
E. Whitaker
D. Wise
J. A. Wood
C. Yoder

TABLE OF CONTENTS

I. History

Early History of Selenogony

STEPHEN G. BRUSH

Department of History and Institute for Physical Science and Technology, University of Maryland, College Park, MD 20742

Modern theories of the origin of the Moon developed from general schemes for the origin of the solar system and also from detailed analyses of the "secular acceleration" of the Moon. After William Ferrel and C. E. Delaunay had suggested that tidal forces slow the Earth's rotation so that the Moon is actually moving more slowly in her orbit, G. H. Darwin extrapolated the history of the lunar orbit back to a time when the Moon was very close to the Earth. He proposed in 1878 that fission of a previous proto-Earth had been triggered by the sun's action in resonance with free oscillations. The hypothesis that the Pacific Ocean basin is the scar left by the Moon's departure from Earth was added by Osmond Fisher. Alternative selenogonics were proposed by Edouard Roche (condensation from a circumterrestrial ring) and Thomas J. J. See (capture after formation in the outer solar system). Darwin's fission theory was rejected following criticism by Harold Jeffreys in 1930.

Speculations about the origin of the Moon must have begun almost as soon as human consciousness, but the kind of evidence needed to develop a quantitative scientific theory has been available for less than 300 years; the evidence required for a rigorous test of competing theories was obtained only with the manned lunar landings beginning in 1969. The first major conference devoted exclusively to the subject was not held until 1984. Yet the history of selenogony—the study of the origin of the moon—is already so rich and complex that presented here are only the main events and themes up to 1935, with references to the voluminous literature for technical details.

Since Galileo's discovery of four small bodies orbiting the planet Jupiter (in 1610), it has been known that the Earth's Moon is only one of several satellites in the solar system. One might therefore expect that selenogony would be only a special case of the theory of formation of satellites, and indeed one of the most popular theories ("binary accretion" or "sister" hypothesis) treats the Moon this way. But many scientists have thought that the Moon deserves her[1] own special hypothesis:

first, because she is unusual in being a single and relatively large companion of her primary, unlike the Jovian and Saturnian systems, which could be described as miniature planetary systems (it was learned only recently that Pluto's satellite may be comparable to our Moon in this respect); second, because we have much more detailed information about her and thus presumably an opportunity to construct a more reliable quantitative theory.

One of the first attempts to explain the Moon's formation in the framework of the new heliocentric astronomy of the 17th century is found in Descartes' *Le Monde*. This work was apparently written around 1630 but withheld from publication because of the condemnation of the heliocentric system in the notorious Trial of Galileo in 1633; it was published posthumously in 1664 (Descartes, 1664). Descartes imagines a universe filled with pieces of matter of various sizes, shapes, and motions, evolving into a system of numerous vortices rotating around stars. Large pieces of matter—planets—can move in orbits at definite distances from the central star, depending on the "force" of their motion (mass or quantity of matter, multiplied by speed); they are kept in dynamic equilibrium at those distances by collisions with smaller particles. Each planet then develops its own vortex of these small particles, revolving around it in the same direction that the planet moves around the star. Another planet may move in the same orbit as long as its force is the same, but if it is smaller it will have to move faster and will soon overtake the larger one. It will then be trapped by the vortex of that larger planet and be forced to move around it. Descartes proposed this scheme to account for the Earth-Moon system but left it to his readers to extend it to the other planetary satellites, confessing that "I have not undertaken to explain everything" (Descartes, 1824, pp. 246–288).

Special theories of the Moon's origin developed from attempts to explain observed variations in the lunar orbit. The periodic time of the Moon's revolution around the Earth is gradually increasing (apart from short-term cyclic changes), a fact that implies (according to Kepler's Third Law) that she is slowly retreating from the Earth. If the average Earth-Moon distance has been continually increasing in the past, either at a constant rate or as a result of a force varying with distance in a known manner, we can extrapolate backward in time to reconstruct the earlier history of the Moon's orbit. This leads us to an epoch when the Moon would have been inside the "Roche limit": the distance at which the Earth's tidal force would break up a body held together by gravitational forces. At that point (if not before) the extrapolation becomes invalid and we must introduce a specific hypothesis about the Moon's earlier motion, and indeed about whether she even existed in her present form.

What I have called a "fact" in the previous paragraph was not recognized as such before the middle of the 19th century. On the contrary, the first quantitative studies of the Moon's motion indicated that the periodic time is *decreasing*, hence the term "secular acceleration" has traditionally been applied to this effect.[2] In 1749 the British astronomer Richard Dunthorne reviewed ancient and modern records of eclipses in order to test Edmund Halley's suggestion (Halley, 1695) that the Moon has gradually been moving faster in her orbit. He concluded that there is indeed a secular acceleration, amounting to about 10″ per century, and this estimate was confirmed by other astronomers (Grant, 1852, pp. 60–64; Forbes, 1972, pp. 11, 20–21, 76–77). Modern values are substantially higher (30″ to 50″ per century).

Laplace attempted to explain the secular acceleration as a combination of the action of the sun on the Moon and the secular variation of the eccentricity of the Earth's orbit. Perturbations of other planets are slowly decreasing this eccentricity, and since the sun's action on the Moon's orbit is greatest when the Earth is at perihelion, his effect on her motion will also be slowly diminishing. Taking into account directly only the radial component of the sun's action, Laplace found that it decreases the angular velocity of the Moon at perihelion, so the decrease in eccentricity of the Earth's orbit will be accompanied by an acceleration of the Moon.[3]

If this explanation is sufficient to account for the entire effect, it implies that the Moon's acceleration is not truly "secular" (changing always in the same direction) but will eventually be reversed, since the change in the eccentricity of the Earth's orbit is itself a cyclic effect. (Indeed, this cycle may be partly responsible for periodic "ice ages."[4]) This cyclic character is consistent with the expectation that gravitational forces should not produce any irreversible effects in a conservative dynamical system.

Since Laplace's theoretical calculations gave a secular acceleration in good agreement with the observed value, it was generally assumed that he had solved the problem. But the British astronomer John Couch Adams, looking into lunar theory half a century later, noticed an error in Laplace's calculation: the tangential component of the sun's action does produce a significant effect. When higher-order terms were computed, it appeared that the Laplace mechanism could account for only about half of the observed acceleration. This result was announced in the same year as the publication of Charles Darwin's *Origin of Species*.[5]

The explanation for this newly discovered discrepancy was already at hand, although it took a few years before it was recognized by astronomers. The German philosopher Immanuel Kant had pointed out as early as 1754 that tidal dissipation should retard the Earth's rotation.[6] The German physician J. Robert Mayer and the American scientist William Ferrel both attempted quantitative determinations of this effect. However, their rough estimates came out much too large, so they assumed that the tidal effect was cancelled by the gradual cooling and contraction of the Earth (which should *increase* its rotation speed, if angular momentum is conserved).[7,8] Mayer's and Ferrel's papers did not attract much attention at the time, but Hermann von Helmholtz emphasized the significance of tidal dissipation in his famous Koenigsberg lecture of 1854.[9] All of these scientists recognized that similar tidal forces acting on the Moon could have increased her rotation period to synchrony with her revolution period, thus explaining why she always keeps the same face to the Earth.

In 1865, the French astronomer Charles-Eugene Delaunay, who had confirmed Adams' calculation of the secular acceleration, proposed that part of the apparent secular acceleration of the Moon is due to a deceleration of the Earth's rotation produced by tidal dissipation (Delaunay, 1865). (The rest would still be attributed to Laplace's mechanism as described above.) Ferrel, strictly speaking, deserves priority for this proposal since he made it at a meeting of the American Academy of Arts and Sciences in Boston in December 1864, a few weeks before Delaunay read his paper to the Academie des Sciences in Paris, but it was Delaunay's reputation that persuaded astronomers to adopt it. G. B. Airy, the British Astronomer Royal, was initially skeptical of Delaunay's claim but eventually accepted it after doing his own calculations.[10]

Although Mayer had noted in 1848 that tidal action would increase the Moon's distance from Earth[11], no one seems to have pursued the cosmogonic implications of this fact for more than two decades. It has not yet been determined what led George Howard Darwin to the idea that the history of the lunar orbit should be traced backwards to a time when the Moon and Earth were in contact. His papers published in 1877–1878 (Darwin, 1877a,b, 1878) suggest that contemporary discussion of the cause of ice ages inspired him to test the hypothesis, proposed by geologist John Evans, that major disturbances in the Earth's surface could change its axis of rotation. This led him to inquire whether the planets could have acquired their present obliquities at a time when they were large, gaseous, rapidly-spinning bodies with pronounced equatorial bulges; satellites might have been spun off at this stage. Having become interested in the process of satellite formation, he realized that mathematical difficulties would prevent him from drawing definite conclusions about the behavior of the system just before separation into two bodies; instead, he decided to follow the two-body system backwards in time to the point just after separation. Thus the son of the biological evolutionist became the first student of the evolution of the Moon's orbit, and the first to work out the quantitative details of the later stages of a specific process by which Mother Earth might have given birth to her satellite.

Darwin announced his theory in 1878 and published the details in a long memoir the next year.[12] He treated the Earth as a homogeneous rotating viscous spheroid, and assumed that the Moon moves in a circular orbit in the plane of the ecliptic. Taking the viscosity of the Earth large enough to give the observed (apparent) secular acceleration, he could work back to a state in which the Moon moved around the Earth as if rigidly fixed to it in a period of 5 hours 36 minutes. This would have been at least 54 million years ago, and the Moon's center would have been no more than 6000 miles from the Earth's surface; both period and distance would be smaller if the Earth were not homogeneous.

"These results point strongly to the conclusion," Darwin declared, "that if the Moon and Earth were ever molten viscous masses, then they once formed parts of a common mass" (Darwin, 1879, p. 536). But how could this mass have broken up? A system rotating in 5 hours, with the combined mass of the present Earth and Moon, would *not* be rotationally unstable. However, Darwin's senior colleague (and his father's nemesis[13]), Sir William Thomson, later known as Lord Kelvin, had shown that a fluid spheroid of the same density as the Earth would have a period of free oscillation of about 1 hour 34 minutes. A less dense body would have a longer period. Darwin could then invoke the sun's tidal action to trigger fission: he proposed that the solar semidiurnal tide, reaching a maximum every 2½ hours at a given place on the Earth's surface, might be in resonance with the free oscillations, thus producing enormous distortion sufficient to disrupt the body.

Darwin undertook further calculations to apply his theory to other possible models, including finite eccentricity and inclination of the Moon's orbit. He was cautious about stating his conclusions on the earliest stages of the Earth-Moon system, recognizing that when one approaches the point of contact an infinitesimal disturbance may cause an irreversible finite change. He thought it likely that most of the tidal dissipation at present is due to ocean tides rather than body tides in the solid Earth, but argued that the early Earth was hotter and more plastic, so body tides would have been more important (Darwin, 1880, pp. 32, 713).

Darwin's theory was quickly popularized by Robert S. Ball, the Royal Astronomer of Ireland, in a lecture at Birmingham, published in *Nature* in 1881 (Ball, 1881). Ball stressed the possible geological effects of the enormous tides that should have been present in the Earth's early history because of the Moon's proximity. Alluding to William Thomson's arguments that the age of the Earth is much less than geologists had assumed[13], he pointed out that to compensate for limiting the time available for geological processes the mathematicians have given geologists a "new and stupendous tidal grinding engine." On the other hand, the social reformers who are attempting to reduce the working day may find their efforts nullified in the long run by the Moon's action in slowing the Earth's rotation—"where will the nine-hour's movement be when the day has increased to 1400 hours?" (Ball, 1881, pp. 79, 103, 1882a,b, 1889).

Ball's speculations stimulated a round of letters to the editor of *Nature*, the most substantial being a proposal by the geologist Osmond Fisher (1882) that the scar left by the Moon's separation did not completely heal. Fisher suggested that the ocean basins are the holes left in the Earth's crust after some flow of the remaining solid toward the original cavity. In this way the birth of the Moon would have resulted in both the Pacific Ocean basin and the separation of the American continent from Europe and Africa.

The major alternative to Darwin's fission theory in the 19th century was the explanation based on Laplace's nebular hypothesis: the Moon condensed from a ring spun off from the rotating gaseous proto-Earth, just as the Earth itself condensed from a ring spun off from the rotating solar nebula.[14] Although this might have been a satisfactory explanation for satellites of the giant planets, there seemed to be some difficulty in accounting for the relatively large mass and orbit of the Moon. The best defense of the Laplacian theory of the origin of the Moon, now known as the "sister" (or "binary planet" or "coaccretion") hypothesis in contrast to Darwin's "daughter" theory, was offered in 1873 by Edouard Roche. Roche, whose 1848 formula for the tidal stability limit of a satellite plays a crucial role in many modern theories, corrected Laplace's calculation of the extent of the Earth's rotating atmosphere. Roche's results showed that the sister theory was tenable, though not compelling.

A variation of the sister theory was the hypothesis proposed by the American geologist Grove Karl Gilbert, that the Moon formed from a ring of small solid particles; the final stage of the process would produce the craters on the Moon's face.[15]

Gilbert's hypothesis, though introduced without specific reference to Darwin's theory, could have been combined with it. Indeed, the first major criticism of Darwin's theory, published by James Nolan (1885, 1886, 1887, 1895) in Australia, was that the material spun off from the Earth would not be able to remain intact in a close orbit, but would immediately be torn apart by tidal forces and form a ring of particles. This is because it would initially be inside the "Roche limit" although Nolan did not refer to Roche's theory. Darwin (1886) was forced to admit that the Moon must have been broken into a flock of meteorites as soon as she escaped from the Earth, but he insisted (contrary to Nolan) that this flock could still exert tidal forces on the Earth. Even a symmetrical ring of fragments would raise tides and the resulting dissipation would expand its orbit beyond the Roche limit, whereupon the fragments could recombine into a single satellite.

Nolan's forgotten pamphlet (1885) also suggested an additional hypothesis that might be used to assist the fission theory:

> . . . the earth, which was supposed to have acquired the rapid rotation which caused the moon to separate, from the process of contraction, could only acquire that condition by the contraction of a denser nucleus. (p. 5)

At the beginning of the 20th century Darwin's theory was widely accepted. Lingering doubts about the excessive time required for tidal evolution, as compared to Lord Kelvin's later estimates of only 20 m.y. for the age of the Earth[13], were dispelled by the discovery of radioactivity and the resulting multibillion-year estimates.[16] Henri Poincaré's mathematical studies of the equilibrium figures of rotating fluids seemed to offer a new and more respectable basis for the theory, and Darwin enthusiastically cooperated with Poincaré in working out the details.[17]

Fisher's hypothesis that the Moon came from the Pacific Ocean basin, revived two decades later by the American astronomer W. H. Pickering (1903, p. 7, 1907), eventually became a standard addition to Darwin's theory. For some geologists it provided a satisfactory catastrophic explanation of the same geographical features that Alfred Wegener's continental drift hypothesis claimed to interpret in a gradualist fashion.[18]

By 1936 the Darwin-Fisher theory had been translated into popular mythology, as illustrated in the following excerpt from a script prepared by the U.S. Office of Education for broadcast as a children's radio program:

(Start with "weird mysterioso fanfare")

FRIENDLY GUIDE.   Have you heard that the Moon once occupied the space now filled by the Pacific Ocean? Once upon a time—a billion or so years ago—when the earth was still young—a remarkable romance developed between the earth and the sun—according to some of our ablest scientists. . . In those days the earth was a spirited maiden who danced about the princely Sun—was charmed by him—yielded to his attraction, and became his bride. . . The Sun's attraction raised great tides upon the earth's surface. . . the huge crest of a bulge broke away with such momentum that it could not return to the body of mother Earth. And this is the way the Moon was born!

GIRL. How exciting! . . .

In 1909 the fission theory was attacked by two American astronomers. Forest Ray Moulton, noting that the Russian mathematician A. M. Lyapunov had disproved Poincaré's conjectures about the stability of his rotating fluids, argued that the Earth-Moon system could not have been produced by Darwin's mechanism.[19] Thomas Jefferson Jackson See proposed that the Moon, like other satellites, was captured through the action of a resisting medium. (Capture had been proposed by other writers but this suggestion was generally ignored.[20])

According to See, the Moon was originally formed in the outer part of the solar system, near the present orbit of Neptune. Following Leonhard Euler, he argued that all the planetary orbits have been gradually shrunk and their eccentricities reduced by the resisting medium. The discovery of retrograde satellites of Saturn and Jupiter suggested that at least some satellites must have been captured; hence we must follow Newton's second rule of reasoning and assign the same causes to the same effects whenever possible—i.e., we must assume that *all* satellites were captured. See claimed that the Moon is approaching the Earth, and stated that there is no direct evidence that the Earth ever rotated more rapidly than at present; thus he rejected the major conclusions of Darwin's theory of tidal evolution of the lunar orbit.[21]

These attacks did not lead to the rejection of Darwin's theory. See was rapidly losing his earlier scientific reputation because of his eccentric behavior, and no one seems to have taken his capture theory very seriously.[22] Harold Jeffreys, in England, came to the defense of the fission theory in 1917, pointing out that the different moments of inertia of the Moon indicate that she must have solidified at a time when she was much closer to the Earth. Jeffreys suggested that Moulton's objections could be avoided by taking account of the heterogeneity of rotating fluid; resonance could then produce fission even for fairly slow rotation. The experiments of A. A. Michelson and others on the Earth's body tides indicated that it is highly elastic rather than viscous, thereby throwing doubt on the tidal dissipation mechanism used by Darwin.[23] But Jeffreys, following G. I. Taylor, estimated that friction in shallow seas could dissipate enough energy to account for the secular acceleration of the Moon, thus reinstating the Darwinian principle for tracing its orbital evolution.[24]

Having become one of the principal advocates of the fission theory, Jeffreys was able to deprive it of most of its support when he rejected it in 1930 (Jeffreys, 1930). His objection—that viscosity in the Earth's mantle would dampen the motions required to build up a resonant vibration and thereby prevent fission—was considered conclusive by later researchers, though I find it unconvincing.[25]

During the next 25 years there seems to have been neither any major progress in developing theories of lunar origin nor any clear agreement on adopting one of the previous theories. The most important writings on the subject were those of the German astronomer F. Nölke, who advocated a modified Laplacian hypothesis in which the Moon condensed from the outer parts of the Earth's atmosphere (Nölke, 1922, 1930, pp. 294–295, 1932, 1934). The revival of selenogony came only in the 1950's with the capture theory of Horst Gerstenkorn and the application of the physico-chemical approach to planetary science by Harold Urey, which I have reviewed elsewhere (Brush, 1981, 1982).

Acknowledgments. My research on the history of planetary science has been sponsored by the History and Philosophy of Science Program of the National Science Foundation and by the Alfred P. Sloan Foundation. This paper was written while I was a visitor at the Institute of Geophysics and Planetary Physics, University of California, Los Angeles. It is based on a longer account that will be included in a book on the history of theories of the origin of the solar system; I have profited from discussing the development of recent theories with many of the participants at the Kona meeting, and their assistance will be noted specifically in the forthcoming book.

Notes

1. In keeping with historical tradition as well as for grammatical convenience, I refer to the Moon as female; for further discussion of this point see Mitroff (1974a, pp. 2107–2119, 1974b, p. 102).

2. Astronomers use the word "secular" to refer to a relatively long-term monotonic variation, as distinct from short-term "cyclic" variations. A major goal of 18th-century celestial mechanics was to explain all observed "inequalities" (deviations from Keplerian motion) in terms of gravitational or other perturbations. Lagrange and Laplace concluded that a mechanical system with purely Newtonian gravitational forces should have no true secular inequalities, but only cyclic ones. This confirmed the notion of the "clockwork universe" or "Newtonian world machine" that, once set in motion with a fixed amount of matter and motion, never runs down. Whereas Newton himself, accepting the existence of secular inequalities, thought divine intervention would be needed to keep the solar system from collapsing, Laplace supposedly told Napoleon, "I have no need of that hypothesis."

3. For an explanation of Laplace's contributions to this problem see de la Rue (1866) and Gillispie (1978, p. 332).

4. Milankovich (1920); Imbrie and Imbrie (1979); Imbrie (1982).

5. Adams (1853, 1859).

6. Kant (1754a,b), reprinted in Kant (1910, pp. 183–191); English translation, Kant (1969, pp. 1–11). See also Wackerbath (1867).

7. Mayer (1848, 1851); Lindsay (1973, pp. 176–195).

8. Ferrel (1853, 1895, pp. 294–297); Burstyn (1971).

9. Helmholtz (1854), English translation in Helmholtz (1862, pp. 59–92).

10. Delaunay (1865, 1866a–e); Ferrel (1866); memoir of Ferrel in National Academy of Sciences *Biographical Memoirs* (1895); Airy (1866); de la Rue (1866); Berry (1898, pp. 308–309, 368–370); Dorling (1979); Bertrand (1866).

11. See note 7; also Lindsay (1973, p. 183).

12. Darwin (1878, 1879). Most of his papers are reprinted in Darwin (1907–1916). A nonmathematical survey, originally published in *Atlantic Monthly* (Darwin, 1898a), may be found in Chapters XV and XVI of his *The Tides* (Darwin, 1898b).

13. See Burchfield (1975); Brush (1978, Chapter 3, pp. 29–44).

14. Laplace (1796), see Note VII at end of Vol. 2. For further details see Jaki (1976).

15. Gilbert (1893); see Hoyt (1982). Harold Urey, who noted that the paper was written between the time of his own conception and birth, admonished other selenologists for ignoring it (Urey, 1959).

16. Darwin, *Scientific Papers*, Vol. II (1908), p. 1v; Eddington (1906). Kelvin himself accepted Darwin's theory (Lord Kelvin, 1908).

17. On Poincaré's ideas about this topic see Brush (1980a).

18. Patterson (1909); Bowie (1929, 1930); Gutenberg (1930); Wegener (1912). For a negative view see Barrell (1907).

19. Moulton (1909a,b); Lyapunov (1905, 1980); see also Schwarzschild (1898, p. 231).

20. See (1909a, pp. 33, 365, 1909b, pp. 387, 481, 534, 634, 1910a, pp. 24, 106, 155, 1910b, Chapter 11); Mackey (1825); Taylor (1898, p. 29).

21. See (1909c, p. 380, 1909d, 1915). He also insisted that lunar craters are due to meteorite bombardment rather than volcanism: "the Moon's surface can be nothing but fragments of rock filled with finer dust; and it is evident that it has never been molten as a whole and has never shown true volcanic activity" (See, 1910a, p. 19). Cf. Gold (1955), Urey (1966).

22. Obituary of See in *New York Times*, 5 July 1962, p. 23; J. Ashbrook (1962); John Lankford (unpublished work, 1980).

23. Michelson (1914); Michelson and Gale (1919); Moulton (1914a,b, 1915); Nölke (1924). For further discussion of this topic see Brush (1980b).

24. Jeffreys (1920); Taylor (1919).

25. Jeffreys estimated the frictional force for the mantle flowing over the (presumably liquid) core from a formula that appears to pertain to liquids flowing over solids; he does not explain why it would be valid in this case. McKinnon and Mueller (1984) have reexamined the effect of the solar resonance and concluded that Jeffrey's conclusion was correct even though based on oversimplified assumptions.

References

Adams J. C. (1853) On the secular variation of the Moon's mean motion. *Phil. Trans. Roy. Soc. London, 143*, 397–406.

Adams J. C. (1859) On the secular variation of the eccentricity and inclination of the Moon's orbit. *Mon. Not. Roy. Astr. Soc., 19*, 206–208.

Airy G. B. (1866) On the supposed effect of friction in the tides, influencing the apparent acceleration of the Moon's mean motion in longitude. *Mon. Not. Roy. Astr. Soc., 26*, 221–235.

Ashbrook J. (1962) Astronomical scrapbook: The sage of Mare Island [T. J. J. See]. *Sky and Telescope, 24*, 193–202.

Ball R. S. (1881) A glimpse through the corridors of time. Origin and development of the Moon from the data of tidal evolution. *Nature, 25*, 79–82, 103–107.

Ball R. S. (1882a) Birth of the Moon by tidal evolution. *Knowledge, 1*, 331–332, 352–353.

Ball R. S. (1882b) On the occurrence of great tides since the commencement of the geological epoch. *Nature, 27,* 201–203.

Ball R. S. (1889) *Time and Tide, a Romance of the Moon.* Society for Promoting Christian Knowledge, London. 192 pp.

Barrell J. (1907) (Review of) The place of origin of the Moon—The volcanic problem, by William H. Pickering [Pickering (1907), reprinted in *J. Geol., 15,* 274–287 (1907)]. *J. Geol, 15,* 503–507.

Berry A. (1898) *A Short History of Astronomy from Earliest Times Through the Nineteenth Century.* Murray, London. Reprinted by Dover Publications, New York (1961). 440 pp.

Bertrand J. (1866) Note sur la variation du moyen mouvement de la Lune. *Compt. Rend. Acad. Sci. Paris, 62,* 162–164.

Bowie W. (1929) Possible origin of oceans and continents. *Gerlands Beitr. Geophysik, 21,* 178–182.

Bowie W. (1930) Crustal changes due to Moon's formation. *Gerlands Beitr. Geophysik, 25,* 137–144.

Brush S. G. (1978) *The Temperature of History: Phases of Science and Culture in the Nineteenth Century.* Burt Franklin, New York. 210 pp.

Brush S. G. (1980a) Poincaré and cosmic evolution. *Physics Today, 33,* no. 3, 42–49.

Brush S. G. (1980b) Discovery of the Earth's core. *Amer. J. Phys., 48,* 705–724.

Brush S. G. (1981) From bump to clump: Theories of the origin of the solar system 1900–1960. In *Space Science Comes of Age: Perspectives in the History of the Space Sciences* (P. A. Hanle and V. D. Chamberlain, eds.), pp. 78–100. Smithsonian Institution Press, Washington, D.C.

Brush S. G. (1982) Nickel for your thoughts: Urey and the origin of the Moon. *Science, 217,* 891–898.

Burchfield J. D. (1975) *Lord Kelvin and the Age of the Earth.* Science History Publications, New York. 260 pp.

Burstyn H. L. (1971) Ferrel, William. *Dictionary of Scientific Biography, 4,* 590–593.

Darwin G. H. (1877a) On a suggested explanation of the obliquity of planets to their orbits. *Phil. Mag.,* series 5, *3,* 188–192.

Darwin G. H. (1877b) The nebular hypothesis, and the obliquity of the axes of planets to their orbits. *Observatory, 1,* 13–17.

Darwin G. H. (1878) On the precession of a viscous spheroid. *Nature, 18,* 580–582.

Darwin G. H. (1879) On the precession of a viscous spheroid and on the remote history of the Earth. *Phil. Trans. Roy. Soc. London, 170,* 447–538.

Darwin G. H. (1880) On the secular changes in the elements of the orbit of a satellite revolving about a tidally distorted planet. *Phil. Trans. Roy. Soc. London, A171,* 713–891. Summary in *Nature, 21,* 235–237, and in *Proc. Roy. Soc. London, 30,* 1–10.

Darwin G. H. (1886) Tidal friction and the evolution of a satellite. *Nature, 33,* 367–368 and *34,* 287–288.

Darwin G. H. (1898a) The evolution of satellites. *Atlantic Monthly, 81,* 444–455. Reprinted in *Ann. Rept. Smithsonian Inst.* (1897), 109–124.

Darwin G. H. (1898b) *The Tides and Kindred Phenomena in the Solar System.* Houghton Mifflin, Boston. 3rd ed., 1911. Reprinted by Freeman, San Francisco, 1962. 378 pp.

Darwin G. H. (1907–1916) *Scientific Papers.* Cambridge University Press, Cambridge.

de la Rue W. (1866) Address delivered by the President, Warren de la Rue, Esq., on presenting the Gold Medal of the Society to Professor J. C. Adams, Director of the Cambridge Observatory. *Mon. Not. Roy. Astr. Soc., 26,* 157–189.

Delaunay C. E. (1865) Sur l'existence d'une cause nouvelle ayant un action sensible sur la valeur de l'equation seculaire de la Lune. *Compt. Rend. Acad. Sci. Paris, 61,* 1023–1032. English translation in *Mon. Not. Roy. Astr. Soc., 26* (1866), 85–91, 148.

Delaunay C. E. (1866a) *Remarques . . . à l'occasion de cette comunication* [Bertrand (1866)]. *Compt. Rend. Acad. Sci. Paris, 62,* 165–166.

Delaunay C. E. (1866b) Sur l'accélération apparente du moyen mouvement de la Lune due aux actions du Soleil et de la Lune sur les eaux de la mer. *Compt. Rend. Acad. Sci. Paris, 62*, 197–200.

Delaunay C. E. (1866c) Réponse a la Note de M. Allégret insérée au Compte Rendu de la Séance du 26 fèvrier. *Compt. Rend. Acad. Sci. Paris, 62*, 575–579.

Delaunay C. E. (1866d) Sur la controverse relative a l'equation seculaire de la Lune. *Compt. Rend. Acad. Sci. Paris, 62*, 704–707.

Delaunay C. E. (1866e) Note sur la question du ralentissement de la rotation de la Terre. *Compt. Rend. Acad. Sci. Paris, 62*, 1107.

Descartes R. (1664) *Le Monde, ou Traité de la Lumière*. Paris. Reprinted in Descartes (1824, vol. 4, pp. 215–332).

Descartes R. (1824) *Oeuvres* (V. Cousin, ed.). Chez F. G. Levrault, Librairie.

Dorling J. (1979) Bayesian personalism, the methodology of scientific research programmes, and Duhem's problem. *Stud. Hist. Phil. Sci., 10*, 177–187.

Dunthorne R. (1749) A letter. . .concerning the acceleration of the Moon. *Phil. Trans. Roy. Soc. London, 46*, 162–172.

Eddington A. S. (1906) A criticism of Sir George Darwin's theories. *Observatory, 29*, 179–181.

Ferrel W. (1853) On the effect of the Sun and Moon upon the rotatory motion of the Earth. *Astron. J., 3*, 138–141.

Ferrel W. (1866) Note on the influence of the tides in causing an apparent secular acceleration of the Moon's mean motion. *Proc. Amer. Acad. Arts Sci., 6*, 379–383, 390–393.

Ferrel W. (1895) Autobiographical sketch. *Biog. Mem. Nat. Acad. Sci. USA, 3*, 287–299.

Fisher O. (1882) On the physical cause of the ocean basins. *Nature, 25*, 243–244.

Forbes E. G. (1972) *The Euler-Mayer Correspondence (1751–1755): A New Perspective on Eighteenth Century Advances in the Lunar Theory*. Elsevier, New York. 118 pp.

Gilbert G. K. (1893) The Moon's face; a study of the origin of its features. *Bull. Wash. Phil. Soc., 12*, 241–292. Abstract in *Trans. N.Y. Acad. Sci., 12*, 93–95.

Gillispie C. C. (1978) Laplace, Pierre-Simon, Marquis de. *Dict. Sci. Biog., 15*, 273–403.

Gold T. (1955) The lunar surface. *Nature, 115*, 585–604.

Grant R. (1852) *History of Physical Astronomy*. Bohn, London. Reprinted by Johnson Reprint Corp., New York, 1966. 638 pp.

Gutenberg B. (1930) Hypotheses on the development of the Earth. *J. Washington. Acad. Sci., 20*, 17–25.

Halley E. (1695) Some account of the ancient state of the City of Palmyra, with short remarks upon the inscriptions found there. *Phil. Trans. Roy. Soc. London, 19*, 160–175.

Helmholtz H. von (1854) *Ueber die Wechselwirkung der Naturkrafte und die darauf bezuglichen neuesten Ermittelungen der Physik*. Grafe and Unzer, Konigsberg. English translation in Helmholtz (1962, pp. 59–92).

Helmholtz H. von (1962) *Popular Scientific Lectures*. Dover, New York. 286 pp.

Hoyt W. G. (1982) W. K. Gilbert's contribution to selenology. *J. Hist. Astron., 13*, 155–167.

Imbrie J. (1982) Astronomical theory of the Pleistocene Ice Ages: A brief historical review. *Icarus, 50*, 408–422.

Imbrie J. and Imbrie K. P. (1979) *Ice Ages: Solving the Mystery*. Enslow, Short Hills, New Jersey, 224 pp.

Jaki S. L. (1976) The five forms of Laplace's cosmogony. *Amer. J. Phys., 44*, 4–11.

Jeffreys H. (1920) Tidal friction in shallow seas. *Phil. Trans. Roy. Soc. London, A221*, 239–264.

Jeffreys H. (1924) *The Earth: Its Origin, History and Physical Constitution*. University, Cambridge. 278 pp.

Jeffreys H. (1930) The resonance theory of the origin of the Moon (second paper). *Mon. Not. Roy. Astr. Soc., 91*, 169–173.

Kant I. (1754a) *Untersuchung der Frage, ob die Erde in ihrer Umdrehung um die Achse, wodurch sie die Abwechselung des Tages und der Nacht hervorbringt, einige Veränderung seit den ersten Zeiten ihres Ursprungs erlitten habe, und woraus man sich ihrere Versichern könne, welche von der königl. Akademie der Wissenschaften zu Berlin zum Preise für das jetztlaufende Jahr aufgegeben worden.* Reprinted in Kant (1910, vol. 1, pp. 183–191). English translation in Kant (1969, pp. 1–11).

Kant I. (1754b) *Die Frage: Ob die Erde veralte? physikalisch erwogen.* Reprinted in Kant (1910, vol. 1, pp. 193–213).

Kant I. (1910) *Gesammelte Schriften.* Reimer, Berlin.

Kant I. (1969) *Universal Natural History and Theory of the Heavens.* Univ. of Michigan Press, Ann Arbor. 180 pp.

Kelvin Lord [William Thomson] (1908) On the formation of concrete matter from atomic origins. *Phil. Mag.,* series 6, *15,* 397–413.

Laplace, P. S. (1796) *Exposition du Système du Monde.* Paris.

Lindsay R. B. (1973) *Julius Robert Mayer: Prophet of Energy.* Pergamon, Oxford.

Lyapunov A. (1905) Sur un problème de Tchebychef. *Mem. Acad. Imp. Sci. St. Petersbourg,* series 8, *17,* no. 3, 1–32.

Lyapunov A. (1908) Problème de minimum dans une question de stabilité des figures d'équilibre d'une masse fluide en rotation. *Mem. Acad. Imp. Sci. St. Petersbourg,* series 8, *22,* no. 5.

Mackey S. A. (1825) *A New Theory of the Earth and of Planetary Motion; In Which it is Demonstrated that the Sun is Viceregent of his own System.* R. Walker, Norwich.

Mayer J. R. (1848) *Beiträge zur Dynamik des Himmels in populärer Darstellung.* Verlag von Johann Ulrich Landherr, Heilbronn. English translation in Lindsay (1973, pp. 148–196).

Mayer J. R. (1851) De l'influence des marées su la rotation de la terre. Submitted to the Academie des Sciences, Paris; published in Mayer (1893, pp. 282–285).

Mayer (J.) R. (1893) *Kleinere Schriften und Briefe von Robert Mayer, nebst Mittheilungen aus seinem Leben* (J. J. Weyrauch, ed.). Verlag der J. G. Cotta'schen Buchhandlung, Stuttgart.

McKinnon W. B. and Mueller S. W. (1984) A reappraisal of Darwin's fission hypothesis and a possible limit to the primordial angular momentum of the Earth (abstract). In *Papers Presented to the Conference on the Origin of the Moon,* p. 34. Lunar and Planetary Institute, Houston.

Michelson A. A. (1914) Preliminary results of measurements of the rigidity of the Earth. *Astrophys. J., 39,* 105–138; *J. Geol., 22,* 97–130.

Michelson A. A. and Gale H. G. (1919) The rigidity of the Earth. *J. Geol., 27,* 585–601; *Astrophys. J., 50,* 330–345.

Milankovich M. (1920) *Theorie mathematique des phénomènes thermiques produits par la radiation solaire.* Gauthier-Villars, Paris. 338 pp.

Mitroff I. I. (1974a) *The Subjective Side of Science.* Elsevier, New York. 330 pp.

Mitroff I. I. (1974b) Science's apollonic moon: A study in the psychodynamics of modern science. *Spring, An Annual of Jungian Psychology,* pp. 102–112.

Moulton F. R. (1909a) In *The Tidal and Other Problems* by T. C. Chamberlin et al. Carnegie Institution of Washington, Publication 107, pp. 135–160.

Moulton F. R. (1909b) Notes on the possibility of fission of a contracting rotating fluid mass. *Astrophys. J., 29,* 1–13.

Moulton F. R. (1914a) Letter to W. W. Campbell, 10 March. Lick Observatory Archives.

Moulton F. R. (1914b) Letter to W. W. Campbell, 23 December. Lick Observatory Archives.

Moulton F. R. (1915) Letter to W. W. Campbell, 30 April. Lick Observatory Archives.

New York Times (1962) Obituary for Capt. T. J. J. See, Astronomer. July 5, 1962, p. 23.

Nölke F. (1922) Ueber die Entstehung der Oberflächenformationen des Mondes. *Astr. Nachr., 215,* 217–228.

Nölke F. (1924) Muss die *Darwinsche* Erklärung der Entwicklung des Erdmondes aufgegeben werden? *Astr. Nachr., 220*, 269–272.

Nölke F. (1930) *Der Entwicklungsgang unseres Planetensystems: Eine kritische Studie.* F. Dümmler, Berlin. 359 pp.

Nölke F. (1932) Die vorgeologische Entwicklung der Erde als Schlüssel zum Verständnis der geologischen Entwicklung. *Gerlands Beitr. Geophysik, 37*, 252–270.

Nölke F. (1934) Der Ursprung des Mondes. *Gerlands Beitr. Geophysik, 41*, 86–91.

Nolan J. (1885) *Darwin's Theory of the Genesis of the Moon.* Robertson, Melbourne. 16 pp.

Nolan J. (1886) Tidal friction and the evolution of a satellite. *Nature, 34*, 286–287.

Nolan J. (1887) Tidal friction and the evolution of a satellite. *Nature, 35*, 75.

Nolan J. (1895) *Satellite Evolution. The Evident Scope of Tidal Friction. The Meaning of Saturn's Rings.* Robertson, Melbourne. 114 pp.

Patterson A. H. (1909) The origin of the Moon. *Science, 29*, 936–937.

Pickering W. H. (1903) *The Moon: A Summary of the Existing Knowledge of our Satellite.* Doubleday, Page and Co., New York. 103 pp.

Pickering W. H. (1907) The place of origin of the Moon. *Pop. Astr., 15*, 274–287.

Roche E. (1873) Essai sur la constitution e l'origine due système solaire. *Mem. Acad. Sci. Lett. Montpellier. Sec. Sci, 8*, 235–324.

Schwarzschild K. (1898) Die Poincarésche Theorie des Gleichgewichtes einer homogenen rotierenden Flüssigkeitsmasse. *Neue Ann. Kgl. Sternwarte München, 3*, 231–299.

See T. J. J. (1909a) Dynamical theory of the capture of satellites and of the division of nebulae under the secular action of a resisting medium. *Astr. Nachr., 181*, 333–350; *Pop. Astr., 17*, 481–494, 534–544.

See T. J. J. (1909b) Origin of the lunar terrestrial system by capture with further considerations on the theory of satellites and on the physical cause which has determined the directions of the rotations of the planets about their axes. *Astr. Nachr., 181*, 365–386; *Pop. Astr., 17*, 634–640, *18*, 24–31, 106–110, 155–161.

See T. J. J. (1909c) The origin of the satellites. *Sci. Amer. Suppl., 68*, 191.

See T. J. J. (1909d) The terrestrial origin of the Moon—A protest. *Sci. Amer., 101*, 91.

See T. J. J. (1910a) The origin of the so-called craters on the Moon by the impact of satellites, and the relation of these satellite indentations to the obliquities of the planets. *Pub. Astr. Soc. Pac., 22*, 13–20.

See T. J. J. (1910b) *Researches on the Evolution of the Stellar Systems*, volume 2: *The Capture Theory of Cosmical Evolution.* Thomas P. Nichols and Sons, Lynn, Massachusetts. 734 pp.

See T. J. J. (1915) The origin of the Moon. *J. Brit. Astr. Assoc., 25*, 282–284.

Taylor F. B. (1898) *An Endogenous Planetary System.* Archer, Ft. Lane, Indiana. 40 pp.

Taylor G. I. (1919) Tidal friction in the Irish Sea. *Phil. Trans. Roy. Soc. London, A220*, 1–33.

Urey H. C. (1959) The face of the Moon: Two-thirds of a century later. Lecture at Philosophical Society, Washington, D.C. (Copy in Urey papers at UCSD library, La Jolla, CA.)

Urey H. C. (1966) "Dust" on the Moon. *Science, 153*, 1419–1420.

Wackerbath A. D. (1867) On an astronomical presentiment of Immanuel Kant relative to the constancy of the Earth's sidereal period of rotation on its axis. *Mon. Not. Roy. Astr. Soc., 27*, 200.

Wegener A. (1912) Die Entstehung der Kontinente. *Petermanns Geogr. Mitt., 58*, 185–195, 253–256, 305–309.

Moon Over Mauna Loa: A Review of Hypotheses of Formation of Earth's Moon

JOHN A. WOOD

Harvard-Smithsonian Center for Astrophysics, 60 Garden Street, Cambridge MA 02138

Important constraints that a model of lunar formation must satisfy are first reviewed: the large mass of the Moon and the substantial prograde angular momentum of the Earth-Moon system; the Moon's depletion in volatile elements and iron (the Moon contains about one-fourth of its cosmic complement of Fe); the correspondence of oxygen isotope signatures in Earth and Moon; and the lunar magma ocean. Some similarities in elemental abundance patterns between the mantles of Earth and Moon have been noted, but there is disagreement as to how and whether this should be applied as a constraint.

Five major models of lunar formation are then reviewed: (1) capture from an independent heliocentric orbit; (2) coaccretion from a swarm of planetesimals in geocentric orbit; (3) fission from a rapidly rotating Earth; (4) collisional ejection, in which a Mars-size planetesimal impacts the Earth, resulting in an orbiting debris disk from which the Moon accretes; and (5) disintegrative capture, wherein a large planetesimal passing through Earth's Roche zone is disrupted by tidal forces and the debris forms an orbiting disk from which the Moon accumulates. The coaccretion model, which has probably had the widest (if most passive) following in the past, is found to have a severe problem: It seems impossible to account for the angular momentum of the Earth-Moon system if it formed by this means. In the light of our present knowledge, the collisional ejection model appears to be the most plausible alternative.

1. Introduction

Considering how conspicuous and starkly beautiful Earth's Moon is, and what a unique object it is in the solar system, it is surprising how little scientific thought was given to its origin before the present epoch. The Moon appears only as a minor detail in the grand World Systems of authors such as Immanuel Kant (1755): Typically these writers considered that satellite systems, including Earth's Moon, are small-scale analogues of the planetary system, and that they formed about their primary planets just as the planets formed about the sun. The only widely cited reference on lunar origin from this period is George Darwin's (1879) treatise on fission of the Moon from the Earth. Early discussions of lunar origin are reviewed in this volume by Brush (1986).

It appears that the first person who really cared about the origin of the Moon was Harold Urey (1952). The Moon was a crucial element of Urey's World System, not a detail in it. He believed the Moon to be a "primary object," probably the only survivor in the inner solar system of a primordial population of bodies that accreted from presolar interstellar dust when the solar system formed. Collisions among the other primary objects, and complex fractionation processes that subsequently affected the debris, produced planets and meteorites with evolved, noncosmic chemical compositions. Urey's great interest in the Moon and his scientific prestige made him a powerful advocate of lunar exploration in the period when NASA was founded and potential space missions were being defined.

By 1964, with four different U.S. programs of lunar exploration at various stages of execution, scientific interest in the Moon had become intense. During January 20-21 of that year, just five months before the first successful Ranger flight to the Moon, a conference on "The Dynamics of the Earth-Moon System" was held at the Institute for Space Studies (New York City) of the NASA Goddard Space Flight Center. Discussions centered on the origin of the Moon via Earth fission or capture, and the subsequent orbital evolution of the Moon. This was (to my knowledge) the first conference that focused on the question of the origin of the Moon. A proceedings volume was published (*The Earth-Moon System*, B. G. Marsden and A. G. W. Cameron, eds., Plenum, New York, 1966).

There was widespread expectation that the Apollo exploration of the Moon would settle the question of its origin; this had been cited frequently as one of the scientific goals of the Apollo program. However, the Moon turned out to be a highly differentiated body, one that preserved a priceless record of the earliest igneous activity and geochemical fractionations in a small planet, but not of how the body formed. The only clear reading on lunar origin to come from Apollo data was that Urey's concept of a cold, primitive Moon had been wrong. As Apollo scientists set to work studying the geological evolution of the Moon, the question of its origin receded into the background.

Recently the question has been reopened by another conference on "The Origin of the Moon," this one held at Kona, Hawaii (October 14-16, 1984) and cosponsored by the Division of Planetary Seiences of the AAS, the Lunar and Planetary Institute (Houston), and NASA. It was the first conference on lunar origin since Apollo; the first, in fact, since the Institute for Space Studies meeting 20 years earlier. The conference revealed that more progress has been made in understanding lunar origin than most of us suspected. Apollo science contributed to the advance, but in relatively indirect ways. Most of the gain in understanding (if it is not illusory) has come from more sophisticated dynamical studies and from a more mature appreciation of the context of the problem, the larger question of the origin of the planetary system. A major share of credit probably goes to digital computers, which have come into widespread use in the years since the Goddard Institute for Space Studies meeting.

The present paper describes and critiques five models of lunar formation in the light of the discussions at Kona. I begin by reviewing those properties of the Moon that seem most likely to preserve information about its mode of origin. A successful model of lunar origin must account for these properties, so they can be thought of as constraints as well as clues.

2. Clues and Constraints

2.1. Mass of the Moon

The Moon is far more massive, relative to its primary, than is the satellite or satellite system of any other planet, with the probable exception of Charon (Table 1). This indicates rather clearly that the event or process that created Earth's Moon was an unusual one, and suggests that some relaxation of the normal aversion to invoking *ad hoc* circumstances might be permissible in this problem.

TABLE 1.   Mass Ratios, Satellite(s)/
Primary Planet.

Mercury	0.
Venus	0.
Earth	0.0122
Mars	0.00000002
Jupiter	0.00021
Saturn	0.00025
Uranus	0.00017
Neptune	0.0013
Pluto	~0.04

2.2. Angular momentum of the Earth-Moon system

The Moon moves in its geocentric orbit in the same prograde sense that the Earth rotates. The amount of angular momentum in the Earth-Moon system is substantial, 3.45×10^{41} rad g cm^2/sec. This is the sum of the rotational angular momentum of the Earth and the orbital angular momentum of the Moon. Figure 1 shows a relationship between angular momentum density and planetary mass, noted by MacDonald (1966); this suggests that the Earth-Moon system contains an anomalously large amount of angular momentum. However, it is questionable whether the curve of Fig. 1, which is established largely by the momenta of the Jovian planets, can be validly applied to the terrestrial planets, which may well

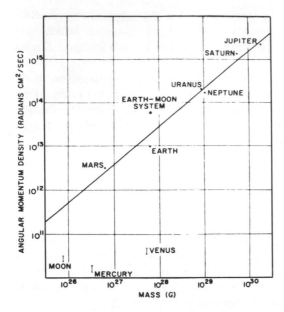

Fig. 1. The angular momentum density of planets (and the Earth-Moon system) as a function of their masses. Figure from MacDonald (1966).

have formed by a fundamentally different mechanism than the Jovian planets (accretion of solids vs. gravitational instability of nebular gases). Only one of the four terrestrial planets, Mars, falls on the curve. Hartmann and Larson (1967) and Burns (1975) have redefined the curve by including data for a number of asteroids at the low-mass end of the plot. This makes the Earth-Moon system fall on the curve, and leaves Mars somewhat deficient in angular momentum. But the value of this generalization is also questionable, since the data embrace objects formed by not two but three mechanisms (most of the asteroids are collisional debris).

That the Earth-Moon system is turning may seem a trivial point to use as a constraint, but as will be seen one of the long-established models of lunar formation has trouble accounting for this basic property. There has, of course, been major exchange of angular momentum between the Earth's rotation and the Moon's orbit. The Moon was once much closer to the Earth than it is now, possibly only a few Earth radii away from it, and tidal interactions have decelerated the rotation of the Earth while they accelerated the Moon and expanded its orbit. There is an extensive literature of theoretical studies of lunar orbital evolution (e.g., Burns, 1977). These considerations, and in particular the angular difference between the inclination of the lunar orbit and the obliquity of Earth's rotation axis, which figures importantly in orbital evolution, are often cited as constraints on lunar orgin (e.g., MacDonald, 1966). This article does not include them among constraints: In the first place, uncertainties in the orbital evolution calculations make it impossible to reliably project back to the earliest epochs of orbital history (e.g., Boss and Peale,

1986), and in the second place, the crucial angular difference between Earth's equator and the lunar orbital plane can have been arbitrarily changed at an early time by the impact of a single relatively small planetesimal at a high latitude on Earth, or on the Moon.

2.3. Volatile element depletion

One of the first observations made in the study of Apollo samples was that the lunar rocks and soils are markedly depleted in volatile elements. The degree of the volatile depletion is indicated in Fig. 2, which compares elemental abundances in the lunar mantle (as deduced from abundances in low-Ti mare basalts) with abundances in undifferentiated planetary material (C1 carbonaceous chondrites).

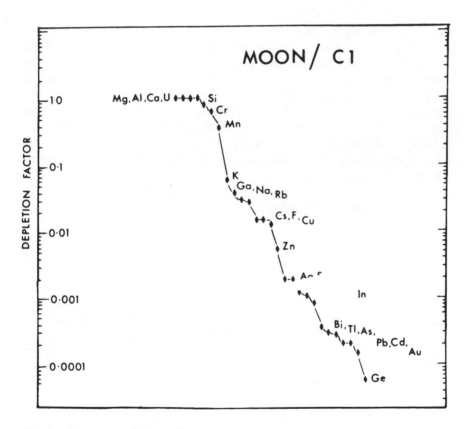

Fig. 2. Comparison of elemental abundances in the mantle source region of low-Ti mare basalts with abundances in C1 chondrites. Volatility of the elements tends to increase to the right. Figure from Ringwood and Kesson (1977).

It should be kept in mind that the Earth is also depleted in volatile elements, and so may be all the terrestrial planets. However, the Moon is much more severely depleted in volatiles than the Earth: For example, the ratio K/U in lunar materials is only about one-fifth as great as in terrestrial rocks. (Potassium and U are volatile and involatile elements, respectively, which have little tendency to be fractionated from one another by igneous processes.)

It is widely believed that the missing volatiles were lost from the substance of the Moon during some highly energetic process of lunar formation that heated the material to very high temperatures. The volatile elements were vaporized at that time, for the most part, and dispersed in space. All the models of lunar formation to be discussed are high-energy processes with the potential for volatile loss.

I-Pu-Xe dating shows that the last lunar volatile element fractionation occurred 20–100 m.y. after the formation of the most primitive meteorites (Swindle et al., 1986), a time consistent with any of the models of lunar formation reviewed in this paper. However, the similarity of initial $^{87}Sr/^{86}Sr$ values in the Moon (LUNI) and in basaltic achondrites (BABI) (Papanastassiou and Wasserburg, 1976) shows that the lunar material was depleted in the volatile element Rb very early, before the canonical $\sim 10^8$ yr of planetary accretion had elapsed (Taylor, 1986). Since all the models of lunar formation that are reviewed are tied to the time scale of planetary formation, this means none of them can be held fully responsible for the volatile depletion of lunar material, if the 10^8 yr figure is correct. The Moon would have to be made of material already depleted in volatiles (as the Earth was). This does not, of course, *exclude* a high-temperature process of lunar formation with additional volatile losses.

2.4. Elevated abundances of refractory elements

The Moon's depletion in volatile elements and the unexpected abundance of the mineral plagioclase (mostly $CaAl_2Si_2O_8$) in samples from the lunar crust also led to a conclusion, early in the Apollo program, that levels of refractory elements in the Moon are greatly enhanced. (Ca, Al, Ti, Ba, Sr, Sc, Y, the rare Earth elements, Zr, Hf, Th, V, Nb, Ta, Mo, W, U, Re, Ru, Os, Rh, Ir, and Pt are the refractory elements. They were believed to be concentrated in the Moon to levels far above the cosmic relative to Mg, Si, and Fe, the major elements of intermediate volatility.) This led to the proposal of several models of lunar origin (Anderson, 1972; Cameron, 1972) that involved incorporation in the Moon of large amounts of refractory Ca,Al-rich material of the type that has been studied extensively in the Allende meteorite.

However, as more geochemical and petrological evidence accumulated, the degree of the hypothetical refractory element enhancement diminished. Current estimates of refractory element abundances in the Moon (e.g., Ca and Al; expressed as oxides in Table 2) do not exceed the cosmic levels by factors large enough to affect ideas about lunar origin. Furthermore, the observations on which these estimates are based

	CaO	Al_2O_3	MgO
Morgan et al (1978)	6.1%	7.6%	29.1%
Taylor (1982)	4.5	6.0	32.0
Wänke et al (1977)	3.8	4.6	32.4
Ringwood (1979)	3.7	4.2	32.7
Cosmic			
C1 chondrites	2.7	3.3	(32.0)

*C1 chondrite levels adjusted to MgO = 32.0.

(Apollo heat-flow measurements at two sites, from which levels of the refractory element U are inferred, and measurements made by orbital geochemical sensors of the composition of the surface layer of the lunar highlands, from which the composition of the entire lunar crust has been inferred) have too much uncertainty associated with them to demonstrate conclusively that refractory elements are concentrated to any degree at all above cosmic in the Moon.

2.5.   Depletion of iron in the Moon

The Moon's mass density (3.344 ± 0.002 g/cm³; Bills and Ferrari, 1977) is much less than the uncompressed densities of the terrestrial planets (~3.7–5.4 g/cm³), which means the Moon must contain a smaller proportion of some heavy major element than the planets do. Only Fe is heavy enough and abundant enough to make the difference. It was recognized before Apollo that the Moon must be deficient in Fe relative to Earth. What is generally pictured is that the dense metallic Fe core that contributes so importantly to the Earth's density is largely missing from the Moon; and indeed, geophysical measurements constrain the Moon's core, if it has one at all, to constitute no more than ~6% of the lunar mass (Hood, 1986), in contrast to a terrestrial core that makes up 32% of the Earth's mass. In fact, however, a broader statement than this can be made: The Moon must be deficient in *elemental* Fe—Fe in any form—not just metallic Fe. The low lunar density cannot be rationalized by postulating that the Moon contains Fe in the cosmic proportion, but all of it is oxidized to FeO and incorporated in silicate minerals.

This can be shown by calculating the mineral norm of a simple Moon consisting only of Si, Mg, and Fe in their cosmic proportions, with enough O to oxidize them fully (the Fe to FeO). The norm consists of 92.3 wt % olivine, 7.7% orthopyroxene; Mg/(Mg + Fe) = 0.56. Using the mineral density data of Deer et al (1962), the net specific gravity of this assemblage is found to be 3.69 g/cm³, much greater than the lunar density. To make this simple system reproduce the lunar density, its content of Fe must be reduced to ~0.25 of the cosmic abundance. In this case the norm works out to 34% olivine, 66% orthopyroxene, Mg/(Mg + Fe) = 0.84, density $= 3.34$ g/cm³. This Mg/(Mg + Fe) ratio is close to the values indicated for the lunar mantle by mare basalt petrology (0.75–0.80; Ringwood, 1979) and geophysical constraints (0.75–0.85; Hood, 1986).

Thus the Moon failed to incorporate three-quarters of its cosmic complement of Fe when it formed. This is one of the strongest constraints that bear on the question of lunar origin. Most models of lunar origin dispose of the unwanted Fe mechanically: Much of the Fe is seen to exist as a metal phase apart from the silicate phases that will eventually join the Moon, and differences in physical properties (density, strength) of the metal and silicate phases cause them to be mechanically fractionated. However, there is another possibility that should also be considered. Hashimoto (1983) has experimentally evaporated oxide mixtures at high temperatures in a vacuum; his work shows that under these circumstances FeO is a relatively volatile component that is lost from the silicate melt more readily than are MgO, SiO_2, CaO, and Al_2O_3 (Fig. 3). It is possible that the Moon's Fe deficiency is simply an aspect of the general pattern of depletion of volatile elements, and that all these elements were distilled away at a time when the substance of the Moon was dispersed and very hot.

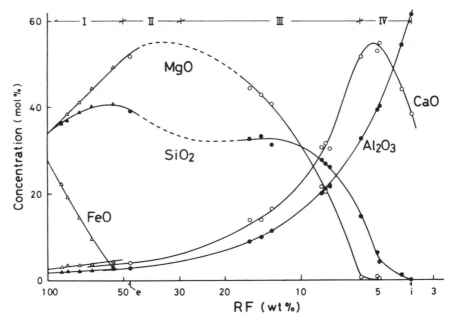

Fig. 3. Change of composition of residual molten oxide material as a charge is vaporized away into a vacuum at 2073 K. RF is the wt % of the original charge remaining. Initial charges contained solar proportions of the oxides shown. Figure from Hashimoto (1983).

2.6. Oxygen isotopes

The oxygen isotope signature of lunar materials collected by the Apollo missions is identical to that of the Earth to within measurement error (Clayton and Mayeda, 1975). Oxygen isotope compositions seem to vary with position in the solar system, since most types of meteorites differ in this property from the Earth and Moon (Clayton et al., 1976). Therefore the similarity of oxygen in the Earth and Moon strongly suggests that both bodies were made from the same batch of protoplanetary material, at the same radial distance from the sun.

2.7. Element abundance patterns in the lunar and terrestrial mantles

Siderophile trace elements are highly depleted in the Earth's mantle. Presumably they were absorbed in drops of molten metal as the latter sank to the core during the primary differentiation of the Earth. Only minor aliquots of these elements, amounts defined by their respective metal-silicate melt partition coefficients, were left in the silicate minerals of the mantle. The mantle of the Moon is also depleted in these elements. The abundance patterns of many of the siderophile elements residual in the mantles of Earth and Moon are similar. Does this mean the two mantles had a common origin, and therefore the Moon was created by some form of fission from the Earth?

A. E. Ringwood and coworkers (Ringwood and Kesson, 1977; Delano and Ringwood, 1978a,b; Ringwood, 1979, 1986) have argued with some conviction that it does. They point out that abundances of the siderophile elements Co, Ni, W, P, S, Se, and Te are the same to within a factor of about two in the terrestrial and lunar mantles, and argue that two separate core-forming fractionations operating in an Earth and Moon that were always independent of one another could not have produced such a correspondence: The partitioning of these elements depends upon pressure, temperature, fO_2, and fS_2 in the interiors of the fractionating bodies, and these conditions are known to have been very different in the early Earth and Moon.

H. Wänke and colleagues (Rammensee and Wänke, 1977; Wänke et al., 1978; Wänke and Dreibus, 1986) share the conviction of Ringwood and his coworkers. They draw attention to the depleted but remarkably similar concentrations of the lithophile elements V, Cr, and Mn in the terrestrial and lunar mantles. Again, this could not be a coincidence; these workers propose that fractions of the elements named joined the terrestrial core as sulfides or oxides before the substance of the Moon was separated from the Earth's mantle, leaving the observed residual amounts in the two mantles.

Drake (1983) has pointed out that for the set of siderophile elements P, Mo, Ge, and Re, mantle concentrations are not similar in Earth and Moon: These elements are more depleted in the Moon, and the amount of depletion correlates with the degree of siderophilic behavior of the element (i.e., its metal/silicate partition coefficient; Fig. 4). Thus if the Moon did separate from the Earth's mantle, an additional amount of core formation had to occur in the Moon but not the Earth, after the separation, to additionally deplete these elements only in the Moon. This hypothetical secondary depletion was earlier proposed by Wänke *et al.* (1978), and more recently has been endorsed by Ringwood and Seifert (1986).

Drake (1983) and also Newsom (1986) conclude that the lunar siderophile pattern is consistent with formation either in the Earth or as an independent system. Morgan *et al.* (1978) and Anders (1978) argue that the fractionation did not occur in the Earth. Kreutzberger *et al.* (1985) show that the patterns of alkali volatile trace elements in Earth and Moon are inconsistent with derivation of all the substance of the Moon from the Earth. The topic of mantle trace element abundances as a key to the origin of the Moon is complex and arcane, and an area of little consensus. A detailed discussion is outside the scope of this paper; the interested reader is referred to reviews by Newsom (1984) and Drake (1986). The present author agrees with Taylor (1982) that too many factors cloud the picture to allow mantle compositions to be used as a convincing argument either for or against formation

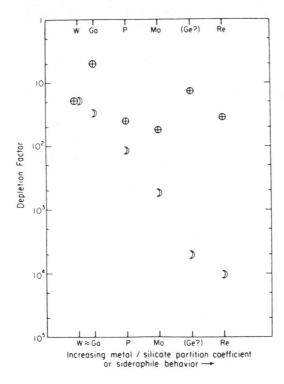

Fig. 4. The depletion of six siderophile elements in mantles of the Earth and Moon, as inferred from the compositions of basalts on both bodies. "Depletion factor" is the ratio (abundance in mantle)/ (abundances in C1 chondrites). The "?" associated with Ge refers to a paucity of data and also the fact that the depletion is partly due to the volatile nature of Ge. These elements are seen to be much more depleted in the Moon than in the Earth, and the degree of depletion correlates with the metal/silicate partition coefficient. Figure from Drake (1983).

of the Moon from the Earth. Consider the number of ways in which the lunar and terrestrial trace element signatures would have been changed or obscured if the Moon *had* formed from the Earth's mantle:

1. The partial vaporization of protolunar material during formation of the Moon that probably contributed to the depletions of volatile elements observed in lunar materials (noted above) would also deplete siderophile elements selectively.

2. By far the most plausible mechanism of Earth fission involves the glancing impact of a large planetesimal that ejected material from the mantle of the Earth (as discussed below). A major amount of debris from the impacting planetesimal inevitably would be mixed with the mantle debris, imparting a composite trace element signature to the material that ultimately accreted into a Moon. Cameron (1985, 1986) concludes that *most* of the material placed in orbit by such an event would be derived from the impactor.

3. Additional core formation and fractionation of mantle siderophiles is likely to have occurred in the Earth after the fission event.

4. Additional core formation and fractionation is also likely to have occurred in the Moon after it accreted; indeed, this seems required. The nature and degree of fractionation is bound to have differed in the two bodies.

5. The Earth probably accreted additional material from heliocentric orbit after fission, and to some degree incorporated its siderophile elements in the terrestrial mantle.

6. The Moon also accreted additional material after its formation. The amounts of new material gained and the nature of its incorporation in the bodies of the Moon and Earth undoubtedly differed.

7. It is not straightforward to infer the composition of the lunar mantle from mare basalt and highlands rock samples. As Taylor (1982) points out, no lunar basalts can be identified as coming from primitive unfractionated material of bulk lunar composition.

8. Similarly, we have no assurance that present-day estimates of the terrestrial mantle composition (based on basalt compositions and xenoliths) are representative of the composition of the mantle at the time of the hypothetical lunar fission event.

2.8. The lunar magma ocean

Apollo samples from the highlands (terrae) of the Moon were found to be enriched in calcic plagioclase feldspar. Some of the samples are true anorthosites (~95% plagioclase); the mean composition of surface material in the highlands all the way around the Moon, as measured by orbiting geochemical sensors, corresponds to that of an anorthositic gabbro containing ~75% plagioclase (Warren, 1985). Magmas produced by the partial melting of planetary interiors are constrained by phase equilibria to contain less than ~55% plagioclase, so some other igneous process must be held responsible for the concentration of plagioclase at the lunar surface. Crystal fractionation is the universally accepted mechanism: Plagioclase is a relatively low-density phase, and would tend to float up to the top of lunar magma bodies and concentrate there.

The amount of plagioclase in the lunar crust is very great. Gravity and topography measurements provide a way of estimating the amount. The highlands are "high" because they are made of relatively low-density rock that "floats" in denser lunar mantle material. The highland rock is light because of its content of plagioclase, the only abundant low-density lunar mineral. For a given crustal density, the total thickness of floating crust needed to buoy the highlands to their observed mean altitude above the maria can be simply calculated. The formula appears in Fig. 5, along with gravity and topographic parameters applicable to the lunar near side (Kaula *et al.*, 1974); these are conservative values. For a hypothetical crust of pure anorthite plagioclase, the crustal thickness t works out to 20.4 km. To collect this much plagioclase from a Moon originally of chondritic composition (minus metal,

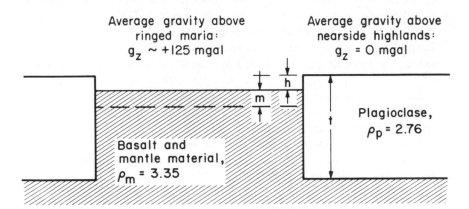

h = mean altitude difference, ~ 2.6 km

m = thickness of mare basalt that would have to be removed to eliminate +125 mgal gravity anomaly, leaving highlands and ringed maria in isostatic equilibrium, ~ 1 km

t = mean crustal thickness

$$\rho_p t = \rho_m \left[t - (h + m) \right]$$

Fig. 5. Summary of gravity and topography parameters on the lunar nearside, and the formula relating them to the mean crustal thickness t (Kaula et al., 1974). Gravity in the highlands is arbitrarily set at 0 mgal; the 125 mgal mean positive gravity anomalies (mascons) in the ringed maria are relative to this value.

sulfide, alkalis) requires that ~30% of its volume (i.e., to a depth of ~200 km) be partially melted, and the plagioclase components extracted. This assumes the plagioclase components are extracted and sent to the lunar crust with perfect efficiency; otherwise a larger fraction of the lunar volume would have to be extracted. Wood *et al.* (1970) and Smith *et al.* (1970) concluded that the outer layers of the Moon, to a depth of several hundred kilometers, were melted when the Moon formed. An integral magma ocean seemed required in order to give plagioclase crystals the freedom to float up from great depths and join the crust. This concept, modified in various ways, has gained wide acceptance. The topic is competently reviewed by Warren (1985).

The lunar crust does not actually consist of pure plagioclase, of course, but of rocks that contain various proportions of other minerals (mostly pyroxenes and olivine). The gravity/topography criterion informs us of the total amount of low-density plagioclase in the crust, but not its degree of dilution by denser minerals. The crust below the plagioclase-rich surface might consist entirely of gabbroic rock containing only ~50% plagioclase; of course, such a crust would have to be thicker than the 20.4 km calculated for pure plagioclase in order to buoy the highlands. These gabbros could have been produced in the Moon by partial melting over a protracted period of time and delivered to the surface little by little ("serial magmatism"); then there would be no need for an integral magma ocean to permit plagioclase crystal flotation. Wetherill (1975) and Walker (1983) have proposed effectively this, dispensing with the magma ocean. In their models the plagioclase-rich rocks characteristic of the near-surface highlands were produced by crystal flotation in gabbroic magma chambers of limited scale, and/or basaltic flow units.

However, these more prosaic models of lunar crust formation have a severe problem. Crystal fractionation in a gabbroic magma chamber would produce ultramafic cumulate rocks (by crystal sinking) as well as plagioclase cumulates (by crystal flotation). If the scale of the magma chambers was smaller than the depth of the impact craters that have excavated and scattered crustal material over the lunar surface, then fragments of ultramafic cumulate rock as well as anorthosite should be abundant on the lunar surface. However, ultramafic rocks are conspicuously absent from lunar highlands regions. Only one such sample, dunite 72415, was among the individual rock samples collected by the Apollo astronauts. The ejection depth from lunar mare basins is estimated to be 30–80 km (Grieve, 1980; Spudis, 1983), i.e., most of the thickness of the lunar crust. Therefore plagioclase fractionation must have consistently occurred in magma bodies of more than 50–120 km vertical extent in order to bury the ultramafic cumulates deeply enough to prevent them from being excavated during the subsequent cratering history of the Moon. Such bodies approach the vertical scale of the hypothetical magma ocean. It does not seem possible to avoid invoking very large-scale surface magmatism on the Moon, though it is arguable whether a surface shell was molten everywhere on the Moon simultaneously, as in the classic magma ocean.

The principal motivation for discounting evidence of a magma ocean has been the difficulty in accounting for a source of heat to melt it. Traditionally, accretional energy has been invoked for this purpose, but Ruskol (1973), Wetherill (1975), and Kaula (1979) have argued that the heat released during formation of the Moon by coaccretion in a geocentric planetesimal swarm is marginal to inadequate to melt a magma ocean. However, the problem may be with the coaccretion model rather than with the magma ocean. As will be seen, other models have no trouble melting a magma ocean during lunar formation.

Like the volatile-poor character of the lunar samples, the evidence for something like a magma ocean early in the history of the Moon requires the mode of origin of the Moon to be a high-energy process. However, where the volatile depletion seems to require that the substance of the Moon was hot while it was in a dispersed state (which facilitates devolatilization), the magma ocean requires high temperature after the Moon accreted.

3. Models of Lunar Origin

Traditionally, the debate about lunar origin has focused on three possibilities: (1) that the Moon was formed in heliocentric orbit elsewhere in the solar system, then captured intact into geocentric orbit; (2) that it accreted as a companion to the Earth from a disk of planetesimals and particles that had been captured into geocentric orbit; and (3) that it fissioned from an early Earth that was rotating so rapidly as to be unstable. More recently, two hybrid models have been added to the lineup: An off-center impact by a very large planetesimal (more properly a planet) threw debris into a geocentrically-orbiting disk from which the Moon accreted; or, a similarly large object that barely missed the Earth was tidally disrupted as it passed through the Roche zone; debris from the Earth-facing side of the object was retained in geocentric orbit, and the Moon accreted from it. These five models will be discussed in turn.

3.1. Intact capture

The intact capture model was much favored in the 1960s (Urey, 1966; MacDonald, 1966; Singer, 1968), but interest in it seems to have waned. Most recently it has been defended by Turcotte and Kellogg (1986) and Singer (1986).

Originally, the most attractive aspect of the capture model was its potential for rationalizing the major compositional differences (especially Fe) between Earth and Moon. If two objects were made in widely separated parts of the solar system and then joined by capture, a large compositional difference would not be hard to understand (e.g., Cameron, 1972). However, it has come to be appreciated that the Moon cannot have been captured from a remote region of the solar system. First, this entails motion of the proto-Moon in an eccentric orbit stretching from its place of origin to the Earth's orbit, which means the encounter would occur at relatively high velocity, and as will be seen this makes capture virtually impossible. Horedt (1976) estimates that a proto-Moon would have to be formed at a mean distance in the range 0.95–1.05 AU and with eccentricity very approximately 0.04 to be eligible for capture. Second, the essentially identical isotopic compositions of lunar and terrestrial oxygen appears to require that the substance of both bodies came from the same part of the solar system. Since the Moon cannot be "exotic," the capture model is deprived of its main selling point.

The dynamics of capture are best explored in the framework of the restricted three-body problem of celestial mechanics. Motions are considered in a reference frame that moves with the Earth and remains aligned with the sun-Earth line. In such a frame, the positions of critical boundary surfaces (zero relative velocity, or Hill, surfaces) can be calculated by use of Jacobi's integral. These surfaces define regions of space where the proto-Moon can and cannot move. The positions of the surfaces depend upon the amount of mechanical energy associated with the proto-Moon's motion relative to the Earth, and the masses of the sun and Earth. In a very low-energy situation, which corresponds to a small proto-Moon velocity relative to the Earth, a zero-velocity surface surrounds the Earth. An object moving around the Earth inside this boundary cannot cross it; the object is permanently locked in geocentric orbit. However, a proto-Moon from elsewhere in the solar system also cannot be captured into this situation, because if its energy is low enough for its Jacobi integral to predict a closed zero-velocity surface around the Earth, it cannot cross the intervening forbidden region and get inside this surface.

On the other hand, if the proto-Moon's energy is high, the zero-velocity surfaces expand, open out, and come to include the sun and a vast region of solar system space. An exotic proto-Moon has no trouble approaching the Earth, but there is nothing to hold it there; it invariably swings back out into distant reaches of space.

The situation is most interesting for intermediate values of proto-Moon energy. Then the Earth can be almost but not quite completely surrounded by a zero velocity surface: A door is left open at the l_1 libration point, on the side of the Earth toward the sun. The lower the energy, the smaller the door. An object from elsewhere in the solar system, having an amount of energy appropriate for this zero velocity surface, can wander in the door and become temporarily trapped. It circulates around and around the Earth until it finally finds its way out again (Fig. 6). Actually the zero velocity surface defined by the present Earth-Moon system has a tiny door; technically the Moon could escape the Earth and go into heliocentric orbit, and conversely there is a very small but nonzero dynamic probability that it was captured from heliocentric orbit without the assistance of energy-dissipating mechanisms (Kopal and Lyttleton, 1963; Szebehely and McKenzie, 1977).

Most capture models rely on the premise, whether explicitly stated or not, that the proto-Moon approached the Earth through a fairly large door in the zero velocity surface surrounding it; some dissipative effect then decreased the proto-Moon's energy,

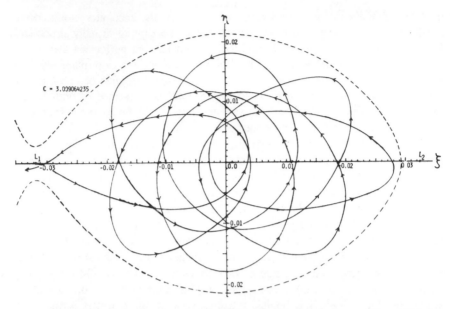

Fig. 6. Example of temporary capture of a satellite in the reference frame of the restricted three-body problem. The capturing planet is at the origin, the sun on the x-axis, far to the left out of the figure. The dashed line is the zero velocity surface for a satellite with a particular small amount of energy. Note that there is a door at the l_1 point to the left of the figure. The satellite for which this zero velocity surface is applicable enters the door and describes seven prograde orbits about the planet before it finds its way out the door again. Figure from Heppenheimer and Porco (1977).

which had the effect of pulling the zero velocity surface in and shrinking its door, so the Moon couldn't find its way out again. The mode of energy dissipation most often invoked is tidal interactions (MacDonald, 1966; Singer, 1968; Öpik, 1972; Winters and Malcuit, 1977), but there are several other possibilities. Kaula and Harris (1973) appeal to collisions with a circumterrestrial swarm of planetesimals; Nakazawa *et al.* (1983) appeal to gas drag in a massive primordial Earth's atmosphere that extended to the zero velocity surface. (In the latter case the timing must be exquisite: The atmosphere must be in the process of dissipation from the Earth at the time when the Moon begins to be decelerated by it, or the Moon will spiral into the Earth.) Clark *et al.* (1975) rely on a fortuitous sudden increase in the mass of the Earth (as by the accretion of a single major planetesimal), rather than energy dissipation, to close the zero velocity surface while the proto-Moon was passing the Earth.

The importance of a zero velocity surface with a small door is that once a proto-Moon is inside, it continues to circulate around the Earth in the same direction

(Fig. 6) and the energy-dissipating mechanism, whatever it may be, can operate repeatedly and with additive effect. The capture of an exotic object that approaches the Earth with high velocity and large energy is ruled out because its zero velocity surface would be too open to trap it, so it could make only one pass by the Earth; and it would have far too much energy to be dissipated in one pass.

Lunar capture is inherently improbable because the range of approach velocities and trajectories that corresponds to a zero velocity surface favorable for entrapment is extremely narrow. Most discouraging, however, is the absence of an explanation for the compositional difference between Earth and Moon (since exotic objects are ruled out). If the two bodies formed independently by similar processes of accretion at the same radial distance in the solar system and from the same reservoir of primordial material, it is reasonable to expect that they would end up with similar inventories of chemical elements. Other models involve more or less different processes of formation for the Earth and Moon, and their advocates take advantage of this to try to rationalize the compositional differences between the two bodies.

3.2. Coaccretion

The idea that the accreting Earth accumulated a disk or swarm of orbiting solid particles from which the Moon then accreted was first articulated by Schmidt (1959), and later developed by Ruskol (1960, 1963, 1972a, 1973), Harris and Kaula (1975), and Harris (1978). Most recently, the concept has been defended by Weidenschilling *et al.* (1986).

The great virtue of this model is that it does not invoke some special, low-probability event. The underlying premise is that the Earth accreted and the protolunar swarm accumulated in a natural way from untold numbers of small heliocentrically orbiting objects, and the dynamic and material properties of the Earth and Moon result inevitably from an averaging of the approach trajectories and behavior upon mutual interaction of these many objects. Coaccretion has probably gained wider acceptance than any of the other lunar origin hypotheses among lunar scientists, who were trained to shun *ad hoc* assumptions. However, the model has difficulty accounting for the compositional differences between Earth and Moon, the melting of the magma ocean, and the angular momentum content of the Earth-Moon system.

Heliocentrically orbiting planetesimals that approached the growing Earth but did not collide with it would have described quasihyperbolic trajectories about Earth's center of mass. Some planetesimals would have passed the Earth in a prograde direction, others in a retrograde sense. Collisions between these two populations would cancel some of the angular momentum of the planetesimals, relative to the Earth; a portion of the collision debris would fall to Earth, and some would be decelerated just enough to leave it in geocentric orbit. It has generally been understood (but see below) that most planetesimals approached the growing Earth-Moon system in a prograde sense—hence the Earth's prograde rotation—so a disk or protolunar swarm (PLS) of decelerated collision debris with net prograde rotation should have built up in orbit around the proto-Earth. Once the beginnings of a disk were established, it would have greatly accelerated the capture of additional disk material, since the collision probability for an incoming heliocentric planetesimal with accumulated disk material was far larger than the likelihood that it would encounter another heliocentric planetesimal. This mechanism of PLS formation is accepted by all the authors cited above. In addition, Weidenschilling *et al.* (1986) raise the possibility that a giant Earth impact of the sort that is central to the collisional ejection model (to be discussed) threw material into geocentric orbit, making a major contribution to the PLS.

Proponents of the coaccretion model explain the volatile element deficiency of the Moon as resulting from devolatilization of PLS material during the many high-energy collisions that established the disk, or as a property of the material that predated its incorporation in the PLS. Ruskol (1972b) proposed that the iron deficiency of the Moon is a consequence of the difference in strength and ductility between silicate minerals and metallic iron. Collisional fragmentation of protoplanetary material, both prior to and during encounter with the PLS, would reduce the relatively weak, brittle silicates to systematically smaller and lighter fragments than Fe metal. The silicate objects, with their lower momentum content, were more readily decelerated and captured by the PLS than the Fe fragments, so the PLS took on an Fe-depleted composition. Thus the PLS operated as a "compositional filter" (Weidenschilling *et al.*, 1986). The process would be particularly effective if, as is generally assumed, most of the protoplanetary material had already had an existence in small bodies that melted and differentiated (as the achondrite and iron meteorite parent bodies did), after which collisions among the bodies reduced many of them to debris fragments. Large Fe-metal core fragments would be especially difficult for the PLS to decelerate.

A crucial aspect of the process is the timing of the nucleation of the Moon from the PLS. Gravitational instability would cause an orbiting swarm to coagulate into one or more proto-Moons in a relatively short time, but Ruskol (1972a) argues that lunar embryos, which formed too early, while the space density of heliocentric planetesimals was still high, could not survive; they would be disrupted by high-velocity collisions with the abundant heliocentric planetesimals. She estimates that the embryos formed or were captured when the Earth had attained about half its present mass, and that there were probably two or three of them. They grew to ~1000 km radius by sweeping up low-velocity PLS material; thereafter they were large enough to accumulate high-velocity heliocentric planetesimal material, as well as additional PLS material. An advantage of several lunar embryos is that it increases the capture cross-section for material in heliocentric orbit, and hence the net accretion rate. Ultimately these moonlets merged to form the Moon.

Harris and Kaula (1975), assuming that once the lunar embryo formed all subsequent accretion was from heliocentric orbits, assessed the relationship between nucleation time and the final Moon/Earth mass ratio, and concluded that the lunar embryo formed relatively early, when the Earth had only ~0.1 its present mass. A difficulty with this picture and also Ruskol's, however, is that direct accretion of heliocentric material would not discriminate between silicate and iron objects; the compositional filter is turned off. Harris (1978), in a reanalysis of the situation, included the effect of accretion to the Moon of material that continued to join the PLS outside the orbit of the lunar embryo. This material is gradually decelerated by "accretion drag" (its encounter with heliocentric particles that have less angular momentum relative to Earth than it has), spirals inward, and is swept up by the Moon. For this model, Harris found that the lunar embryo can have nucleated when the Earth had attained ~0.4 of its final mass, and that one-third to one-half of its mass can have come from the PLS rather than directly from heliocentic orbit. (Even this does not rationalize the lunar composition, however, since, as noted, approximately three-quarters of the Moon's cosmic complement of Fe has to be withheld from it.)

The problem with melting a magma ocean during coaccretion of the Moon is that the heat source which is traditionally invoked, namely the gravitational potential energy that is converted to heat during the accretion of the Moon, may be released too gradually to be effective. Ruskol (1973) argues that the accretion of the Moon would not have ended until planetary accretion did, since protoplanetary material continued to be available for accretion until that time, and the duration of planetary accretion was ~10^8 yr. Another advantage of Ruskol's (1973) plan to make the Moon from several submoons is that a major fraction of the potential energy of accretion would be converted to heat in the ~1 hr during which two submoons of comparable size coalesced; this goes a long way toward providing the heat needed to melt the magma ocean. The melting of a magma ocean is also more plausible if lunar accretion was suppressed until late in the ~10^8 yr interval of planetary accretion, after which it proceeded very rapidly (Weidenschilling et al., 1986), and/or if the act of accretion threw up a transient lunar atmosphere that retarded radiative heat loss (Matsui and Abe, 1986).

The most serious problem for the coaccretion model appears to be accounting for the angular momentum of the Earth-Moon system. Only as recently as the Kona Conference was this problem widely appreciated. It has always been understood, and occasionally proven in rather broad terms (e.g., Schmidt, 1959; Lyttleton, 1972), that the accretion of a planet from dispersed, heliocentrically orbiting material would naturally transform some of the prograde orbital angular momentum of the raw material into rotational angular momentum of the planet. As a corollary, material captured into Earth's PLS would also have net prograde angular momentum. Giuli (1968) attempted to confirm this by simulating the accretion of the Earth on a planetesimal-by-planetesimal basis, using a computer to follow the trajectories and assess the angular momentum inputs of discrete objects. (For the moment I will leave the PLS aside, and concentrate on the Earth.) Giuli investigated a number of sets of a, e for planetesimals, which define orbits in two dimensions that are capable of accreting the planetesimals to the proto-Earth, if the latter is in the right place at the right time in its orbit ($a = 1$ AU, $e = 0$). For each planetesimal a and e, there is a limited range of the angle θ (see Fig. 7) for which accretion can occur. Giuli refers to all the orbits in one of these ranges of θ, having a common a and e, as a "band." He found that for most bands investigated, one end of the angular range corresponds to a trajectory that barely skims the limb of the Earth in a prograde sense; the other end of the range corresponds to skimming the retrograde limb of the Earth, and trajectories in between impact symmetrically across the body of the Earth, half imparting prograde and half retrograde angular momentum (Fig. 8). Thus accretion of a large number of planetesimals from these bands would impart no net angular momentum to the Earth; their effects would cancel each other.

Some of Giuli's bands did not impact the Earth symmetrically. He found that bands originating in circular orbits contributed net negative angular momentum to the Earth. On the other hand, bands originating in eccentric orbits, for which Earth

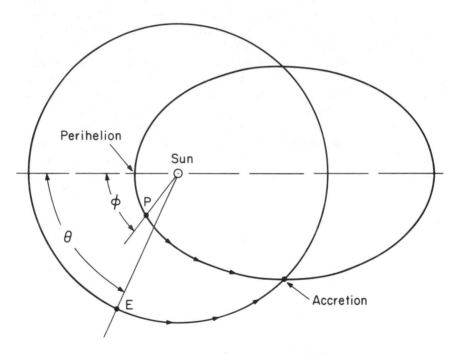

Fig. 7. Trajectories followed by Earth (started at E) and a planetesimal (started at P) during simulated Earth accretion. Variables in the situation are ₐ and ₑ for the planetesimal orbit and also the angle Θ, which describes the orientation of the planetesimal orbit relative to the starting position of Earth. In general there is a small range of values of Θ (φ held constant) that result in Earth impact. Exactly where the planetesimal strikes the surface of the Earth, and how much angular momentum it imparts to the Earth, vary with the value of Θ in this range; see Fig. 8.

impact occurred at the lowest possible velocity (the escape velocity), produced net positive angular momentum. These corresponded to a very narrow range of *a* and *e*. Giuli's conclusion was that an accretion scenario can be postulated that would produce an Earth with a prograde rotation period of 15 hours. However, to do this requires the planetesimal feedstock for the Earth to come from orbits with an implausibly limited range of orbital elements: *a*, 0.958–0.973 or 1.029–1.046 AU; *e*, 0.025–0.039. These values hold for an Earth with its present mass; for an earlier, smaller proto-Earth, values of *a* and *e* even closer to 1 and 0 are required.

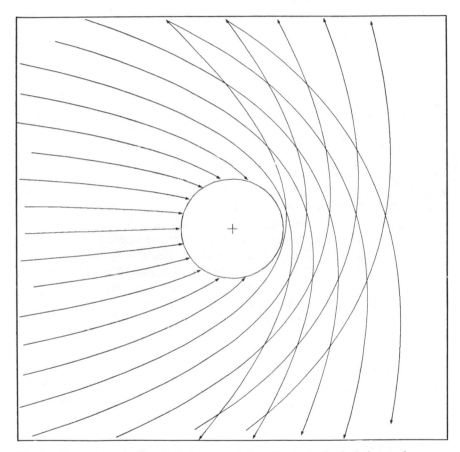

Fig. 8. Trajectories of a "band" of planetesimals accreting to the Earth. Reference frame fixed in Earth; sun-Earth line maintains alignment. Earth has its present radius. Planetesimals shown originated in orbits having the same a (1 AU) and e (0.05), but with orientations of the line of apsides (Θ) rotated by equal small increments. Impacts that impart positive and negative angular momentum to the Earth are evenly balanced. Trajectories computed by the author (see Appendix).

Harris (1977) noted that the bands of orbits that contribute positive angular momentum to Giuli's proto-Earth are just tangent to Earth's orbit at perihelion or aphelion. In fact, when Earth is not in the vicinity they don't quite cross Earth's orbit, but the presence of Earth's gravitational field enlarges their orbits to the point that the planetesimals can impact one (but only one) hemisphere of the Earth—the inner or sun-facing hemisphere if the planetesimal is at aphelion, the outer hemisphere for planetesimals at perihelion. Since eccentrically orbiting planetesimals at aphelion move more slowly than an object in circular orbit at the same radial distance, they are overtaken by the Earth; striking its sun-facing hemisphere, they impart prograde rotation to it. For analogous reasons the planetesimals at perihelion also impart prograde rotation to the Earth when they overtake it. In an analytical treatment of accretion that takes this effect into account, Harris (1977) confirmed that a suitably rotating Earth can be accreted from planetesimals if their range of orbital elements is sufficiently narrowly constrained. However, he observed that the value and range of e required are implausibly small, and an Earth accreted from planetesimals with a more credible range of orbital elements would have only ~10% of the required angular momentum.

Not surprisingly, it turns out that the same angular momentum problem is associated with the capture of a PLS from heliocentric orbits. Weidenschilling et al. (1986) studied the near-Earth trajectories of planetesimals started in a wide range of heliocentric orbits, and found it impossible to capture a PLS with an adequate amount of prograde angular momentum via planetesimal collisions in near-Earth space unless the source population of heliocentric planetesimals included very few objects in Earth-crossing orbit. This is equivalent to the orbital constraints Giuli (1968) invoked in order to explain the Earth's rotation. Weidenschilling et al. consider this to be a possible state of affairs, but their opinion is not widely shared. A commonly held view is that planetesimal eccentricities would have been large enough during the late stages of accretion (because of orbit perturbations by numerous growing subplanets in the system; e.g., Wetherill, 1986) to cause the great majority of objects entering near-Earth space to arrive in Earth-crossing orbits.

A simulation of Earth accretion and PLS accumulation (via disintegrative capture) carried out several years ago by the author (unpublished; reviewed in an Appendix to this article) confirmed the difficulty of accreting an Earth-Moon system with the observed content of angular momentum. This problem with the Earth's angular momentum forces a reexamination of the basic premise of coaccretion, that it consisted of the accumulation of a statistically large number of small planetesimals. The alternative one is driven to is a more hierarchical form of accumulation, involving a few very large planetesimals in the late stages. The fortuitous off-center impact of one large planetesimal could account for the rotation of the Earth, and perhaps also for the formation of the PLS. The absence of such an event might be the reason for the small, negative rotation rate of Venus.

3.3. Earth fission

In Darwin's (1879) original proposal of the fission origin of the Moon, rotational instability was not the cause of fission. Darwin assumed a prefission Earth with the same angular momentum as the present Earth-Moon system, rotating with a period of ~4 hours. He noted that the period of the slowest free oscillation of a fluid mass having the Earth's size and mean density is ~2 hours, and concluded that a resonance between this period and the period of tides raised on the Earth by the sun would cause the tidal bulges to grow higher and higher, until the tip of a highly elongated Earth broke off to become the Moon. The concept was refuted by Jeffreys (1930), who showed that frictional damping in the Earth, which increases as the cube of the tidal distortion, would prevent the amplitude of the distortion from growing large enough to result in fission.

The fission hypothesis was resurrected by Ringwood (1960) and Wise (1963), this time with rotational instability as the cause of the event. In this model the Earth was formed spinning at the verge of rotational instability, with a period of ~2.6 hours. Initially it had uniform density; segregation of its core decreased its moment of inertia, which caused it to spin faster (~2.1 hours), making it rotationally unstable. The Earth responded to this instability by adopting increasingly distorted geometrical forms (Fig. 9), until its most tenuously connected end broke off. The remaining Earth relaxed into a less distorted form and, rotating faster than the orbital period of the new Moon, began to tidally accelerate the motion and expand the orbit of the Moon. Jeffrey's objection does not apply in this case, because there is no differential rotation between the form and the substance of the prefission Earth. This model has been promoted by Wise (1969), O'Keefe (1966, 1969, 1972), O'Keefe and Sullivan (1978), and Binder (1974, 1978, 1986).

Recently, numerical hydrodynamic simulations of the fission of a rapidly spinning Earth have been carried out (Durisen and Scott, 1984; Durisen and Gingold, 1986; Boss and Peale, 1986) that represent a great advance over the crude understanding of the physical nature of the event that existed in the 1960s. These have shown that fission would not separate an intact Moon, as shown in Fig. 9, but instead dispersed material would be shed from the ends of the spinning elongated figure, building up a particulate disk or PLS. In some ways this makes the fission model more attractive, since hot, dispersed material shed by the Earth would be devolatilized more efficiently than a Moon fissioned intact, no matter how hot it was.

Advantages of the fission hypothesis are that it explains the Fe-poor character of the Moon, if the fission occurred after core formation in the Earth; and it also explains the common oxygen isotope signatures of the two bodies, the similarities that have been reported in trace element abundance patterns in the mantles of Earth and Moon, and the geochemical evidence that a large amount of metallic iron was once in contact with the substance of the Moon in spite of the small size of the

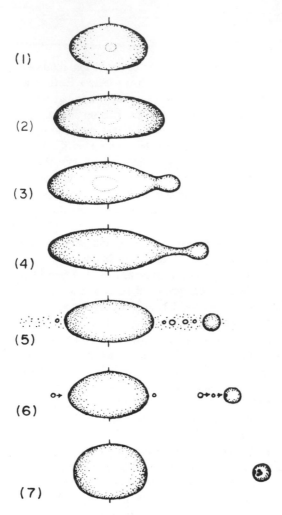

(1)

(2)

(3)

(4)

(5)

(6)

(7)

Fig. 9. Sequence of forms adopted by a rapidly spinning Earth after it becomes rotationally unstable. (1) Oblate Maclaurin spheroid; (2) Jacobian triaxial ellipsoid; (3) Poincare figure. Figure from Wise (1963).

present lunar core (Rammensee and Wänke, 1977; Ringwood, 1979). [But there is an uncomfortable aspect of this last triumph. Terrestrial core separation needs to be ~97% complete before fission occurs, to account for lunar siderophile element depletions. As O'Keefe and Sullivan (1978) point out, if fission is caused by core formation, it is unlikely that the process would go so nearly to completion before it caused destabilization.]

However, the fission model has very severe dynamic problems: In order to fission, the Earth had to have about four times as much angular momentum as the Earth-Moon system now has. There is no good explanation of why the Earth had such an excess of angular momentum in the first place, or where the surplus angular momentum went after fission occurred.

Why would a prefission Earth have so much angular momentum? Fission advocates rarely address this question. The previous section discussed at some length the evidence from a number of independent studies that accretion of the Earth from small heliocentrically orbiting planetesimals cannot even account for the amount of angular momentum the Earth-Moon system has, let alone four times this much. A series of major planetesimal impacts, all fortuitously off-center in a prograde sense, could do so, but lunar formation by this mechanism would have more in common with the collisional ejection model (discussed below) than with classic fission. Binder (1978, 1986) draws an analogy with contact binary stars, which have too much angular momentum to be stable as a single star. However, this has no relevance to the rotational state of planets and satellites. Stars are formed by the contraction of masses of interstellar gas to $\sim 10^{-7}$ of their original dimension. Conservation of even a fraction of the original angular momentum of the interstellar gas guarantees that the resulting stellar system will be rotating very rapidly, in many cases too rapidly to collapse, all the way to stellar dimensions as an integral object. No one has proposed that the terrestrial planets formed by an analogous process. G. P. Kuiper, H. C. Urey, and A. G. W. Cameron have advocated planet formation in gaseous protoplanets, but no element of shrinkage through many orders of magnitude of dimension with angular momentum conserved is involved. In these models the planets form by a rainout of refractory solids/liquids to the centers of the protoplanets, and it is to be expected that viscous interaction of the particles with protoplanet gases on the way down would remove most of their angular momentum, so that a planet which accumulated from them would be rotating at not much more than the moderate rate of the gaseous protoplanet itself.

Where could the excess angular momentum have gone after fission? Only very speculative answers have been offered. Wise (1969) suggests that the postfission Earth, at a very high temperature after the release of energy associated with accretion, core formation, rotational distortion, and tidal interaction with the fissioned Moon, boiled off a transient atmosphere of silicate vapors at $\sim 5000K$. If such an atmosphere extended to an altitude of one (additional) Earth radius, and rotated synchronously with the Earth beneath, then only 3.7% of the mass of the Earth would have to be lost by thermal escape from the top of the atmosphere to bring the angular momentum of the system down to its present value. (It is doubtful, however, that mechanical coupling between the Earth and its upper atmosphere would be good enough to brake the rotation of the Earth by the required amount.) O'Keefe (1972) proposes that the Moon that originally fissioned was much larger than the present Moon, ~ 0.1 Earth mass; the very high temperatures in that epoch boiled away 95% of the mass of the Moon. This mass and the angular momentum associated with it were lost to space, reducing the system's angular momentum by the required amount. (But the distillation away of such a large fraction of the substance of the

Moon should have left a residual body of very specialized composition: see Fig. 3. The Apollo samples do not bear this out.)

3.4. Collision ejection

The idea that the Moon was assembled from a disk of material that had been ejected from the Earth by a catastrophic collision was late in coming because (1) invoking the services of a sufficiently large planetesimal seemed uncomfortably *ad hoc*: Safronov (1966), always an advocate of a hierarchical process of accumulation of the planets, nonetheless estimated the size of the largest objects impacting the Earth toward the end of its accretion to be only $\sim 10^{-3}$ Earth mass, which is not large enough; and (2) it always seemed obvious that debris spalled off the Earth at less than the escape velocity would, after executing one geocentric orbit, return to its original geocentric radial distance, i.e., the surface of the Earth, and reaccrete.

Recently Hartmann and Davis (1975), Wetherill (1976, 1986), Greenberg (1979), and Hartmann (1986) have argued that the upper end of the size distribution of planetesimals during planetary accretion included much larger objects than those visualized by Safronov. According to this analysis the addition of an object of ~ 0.1 Earth mass to the Earth late in its accretion is to be expected, and is not *ad hoc* at all. Hartmann and Davis (1975) further pointed out that such a large collision might eject more than enough debris from the Earth to make a Moon, most of it Fe-depleted mantle material.

However, Hartmann and Davis did not address problem (2) above. Cameron and Ward (1976) first suggested a way out: An impact of the scale contemplated, between two bodies already hot and partially molten from the release of accretional and core-forming energy, would eject debris mostly in the form of vapor, not solids. The vapor would expand as it receded from the Earth, and gas pressure effects would continue to accelerate elements of the gas and also any solids entrained in them for some time after the material was initially dispatched from the point of impact. Because of this continued acceleration, the motion of the debris would not be strictly ballistic; it is to be expected that some of the debris would be accelerated into orbits whose perigees cleared the Earth. Cameron and Ward (1976) postulated an off-axis collision that would impart much of the early Earth's prograde angular momentum to it (Fig. 10), and most of the debris from such an impact would also move in a prograde sense. Thus after collisions and viscous interaction between elements of the ejecta placed in orbit had cancelled out most of its random motions, a disk of material would be left in prograde orbit about the Earth.

This theme has been developed further by Ward and Cameron (1978), Ringwood (1979), Cameron (1983, 1985, 1986), Thompson and Stevenson (1983), and Stevenson (1984, 1985). Major topics of discussion are the formation of the orbiting disk, its nature and evolution, and the source of observed similarities and differences in composition between Earth and Moon.

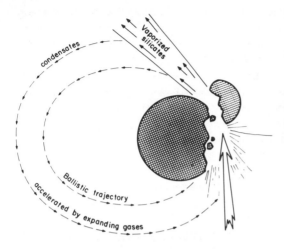

Fig. 10. The collisional ejection model: Expansion of ejected silicate vapor accelerates condensates into orbits with perigees that clear the Earth's surface (in contrast to the trajectories of solid debris knocked into simple ballistic orbits, which would reaccrete to the Earth).

Would the expanding-vapor model just sketched really place enough debris and enough angular momentum in geocentric orbit to make a Moon of? On the basis of a crude numerical simulation by Cameron (1985, 1986) and simple physical considerations cited by Stevenson (1984), the answer is yes. Success is favored by a large planetesimal (>0.1 Earth mass, i.e., Mars-size); a prograde off-axis impact point; and an impact velocity somewhat (~2 km/sec) above the escape velocity, which means a moderately eccentric (e ~0.2) preterrestrial orbit for the planetesimal. Cameron finds that most of the debris launched into geocentric orbit originated in the planetesimal, not the mantle of the Earth. Stevenson lists two physical effects in addition to vapor expansion that might have helped put a portion of the impact debris in orbit: the fact that the center of mass of the colliding system is significantly displaced from the center of mass of the Earth at the time of the collision; and the viscous transfer of energy and momentum in a broad jet of debris emanating from the impact.

Cameron and Ward (1976) assume the expanding vapor would begin to condense into solids soon after ejection, but Thompson and Stevenson (1983) argue that energy-dissipative processes (noted below) in the disk would keep it hot and largely vaporized during most of its existence. The disk is expected to behave qualitatively like the solar nebula: viscous interaction between differentially rotating annular elements of the disk cause it to spread, with angular momentum and some mass being transferred outward, the remaining mass inward. The spreading rate depends upon the effective viscosity of the disk, which controls the rate at which mechanical energy can be dissipated as heat. Ward and Cameron (1978) conclude that particulate disk material inside the Roche limit would try repeatedly to coalesce via gravitational instability, only to be torn apart by tidal forces. The "viscosity" of the disk would arise from the noncircular components of motion of particles as they moved into and out of these unsuccessful gravitational associations; this would determine the rate of energy dissipation in the disk. Normally a hot gaseous disk, such as Thompson and Stevenson (1983) envisage, would not be subject to such gravitational instability effects, because the tendency to instability is inversely related to the speed of sound in the gas, and the sound speed in hot gases is relatively high. However, Thompson and Stevenson point out that the sound speed in vapor-plus-solid or vapor-plus-liquid mixtures can be much lower than the value for pure vapors at the same temperature: As a result, the disk would have stayed hot, because wherever it became cool enough to condense extensively gravitational instability set in, the rate of viscous energy dissipation was turned up, and additional heat was generated. Consequently the disk could not have cooled and condensed fully until late in its spreading history.

As a result of this efficient energy-dissipation process and the high effective viscosity of the disk material, the disk is expected to expand beyond Earth's Roche limit and cool in ~100 yr. Once an important fraction of the disk material is outside the Roche limit, gravitational instability very promptly collects the condensed matter into Moonlets, and these soon coalesce to form the Moon. Because of the short time-scale of lunar accretion and the relatively high temperature of the solids being accreted, the Moon should be partly or wholly molten when it forms (Thompson and Stevenson, 1983).

This hypothesis is so new that its weaknesses have not yet become apparent. Much work needs to be done toward modelling the mechanical and thermal behavior of the hypothetical vapor/debris disk. No attempt has been made to understand the chemical fractionations that would result from formation of the Moon by collisional ejection. If Earth and the impactor had already separated cores, and if mostly mantle material was ejected and vaporized in the collision, this could account for the Fe-poor lunar composition. Also, Cameron (1985, 1986) argues that the Fe metal core of the impactor would be less decelerated during the collision than mantle material, because of its greater density; therefore the metal was less vaporized than silicates and was correspondingly underrepresented in the vapor disk. Another possibility is that the relatively high volatility of Fe (Fig. 3) prevented most of that element from being incorporated in the Moon along with the other volatile elements, even though it was present in the disk of vaporized debris. If the relatively eccentric planetesimal orbit favored by Cameron (1985, 1986) means that the planetesimal formed at a radial distance much different than 1 AU, we should worry about the close match in oxygen isotope compositions between Earth and Moon. Cameron postulates that the planetesimal was formed at ~1 AU and gravitationally perturbed into an eccentric orbit.

Undoubtedly collisional ejection would fractionate elements according to their volatility, which in principle would account for the volatile-depleted character of the Moon. However, there are limits on the degree to which volatility fractionation in the circumterrestrial disk can be invoked to rationalize the lunar composition, set by the fact that the rare Earth elements in the Moon have not been fractionated according to volatility (Taylor, 1986). Vapor fractionation appears to be limited to elements with condensation temperatures less than ~1100K. Kreutzberger *et al.* (1985) have shown that even at <1100 K the fractionation pattern is very difficult to understand.

3.5. Disintegrative capture

Öpik (1972) first showed that if an inviscid planetesimal in independent heliocentric orbit passed close enough to the Earth to be significantly decelerated by the tidal interaction (and hence potentially captured during a single passage), it would move inside Earth's Roche limit and would be tidally disrupted in the process. An interesting aspect of such an event, as Öpik pointed out, is that for a range of low-velocity encounters it would lead to geocentric capture of a portion of the debris of the disrupted planetesimal, even if tidal deceleration effects are ignored. Fragments derived from the inner (Earth-facing) side of the planetesimal would have been moving with the same velocity and energy as the center of mass of the planetesimal since (before disruption) they were bound together, yet the fact that they were closer to the center of the Earth would mean that the geocentric escape velocity applicable to them would be larger than the escape velocity for fragments derived from the center of mass or the outermost face of the planetesimal. For low-velocity encounters (asymptotic velocity <2 km/sec) this results in some debris from the inner side of the planetesimal being captured into geocentric orbits (Fig. 11). As much as 50% of the planetesimal debris can be captured, in the case of encounter at the parabolic velocity. It should be stressed that this material is captured because its position in the predisruption planetesimal constrains it to approach the Earth at an unnaturally low velocity, not because its motion is retarded by tides or any other dissipative process.

This concept was adopted by Wood and Mitler (1974) and Smith (1974), with the added observation that such a process would capture mostly crustal and mantle material from the inner face of a differentiated planetesimal; the bulk of the core material from such a system would be moving too fast to be captured, and would remain in heliocentric orbits. Thus the Fe-poor character of the Moon might be accounted for. Wood and Mitler (1974) proposed that many small planetesimals suffered this fate at the time when other such planetesimals were accreting to form the Earth, and that the integrated effect of many such disintegrative captures was to establish an Fe-poor PLS about the Earth. But this model suffers from the same angular momentum problem as coaccretion: There is no reason to expect substantially more planetesimals to be disrupted and partially captured while passing Earth in a prograde sense than in a retrograde sense.

Smith (1974) and Mitler (1975) postulated the disintegrative capture of just one planetesimal (again, Mars-size), which fortuitously encountered the Earth in a prograde direction. Clearly the conditions favorable for this disintegrative capture model and for collisional ejection (previous section) are almost identical. Smith (1974) and Mitler (1975) did not address the question of the rotation of the Earth itself. It may have been fortuitously the same as that of the PLS, but another possibility is that evolution of the PLS along the lines discussed in the last section caused

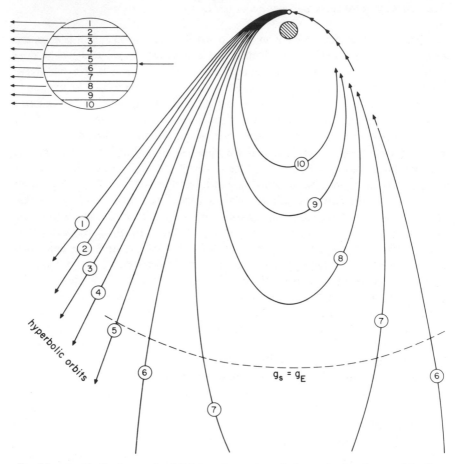

Fig. 11.   An idealized example of disintegrative capture. A planetesimal at the top of the fig-ure is tidally disrupted during passage at parabolic velocity through Earth's Roche zone. The subsequent trajectories of material from ten parallel slabs in the original planetesimal (inset, upper left) are followed. Debris from slabs 1–5 (farthest from Earth) is returned to heliocentric orbits; debris from slabs 6–10 is captured in geocentric orbits. In the case of encounters at greater than the parabolic velocity, a smaller number of slabs, those that pass closest to Earth, are retained in geocentric orbits. These would contain a disproportionately large amount of crustal and mantle material if a differentiated planetesimal was disrupted; most of the core material would return to heliocentric orbits. Figure from Wood and Mitler (1974).

a portion of it to lose angular momentum and migrate inward until it joined the Earth at Earth-skimming orbital velocity. This would spin up the Earth, and bring its rotation axis closer to that of the PLS. Whether it is possible for such a process to account for the early Earth's rotation rate is not known.

Advantages of the model are that it has the potential of accounting for the compositional differences between Earth and Moon, the Moon's prograde orbit, and the lunar magma ocean (if disintegrative capture occurred when planetary accretion was nearly completed, and the PLS promptly reconstituted itself as the Moon). The encounter postulated is as plausible as the one that would lead to collisional ejection; perhaps slightly more so, since the cross section for a disintegrative capture interaction (if it works at all) is probably larger than the cross section for collisional ejection.

The most serious liability of disintegrative capture is that it is not clear it would work. The process has been investigated in only a very approximate way. All the authors cited above except Öpik make the simplifying assumption that the passing planetesimal remains an integral sphere until it reaches perigee, whereupon it disrupts and fragments follow the trajectories dictated by their positions in the system at the time of disruption. In fact, the planetesimal would have been markedly distorted as it approached the Roche zone (Öpik, 1972; Kaula and Beachey, 1986), and disruption can have occurred before or after perigee passage, depending upon the circumstances of encounter. The above authors also ignore the continuing gravitational influence of the substance of the planetesimal after its disruption on the debris fragments as they recede from one another.

An apparently fatal objection to the model has been registered by Mizuno and Boss (1985), who numerically simulated the passage of viscous fluid planetesimals through Earth's Roche zone. These authors concluded that disruption would not occur at all, because the effective viscosity of the material of the passing planetesimal is so great and the tidal stresses are applied so briefly (< 2 hours) that the planetesimal would not have time to deform enough to disrupt.

4. A Summing Up

How the Earth's Moon was formed is still not known. Perhaps it will never be. But it became clear at the Kona conference that a major shift of confidence has occurred among lunar scientists toward the collisional ejection model. This did not occur because strong evidence was presented that the Moon was formed by this means, or even that it could have been: It happened because several independent investigators showed that coaccretion, the model that had been most widely accepted by lunar scientists (at least at a subconscious level), could not account for the angular momentum content of the Earth-Moon system. With this alternative seemingly removed, many in the lunar community turned to collisional ejection as the model that appeared most plausible among those remaining.

I have attempted to review the five major models of lunar formation in this paper. My personal editorial comments and judgments have crept in from time to time, as the reader will have noticed. By way of summarizing what 1 see as the strengths and weaknesses of the five models, I present a "Report Card" (Table 3). The line in parentheses symbolizes my doubt that similarities in trace element abundance patterns between the mantles of Earth and Moon are a reliable constraint. Perhaps the question of lunar origin should be systematically reexamined in another twenty years. An appropriate conference site would be the Moon.

TABLE 3.   A "Report Card"*: The Author's Opinion of How Well Five Models
of Lunar Origin Satisfy Seven Constraints.

	Intact Capture	Coaccretion	Earth Fission	Collisional Ejection	Disintegrative Capture
Lunar mass	B	B	D	I	B
Earth-Moon angular momentum	C	F	F	B	C
Volatile element depletion	C	C	B	B	C
Fe depletion	F	D	A	I	B
Oxygen isotopes	B	A	A	B	B
(Similarity of mantle trace element patterns)	(C)	(D)	(A)	(C)	(C)
Magma ocean	D	C	A	A	B
Physical plausibility	D–	C	F	I	F

*For readers unfamiliar with the U.S. educational system: A is the best grade; F (failing) is the worst; I (incomplete) means all assignments have not been completed and a grade cannot yet be awarded.

5.   Appendix

Computer simulation of earth accretion and capture of a protolunar swarm

In 1981, the author developed a computer program to simulate the accretion of the Earth and the accumulation of a protolunar swarm from a population of heliocentrically orbiting planetesimals. The model was based on the premise that the systematic accretion of a large number of relatively small planetesimals was involved, and no one of them affected the outcome greatly.

A characteristic eccentricity for the planetesimals was entered as input, as well as a size distribution and maximum mass for the planetesimals An embryonic Earth some fraction of its present mass was assumed at the outset. A series of planetesimals were then allowed to interact with the Earth until it grew to its present mass. Planetesimals that missed the Earth but passed through its Roche zone were assumed to have been tidally disrupted if they were larger than the critical mass beyond which their material strength was comparable to their gravitational cohesion, and the fraction of their mass and angular momentum that would have been captured in this situation was added to the PLS. A record was kept of the proportions of core and mantle material captured from hypothetically differentiated planetesimals, in the forlorn hope that the disintegrative capture model of lunar formation would be found to work. When material was disintegratively captured into the PLS in a sense opposite to its net rotation, the new mass was assumed to collide with a like amount of PLS material, cancelling the angular momenta of both and adding their mass to the Earth.

The simulated accretion proceeded in a series of cycles consisting of two stages: First, values of e and a were chosen; then a number of planetesimals were released with this e and a, but different values of the angle θ (Fig. 7). The simulation was carried out in two dimensions, but the results were corrected for the effect of planetesimals whose inclinations caused them to approach the Earth-Moon system out of the plane of the ecliptic.

The eccentricity for each set (band) of planetesimals was randomly chosen from a range of values centered on and weighted in favor of the value of characteristic eccentricity that was entered as input. The range of values of mean distance for which an object with this e would cross Earth's orbit ($a = 1$ AU, $e = 0$) was reckoned, and a was randomly chosen from this range ± 0.05 AU. The extra ± 0.05 AU of range was added to include the effect of planetesimal orbits that do not normally intersect Earth's orbit, but which under the gravitational influence of the Earth might be perturbed into interaction with it. For orbits that intersect Earth's without perturbation, a random choice was made of whether the encounter would occur as the planetesimal was crossing into or out of Earth's orbit.

A series of test points were released suitably far away from the Earth in orbits having these values of a and e but different angles θ between their lines of apsides and the position of the Earth at the time of release, and by a process of trial and error the range in θ was found that would result in collision with the Earth or capture into its PLS. A set of planetesimals was then released with the established values of a and e, and with values of θ randomly chosen in this fertile range. The number of planetesimals in the set was proportional to the angular width of the allowed range in θ. The planetesimal size was randomly chosen within the weighted size distribution previously specified. Planetesimal trajectories were integrated in double precision, using an Adams-Bashforth-Moulton predictor/corrector technique.

Several dozen runs were carried out using different input parameters, experimental biasing factors, and increasingly complex program modifications that were intended to make the simulation more realistic. The results were surprising. None of the runs created an Earth and PLS with as much as half the angular momentum of the present Earth-Moon system, and about half of the runs produced Earth-Moon systems with retrograde rotatation.

The most successful run for which data survives had a characteristic eccentricity of 0.04. It accreted the Earth from a 0.25-Earth-mass nucleus to its present mass, producing an Earth and PLS with 1.37×10^{41} rad g cm^2/sec of prograde angular momentum, 40% of the value for the present Earth-Moon system. The mass of material in the PLS was 3.6×10^{25} g, 49% of the present lunar mass. The record of growth of the system is shown in Fig. A1. 1366 planetesimals, having masses 10^{-5}–10^{-3} the current mass of the system, joined the Earth and PLS in 388 sets. The simulation had to be biased to achieve even this much success: All planetesimals passing through the Roche zone, regardless of size, were assumed to be disrupted and partially captured; this maximizes the mass of the PLS and hence the angular momentum of the system.

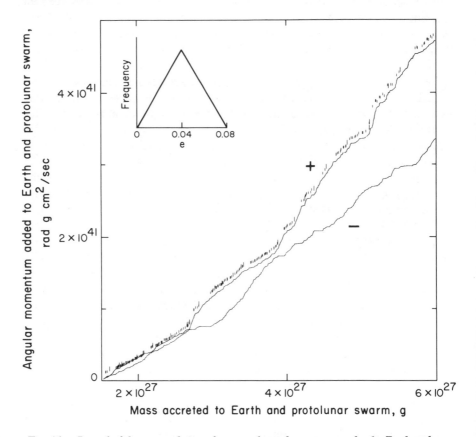

Fig. A1. Record of the accumulation of mass and angular momentum by the Earth and pro-tolunar swarm in the simulation described in the text. Assumed distribution of eccentricities among arriving planetesimals is shown at upper left. Two curves are shown, one for positive and one for negative angular momentum accumulated; net angular momentum of the system is the vertical difference between the curves. Small tick marks divide the mass increments due to the 388 sets (bands) of planetesimals accreted and captured.

Inspection of Fig. A1 suggests that the Earth-Moon system in this run acquired its record amount of angular momentum because the roll of the dice favored it this time, not because some systematic physical effect inexorably converted the planetesimals' orbital angular momentum into rotational angular momentum of the Earth-Moon system. The curves of positive and negative angular momentum do not diverge steadily: For the first 20% of their lengths, slightly more negative than positive angular momentum is being acquired by the system; at 40% of their lengths, the curves almost touch or cross again; beyond this they diverge, but along paths so erratic as to suggest they might rejoin if the system continued to grow.

I abandoned this effort when it began to seem clear that, as Safronov (1966) has argued all along, the rotation (and obliquity) of the Earth must be due to the off-center accretion of a very small number of larger planetesimals, instead of some systematic effect associated with the accretion of a large number of small planetesimals.

Acknowledgments. I am grateful to a number of people for their critical reading of this paper, and the helpful suggestions they made: A. G. W. Cameron, W. Cameron, C. R. Chapman, M. J. Drake, R. J. Greenberg, A. W. Harris, L. L. Hood, W. M. Kaula, J. Longhi, H. E. Newsom, J. A. O'Keefe, G. Ryder, S. F. Singer, and S. R. Taylor.

6. References

Anders E. (1978) Procrustean science: Indigenous siderophiles in the lunar highlands, according to Delano and Ringwood. *Proc. Lunar Planet. Sci. Conf. 9th,* pp. 161–184.

Anderson D. L. (1972) The origin of the moon. *Nature, 239,* 263–265.

Bills B. G. and Ferrari A. J. (1977) A harmonic analysis of lunar topography. *Icarus, 31,* 244–259.

Binder A. B. (1974) On the origin of the moon by rotational fission. *The Moon, 11,* 53–76.

Binder A. B. (1978) On fission and the devolatilization of a moon of fission origin. *Earth Planet. Sci. Lett., 41,* 381–385.

Binder A. B. (1986) The binary fission origin of the Moon, this volume.

Boss A. P. and Peale S. J. (1986) Dynamical constraints on the origin of the Moon, this volume.

Brush S. G. (1986) The early history of selenogony, this volume.

Burns J. A. (1975) The angular momenta of solar system bodies: implications for asteroid strength. *Icarus, 25,* 545–554.

Burns J. A. (1977) Orbital evolution. In *Planetary Satellites* (J. A. Burns, ed.), pp. 113–156. Univ. of Arizona, Tucson.

Cameron A. G. W. (1972) Orbital eccentricity of Mercury and the origin of the moon. *Nature, 240,* 299–300.

Cameron A. G. W. (1983) Origin of the atmospheres of the terrestrial planets. *Icarus, 56,* 195–201.

Cameron A. G. W. (1985) Formation of the prelunar accretion disk. *Icarus,* in press.

Cameron A. G. W. (1986) The impact theory for origin of the Moon, this volume.

Cameron A. G. W. and Ward W. R. (1976) The origin of the moon (abstract). In *Lunar Science VII,* pp. 120–122. The Lunar Science Institute, Houston.

Clark S. P., Jr., Turcotte D. L., and Nordmann J. C. (1975) Accretional capture of the moon. *Nature, 258,* 219–220.

Clayton R. N. and Mayeda T. K. (1975) Genetic relations between the moon and meteorites. *Proc. Lunar Sci. Conf. 6th,* pp. 1761–1769.

Clayton R. N., Onuma N., and Mayeda T. K. (1976) A classification of meteorites based on oxygen isotopes. *Earth Planet. Sci. Lett., 30*, 10–18.

Darwin G. H. (1879) On the precession of a viscous spheroid and on the remote history of the earth. *Phil. Trans. Roy. Soc. Part II, 170*, 447–530.

Deer W. A., Howie R. A., and Zussman J. (1962) *Rock-forming Minerals*, vols. 1 and 2. Longman, London. 333 pp. and 379 pp.

Delano J. W. and Ringwood A. E. (1978a) Indigenous abundances of siderophile elements in the lunar highlands: Implications for the origin of the moon. *Moon and Planets, 18*, 385–425.

Delano J. W. and Ringwood A. E. (1978b) Siderophile elements in the lunar highlands: Nature of the indigenous component and implications for the origin of the moon. *Proc. Lunar Planet. Sci. Conf. 9th*, pp. 111–159.

Drake M. J. (1983) Geochemical constraints on the origin of the moon. *Geochim. Cosmochim. Acta, 47*, 1759–1767.

Drake M. J. (1986) Is lunar bulk material similar to Earth's mantle?, this volume.

Durisen R. H. and Gingold R. A. (1986) Numerical simulations of fission, this volume.

Durisen R. H. and Scott E. H. (1984) Implications of recent numerical calculations for the fission theory of the origin of the moon. *Icarus, 58*, 153–158.

Giuli R. T. (1968) On the rotation of the earth produced by gravitational accretion of particles. *Icarus, 8*, 301–323.

Greenberg R. (1979) Growth of large, late-stage planetesimals. *Icarus, 39*, 141–150.

Grieve R. A. F. (1980) Cratering in the lunar highlands: Some problems with the process, record and effects. In *Proceedings of the Conference on the Lunar Highlands Crust* (J. J. Papike and R. B. Merrill, eds.), pp. 173–196. Lunar and Planetary Institute, Houston.

Harris A. W. (1977) An analytical theory of planetary rotation rates. *Icarus, 31*, 168–174.

Harris A. W. (1978) Satellite formation, II. *Icarus, 34*, 128–145.

Harris A. W. and Kaula W. M. (1975) A co-accretional model of satellite formation. *Icarus, 24*, 516–524.

Hartmann W. K. (1986) Moon origin: The impact-trigger hypothesis, this volume.

Hartmann W. K. and Davis D. R. (1975) Satellite-sized planetesimals and lunar origin. *Icarus, 24*, 504–515.

Hartmann W. K. and Larson S. M. (1967) Angular momenta of planetary bodies. *Icarus, 7*, 257–260.

Hashimoto A. (1983) Evaporation metamorphism in the early solar nebula—evaporation experiments on the melt $FeO-MgO-SiO_2-CaO-Al_2O_3$ and chemical fractionations of primitive materials. *Geochem. J., 17*, 111–145.

Heppenheimer T. A. and Porco C. (1977) New contributions to the theory of capture. *Icarus, 30*, 385–401.

Hood L. L. (1986) Geophysical constraints on the lunar interior, this volume.

Horedt Gp. (1976) Pre-capture orbits of the moon. *Moon and Planets, 15*, 439–443.

Jeffreys H. (1930) The resonance theory of the origin of the moon. *Mon. Not. Roy. Astron. Soc., 91*, 169–174.

Kant I. (1755) *Allgemeine Naturgeschichte und Theorie des Himmels*.

Kaula W. M. (1979) Thermal evolution of Earth and Moon growing by planetesimal impacts. *J. Geophys. Res., 84*, 999–1008.

Kaula W. M. and Beachey A. E. (1986) Mechanical models of close approaches and collisions of large protoplanets, this volume.

Kaula W. M. and Harris A. W. (1973) Dynamically plausible hypotheses of lunar origin. *Nature, 245*, 367–369.

Kaula W. M., Schubert G., Lingenfelter R. E., Sjogren W. L., and Wollenhaupt W. R. (1974) Apollo laser altimetry and inferences as to lunar structure. *Proc. Lunar Sci. Conf. 5th*, pp. 3049–3058.

Kopal Z. and Lyttleton R. A. (1963) On the elliptic case of the restricted problem of three bodies and the remote history of the earth-moon system. *Icarus, 1*, 455–458.

Kreutzberger M. E., Drake M. J., and Jones J. H. (1985) Origin of the earth's moon: constraints from alkali volatile trace elements. *Geochim. Cosmochim. Acta*, in press.

Lyttleton R. A. (1972) On the formation of planets from a solar nebula. *Mon. Not. Roy. Astron. Soc., 158*, 463–483.

MacDonald G. J. F. (1966) Origin of the moon: dynamical considerations. In *The Earth-Moon System* (B. G. Marsden and A. G. W. Cameron, eds.), pp. 165–209. Plenum, New York.

Matsui T. and Abe Y. (1986) Origin of the moon and its early thermal evolution, this volume.

Mitler H. E. (1975) Formation of an iron-poor moon by partial capture, or: Yet another exotic theory of lunar origin. *Icarus, 24*, 256–268.

Mizuno H. and Boss A. P. (1985) Tidal disruption of dissipative planetesimals. *Icarus, 63*, 109–133.

Morgan J. W., Hertogen J., and Anders E. (1978) The Moon: composition determined by nebular processes. *Moon and Planets, 18*, 465–478.

Nakazawa K., Komuro T., and Hayashi C. (1983) Origin of the moon—capture by gas drag of the earth's primordial atmosphere. *Moon and Planets, 28*, 311–327.

Newsom H. E. (1984) The lunar core and the origin of the moon. *EOS (Trans. Amer. Geophys. Union), 65*, 369–370.

Newsom H. E. (1986) Constraints on the origin of the Moon from the abundance of molybdenum and other siderophile elements, this volume.

O'Keefe J. A. (1966) The origin of the moon and the core of the earth. In *The Earth-Moon System* (B. G. Marsden and A. G. W. Cameron, eds.), pp. 224–233. Plenum, New York.

O'Keefe J. A. (1969) Origin of the moon. *J. Geophys. Res., 74*, 2758–2767.

O'Keefe J. A. (1972) The origin of the moon: theories involving joint formation with the earth. *Astrophys. Space Sci., 16*, 201–211.

O'Keefe J. A. and Sullivan E. C. (1978) Fission origin of the moon: cause and timing. *Icarus, 35*, 272–283.

Öpik E. J. (1972) Comments on lunar origin. *Irish Astron. J., 10*, 190–238.

Papanastassiou D. A. and Wasserburg G. J. (1976) Early lunar differentiates and lunar initial [87]Sr/[86]Sr (abstract). In *Lunar Science VII*, pp. 665–667. The Lunar Science Institute, Houston.

Rammensee W. and Wänke H. (1977) On the partition coefficient of tungsten between metal and silicate and its bearing on the origin of the moon. *Proc. Lunar Sci. Conf. 8th*, pp. 399–409.

Ringwood A. E. (1960) Some aspects of the thermal evolution of the earth. *Geochim. Cosmochim. Acta, 20*, 241–259.

Ringwood A. E. (1979) *Origin of the Earth and Moon*. Springer-Verlag, New York. 295 pp.

Ringwood A. E. (1986) Composition and origin of the Moon, this volume.

Ringwood A. E. and Kesson S. E. (1977) Basaltic magmatism and the bulk composition of the moon, II. Siderophile and volatile elements in the moon, earth, and chondrites: implications for lunar origin. *The Moon, 16*, 425–464.

Ringwood A. E. and Seifert S. (1986) Nickel-cobalt abundance systematics and their bearing on lunar origin, this volume.

Ruskol E. L. (1960) The origin of the moon. I. Formation of a swarm of bodies around the earth. *Soviet Astronomy–AJ, 4*, 657–668.

Ruskol E. L. (1963) On the origin of the moon. II. The growth of the moon in the circumterrestrial swarm of satellites. *Soviet Astronomy–AJ, 7*, 221–227.

Ruskol E. L. (1972a) The origin of the moon. III. Some aspects of the dynamics of the circumterrestrial swarm. *Soviet Astronomy–AJ, 15*, 646–654.

Ruskol E. L. (1972b) On the possible differences in the bulk chemical composition of the earth and the moon forming in the circumterrestrial swarm. In *The Moon* (S. K. Runcorn and H. C. Urey, eds.), pp. 426–428. D. Reidel, Dordrecht.

Ruskol E. L. (1973) On the model of the accumulation of the moon compatible with the data on the composition and the age of lunar rocks. *The Moon, 6,* 190–201.

Safronov V. S. (1966) Sizes of the largest bodies falling onto the planets during their formation. *Soviet Astronomy–AJ, 9,* 987–991.

Schmidt O. Yu. (1959) *A Theory of the Origin of the Earth.* Lawrence and Wishart, London. 139 pp.

Singer S. F. (1968) The origin of the moon and geophysical consequences. *Geophys. J. Roy. Astron. Soc., 15,* 205–226.

Singer S. F. (1986) Origin of the Moon by capture, this volume.

Smith J. V. (1974) Origin of the moon by disintegrative capture with chemical differentiation followed by sequential accretion (abstract). In *Lunar Science V,* pp. 718–720. The Lunar Science Institute, Houston.

Smith J. V., Anderson A. T., Newton R. C., Olsen E. J., Wyllie P. J., Crewe A. V., Isaacson M. S., and Johnson D. (1970) Petrologic history of the moon inferred from petrography, mineralogy, and petrogenesis of Apollo 11 rocks. *Proc. Apollo 11 Lunar Sci. Conf.,* pp. 897–925.

Spudis P. D. (1983) The excavation of lunar multi-ringed basins: Additional results for nearside basins (abstract). In *Lunar and Planetary Science XIV,* pp. 735–736. Lunar and Planetary Institute, Houston.

Stevenson D. J. (1984) Lunar formation from impact on the earth: is it possible? (abstract). In *Papers Presented to the Conference on the Origin of the Moon,* p. 60. Lunar and Planetary Institute, Houston.

Stevenson D. J. (1985) Implications of very large impacts for Earth accretion and lunar formation (abstract). In *Lunar and Planetary Science XVI,* pp. 819–820. Lunar and Planetary Institute, Houston.

Swindle T. D., Caffee M. W., Hohenberg C. M., and Taylor S. R. (1986) I-Pu-Xe dating and the relative ages of the Earth and Moon, this volume.

Szebehely V. and McKenzie R. (1977) Stability of the sun-earth-moon system. *Astron. J., 82,* 303–305.

Taylor S. R. (1982) *Planetary Science: A Lunar Perspective.* Lunar and Planetary Institute, Houston. 481 pp.

Taylor S. R. (1986) The origin of the Moon: Geochemical considerations, this volume.

Thompson A. C. and Stevenson D. J. (1983) Two-phase gravitational instabilities in thin disks with application to the origin of the moon (abstract). In *Lunar and Planetary Science XIV,* pp. 787–788. Lunar and Planetary Institute, Houston.

Turcotte D. L. and Kellogg L. H. (1986) Implications of isotope data for the origin of the Moon, this volume.

Urey H. C. (1952) *The Planets.* Yale University, New Haven. 245 pp.

Urey H. C. (1966) The capture hypothesis of the origin of the moon. In *The Earth-Moon System* (B. G. Marsden and A. G. W. Cameron, eds.), pp. 210–212. Plenum, New York.

Walker D. (1983) Lunar and terrestrial crust formation. *Proc. Lunar Planet. Sci. Conf. 14th,* in *J. Geophys. Res., 88,* B17–B26.

Ward W. R. and Cameron A. G. W. (1978) Disc evolution within the Roche limit (abstract). In *Lunar and Planetary Science IX,* pp. 1205–1207. Lunar and Planetary Institute, Houston.

Wänke H. and Dreibus G. (1986) Geochemical evidence for formation of the Moon by impact induced fission of the proto-Earth, this volume.

Wänke H., Baddenhausen H., Blum K., Cendales M., Dreibus G., Hofmeister H., Kruse H., Jagoutz E., Palme C., Spettel B., Thacker R., and Vilcsek E. (1977) On the chemistry of lunar samples and achondrites. Primary matter in the lunar highlands: a re-evaluation. *Proc. Lunar Sci. Conf. 8th,* pp. 2191–2213.

Wänke H., Dreibus G., and Palme H. (1978) Primary matter in the lunar highlands: The case of the siderophile elements. *Proc. Lunar Planet. Sci. Conf. 9th,* pp. 83–110.

Ward W. R. and Cameron A. G. W. (1978) Disc evolution within the Roche limit (abstract). In *Lunar and Planetary Science IX,* pp. 1205–1207. Lunar and Planetary Institute, Houston.

Warren P. H. (1985) The magma ocean concept and lunar evolution. *Ann. Rev. Earth Planet. Sci.,* *13,* 201–240.

Weidenschilling S. J., Chapman C. R., Herbert F., Davis D. R., Drake M. J., Jones J., and Hartmann W. K. (1986) Origin of the Moon from a circumterrestrial disk, this volume.

Wetherill G. W. (1975) Possible slow accretion of the moon and its thermal and petrological consequences (abstract). In *Papers Presented to the Conference on Origins of Mare Basalts and their Implications for Lunar Evolution,* pp. 184–188. The Lunar Science Institute, Houston.

Wetherill G. W. (1976) The role of large bodies in the formation of the earth. *Proc. Lunar Sci. Conf. 7th,* pp. 3245–3257.

Wetherill G. W. (1986) Accumulation of the terrestrial planets and implications concerning lunar origin, this volume.

Winters R. R. and Malcuit R. J. (1977) The lunar capture hypothesis revisited. *Moon and Planets,* *17,* 353–358.

Wise D. U. (1963) An origin of the moon by rotational fission during formation of the earth's core. *J. Geophys. Res., 68,* 1547–1554.

Wise D. U. (1969) Origin of the moon from the earth: Some new mechanisms and comparisons. *J. Geophys. Res., 74,* 6034–6045.

Wood J. A. and Mitler H. E. (1974) Origin of the moon by a modified capture mechanism, or: Half a loaf is better than a whole one (abstract). In *Lunar Science V,* pp. 851–853. The Lunar Science Institute, Houston.

Wood J. A., Dickey J. S., Marvin U. B., and Powell B. N. (1970) Lunar anorthosites and a geophysical model of the moon. *Proc. Apollo 11 Lunar Sci. Conf.,* pp. 965–988.

II.   Dynamical Constraints

Dynamical Constraints on the Origin of the Moon

A. P. BOSS

Department of Terrestrial Magnetism, Carnegie Institution of Washington, 5241 Broad Branch Road, N.W., Washington, DC 20015

S. J. PEALE

Department of Physics, University of California, Santa Barbara, Santa Barbara, CA 93106

Dynamical studies dealing with the origin of the Moon are described and are used to try to eliminate dynamically impossible or implausible theories of lunar origin. The origin of the Moon is discussed within the context of the general theory of terrestrial planet formation by accumulation of planetesimals. The past evolution of the lunar orbit is of little use in differentiating between the theories, primarily because of the inherent uncertainty in a number of model parameters and assumptions. The various theories that have been proposed are divided into six categories. Rotational fission and disintegrative capture appear to be dynamically impossible for viscous protoplanets, while precipitation fission (precipitation of Moon-forming material from a hot, extended primordial atmosphere of volatilized silicates), intact capture, and binary accretion appear to be dynamically implausible. Precipitation fission and binary accretion suffer chiefly from having insufficient angular momentum to form the Moon, while intact capture requires forming the Moon very close to the Earth without encountering any perturbations prior to capture. The only mechanism proposed so far that is apparently not ruled out by dynamical constraints and that also seems the most plausible involves formation of the Moon following a giant impact that ejects portions of the differentiated Earth's mantle and parts of the impacting body into circumterrestrial orbit. The Moon must have accreted subsequently from this circumterrestrial disk. The giant impact model contains elements of several of the other models and appears to be dynamically consistent with the absence of major satellites for the other terrestrial planets. While the giant impact mechanism for forming the Moon thus emerges as the theory with the least number of obvious flaws, it should be emphasized that the model is relatively new and has not been extensively developed nor thoroughly criticized. Much further work must be done to learn whether the giant impact mechanism for lunar formation can be made into a rigorous theory.

1. Introduction

G. H. Darwin initiated the dynamical study of the origin of the Moon in a series of papers published beginning in 1879. Darwin's analysis of the evolution of the lunar orbit led him to conclude that the Moon had once been very close to the Earth, and hence he suggested that the Moon formed from the outer layers of the Earth. In the century since Darwin's seminal work, a number of other origins for the Moon have been proposed and debated at length, but seldom have any of the competing theories been considered to be definitively ruled out. Thus the number of competing theories has been a monotonically increasing function of time. On the other hand, some workers have concluded that all theories of lunar origin are implausible. Recent dynamical work has helped to dispel this haze of uncertainty surrounding the Moon's formation by suggesting a promising new variant on the older theories of origin and by adding decisive reasons for dismissing some of the older theories. The result is that only one of the possibilities for lunar origin proposed so far—that the Moon resulted from a giant impact on the early Earth—appears to satisfy the criteria of dynamical possibility and plausibility. However, the reader is cautioned that the theory of this scenario is embryonic and may not survive the scrutiny it will receive in the next few years. Previous reviews of dynamical (Kaula, 1971; Kaula and Harris, 1975) and other aspects (Kaula, 1977; Wood, 1977; Ringwood, 1979; Taylor, 1982; Brush, 1982) of lunar formation have reached conclusions that are similar but not identical to the conclusion of this review.

We will describe the dynamical constraints on the various theories of lunar origin proposed to date. We interpret this topic to include studies of orbital evolution both prior to and after lunar formation, rotational evolution of protoplanetary bodies, and the orbital and fluid dynamical processes involved in the formation of the terrestrial planets in general. The results of these dynamical studies may properly be termed "constraints" because of the mathematical certainty that usually accompanies such work, provided that the assumptions used in constructing the solution are appropriate. Geophysical and geochemical evidence, while potentially of equal or even greater importance, will not be considered here. We shall see that the dynamical constraints alone appear successful at eliminating most of the proposed models of lunar origin.

Two temporally different approaches may be taken in considering the formation of the Moon. In the most general of the two approaches, the formation of the terrestrial planets is considered as an initial value problem, where plausible but assumed initial conditions are evolved forward in time to learn if the terrestrial planets, including the Moon, result. Various subproblems related more directly to the formation of the Moon may then be studied within the framework of the general theory. The other approach is more specific to the Earth-Moon system and involves integrating the lunar orbit backward in time to learn as much as possible about where the Moon may have been early in its history. These two approaches are described in Sections 2 and 3 of this review. The history of the lunar orbit will be seen to be of little use in sorting out the possibilities for lunar formation, whereas the general theory of planetary formation seems to provide the necessary framework in which we can hope to understand the Moon's origin.

In Sections 4–6, the detailed models for lunar formation are described along with the dynamical studies that in most cases eliminate each model. The different models of lunar origin have been grouped into six general categories in order to accentuate their common features and provide a logical structure to this review. The association of a particular model with just one of these categories is a convenient simplification; some models could fit into several categories. The six categories are in essence subcategories of the three classical models (fission, capture, and binary accretion): (1) rotational fission, (2) precipitation fission, (3) intact capture, (4) disintegrative capture, (5) binary accretion, and (6) giant impact accretion.

2. Terrestrial Planet Formation

Safronov (1969) and Wetherill (1980) have provided detailed reviews of the general theory of the formation of the terrestrial planets. In this section we will give a short sketch of the theory.

The most fundamental assumption of terrestrial planet formation is that it occurred as a natural part of the formation of the sun and the rest of the solar system. A rotating, cold, dense interstellar cloud began about 4.6×10^9 years ago to contract and collapse because of self-gravity. The low angular momentum matter in the cloud collapsed within about 10^5 years to form a single, central protosun, while the high angular momentum matter collapsed to form a bar-like disk surrounding the protosun (Boss, 1985a). While the subsequent evolution of the protosun and surrounding nebula in this phase is not yet known, the protosun must have contracted to a relatively small size and entered a phase of pre-main-sequence evolution similar to other stars with the same mass. The sun is thereafter considered to be effectively a point mass gravitationally stabilizing the surrounding disk of gas and dust grains. This disk constitutes the solar nebula from which the planets were formed. The solar nebula evolves essentially independently of the sun, except for the possibility of the sun having experienced a phase with a violent stellar wind (T Tauri wind) that may have been responsible for sweeping residual gas out of the solar nebula. The timing of this putative event is uncertain, but, based on estimated ages of T Tauri stars, probably occurred during the first 10^6 years of pre-main-sequence evolution.

We will further assume that the terrestrial planets formed by accumulation of dust grains rather than by gravitational collapse of the gas in the solar nebula. Once a quasiequilibrium solar nebula is formed the gas and dust grains begin to separate, because while the gas is supported by thermal pressure and forms a relatively thick disk, the dust grains are largely unaffected by the gas pressure gradient and begin to sediment to the midplane of the nebula, at least in the later stages of nebular mass accumulation, when convectively driven turbulence ceases (Weidenschilling,

1984c). At the same time the dust grains begin the process of planetary accumulation by growing through collisions that result in coagulation of the grains. At this stage the grains are held together by electromagnetic (Van der Waals) forces between the molecules. In regions of the nebula that are not turbulent, the dust grains can grow and sediment to form a thin dust layer within about 10^3 years (e.g., Nakagawa et al., 1981).

When the dust layer becomes sufficiently compact, it becomes gravitationally unstable to radial wave perturbations (e.g., Safronov, 1969) and very rapidly forms into bodies up to 0.1 km in diameter. This first generation of objects is also gravitationally unstable and accumulates into clusters that collapse on time scales of a few thousand years, determined by the rate of gas drag dissipation of internal rotation and random kinetic energies. The second generation planetesimals are of the order of 5 km in size (Goldreich and Ward, 1973). These bodies are relatively closely packed in the midplane of the solar nebula and hence their rate of collisions resembles that in a very cold, ideal gas. Because of the small eccentricity of the planetesimals during this phase and their relatively close packing, the accumulation process can be modeled by a statistical theory of particles in a box (Safronov, 1969). The collisions lead to accumulation, because the kilometer-sized bodies are so massive that gravitational forces are large enough to hold bodies together following gentle collisions.

Safronov (1969) found that accumulation led to a mass distribution with most of the mass concentrated in the largest bodies and with a very sharp dropoff at the high mass end. Safronov's calculations assumed that these larger planetesimals are unaffected by gas drag; Nakagawa et al. (1983) studied accumulation in the presence of the gaseous component of the solar nebula, assuming that it had not yet been dissipated by a T Tauri phase of the sun, and found results similar to those of Safronov. A different result was found by Greenberg et al. (1978), who proposed that runaway growth of one planetesimal could occur. Starting with a swarm of 10^{12} bodies 1 km in diameter, within about 10^4 years bodies up to 10^{24} g (500 km radius) accumulate by this process, with the bulk of the mass residing in <10-km-size bodies (Greenberg et al., 1978). Calculations by Greenberg (1980) and Weidenschilling suggest that the Greenberg et al. (1978) scenario is relatively unaffected by the inclusion of gas drag. Whether or not runaway accretion occurs depends in large part on the validity of the assumed gravitational cross-sections as a function of relative velocity, and subsequent work suggests that the gravitational cross-section is insufficient to result in runaway growth to planet-sized bodies (Wetherill and Cox, 1982, 1984; Wetherill, 1986). At any rate, this phase terminates when all the particles in the "box" are accumulated. If the phase studied by Greenberg et al. (1978) terminates with the formation of a 10^{24} g body in each box, each box would only be about 10^{-4} AU in radial extent. Larger boxes would produce larger bodies. Roughly 10^4 bodies of mass 10^{24} g would be sufficient to form the terrestrial planets.

In the subsequent stages of the accumulation process, the fact that the planetesimals are in orbit around the sun must be included, because at this point the largest bodies are on isolated, nearly circular orbits and require significant variations in radius in order to collide (Cox and Lewis, 1980; Wetherill, 1980). In this relatively loosely packed stage, the mutual gravitational perturbations due to close encounters between the planetesimals on heliocentric orbits result in increasingly eccentric orbits, radial variations, and hence further collisions and growth. Calculations of the evolution in this phase, assuming the absence of significant gas drag, show that a swarm of 500 bodies of mass 2.5×10^{25} g, or a range between 5.7×10^{24} to 1.1×10^{26} g based on the results of Safronov (1969) and Nakagawa et al. (1983), initially distributed in semi-major axis so as to contain orbital angular momentum equal to that of the terrestrial planets, will indeed evolve to produce a small number of planets with a strong resemblance to the terrestrial planets of our solar system (Wetherill, 1985, 1986).

A key factor in the success of the accumulation process is that the relative velocities (or equivalently the orbital eccentricities) between planetesimals must not become so large as to result in fragmentation following collisions rather than accumulation, or so small that the orbits no longer intersect and the accumulation process is halted prematurely. Safronov (1969) and Wetherill (1980) indicate that the dynamics of the accumulation process regulate the relative velocities of the swarm to intermediate values that avoid both of these extremes.

The final spin angular momenta of the terrestrial planets result from the sum of contributions of the constituent impacting objects. The angular momentum of the impacting planetesimal is shared by the spin and orbital angular momenta of the accumulating planet, the partition for each planetesimal being determined by the details of the collision. Hence, the spin magnitude and direction both do a random walk as the planet accumulates (Safronov, 1969). For eccentric orbits of the colliding planetesimals a prograde spin of the planets results (Giull, 1968). Even the specific angular momentum of the Earth-Moon system can be attained as a maximum, but only with such small planetesimal eccentricities that only a relatively small zone around the embryo planet could be accreted (Harris, 1977b; Harris and Ward, 1982). The difficulty in obtaining sufficient spin angular momentum from the ordered accretion of small planetesimals leads to the supposition that the primordial spins of the terrestrial planets are essentially the stochastic result of the last few large impacts—perhaps deminated by the last such impact. As pointed out by Hartmann (1984) and others, such stochastic events must have occurred in the formation of the terrestrial planets, and hence their inclusion should not be considered a liability to the theory.

The final stage of growth requires on the order of 10^7–10^8 years to produce bodies the size of the terrestrial planets, regardless of whether or not a dynamically important amount of gas is present (Safronov, 1969; Wetherill, 1980; Nakagawa et al., 1983). Because this time scale is significantly longer than the inferred ages of T Tauri stars, if a strong solar wind was responsible for the removal of the residual gas from the solar nebula, this process probably occurred prior to the final stages of the accumulation of the terrestrial planets, so that the theory of gas-free accumulation should be appropriate.

Cameron (1985a) has proposed a very different early history of the inner solar nebula where much of the mass of a terrestrial planet results as a core of a giant gaseous protoplanet whose gaseous envelope is evaporated during a hot, dissipative phase in the nebular gas. However, the later stages of accretion in Cameron's model are essentially identical to that described above, where relatively few large bodies are accreted in a cold, gas-free environment.

3. Evolution of the Lunar Orbit

Darwin (1879, 1880) pioneered the quantitative study of the lunar orbit. The mechanism that drives the evolution is illustrated in Fig. 1. The gravitational forces of the Moon raise tidal bulges on the Earth that in the absence of rotation would be aligned with the Moon and the center of the Earth. This occurs because the gravitational acceleration of the Moon on the Earth falls off as the square of the distance, so that the nearside of the Earth is accelerated more toward the Moon than the center, while the center is accelerated more than the farside of the Earth. The net result is to produce a differential acceleration across the Earth and for

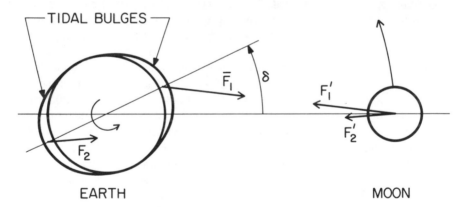

Fig. 1. Schematic of tidal bulges raised on the Earth by the differential gravitational acceleration of the Moon. Because the Earth rotates faster than the Moon revolves and dissipation leads to a time lag in the response of the Earth to the tide raising potential (i.e., high tide occurs after the Moon has passed overhead), the tidal bulge leads the Moon by a phase angle δ. F_1 and F_2 are the lunar attractions for the near and far bulges, respectively, and F_1' and F_2' are the corresponding equal and opposite reactions on the Moon. Since $|F_1| > |F_2|$ a net torque retards the Earth's spin and increases the orbital angular momentum of the Moon, thereby increasing the Earth-Moon separation.

the Earth to respond by deforming. Because the Moon orbits around the Earth at a rate slower than the Earth rotates, the Moon falls behind as the Earth rotates underneath it. If the Earth were perfectly elastic, the tidal bulges would respond to the changing location of the Moon by precisely following it across the sky. Then the phase angle δ in Fig. 1 would be zero. Because the Earth is not perfectly elastic, but instead dissipates energy as the tidal bulges try to follow the Moon, the bulges fail to keep up with the Moon, and the phase angle is nonzero (δ is about $0.6°$ for the solid part of the Earth; Lambeck, 1980). Because the closer displaced tidal bulge will be pulled stronger than the farther bulge, a net torque results that reduces the spin rate of the Earth as shown by Fig. 1. By conservation of angular momentum, this same gravitational interaction must increase the orbital angular momentum of the Moon, which drives it further from the Earth. This simple picture is considerably complicated on the real Earth by the fluid oceans whose periods of natural oscillation in their basins are comparable to the fundamental tidal period near 12 hours. The net effect is qualitatively the same, however, and angular momentum is secularly transferred from the Earth's spin to the orbital motion of the Moon. Other combinations of spin and orbital periods and planet/satellite mass ratios can result in a satellite evolving closer to the planet instead, or becoming trapped in a stable spin-orbital state (Goldreich and Peale, 1966; Counselman, 1973). In any case, the evolution of the Earth-Moon system has been one of increasing separation with time, and consequently the Moon must have been closer to a more rapidly rotating Earth in the past.

The essence of the problem of the lunar orbit is to determine this past evolution as accurately as possible. Clearly if one could trace the position of the Moon backward in time sufficiently far, one would have a potentially powerful means to differentiate between various theories of lunar origin. For example, if the orbit should begin to diverge from the Earth, perhaps at high inclination with respect to the Earth, or even in a retrograde sense of orbital motion, then a strong case for a capture origin could be made. In practice, the accuracy of calculations of the past history of the lunar orbit is limited by the uncertainties in many model assumptions and a complete breakdown of the simple tidal models when the Moon is close.

Using a particular model for tidal dissipation in the Earth, Darwin (1880) surmised that the Moon separated from the Earth on the order of 10^8 years ago. Estimates of the time scale for the evolution of the lunar orbit have necessarily increased as measurements of the rate of the Moon's recession have improved. In this century interest in the lunar orbit evolution was revived by Gerstenkorn (1955). MacDonald (1964) and Gerstenkorn (1967) estimated a time interval of about 2×10^9 years since the minimum Earth-Moon separation by assuming that the effective δ has always been equal to its present value. This time is considerably less than the accepted age of $4.5–4.6 \times 10^9$ years for the Earth and Moon; attempts have been made to associate geological events such as lunar mare volcanism with such a period of extreme tidal interaction (e.g., Turcotte et al., 1977). Peale and Cassen (1978), however, showed that the evolution away from small separations is so rapid that negligible tidal heating would have occurred in the Moon. As most of the tidal dissipation in the Earth that accounts for the measured secular acceleration of the Moon's mean motion of -25 arcseconds/century^2 (Yoder *et al.,* 1984) can be attributed to the oceans alone (Lambeck, 1975, 1980), and the configuration of the ocean basins has changed dramatically even over the last 2×10^8 years, the assumption of a constant effective δ over several billion years in determining the time scale of the lunar orbit evolution must be incorrect. Hence the precise time scale associated with lunar orbit evolution cannot be specified from observation and is most likely the age of the Earth itself. Modern students of lunar orbital history (e.g., Goldreich, 1966) therefore ignore the time scale in determining the geometric evolution of the system.

The dynamical history of the Earth-Moon system is usually determined by integrating the Lagrange planetary equations or their equivalent backward in time from the current configuration. The perturbation is that of the simple tidal distribution of mass pictured in Fig. 1 generalized to inclined orbits, where the extent of the variation in tidal models is limited to the amplitude and frequency dependence of the phase angle δ or more appropriately to the assumed time phase lag of each simply periodic term in a Fourier series expansion of the gravitational potential corresponding to the Earth tide (e.g., Kaula, 1964). Ideally the orbital eccentricity, semi-major axis, and inclination of the orbit relative to the ecliptic and to Earth's equator are followed back in time, including precessional dynamics and the effect of tides raised by the sun on the Earth.

The perturbing potential due to the tidal distribution of mass varies as $1/r^6$, which follows from the amplitude of the tide being the result of the differential of a $1/r^2$ force and that tide in turn being a second harmonic distribution of mass that must fall off as $1/r^3$ (Love, 1927). The torques derived from this potential also vary as $1/r^6$, which can be seen from Fig. 1 where the amplitude of the tide varies as $1/r^3$ and the torque on the bulge is proportional to the differential of a force varying as $1/r^2$. The tidal potential also varies as the fifth power of the radius of the tidally deformed body, which along with the $1/r^6$ dependence means that dissipation in the Moon can be ignored as long as the Earth-Moon separation exceeds several Earth radii (MacDonald, 1964; see Goldreich, 1966, for a discussion of the relative importance of lunar dissipation).

As described above, one major effect of the tidal perturbations is to increase the semi-major axis of the Moon's orbit with time. The tidal perturbations also tend to increase the eccentricity of the orbit with time. Because of the strong dependence of the tidal torque on Earth-Moon separation, the torque will be greatest at perigee, and the stronger tug on the Moon at perigee tends to sling it farther away at apogee, thereby increasing the orbital eccentricity. Because the current eccentricity of the lunar orbit is small (e = 0.055) and is currently increasing, the eccentricity was probably small throughout most of the lunar orbit evolution.

The Earth's equatorial bulge causes the Moon's orbit to precess with a precessional angular velocity that is perpendicular to the Earth's equator, whereas the effect of the sun on the orbit is to induce a precessional angular velocity that is perpendicular to the ecliptic plane. The resultant precessional angular velocity is perpendicular to a third plane called the Laplacian plane, with which the Moon's orbit maintains an approximately constant inclination as it precesses. The Laplacian plane approaches the Earth's equator when the Moon is close to the Earth (Earth caused precession dominating that of the sun), and it approaches the ecliptic when the Moon is far away. Hence the Moon currently maintains about a 5° inclination relative to the ecliptic as it precesses with about an 18-year period. When the Moon is close to the Earth, the tides tend to decrease the orbital inclination relative to the Earth's equator (Laplacian plane). In an inclined orbit the Earth's rotation carries the tidal bulge out of the plane of the orbit leading to components of the torques that drive the spin angular momentum of the Earth and the lunar orbital angular momentum toward alignment. With the current Earth-Moon separation, precessional averaging essentially places the Moon in the ecliptic and both lunar and solar tides tend to increase the Earth's obliquity, albeit on the long time scale of slowing the Earth's spin. Currently the lunar orbit inclination relative to the ecliptic plane is slowly decreasing, since the Earth tides increase that component of orbital angular momentum perpendicular to the ecliptic while leaving the component in the ecliptic essentially untouched.

The evolution of the orbital inclination is the most important result of Goldreich's (1966) calculations in which he carried the Moon back to a separation of about 3 R_E, which is roughly the Roche limit of the Earth (see Section 7 for a discussion of the Roche limit). These calculations improved on several deficiencies present in previous calculations (discussed by Ruskol, 1966), but were restricted to orbits of zero eccentricity. Darwin had modeled the tidal torque by assuming that the Earth responded as a viscous fluid with a phase angle (rate of dissipation) that depended on the frequency of the disturbance, whereas MacDonald (1964) assumed that the phase angle was independent of frequency. Goldreich considered the cases of temporally constant phase angle for the MacDonald tidal model and the case of constant and equal phase angles for different components of the Darwin tidal model. Goldreich found that the inclination of the Moon's orbit to the Earth's equator always reached a minimum value near 10° when the Moon was inside 10 R_E and became larger for smaller separation, regardless of whether Darwin's or MacDonald's tidal model was used, and whether or not the influence of solar tides was included. Goldreich therefore concluded that the Moon could not have been formed by accretion in the Earth's equatorial plane within about 10 R_E. He also excluded accretional formation outside 30 R_E because the Moon would now lie in the ecliptic plane as would be the case for equatorial origin inside 10 R_E. The calculations were not evolved farther backward in time because at separations less than 3 R_E the eccentricity would have been significant, and the calculations were limited to circular orbits. Goldreich's conclusions have withstood subsequent scrutiny (Gerstenkorn, 1967, 1968, 1969; Goldreich, 1968; Mignard, 1981; Conway, 1982).

O'Keefe (1972a) argued that even if the Moon was once in the Earth's equatorial plane, the inclination might have increased to the present value, rather than remain small as predicted by Goldreich (1966). O'Keefe based his argument on Darwin's investigation of different models of tidal dissipation in the Earth, which assumed that the Earth was a very viscous fluid (with a viscosity greater than about 10^{15} poise) during a time when the Moon was at 3.83 R_E, where the Moon's orbital period would have been twice the Earth's rotational period. In Darwin's model the phase angles vary for different components of the tide and, because of the large viscosity, could have been as large as 90°, meaning that their amplitudes would have been near zero. In that case other components of the tide would dominate. O'Keefe pointed out that this resonant situation would lead to rapid growth of an initially zero inclination. This proposal was investigated in greater detail by Rubincam (1975), who concluded that the inclination to the equator could indeed grow to the 10° necessary to evolve to the present configuration, but only if the initial inclination was at least 3° and the viscosity of the Earth was greater than 10^{18} poise. Goldreich (1966) also investigated the evolution when the phase angles depended on frequency, but with his models of the tidal dissipation in the Earth the effect of frequency dependent phase angles made little difference.

As pointed out by Kaula and Harris (1975), the analysis used by Darwin, O'Keefe, and Rubincam is dependent on assuming a linear model of viscous dissipation in the Earth, whereas for viscous dissipation in an Earth subjected to the tidal forces of a nearby Moon, the strains must have been large enough to result in a nonlinear viscosity, formally invalidating the theoretical analysis. However, the same sort of criticism can be applied to all models of the lunar orbit, simply because of the absence of knowledge about tidal dissipation in the extreme conditions associated with small separations. Even if the proper nonlinear viscoelastic model for the Earth were known, a nonlinear theory of the tidal distortions has not been developed, so we would still be unable to definitively follow the evolution of the Earth-Moon system at small separations.

Because of the fundamental uncertainties in both the past history of the tidal dissipation in the Earth-Moon system and in the tidal torques that would occur when the Earth-Moon separation was on the order of a few R_E, we must conclude that the evolution of the Moon's orbit cannot be reliably determined indefinitely backward in time. The invariance of computational results to changes in the tidal model implies that it is likely that roughly 4.5×10^9 years ago, the Moon was in a prograde orbit about the Earth, at a separation of 10 R_E, with an inclination of about 10°. An inclination of 10° at that time does not necessarily rule out an origin of the Moon in the Earth's equatorial plane, because special circumstances that cannot be ruled out could increase the inclination, such as a major impact on the Earth or Moon, the possibility of resonances that result in inclination growth, or even the possibility of orbital resonances with asymmetric distributions of mass on the surface of the Earth.

4. Rotational Fission

We now begin the discussion of the detailed dynamics of the various proposals for lunar origin. We will begin appropriately with Darwin's (1879, 1880) favored mechanism, namely the fission of a rapidly rotating proto-Earth into two bodies, the Earth and the Moon, as a consequence of a dynamic fission instability.

The problem of the equilibrium of a rotating, self-gravitating fluid body captured the attention of many of the great classical scientists. Comprehensive reviews of the subject have been given by Lyttleton (1953), Chandrasekhar (1969), and Tassoul (1978). These reviews have been brought up to date by the review of Durisen and Tohline (1985).

Chandrasekhar (1969) gave a historical introduction to this subject. Newton in 1687 was the first to consider the equilibrium of a uniform density, uniformly rotating fluid body restricted to a rotation rate small enough that the departures from a spherical shape could be considered as small perturbations. Maclaurin was able to solve for the equilibrium of more rapidly rotating bodies, for which the figures of equilibrium are termed oblate spheroids. Spheroids have surfaces that are ellipses rotated about the axis of rotation, making them axisymmetric. Jacobi was later able to find equilibrium figures for ellipsoids, which are nonaxisymmetric bodies with elliptical cross-sections perpendicular to each coordinate axis. Spheroids have two semi-major axes that are equal (a = b ≠ c); ellipsoids have three different semi-major axes (a ≠ b ≠ c). Both of these types of bodies are restricted to solid body rotation. Other forms of ellipsoidal equilibrium that involve internal motion (vorticity) were found by Dedekind and Rieman. Dedekind ellipsoids have a figure that is stationary in an inertial frame, while the Riemann-S ellipsoids involve both uniform vorticity and uniform rotation about parallel axes. While all four of these types of bodies are also restricted to uniform density (incompressible), qualitatively similar equilibria occur for nonuniform density (compressible) and nonuniformly rotating bodies (Tassoul, 1978; Durisen and Tohline, 1985). The permitted values of the semi-major axis for each of the incompressible equilibria are illustrated in Fig. 2.

Each of these types of configurations can be thought of as a sequence of forms, where the shape varies as the angular momentum increases. The Maclaurin spheroids, for example, constitute a sequence starting with spheres (zero rotation) that become progressively more oblate as the angular momentum increases. The Jacobi ellipsoid sequence begins with an equilibrium configuration that is part of the Maclaurin spheroid sequence. This phenomenon is termed a bifurcation, and it occurs at a well-defined amount of angular momentum. It is convenient to quantify points of bifurcation by the ratio of rotational energy to the absolute value of the gravitational

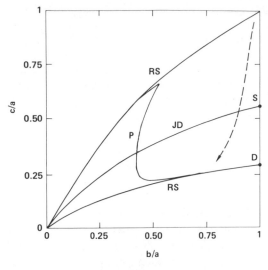

Fig. 2.   Classification of four different types of equilibria for incompressible, self-gravitating, rotating fluid bodies according to the ratios of their semi-major axes a, b, and c (after Durisen and Tohline, 1985). The upper right-hand corner corresponds to a non-rotating sphere, while the right-hand border is the Maclaurin sequence (angular momentum increases downward and to the left in this plot). JD labels the Jacobi and Dedekind ellipsoid sequences, while the two curves labelled RS denote the region within which the Riemann S-type ellipsoids occur. The Maclaurin spheroids are secularly or dynamically unstable to bar-shaped perturbations at and below the points marked S and D, respectively, while the ellipsoidal equilibria are dynamically unstable to pear-shaped perturbations at and to the left of the curve marked P. The dashed curve represents a possible trajectory for a viscous object whose rotation rate is increased on a time scale shorter than that for secular instability, resulting in evolution toward a dynamic pear-shaped instability.

energy $(T/|W|)$; $T/|W|$ equals 0.138 where the Jacobi sequence bifurcates from the Maclaurin sequence. Other points of bifurcation exist as well. Poincaré found pear-shaped equilibria that bifurcate from the Jacobi sequence at $T/|W| = 0.163$; higher order figures bifurcate at even higher rotation rates.

The presence of two possible uniformly rotating equilibria for $T/|W| > 0.138$ implies that thereafter a Maclaurin spheroid could be deformed into a Jacobi ellipsoid, which is also a lower total energy configuration, through a suitable dissipative mechanism. It is also possible to directly study the stability of the equilibria to small perturbations, which can grow on either a long (secular) or short (dynamic) time scale. A secular instability proceeds on a time scale determined by the dissipative mechanism, while a dynamic instability occurs within a few rotational periods. For $T/|W| > 0.138$, Maclaurin spheroids are secularly unstable to the growth of perturbations (with two-fold symmetry) that deform them into Jacobi ellipsoids, while for $T/|W| > 0.274$, the Maclaurin spheroids become dynamically unstable to the same modes (e.g., Chandrasekhar, 1969). The Jacobi ellipsoids are themselves dynamically unstable to pear-shaped perturbations (three-fold symmetry) for $T/|W| > 0.163$. The points of instability are also depicted in Fig. 2.

One can imagine then that as a dissipative, uniformly rotating, incompressible fluid body is spun progressively faster, it proceeds through a series of equilibria, leading from the Maclaurin spheroids through the Jacobi ellipsoids. Poincare hypothesized that the pear-shaped sequence that bifurcates from the Jacobi ellipsoids might lead stably and continuously to the equilibrium of two bodies, one larger than the other. This conjecture led Darwin to hypothesize that such a "fission" mechanism could explain the origin of binary stars. However, Liapounoff (1905) found that the entire pear-shaped sequence was dynamically unstable, which eliminates the possibility of a gradual deformation into two bodies. Jeans (1919) then proposed that the dynamical instability of a pear-shaped configuration would result in fission into two bodies, a configuration already described by Darwin's equilibria for tidally distorted, binary objects.

Darwin of course also considered fission to be the means by which the Moon formed from the outer layers of the Earth. Darwin (1879) first suggested that since the fundamental resonant period of the proto-Earth may have been about 2.5 hours, if the proto-Earth had at one time a rotational period of 5 hours, then the solar semi-diurnal tides might resonantly excite the proto-Earth into fissioning. However, Jeffreys (1930) showed that the dissipation accompanying the motion of a fluid mantle over a core damps any such tidally induced instability (see also McKinnon and Mueller, 1984). Darwin (1880) also hypothesized a purely rotationally driven fission origin of the Earth-Moon system, starting from an ellipsoidal proto-Earth with a period of about 2 to 4 hours. It is this possibility that has motivated the work on the fission origin of the Moon in the last several decades, and coupled with the interest in lunar origin associated with the Apollo flights, has led many to favor a fission origin (Ringwood, 1960; Wise, 1963, 1969; O'Keefe, 1969, 1970a,b, 1972a,b, 1974, 1984; O'Keefe and Sullivan, 1978; Binder, 1974, 1978, 1980, 1984; Schofield, 1981; Andrews, 1982).

One obvious problem with a fission origin of the Moon is the necessity to take an otherwise equilibrium proto-Earth and spin it up beyond the dynamical instability limit. While the quasiequilibrium contraction phase of a star with constant total angular momentum provides a reasonable means to accomplish this in the stellar context, no equivalent phase of terrestrial planet formation is thought to exist (Section 2). Ringwood (1960) and Wise (1963) helped to revive interest in the fission origin of the Moon in this century by proposing that the proto-Earth may have formed very close to the dynamical limit, and that the decrease in the moment of inertia accompanying rapid core formation may have triggered dynamical instability. This effect is relatively small (O'Keefe, 1970a), so the protoplanet would have to be formed very close to instability. However, it is impossible to form a terrestrial planet rotating at the dynamical limit through the accumulation of a large number of planetesimals (Section 2; see also Weidenschilling, 1984b); large spin rates seem to require the tangential impact of a relatively large (i.e., a mass on the order of

0.1 that of the protoplanet) planetesimal. The latter situation forms the basis of the giant impact model to be discussed in Section 9. Because the proto-Earth could not have been brought to dynamical instability except through a giant impact, which may itself be capable of producing the Moon, the fission hypothesis would be obviated in this case.

A rotational period of about 2.5 hours is generally thought necessary to result in fission (e.g., O'Keefe and Sullivan, 1978). Moulton (1909) was the first to raise another important argument against the fission origin of the Moon by pointing out that the amount of angular momentum in the present Earth-Moon system is insufficient to produce a proto-Earth rotating above the dynamical instability limit. The present system contains only about one-fourth the angular momentum necessary for a rotational period of 2.5 hours and implies a proto-Earth with a period of about 5 hours. Wise (1969) argued that Moulton's criticism could be avoided if the excess angular momentum was lost along with part of an atmosphere following the fission event. O'Keefe (1969, 1970,a,b, 1972a,b, 1974) similarly argued that tidal dissipation during asynchronous rotation immediately following fission could lead to vaporization and removal of enough mass and angular momentum, but this is energetically impossible since only the proto-Moon's rotational energy (2.5 hours rotation period, 0.1 M_E) is available to tidally heat the proto-Moon, and this energy is insufficient to vaporize the silicates by a factor of over 10. Binder (1980) proposed that mass and angular momentum loss resulted from the catastrophic boiling that would occur if matter with temperatures of 3000° to 4000°K became a part of a low-pressure body like the Moon. In both cases, a quantitative estimate of the amount of mass and angular momentum loss, even assuming the extreme situations envisioned, has not been made. If one assumes that about 0.1 of the mass of the Earth and the excess initial angular momentum is carried away following fission, then the energy required to move the matter out of the Earth's gravitational potential is about 0.1 of the total energy of the Earth-Moon system, including the gravitational binding energy. Any viable fission scenario must provide a physical process following formation of the Earth-Moon system that could provide the energy required for loss of the excess angular momentum.

Because of the inability of linearized analytical methods to calculate the evolution of a nonlinear, dynamical instability, the classical workers could only hypothesize about the actual outcome of a fission instability. Recently numerical techniques have been used to study the dynamical evolution of unstable, rapidly rotating, compressible fluid bodies (see Durisen and Tohline, 1985), primarily in order to learn the implications for the formation of binary stars. The dynamical studies have clearly shown that the dynamic instability of a Maclaurin spheroid-like body (compressible and differentially rotating) to a bar-shaped perturbation does not result in fission into a binary system, contrary to the classical hypotheses. Instead, any binary system that begins to form is aborted by the loss of its orbital angular momentum to spiral arms that grow coevally and transfer angular momentum outward (Boss, 1984). The system then merges back into a single central body rotating below the limit for dynamic instability. The excess angular momentum is deposited in an expanding disk of gas in the equatorial plane of the object; even if this matter could be later assembled into a single body rather than reaccreted or lost from the system, the mass of the secondary would be much less than that of the central body. The fission of a rapidly rotating protostar into a close binary has been cited as a model for a fission origin of the Moon (Binder, 1978, 1984). But fission cannot account for the formation of close binary stars, because these stars are observed to have nearly equal mass members (Lucy and Ricco, 1979; van't Veer, 1981). Instead, a collapsing cloud with excess angular momentum fragments into two orbiting protostars of comparable mass before a single protostar can form (Lucy and Ricco, 1979; Boss, 1984, 1985a, and references therein).

Considering the mass ratio of the Earth-Moon system, a fission origin for the Moon cannot be similarly dismissed. Indeed, Durisen and Scott (1984) proposed that inefficient formation of the Moon from a circumterrestrial ring (with a mass of about 0.1 M_E) might well explain the necessary loss of mass and angular momentum. In particular, the rotational energy of the still rapidly rotating proto-Earth is available for powering the needed mass and angular momentum loss, and a physical mechanism (gravitational torques between a nonaxisymmetric proto-Earth and the ring matter) exists that potentially could transfer the required energy to the ring. However, the calculations of Durisen *et al* (1985), which are the basis for the proposals by Durisen and Scott (1984), simulate stellar bodies and are not directly applicable to the protoplanetary fission scenario; the fission instability of a rapidly rotating, slightly compressible, viscoelastic protoplanet was not considered. Formation of the Moon following the fission of a gaseous proto-Earth has been proposed by Petersons (1984), but the general theory of terrestrial planet formation (Section 2) disallows a gaseous proto-Earth (but see Cameron, 1985a). While Durisen and Scott (1984) and Durisen and Gingold (1986) pointed out the differences in compressibility between the stellar fission models and protoplanets, Boss and Mizuno (1984, 1985) found the most important difference to be viscous damping, which is negligible for stellar models.

The proto-Earth would have had to be completely molten in order to have a viscosity low enough to validate the classical and contemporary work on fission in inviscid fluid bodies. Roberts and Stewartson (1963) and Rosenkilde (1967) extended the classical work on equilibria to include the effects of very small viscosity, and found the Maclaurin spheroids to continue to be secularly unstable for $T/|W| > 0.138$. O'Keefe and Sullivan (1978) proposed that the proto-Earth was initially totally molten with a Maclaurin spheroid configuration and rotating with $T/|W| > 0.138$. A small amount of viscosity might then transform the proto-Earth into a Riemann S-type ellipsoid, which eventually would become unstable to the pear-shaped instability. Riemann S-type ellipsoids are inviscid fluid bodies with internal motions, so O'Keefe and Sullivan's mechanism could only have occurred for a totally molten, nearly inviscid proto-Earth. Wise (1963) perceptively noted that the neglect of viscosity in models of fission is not justifiable.

While the material that formed the terrestrial planets may have been initially "cold" because of the low temperatures in the presolar nebula (e.g., Boss, 1985a), substantial melting may have resulted from dissipation of the kinetic energy of accretion as the bodies grew. Larger bodies have deeper gravitational potential wells, and hence more energy per unit mass of incoming matter is released. Bodies that are larger than a critical size will have collisions violent enough to result in at least local melting. The critical size is determined by requiring that the self-gravitational energy exceed the energy needed for melting the entire body. For terrestrial matter, the critical mass is about 10^{25} g (1000 km radius). An unknown fraction of the energy released by accretion will be lost by radiation and kinetic energy of ejecta during the impact (e.g., Kaula, 1979, 1980), reducing the energy available for melting the planet. More importantly, convection in large, partially molten bodies is quite efficient at removing heat from the interior of the body, and estimates of the minimum viscosity encountered in the evolution of Earth's mantle start at about 10^{15} poise (Turcotte et al., 1979; Schubert et al., 1980). These viscosities correspond to mantles that are partially molten; terrestrial magmas have viscosities in the range of 10^{2} to 10^{6} poise.

The effective viscosity of Maxwell and Kelvin-Voigt viscoelastic models is related to the specific dissipation function Q and the dynamic time scale (e.g., Poirier et al., 1983). For seismic waves in the present Earth, the effective Q is about 100 (Knopoff, 1964). The effective Q of a partially molten protoplanet may be much lower, because Q depends on temperature and on the frequency and amplitude of the motions being dissipated. An effective viscosity of about 10^{15} poise results for both Maxwell and Kelvin-Voigt viscoelastic models when Q = 1 and the dynamic time scale is a few hours.

Boss and Mizuno (1984, 1985) simulated the effects of an effective viscosity of about 10^{15} poise on the dynamic instability of rapidly rotating proto-Earth models with $T/|W| > 0.274$. They found that a rapidly rotating, dissipative proto-Earth subjected to a large bar-shaped perturbation damps out the nascent perturbation and returns to its initially axisymmetric configuration; no tendency toward fission was found. This result is independent of concerns about the numerical techniques, because the same numerical code was used in the inviscid, stellar fission calculations of Durisen et al. (1985). The numerical result was confirmed by an analysis of the linear stability of a highly dissipative protoplanet (Boss and Mizuno, 1985), which showed that in the presence of a very large simulated viscosity, the dynamic instability disappears, though a secular instability remains, with a time scale long compared to the rotational period.

If the proto-Earth was viscous, then it must have been very nearly in solid body rotation, because viscous shear will damp out any differential rotation. However, an axisymmetrically stable model of a uniformly rotating polytrope with the compressibility of the Earth (n = 1/2) can only be constructed if $T/|W| < 0.17$ (Tassoul, 1978). Hence a viscous proto-Earth cannot be spun fast enough to undergo dynamical fission ($T/|W| > 0.274$). Instead, a viscous protoplanet that is spun faster and faster must simply begin to lose matter from its equatorial regions (Boss, 1985b).

Because fission must occur through a dynamic instability, the fission instability cannot occur in a viscous, partially molten or solid protoplanet.

5. Precipitation Fission

The second mechanism for the origin of the Moon involves formation of the Moon from planetesimals formed by the precipitation of matter derived from a massive proto-Earth atmosphere. As originally proposed by Ringwood (1970), the mechanism did not involve rotational instability, but was chemically similar to fission in that the Moon was formed from matter ultimately derived from the mantle of the Earth. In a later version of the theory, Ringwood (1972) explicitly involved a rotational fission instability in the process.

Ringwood's scenario assumes that the Earth formed directly from the accretion of planetesimals no larger than 10 km over a time period of less than 10^6 years. Because the gravitational energy liberated by the accreting planetesimals and by core formation is energetically sufficient to produce surface temperatures of $1000°-2000°C$ on the proto-Earth, Ringwood then supposed that silicates would be preferentially volatilized and would enter a massive ($0.2–0.5\ M_E$) atmosphere composed primarily of hydrogen. The atmosphere would later be dissipated by a T Tauri wind phase of the sun, and as the atmosphere was dissipated and cooled, precipitation of the metals and oxides would produce a circumterrestrial ring from which the Moon could later form. This precipitation hypothesis received early support from Cameron's (1970) models of the primitive solar nebula, which had predicted temperatures in the terrestrial planet region sufficient to vaporize metals and silicates and implied a gaseous proto-Earth. But it is not clear why any precipitates would not have either sedimented onto the Earth prior to the T Tauri wind, or else been carried away with the rest of the atmosphere by the wind once it began. Even more crucial is the absence of sufficient net angular momentum in such a massive atmosphere; precipitation could only lead to sedimentation back onto the proto-Earth, not to the formation of a disk at several R_E.

Cameron (1970) and Ringwood (1972) realized that the proto-Earth would have to be extremely flattened by rotation in order to possibly produce a disk by precipitation. Ringwood envisioned that a rapidly rotating, massive proto-Earth atmosphere would eject matter in the equatorial plane in the form of rings, and hinted that magnetic fields generated by an early terrestrial dynamo might aid the process of transferring angular momentum to the disk. The gas hydrodynamics and magnetohydrodynamics involved in this scenario have not been investigated, so the details of the process, especially the angular momentum transfer necessary to form the disk, must remain very speculative. Ringwood (1979) concluded that the most serious problems associated with the precipitation hypothesis are the difficulties in completely removing the massive hydrogen atmosphere and in producing sufficiently high temperatures in the proto-Earth mantle to evaporate silicates.

Two of the arguments against fission discussed in Section 4 are also applicable to the precipitation fission hypothesis, namely the inability to form a proto-Earth rotating close to dynamical instability through the impact of a large number of planetesimals, and the question of what mechanism was able to dispose of the excess angular momentum of the proto-Earth without also destroying the prelunar disk. We conclude that the total of all the problems for precipitation fission is fatal for the hypothesis.

6. Intact Capture

The third mechanism of lunar formation is accumulation on a heliocentric orbit in a manner similar to that of the other terrestrial planets, followed by a close encounter with the Earth that results in capture of the Moon into Earth orbit. The formation of the Moon in heliocentric orbit through the same processes (Section 2) that formed the terrestrial planets is completely reasonable. It is also reasonable for a proto-Moon so formed to have its orbital eccentricity increased through gravitational perturbations a sufficient amount to yield an orbit that crossed that of the Earth (e.g., Wetherill, 1985, 1986). In addition to the problem of matching the Moon's chemistry, the dynamical problem with this mode of origin lies solely with capturing the proto-Moon in Earth orbit. We will discuss three different schemes for achieving this capture, based on the three-body problem: dissipation of energy by tides, collision and gas drag in circumterrestrial orbit, and altering the masses of the Earth or the sun.

First consider the three-body problem consisting of the dynamical evolution of the Earth, Moon, and sun. In order to make progress, one must assume that the mass of the Moon is infinitesimal and that the Earth moves on a circular orbit about the sun (restricted three-body problem). These are reasonable assumptions for the Earth-Moon-sun system ($M_E/M_M = 81$, $e_E = 0.017$). In this case the energy of the infinitesimal mass (hereafter the Moon) is related to a quantity C, termed the Jacobi constant ($C > 0$). The Jacobi constant involves the position of the Moon relative to the Earth-sun system, and hence the value of C defines surfaces in space accessible to the Moon. The Jacobi constant also involves the square of the relative velocity of the Moon, which clearly must always be positive, and the Jacobi constant with the velocity set equal to zero then restricts the possible motions of the Moon. The surfaces of zero velocity enclose regions of permitted trajectories for the Moon, with the Moon's orbit being prohibited from evolving from one region to another across a zero velocity surface. For relatively large values of C, permitted orbits for the Moon consist of orbits that either encircle the combined Earth-sun system at a large distance, or else are restricted to small volumes around either the earth or the sun. For relatively small values of C, permitted orbits exist throughout the system, but cannot pass through small regions equidistant from the Earth and sun. Intermediate values correspond to orbits that encircle both the sun and the Earth. Suppose then that the Moon initially had an intermediate value of C, corresponding to a heliocentric orbit intersecting that of the Earth. If C was increased sufficiently during the time that the Moon was close to the Earth, then the Moon might become trapped in a stable orbit about the Earth. Increasing C could result from decreasing the kinetic energy of the Moon. For example, Bailey (1969) pointed out that changes in the eccentricity of the Earth's orbit could gravitationally perturb the Moon and result in its capture, but only if the Moon was on an orbit very close to that of the Earth, with a relative velocity on the order of 1 km s^{-1}. The key point is that the orbital energy of the Moon with respect to the Earth must be reduced from a positive value to a negative value, i.e., energy must be lost from the Moon. Even for a temporary capture, the small range of C compatible with Earth orbit makes capture at best a low probability event.

One possible means of dissipating sufficient energy for capture is through tidal dissipation during a close encounter with the Earth (Gerstenkorn, 1955, 1969; Singer, 1968, 1970, 1972; Singer and Bandermann, 1970; Alfvén and Arrhenius, 1969, 1972, 1976; Winters and Malcuit, 1977; Malcuit et al., 1977; Conway, 1984).

Gerstenkorn (1955) originally proposed that the Moon was captured into a retrograde orbit (i > 90°) about the Earth, consistent with his calculation of the evolution of the lunar orbit implying high inclinations when the Moon was very close to the Earth. Gerstenkorn (1969) investigated tidal dissipation in a Moon that spent several years passing within about 3 R_E (the Roche limit, defined in Section 7). Using relatively small tidal dissipation rates, Gerstenkorn concluded that sufficient tidal dissipation would have occurred to heat by 1000°K and trap a Moon that had a relative velocity at infinity of about 1 km s^{-1}. The important point here, however, is the capture of the Moon on its first pass, not the dissipation resulting from repeated passes inside the Roche limit.

Singer (1968) sought to improve on Gerstenkorn's approach by hypothesizing capture from an initially prograde orbit, so that less rotational kinetic energy would have to be dissipated compared to the retrograde rotation case. Using a Darwin model for tidal dissipation, Singer (1970) estimated that the Moon could be captured at a separation of about 5 R_E, but that subsequently the Moon would pass within the Roche limlt before receding out to its present location. The evolution obtained by Singer (1970) is shown in Fig. 3. Singer's calculation was actually a backward evolution of the lunar orbit in time terminating in a parabolic orbit; hence Singer did not address the question of the probability of obtaining the parabolic orbit. Consideration of the discussion in Section 3 shows that extrapolating the evolution of the lunar orbit this far back in time is unwarranted. Singer (1972) concluded that the Moon could only be captured if it was originally formed very close to

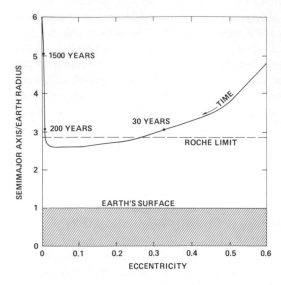

Fig. 3. Evolution of the semi-major axis and eccentricity of the Moon in the capture scenario of Singer (1970). The basic calculation involved the evolution of the present lunar orbit backward in time until a parabolic orbit was achieved. This parabolic orbit defines zero time in this plot, when the Moon was roughly 5 R_E distant. The Moon at first evolves closer to the Earth, spends on the order of 100 years within the Roche limit, and then moves outward to its present separation from the Earth.

the Earth with very little eccentricity (again, with very low relative velocity). Singer (1972) noted that in such a situation, it would be more likely for the Moon to impact the Earth than to be captured.

During the capture of an intact Moon, most of the tidal dissipation occurs within the Moon from the radial tide. This dissipation can be estimated as a fraction of the work done on the Moon by tidal forces during the single pass. Although based on a linear theory of the tides, the value of 1.5×10^{33} ergs obtained by Kaula and Harris (1973), where 0.1 for the fraction of tidal energy dissipated was assumed, is probably a reasonable estimate. Malcuit *et al.* (1977) estimate that in a close encounter with a perigee of 3 R_E about 10^{34} ergs could be dissipated, corresponding to a Q of about 1 in the linear theory, which may be reasonable for a recently formed object (Section 4). With the larger figure, the Moon's relative velocity before encounter could then be no greater than about 0.1 km s^{-1}, which is already an order of magnitude below Gerstenkorn's 1 km s^{-1}, and it represents an exceedingly low relative velocity, considering that the orbital velocity of the Earth is roughly 30 km s^{-1}.

Multiple close encounters with the Earth have also been hypothesized in order to tidally dissipate more energy and thus allow larger relative velocities (e.g., Malcuit *et al.*, 1977). But each close encounter is much more likely to increase the relative velocity on the next pass rather than to reduce it, and the Moon has to avoid colliding with the Earth as well. The probability of a successful capture of an intact Moon by tidal friction is negligibly small even if a small relative velocity could be contrived beforehand.

The dissipation of kinetic energy necessary for capture could also be achieved by having the Moon collide with objects already in Earth orbit (Dyczmons, 1978) or through gas drag in a massive primordial terrestrial atmosphere (Horedt, 1976; Nakazawa et al., 1983). The mechanism of Ruskol (1972a) can be used to place objects in Earth orbit during the accretion process where collisions between planetesimals while in close proximity to the forming Earth leave one of the pair or fragments thereof in orbit. However, the orbiting material does not survive long because of angular momentum loss through collisions with other infalling planetesimals (Weidenschilling, 1984a,b), and it is not clear that a swarm of sufficient density could exist to provide a significant collision cross-section for capturing the Moon. While the general theory of gas-free accretion discussed in Section 2 does not predict that the terrestrial planets had massive atmospheres, such atmospheres are a natural consequence of the theory of planetary formation in the presence of the gaseous component of the solar nebula, which has been developed by the Kyoto group founded by C. Hayashi, and they follow from Cameron's (1985a) theory involving giant gaseous protoplanets. Nakazawa et al. (1983) have shown that a Moon, again initially with little relative velocity, can be slowed down within about 100 years and captured by gas drag occurring primarily at perigee in the primordial atmosphere. Here again sufficient energy must be dissipated in the first pass to effect capture. The principal problem associated with this mechanism is the need for having a T Tauri wind phase of the sun precisely coincide with the lunar capture, thereby stripping away the primordial atmosphere. Otherwise, the Moon would rapidly spiral further in through gas drag and eventually impact the Earth.

The third possibility for intact capture involves a change in the mass of the Earth or the sun during the close encounter with the Moon. Szebehely and Evans (1980) found that if the mass of the sun was suddenly decreased by about one-third, then the increased importance of the Earth's gravitational field could trap the Moon. The requirement of a drastic loss of mass from the sun rules out this possibility. Lyttleton (1967) considered the effect of adding mass to the Earth during a close encounter, and estimated that about one-sixth of the mass of the Moon would have to be added in about 10 years in order to achieve capture. But adding mass over any extended interval of time will not work since the increment in mass must occur while the Moon is in close proximity with the Earth. Clark et al. (1975) and Nordmann and Turcotte (1977) also studied the effects of increasing the Earth's mass to achieve capture and showed that the chances for capture increase considerably if the added mass is a large fraction (e.g., 0.1) of the Earth's mass. Clearly the timing required for success in the case of a large impact, given the occurrence of the hypothesized events, is such as to make this as well as many of the other schemes for capture verge on the miraculous. Given the miracle, capture by this means still depends on having the Moon initially on an orbit with low relative velocity, very close to the Earth.

The capture origin has often been advocated as a means of accounting for chemical differences between the Earth and the Moon, with its original orbital position ranging from inside the orbit of Mercury (Cameron, 1972, 1973) to the asteroid belt (Malcuit *et al.*, 1984). The implicit assumption behind this motivation is that the temperature gradient in a hot solar nebula led to an equilibrium condensation sequence as the nebula cooled, which produced an orderly gradation of compositionally different planetesimals or terrestrial planets. Recent work by Wetherill (1985, 1986) implies that it is possible to get planetesimals formed in different thermal and chemical environments into Earth-crossing orbits, as such planetesimals undergo great variations in semi-major axis in the process of accumulating into the terrestrial planets. However, this mixing of different chemistries makes it difficult to accumulate a body as large as the Moon with such a distinct chemistry. Finally, even if such a Moon could be formed elsewhere in the primordial nebula, it would necessarily arrive at the Earth with a velocity considerably larger than the 0.1 km s^{-1} needed to effect capture by any of the proposed mechanisms. An origin at high inclination in the solar nebula (Anderson, 1972; Hanks and Anderson, 1972) is also extremely unlikely because the solid matter sediments to the midplane of the nebula (i = 0) faster than it could possibly accumulate at high inclination (Fremlin, 1973).

We have seen that in order for any capture scenario to have a nonvanishing probability of occurrence, the precapture orbit of the Moon must have been very close to that of the Earth. However, it is then difficult to see how the Moon could have completely accumulated so close to the Earth without having been swept up by the Earth, or else perturbed to a more distant eccentric orbit at an earlier time. This problem together with the low probabilities and other problems discussed above forces us to abandon capture of an intact Moon as a viable alternative origin.

7. Disintegrative Capture

The fourth mechanism for lunar origin is a variation on the capture mechanism that involves the tidal disruption of a proto-Moon during a close encounter with the proto-Earth. One motivation for this mechanism is that the cross-section for tidal disruption was thought to be much larger than that for intact capture. The fragments resulting from tidal disruption might be more easily trapped in circumterrestrial orbit, where they could later reaccrete into the Moon. This mechanism clearly retains the advantages of the original capture hypothesis, while providing what seems to be a more efficient means for accomplishing the actual capture. The disintegrative capture model was proposed by Alfvén (1963), Alfvén and Arrhenius (1969), and Öpik (1969, 1971), and is based on the concept of instability to tidal forces within a certain radius of the proto-Earth, termed the Roche limit, which is equal to about 3 R_E.

In 1847 Roche began the study of the stability of bodies subjected to tidal forces (see Chandrasekhar, 1969). Roche considered an incompressible, inviscid, self-gravitating liquid satellite in circular orbit about a spherical planet, with the satellite spinning synchronously with the orbital motion. Roche found ellipsoidal equilibria for the tidally accelerated satellite to exist for radii greater than a critical value, termed the Roche limit. The absence of equilibria inside the Roche limit means that a liquid satellite with a circular orbit inside the limit has insufficient self-gravity to resist being destroyed by tidal forces.

Later workers extended Roche's analysis to solid satellites. Jeffreys (1947) and Öpik (1950) compared the tidal stresses with tensile strengths of solid materials and derived a criterion for disruption of a solid satellite when the former exceed the latter. However, this work ignored the self-gravity of the satellite, which was included by Sekiguchi (1970). Sekiguchi calculated the stress in a rigid body on a linear trajectory toward the planet, and Aggarwal and Oberbeck (1974) considered nonrigid satellites on circular orbits. A criterion for tidal disruption similar to the Roche limit was found in each case. It is now realized that Phobos, the larger satellite of Mars, is *inside* the Roche limit, yet the acceleration of a particle on its surface is everywhere toward the satellite's interior (Dobrovolskis, 1982). This is due to Phobos' compressive strength, which prevents its relaxation to sufficient elongation for disruption. Phobos would be stable even if it were a pile of unconsolidated debris (Harris, 1977c)! This resistance to deformation is a fundamental property in the discussion of tidal disintegration below.

The general theory of terrestrial planet accumulation (Section 2) that seems best supported at the present time predicts that the proto-Earth was formed primarily by the impact of bodies similar to the Moon in size. For each planetesimal that impacts, a larger number must pass close to the proto-Earth without suffering an impact, because the cross-sectional area of the Roche limit is larger than that of the Earth. Öpik (1971) reasoned that because a proto-Moon that passes close to the proto-Earth is within the Roche limit for a time on the order of the fundamental oscillation period of the proto-Moon, the tidal forces should have sufficient time for complete disruption.

Öpik (1971) envisioned that about one-half of the disintegrated planetesimal would end up in circumterrestrial orbit, and the remainder would continue on a heliocentric orbit. Malcuit *et al.* (1984) associated the large circular lunar maria with the impacts of remnants of an incomplete tidal disintegration. Smith (1974, 1979) hypothesized that because the Roche limit depends on the inverse one-third power of the density, the less dense silicate layers of a differentiated planetesimal might be preferentially disrupted, leaving the denser iron core to pass on by. Wood and Mitler (1974) and Mitler (1975) considered the orbital dynamics of a disintegrated planetesimal and found that in the most favorable cases, the fraction that is trapped in Earth orbit is less than one-half, but that it would preferentially be drawn from the outer layers, which would be significant if the planetesimal had already differentiated. The mass of the planetesimal would then have to be many times the mass of the final Moon, especially if the subsequent reaccretion process loses substantial matter to the Earth. Also, Wood and Mitler found that the planetesimal must not approach the proto-Earth at a high relative velocity if mass is to be captured; the velocity at infinity must be about 1 km s^{-1} or less, which restricts the planetesimal to heliocentric orbits close to the Earth. Harris (1975) pointed out that if the proto-Moon was tidally disrupted, subsequent collisions among the fragments would further reduce the fragment size, leading to a time scale for tidal evolution to orbital radii beyond the Roche limit that is incompatible with the age of the Moon. The orbital dynamics of a disintegrated planetesimal were also considered by Buck (1982) and Kaula and Beachey (1986).

The fundamental problem with disintegrative capture, however, is the assumption of tidal disruption of the planetesimal. Mizuno and Boss (1984, 1985) studied the tidal disruption of a planetesimal with a simulated viscosity of about 10^{14} poise, which is lower than the effective viscosity of solid or partially molten planetesimals (Section 4). They studied the dynamical evolution of the nonlinear distortion of a three-dimensional planetesimal on a parabolic trajectory past the proto-Earth (Fig. 4). Mizuno and Boss found that complete tidal disruption does not occur, even for perigees of grazing incidence with the proto-Earth; at most, only a small fraction of the mass is lost from the planetesimal. This result was confirmed by the work of Kaula and Beachey (1986). A dissipative planetesimal cannot be disrupted in the short period of time involved in a close encounter. Like Phobos, the planestesimal simply cannot deform sufficiently for disruption. Unless the proto-Moon is totally molten and nearly inviscid, the disintegrative capture mechanism cannot work.

The presence of binary craters on terrestrial bodies (e.g., Sekiguchi, 1970; Aggarwal and Oberbeck, 1974) is not inconsistent with the finding that tidal disruption does not occur for viscous bodies (Mizuno and Boss, 1985). Because most craters are single, most impacting bodies did not disrupt, yet all impacting bodies must pass through the same tidal field. Hence whether or not a binary crater is formed must depend on the physical condition of the incoming object. Binary craters could then result from binary impactors (e.g., Weidenschilling, 1980).

If tidal disruption was very efficient, the accretion process ,would have to more closely resemble the runaway growth scenario (e.g., Greenberg et al., 1978), where a few large planets sweep up greatly smaller planetesimals that have been repeatedly tidally disrupted on previous close encounters. However, tidal disruption will not occur unless the planetesimals are primarily molten at the time of the tidal encounter. The absence of tidal disruption during the planetary accumulation process can be expected to favor a mass spectrum with most of the mass residing in the larger bodies. In such a situation, the accumulation process will include the stochastic impacts of large bodies (e.g., Hartmann, 1984).

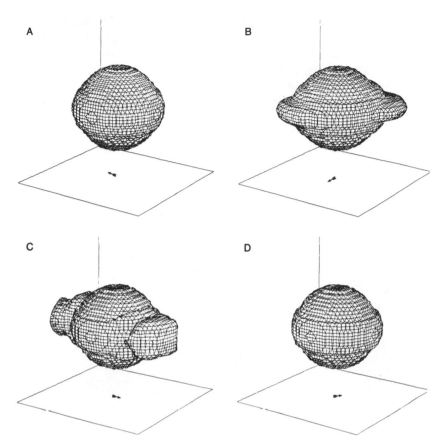

Fig. 4. *Time evolution of the attempted tidal disruption of a viscous planetesimal (radius 1000 km) on a parabolic orbit that brings it within grazing incidence of the Earth (perigee of 1.2 R_E = 7700 km), calculated by Mizuno and Boss (1985). The three-dimensional shape of the tidally distorted atmosphere of the planetesimal is represented by three-dimensional isodensity contours. The density plotted is 0.01 g cm^{-3}, which lies well outside the solid surface defined as 3 g cm^{-3}. Whereas the solid surface is largely undistorted during the close encounter, the atmosphere undergoes a more dramatic evolution. The arrows point toward the location of the tidally-distorting Earth, which appears to move past the planetesimal in this frame of reference. The Earth-planetesimal separation varies through the sequence: A—2.8 R_E, B—1.2 R_E, C—2.7 R_E, D—6.6 R_E. A few percent of the planetesimal mass is lost during the encounter, but the planetesimal easily survives.*

8. Binary Accretion

The fifth mechanism for the formation of the Moon is binary accretion, whereby the Moon forms coevally in orbit with the Earth. This mechanism was first studied by Ruskol (1961, 1963, 1972a,b,c, 1975). Ruskol envisioned that as the Earth accumulated from a range of different sized planetesimals, a swarm of particles up to 100 km in size would form about the Earth. Initially spherical, the swarm would evolve through collisions into a thin circumterrestrial ring (see Fig. 5). Once such a ring formed, it would grow by trapping further heliocentric planetesimals through collisions. The circumterrestrial swarm would eventually become dense enough to be gravitationally unstable (a process similar to the instability of the solar nebula as a whole; see Section 2) and then would begin to accumulate into larger and larger bodies. Ruskol (1972a) predicted that in order to form the Moon through

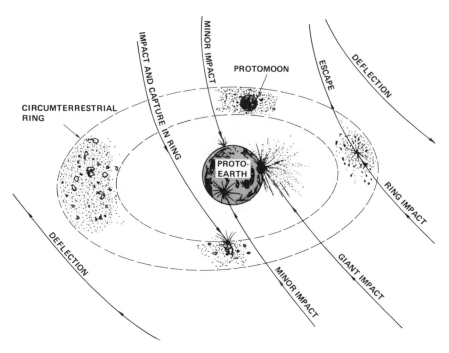

Fig. 5. Schematic of possible dynamical processes involved in the formation of the Moon in either the binary accretion or the giant impact accretion models (the velocities of the heliocentric planetesimals are shown in a frame of reference moving with the Earth). In the former, a circumterrestrial ring or disk of debris is formed by the collisions of unbound planetesimals in near-Earth orbit that result in their capture by the Earth. The disk subsequently traps more heliocentric planetesimals directly by collisions between bound and unbound particles; however, it is unlikely that this mechanism can produce sufficient angular momentum to initially form or maintain the circumterrestrial ring. In the giant impact model, debris from a single giant, tangential impact is ejected and placed in Earth orbit, thereby accounting for the necessary angular momentum to form the ring. In either model, the Moon is expected to form relatively rapidly from accumulation of the ring matter.

this process, it would be necessary for a 1400-km-radius proto-Moon to exist (with a mass one-half that of the present Moon), or else a small number of proto-Moons with 1000-km-radii would be needed to collide and produce a single Moon. The origin of these proto-Moons was then problematical; they must have either formed out of the swarm, or been captured intact (Section 6).

The binary accretion model was further developed by Harris and Kaula (1975), who also studied the growth of the Moon in the framework of Safronov's (1969) model of terrestrial planet formation. Harris and Kaula found that in order to produce the Moon, a proto-Moon embryo has to begin to grow when the Earth has accumulated about 0.1 of its final mass. The proto-Moon initially has a mass of about 0.01 of its final mass in this model; the proto-Moon embryo forms from the gravitational instability of the circumterrestrial ring. Thereafter both the proto-Earth and the proto-Moon grow by accretion from the planetesimal swarm. The Moon grows relatively faster than the Earth, because both collect material at a rate roughly proportional to their cross-sectional area (with the Moon having a larger area to mass ratio) and both are subjected to the about the same flux density, which is dominated by the gravitational focusing produced by the Earth (Harris and Kaula, 1975; see also Cox, 1984). Harris and Kaula (1975) and Harris (1977a, 1978) also pointed out the importance of tidal torques for preventing the spiralling inward of the proto-Moon that would otherwise occur through planetesimal drag and growth of the Earth; the Moon then remains at a distance of about 10 R_E from the Earth throughout its formation phase.

In the models of Harris and Kaula, the Moon would have accreted from a combination of planetesimals on both heliocentric and circumterrestrial orbits. In Ruskol's (1972a, 1973) version, the Moon formed primarily from two or three proto-Moons that accumulated in Earth orbit; the energy released in the final coalescence could have heated the Moon by up to 1000°K, resulting in substantial melting (Ruskol, 1973, 1979; Safronov, 1975). Wetherill (1976a,b) pointed out that the runaway growth predicted in Safronov's theory, where the ratio of the most massive body to the next most massive body becomes large, only occurs in the limit of infinite mass and for restricted values of the relative velocity. Instead, the dominant planetesimal may have emerged through attrition, so that the next largest size body (about 10^{26} g) might have been 10 times larger than Safronov's (1969) prediction of about 10^{25} g. Wetherill (1976b) found that the impacts of bodies that large still would not produce a Moon with a totally molten outer layer, although the amount of partial melt is quite uncertain. This result is consistent with the rheological assumptions of Section 4, where it was assumed that the terrestrial protoplanets were at most partially molten.

In recognition of the bulk chemical differences between the whole Earth and the Moon, Ruskol (1972b) and Kaula (1975) proposed that the circumterrestrial swarm of debris had acted as a compositional filter. Because silicate planetesimals are more easily fragmented than iron planetesimals, the smaller planetesimals should be more silicate-rich than the larger planetesimals. The smaller planestesimals are more likely to be trapped in Earth orbit by collisions and gas drag in the circumterrestrial ring. Furthermore, for the large, previously differentiated planetesimals that suffer fragmentary collisions, the silicate fragments of the planetesimal mantle are more likely to be trapped than the dense iron cores, which will pass through the debris ring unimpeded (Greenberg *et al.*, 1984; Chapman and Greenberg, 1984). One problem with the compositional filter mechanism is that by the time that the Moon was forming, the Earth was nearly complete, and recent calculations (Wetherill, 1985, 1986) suggest that at that time the great majority of the mass would have been in very large bodies (i.e., 10^{26} g or more), not in the small bodies for which the compositional filter may be effective. The compositional filter can only have been operative if a means can be found to maintain a significant distribution of relatively small-sized planetesimals rather late in the accumulation process. While runaway accretion would result in a mass distribution suitable for filtering (Weidenschilling *et al.*, 1986), runaway accretion is unlikely to have occurred to the necessary extent (Wetherill and Cox, 1982, 1984; Wetherill, 1986).

Another problem with the binary accretion hypothesis is the need to keep the material in the circumterrestrial swarm from prematurely accumulating into a few large bodies, because once this happens, the swarm again loses its ability to act as a compositional filter. Harris and Kaula (1975) had found that the ring would begin to accumulate by gravitational instability once its mass was on the order of 0.01 M_M. Because planetesimals in near-Earth orbit have periods of a few hours, whereas those in heliocentric orbit have periods of years, the accumulation process generally must occur on a much shorter time scale in Earth orbit than in heliocentric orbit. Chapman and Greenberg (1984) estimated the time scale for accretion in Earth orbit to be on the order of a year, while new material would be added to the circumterrestrial swarm only on times of the order of 10^7 years. Hence, in order to retain the compositional filter, one must find a way to prevent accumulation in the circumterrestrial swarm in order to delay the formation of the proto-Moon and allow time for further iron/silicate differentiation. Weidenschilling *et al.* (1986) have proposed that premature formation of the Moon could be prevented by repeated shattering following impacts with incoming planetesimals, but considering that on the order of 10^6 successive high velocity impacts are necessary (once about every 10 years for about 10^7 years), this seems very unlikely.

The most important dynamical problem associated with the binary accretion model is the origin of the angular momentum necessary for creation and maintenance of the circumterrestrial swarm. The swarm is postulated to have originally formed through inelastic collisions near Earth between planetesimals on heliocentric orbit, resulting in decreased kinetic energy, increased Jacobi constants (Section 6), and hence capture into Earth orbit; subsequent mass is added through further collisions within the heliocentric swarm and between members of the geocentric and heliocentric swarms. The question becomes whether or not adding mass in this manner can also result in the addition of sufficient angular momentum to (1) counteract the collisional drag causing the geocentric particles to spiral into the Earth and (2) deposit the current angular momentum of the Earth-Moon system.

Ruskol (1972c) estimated that the incoming planetesimals would carry sufficient angular momentum for the Moon to be formed around 10–30 R_E, with about one-half the total angular momentum of the Earth-Moon system; the remaining one-half would have to have been accumulated by the Earth. Harris and Kaula (1975) favored forming the Moon at 10 R_E. These distances are consistent with calculations of the evolution of the lunar orbit (Section 3). However, as noted in Section 2, studies by Giuli (1968) and Harris (1977b) found that the Earth could not have received the majority of its angular momentum through the accumulation of a large number of planetesimals on moderately eccentric (e > 0.03) orbits (see also Harris, 1978). The same concerns apply to the circumterrestrial swarm; the net effect of adding together the angular momentum from a large number of planetesimals on eccentric orbits is only a small amount of prograde angular momentum. This was confirmed by Weidenschilling (1984a), whose numerical integrations of planetesimal orbits in near-Earth space found little preference for collisions between initially unbound or between bound and unbound planetesimals to transfer net angular momentum to the circumterrestrial swarm. Similarly, Herbert and Davis (1984) found that the addition of heliocentric planetesimals to a preexisting circumterrestrial swarm results in increased mass but decreased total angular momentum for the swarm, so that collisions between the swarm particles and incoming planetesimals tend to result in either infall onto the Earth or else ejection from Earth orbit. Only specialized distributions of planetesimals, for example, depleted in Earth-crossing orbits, result in appreciable additional angular momentum for the circumterrestrial swarm (Herbert et al., 1985). Such special cases can be excluded, because the time scale for repopulating the Earth-crossing orbits (depleted by impacts with the Earth-swarm system) with nearby planetesimals must be less than the time scale for the nearby planetesimals to impact the Earth-swarm system themselves.

Even if the circumterrestrial ring was once formed by some means (such as impact of two large bodies within Earth's sphere of influence, suggested by Weidenschilling et al., 1986), the subsequent flux of heliocentric planetesimals necessary for compositional filtering and mass gain dooms the ring to decay onto the Earth. Except for an unrealistically small mean eccentricity of the planetesimal orbits (Harris, 1977b) or special initial planetesimal distributions (Herbert et al., 1985), the angular momentum necessary to form the Earth-Moon system apparently cannot be plausibly obtained by accumulation of a large number of small planetesimals in this general model of the formation of the terrestrial planets.

9. Giant Impacts

The sixth method of forming the Moon is through the impact of a massive protoplanet on the surface of the Earth during the late stages of accretion. In the resulting explosion, debris from the Earth's mantle (presumably already differentiated) and from the impacting object will be blown off of the Earth. If a sufficient amount of the ejected matter enters Earth orbit, then the Moon may accumulate from the resulting circumterrestrial ring. The giant impact accretion model contains some of the elements of the precipitation fission, intact and disintegrative capture, and binary accretion models, while retaining distinct differences (Fig. 5). For example, the circumterrestrial disk formed in the giant impact model is not subject to the danger encountered by the disk in the binary accretion model of decay onto the Earth following collisions with a flux of heliocentric planetesimals. This is simply because there need no longer be a large flux of small planetesimals at the time of the giant impact; the final growth of the Earth may occur through collisions with just a few more very large planetesimals.

Öpik (1971) hinted that a near-grazing collision would be an attractive means of forming the Moon. Hartmann and Davis (1975) were the first to seriously consider the collision of a large body with the Earth as an event that could form the Moon. Hartmann and Davis's (1975) study of the formation of the terrestrial planets suggested that the largest bodies available to impact the Earth would have had a radius up to 3000 km (roughly Mars size). Soon thereafter, Cameron and Ward (1976) noted that if a Mars-sized impactor struck the Earth near grazing incidence with a relative velocity of 11 km s^{-1}, it could impart the current angular momentum of the Earth-Moon system. Cameron and Ward also pointed out that much of the debris from the impact would be vaporized, and the extra accelerations resulting from pressure gradients in the expanding, cooling vapor would be important for injecting the debris into circumterrestrial orbit. Matter ejected purely on ballistic trajectories from the impact must either leave the Earth completely or eventually reimpact the Earth, because its perigee would be effectively below the Earth's surface. Ringwood and Kesson (1977) envisioned a smaller body with a mass of 0.2 M_M, impacting a hot Earth at 13 km s^{-1}, leading to the vaporization of 2 M_M of matter from the mantle; the protolunar material would receive its angular momentum from the Earth through magnetic fields or viscous coupling. However, this variation raises again the problem of the initial rapid rotation of the Earth that is more simply solved by having the giant impact occur tangentially. Ringwood (1978, 1984) also proposed that instead of having a single giant impact, several large impacts occurred, but this variation encounters a familiar difficulty: The net angular momentum from the random impacts of several bodies is less likely to be sufficient to produce the angular momentum of the Earth-Moon system.

It has been thought that the accumulation of the terrestrial planets proceeded through the accretion of relatively small bodies (e.g., Baldwin, 1974; Greenberg *et al.,* 1978). However, more recent accretional models (Wetherill, 1985, 1986) indicate that much of the Earth most likely accumulated from a number of relatively massive planetesimals, with masses of at least 10^{24} g. It appears reasonable for the Earth to have been struck by at least one protoplanet with a mass at least as large as Mars (Wetherill, 1985, 1986; Fig. 6). The impacting body might even have been as large as three times the mass of Mars, and the relative velocity on the order of 10 km s^{-1}. If we assume the impacting body has a mass of 0.1 M_E, a relative velocity of 10 km s^{-1}, and a radius of 3000 km, it must have an impact parameter b $>$ 5000 km relative to a proto-Earth of 6000 km radius to have an angular momentum greater than that of the present Earth-Moon system. The body misses altogether or contributes little to the proto-Earth if b $>$ 7000 km. Given that an impact occurs with b $<$ 7000 km, the probability is about 5% that it will strike with impact parameter between 5000 and 7000 km and within a sector of about 40° spanning the solar system plane on the prograde side. Within this range a major fraction of the angular momentum should remain in the system after impact. This probability is small, but it is enormous compared to those discussed earlier for intact capture.

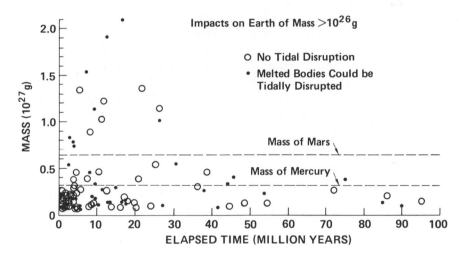

Fig. 6.   The mass distribution of large bodies impacting the proto-Earth as a function of time during 10 different calculations of the final stages of accumulation of the terrestrial planets (from Wetherill, 1985). Only impacts of bodies greater than 10^{26} g are depicted, and the two types of symbols correspond to two types of calculations, where tidal disruption during close encounters was either prohibited or was allowed provided that the bodies had previously been subjected to sufficient collisions to produce complete melting (in the absence of energy loss). In either case, it can be seen that a giant impact of a body with a mass on the order of Mars or more is likely to have occurred during the accumulation of the Earth.

Early work on the effects of high-velocity impacts of iron or silicate bodies into a silicate mantle led to the conclusion that little vaporization would occur for impact velocities of 10 km s^{-1} (Jakosky and Ahrens, 1978; Rigden and Ahrens, 1981). However, improved equations of state have since led to predictions of substantially increased amounts of vaporization (Boslough and Ahrens, 1983). Boslough and Ahrens found that a body as small as 0.2 M_M impacting at 15 km s^{-1} could vaporize one lunar mass of silicates. Even larger bodies can vaporize more matter at lower impact velocities (Jakosky and Ahrens, 1978), so it appears reasonable for a body the mass of Mars (about 8 M_M) with an impact velocity of 10 km s^{-1} to vaporize several times the mass of the Moon, thereby allowing the subsequent processes of orbital injection and lunar accumulation to be comfortably less than 100% efficient. Melosh (1985) and Melosh and Sonnett (1986) have used theoretical models based on the steady-state structure of jets produced during high-velocity impacts to predict the amount of matter blown outward following the impact of different sized bodies with the Earth. They find that for impact parameters close to tangential impacts, a Mars-sized body will result in the jetting outward of up to almost half the projectile mass, with about half the ejected mass being contributed by the impactor and half by the Earth.

Cameron (1984, 1985b) has simulated the ejection of the debris following the impact and explosion. He found that if the vaporized matter quickly recondenses into small (cm) particles, the resulting particle cloud could still act like a compressible fluid, and the pressure gradients within could then effect the process of orbital injection. However, Cameron found that because the particle cloud would be effectively "cold," the desired accelerations would be minimal, and little matter would be emplaced beyond the Roche limit. He then suggested that the self-gravity of the ejected matter or stirring by remnants from the impacting iron core could further accelerate the prelunar matter. Stevensen (1984) also considered mechanisms for achieving orbital injection and found that if turbulently viscous jets are formed in the explosion, the viscous stresses can inject a portion of the jet into Earth orbit. Stevenson further proposed another variation—that a giant impact could have created a rapidly rotating, massive atmosphere of silicate vapor and liquid, which could then cool and leave behind matter outside the Roche limit; this scenario is very reminiscent of the precipitation model (Section 5). Boss (1985b) found that mass shedding from a rapidly rotating proto-Earth spun up following a giant impact could also contribute mass and angular momentum to a prelunar disk.

Ward and Cameron (1978) studied the evolution of a prelunar disk that might have been formed in the giant impact model. Assuming the disk to be composed of particles rather than gas, they investigated the portions of the disk within the Roche limit of the Earth, where tidal forces would have prohibited large bodies from forming through gravitational instability. Instead, the disk would have repeatedly tried to gravitationally fragment, only to have the nascent bodies broken up by tidal stresses on rotational period time scales. The process results in dissipation of energy and transfer of mass inward and outward through the action of an effective viscosity. The prelunar mass must first be moved outside the Roche limit if accumulation is to occur. Ward and Cameron (1978) estimated that a "cold" disk could spread out beyond the Roche limit within a few years by this process. However, Thompson and Stevenson (1983) pointed out that the energy dissipation rate estimated by Ward and Cameron would result in heating the disk to 6000°K, well above the vaporization point. Cameron (1983) then suggested that the disk would be regulated by vaporization to a temperature closer to 2000°K within the Roche limit. The presence of substantial amounts of hot, vaporized gas must modify the original assumption of a cold, particulate disk that was employed in deriving the dispersion relation for gravitational instability; thus this scenario requires further study to self-consistently determine the evolution of a "cold" prelunar disk.

Thompson and Stevenson (1983) considered another source of gravitational instability in the prelunar disk. They showed that for a mixed two-phase disk composed of liquid silicates and bubbles of vaporized silicates, the sound speed can be much lower than in either phase in a pure state. Because the sound speed measures the ability of thermal pressure to resist gravitational instability, a two-phase disk with a very low sound speed (around 0.01 km s^{-1}) can be gravitationally unstable. Thompson and Stevensen found that even for a hot (2800°K) disk with a mass only twice that of the Moon, gravitational instability can occur inside and outside the Roche limit. Assuming that the instability results in an effective viscosity as proposed by Ward and Cameron (1978), Thompson and Stevenson find that the disk should expand on a time scale of about 50 years, which is slightly less than the estimated cooling time of their disk of about 100 years. It is then possible in Thompson and Stevenson's (1983) model, but as yet not demonstrated, that the Moon formed as a molten body outside the Roche limit through processes initiated by gravitational instability of a "hot" prelunar disk.

Many uncertainties remain; we note only a few here. The nonlinear development of the gravitational instability in the disk is unknown; the analysis to date applies to the linear regime. The two-phase process of Thompson and Stevenson depends on having a very large mass ratio (200:1) of liquid to gas in the prelunar disk in order to obtain the anomalously low sound speed, and it remains to be seen whether or not the prelunar disk would evolve toward the critical ratio. This scenario would almost certainly lead to formation of the Moon in the equatorial plane, since the time required to move material outside the Roche limit would allow damping of the material to the equatorial plane before the Moon could accumulate. Hence, if the 10° inclination of the lunar orbit to the Earth's equator when the Moon was near 10 R_E is considered a firm constraint, another event such as another large impact on the Earth or on the proto-Moon is necessary to alter the relative inclination. A subsequent impact on the Earth is preferred in order to retain the chemical composition of mantle-like material for the Moon, unless the impacting body was a silicate fragment from the prior disruption of a differentiated planetesimal.

In our opinion, this sixth mechanism of lunar formation has the least obvious flaws, although it is a series of events of perhaps somewhat low probability. Still, the Earth-Moon system is unique in the solar system, so it is not outrageous to invoke a rare event if the probability for that event is not vanishingly small. We shall qualify the support for this model in our conclusions, which follow in the next section.

10. Conclusions

We have considered six different categories of models for the formation of the Moon, using the general theory of the accumulation of the terrestrial planets to define the processes occurring throughout the terrestrial planet region during lunar formation. Of the six categories, five can apparently be ruled out on the basis of dynamical impossibility or implausibility; only a single candidate mechanism remains. Rotational fission and disintegrative capture are dynamically impossible for the viscous protoplanets that form by accumulation of planetesimals; this conclusion can only be refuted if it is shown that the protoplanets were completely molten and hence inviscid at the time of the hypothesized events. Precipitation fission is dynamically implausible because a proto-Earth forming by the accumulation of a large number of planetesimals has a negligible chance of accumulating sufficient angular momentum to result in rotational fission of a massive primordial atmosphere. This objection also applies to rotational fission and adds to the previous argument against the mechanism. Intact capture requires dissipation of the excess kinetic energy between heliocentric and circumterrestrial orbit. Because the possible mechanisms for achieving this dissipation are relatively inefficient, the Moon could only have been captured from a orbit that was very close to the orbit of the Earth. This would require the Moon to have grown to its present size without having suffered any appreciable close encounters with the Earth prior to the capture event; otherwise the Moon would have been either swept up by the Earth, captured as a much smaller body, or else perturbed to another orbit. Thus intact capture is dynamically implausible. Binary accretion is also dynamically implausible, in part for the same reason given for precipitation fission: The net angular momentum obtained from a swarm of planetesimals trapped in Earth orbit is insufficient to form or maintain a circumterrestrial disk.

The only remaining model is formation of the Moon following the impact of a massive (Mars-size or larger) protoplanet into the silicate mantle of a differentiated proto-Earth. Such impacts are likely to have occurred during the latter phases of the accumulation of the terrestrial planets, with a probability on the order of unity. If one of these giant impacts happened to strike the Earth nearly tangentially, the resulting explosion could eject sufficient mantle and impactor material into circumterrestrial orbit to later result in the formation of the Moon. Preliminary work on the evolution of the ejected debris has suggested several mechanisms for ensuring that the mass in the prelunar accretion disk is spread outside the Earth's Roche limit, thereby allowing the Moon to safely accumulate.

The giant impact model of lunar origin is not inconsistent with the absence of major satellites for the other terrestrial planets. Given that the Earth experienced at least one giant impact, the probability may be about 5% that the impact has the proper impact parameter and orientation to leave the bound mass with the angular momentum of the Earth-Moon system. If one considers that an impact could just as well have produced a retrograde Earth-Moon system, or one with arbitrary inclination, then the probability of producing a generic Earth-Moon system is about 50% *for each impact*. If this estimate is correct, and each terrestrial planet encountered at least one giant impact, then it is likely that more than one terrestrial planet should have obtained a giant satellite through this mechanism, and furthermore, that the spins of the terrestrial planets may be the result of stochastic giant impacts. Because the other terrestrial planets may have lost major satellites through tidal evolution (Ward and Reid, 1973; Counselman, 1973; Harris, 1978; O'Keefe, 1984), the absence of such satellites does not provide an observational constraint.

Whereas the giant impact model appears to meet the dual criteria of dynamic possibility and plausibility, we must caution the reader not to jump on this bandwagon prematurely. The model is new and several parts of the scenario have been asserted without many key processes being worked out in detail. A completely self-consistent theory of lunar origin is far from being constructed. It is possible (or even likely?) that thorough exploration of the various processes will reveal a fatal flaw in the hypothesis. One constraint imposed by the model in its present form is that the Moon was essentially completely molten during some phase of its formation. A molten state is desirable for the devolatilization, but the lack of major thrust faults on the lunar surface that would have formed in a solid crust as the interior shrank upon cooling is not consistent with such an initial molten state after consolidation (Solomon and Chaiken, 1976; Solomon, 1977). A more constant lunar radius is implied, which requires forming the Moon from solid matter that perhaps formed during the cooling of the lunar disk. The details of an impact explosion of a scale well beyond any so far considered are essentially unknown. Will the result of near-grazing incidence impacts be the shattering of the projectile planetesimal into many smaller escaping bodies, so that collisional disintegration rules and removes from the mass spectrum most of the relatively large bodies needed for the giant impact scenario? Will the partition of the energy into heating, melting, vaporization, and mechanical kinetic energy really be favorable for the insertion of sufficient material into circumterrestrial orbits? What are the constraints on the impact (if such can be found) that could preserve a presumably molten iron core of the impacting body in intact globs to be later accreted by the Earth while much of the planetesimal's silicate mantle and a lot of the Earth's mantle are vaporized and ejected into orbit by pressure gradients? Can it really happen this way? What becomes of the material once it is inserted in orbit? Will turbulent viscosity (a notoriously "hand-waved" phenomenon) be capable of pushing sufficient material beyond the Roche limit to form the Moon or can interactions induced by gravitational instabilities effect the transfer of angular momentum? Can gravitational instabilities occur in the hot gas-liquid phase or must they wait for the formation of the thin disk of solid particles? If the latter is the case, is most of the material left inside the Roche limit?

Several or perhaps even most of these questions may defy definitive answers. On the other hand, the giant impact scenario for the origin of the Moon is the only hypothesis so far offered that does not appear to be impossible or so improbable as to currently warrant rejection. The problems for the hypothesis are reasonably well defined and we hope they are vigorously and objectively attacked in the next few years. It will be interesting to watch this model evolve as the various processes so far invoked for its success are investigated in detail. Have we found the mechanism of lunar origin?

Acknowledgments. We thank Pat Cassen, Don Davis, Richard Durisen, Richard Greenberg, Alan Harris, William Kaula, and George Wetherill for their comments on this review. A.P.B. has been indirectly supported by the Innovative Research Program of the National Aeronautics and Space Administration under grant NAGW-398. S.J.P. has been supported by the Planetary Geology and Geophysics Program of the National Aeronautics and Space Administration under grant NGR 05-010-062.

11. References

Aggarwal H. R. and Oberbeck V. R. (1974) Roche limit of a solid body. *Astrophys. J., 239*, 577–588.

Alfvén H. (1963) The early history of the Moon and the Earth. *Icarus, 1*, 357–363.

Alfvén H. and Arrhenius G. (1969) Two alternatives for the history of the Moon. *Science, 165*, 11–17.

Alfvén H. and Arrhenius G. (1972) Origin and evolution of the Earth-Moon system. *The Moon, 5*, 210–230.

Alfvén H. and Arrhenius G. (1976) *Evolution of the Solar System.* NASA SP-345, Scientific and Technical Information Office, NASA, Washington, D.C.

Anderson D. L. (1972) The origin of the Moon. *Nature, 239*, 263–265.

Andrews D. J. (1982) Could the Earth's core and Moon have formed at the same time? *Geophys. Res. Lett., 9*, 1259–1262.

Bailey J. M. (1969) The Moon may be a former planet. *Nature, 223*, 251–253.

Baldwin R. B. (1974) On the accretion of the Earth and the Moon. *Icarus, 23*, 97–107.

Binder A. B. (1974) On the origin of the Moon by rotational fission. *The Moon, 11*, 53–76.

Binder A. B. (1978) On fission and the devolatilization of a Moon of fission origin. *Earth Planet. Sci. Lett., 41*, 381–385.

Binder A. B. (1980) The first few hundred years of evolution of a Moon of fission origin. *Proc. Lunar Planet. Sci. Conf. 11th*, pp. 1931–1939.

Binder A. B. (1984) On the origin of the Moon by rotational fission (abstract). In *Papers Presented to the Conference on the Origin of the Moon*, p. 47. Lunar and Planetary Institute, Houston.

Boslough M. B. and Ahrens T. J. (1983) Shock-melting and vaporization of anorthosite and implications for an impact-origin of the Moon (abstract). In *Lunar and Planetary Science XIV*, pp. 63–64. Lunar and Planetary Institute, Houston.

Boss A. P. (1984) Angular momentum transfer by gravitational torques and the evolution of binary protostars. *Mon. Not. Roy. Astron. Soc., 209,* 543–567.

Boss A. P. (1985a) Three dimensional calculations of the formation of the presolar nebula from a slowly rotating cloud. *Icarus, 61,* 3–9.

Boss A. P. (1985b) Protoearth mass shedding and the origin of the Moon (abstract). *Bull. Amer. Astron. Soc., 17.*

Boss A. P. and Mizuno H. (1984) The dynamic fission instability and the origin of the Moon (abstract). In *Papers Presented to the Conference on the Origin of the Moon,* p. 36. Lunar and Planetary Institute, Houston.

Boss A. P. and Mizuno H. (1985) Dynamic fission instability of dissipative protoplanets. *Icarus, 63,* 134.

Brush S. G. (1982) Nickel for your thoughts: Urey and the origin of the Moon. *Science, 217,* 891–898.

Buck W. R. (1982) Lunar breakup and capture close to the Earth (abstract). In *Lunar and Planetary Science XIII,* pp. 73–74. Lunar and Planetary Institute, Houston.

Cameron A. G. W. (1970) Formation of the Earth-Moon system. *EOS (Trans. Amer. Geophys. Union), 51,* 628–633.

Cameron A. G. W. (1972) Orbital eccentricity of Mercury and the origin of the Moon. *Nature, 240,* 299–300.

Cameron A. G. W. (1973) Properties of the solar nebula and the origin of the Moon. *The Moon, 7,* 377–383.

Cameron A. G. W. (1983) Origin of the atmospheres of the terrestrial planets. *Icarus, 56,* 195–201.

Cameron A. G. W. (1984) Formation of the prelunar accretion disk (abstract). In *Papers Presented to the Conference on the Origin of the Moon,* p. 58. Lunar and Planetary Institute, Houston.

Cameron A. G. W. (1985a) Formation and evolution of the primitive solar nebula. In *Protostars and Planets* (D. C. Black and M. S. Mathews, eds.), Univ. of Arizona Press, Tucson, in press.

Cameron A. G. W. (1985b) Formation of the prelunar accretion disk. *Icarus, 62,* 319–327.

Cameron A. G. W. and Ward W. R. (1976) The origin of the Moon (abstract). In *Lunar Science VII,* pp. 120–122. The Lunar Science Institute, Houston.

Chandrasekhar S. (1969) *Ellipsoidal Figures of Equilibrium.* Yale University Press, New Haven. 253 pp.

Chapman C. R. and Greenberg R. (1984) A circumterrestrial compositional filter. In *Papers Presented to the Conference on the Origin of the Moon,* p. 56. Lunar and Planetary Institute, Houston.

Clark S. R., Turcotte D. L., and Nordmann J. C. (1975) Accretional capture of the Moon. *Nature, 258,* 219–220.

Conway B. A. (1982) On the history of the lunar orbit. *Icarus, 51,* 610–622.

Conway B. A. (1984) The Moon's orbit history and inferences on its origin (abstract). In *Papers Presented to the Conference on the Origin of the Moon,* p. 33. Lunar and Planetary Institute, Houston.

Counselman C. C. (1973) Outcomes of tidal evolution. *Astrophys. J., 180,* 307–314.

Cox L. P. (1984) A numerical investigation of planetesimal collision trajectories with a Moon accumulating in Earth orbit (abstract). In *Papers Presented to the Conference on the Origin of the Moon,* p. 38. Lunar and Planetary Institute, Houston.

Cox L. P. and Lewis J. S. (1980) Numerical simulation of the final stages of terrestrial planet formation. *Icarus, 44,* 706–721.

Darwin G. H. (1879) On the precession of a viscous spheroid, and on the remote history of the Earth. *Philos. Trans. Roy. Soc., 170,* 447–530.

Darwin G. H. (1880) On the secular changes in the elements of the orbit of a satellite revolving about a tidally distorted planet. *Philos. Trans. Roy. Soc., 171,* 713–891.

Dobrovolskis A. R. (1982) Internal stresses in Phobos and other triaxial bodies. *Icarus, 52,* 136–148.

Durisen R. H. and Gingold R. A. (1986) Numerical simulations of fission, this volume.

Durisen R. H. and Scott E. H. (1984) Implications of recent numerical calculations for the fission theory of the origin of the Moon. *Icarus, 58*, 153–158.

Durisen R. H. and Tohline J. E. (1985) Fission of rapidly rotating fluid systems. In *Protostars and Planets* (D. C. Black and M. S. Mathews, eds), Univ. of Arizona Press, Tucson, in press.

Durisen R. H., Gingold R. A., Tohline J. E., and Boss A. P. (1985) The binary fission hypothesis: a comparison of results from different numerical codes. *Astrophys. J.*, in press.

Dyczmons V. (1978) Formation of planetary and satellite systems. *Astrophys. Space Sci., 58*, 521–523.

Fremlin J. H. (1973) The origin of the Moon. *Nature, 242*, 317–318.

Gerstenkorn H. (1955) Uber Gezeitenreibung beim Zweikorperproblem. *Z. Astrophys., 36*, 245–274.

Gerstenkorn H. (1967) On the controversy over the effect of tidal friction upon the history of the Earth-Moon system. *Icarus, 7*, 160–167.

Gerstenkorn H. (1968) A reply to Goldreich. *Icarus, 9*, 394–397.

Gerstenkorn H. (1969) The earliest past of the Earth-Moon system. *Icarus, 11*, 189–207.

Giuli R. T. (1968) Gravitational accretion of small masses attracted from large distances as a mechanism for planetary rotation. *Icarus, 9*, 186–190.

Goldreich P. (1966) History of the lunar orbit. *Rev. Geophys., 4*, 411–439.

Goldreich P. (1968) On the controversy over the effect of tidal friction upon the history of the Earth-Moon system: a reply to comments by H. Gerstenkorn. *Icarus, 9*, 391–393.

Goldreich P. and Peale S. J. (1966) Spin-orbit coupling in the solar system. *Astron. J., 71*, 425–438.

Goldreich P. and Ward W. R. (1973) The formation of planetesimals. *Astrophys. J., 183*, 1051–1061.

Greenberg R. (1980) Numerical simulation of planet growth: early runaway accretion (abstract). In *Lunar and Planetary Science XI*, pp. 365–367. Lunar and Planetary Institute, Houston.

Greenberg R., Chapman C. R., Davis D. R., Drake M. J., Hartmann W. K., Herbert F. L., Jones J., and Weidenschilling S. J. (1984) An integrated dynamical and geochemical approach to lunar origin modelling (abstract). In *Papers Presented to the Conference on the Origin of the Moon*, p. 51. Lunar and Planetary Institute, Houston.

Greenberg R., Wacker J. F., Hartmann W. K., and Chapman C. R. (1978) Planetesimals to planets: numerical simulation of collisional evolution. *Icarus, 35*, 1–26.

Hanks T. C. and Anderson D. L. (1972) Origin, evolution and present thermal state of the Moon. *Phys. Earth Planet. Inter., 5*, 409–425.

Harris A. W. (1975) Collisional breakup of particles in a planetary ring. *Icarus, 24*, 190–192.

Harris A. W. (1977a) The effect of tidal friction on the origin and thermal evolution of the Moon (abstract). In *Lunar Science VIII*, pp. 401–402. The Lunar Science Institute, Houston.

Harris A. W. (1977b) An analytical theory of planetary rotation rates. *Icarus, 31*, 168–174.

Harris A. W. (1977c) On the origin of the satellites of Mars. *Bull. Amer. Astron. Soc., 9*, 519.

Harris A. W. (1978) Satellite formation, II. *Icarus, 34*, 128–145.

Harris A. W. and Kaula W. M. (1975) A co-accretional model of satellite formation. *Icarus, 24*, 516–524.

Harris A. W. and Ward R. (1982) Dynamical constraints on the formation and evolution of planetary bodies. *Ann. Rev. Earth Planet. Sci., 10*, 61–108.

Hartmann W. K. (1984) Stochastic $\neq$ ad hoc. In *Papers Presented to the Conference on the Origin of the Moon*, p. 39. Lunar and Planetary Institute, Houston.

Hartmann W. K. and Davis D. R. (1975) Satellite-sized planetesimals and lunar origin. *Icarus, 24*, 504–515.

Herbert F. and Davis D. R. (1984) Models of angular momentum input to a circumterrestrial swarm from encounters with heliocentric planetesimals (abstract). In *Papers Presented to the Conference on the Origin of the Moon*, p. 53. Lunar and Planetary Institute, Houston.

Herbert F., Davis D. R., and Weidenschilling S. J. (1985) On forming the Moon in geocentric orbit; dynamical evolution of a circumterrestrial swarm (abstract). In *Lunar and Planetary Science XVI*, pp. 341–342. Lunar and Planetary Institute, Houston.

Horedt G. P. (1976) Pre-capture orbits of the Moon. *The Moon, 15*, 439–443.

Jakosky B. M. and Ahrens T. J. (1978) Constraints on lunar formation by impact vaporization (abstract). In *Lunar and Planetary Science IX*, pp. 582–584. Lunar and Planetary Institute, Houston.

Jeans J. H. (1919) *Problems of Cosmogony and Stellar Dynamics.* Cambridge University Press, Cambridge. 293 pp.

Jeffreys H. (1930) Amplitude of tidal resonances. *Mon. Not. Roy. Astron. Soc., 91*, 169–173.

Jeffreys H. (1947) The relation of cohesion to Roche's limit. *Mon. Not. Roy. Astron. Soc., 107*, 260–272.

Kaula W. M. (1964) Tidal dissipation by solid friction and the resulting orbital evolution. *Rev. Geophys., 2*, 661–685.

Kaula W. M. (1971) Dynamical aspects of lunar origin. *Rev. Geophys. Space Phys., 9*, 217–238.

Kaula W. M. (1975) Mechanical processes affecting differentiation of protolunar material. In *Proc. Soviet-American Conference on Cosmochemistry of the Moon and Planets*, pp. 630–637, Nauka, Moscow, translated in English in NASA SP-370 (1977), pp. 805–813.

Kaula W. M. (1977) On the origin of the Moon, with emphasis on bulk composition. *Proc. Lunar Sci. Conf. 8th*, pp. 321–331.

Kaula W. M. (1979) Thermal evolution of Earth and Moon growing by planetesimal impacts. *J. Geophys. Res., 84*, 999–1008.

Kaula W. M. (1980) The beginning of the Earth's thermal evolution. *The Continental Crust and its Mineral Deposits* (D. W. Strangway, ed.), pp. 25–34. Geological Association of Canada Special Paper 20.

Kaula W. M. and Beachey A. E. (1986) Mechanical models of close approaches and collisions of large protoplanets, this volume.

Kaula W. M. and Harris A. W. (1973) Dynamically plausible hypotheses of lunar origin. *Nature, 245*, 367–369.

Kaula W. M. and Harris A. W. (1975) Dynamics of lunar origin and orbital evolution. *Rev. Geophys. Space Phys., 13*, 363–371.

Knopoff L. (1964) Q. *Rev. Geophys., 2*, 625–660.

Lambeck K. (1975) Effects of tidal dissipation in the oceans on the Moon's orbit and the Earth's rotation. *J. Geophys. Res., 80*, 2917–2925.

Lambeck K. (1980) *The Earth's Variable Rotation: Geophysical Causes and Consequences.* Cambridge University Press, Cambridge. 449 pp.

Liapounoff A. M. (1905) Sur un probleme de Tchebychef. *Mem. Acad. St. Petersburg, XVIII*, 3.

Love A. E. H. (1927) *A Treatise on the Mathematical Theory of Elasticity.* Dover (1944), New York.

Lucy L. B. and Ricco E. (1979) The significance of binaries with nearly identical components. *Astron. J., 84*, 401–412.

Lyttleton R. A. (1953) *The Stability of Rotating Liquid Masses.* Cambridge University Press, Cambridge. 150 pp.

Lyttleton R. A. (1967) Dynamical capture of the Moon by the Earth. *Proc. Roy. Soc. London, A296*, 285–292.

MacDonald G. J. F. (1964) Tidal friction. *Rev. Geophys., 2*, 467–541.

Malcuit R. J., Winters R. R., and Mickelson M. E. (1977) Is the Moon a captured body? (abstract). In *Lunar Science VIII*, pp. 608–609. The Lunar Science Institute, Houston.

Malcuit R. J., Winters R. R., and Mickelson M. E. (1984) A testable gravitational capture model for the origin of the Earth's Moon (abstract). In *Papers Presented to the Conference on the Origin of the Moon*, p. 43. Lunar and Planetary Institute, Houston.

Melosh H. J. (1985) When worlds collide: Jetted vapor plumes and the Moon's origin (abstract). In *Lunar and Planetary Science XVI*, pp. 552–553. Lunar and Planetary Institute, Houston.

Melosh H. J. and Sonnett C. P. (1986) When worlds collide: Jetted vapor plumes and the Moon's origin, this volume.

McKinnon W. B. and Mueller S. W. (1984) A reappraisal of Darwin's fission hypothesis and a possible limit to the primordial angular momentum of the Earth (abstract). In *Papers Presented to the Conference on the Origin of the Moon*, p. 34. Lunar and Planetary Institute, Houston.

Mignard F. (1981) The lunar orbit revisited, III. *Moon and Planets, 24*, 189–207.

Mitler H. E. (1975) Formation of an iron-poor Moon by partial capture, or: yet another exotic theory of lunar origin. *Icarus, 24*, 256–268.

Mizuno H. and Boss A. P. (1984) Tidal disruption and the origin of the Moon (abstract). In *Papers Presented to the Conference on the Origin of the Moon*, p. 37. Lunar and Planetary Institute, Houston.

Mizuno H. and Boss A. P. (1985) Tidal disruption of dissipative planetesimals. *Icarus, 63*, 109.

Moulton F. R. (1909) Notes on the possibility of fission of a contracting rotating fluid mass. *Astrophys. J., 29*, 1–13.

Nakagawa Y., Hayashi C., and Nakazawa K. (1983) Accumulation of planetesimals in the solar nebula. *Icarus, 54*, 361–376.

Nakagawa Y., Nakazawa K., and Hayashi C. (1981) Growth and sedimentation of dust grains in the primordial solar nebula. *Icarus, 45*, 517–528.

Nakazawa K., Komuro T., and Hayashi C. (1983) Origin of the Moon—Capture by gas drag of the Earth's primordial atmosphere. *Moon and Planets, 28*, 311–327.

Nordmann J. C. and Turcotte D. L. (1977) Numerical calculations of the cross-section for the accretional capture of the Moon by the Earth. *Proc. Lunar Sci. Conf. 8th*, pp. 57–65.

O'Keefe J. A. (1969) Origin of the Moon. *J. Geophys. Res., 74*, 2758–2767.

O'Keefe J. A. (1970a) The origin of the Moon. *J. Geophys. Res., 75*, 6565–6574.

O'Keefe J. A. (1970b) Apollo 11: implications for the early history of the solar system. *EOS (Trans. Amer. Geophys. Union), 51*, 633–636.

O'Keefe J. A. (1972a) Inclination of the Moon's orbit: the early history. *Irish Astron. J., 10*, 241–250.

O'Keefe J. A. (1972b) The origin of the Moon: theories involving joint formation with the Earth. *Astrophys. Space Sci., 16*, 201–211.

O'Keefe J. A. (1974) The formation of the Moon. *The Moon, 9*, 219–225.

O'Keefe J. A. (1984) Fission of Venus? (abstract). In *Lunar and Planetary Science XV*, pp. 615–616. Lunar and Planetary Institute, Houston.

O'Keefe J. A. and Sullivan E. C. (1978) Fission origin of the Moon: cause and timing. *Icarus, 35*, 272–283.

Öpik E. J. (1950) Roche's limit; rings of Saturn. *Irish Astron. J., 1*, 25–26.

Öpik E. J. (1969) The Moon's surface. *Ann. Rev. Astron. Astrophys., 7*, 473–526.

Öpik E. J. (1971) Comments on lunar origin. *Irish Astron. J., 10*, 190–238.

Peale S. J. and Cassen P. (1978) Contribution of tidal dissipation to lunar thermal history. *Icarus, 36*, 245–269.

Petersons H. F. (1984) Possible conditions for formation of satellite from protoearth by ejection of material. *Earth, Moon and Planets, 31*, 15–24.

Poirier J. P., Boloh L., and Chambon P. (183) Tidal dissipation in small viscoelastic ice Moons: the case of Enceladus. *Icarus, 55*, 218–230.

Rigden S. M. and Ahrens T. J. (1981) Impact vaporization and lunar origin (abstract). In *Lunar and Planetary Science XII*, pp. 885–887. Lunar and Planetary Institute, Houston.

Ringwood A. E. (1960) Some aspects of the thermal evolution of the Earth. *Geochim. Cosmochim. Acta, 20*, 241–249.

Ringwood A. E. (1970) Origin of the Moon: the precipitation hypothesis. *Earth Planet. Sci. Lett., 8*, 131–140.

Ringwood A. E. (1972) Some comparative aspects of lunar origin. *Phys. Earth Planet. Inter., 6*, 366–376.

Ringwood A. E. (1978) Origin of the Moon (abstract). In *Lunar and Planetary Science IX*, pp. 961–963. Lunar and Planetary Institute, Houston.

Ringwood A. E. (1979) *Origin of the Earth and Moon.* Springer-Verlag, New York. 295 pp.

Ringwood A. E. (1984) Origin of the Moon (abstract). In *Papers Presented to the Conference on the Origin of the Moon*, p. 46. Lunar and Planetary Institute, Houston.

Ringwood A. E. and Kesson S. E. (1977) Composition and origin of the Moon. *Proc. Lunar Sci. Conf. 8th*, pp. 371–398.

Roberts P. H. and Stewartson K. (1963) On the stability of a Maclaurin spheroid of small viscosity. *Astrophys. J., 137*, 777–790.

Rosenkilde C. E. (1967) The tensor virial-theorem including viscous stress and the oscillations of a Maclaurin spheroid. *Astrophys. J., 148*, 825–832.

Rubincam D. P. (1975) Tidal friction and the early history of the Moon's orbit. *J. Geophys. Res., 80*, 1537–1548.

Ruskol E. L. (1961) The origin of the Moon—I. *Soviet Astronomy-AJ, 4*, 657–668.

Ruskol E. L. (1963) The origin of the Moon—II. *Soviet Astronomy-AJ, 7*, 221–227.

Ruskol E. L. (1966) On the past history of the Earth-Moon system. *Icarus, 5*, 221–227.

Ruskol E. L. (1972a) The origin of the Moon—III. Some aspects of the dynamics of the circumterrestrial swarm. *Soviet Astronomy-AJ, 15*, 646–654.

Ruskol E. L. (1972b) Possible differences in the chemical composition of the Earth and Moon, for a Moon formed in the circumterrestrial swarm. *Soviet Astronomy-AJ, 15*, 1061–1063.

Ruskol E. L. (1972c) On the initial distance of the Moon forming in the circumterrestrial swarm. *The Moon, 5*, 402–404.

Ruskol E. L. (1973) On the model of the accumulation of the Moon compatible with the data on the composition and the age of lunar rocks. *The Moon, 6*, 190–201.

Ruskol E. L. (1975) The origin of the Moon. In *Proc. Soviet-American Conference on Cosmochemistry of the Moon and Planets*, pp. 638–644. Nauka, Moscow, translated in English in NASA SP-370 (1977), pp. 815–822.

Ruskol E. L. (1979) Thermal effect of the collision of two massive bodies and the initial temperature on the Moon (abstract). In *Lunar and Planetary Science X*, pp. 1042–1044. Lunar and Planetary Institute, Houston.

Safronov V. S. (1969) *Evolution of the Protoplanetary Cloud and Formation of the Earth and the Planets.* Nauka, Moscow. Translated by the Israel Program for Scientific Translation (1972).

Safronov V. S. (1975) Time scale for the formation of the Earth and planets and its role in their geochemical evolution. In *Proc. Soviet-American Conference on Cosmochemistry of the Moon and Planets*, pp. 624–629. Nauka, Moscow, translated in English in NASA SP-370 (1977), pp. 797–803.

Schofield N. (1981) On the formation of the Earth and Moon by gravitational accretion in a dust disc. *Mon. Not. Roy. Astron. Soc., 197*, 1031–1047.

Schubert G., Stevenson D., and Cassen P. (1980) Whole planet cooling and the radiogenic heat source contents of the Earth and Moon. *J. Geophys. Res., 85*, 2531–2538.

Sekiguchi N. (1970) On the fissions of a solid body under influence of tidal force. *The Moon, 1*, 429–439.

Singer S. F. (1968) The origin of the Moon and geophysical consequences. *Geophys. J. Roy. Astron. Soc., 15*, 205–226.

Singer S. F. (1970) Origin of the Moon by capture and its consequences. *EOS (Trans. Amer. Geophys. Union), 51*, 637–641.

Singer S. F. (1972) Origin of the Moon by tidal capture and some geophysical consequences. *The Moon, 5*, 206–209.

Singer S. F. and Bandermann L. W. (1970) Where was the Moon formed? *Science, 170*, 438–439.

Smith J. V. (1974) Origin of Moon by disintegrative capture with chemical differentiation followed by sequential accretion (abstract). In *Lunar Science V*, pp. 718–720. The Lunar Science Institute, Houston.

Smith J. V. (1979) A new heterogeneous accretion model for the inner planets, especially the Earth (abstract). In *Lunar and Planetary Science X*, pp. 1131–1133. Lunar and Planetary Institute, Houston.

Solomon S. (1977) The relationship between crustal tectonics and internal evolution of the Moon and Mercury. *Phys. Earth Planet. Inter., 15*, 135–145.

Solomon S. and Chaiken J. (1976) Thermal expansion and thermal stress in the Moon and terrestrial planets: clues to early thermal history. *Proc. Lunar Sci. Conf. 7th*, pp. 3229–3243.

Stevenson D. J. (1984) Lunar origin from impact on the Earth: Is it possible? (abstract). In *Papers Presented to the Conference on the Origin of the Moon*, p. 60. Lunar and Planetary Institute, Houston.

Szebehely V. and Evans R. T. (1980) On the capture of the Moon. *Celestial Mechanics, 21*, 259–264.

Tassoul J.-T. (1978) *Theory of Rotating Stars*. Princeton University Press, Princeton, N.J.

Taylor S. R. (1982) *Planetary Science: A Lunar Perspective*. Lunar and Planetary Institute, Houston. 481 pp.

Thompson A. C. and Stevenson D. J. (1983) Two-phase gravitational instabilities in thin disks with application to the origin of the Moon (abstract). In *Lunar and Planetary Science XIV*, pp. 787–788. Lunar and Planetary Institute, Houston.

Turcotte D. L., Cisne J. L., and Nordmann J. C. (1977) On the evolution of the lunar orbit. *Icarus, 30*, 254–266.

Turcotte D. L., Cooke F. A., and Willeman R. J. (1979) Parameterized convection within the Moon and the terrestrial planets. *Proc. Lunar Planet. Sci. Conf. 10th*, pp. 2375–2392.

van't Veer F. (1981) The initial mass ratio of solar type contact binaries. *Astron. Astrophys., 98*, 213–217.

Ward W. R. and Cameron A. G. W. (1978) Disc evolution within the Roche limit (abstract). In *Lunar and Planetary Science IX*, pp. 1205–1207. Lunar and Planetary Institute, Houston.

Ward W. R. and Reid M. J. (1973) Solar tidal friction and satellite loss. *Mon. Not. Roy. Astron. Soc., 164*, 21–32.

Weidenschilling S. J. (1980) Hektor: nature and origin of a binary asteroid. *Icarus, 44*, 807–809.

Weidenschilling S. J. (1984a) Capture of planetesimals into a circumterrestrial swarm (abstract). In *Papers Presented to the Conference on the Origin of the Moon*, p. 54. Lunar and Planetary Institute, Houston.

Weidenschilling S. J. (1984b) The lunar angular momentum problem (abstract). In *Papers Presented to the Conference on the Origin of the Moon*, p. 55. Lunar and Planetary Institute, Houston.

Weidenschilling S. J. (1984c) Evolution of grains in a turbulent solar nebula. *Icarus, 60*, 553–567.

Weidenschilling S. J., Greenberg R., Chapman C. R., Herbert F., Davis D. R., Drake M. J., Jones J., and Hartmann W. K. (1986) Origin of the Moon from a circumterrestrial disk, this volume.

Wetherill G. W. (1976a) The role of large impacts in the formation of the Earth and Moon (abstract). In *Lunar Science VII*, pp. 930–932. The Lunar Science Institute, Houston.

Wetherill G. W. (1976b) The role of large bodies in the formation of the Earth and Moon. *Proc. Lunar Sci. Conf. 7th*, pp. 3245–3257.

Wetherill G. W. (1980) Formation of the terrestrial planets. *Ann. Rev. Astron. Astrophys., 18*, 77–113.

Wetherill G. W. (1985) Occurrence of giant impacts during the growth of the terrestrial planets. *Science, 228*, 877–879.

Wetherill G. W. (1986) Accumulation of the terrestrial planets and implications concerning lunar origin, this volume.

Wetherill G. W. and Cox L. P. (1982) Gravitational cross sections and "runaway accretion" (abstract). In *Lunar and Planetary Science XIII*, pp. 855–856. Lunar and Planetary Institute, Houston.

Wetherill G. W. and Cox L. P. (1984) The range of validity of the two-body approximation in models of terrestrial planet accumulation. *Icarus, 60*, 40–55.

Winters R. R. and Malcuit R. J. (1977) The lunar capture hypothesis revisited. *The Moon, 17*, 353–358.

Wise D. U. (1963) An origin of the Moon by rotational fission during formation of the Earth's core. *J. Geophys. Res., 68*, 1547–1554.

Wise D. U. (1969) Origin of the Moon from the Earth: some new mechanisms and comparisons. *J. Geophys. Res., 74*, 6034–6045.

Wood J. A. (1977) Origin of Earth's Moon. In *Planetary Satellites* (J. A. Burns, ed.), pp. 513–529. Univ. of Arizona Press, Tucson.

Wood J. A. and Mitler H. E. (1974) Origin of the Moon by a modified capture mechanism, *or* half a loaf is better than a whole one (abstract). In *Lunar Science V*, pp. 851–853. The Lunar Science Institute, Houston.

Yoder C. F., Williams J. G., Dickey J. O., and Newhall X. X. (1984) Tidal dissipation in the Earth and Moon from lunar laser ranging (abstract). In *Papers Presented to the Conference on the Origin of the Moon*, p. 31. Lunar and Planetary Institute, Houston.

III. *Geochemical Constraints*

Is Lunar Bulk Material Similar to Earth's Mantle?

MICHAEL J. DRAKE

Lunar and Planetary Laboratory, University of Arizona, Tucson, AZ   85721

The Earth and Moon have the same ratios of $^{18}O/^{17}O/^{16}O$. The Moon is impoverished in metal compared to Earth. Siderophile trace element concentrations in the Moon are generally lower than in the upper mantle of the Earth, implying separation of metal from silicate following lunar assembly. Refractory elements may be enriched in the Moon compared to the upper mantle of the Earth, but identical concentrations cannot be rigorously excluded. The Moon is depleted in volatile elements compared to Earth, but detailed examination of volatile element ratios suggest that this depletion is not simply related to high temperature processes during lunar assembly. The mg# of the lunar mantle appears to be lower than the upper mantle of the Earth, although extreme lunar estimates overlap the terrestrial value. The Fe/Mn ratio of the lunar mantle appears to be higher than the upper mantle of the Earth, although Mn concentrations in both objects are low compared to eucrites, shergottites, and CI chondrites. The upper mantle of the Earth is presently more oxidized than the Moon. On the basis of these observations the Rotational Fission hypothesis of lunar origin may be rigorously excluded. The Intact Capture and Disintegrative Capture hypotheses are not readily testable by geochemical criteria. Both the Collisional Ejection hypothesis and the Binary Accretion hypothesis remain viable in their present, imprecisely defined, variants.

1. Introduction

The answer to the question posed for me by the organizers of the Conference on the Origin of the Moon, "Is lunar bulk material similar to Earth's mantle?," is "yes." Lunar bulk material is not identical in composition to the upper mantle of the Earth, however. It is not possible to reconstruct with any certainty the bulk compositions of the Moon and of the Earth, although nodules entrained in alkalic basalts permit an estimate of the concentrations of most elements except the highly incompatible ones in the upper mantle of the Earth. It is possible to establish within known limits certain compositional characteristics of both planetary bodies. It is the aim of this paper to specify those parameters with appropriate caveats so that they may provide critical tests to be satisfied *simultaneously* by any successful hypothesis of lunar origin. The parameters of interest are oxygen isotopic composition, metal/silicate ratio, siderophile trace element concentrations, refractory lithophile trace element concentrations, volatile element ratios and volatile/refractory element ratios, bulk mg#, Fe/Mn ratios, and redox state. I shall first estimate the values of these parameters and then briefly examine various hypotheses for lunar origin in their light.

2. Oxygen Isotopes

The oxygen isotopic compositions of a variety of solar system materials are summarized in Fig. 1. This diagram is important in that it places constraints on the provenance of lunar material. The cause of variations in oxygen isotopic composition is unresolved (e.g., Thiemens and Heidenreich, 1983; Esat *et al.*, 1985), but, regardless of cause, the Earth and Moon define the same mass fractionation line. The only other known materials to fall on this line are the enstatite chondrites and the enstatite achondrites (not shown). Meteorites such as ordinary (H,L,LL) and carbonaceous (C1,C2,C3) chondrites are plausibly associated with the asteroid belt at 2.1–3.3 AU (Greenberg and Chapman, 1983). The case has been made that the eucrites come from an asteroid, possibly Vesta at 2.4 AU (Consolmagno and Drake, 1977). There is increasing evidence linking the SNC meteorites with

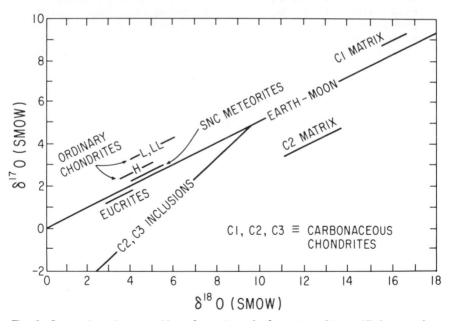

Fig. 1. Oxygen isotopic composition of a variety of solar system objects. All data are from the laboratory of R. N. Clayton, University of Chicago.

Mars at 1.5 AU (e.g., Bogard and Johnson, 1983; Bogard et al., 1984). Thus materials from objects with semi-major axes greater than 1 AU appear to have oxygen isotopic compositions distinct from the Earth and Moon. We have no information concerning the oxygen isotopic composition of materials from objects with semi-major axes less than 1 AU.

3.   Metal/Silicate Ratio

The low mean uncompressed density of the Moon (3.34 gcm^{-3}) compared to the Earth (4.45 gcm^{-3}) has long indicated that the Moon is impoverished in metal compared to Earth. Geophysical arguments including the moment of inertia factor are consistent with the Moon containing quantities of metallic iron ranging from vanishingly small up to a few weight percent (Hood and Jones, 1985). In contrast, the metal core of the Earth constitutes approximately 30% of the mass of the planet.

4.   Siderophile Trace Elements

Estimating siderophile trace element concentrations in the Moon and the upper mantle of the Earth has been a subject of considerable debate (e.g., Anders, 1978; Delano and Ringwood, 1978), but some element abundances are now established within known uncertainties (Drake, 1983; Newsom, 1984). The most successful approach applicable to both Earth and Moon follows Rammensee and Wänke (1977) in estimating depletion factors relative to CI abundances for siderophile trace elements from correlations with a lithophile incompatible trace element. Results are summarized in Table 1 for those elements with concentrations that may be inferred within useful uncertainties. The concentrations of other siderophile elements are considerably more uncertain. It must be stressed that we are sampling only the crusts and upper mantles of the Earth and Moon via basalts and nodules (Earth only). While the crust and upper mantle of the Moon constitute approximately 64% of that body, it must be remembered that the crust and upper mantle of the

TABLE 1.   Siderophile Element Depletions in the Upper Mantles of the Earth and Moon.*

	Earth	Moon
W	22 ± 10	22 ± 7
Ga	4 – 7	20 – 40
P	43 ± 10	115 ± 25
Mo	44 ± 15	1200 ± 750
Re	420	10^5

*After Newsom (1984) and Drake et al. (1984). Note that while metal/silicate partition coefficients vary significantly as a function of intensive variables, they generally increase from values of on the order of 25 for W to on the order of 10^5 for Re.

Earth constitute only 32% of its silicate portion. If convection in the Earth is "whole mantle" in scale this caveat loses its significance, but this issue is currently unresolved among geophysicists (e.g., Davies, 1981).

4.1. Tungsten

Within sampling uncertainty the concentrations of W in the upper mantles of the Moon and the Earth are the same (Rammensee and Wänke, 1977; Newsom and Drake, 1982; Newsom and Palme, 1984). This observation was cited by Ringwood (1979) in support of his Fission hypothesis.

4.2. Gallium

The concentrations of Ga in the upper mantles of the Moon and the Earth are different although considerable uncertainty exists concerning absolute values (Drake et al., 1984). Since Ga is volatile as well as somewhat siderophile, these differences may be the result of volatility considerations alone.

4.3. Phosphorus

The concentration of P in the upper mantle of the Moon is approximately three times lower than in the upper mantle of the Earth (Newsom and Drake, 1983). There is virtually no overlap in the raw data, contrary to the assertion of Ringwood (1979). Because P is somewhat volatile, the low lunar concentration of P *taken in isolation* could be due to volatility considerations above. Taken in concert with refractory Mo below, it is clear that the siderophile nature of P is most important.

4.4. Molybdenum

The concentration of Mo in the upper mantle of the Moon is approximately 30 times lower than in the upper mantle of the Earth (Newsom and Palme, 1984). Because Mo is refractory, this difference must be attributed to the siderophile behavior of Mo during metal/silicate fractionation.

4.5. Rhenium

Rhenium behaves as a somewhat incompatible element during magma genesis and its concentration cannot be inferred precisely from basalts. It appears that the concentration of Re in the upper mantle of the Moon is approximately 25 times lower than in the upper mantle of the Earth, and the abundance of Re in the upper mantle of the Earth is reasonably well constrained from analyses of nodules (Morgan et al., 1981). This difference must be attributed to the siderophile behavior of Re during metal/silicate fractionation. A suggestion by Ringwood *et al.* (1981) that Re behaves as a volatile oxide in the Moon does not appear to be supported by observation (Chou *et al.*, 1983; Drake, 1983).

5. Refractory Trace Elements

Refractory trace elements occur in the sun and in primitive meteorites in approximately constant relative proportions and, hence, are expected to be present in planetary bodies in chondritic relative proportions. It is important to estimate the concentration of refractory elements in the upper mantles of the Earth and Moon because they should be related to each other in a predictable manner in any successful hypothesis of lunar origin. Uranium is a useful refractory element to estimate because its concentration is obtainable from several geochemical and geophysical approaches, as discussed by Taylor (1982).

5.1. Uranium concentration in the Earth

The concentration of U in the Earth may be inferred from three techniques: (1) using known solar system ratios of refractory elements; (2) using known terrestrial ratios of elements, including those constrained by isotope systematics; and (3) from heat flow.

Approaches (1) and (2) first require estimation of the concentrations of elements that may be related to U through element ratios. Table 2 illustrates the two approaches. One approach involves estimating the concentration of elements in a model mantle, in this case "pyrolite" (Ringwood, 1975), by reconstruction from basalt and depleted nodule data. A second approach involves estimating which mantle nodules are most primitive, i.e., which nodules have had the least amount of basaltic melt extracted from them (Jagoutz *et al.*, 1979). (Note that estimated concentrations of refractory Ca and Al, and somewhat more volatile Si, are in good agreement using these approaches.) Using CI abundance ratios, a concentration of approximately 20 ppb U is inferred for the upper mantle of the Earth. The lower concentration derived

TABLE 2. *Estimates of the Concentration of a Refractory Element (U) in the Upper Mantle of the Earth, Using CI Ratios and Pyrolite (Ringwood, 1975), Mantle Nodule (Jagoutz et al., 1979) Approaches.*

CI abundances: Al = 0.862 wt %	Ca = 0.902 wt %	Si = 10.67 wt %
U = 8 ppb	Sr = 7.91 ppm	K = 569 ppm
Ca = 2.2 wt % in pyrolite	→ 19.5 ppb U	
Ca = 2.5 wt % in nodule	→ 22.2 ppb U	
Al = 2.1 wt % in nodule, pyrolite	→ 19.5 ppb U	
Si = 21.1 wt % in nodule, pyrolite	→ 15.8 ppb U	

Note: Al (Earth)/Al(CI) = 2.44, Si (Earth)/(CI) = 1.98, Al = 1.23 when CI and Si normalized.

from Si is consistent with a subchondritic concentration of Si due to its relative volatility, as demonstrated by the CI and Si normalized value of Al = 1.23 (it should be 1.0 if Al and Si have the same volatility and, hence, are present in the Earth in the chondritic ratio). Note, however, that Palme and Nickel (1985) argue that even the most primitive nodules from the present day upper mantle cannot be considered to retain primordial compositions.

Table 3 uses the same estimates of the upper mantle concentration of Ca to yield a U concentration using the CI Ca/Sr ratio, the terrestrial Rb/Sr ratio constrained from considerations of Rb/Sr and Sm/Nd isotope systematics (e.g., O'Nions *et al.*, 1979), and the terrestrial K/Rb ratio. Again a U concentration of approximately 20 ppb is obtained. A weak test of this value is that the inferred K concentration of 180–220 ppm is broadly consistent with the present day atmospheric $^{40}Ar/^{36}Ar$ ratio of 296 (Hart and Hogan, 1978).

TABLE 3. *Estimates of the Concentration of a Refractory Element (U) in the Upper Mantle of the Earth, Using Earth Ratios and Pyrolite (Ringwood, 1975), Mantle Nodule (Jagoutz et al., 1979) Approaches.*

Earth ratios:
$K/U = 10^4$ Rb/Sr = 0.03   K/Rb = 310

Rb/Sr systematics:

Ca = 2.2 wt % in pyrolite	$\rightarrow$ 19.3 ppm	Sr $\rightarrow$ 0.58 ppm	Rb $\rightarrow$ 179 ppm	K $\rightarrow$ 17.9 ppb U
Ca = 2.5 wt % in mantle	$\rightarrow$ 21.9 ppm	Sr $\rightarrow$ 0.66 ppm	Rb $\rightarrow$ 204 ppm	K $\rightarrow$ 20.4 ppb U

K/Ar systematics:
~200 ppm K consistent with atmospheric $^{40}Ar/^{36}Ar$ = 296

Global heat flow may be used to estimate the U concentration in the Earth using known Earth ratios of K/U/Th. If heat production equals heat flow, approximately 40 ppb U is required in the mantle (upper plus lower), assuming U does not enter the core. However, it is now generally agreed that the Earth loses heat with a long time constant, i.e., radioactive heat production does not equal heat flow (e.g., Davies, 1980; McKenzie and Richter, 1981). This conclusion leads to estimated U concentrations consistent with the approximately 20 ppb deduced from geochemical considerations.

In summary, a concentration of 20 ppb U in the mantle of the Earth is consistent with a variety of geochemical and geophysical considerations.

5.2.   Uranium concentration in the Moon

Direct samples of the lunar mantle are not available and there is no generally accepted equivalent of "pyrolite" for the Moon. Thus a different geochemical approach to the estimation of the bulk lunar U concentration is needed. The only available "global" chemical analyses of the lunar surface come from the orbital gamma-ray and X-ray experiments. It should be remembered that these experiments analyzed a relatively small area of the surface entirely contained within ±30° of the lunar equator (e.g., Metzger *et al.*, 1974) and cannot be proven to be representative of the entire lunar highland surface.

The mean highland surface concentration of K is approximately 600 ppm and of Th is approximately 0.9 ppm (Metzger *et al.*, 1977). Our task is to convert these surface estimates into global lunar estimates. This exercise is summarized in Table 4. If the mean crustal thickness is 75 km, the crust corresponds to 12.4% by volume of the Moon. If K and Th are uniformly distributed throughout a 75-km-thick crust, and if K and Th were quantitatively extracted from the mantle into the crust, the bulk Moon concentrations of K and Th are 75 ppm and 112 ppb respectively. Using a lunar K/U ratio of 2500 and a chondritic Th/U ratio of 3.53, we arrive at an estimate of approximately 31 ppb U in the bulk Moon from both K and Th.

TABLE 4. *Estimates of the Concentration of a Refractory Element (U) in the Moon Using Mean K and Th Concentrations in the Terrae From the Apollos 15 and 16 Orbital Experiments (Metzger et al, 1977).*

K in highland surface ~600 ppm from orbital geochemistry
If mean crustal thickness is 75 km, crust is 12.4% of Moon
If no K left in mantle → 75 ppm K in bulk Moon, more if K in mantle
From K/U = 2500 → U = 30 ppb

Th in highland surface ~0.9 ppm from orbital geochemistry
If no mean crustal thickness is 75 km, crust is 12.4% of Moon
If no Th left in mantle → 112 ppb Th in bulk Moon, more if Th in mantle
From Th/U = 3.53 → U = 32 ppb

This value is significantly higher than the value of 20 ppb U for the Earth. However, the following caveats must be noted. The mean thickness of the lunar crust is uncertain. If the mean thickness were as low as 50 km (corresponding to 8.4 vol % of the Moon), keeping all other assumptions the same, the bulk Moon concentrations of K and Th would be 50 ppm and 76 ppb respectively, leading to an estimate of the concentration of U of approximately 20 ppb, the same as estimated for the Earth. If any K and Th remain in the mantle (as must be the case since mare basalts derived from the interior contain K and Th), the estimates of bulk U concentration are correspondingly increased.

If K and Th are not uniformly distributed in the crust, the mean crustal concentrations (and, hence, bulk lunar concentrations) must be appropriately adjusted. Ryder and Wood (1977), on the basis of sample studies, and Andre and Strain (1983), on the basis of Apollo orbital geochemical measurements, have argued that the crust is heterogeneous and layered. Details of the distribution of K and Th with depth await flight of a spacecraft such as the proposed Lunar Geoscience Orbiter. Although orbital gamma-ray experiments analyze only the outer few centimeters of the lunar surface, depth information may be obtained from analyses of large basin ejecta. For example, the Gargantuan Basin (3000 km diameter) and the Big Backside Basin (1900 km diameter) may have penetrated to the lunar mantle (Croft, 1981). Ejecta close to the basin rims represent excavated, deep-seated rocks. Until global orbital geochemical maps are created, considerable uncertainty will exist concerning crustal heterogeneity and hence global lunar U concentration.

The global lunar U concentration may also be estimated from heat flow measurements. The Apollo 15 and 17 heat flow probes returned measurements of 21 erg s^{-1} cm^{-2} and 14 erg s^1 cm^{-2} respectively (Langseth et al., 1976), from which Keihm and Langseth (1977) estimated a mean heat flow of 14-18 erg s^{-1} cm^{-2}. This range of heat flow corresponds approximately to a global lunar U concentration of 33–44 ppb if heat production equals heat flow. This value is higher than the estimate of 20 ppb for the Earth. However, both heat flow probes were located at mare/terrae boundaries and several authors (most recently Rasmussen and Warren, 1985) have argued that this geological setting will lead to an overestimate of global U concentration. After taking into account the variable thickness of porous, low thermal-conductivity megaregolith, the thickness of the lunar lithosphere, and the composition of the crust at Apollo 15 and 17, Rasmussen and Warren (1985) have revised the estimate of global U concentration to 19 ppb, indistinguishable from the terrestrial value.

5.3. Summary

"Best" estimates of global U concentration are approximately 20 ppb for Earth and approximately 30 ppb for the Moon. There is, however, considerable uncertainty in the lunar value in particular. One cannot exclude the possibility that the global concentrations of U in the Earth and Moon are both approximately 20 ppb.

6. Volatile Trace Elements

It has been known conclusively since the return of the Apollo 11 samples that the Moon is impoverished in volatile elements compared to the Earth. For example, there is no trace of water on the Moon. This depletion of volatile elements is well illustrated by volatile/refractory element ratios, e.g., the K/U ratio for the Earth is 10^4 but is only 2500 for the Moon (Tables 3 and 4). Such simple depletions argue for high-temperature processes controlling concentrations in proportion to volatility. It seems likely that this deduction is too simplistic, however.

Figure 2 (after Kreutzberger et al., 1985) illustrates the convariance of the volatile alkali metals Rb and Cs in the Earth, Moon, eucrites, and CI chondrites. Mare basalts define an array of points of slope unity with Cs depleted relative to Rb by a factor of three compared to CI chondrites. Lunar terrae rocks are consistent

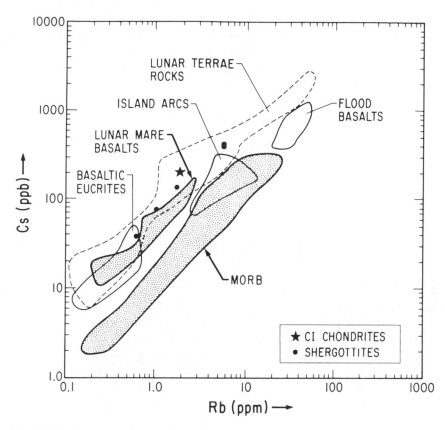

Fig. 2. Rubidium versus Cs in terrestrial, lunar, and meteoritic samples (after Kreutzberger et al., 1985).

with this ratio. Young, unaltered terrestrial basalts [mid-ocean ridge basalts (MORB), intraplate ocean island basalts, and island arc basalts] also define an array of points of slope unity, but with Cs depleted relative to Rb by a factor of ten. Older basaltic rocks that presumably have suffered alteration of these readily remobilized elements show much greater scatter. It is concluded by Hofman and White (1983) that these Cs/Rb ratios represent planetary values. Kreutzberger *et al.* (1985) concur and note in particular that in the case of the Earth there is no evidence for a secular decrease of the Cs/Rb ratio due to either metallization of Cs and extraction into the core or to differential geochemical behavior of Cs and Rb at more modest pressures. The higher Cs/Rb ratio in upper continental crust is likely to reflect intracrustal processes.

It should be noted that Cs/Rb ratios for the Earth and Moon are not in accord with predictions based on considerations of volatility. The volatility of Cs is somewhat greater than Rb in elemental and oxide form, and in silicate melts (Kreutzberger *et al.*, 1985). With the Moon being depleted in volatile elements, one would predict that its Cs/Rb ratio should fall below that of the Earth in Fig. 2, rather than between the Earth and CI chondrites. This apparent paradox must be satisfied by any successful hypothesis of lunar origin.

7. Mg/(Mg + Fe) Ratio

Iron and Mg are the most abundant elements in the Earth and Moon after O and Si, and their abundances vary considerably during magmatic events. The Mg/(Mg + Fe) atomic ratio, the "mg#," is thus a significant parameter.

7.1. mg# of the Earth

The mg# of the upper mantle of the Earth appears to be reasonably well established at approximately 0.89 from studies of mantle nodules (see Jagoutz *et al.*, 1979 and Palme and Nickel, 1985 for recent discussions) and from model constructs such as "pyrolite" (Ringwood, 1975). Several cautions must be observed. First, we have no direct information about the lower mantle (which corresponds to 68% of the volume of the mantle) unless convection is "whole mantle" in scale (e.g., Davies, 1981). Second, it has been proposed that the mg# of the Earth has undergone a secular increase due to loss of Fe to the core by disproportionation (Jagoutz and Wänke, 1982) although there is no direct evidence for continued growth of the core in the last 3.5×10^9 years of geologic time (Newsom and Palme, 1984; Newsom *et al.*, 1985). Third, most mantle nodules are erupted in alkalic basalts on continents and could represent a biased sample.

7.2. mg# of the Moon

The mg# of the Moon is much more difficult to determine, partly because we lack direct samples of the lunar mantle. Estimates are based on geophysical and petrological approaches.

The geophysical approach is typified by Buck and Toksöz (1980). These authors used the lunar structure of Goins *et al.* (1978) to infer P- and S-wave velocities for phase assemblages computed using a normative calculation approach from various estimates of lunar bulk composition. The best fit for the upper mantle of the Moon was for a mg# of 0.79 and a subchondritic Mg/Si ratio of approximately 0.77 (compared to the CI ratio 1.06). Note, however, that a consistent fit for the lower mantle was arbitrarily forced by adding 2 wt % Fe-FeS. Nakamura (1983) inferred a much higher S-wave velocity for the lower mantle, implying a higher mg# or a more aluminous composition for that part of the Moon [see Hood and Jones (1985) for a detailed discussion]. Thus the computed mg# of 0.79 based on the

---

upper mantle may require revision upwards. Note that upward revision of the mg# *requires* the presence of a 300–500-km-radius iron core, a core size inconsistent with derivation of the bulk of lunar matter from the Earth's mantle (Newsom, 1984). Alternatively, a more aluminous lower mantle implies an upward revision of estimates of refractory element concentrations.

Three petrological estimates are given for purposes of illustrating the assumptions implicit or explicit in each technique. These estimates are by Ringwood (1979), Delano and Lindsley (1983), and Warren (1986).

Ringwood (1979) noted that certain low-Ti mare basalts (12009, Apollo 15 green glass, etc.) exhibit characteristics that may indicate they are *primary*, i.e., that they have travelled as magmas from the interior of the Moon to the lunar surface without modification. If this assumption is correct, phase equilibrium studies can yield the phase composition and assemblage at the pressure and temperature of origin inside the Moon. Effective use of this approach requires that the mare basalts were produced by *equilibrium* partial melting with at least two phases remaining in the source region following melt extraction. Ringwood (1979) concludes that the mg# of the source region is approximately 0.75. If the source region is primitive and if the Moon accreted homogeneously, 0.75 would be the mg# of the bulk Moon. Details of mare basalt petrogenesis remain obscure, but most authors conclude that cumulates from the magma ocean were involved in some way, in which case the source region is not primitive. Ringwood (1979) estimates a bulk lunar mg# of 0.81 by mass balance of an "ultimate source of mare basalts" in the lower mantle with mg# = 0.75 and differentiated upper mantle with mg# = 0.88–0.90.

Delano and Lindsley (1983) and Delano (1984) have used compositional trends in mare glasses inferred to be volcanic to deduce the lunar mg#. Noting that glass compositional arrays terminate near chondritic ratios of refractory elements, Delano and Lindsley (1983) infer the composition of a "primitive liquid" (mg# = 0.74) derived by melting the primordial lunar interior and infer a bulk Moon composition by assuming that olivine is the sole residual mineral in the source region. There appears to be some inherent imprecision in this approach since Delano and Lindsley (1983) infer a bulk lunar mg# of 0.87, while Delano (1984) reports a mg# for the silicate portion of the Moon of 0.79. Delano (personal communication, 1985) argues that the result of Delano and Lindsley (1983) is incorrect due to a mathematical artifact.

Warren (1986) estimates the lunar mg# from considerations of the petrogenesis of nonmare highland samples and from estimating the mg# necessary to produce the value recorded in lunar soils. Warren concludes that the lunar mg# lies between 0.84 and 0.93 and most probably between 0.87 and 0.91, indistinguishable from the Earth's upper mantle. It must be noted that the petrogenesis of the nonmare highland rocks with the most Mg-rich olivines (the so-called Mg-suite) is unknown and the provenance of lunar soils is unclear in detail. Thus relating mg#'s calculated from such samples to the mg# of the bulk Moon is indirect and inevitably imprecise.

This mg# should be viewed as an upper limit but certainly cannot be excluded rigorously as the true lunar mg#.

In summary, most estimates of the mg# of the bulk Moon are lower than the mg# of the upper mantle of the Earth, with only the most extreme lunar estimates overlapping terrestrial values. A lower mg# for the Moon is consistent with the observation that lunar basalts, in general, are significantly more FeO-rich than terrestrial basalts (Wänke *et al.*, 1983).

8. Fe/Mn Ratio

The Fe/Mn ratio is an important parameter because Mn is considerably more volatile than Fe but behaves in an almost identical way in magmatic events (Wänke *et al.*, 1983). The excellent correlation of Fe with Mn in lunar samples was originally demonstrated by Laul *et al.* (1972). The Fe/Mn weight ratio of the lunar mantle is well established at approximately 70, while that of the upper mantle of the Earth is approximately 61. Although these values are different, Wänke *et al.* (1983) emphasize that estimated Mn concentrations for the Earth and Moon are both low compared to eucrites, shergottites, and CI chondrites. The low Mn concentration in the Earth's mantle is postulated to be due to possible chalcophilic behavior of Mn during core formation, although there is no evidence for such behavior at low pressures (Jones and Drake, unpublished data, 1985).

9. Oxidation State

The upper mantle of the Earth is currently considerably more oxidized than the mantle of the Moon. There is evidence for upper mantle heterogeneity in oxidation state (Arculus *et al.*, 1984) but, broadly speaking, it is close to the QFM (quartz-fayalite-magnetite) buffer (i.e., significant amounts of Fe^{3+} are stable). In contrast, the Moon is close to the QFI (quartz-fayalite-iron) buffer with no significant Fe^{3+} being stable. The reason for the present oxidized state of the Earth's upper mantle is unclear (Arculus and Delano, 1981) to the extent that it may be a primordial consequence of inhomogeneous accretion (Wänke, 1981) or it may have evolved through geologic time (Turekian, 1981; Sato, 1978).

10. Discussion

Five principal hypotheses were discussed at the Conference on the Origin of the Moon: Intact Capture, Disintegrative Capture, Fission, Collisional Ejection, and Binary Accretion. Each of these hypotheses comes in a number of variants, and some hypotheses grade into each other as parameters are varied. In addition, chemical processing of lunar precursor materials at high temperatures in the solid, liquid, or gaseous states is a consequence of all hypotheses except Intact Capture. These chemical processes have not been examined in detail by the proponents of these hypotheses, and will be dealt with here only in most general terms. Below, the general predictions of each hypothesis are examined in the light of the compositional constraints. Unless otherwise stated, all conclusions will be based on the most probable values of compositional parameters as outlined in the preceding sections.

10.1. Intact Capture hypothesis

The Intact Capture hypothesis postulates that the Moon was formed elsewhere in the solar system and was captured intact with its present compositional properties. The principal objections to this theory are dynamical (e.g., Hansen, 1982). The only compositional property constraining this hypothesis is the commonality of oxygen isotopic composition. While this observation does not rule out intact capture, it does appear to require that the Moon formed with a semi-major axis of approximately 1 AU or less. It is unclear if an environment existed at this heliocentric radius with the properties appropriate for direct accretion of the Moon.

10.2. Disintegrative Capture hypothesis

As proposed by Wood and Mitler (1974), Mitler (1975), and Smith (1974), the Moon is formed in circumterrestrial orbit by accretion of the silicate mantles and crusts of differentiated lunar-sized bodies. The cores are preferentially lost from the vicinity of 1 AU or are accreted onto the Earth. These lunar-sized bodies were originally in heliocentric orbits and were disrupted during passes through the Roche limit. Depending on the physical state of these bodies during close approach, crustal materials rich in incompatible elements reflecting igneous differentiation might be preferentially accumulated, and partial loss of the more volatile elements might occur if the bodies were partially or wholly molten.

This class of hypotheses may be inconsistent with the oxygen isotope data if heliocentric semi-major axes significantly in excess of 1 AU are involved. The hypothesis clearly can satisfy the density constraint, but it is unlikely that it can satisfy the siderophile trace element data unless the precursor bodies were themselves very poor in metal. The refractory elements, mg#, and Fe/Mn ratios do not constrain the hypothesis. Volatile elements are consistent with the hypothesis in that nature has clearly produced other bodies with not only volatile depletion but also the correct Rb/Cs ratios, i.e., the eucrites. The low lunar oxidation state may be similarly explained. A serious problem is that it does not appear that the early Moon was ever close to the Roche limit, approximately 2.5 Earth radii (Hansen, 1982). In addition, it is unclear if disruption will occur during the short passage through the Roche Zone (Mizuno and Ross, 1984). The hypothesis also suffers from the problem that one must postulate the existence of a large number (50+) of differentiated lunar-sized objects, only some of which are likely to suffer close approaches to Earth, in order to make the Moon. While not impossible, it is unclear if the requisite number of lunar-sized bodies are a natural consequence of the dynamical accumulation of planets.

10.3. Rotational Fission hypothesis

The hypothesis that the Moon was formed by rotational fission of the Earth was originally proposed by Darwin (1880). Current advocates include Binder (1984) and Durisen and Scott (1984). This hypothesis is by far the most readily testable because it predicts that the bulk composition of the Moon should be the same as that of the Earth's upper mantle 4.5×10^9 years ago, at least for nonvolatile elements. More volatile elements are subject to loss during fission because of the likelihood that the rapidly rotating Earth is partially or substantially molten.

This hypothesis clearly satisfies the oxygen isotopic constraint. The low mean density of the Moon is predicted in that metal had already been extracted into the Earth's core prior to fission. Siderophile trace element abundances may be satisfied if the Moon has a trivially small fraction of metal (Newsom, 1984). If refractory element abundances in the Moon are higher than in the Earth's upper mantle, then fission is not consistent with observation. The Moon is depleted in volatile elements as predicted, but should have the same or higher Cs/Rb ratio than Earth because Cs is marginally more volatile than Rb, contrary to observation. The best estimate of the mg# of the Moon is lower than the mg# of the Earth's upper mantle, contrary to the prediction of the same mg#. The Fe/Mn ratios are consistent with the hypothesis as Mn is more volatile than Fe. If the present oxidized state of the upper mantle is a primordial feature it is necessary to reduce fissioned material even though loss of H_2O is likely to be an oxidizing event due to preferential loss of H_2.

Finally, Hansen (1982) concludes that the Moon was never closer to the Earth than 38–53 Earth radii and was at an inclination of $3°$–$22°$ relative to the plane of the Earth's equator 4.5×10^9 years ago, while fission requires that the Moon graze the Earth or be assembled just outside the Roche limit in the plane of the Earth's equator. Note, however, that modeling lunar orbital evolution inside 10 Earth radii is fraught with uncertainties (Burns, personal communication, 1984).

10.4. Collisional Ejection hypothesis

The Collisional Ejection hypothesis, originally proposed by Hartmann and Davis (1975), postulates a collision of a large, perhaps Mars-sized object with the Earth, leading to the ejection of mantle material from both planets. Estimates of the relative contributions of target and projectile to the ejecta vary from predominantly target (Ringwood, 1979; Wänke and Dreibus, 1984) through roughly equal contributions (Melosh and Sonett, 1986) to predominantly projectile (Cameron, 1986). From a geochemical viewpoint, Cameron's (1986) hypothesis is a variant of the Capture hypothesis. Because of the currently unconstrained nature of the hypothesis, both in terms of contributions of target and projectile to the ejecta and in terms of the thermal processing of the ejecta prior to accumulation to form the Moon, this hypothesis may be examined for consistency with compositional constraints only in the most general terms. Note, however, that although ejecta is initially in the vapor state (Melosh and Sonett, 1986; Stevenson, 1985) elemental loss at very high temperatures cannot have been significant because rare earth element and other element patterns do not show evidence for such a process (see, for example, Boynton, 1975).

The Collisional Ejection hypothesis is clearly consistent with the oxygen isotope constraints if the overwhelming contribution to the ejecta comes from the target. If a significant fraction of the ejecta comes from the projectile, the projectile's oxygen isotopic composition must be the same as Earth's or at least differ so little that its signature is not measurable within the limits of analytical precision. The hypothesis is consistent with density constraints if both target and projectile were differentiated into mantle and core prior to collision and the projectile core was accreted into the Earth. The siderophile trace element constraints are met only if a trivial fraction (0.1 wt % to 1.0 wt %; see Newsom, 1984) of metal is separated from silicate subsequent to lunar assembly. Volatile element constraints may be met only if a significant fraction of the projectile is incorporated in the ejecta and the projectile has a Rb/Cs ratio closer to chondritic than the present lunar value. The mg# constraint may be met if a significant fraction of the projectile is incorporated in the ejecta and the projectile has an mg# less than the current best estimate of the lunar value. The Fe/Mn ratio constraint may be met if most of the ejecta comes from the target, or if the projectile had a similar Fe/Mn ratio. The differences in lunar and terrestrial oxidation states must also be explained, either by incorporating still more reduced projectile material into the ejecta or by reduction of upper mantle material in the collision. A key question is whether all of the conditions necessary for the Collisional Ejection hypothesis to meet the compositional constraints may be met simultaneously; this question is currently unanswered.

Finally, it should be noted that most recent lunar orbital evolution calculations (Hansen, 1982) are inconsistent with this hypothesis. Again it must be noted that modeling lunar orbital evolution inside ten Earth radii is nonunique (Burns, personal communication, 1984).

10.5. Binary Accretion hypothesis

The Binary Accretion hypothesis is attributed to Ruskol; see Ruskol (1977) for a general discussion. Recent variants have been advanced by Wasson and Warren (1979, 1984) and Weidenschilling et al. (1986). In this hypothesis differentiated planetesimals from the general vicinity of 1 AU are disrupted and the silicate portions are preferentially captured into geocentric orbit, ultimately to accrete to form the Moon.

This hypothesis satisfies the oxygen isotopic constraints in that all protolunar material originates in the general vicinity of 1 AU. The density constraint and siderophile trace element abundance constraints can be met if metal is efficiently separated from silicate with only silicate being accreted to form the Moon. If refractory element abundances are higher in the Moon than in the Earth, the Binary Accretion hypothesis is only consistent with observation if basaltic crustal material from the differentiated planetesimals with unfractionated incompatible refractory element patterns—an unlikely circumstance—was preferentially accreted. The volatile element constraints may be met in that nature has already provided material with appropriate volatile depletions and Cs/Rb ratio in the form of the eucrites. A similar comment applies to oxidation state. The mg# constraint may be satisfied if the differentiated planetesimals had the appropriate value, although there is no obvious reason why planetesimals that were accreted into the Earth at approximately 1 AU should have a different mean mg# from those accreted into the Moon or, again, if basaltic crustal material were preferentially accreted. The Fe/Mn ratio constraint may be met if the Fe/Mn ratio of the Earth is *not* affected by core formation and if protolunar material had a similar Fe/Mn ratio. In this context it is important to establish whether Mn behaves as a chalcophile element during metallic liquid/silicate liquid separation.

Finally, lunar orbital evolution calculations (Hansen, 1982) are consistent with the Binary Accretion hypothesis, although other dynamical difficulties exist (Harris and Kaula, 1975) that have only been qualitatively addressed (Weidenschilling *et al.*, 1986).

11. Conclusions

The geochemical considerations discussed in the previous section are summarized in a geochemical "truth" table (Table 5). I will resist the temptation to assign a pass/fail/incomplete grade (cf. Wood, 1986). Rather, let me note that on geochemical grounds alone, only "pure" Fission (i.e., lunar matter is derived solely from Earth's mantle) may be ruled out. Intact Capture is unlikely on dynamical grounds.

TABLE 5.   Geochemical "Truth Table."

	Intact Capture	Disintegrative Capture	Fission	Collisional Ejection	Binary Accretion
Oxygen Isotopes	?	?	Y	Y	Y
Density	?	Y	Y	Y	Y
Siderophile Elements	?	?	Y	Y	Y
Refractory Elements	?	?	N	?	?
Volatile Elements	?	Y	N	?	Y
mg#	?	?	N	?	?
Fe/Mn Ratio	?	?	Y	Y	Y
Oxidation State	?	Y	?	?	Y

Y= hypothesis is consistent with the constraint; ? = hypothesis is unconstrained; N = hypothesis is inconsistent with the constraint.

Disintegrative Capture seems unlikely on a variety of grounds. From a geochemical standpoint both Collisional Ejection and Binary Accretion remain viable hypotheses. Neither hypothesis unambiguously satisfies all geochemical constraints. For Collisional Ejection, it remains to be demonstrated that the same conditions— target/projectile ratios, thermal processing, etc.—can simultaneously satisfy all geochemical constraints. For both hypotheses, dynamical questions remain. Although theories for the origin of the Moon have evolved significantly since Apollo 11, an unambiguous solution has yet to be attained. Perhaps, as noted by Weidenschilling *et al.* (1986), elements of all hypotheses may ultimately have played a role in lunar origin.

Acknowledgments. Reviews by G. J. Taylor, J. Larimer, an anonymous reviewer, and colleagues at Kona and in Tucson are gratefully acknowledged. This work was supported by NASA grants NAG 9-39 and NAGW 680.

12. References

Anders E. (1978) Procrustean science: indigenous siderophiles in the lunar highlands according to Delano and Ringwood. *Proc. Lunar Planet. Sci. Conf. 9th*, pp. 161–184.

Andre C. G. and Strain P. L. (1983) The lunar nearside highlands: evidence of resurfacing. *Proc. Lunar Planet. Sci. Conf. 13th*, in *J. Geophys. Res., 88*, A544–A552.

Arculus R. and Delano J. W. (1981) Intrinsic oxygen fugacity measurements: techniques and results for spinels from upper mantle peridotites and megacryst assemblages. *Geochim. Cosmochim. Acta, 45*, 899–913.

Arculus R., Dawson J. B., Mitchell R. H., Gust D. A., and Holmes D. A. (1984) Oxidation states of the upper mantle recorded by megacryst ilmenite in kimberlite and type A and B spinel lherzolites. *Contrib. Mineral. Petrol., 85*, 85–94.

Binder A. B. (1984) On the origin of the Moon by rotational fission (abstract). In *Papers Presented to the Conference on the Origin of the Moon*, p. 47. Lunar and Planetary Institute, Houston.

Bogard D. D. and Johnson P. (1983) Martian gases in an Antarctic meteorite? *Science, 221*, 651–654.

Boynton W. V. (1975) Fractionation in the solar nebula: condensation of yttrium and the rare earth elements. *Geochim. Cosmochim. Acta, 39*, 569–584.

Buck W. R. and Toksöz M. N. (1980) The bulk composition of the Moon based on geophysical constraints. *Proc. Lunar Planet. Sci. Conf. 11th*, pp. 2043–2058.

Cameron A. G. W. (1986) The impact theory for origin of the Moon, this volume.

Chou C.-L, Shaw D. M., and Crocket J. M. (1983) Siderophile trace elements in the Earth's oceanic crust and upper mantle. *Proc. Lunar Planet. Sci. Conf. 13th*, in *J. Geophys. Res., 88*, A509–A518.

Consolmagno G. J. and Drake M. J. (1977) Composition and evolution of the eucrite parent body: evidence from rare earth elements. *Geochim. Cosmochim. Acta, 41*, 1271–1282.

Croft S. K. (1981) The modification stage of basin formation: conditions of ring formation. In *Multi-ring Basins, Proc. Lunar Planet. Sci. 12A* (P. H. Schultz and R. B. Merrill, eds.), pp. 227–257. Pergamon, N.Y.

Davies G. F. (1980) Thermal histories of convective earth models and constraints on radiogenic heat production in the Earth. *J. Geophys. Res., 85*, 2517–2530.

Davies G. F. (1981) Earth's neodymium budget and structure and evolution of the mantle. *Nature, 290*, 208–213.

Darwin G. H. (1880) On the secular changes in the orbit of a satellite revolving around a tidally disturbed planet. *Phil. Trans. Roy. Soc. London, 171*, 713–891.

Delano J. W. (1984) Abundances of Ni, Cr, Co, and major elements in the silicate portion of the Moon: constraints from primary lunar magmas (abstract). In *Papers Presented to the Conference on the Origin of the Moon*, p. 15. Lunar and Planetary Institute, Houston.

Delano J. W. and Lindsley D. H. (1983) Mare glasses from Apollo 17: constraints on the Moon's bulk composition. *Proc. Lunar Planet. Sci. Conf. 14th*, in *J. Geophys. Res., 88*, B3–B16.

Delano J. W. and Ringwood A. E. (1978) Siderophile elements in the lunar highlands: nature of indigenous component and implications for the origin of the Moon. *Proc. Lunar Planet. Sci. Conf. 9th*, pp. 111–159.

Drake M. J. (1983) Geochemical constraints on the origin of the Moon. *Geochim. Cosmochim. Acta, 47*, 1759–1767.

Drake M. J., Newsom H. E., Reed S. J. B., and Enright M. C. (1984) Experimental determination of the partitioning of gallium between solid iron metal and synthetic basaltic melt: electron and ion microprobe study. *Geochim. Cosmochim. Acta, 48*, 1609–1615.

Durisen R. H. and Scott E. H. (1984) Implications of recent numerical calculations for the fission theory of the origin of the Moon. *Icarus, 58*, 153–158.

Esat T. M., Spear R. H., and Taylor S. R. (1985) Anomalous mass fractionation in distillation: implications for the early history of meteorites and the solar system (abstract). In *Lunar and Planetary Science XVI*, pp. 217–218. Lunar and Planetary Institute, Houston.

Goins N. R., Toksöz M. N., and Dainty A. M. (1978) Seismic structure of the lunar mantle: an overview. *Proc. Lunar Planet. Sci. Conf. 9th*, pp. 3575–3588.

Greenberg R. and Chapman C. R. (1983) Asteroids and meteorites: parent bodies and delivered samples. *Icarus, 55*, 455–481.

Hansen K. S. (1982) Secular effects of oceanic tidal dissipation on the Moon's orbit and the Earth's rotation. *Rev. Geophys. Space Phys., 20*, 457–480.

Harris A. W. and Kaula W. M. (1975) A co-accretional model of satellite formation. *Icarus, 24*, 516–524.

Hart R. and Hogan L. (1978) Earth degassing models and the heterogeneous vs. homogenous mantle. In *Terrestrial Rare Gases* (E. C. Alexander and M. Ozima, eds.). Japan Sci. Soc. Press, City.

Hartmann W. K. and Davis D. R. (1975) Satellite-sized planetesimals and lunar origin. *Icarus, 24*, 504–515.

Hofmann A. W. and White W. M. (1983) Ba, Rb, and Cs in the Earth's mantle. *Z. Naturforsch, 38a*, 256–266.

Hood L. L. and Jones J. H. (1985) Lunar density models consistent with mantle seismic velocities and other geophysical constraints (abstract). In *Lunar and Planetary Science XVI*, pp. 360–361. Lunar and Planetary Institute, Houston.

Jagoutz E. and Wänke H. (1982) Has the Earth's core grown over geologic times? (abstract). In *Lunar and Planetary Science XIII*, pp. 358–359. Lunar and Planetary Institute, Houston.

Jagoutz E., Palme H., Baddenhausen H., Blum K., Cendales M., Dreibus G., Spettel B., Lorenz V., and Wänke H. (1979) The abundances of major, minor, and trace elements in the Earth's mantle as derived from primitive ultramafic nodules. *Proc. Lunar Planet. Sci. Conf. 10th*, pp. 2031–2050.

Keihm S. J. and Langseth M. J. (1977) Lunar thermal regime to 300 km. *Proc. Lunar Sci. Conf. 8th*, pp. 499–514.

Kreutzberger M. E., Drake M. J., and Jones J. H. (1985) Origin of the Moon: constraints from volatile elements. *Geochim. Cosmochim. Acta*, in press.

Langseth M. G., Keihm S. J., and Peters K. (1976) Revised lunar heat flow values. *Proc. Lunar Sci. Conf. 7th*, pp. 3143–3171.

Laul J. C., Wakita H., Showalter D. L., Boynton W. V., and Schmitt R. A. (1972) Bulk rare earth, and other trace elements in Apollo 14 and 15 and Luna 16 samples. *Proc. Lunar Sci. Conf. 3rd*, pp. 1181–1200.

McKenzie D. P. and Richter F. M. (1981) Parameterized thermal convection in a layered region and the thermal history of the earth. *J. Geophys. Res., 86*, 11,667–11,680.

Melosh H. J. and Sonett C. P. (1986) When worlds collide: Jetted vapor plumes and the Moon's origin, this volume.

Metzger A. E., Trombka J. I., Reedy R. C., and Arnold J. R. (1974) Elemental concentrations from lunar orbiter gamma-ray measurements. *Proc. Lunar Sci. Conf. 5th*, pp. 1067–1078.

Metzger A. E., Haines E. L., Parker R. E., and Radocinski R. G. (1977) Thorium concentrations in the lunar surface I. Regional values and crustal content. *Proc. Lunar Sci. Conf. 8th*, pp. 949–999.

Mitler H. E. (1975) Formation of an iron-poor moon by partial capture, or: yet another exotic theory of lunar origin. *Icarus, 24*, 256–268.

Mizuno H. and Boss I. P. (1984) Tidal disruption and the origin of the Moon (abstract). In *Papers Presented to the Conference on the Origin of the Moon*, p. 34. Lunar and Planetary Institute, Houston.

Morgan J. W., Wandless G. A., Petrie R. K., and Irving A. J. (1981) Composition of the Earth's upper mantle - I. Siderophile trace element abundances in ultramafic nodules. *Tectonophys., 75*, 47–67.

Nakamura Y. (1983) Seismic velocity structure of the lunar mantle. *J. Geophys. Res., 88*, 677–686.

Newsom H. E. (1984) The lunar core and the origin of the Moon. *EOS (Trans. Am. Geophys. Union), 65*, 369–370.

Newsom H. E. and Drake M. J. (1982) Constraints on the Moon's origin from the partitioning behavior of tungsten. *Nature, 297*, 210–212.

Newsom H. E. and Drake M. J. (1983) Experimental investigation of the partitioning of phosphorus between metal and silicate phases: implications for the Earth, Moon and eucrite parent body. *Geochim. Cosmochim. Acta, 47*, 93–100.

Newsom H. E. and Palme H. (1984) The depletion of siderophile elements in the Earth's mantle: new evidence from molybdenum and tungsten. *Earth Planet. Sci. Lett., 69*, 354–364.

Newsom H. E., White W. M., and Jochum K. P. (1985) Did the Earth's core grow through geological time (abstract)? In *Lunar and Planetary Science XVI*, pp. 616–617. Lunar and Planetary Institute, Houston.

O'Nions R. K., Evenson N. M., and Hamilton P. J. (1979) Geochemical modeling of mantle differentiation and crustal growth. *J. Geophys. Res., 84*, 6091–6101.

Palme H. and Nickel K. G. (1985) Ca/Al ratio and composition of the Earth's upper mantle. *Geochim. Cosmochim. Acta*, in press.

Rammensee W. and Wänke H. (1977) On the partition coefficient of tungsten between metal and silicate and its bearing on the origin of the Moon. *Proc. Lunar Sci. Conf. 8th*, pp. 399–409.

Rasmussen K. L. and Warren P. H. (1985) Megaregolith thickness, heat flow, and the bulk composition of the Moon. *Nature, 313*, 121–124.

Ringwood A. E. (1975) *Composition and Petrology of the Earth's Mantle.* McGraw-Hill, New York. 618 pp.

Ringwood A. E. (1979) *Origin of the Earth and Moon.* Springer-Verlag, New York. 295 pp.

Ringwood A. E., Kesson S. E. and Hibberson W. (1981) Rhenium depletion in mare basalts and redox state of the lunar interior (abstract). In *Lunar and Planetary Science XII*, pp. 891–893. Lunar and Planetary Institute, Houston.

Ruskol Ye. L. (1977) The origin of the Moon. In *The Soviet-American Conference on the Cosmochemistry of the Moon and Planets*, pp. 815–822. NASA SP-370, NASA/Johnson Space Center, Houston.

Ryder G. and Wood J. A. (1977) Serenitatis and Imbrium impact melts: implications for large-scale layering in the lunar crust. *Proc. Lunar Sci. Conf. 8th*, pp. 655–668.

Sato M. (1978) Oxygen fugacity of basaltic magmas and the role of gas-forming elements. *Geophys. Res. Lett., 5*, 447–449.

Smith J. V. (1974) Origin of the Moon by disintegrative capture with chemical differentiation followed by sequential accretion (abstract). In *Lunar Science V*, pp. 718–720. The Lunar Science Institute, Houston.

Stevenson D. J. (1985) Lunar origin from impact on the Earth (abstract)? In *Papers Presented to the Conference on the Origin of the Moon*, p. 60. Lunar and Planetary Institute, Houston.

Taylor S. R. (1982) *Planetary Science: A Lunar Perspective*. Lunar and Planetary Institute, Houston. 481 pp.

Theimans M. H. and Heidenreich J. E., III (1983) The mass-independent fractionation of oxygen: a novel isotope effect and its possible cosmochemical implications. *Science, 219*, 1073–1075.

Turekian K. K. (1981) Origin and evolution of continents and oceans. In *Life in the Universe* (J. Billingham, ed.), pp. 101–110. M.I.T. Press, Cambridge.

Wänke H. (1981) Constitution of terrestrial planets. *Phil. Trans. Roy. Soc. London, A303*, 287–302.

Wänke H. and Dreibus G. (1984) Geochemical evidence for the formation of the Moon by impact induced fission of the proto-Earth (abstract). In *Papers Presented to the Conference on the Origin of the Moon*, p. 48. Lunar and Planetary Institute, Houston.

Wänke H., Dreibus G., Palme H., Rammensee W., and Weckwerth G. (1983) Geochemical evidence for the formation of the Moon from material of the Earth's mantle (abstract). In *Lunar and Planetary Science XIV*, pp. 818–819. Lunar and Planetary Institute, Houston.

Warren P. H. (1986) The bulk-Moon MgO/FeO ratio: A highlands perspective, this volume.

Wasson J. T. and Warren P. H. (1975) Formation of the Moon from differentiated planetesimals of chondritic composition (abstract). In *Lunar and Planetary Science X*, pp. 1310–1312. Lunar and Planetary Institute, Houston.

Wasson J. T. and Warren P. H. (1984) The origin of the Moon (abstract). In *Papers Presented to the Conference on the Origin of the Moon*, p. 57. Lunar and Planetary Institute, Houston.

Weidenschilling S. J., Greenberg R., Chapman C. R., Herbert F. L., Davies D. R., Drake M. J., Jones J. H., and Hartmann W. K. (1986) Origin of the Moon from a circumterrestrial disk, this volume.

Wood J. A. (1986) Moon over Mauna Loa: A review of hypotheses of formation of Earth's Moon, this volume.

Wood J. A. and Mitler H. E. (1974) Origin of the Moon by a modified capture mechanism, or: half a loaf is better than a whole one (abstract). In *Lunar Science V*, pp. 851–853. The Lunar Science Institute, Houston.

The Origin of the Moon: Geochemical Considerations

STUART ROSS TAYLOR

Research School of Earth Sciences, Australian National University, Canberra, Australia, 2601

"Of course I was not there when the solar system originated."
(Harold C. Urey, letter to NASA Planetology Committee, September 20, 1963)

Capture models for the origin of the Moon do not meet chemical, isotopic, or dynamical constraints and have effectively been abandoned. The composition of the bulk Moon differs in several important ways from that of the terrestrial mantle, ruling out fission hypotheses unless substantial element fractionation occurs during or following fission. Large Mars-sized impactor models cut several Gordian knots, since the lunar material is derived mostly from the impactor, whose chemical and isotopic composition become free parameters. Both volatile and siderophile element depletions, observed in the Moon, are common in meteorites (e.g., eucrites) formed at 4.5 aeons, so that the lunar composition is nonunique, and a population of precursor planetesimals of appropriate composition existed. Of the five hypotheses for lunar origin, either the large impactor models or the double-planet models remain as viable candidates. Establishment of the time of volatile depletion, and whether one or more episodes of volatile loss were involved, might distinguish between these two hypotheses. Whole Moon melting follows accretion. Core formation in the Moon depletes the mantle in siderophile elements. Additional Ni and Co are trapped in deep olivine and orthopyroxene phases. Crystallisation results in an anorthositic crust and a differentiated lunar mantle while the final residual liquid (KREEP) invades the crust at 4.3–4.4 aeons.

1. The Apollo Sample Return: Demise of Capture Models

The chemical and isotopic information from the samples returned by the Apollo manned missions immediately provided new limits on theories of lunar origin. Prior to 1969, the composition of the Moon was constrained only by density measurements.

Hence, it was known to be low in iron relative to bulk Earth compositions. From the evidence of massive cratering in the highlands, the age of that surface was supposed to be older than that of any exposed terrestrial surface. Its probable ancient age and the low density of the Moon led to hypotheses that the Moon might represent an essentially primitive object (Urey, 1962). Estimates of the cratering flux by perceptive students gave essentially correct ages of 3–4 aeons for the lunar maria (e.g., Baldwin, 1949; Hartmann, 1965). Although the material filling the maria was correctly identified as basaltic lava by acute observers (e.g., Baldwin, 1949), the anorthositic nature of the highlands eluded investigators until the first sample return.

The Apollo sample results immediately demonstrated that the Moon was formed at least 4.4 aeons ago, that it was highly depleted in volatile elements, and that a large fraction, if not all, had been extensively differentiated (e.g., Taylor, 1975). The similarity of oxygen isotope ratios to those in the Earth indicated a nearby origin, rather than in some remote part of the solar nebula. These observations reinforced the dynamical difficulties inherent in capture models, so that capture of a primitive object became less attractive. Since the hypothesis now possessed few advantages over double planet scenarios, it was effectively abandoned by most workers.

Attention then became focused on double-planet and fission scenarios. Double-planet hypotheses immediately encounter the difficulty of accounting for the density difference, and selective accretion of silicate rather than metallic iron became a feature of these models (Ruskol, 1977; Wood and Mitler, 1974; Smith, 1979, 1982). All such models presuppose prior fractionation of silicate-metal-sulfide phases, in addition to volatile element depletion in precursor planetesimals, before accretion of the Earth and Moon. This important point will be addressed later, but it should be recalled that the Earth is depleted in volatile elements (e.g., low Rb/Sr, K/U relative to CI meteorites), although not to the same extent as the Moon.

2. The Fission Hypothesis

The principal philosophical support for this hypothesis was the similarity between the density of the Moon (3.34 gm cm^{-3}) and the uncompressed density of the terrestrial mantle (3.32 gm cm^{-3}). Given the Apollo data, it became the most readily testable of all hypotheses. A necessary corollary is that core formation in the Earth should precede fission, and hence the Earth has to be close to its present size before fission occurs. Wetherill-Safronov scenarios (e.g., Wetherill, 1980) for the accretion of the terrestrial planets apparently demand sweep-up time scales of 10^8 years. If we assume that core formation was coeval or immediately followed accretion, then the origin of the Moon occurs at about 4.45 aeons, measurably later than the 4.56 aeon age for the formation of the meteorites (and presumably planetesimals). This point will be discussed later, along with the complexity introduced by the difference between lunar and terrestrial mantle FeO contents.

Serious dynamical constraints beset simple fission models. The angular momentum of the Earth-Moon system is insufficient by a factor exceeding three to form a bulge during rotation, from which the Moon might form (Weidenschilling, 1984). Catastrophic core formation might speed up the rotation rate (Wise, 1963), thereby facilitating fission. The precipitation hypothesis (Ringwood, 1970) was an attempt both to overcome the angular momentum problem and to provide for depletion in volatile elements.

2.1. The lunar orbit

The lunar orbit is inclined at 5.1° to the plane of the ecliptic, whereas the axial plane of the Earth is inclined at 23.4°. The inclination of the lunar orbit to the equatorial plane of the Earth thus varies between 18.3° and 28.5°. Simple versions of the fission hypothesis place the orbit of the Moon in the equatorial plane of the Earth. Although most planetary orbits lie close to the plane of the ecliptic, high axial inclinations are common and are generally attributed to the effects of large impacts during accretion consistent with the planetesimal hypothesis (Wetherill, 1980). Thus, the present Earth-Moon orbital arrangement could be due to a late large terrestrial impact following the formation of the Moon.

A more serious constraint comes from orbital calculations showing that the Moon has never been closer to the Earth than about 240,000 km (Hansen, 1982). This is consistent with a double-planet scenario, but not with fission models, or with disintegrative capture models in which a ring of debris enriched in silicates accumulates just beyond the Roche limit (2.89 earth radii or 18,400 km; Wood and Mitler, 1974; Smith, 1979). If Hansen's calculations are correct, all hypotheses for lunar origin except double-planet models are excluded.

2.2. Volatile element depletion

If the fission hypothesis is correct, then the mantle of the Earth should bear a close and identifiable relationship to the composition of the bulk Moon. The composition of the upper mantle of the Earth is constrained by experimental petrology, mantle derived xenoliths, and by basalts derived therefrom by partial melting. The composition of the lower mantle remains uncertain, but serious cosmochemical problems arise if it is the same as the upper mantle, whose Mg/Si ratio is far above that of any meteorite group. The lower mantle is probably dominated by perovskite ($MgSiO_3$) phases (Liu and Bassett, 1985) and hence may be different in composition to the upper mantle. Accordingly, there is no unambiguous evidence that the Earth has other than chondritic ratios for the major lithophile elements.

The volatile trace elements, however, are clearly depleted relative to CI abundances. The terrestrial Rb/Sr ratio is 0.031 compared to CI values of 0.30. The lunar Rb/

Sr ratio of 0.009 is depleted relative to the Earth. This volatile/refractory element depletion is likewise shown by K/U ratios (60,000 for CI, 10,000 for the Earth, and 2500 for the Moon) (Fig. 1) and for many other elements. The most volatile elements (e.g., Bi, Tl) are depleted in the Moon relative to the Earth by factors of about 50 (Wolf and Anders, 1980). This extensive depletion (shown also by total absence of lunar water) is a first-order fact requiring explanation in all hypotheses. However, Earth-Moon differences in volatile element abundances are not always simply related to volatility. Drake (1986) has pointed out that the lunar Cs/Rb ratio is higher than that of the Earth, contrary to predictions based on relative volatility.

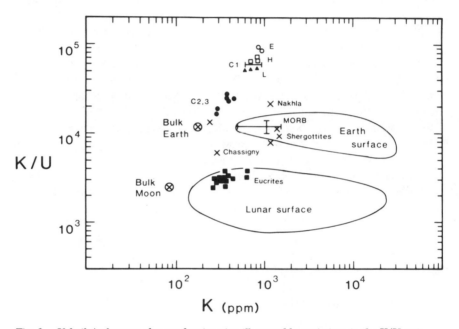

Fig. 1.   Volatile/refractory element fractionation illustrated by variations in the K/U ratio, plotted against potassium abundance, in meteorites, Earth, and Moon. The CI data are taken as being representative of the original solar nebula, on account of their similarity to solar values. Potassium is volatile at about 1000 K and uranium is highly refractory. Widespread depletion in potassium, relative to uranium, is a feature of the inner solar system. Both elements concentrate in residual melts during intraplanetary igneous fractionation processes, so that the K/U ratio in surface rocks approximates to the bulk planetary ratio. If the SNC meteorites (Shergotty, Nakhla, Chassigny) come from Mars, then that body is at least as volatile-rich as the Earth. The K/U value for Nakhla, and the K/La data (Fig. 2) indicate higher K values for the SNC parent body. Uranium is mobile, under oxidising conditions as U^{6+}. This possibility explains the wide range in K/U ratios in the SNC meteorites. Both the Moon and the eucrites are the most depleted in K. (Adapted from Taylor, 1982, Fig. 8.3. Additional data from Smith et al., 1984.)

2.3. Lunar cores and siderophile element depletions

The Mg/(Mg + Fe) ratio of the terrestrial mantle and the Moon should be similar, if the fission hypothesis is correct, since the ratio should not be altered by the fission process. The FeO content of the terrestrial mantle is usually estimated at about 8%. Density considerations limit higher values. The bulk Moon FeO content is estimated by most workers to be about 13%, making the bulk lunar Mg/(Mg + Fe) value = 0.81. If about 2% Fe is removed in a lunar core, the Mg/(Mg + Fe) value for the lunar mantle becomes about 0.84, consistent with the values required for the mare basalt source regions. The Mg/(Mg + Fe) value for the terrestrial upper mantle is 0.90. Warren (1984) argued on the basis of estimates from highland samples that the lunar Mg/(Mg + Fe) ratio could be similar to that of the terrestrial mantle. However, for bulk lunar FeO values of 13%, this would lead to bulk lunar MgO values exceeding 60%. These seem unrealistic. Lunar mare basalts possess very high Fe/Ni ratios and are depleted in Ni by a factor of four and trace siderophile elements by larger factors relative to terrestrial basalts (Taylor, 1979).

The various indirect arguments for the existence of a small (400-km radius) metallic core comprising about 2% of lunar volume have become stronger (Newsom, 1984). The existence of a lunar core has several consequences.

1. It provides a sink to deplete the lunar mantle in Ni and the other trace siderophile elements. The extreme depletion in these elements in the lunar mantle contrasts strongly with their higher abundances in the terrestrial mantle, indicating a lack of core-mantle equilibrium in the Earth in contrast to the Moon.

2. It reduces the lunar mantle Fe content to provide an acceptable Mg/(Mg + Fe) value for the source region of lunar basalts.

3. It implies whole Moon differentiation. Whether the Moon was wholly or partially melted is less of an issue than the evidence that extensive planetary-wide differentiation occurred.

4. It raises a paradox for the fission hypothesis. The extreme depletion of the lunar mantle in siderophile elements is consistent with their removal into a lunar core (Newsom, 1984, 1986). If the Moon was derived from the terrestrial mantle, both would have the same FeO content at that time. Separation of a lunar core should then deplete the mantle of iron. However, the converse is true. The lunar mantle is enriched in FeO compared to the Earth's mantle, although Ni and other trace siderophile elements are severely depleted. One answer to this dilemma for the fission model is to remove the Moon from the terrestrial mantle before core separation is complete, requiring an ad hoc adjustment to the hypothesis.

2.4. Whole Moon differentiation and magma oceans

It is generally agreed that large scale melting and differentiation of the Moon occurred before 4.4 aeons. However, estimates of the depth to which the Moon

---

was affected have varied from shallow (200 km) to whole Moon involvement. In this section, it is concluded that evidence for the latter case is strong. Although alternative scenarios such as serial melting (Walker, 1983) have been proposed, the following lines of evidence support the concept of a magma ocean. These include the presence of a thick plagioclase-rich crust, complementary Eu anomalies in the crust and the mare basalts, the extensive near-surface concentrations of REE, U, Th, and other incompatible elements, the uniform REE and isotopic systematics of KREEP, and the isotopic evidence that this extensive lunar differentiation was completed very early, by about 4.4 aeons. Deep melting is consistent with geochemical balance arguments and by the most recent assessment of the lunar internal structure (Hood and Jones, 1985). This favors models in which $Mg/(Mg + Fe)$ increases at depth, consistent with a deep lunar interior dominated by early Mg-rich olivine and orthopyroxene cumulates to depths of over 1000 km. Such models also require lunar cores (350–500-km radius) to account for the lunar coefficent of moment of inertia (0.3905 ± 0.0023), which needs a density increase in the deep interior, in addition to a low density crust, to offset the low density Mg-rich interior. Such scenarios remove the possibility of retention of unmelted primordial lunar compositions deep in the interior.

Such regions have appealed to some geochemists wishing to derive mare basalts from unfractionated lunar material, which accordingly would provide unique information on its origin (Delano and Ringwood, 1978). However, the most primitive lunar basaltic material available (green glass—15425–15427) comes of course from a source region already fractionated, shown by the presence of a negative europium anomaly. The evidence for a primitive volatile-rich interior deduced by Delano and Lindsley (1983) from volatile-rich rims on volcanic glass spheres is readily explained by coating of such spheres in fire-fountain-type eruptions, with the volatiles (e.g., Zn, Pb) being concentrated in such an environment, following scavenging from their miniscule lunar budget. The presence of primitive Pb isotope ratios, for example, is well explained by the segregation of Pb in sulfide phases in a low μ ($^{238}U/^{204}Pb$) environment during the initial lunar differentiation at about 4.4 aeons. The case for whole Moon differentiation seems reasonably consistent with the geochemical data and assists in explaining the siderophile element depletion, by sequestering these in a metallic core. An additional depletion of Ni will occur by segregation of early crystallising olivine and orthopyroxene in the deep interior. The lack of radioactive heat sources in those deep cumulates, and their depth and refractory nature, will prevent magma formation.

2.5. Refractory element abundances and the lunar crust

A basic question for the fission hypothesis is whether the refractory element abundances in the Moon and the Earth are similar or different. The refractory element abundances in the Earth are constrained by an interlocking set of elemental and isotopic ratios, so that a value of 18 ppb U represents a consensus among geochemists. In this case, heat flow is not equal to heat production and the Earth is losing heat (McKenzie and Richter, 1981). The abundance of uranium is tied to that of other refractory elements (e.g., Al, Ti) on the assumption that refractory elements are not substantially fractionated on a planetary scale in cosmochemical processes. Assuming no fractionation between U and Al, a uranium abundance of 18 ppb indicates an Al_2O_3 content of 3.60% for the bulk Earth, consistent with other estimates.

The lunar uranium abundance is more disputed. However, both Al and U are concentrated in the highland crust of the Moon. The average thickness of this crust is 73.4 km, comprising 12.3% of lunar volume. The Al content of this crust is thus a significant fraction of the lunar Al budget. The variation in crustal thickness (nearside 64 km; farside 86 km) is conventionally held responsible for the centre of figure (CF)/centre of mass (CM) offset, while a low density crust, coupled with an increase in density in the deep interior, accounts for the value of the coefficient of moment of inertia (see above). Thinner crusts must be of lower density (higher Al_2O_3), while a lower Al_2O_3 content for the bulk crust may be accommodated by a thicker crust. It is accordingly difficult to lower the bulk Al_2O_3 content of the crust because of these geophysical constraints. Alternative explanations for the CF/CM offset suggest core or mantle asymmetries. Such explanations seem unlikely for a fully differentiated moon. On the contrary, variations in thickness of a floating aluminous crust are readily explicable in terms of lunar magma oceans. It is difficult to escape from the conclusion that the Moon possesses a thick aluminous crust. Taylor (1982) concluded on geochemical grounds that the average crust contained 24.6% Al_2O_3. Spudis (1984), working from stratigraphic constraints, estimates 25% Al_2O_3. If it averages 24.6% Al_2O_3, this contributes 3.0% Al_2O_3 to the bulk lunar composition. Even if the crustal composition is based on the lowest observable Al_2O_3 content (about 20%), this contributes 2.5% Al_2O_3 to the bulk Moon composition. This exceeds the total Al_2O_3 content (2.4%) for a moon of chondritic composition and comprises a large proportion of a moon with a terrestrial mantle Al_2O_3 abundance of 3.6%.

A further petrological constraint favors high-alumina moons. It is a requirement that plagioclase crystallise early, during magma ocean cooling, not only to account for the thick plagioclase-rich highland crust, but also to account for the universal depletion of the lunar interior in the REE europium. Mare basalts, which sample to depths of 200–400 km (50% of lunar volume), provide a lower limit on the amount of prior plagioclase (and Eu) removal. Nevertheless, sufficient Al_2O_3 is retained in the source regions to produce basalts by partial melting that typically contain 8–14% Al_2O_3 (Taylor, 1982, Tables 6.3a,b). Terrestrial mantle compositions (3.6% Al_2O_3) do not meet these requirements, crystallising pyroxene before plagioclase, and so the bulk Moon must be enriched in Al_2O_3 relative to the terrestrial mantle. A bulk lunar Al_2O_3 content of 6.0% implies a uranium abundance of 30 ppb, a little below the lower heat flow limit. The heat flow estimates provide a range from 33 to 46 ppb (Langseth *et al.,* 1976). This assumes that heat flow is equal to heat production, a reasonable view for the Moon on account of the near-surface concentration of heat-producing elements, its small size, and cessation of volcanic activity about 3 aeons ago. Both heat flow values were obtained at mare basin margins. Rasmussen and Warren (1985) suggest that focusing effects at such locations enhance the heat flow and estimate a bulk lunar uranium abundance of 19 ppb. Several factors suggest that this value is too low. The present basins filled with mare basalt (only a few tens of meters thick at the edges) represent only the final impacts. Modest estimates of cratering rates (Wilhelms, 1985) since crustal solidification at 4.4 aeons allow for at least 80 basins (>300 km diameter) and 10,000 craters in the range 30–300 km diameter. Spudis (1984) suggests that this impact history has created a megaregolith down to the 25-km seismic discontinuity. If the crustal structure is effectively that of a megaregolith 25 km thick, due to the superimposed effects of over 80 basins (plus several thousand craters), then it seems unlikely that any simple focusing mechanism will operate or that heat transport will be affected only by the most recent basin margin effects.

Further constraints may be placed by the abundance of potassium. The bulk lunar K/U ratio is about 2500. The highland crustal average is estimated at 500 ppm, which contributes 60 ppm to the bulk lunar budget. Mare basalts typically contain 500 ppm K (Taylor, 1982, Tables 6.4a,b), which for $D_{L/R}$ values of 10 indicate source region contents of at least 50 ppm. Thus, the bulk lunar K abundance is at least 85 ppm, which gives a bulk lunar U value of 34 ppb for K/U = 2500. The Rasmussen-Warren value of 19 ppb U yields a bulk lunar K value of 48 ppb, less than is apparently concentrated in the highland crust. It is thus difficult to escape from the conclusion that the lunar uranium value is greater than 30 ppb, consistent with the estimate by Drake (1985). In this case, the Moon is enriched in refractory elements relative to the Earth by a factor of greater than 1.5. Such an enrichment cannot, of course, be caused merely by loss of volatile elements during fission, but could be consistent with condensation from a vapor phase, as in the large impactor hypothesis.

2.6. Summary of Earth-Moon differences

Philosophical difficulties beset attempts to compare the chemistry of the Earth and the Moon. Even if a perfect compositional match was obtained, this implies similarity of process, rather than a genetic connection. Thus, the shergottite class of meteorites shows a very close correspondence in trace element chemistry to terrestrial basalts (Stolper and McSween, 1979). The correspondence is much closer than between lunar and terrestrial basalts. However, no one supposes that Mars, the probable source of the shergottites, is derived from the Earth. Likewise, the eucrites are much closer to lunar basalts than are terrestrial basalts, but a genetic connection is not suggested. The overall view of lunar and terrestrial mantle geochemistry reveals more differences than similarities and refutes simple versions of the fission hypothesis (Table 1). The depletion of the volatile elements, the differences in FeO content, siderophile element and refractory element abundances all suggest irreconcilable differences. The high lunar FeO and low siderophile abundances present a paradox.

TABLE 1. *Element Abundances in CI Chondrites (Volatile-free), Primitive Earth Mantle (= Mantle Plus Crust) and Bulk Moon.*

	CI	Earth Mantle +Crust	Bulk Moon
SiO_2	34.2	49.9	43.4
TiO_2	0.11	0.16	0.3
Al_2O_3	2.44	3.64	6.0
FeO	35.8	8.0	13.0
MgO	23.7	35.1	32.0
CaO	1.89	2.89	4.5
Na_2O	0.98	0.34	0.09
K_2O	0.10	0.02	0.01
ϵ	99.2	100.1	99.3
Volatile elements			
K (ppm)	854	180	83
Rb (ppm)	3.45	0.55	0.28
Cs (ppb)	279	18	12
Moderately volatile element			
Mn (ppm)	2940	1000	1200
Moderately refractory element			
Cr (ppm)	3975	3000	4200
Refractory elements			
Sr (ppm)	11.9	17.8	30
U (ppb)	12.2	18	33
La (ppb)	367	551	900
Eu (ppb)	87	131	210
V (ppm)	85	128	150
Siderophile elements			
Ni (ppm)	16500	2000	400
Ir (ppb)	710	3.2	0.01
Mo (ppb)	1380	59	1.4
Ge (ppm)	48.3	1.2	0.0035

Data sources: CI and terrestrial data from Taylor and McLennan (1985, Tables 11.1 and 11.3), lunar data from Taylor (1982, Tables 8.1 and 8.2), except for Ni, Ir, and Mo data calculated from CI/Moon depletion factors given by Newsom (1986), and Ge data from Dickinson and Newsom (1984).

Most attempts to find similarities between the chemistry of the terrestrial mantle and the Moon result in taking the extreme limits of the measured data or in the least plausible explanations. These models require the thinnest lunar crust (with the lowest Al_2O_3 content), that the heat flow data be ignored, that siderophile element abundances be based on highland rock data (particularly liable to meteoritic contributions), that the cause of the CF/CM offset lies deep in the inaccessible interior rather than in the crust, and that fission occur at a time earlier than the establishment of the present mantle-core relationship. Such attempts to match Earth-Moon chemistry have been labeled Procrustean, after the mythological inn-keeper (Anders, 1978). A final assessment is that "if the Moon was derived from the terrestrial mantle by fission, then the chemical evidence for such an event has been destroyed" (Taylor, 1982, p. 425).

3. The Large Impactor Hypothesis

This model states that the Moon was derived from material splashed during a glancing collision of a Mars-sized body with the Earth (Cameron and Ward, 1976; Hartmann and Davis, 1975). In the latest versions of this model (Cameron, 1984; Stevenson, 1984; Kaula and Beachy, 1984), two constraints have appeared. In order for the material to be placed in lunar orbit, it must be derived mostly from the Mars-sized impactor, removing the geochemical problems to another part of the wood. In addition, it must be vaporised, since solid ejecta will return to the Earth. This scenario cuts a number of Gordian knots. It removes the angular momentum problem. A grazing impact might place material in a nonequatorial orbit, accounting for the orbital dilemma. Several provisos may be made:

1. The impactor must form in the general vicinity of the Earth, to account for the similarity in oxygen isotopic signatures between Earth and Moon.

2. Mantle-core separation had already occurred in both Earth- and Mars-sized impactors before impact.

3. Time scales of 10^8 years are required to allow for accretion of Earth and impactor and for mantle-core separation in both bodies.

4. A seductive feature of the model is that it allows for an ad hoc composition for the impactor, within the general constraints imposed by inner planetary compositions. Accordingly, appropriate Fe/Mg ratios, siderophile element contents and Al and U abundances become free parameters, thus bypassing the imbroglio of lunar-terrestrial compositional comparisons.

5. It is apparently a requirement of the model that the material that finished up in the Moon, in addition to being derived mainly from the impactor, is vaporised at temperatures of about 2000 K (Stevenson, 1984). This could account for the extensive depletion of the Moon in volatile elements (e.g., alkalies, Bi, Tl, H_2O, etc.). However, recondensation temperature limits of 1500 K (Boynton, 1983) might be placed by the absence of combined Eu and Yb anomalies in bulk lunar REE

patterns, in contrast to those observed in some Allende inclusions. This constraint could be overcome by postulating simple evaporation and total recondensation for the REE. Since, however, the model wishes to use this mechanism to deplete the Moon in the other volatile elements, special pleading must be invoked and the shadow of William of Ockham appears. Evaporation of MgO and SiO_2 might also be expected at these temperatures, but such effects are not obvious. Extensive processing of material through a vapor phase might also produce some isotopic anomalies, although none have been detected. Depletion of volatile elements in the Moon appears to be limited to those elements with condensation temperatures below about 1100 K, which limits thermal processing of protolunar material to such temperatures or to short time intervals.

4. Early Volatile and Siderophile Element Depletion

Volatile element depletion is not restricted to the Moon but is common in the solar system, as is shown by the widespread variations in K/U and K/La ratios in meteorites and planets (Figs. 1 and 2). The terrestrial planets exhibit variations in volatile-refractory element ratios and in density consistent with metal-silicate fractionation. Since planets of Earth's size cannot lose elements of high atomic weight, three basic questions arise: When does the volatile-refractory element separation occur, where is the location, and what is the mechanism responsible? The meteorites provide some basic clues. They display much variation in metal-silicate and volatile-refractory ratios. The production of metal is apparently separate from and unrelated to the volatile depletion. The type 1 enstatite chondrites provide this interesting information, since they contain all their complement of iron as metal but retain the primitive levels of volatile element abundances. The H, L, and LL chondrites display both variations in metal and volatile elements.

4.1. The eucrite evidence

The lunar composition is clearly highly fractionated, relative to primitive solar nebula abundances. Does it represent an exceptional case among planetary compositions? Although there is a strong temptation to regard the Moon as a special case, volatile depletion is not restricted to that body, for very similar depletions (in K/U, Bi, Tl, H_2O, lack of Fe^{3+}, etc.) are observed in the eucrites, leading to the comment that "the lunar deficiency of alkalies and volatiles is not unique, and does not necessarily imply special processes in its evolution as a satellite" (Morgan et al., 1978, p. 35).

The eucrites are derived by partial melting from the mantle of a small asteroid. In addition to the volatile element depletion (K/U = 3100), they are depleted in Ni (~10 ppm) and the other siderophile elements. Metal was absent in the source regions (Stolper, 1977). Samarium-neodymium and $^{207}Pb/^{206}Pb$ data for eucrites indicate crystallisation ages of 4.54–4.56 aeons. Two possibilities exist: (1) The eucrite parent body was accreted from metal-free volatile-depleted precursors. (2) The eucrites were derived from the mantle of an asteroid that had already formed a metallic core. In either alternative, K and Rb had already been depleted relative to CI abundances of U and Sr. The refractory element abundances in the parent body

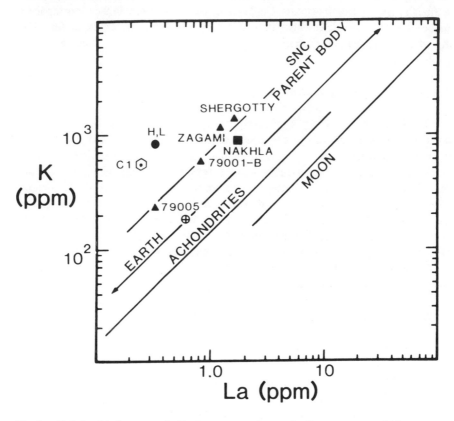

Fig. 2.   Relationship between volatile element potassium and refractory rare earth element lanthanum for CI, H, and L chondrites, Earth, Moon, achondrites, and SNC parent body (= Mars?). All planetary bodies are depleted in potassium (volatile at about 1000 K), relative to La, illustrating the large scale of depletion of volatile elements in the inner portions of the solar nebula. Both K and La are incompatible elements in intraplanetary igneous differentiation processes, concentrating in residual melts during crystallisation of silicate melts. The K/La ratio of surface rocks thus approximates to the bulk planetary ratio. Mars, if it is the source of the SNC meteorites, is thus more volatile-rich than the Earth. The K/U (Fig. 1) and K/La plots show slight differences, attributable to differences in behaviour of La and U in intraplanetary differentiation processes. (Based on Smith et al., 1984.)

are parallel to those of CI chondrites, showing no anomalies in the abundances of the more volatile REE, Eu, and Yb.

The similarity between eucrites and lunar mare basalts is remarkable. Were it not for the differences in age and oxygen isotope signature, it might be difficult to distinguish them on petrological or geochemical grounds. The similar levels of depletion in the highly volatile elements (e.g., Bi, Tl) and the absence of water is particularly significant. It demonstrates that it is unnecessary to appeal to high-temperature processing of terrestrial material following fission to cause the high lunar depletion in volatiles.

The fact that both volatile and siderophile element depletion occurred at 4.5 aeons and that basalts were generated by partial melting of a planetesimal mantle similar to that of the lunar mantle must give pause to those who advocate unique conditions for the origin of the Moon.

4.2. Timing of the lunar volatile depletion

Volatile depletion, in the fission and large impactor models for lunar origin, occurs during the fission or impact event. In the double-planet scenario, the Moon accretes from planetesimals in which volatile depletion has already occurred. Can we establish the time at which the lunar depletion in volatile elements occurred? The eucrites, and other meteorites, provide direct evidence of volatile loss at about T_o (4.56 aeons). We also know that the Earth accreted from volatile-depleted planetesimals that had average K/U ratios of 10,000 and Rb/Sr = 0.031 and that volatile/refractory element fractionation was widespread in the early solar system (Figs. 1 and 2).

In principle the Rb-Sr isotopic systematics could provide evidence for the time of separation of volatile Rb from refractory Sr (and hence also K from U). The best constraints on the lunar initial $^{87}Sr/^{86}Sr$ (LUNI) come from measurements on the low Rb/Sr anorthosites. For example, sample 60025 has a measured $^{87}Sr/^{86}Sr$ ratio of 0.69895 ± 3, which constrains LUNI to be equal to or less than that value (Nyquist, 1977). This is very primitive and close to BABI (0.69898 ± 3), implying an early origin for the Moon. We do not know the terrestrial initial $^{87}Sr/^{86}Sr$, but if the Earth formed in 10^8 years by accretion from a suite of planetesimals with bulk Rb/Sr = 0.031, then the terrestrial initial $^{87}Sr/^{86}Sr$ ratio must be very low, closer to ALL (0.69877 ± 2) than to BABI, if we wish to derive the Moon from the terrestrial mantle. This evidence is consistent with models that establish the basic differences in volatile/refractory element ratios between the Earth and the Moon in precursor phases prior to their formation and accrete the two bodies from a distinct set of planetesimals, as in the double-planet scenario.

In the impact-induced fission model, the absence of knowledge of the terrestrial initial $^{87}Sr/^{86}Sr$ ratio is not critical since terrestrial mantle involvement in lunar composition is minimal. The impactor $^{87}Sr/^{86}Sr$ initial ratio is a free parameter.

However, in the 10^8 years or so required to accrete both Earth and impactor, a measurable increase in $^{87}Sr/^{86}Sr$ will occur so that the impactor initial $^{87}Sr/^{86}Sr$ must be very low. The very primitive $^{87}Sr/^{86}Sr$ ratio of the Moon presents some difficulties for these scenarios.

In summary, either the Earth or the impactor have initial $^{87}Sr/^{86}Sr$ much lower than BABI, or the Wetherill-Safronov times for planetesimal sweep-up of 10^8 years are too long, or the fission and impact-induced fission models are incorrect. Further constraints are derived from the I-Pu-Xe dates (Swindle et al., 1984, 1986) that indicate that the Moon appears to be somewhat older than the Earth. A full discussion is given in Swindle et al. (1986).

4.3. Mechanisms for volatile depletion and element fractionation

Depletion of the volatile elements occurs both in the early solar nebula and in later collisional episodes. The mechanisms responsible for the early volatile depletion in the nebula are speculative, but some constraints may be noted. Temperatures of the order of 1000–1100 K seem required to account for the observed depletions. The separation of volatile (e.g., K) from refractory (e.g., U) elements occurs on a large scale, so that the Earth, Moon, meteorites, the eucrite parent body, and other planetesimals acquired differing K/U ratios (Figs. 1 and 2). However, heavy elements cannot be lost by heating from condensed bodies of planetesimal size, and the separation must take place in dispersed phases before accretion or during collisions between small bodies. Could this separation have occurred during condensation of an initially hot solar nebula? There is little evidence that the primordial solar nebula was hot, and there is considerable evidence from the isotopic heterogeneities (Clayton, 1978), and supported by astrophysical observations and theory, that it was initially cold (Gehrels, 1978). However, heating of the inner nebula (in the region now occupied by the terrestrial planets), either during collapse of the central region to form the sun or due to early solar activity, has occurred. A preferred scenario is heating of the inner nebula by early intense solar activity.

If volatile elements, along with the rare gases, are removed from the inner solar system by early strong solar winds and flares, then the astrophysical evidence provides stringent time constraints. Nebulae surrounding young stars clear on time scales of 10^6 years. Strong T Tauri type solar winds occur on similarly short time scales (Gehrels, 1978). In order to sweep out gases and volatiles by such a mechanism, then the mechanism operates on times one or two orders of magnitude shorter than the planetesimal sweep-up times of 10^8 years. Accordingly, volatile-refractory element separation, if accomplished by such means, can only occur within a few million years of the sun arriving on the main sequence. The evidence from the meteorites that such fractionation occurs on short time scales at about 4.5 aeons may thus be consistent with such scenarios.

The timing is crucial, since not all volatile elements are removed from the inner solar system. Enough volatile material remains to account for the volatile components in the inner planets. Accordingly, by the time the early sun is radiating, some material must already be in planetesimals capable of surviving heating and dispersion by early strong solar activity.

Large impact events also provide a mechanism for depleting the parent material in volatile elements. Small-scale examples of depletion of very volatile elements (e.g., Bi, Cs, Tl, Pb) occur in tektite-forming events. Impact glasses (e.g., Henbury, Zhamanshin), in contrast, are subjected to less heating and retain their complement of such elements (Taylor and McLennan, 1979). Such mechanisms can operate either during planetesimal breakup, in disintegrative capture processes, in processing through a circumterrestrial disk, or during large-scale impactor events. In the latter case, the material that goes into the Moon is postulated to have condensed from a vapor phase. There are accordingly several possible mechanisms for depleting the lunar material in volatile elements. Temperatures of recondensation seem to have been about 1000–1100 K, since only elements whose condensation temperatures are less than this are depleted.

5. Implications for Terrestrial Planetary Formation

Current theories for the origin of the inner planets call for their assembly from a diverse suite of planetesimals, probably in a gas-free environment (Wetherill, 1980). The time scale for the accretion of the inner planets from a large population of planetesimals is of the order of 10^8 years, although much of the accretion is completed in about half that time (Wetherill, 1980).

If the terrestrial planets accrete from planetesimals, with sweep-up times of about 10^8 years, then the inner planets were not complete until accretion ended about 4.45 aeons, about 100 m.y. later than the time scales for element fractionation indicated by the meteorites. It is a consequence of such models that the inner planets thus accrete from a heterogeneous swarm of already differentiated planetesimals, with varying contents of metal and volatile elements.

The implications of preaccretion metal-silicate and refractory-volatile element fractionation for the compositions of the Earth, Moon, and the inner planets are considerable. The metal now in the Earth's core, and the silicates in the mantle are accreted as separate phases (but not necessarily in separate planetesimals) whose equilibration was established in precursor events. Thus, there is no necessary close relationship between, for example, the Ni, Co, and siderophile element contents of the mantle and the core. Melting and segregation of metal, sulfide, and silicate (i.e., core formation) most likely occurred during or shortly after accretion, but new metal-silicate equilibria was not necessarily established, and the bulk $Mg/(Mg + Fe)$ ratio of the mantle is only indirectly related to core compositions. Partitioning of K, U, or Th into the core is thus unlikely. The light element in the core is probably mainly sulfur (Ahrens, 1979). If all element fractionation processes occurred at low pressures in small bodies or dispersed phases in the inner solar system close to 4.56 aeons, then the meteorites provide the classical Goldschmidtian evidence of siderophile-chalcophile-lithophile (plus volatile-refractory) element fractionation, which points to sulfur accreted as troilite or oldhamite as the most viable candidate. Limited high-pressure reaction between metal and silicate is not precluded but is not extensive in this scenario, in which the preexisting metal and sulfide fall out to form the core during melting concomitant with accretion.

Some further consequences follow. The satellites of the outer planets have low densities, although Ganymede, Callisto, Titan, and Triton are about the size of Mercury. Accordingly, it appears that free metal was not available in large amounts beyond the asteroid belt. Probably volatile-refractory element fractionation, which may be linked to early solar flare-ups in the inner solar system, was not effective at and beyond 5 AU (as is shown, inter alia, by the large amounts of condensed water/ice in the Jovian and Saturnian satellites). Accordingly, it may be predicted that the volatile-refractory element ratios (e.g., K/U ratios) in these satellites are about the same as those in CI meteorites. This would have the consequence that radioactive heat generation is generally higher in these bodies than for the inner planets. The decrease in density of the Galilean satellites with distance from Jupiter is probably due to mild warming of a proto-Jupiter disk, changing the rock/ice ratio. This scenario provides a low-energy analogue of events close to the sun in the early solar nebula.

6. Conclusions

The current state of geochemical knowledge for the Moon, Earth, and meteorites clarifies several points with respect to the origin of the Moon.

1. Classical capture models can be excluded.

2. The composition of the bulk Moon differs significantly from that of the terrestrial mantle. Fission models can be ruled out, unless very complex fractionation processes occur following fission but preceding accretion of the Moon.

3. Large Mars-sized impactor models, in which the material making up the Moon is both derived from the impactor and is vaporised during the event, become viable. They cut the Gordian knot of Earth-Moon comparisons at the cost of making the chemical and isotopic composition of the impactor free parameters.

4. Double-planet scenarios remain viable (Taylor, 1982) since the meteorite data predict a population of precursor planetesimals depleted in siderophile and volatile elements. The problems of accreting these to the Moon remains an outstanding problem, possibly answered by Ruskol-type scenarios.

5. The origin of the Moon is complex and elements of several theories are needed to account for its composition. This, however, is not unique, and geochemical processes

operating at the beginning of the solar system produced similar depletions in siderophile and volatile elements to those observed in the Moon.

6. Hypotheses in science should be testable. A test to distinguish between the two viable hypotheses of lunar origin, large impactor and double planet, can be made in principle by dating the volatile depletion event. In the simplest case, this depletion occurs at T_0 for the double-planet hypothesis, with the Moon then accreting in a circumterrestrial disk from fractionated planetesimals. The Mars-sized impactor hypothesis predicts that major volatile depletion occurs in a vapor phase as a consequence of the impact. However, this scenario is complicated by the documented meteoritic evidence of volatile depletion at T_0, so that we need to identify the signature of a second volatile depletion event.

Acknowledgments. I thank James R. Arnold for many interesting lunchtime discussions about the early solar system and for comments on this paper. Scott McLennan assisted in several ways. John Larimer and an anonymous reviewer made many perceptive comments. Jeff Taylor made a major contribution by pointing out that the original paper was too terse. The efforts of all these people have greatly improved the paper. I thank Karen Buckley and Julie Stringer for secretarial assistance.

7. References

Anders E. (1978) Procrustean science: Indigenous siderophiles in the lunar highlands, according to Delano and Ringwood. *Proc. Lunar Planet. Sci. Conf. 9th*, pp. 161–184.

Ahrens T. J. (1979) Equation of state of iron sulfides and constraints on the sulfur content of the Earth. *J. Geophys. Res., 84*, 985–998.

Baldwin R. B. (1949) *The Face of the Moon.* University of Chicago Press, Chicago. 239 pp.

Boynton W. V. (1983) Cosmochemistry of the rare earth elements: meteorite studies. In *Rare Earth Element Geochemistry* (P. Henderson, ed.), pp. 63–114. Elsevier, Dordrecht.

Cameron A. G. W. (1984) Formation of the prelunar accretion disk (abstract). In *Papers Presented to the Conference on the Origin of the Moon,* p. 58. Lunar and Planetary Institute, Houston.

Cameron A. G. W. and Ward W. R. (1976) The origin of the moon (abstract). In *Lunar Science VII*, pp. 120–122. The Lunar Science Institute, Houston.

Clayton R. N. (1978) Isotopic anomalies in the early solar system. *Annu. Rev. Nuclear Sci., 28*, 501–522.

Delano J. W. and Lindsley D. H. (1983) Mare glasses from Apollo 17: Constraints on the Moon's bulk composition. *Proc. Lunar Planet. Sci. Conf. 14th*, in *J. Geophys. Res., 88*, B3–B16.

Delano J. W. and Ringwood A. E. (1978) Siderophile elements in the lunar highlands: nature of the indigenous component and implications for the origin of the moon. *Proc. Lunar Planet. Sci. Conf. 9th*, pp. 111–159.

Dickinson T. and Newsom H. (1984) Ge abundances in the lunar mantle and implications for the origin of the moon (abstract). In *Papers Presented to the Conference on the Origin of the Moon,* p. 16. Lunar and Planetary Institute, Houston.

Drake M. J. (1986) Is lunar bulk material similar to Earth's mantle?, this volume.

Gehrels T. (ed.) (1978) *Protostars and Planets.* University of Arizona Press, Tucson. 756 pp.

Haines E. L. and Metzger A. E. (1980) Lunar highland crustal models based on iron concentrations: Isostasy and center-of-mass displacement. *Proc. Lunar Planet. Sci. Conf. 11th*, pp. 689–718.

Hansen K. S. (1982) Secular effects of oceanic tidal dissipation on the moon's orbit and the Earth's rotation. *Rev. Geophys. Space Phys., 20*, 457–480.

Hartmann W. K. (1965) Terrestrial and lunar flux of large meteorites in the last two billion years. *Icarus, 4*, 157–165.

Hartmann W. K. and Davis D. R. (1975) Satellite-sized planetesimals and lunar origin. *Icarus, 24*, 504–515.

Hood L. L. and Jones J. (1985) Lunar density models consistent with mantle seismic velocities and other geophysical constraints (abstract). In *Lunar and Planetary Science XVI*, pp. 360–361. Lunar and Planetary Institute, Houston.

Kaula W. M. and Beachy A. E. (1984) Mechanical models of close approaches and collisions of large protoplanets (abstract). In *Papers Presented to the Conference on the Origin of the Moon*, p. 59. Lunar and Planetary Institute, Houston.

Langseth M. G., Keihm S. J., and Peters K. (1976) Revised lunar heat flow values. *Proc. Lunar Planet. Sci. Conf. 7th*, pp. 3143–3171.

Liu L. and Bassett W. A. (1985) *Elements, oxides and silicates: High pressure phases.* Oxford University Press, in press.

Mason B. and Taylor S. R. (1982) Inclusions in the Allende meteorite. *Smithson. Contrib. Earth Sci., 25*, 1–30.

McKenzie D. and Richter F. M. (1981) Parameterized thermal convection in a layered region and thermal history of the earth. *J. Geophys. Res., 86*, 11667–11680.

Morgan J. W., Higuchi H., Takahashi H., and Hertogen J. (1978) A "chondritic" eucrite parent body: Inference from trace elements. *Geochim. Cosmochim. Acta, 42*, 27–38.

Newsom H. E. (1984) The lunar core and the origin of the moon. *EOS (Trans. Am. Geophys. Union,) 65*, 369–370.

Newsom H. E. (1986) Constraints on the origin of the moon from the abundance of molybdenum and other siderophile elements, this volume.

Nyquist L. E. (1977) Lunar Rb-Sr chronology. *Phys. Chem. Earth, 10*, 103–142.

Rasmussen K. L. and Warren P. H. (1985) Megaregolith thickness, heat flow and the bulk composition of the Moon. *Nature, 313*, 121–124.

Ringwood A. E. (1970) Origin of the moon: The precipitation hypothesis. *Earth Planet. Sci. Lett., 8*, 131–140.

Ruskol El. Y. (1977) The origin of the moon. *NASA SP 370*, p. 815–822.

Smith J. V. (1979). Mineralogy of the planets: A voyage in space and time. *Mineral. Mag., 43*, 1–89.

Smith J. V. (1982) Heterogeneous growth of meteorites and planets, especially the earth and moon. *J. Geol., 90*, 1–125.

Smith M. R., Laul J. C., Ma M. -S., Huston T., Verkouteren R. M., Lipschutz M. E., and Schmitt R. A. (1984) Petrogenesis of the SNC meteorites: Implications for their origin from a large dynamic planet, possibly Mars. *Proc. Lunar Planet. Sci. Conf. 14th*, in *J. Geophys. Res., 89*, B612–B630.

Spudis P. D. (1984) Apollo 16 site geology and impact melts: Implications for the geologic history of the lunar highlands. *Proc. Lunar Planet. Sci. Conf. 15th*, in *J. Geophys. Res., 89*, C95–C107.

Stevenson D. J. (1984) Lunar origin from impact on the Earth (abstract). In *Papers Presented to the Conference on the Origin of the Moon*, p. 60. Lunar and Planetary Institute, Houston.

Stolper E. M. (1977) Experimental petrology of the eucritic meteorites. *Geochim. Cosmochim. Acta, 41*, 587–611.

Stolper E. M. and McSween H. Y. (1979) Petrology and origin of the shergottite meteorites. *Geochim. Cosmochim. Acta, 43*, 1475–1498.

Swindle T. D., Caffee M. W., Hohenberg C. M., Hudson G. B., Laul J. C., Simon S. B., and Papike J. J. (1984) Noble gas component organisation in Apollo 14 breccia 14318; ^{129}I and ^{244}Pu regolith chronology. *J. Geophys. Res.*, in press.

Swindle T. D., Caffee M. W., Hohenberg C. M., and Taylor, S. R. (1986) I-Pu-Xe dating and the relative ages of the Earth and Moon, this volume.

Taylor S. R. (1975) *Lunar Science: A Post-Apollo View*. Pergamon, New York. 372 pp.

Taylor S. R. (1979) Nickel abundances in lunar mare basalts (abstract). In *Lunar and Planetary Science X*, pp. 1215–1216. Lunar and Planetary Institute, Houston.

Taylor S. R. (1982) *Planetary Science: A Lunar Perspective*. Lunar and Planetary Institute, Houston. 481 pp.

Taylor S. R. and McLennan S. M. (1979) Chemical relationships among irghizites, zhamanshinites, Australasian tektites and Henbury impact glasses. *Geochim. Cosmochim. Acta, 43*, 1551–1565.

Taylor S. R. and McLennan S. M. (1985) *The Continental Crust: Its Composition and Evolution*. Blackwell, Oxford. 312 pp.

Urey H. C. (1962) Origin and history of the moon. In *Physics and Astronomy of the Moon* (Z. Kopal, ed.), pp. 481–523. Academic, New York.

Walker D. (1983) Lunar and terrestrial crust formation. *Proc. Lunar Planet. Sci. Conf. 14th*, in *J. Geophys. Res, 88*, B17–B25.

Warren P. H. (1984a) Megaregolith thickness, heat flow and the bulk composition of the moon (abstract). In *Papers Presented to the Conference on the Origin of the Moon*, p. 18. Lunar and Planetary Institute, Houston.

Warren P. H. (1984b) The bulk-moon MgO/FeO ratio: A highlands perspective (abstract). In *Papers Presented to the Conference on the Origin of the Moon*, p. 19. Lunar and Planetary Institute, Houston.

Weidenschilling S. J. (1984) The lunar angular momentum problem (abstract). In *Papers Presented to the Conference on the Origin of the Moon*, p. 55. Lunar and Planetary Institute, Houston.

Wetherill G. W. (1980) Formation of the terrestrial planets. *Annu. Rev. Astron. Astrophys., 18*, 77–113.

Wilhelms D. E. (1985) Lunar impact rates reconsidered (abstract). In *Lunar and Planetary Science XVI*, pp. 904–905. Lunar and Planetary Institute, Houston.

Wise D. U. (1963) An origin of the moon by rotational fission during formation of the earth's core. *J. Geophys. Res., 68*, 1547–1554.

Wolf R. and Anders E. (1980). Moon and Earth: Compositional differences inferred from siderophiles, volatiles and alkalis in basalts. *Geochim. Cosmochim. Acta, 44*, 2111–2124.

Wood J. A. and Mitler H. E. (1974) Origin of the moon by a modified capture mechanism (abstract). In *Lunar Science V*, pp. 851–853. The Lunar Science Institute, Houston.

Nebular Chemistry and Theories of Lunar Origin

JOHN W. LARIMER

Department of Geology and Center for Meteorite Studies, Arizona State University, Tempe, AZ 85281

The origin of the Moon is regarded as one stage in the history of planetary matter, the long sequence of events that determines the final elemental and isotopic composition of the rocks. Lunar and terrestrial rocks have passed through the entire sequence of events, making the record difficult to decipher. Other, less extensively processed materials fortunately exist, such as the sun and primitive meteorites, that have escaped the planetary processes of melting and volcanism. The isotopic anomalies observed in meteorites indicate that some interstellar dust, still bearing its nucleogenetic signature, survived nebular processing. But the distribution of this dust in the nebula and its bearing on the origin of the Moon is not yet understood. The elemental compositions of the sun and chondritic meteorites are very similar, suggesting that the chemical inhomogeneities in the nebula implied by the isotopic anomalies were localized and superimposed on an elementally homogeneous nebula. Chondrites contain a number of chemically and mineralogically distinct components that appear to have formed in the nebula within specific temperature intervals. Each component contains a suite of elements in solar proportions and each class of chondrite contains its own unique blend of the various components. This mixing gives rise to a characteristic elemental fractionation pattern where the concentrations of groups of elements appear to increase or decrease in unison. Elemental concentrations in the bulk Earth and Moon are constrained using simple geochemical principles and mass balance relations. The patterns in these abundance estimates are similar to those observed in chondrites, suggesting the presence of the same set of components blended in different proportions. Relative to Type I chondrite composition, the Earth is strongly depleted in volatile elements and slightly enriched in metallic elements. The Moon's unique blend of nebular components falls into an interesting pattern: normalized to the refractory elements, the Moon is depleted in Mg-silicates by a factor of 2–3, in metallic elements by a factor of 10, and in the volatile elements by a factor of 30, relative to the Earth. Evidently the depletion factors are related to the temperature at which the components become stable, with the extent of depletion increasing as the temperature decreases. The implication is that accretion became progressively less efficient with decreasing temperature, suggesting that the Moon may have been competing for the components and systematically losing more of its share as accretion progressed. A possible interpretation is that the Earth and Moon accreted from the same mix of components as the proto-Moon orbited the proto-Earth, with the Earth winning, and the Moon progressively losing, its solar complement of the components.

1. Introduction

Lunar rocks, as well as all other samples of planetary material, are the end product of their cosmic history, the long sequence of events that determine the final elemental and isotopic composition. From this broad perspective, the origin of the Moon is one step in a series of events in the chemical history of planetary matter. The history begins with nucleosynthesis, continues with a nebular, star-forming stage, and culminates in accretion. The scene is then set for planetary differentiation: the formation of a core, mantle, crust, and atmosphere. Each of these events leaves its signature, although the more ancient signatures are often obscured by the more recent events. The challenge is to decipher this history in order to understand the elemental and isotopic changes that occur at each stage. Lunar and terrestrial rocks, having passed through the full sequence of processing, are the most complex to trace. Fortunately, other materials exist that are processed less extensively: both the sun and primitive meteorites escaped planetary differentiation. A large body of observational, analytical, and experimental data provides additional constraints and insights.

Some theories of lunar origin reverse the historic sequence by suggesting that the Moon accreted from material that had already passed through the planetary differentiation stage. There are several currently popular ideas: fission from the Earth, a gargantuan terrestrial impact, and disintegrative capture. In these theories the Moon accretes from the debris of preexisting, differentiated bodies; the debris originates from the Earth's mantle either by fission or impact, from a differentiated giant impactor, or from a swarm of differentiated planetesimals that break up as they approach the Earth. An important stimulus for these theories is that they explain the Moon's low metallic Fe content by a known process: mantles of differentiated bodies are metal-deficient because the metal has been sequestered into a core. The violent, energetic nature of these events introduces other complexities.

It is also possible, however, that the Moon simply accreted from planetary material in orbit around the Earth. Such an origin is generally accepted for most satellites in the solar system. Reminiscent of the Earth-Moon system, many of these satellites also have densities that differ from the planet they orbit, an indication of compositional differences. In addition, density variations within satellite systems, such as Jupiter's Galilean satellites, imply different compositions within a system. Although the casual relations are not understood, it appears that accretion in planetary orbit can produce chemical fractionations.

In the following discussion, the mystery of lunar origin will be approached by tracing the cosmic history of planetary matter from nucleosynthesis through accretion. This permits specific predictions to be made on the probable range of bulk compositions at the time of origin. Comparing these predictions to estimates of the Moon's compositions will reveal any unique compositional features that might suggest a special event, or perhaps a reversal in the expected sequence of events. Obviously, complete coverage of all these topics is beyond the scope of this paper. For the sake of brevity, only a few of the more recent, pertinent discoveries that bear most heavily on the problem of lunar origin will be highlighted.

2. Cosmic History

The focus here is on those events and processes that lead to important, observable changes in the composition of planetary material. Nucleosynthesis clearly qualifies as an important event, though it seems far removed in space and time from events associated with the origin of the moon. Nebular processes have been explored extensively and their effects must be considered in any theory of lunar origin. Planetary differentiation and volcanism determine the final character of lunar rocks. Moreover, if these most recent processes can be understood, they may also provide information on the composition of the lunar interior.

Prenebular history

Isotopic data. The discovery of certain isotopic anomalies in primitive meteorites has led to a change in perspective regarding both the prenebular and nebular history of planetary material. The first such anomaly discovered, and also the most widespread and potentially most important to the question of lunar origin, is the variation in ^{16}O (Clayton *et al.*, 1973). Unfortunately, the present ^{16}O data provide few constraints on the various theories of lunar origin; they do, however, provide constraints on the earlier history of planetary material.

Before the discovery of anomalous oxygen, all available evidence indicated that the nebula was chemically homogeneous. Most variations in elemental and isotopic composition could safely be attributed to chemical processes, radioactive decay, or expected exposure to energetic particles. But with the discovery of anomalous isotopes, especially the relatively abundant ^{16}O, the idea of a perfectly homogeneous nebula faded quickly. As Clayton *et al.* (1973) noted in reporting their original discovery, the most plausible explanation for anomalous ^{16}O is that the solar system formed from an incompletely homogenized mixture of interstellar dust and gas derived from several nucleogenetic sources. The implications are profound. The presence of anomalous O implies that interstellar dust still carrying a nucleogenetic signature might be preserved in meteorites. Different bodies evidently contain different amounts of ^{16}O, suggesting that each object in the solar system could have accreted from its own unique blend of interstellar and locally formed nebular dust. Furthermore, the preservation of interstellar dust implies incomplete volatilization of planetary material during the nebular stage.

What makes oxygen isotopes particularly interesting is their usefulness as a cosmic fingerprint. Chemical processes, in contrast to nebular processes, change the proportions of the three isotopes of oxygen in a predictable way. The change in the $^{18}O/^{16}O$ ratio is always twice that of the $^{17}O/^{16}O$ ratio because the mass difference of

$^{18}O/^{16}O$ is twice that of $^{17}O/^{16}O$. For this reason, data from various planetary samples are plotted on $^{17}O/^{16}O$ vs. $^{18}O/^{16}O$ diagrams along with lines of slope = 1/2. All samples that plot on any one line (with slope = 1/2) can, in principle, be derived from the same oxygen reservoir. All samples plotting off that line must be derived from some other, isotopically distinct reservoir. "Chemical processes" in this context includes all reactions that occur during planetary differentiation, volcanism, metamorphism, and weathering: in effect, all processes that occur after accretion. Thus a single measurement on any sample from a particular body is sufficient to determine the relative proportions of the oxygen isotopes in that body. (The exceptions to this generalization are those meteorites, like Allende, in which the original mix of isotopes has never homogenized).

Data collected to date indicate that each of the various groups of meteorites is derived from its own oxygen reservoir, implying a separate parent body for each group. All lunar and terrestrial samples, on the other hand, plot on a single line, to within experimental limits, suggesting an origin from the same or very similar reservoirs. But the significance of this similarity is unclear. The isotopic composition of the Earth and Moon falls in the middle of the range defined by the meteorites. A few meteorite groups have virtually the same composition as the Earth and Moon, but most do not. However, no pattern is evident: relative to the Earth and Moon, some meteorite parent bodies are more enriched and others more depleted in ^{16}O. Thus while the data appear to suggest a common origin for the Earth and Moon, they would not be inconsistent with a theory in which lunar material originates as far away as the asteroid belt.

Elemental data. The elemental composition of the sun, which dominates the solar system, appears very similar to other stars. It is also similar to the elemental composition predicted from nucleogenetic theory. This suggests that isotopically anomalous material, while important locally, was not sufficiently abundant to produce a chemically unique solar system. The sun evidently formed from a well-mixed collection of stellar debris.

Furthermore, primitive meteorites known as Type I carbonaceous chondrites contain the nonvolatile elements in proportions nearly identical to those in the sun. A recent comparison (Anders and Ebihara, 1982) of the chondritic composition with that of the sun indicates that for over 60 elements the two are identical to within a factor of 2. Moreover, this uncertainty factor of 2 results mainly from the difficulties inherent in determining elemental concentrations in the sun. The proportions of the oxygen isotopes in these meteorites is very similar, but not quite identical, to samples of lunar and terrestrial material (Clayton and Mayeda, 1984).

These observations suggest that the stellar debris that accumulated to form the solar system must have acquired a uniform elemental and, to a large extent, isotopic composition over the inner solar system. If this inference of a generally uniform initial composition is correct, then any deviations from solar composition in an object must have developed during the nebular or postnebular stages of cosmic history.

Since major changes in bulk compositions are not likely to occur after accretion, this leaves the nebula as the most likely setting. Of course, this reasoning does not include those theories in which the Moon forms from the debris of a preexisting body that had already differentiated into a core, mantle, and crust. One test of such theories is to compare the Moon's composition with that of planetary material at the end of the nebular stage to see whether an additional processing step is required.

Nebular history

Many features of chondritic meteorites appear to be the product of processes that occurred prior to or during accretion. For this reason a common exercise is to compare the composition of a sample, or an estimate of the bulk composition of a body such as the Earth or Moon, with chondritic abundances. Any deviation from chondritic composition is then considered to be an indication of postaccretion processing.

What is often overlooked is that chondrites themselves differ in composition. In fact, the unique compositional variations among different types of chondrites are some of the best documented primitive features. Type I carbonaceous chondrites are remarkably similar to the sun, as just discussed, but in all other chondrites, at least a few elements occur in different proportions. These differences fall into patterns that clearly are not the result of planetary igneous, metamorphic, or sedimentary processes, but instead suggest more primitive, nebular processes.

A characteristic feature of chondritic fractionation is that groups of elements appear to move in unison. Each class of chondrites appears to contain its own blend of these groups of elements. If one element in a group is enriched or depleted, then all other elements are enriched or depleted as well, and by nearly the same factor. One well known group of elements is the more noble metals, which are usually alloyed in the FeNi phase. In most groups, however, the only property shared by all elements is volatility. Four groups of elements with similar volatilities are recognized; they are classified on the basis of their fractionation behavior: refractories ($T > 1300°K$), silicates ($1300° > T > 1000°K$), normally depleted ($1000° > T > 600°K$), and strongly depleted ($T < 600°K$).

The temperature ranges given in parentheses are those in which elements belonging to the group are predicted to condense from a cooling gas with solar composition at a pressure $P = 10^{-4}$ atm. These temperatures reflect the cosmic volatility of an element and depend on several factors: elemental vapor pressures; an element's reactivity, which determines whether it occurs as a metal or compound; elemental abundances, which govern the partial pressures in a cosmic gas; and the total pressure of the system.

As might be expected given the diverse chemical properties of the elements, there are some that do not fit neatly into this or any other classification scheme. For example, some elements are metallic as well as refractory (e.g., Ir and Pt) and appear to belong to two groups. In addition, a few elements seem to behave erratically by occasionally switching groups or displaying different fractionation factors relative to other members of their group. Such elements inevitably are ones predicted to condense near a temperature limit between two groups. Once this is realized, their behavior no longer seems anomalous: it reflects nothing more complicated than our inability to develop a simple, yet comprehensible classification scheme. Among the elements that display such behavior, perhaps the most interesting are the alkali metals (Na, K, Rb, and Cs) and Mn.

The interpretation of the chondritic fractionations is based on detailed studies of the condensation sequence of the elements in a cooling gas of solar composition. Specific predictions regarding the composition and mineralogy of the condensate can be made and compared to observations drawn from the meteorites. The results are summarized in Fig. 1. Note that this presentation of the thermodynamic calculations

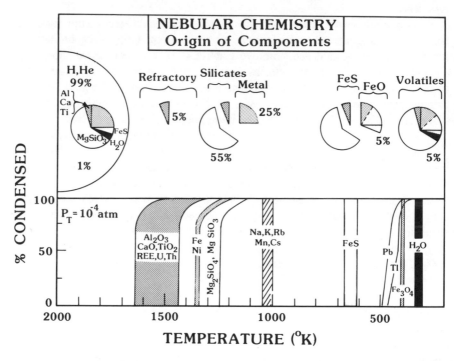

Fig. 1. *Nonvolatile planetary material represents about 1 wt % of the total mass of the sun, or solar system. In primitive meteorites, the nonvolatile elements are observed to be distributed among several components, descriptively referred to as: refractory (-rich), metal, silicate, FeS and volatile (-rich). Iron occurs as metal, sulfide, and oxide. Each component contains a group of elements and compounds that condense from a gas of solar composition over a restricted temperature range.*

---

is not meant to imply that all the material in the nebula formed during a single, all-inclusive vaporization event followed by condensation; it is simply a means of displaying the temperature ranges over which the various solids would be in equilibrium with a gas of solar composition.

In addition to predicting temperatures at which elements will condense or vaporize, these studies examine a variety of other pertinent reactions. The chemistry of Fe is an illustrative example. Consider the reactions that occur with decreasing temperatures. First, gaseous Fe condenses to FeNi grains, whose composition and condensation temperature is pressure dependent. As the metal grains cool, they acquire ever increasing amounts of the more volatile trace metals such as Au and Cu. At $670°$ K the metal reacts with H_2S in the gas to form FeS, which is a pressure-independent reaction. From the onset of condensation, a small but continually increasing amount of Fe reacts with H_2O in the gas to form FeO, which gradually is incorporated into the Mg-silicates. The FeO content increases markedly between about $600°$ and $500°$ K. However, several factors could affect the extent to which Fe is oxidized: the rate at which CO and H_2 react to form H_2O and CH_4, the C/O ratio (Larimer and Bartholomay, 1979), the gas/dust ratio (Wood, 1984), and the rate at which FeO reacts with the silicates. Any metallic Fe left unreacted when the temperature falls to $400°$ K will combine with gaseous H_2O to form Fe_3O_4, another pressure-independent reaction.

These predictions are then compared to observations drawn from chondrites. The key observation is that chondrites consist of a number of discrete components that can be studied microscopically, separated and analyzed. Each component carries its own charateristic group of elements. Significantly this grouping of the elements is generally similar to the grouping based on fractionation behavior. Each component also displays a characteristic mineralogy. For example, the component that carries the refractory elements is comprised of a suite of minerals predicted to be the only stable solid phases that can exist in a gas/dust mixture with solar composition at high temperatures, $T > 1300°$ K. In all cases, the mineralogy and elemental abundances of the various components are generally consistent with each other and with the formation temperatures inferred from the computed nebular chemistry (Grossman and Larimer, 1974).

The importance of these components lies in the information they contain regarding conditions in the nebula and the nature of the accretion process. Each chondrite class apparently has acquired its own characteristic mix of components, with the proportions varying from class to class. The relative proportions of the various components, as they would occur in an unfractionated nebula, are also shown in Fig. 1. Type I carbonaceous chondrites are the only samples of planetary material in which all the components appear in their solar proportions. All other chondrite classes acquired the components in distinctly nonsolar proportions.

If chondrites are derived from bodies that accreted as mixtures of nebular components, it naturally follows that other bodies may have formed from the same set of materials. This idea was first applied to the Earth and Moon in an attempt to explain the similarities and differences in bulk composition (Ganapathy and Anders, 1974). It has since been modified and extended to Mercury, Venus, Mars, and the eucrite parent body (Anders, 1977; Morgan and Anders, 1979, 1980).

In this model, planetary bodies are assumed to consist of seven components: (1) refractory-rich material, (2) melted metal, (3) melted silicate, (4) unmelted metal, (5) unmelted silicate, (6) FeS, and (7) volatile-rich material. All but one of these components can be readily observed in some chondrites, such as Allende. The Ca-Al-rich inclusions, enriched in all refractory elements, represent the refractory-rich component. Chondrules consist of melted and partially devolatilized metal and silicates. The fine-grained matrix contains unmelted metal and silicates as well as FeS. There is no isolated volatile-rich component; instead, the volatile elements and compounds appear to be dispersed throughout the matrix material: H_2O as hydrated silicates; C as organic compounds, graphitic material, or carbonates; noble gases trapped largely in the graphitic material; and Bi, In, Pb, Tl, etc., as dissolved trace components in the more abundant metal, silicate, or sulfide hosts. It is only for bookkeeping purposes that the volatile elements are lumped together in a single component.

On the Earth and Moon, however, these components have been obliterated by extensive postaccretion processing and are no longer recognizable. The amount of each component present must be inferred from the physical and chemical properties of the body. Useful physical properties include the body's density, which constrains the Fe/Si ratio, and its surface heat-flow, which constrains the radioactive Th and U contents. The geochemical behavior of a number of elements results in a characteristic distribution that can also be used to constrain the proportions of the various components in a body, as will be discussed in detail in the next section.

This multicomponent model provides estimates on the concentrations of nearly all elements in the body. Since each component contains a known group of elements in fixed proportions, then a constraint on the concentration of any one element automatically constrains all the others (Anders, 1977). The results of this approach are presented in Fig. 2, where the proportions of the components in the Earth and Moon are compared to those in various chondritic meteorites. This comparison suggests that the Earth is slightly enriched in metal but is otherwise similar to H-group chondrites. The Moon, in contrast, is poorer in metal and enriched in refractory elements relative to the Earth or any class of chondrite.

This cosmochemical approach to the origin and composition of the Moon has an advantage over other models; it provides an elegant, heuristic theoretical framework. Its main disadvantage is the sweeping nature of the assumptions involved. There are some simple tests of the assumptions, however. For example, if the ratio of one pair of elements is used to establish the proportions of two components in a body, then the ratios of all other pairs of elements belonging to those two components should display the same ratio. Based on the data available, such tests are generally consistent (Ganapathy and Anders, 1974).

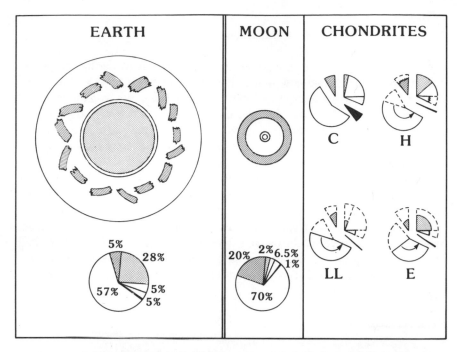

Fig. 2. The bulk composition of the Earth and Moon considered in terms of the nebular components observed in chondrites. In both bodies the elements have been extensively redistributed and the components can no longer be recognized. On Earth, the refractory elements are distributed throughout the crust and mantle, while on the Moon they are concentrated in the crust. Metal and sulfide, if present, are concentrated in the cores. The volatile content of the Earth and Moon is small, proportionately less than the line thickness in the figure. Different classes of chondrites (C, H, LL, and E) contain different amounts of the components. The patterns used to designate components are the same ones used in Fig. 1.

Here we will examine the implications of the model by considering a much broader data base. The extensive studies of trace element abundance and distribution in lunar rocks has stimulated a similarly extensive study of terrestrial material. The result of these efforts is a greatly expanded data base that has yet to be fully utilized. In fact, as the quality and quantity of the data have improved, models of lunar origin seem to have become less specific chemically. Some variants of the giant impactor and disintegrative capture models appear to leave chemical composition as a free parameter. From the chemical point of view such models cannot be tested directly; their viability hinges on the extent to which other models fail to account for the data. Many of the data can be used to test the multicomponent model, however, and to quantify the fractionation processes that produced bodies of different composition.

3. Comparative Chemistry

At this juncture, we will change our approach. Instead of tracing the history from beginning to end, we now turn to the end product, the lunar and terrestrial rocks, and use their composition of the Moon and Earth. This means working backward from the present to the time of origin, using the composition of today's crustal rocks to constrain the material that accreted 4.5×10^9 years ago. The final and most interesting step will be to compare the bulk compositions obtained in this way with that predicted for nebular materials.

The first problem is to find an informative method of comparing large numbers of elemental abundances in the Earth and Moon with solar or chondritic abundances. The method adopted is to first group the elements according to their cosmochemical behavior, normalize to solar system abundances, and then check for fractionations. By grouping the elements at the outset, it becomes possible to test for fractionations within each group as well as between the groups. This is the method commonly used to compare solar and Type I chondritic compositions (Anders and Ebihara, 1982).

Elemental ratios

The bulk composition of a body can be constrained in either a relative or an absolute sense. An absolute abundance is the average concentration of an element in a body per unit mass, while relative abundances are simply ratios between two or more elements. Obviously, if absolute abundances are known, then relative abundances are redundant. However, absolute abundances are more difficult to obtain. In undifferentiated chondrite parent bodies, where all elements occur in the same proportions throughout, analysis of any sample yields absolute abundances. But in differentiated bodies, such as the Earth and Moon, where the elements have been extensively and unevenly redistributed, no representative, average sample exists. Crustal rocks, which display huge enrichments or depletions of certain elements, are clearly not representative, average samples. Nonetheless, these unrepresentative crustal rocks can still provide good estimates, or definitive upper and lower limits, on bulk elemental ratios for numerous elements.

Absolute abundances are of secondary importance for our purposes. If it can be shown that a group of elements, ideally one of the groups observed in chondrites, are all present in solar proportions, then the absolute abundance of only a few elements constrains all the others. The first step is to determine whether the elemental abundances in the Earth and Moon parallel the chondritic pattern. If all elements within each group occur in the same proportions, then the same set of components must be present. But if an altogether different pattern emerges, then some other mode of origin is indicated.

For many elements, good estimates of their bulk ratios can be obtained by analyzing any crustal igneous rock. The rationale is straightforward. The rocky fraction of planetary material consists largely of MgFe-silicate with minor amounts of Ca-Al silicate. These materials dominate the mineralogy of mantles and control the evolution of the mantle-crust systems. Crustal igneous rocks are rich in all the elements that are not readily accommodated in these phases. Elements such as U and Th, which occur as unusually large ions with +4 charges, cannot readily occupy mineral lattice sites that are normally reserved for much smaller ions of Ca, Fe, and Mg with +2 charges. During planetary differentiation, when the mantle partially melts, these incompatible elements are excluded from the solid minerals and become concentrated in the magmatic liquids. The liquids, rich in incompatible elements, ascend toward the surface where they solidify into a suite of plutonic and volcanic igneous rocks. These rocks display variable major element contents, depending on the extent of melting and other complicating factors. Nonetheless, many of the most incompatible trace elements occur in nearly constant proportions that appear to reflect the body's bulk ratio. Many examples have been reported in studies of differentiated meteorites, lunar and terrestrial rocks; K/U, K/La, Tl/U, and W/La are among the best studied pairs of elements (Taylor, 1982).

In addition, there are other methods to constrain the bulk elemental ratios when the element pair does not always display the same ratio in crustal rocks. Radioactive decay produces systematic changes in the isotopic ratios of several elements and provides a useful tool for constraining the U/Pb, Rb/Sr, Lu/Hf, and Sm/Nd ratios in a body. Radioactive decay of a parent isotope produces a predictable increase in the abundance of a daughter isotope; the extent of the increase depends on the time elapsed and the parent/daughter ratio. The parent/daughter ratio and amount of daughter isotope in a rock or mineral can be measured and used to determine both the age and the amount of daughter isotope present when the rock formed. The amount of a daughter isotope present at the time of formation reflects the parent/daughter ratio in the parent material from which the magma was derived. In cases where the parent magma can be shown to have originated in the mantle, the parent/daughter ratio of the mantle, a major reservoir of trace elements, is constrained.

Useful limits on the bulk ratios of many elements can also be obtained from an understanding of the forces that control elemental redistribution. The rare earth elements (REE) provide a good illustrative example. All 14 REE are remarkably similar in their geochemical behavior. They are incompatible in the common mantle minerals, normally occurring as +3 ions, with the occasional exception of Eu, which occurs as +2 ions in the reducing lunar environment. The one difference between the REE is the systematic decrease in ionic size with increasing atomic number. Light rare earth elements (LREE) form larger ions than the heavier rare earth elements (HREE). The LREE are therefore less compatible, with the result that during the melting of planetary mantles the LREE are more highly concentrated in magmatic liquids than the HREE, whereas the mantle displays the complementary pattern. The implication is that the crustal LREE/HREE ratio must be greater than that for the body as a whole. This implication, expressed algebraically, is:

$$(LREE/HREE)_C > (LREE/HREE)_B$$

where the subscripts C and B are used to differentiate crustal and bulk ratios.

The reasoning can be extended to obtain limits on absolute abundances. The ratio on the left hand side of the inequality is the crustal ratio, a determinable quantity, which is always greater than the bulk ratio. And if the ratio of the two elements is constrained, then any independent constraint on the bulk content of one element automatically constrains the other. If the HREE can be constrained at some maximum permissible value, the LREE is also constrained:

$$(LREE)_B < (LREE/HREE)_C \, (HREE)_{MAX}$$

Any exception violates the inequality and, in geochemical terms, implies that the HREE is more enriched in the crust than the LREE, which is not reasonable. Alternatively, if a minimum value for a LREE can be established for the body, then a minimum value for a HREE can be obtained:

$$(HREE)_B > (HREE/LREE)_C \, (LREE)_{MIN}$$

Similar reasoning can be applied to other groups of elements: Mg, Ca, Sr, Ba; Na, K, Rb, Cs; etc. It is also possible to extend this reasoning to constrain ratios across these groups of elements when the effects of ionic charge are taken into account: $(K^{+1}/Ca^{+2}) > (K^{+1}/Ca^{+2})$, $(Rb^{+1}/Sr^{+2}) > (Rb^{+1}/Sr^{+2})$, etc. These simple geochemical arguments obviously provide a powerful tool for constraining bulk compositions.

To illustrate the strengths and weaknesses of this approach, let us consider a fictitious body (Table 1). Studies of the crustal rocks on this body reveal a REE pattern similar to that observed on other bodies: all REE are enriched relative to solar system abundances and there is a logarithmic decrease in the abundances with increasing atomic number. For the bulk body, we will use a lower limit of 1000 ppb for La, the lightest REE, and an upper limit of 100 ppb for Lu, the heaviest REE, on the assumption that these values can be obtained independently. Upper and lower limits for the remaining REE are calculated according to the method just discussed. Only the odd numbered REE data are presented for simplicity. The most probable concentrations must lie closer to the center of the ranges than to

TABLE 1. Constraints on the REE Concentrations in a Body.

Element	La	Pr	Eu	Tb	Ho	Tm	Lu
			Concentrations in ppb				
CRUST	10200	3420	1970	1210	1675	595	500
MAX	2040	727	394	242	335	119	100
MIN	1000	356	193	119	165	59	50
AVG	1520	542	294	180	250	89	75
C-I	243	90	54	35	54	22	24
f	6.25	6.02	5.44	5.14	4.63	4.05	3.13

Notes: The MAX, MIN, and AVG values are calculated on the assumptions that an upper limit (MAX) for Lu of 100 ppb and a lower limit (MIN) for La of 1000 ppb have been established independently. The (chondrite normalized) REE pattern in the crust (CRU) of this fictitious body is similar to that observed in basaltic achondrites, lunar, and terrestrial rocks: a logarithmic decrease with atomic number. The averages (AVG) of the upper and lower limits are normalized to Type I chondrites (C-I) to obtain the enrichment factors (f) for the body.

the limits but it is impossible to be more precise. The average of the limits therefore is adopted as the best estimate of the bulk concentration of the elements in the body. These averaged values are compared to Type I chondrite concentrations to ascertain the extent of fractionation relative to solar system abundances. The obvious strength of this approach is that many element ratios in a body can be constrained in a simple, useful manner.

One weakness is also evident. The variation in f values implies that the REE are fractionated relative to solar abundances. However, since we put the REE into this imaginary body in their solar proportions, the fractionation is not real, but merely an artifact of the computations. As noted above, the best estimate of the bulk concentrations is simply an average of the upper and lower limits; since these limits will vary in a regular manner, the averages of the limits will also vary, giving rise to the apparent fractionation. Therefore, any slight systematic fractionation should be regarded as a weakness in the model rather than a characteristic feature of the body. Moreover, for real bodies like the Earth and Moon additional constraints exist: Sm/Nd and Lu/Hf isotope systematics, an Eu anomaly, and K/La ratios can be used for midcourse corrections and tightening of the end points.

The point to be stressed before considering the controversial question of absolute abundances is that for either a fictitious body or for the Earth and Moon, many elemental ratios can be constrained independent of reliable constraints on absolute abundances. If we had expanded our study of the imaginary body to include the entire group of refractory elements, and if all of their f values were similar to those calculated for the REE, then the presence of a refractory component would be strongly suggested. Conversely, large differences in elemental ratios or f values would rule out a refractory component, or at least rule out the one found in chondrites.

Absolute abundances

The controversial nature of estimating absolute abundances can be illustrated by considered a pertinent example. When Ganapathy and Anders (1974) first estimated the bulk compositions of the Earth and Moon in terms of the multicomponent model, the accepted bulk U content of the Moon was 60–75 ppb. There was one much lower estimate, which was widely ignored, of 30–40 ppb (Larimer, 1971). Today, the pendulum has swung in the other direction; now bulk U concentrations of 15–20 ppb are considered possible (Rasmussen and Warren, 1984). The most widely accepted estimates at present, however, fall between 30 and 40 ppb (Drake, 1986; Taylor, 1986).

The bulk U content of the Moon is of special interest because it can be used to test and develop various models of origin. A lunar value of 15–20 ppb would be similar to the value inferred for the Earth's mantle, and would therefore be consistent with the fission model, whereas the much higher value would seem to virtually eliminate the model. The best current estimate of the bulk U content is also used to estimate the refractory element content of the Moon in the multicomponent model. Here, rather than rely on the best estimate for a single element, we will establish constraints on the bulk concentrations of a large number of elements. Several simple, basic relations will be used. Upper and lower limits on the bulk elemental concentrations are established first, based on a few simplifying assumptions. Then the most important sources of potential error are discussed.

A simple proposition serves to fix a firm lower limit on the amount of any element in a body: the body must contain at least the amount found in the crust. The only data required to establish this limit are the average crustal concentrations of the elements and the mass of the crust. A lower limit obtained in this way is most useful if a large fraction of the element is concentrated in the crust. For many elements, the fraction concentrated in the crust is too small to provide a useful limit.

Such a simple concept would not be controversial if the mass of the lunar crust and the extent to which it is chemically homogeneous could be agreed upon. The mass of the Earth's crust is well constrained and, while it obviously is chemically inhomogeneous, geochemists generally agree on how to obtain reliable average crustal concentrations. The uncertainties regarding the Moon will be discussed below; for now we will simply assume that the lunar crust is chemically homogeneous, with a mass equal to 10% of the global mass. Note that if these assumptions prove to be invalid, the general effect will be to shift all of the upper and lower limits by some common factor. Whole groups of elements will move together while retaining their relative positions.

Lanthanum is a good example of an element whose minimum bulk concentration in the Earth and Moon can be set by considering its crustal concentrations. Taylor (1982) estimates average crustal concentrations of 19 and 5.3 ppm for the Earth and Moon, respectively. The massive lunar crust, which comprises about 10% of the Moon's mass, contrasts with the terrestrial crust, which comprises only 0.4% of the Earth's mass. This large difference in mass fraction of the crusts is largely offset by the much greater enrichments of incompatible elements in the terrestrial crust. The concentrations obtained by multiplying the mass fraction of the crusts by the crustal concentrations yields minimum La values of 0.8 and 0.53 ppm for the Earth and Moon.

These limits are unrealistically low, however, because some portion of the body's total content of each element is almost certainly present in the mantle or core. A more realistic lower limit can be obtained by considering the analytical and experimental studies on trace element distribution during partial melting of mantle minerals. These studies, in which the analytical data on mantle-derived basalts are used to infer the trace element contents in the mantle, indicate that at least 20%, and more likely 40–50%, of the total La content in both the Earth and Moon still resides in the mantle. In other words, the crust contains no more than 80% of the total La. A more realistic lower limit for La can therefore be obtained by dividing the average crustal concentrations by 0.8, resulting in values of 0.1 and 0.66 ppm for the Earth and Moon, respectively.

Once lower limits for a few elements in a body are established, the remainder can be obtained by using this method or one of the others discussed earlier. For example, one can use isotopic and elemental ratios or differential crustal enrichments that result from differences in ionic size or charge. Since the intent is to acquire the maximum permissible lower limit, when two or more methods can be used the one that yields the largest bulk concentration is adopted.

Upper limits are obtained in a similar manner, the single difference being that the crustal concentrations must now be converted to maximum permissible bulk concentrations. For this purpose, a few elements are selected that are known to be concentrated in the crust, but not as concentrated as those used to set the lower limits. An example is Lu, the least enriched of the REE. The estimated crustal concentrations are 0.21 ppm for the Moon and 0.3 ppm for the Earth. If Lu is not enriched in crustal rocks relative to the bulk concentration (which is unreasonable to expect), then these values are firm upper limits on the bulk concentrations. But trace element distribution studies indicate that Lu is enriched by a factor of 2–3 in the lunar crust and by a factor of 10–20 in the Earth's crust, relative to the mantle source regions. Somewhat more realistic, though still extreme, upper limits of 0.14 and 0.05 ppm for the Moon and Earth have therefore been set on the assumption of minimum enrichment factors of 1.5 and 5, much less than the probable values of 2–3 and 10–20.

Once upper limits are fixed for a few elements, elemental and isotopic ratios can be used together with established differential enrichment patterns to fix other upper limits. There are also independent means to establish upper limits; a good example is the lunar U concentration of 47 ppb, based on the lunar heat flow data (Langseth *et al.*, 1976). Interestingly, this value is about a factor of 6 higher than the solar abundance, which is similar to the upper limit inferred for Lu, as well as for the other REE and a large number of refractory elements.

As extreme as these limits may seem, they nonetheless constrain the composition of the Earth and Moon in a useful way. The computed values, based on Taylor's (1982) summary of crustal data, are summarized in Table 2 and Figs. 3–6. Of the 37 elements considered, the upper limits of only 3 (Al, Ca, and V) are based on the common *ad hoc* assumption of a chondritic ratio. The concentrations of

TABLE 2. *Estimated Concentration of Selected Elements in Bulk Moon and Earth.*

Element	Moon	Basis of Estimate	Earth	Basis of Estimate	C1†
		Refractories > 1300° K			
Al %	3.2 ± 1.6	CR,CI	1.1 ± 0.6	R,CI	0.862
CA %	3.4 ± 1.6	CR,LI	1.2 ± 0.6	R,CI	0.902
Sc ppm	23 ± 14	CR,C	8 ± 5	CR,R	5.76
Ti ppm	1600 ± 800	CR,R	540 ± 200	CR,R	436
V ppm	190 ± 120	R,CI	80 ± 40	R,CI	57
Sr ppm	30 ± 10	CR,R,I	12 ± 4	CR,R,I	7.9
Y ppm	4.8 ± 3.0	CR,R	1.8 ± 1	CR,R	1.5
Zr ppm	13 ± 4	CR,C	8 ± 5	CR,C	3.69
Nb ppm	0.8 ± 0.2	CR,R	0.45 ± .20	CR,R	0.25
Ba ppm	11 ± 3	CR,R	3.5 ± 1.5	CR,R	2.27
La ppb	990 ± 330	CR,R	300 ± 100	CR,R	243
Ce ppb	2200 ± 700	CR,R	900 ± 500	CR,R	619
Pr ppb	300 ± 90	CR,R	190 ± 140	CR,R	90
Nd ppb	1350 ± 400	CR,R	950 ± 400	CR,R,I	462
Sm ppb	460 ± 200	CR,R	200 ± 90	CR,R,I	142
Eu ppb	300 ± 160	CR,R	10 ± 80	CR,R	54
Gd ppb	900 ± 550	CR,R	370 ± 260	CR,R	196
Tb ppb	160 ± 100	CR,R	65 ± 40	CR,R	35
Dy ppb	1050 ± 650	CR,R	380 ± 270	CR,R	242
Ho ppb	220 ± 120	CR,R	85 ± 60	CR,R	54
Er ppb	620 ± 350	CR,R	230 ± 170	CR,R	160
Tm ppb	90 ± 50	CR,R	33 ± 23	CR,R	22
Yb ppb	570 ± 300	CR,R	230 ± 160	CR,R	166
Lu ppb	90 ± 50	CR,R	3.0 ± 20	CR,R	24
Hf ppb	440 ± 200	CR,C	240 ± 180	CR,R	119
Th ppb	140 ± 40	CR,H	40 ± 20	CR,R	29
U ppb	37 ± 10	CR,H	12 ± 6	CR,R	8
		Volatiles 1300–600° K			
Na ppm	700 ± 300	CR,CI	700 ± 300	CR,CI	4830
K ppm	85 ± 22	CR,R,C	135 ± 45	CR,I,O	570
Rb ppm	0.27 ± 0.07	CR,R,I	0.42 ± 0.13	CR,R,I	2.3
Cs ppb	12 ± 3	CR,R	9 ± 4	CR,R	186
Mn ppm	1050 ± 300	C	650 ± 30	C	1960

TABLE 2. *Continued.*

Element	Moon	Basis of Estimate	Earth	Basis of Estimate	Cl†
		Volatiles < 600° K			
H₂O%			0.0375 ± 0.0125	CR,O	20(%)
³⁶Ar ppb			> 0.04	CR	1.21
Cd ppb	0.55 ± 0.15	R	18 ± 14	R	6.73
In ppb			> 0.07	CR	78
Tl ppb	0.35 ± 0.25	R	3.24 ± 1.72	R	143
Pb²⁰⁴ ppb	0.059 ± 0.016	I	1.33 ± 0.67	I	47
Bi ppb	0.064 ± 0.020	R	0.8 ± 0.4	CR,R	111
		Major Elements			
Mg%	19 ± 2	O	15 ± 1.5	O	9.55
Si%	20 ± 2	O	16.5 ± 1.5	O	10.67
Fe%	9.5 ± 2	O	30 ± 2	O	18.51

*Estimates based on: crustal abundances (CR), crustal ratios (R), element correlations (C), isotopic systematics (I), heat flow (H), chondritic ratio (CI), other (O).

†Anders and Ebihara (1982).

Fe, Mg, and Si are based on the requirement of mass balance, allowing for the mass of the Earth's core, an assumed lunar core comprising 2% of the mass of the Moon, and an assumed solar Mg/Si ratio of 1. A recent estimate of the Earth's bulk Mg and Si content, based on studies of the least fractionated, mantle-derived xenoliths is virtually identical to the estimates in Table 2 (Palme and Nickel, 1985).

One additional, clarifying remark should be inserted here. The terrestrial concentrations given in Table 2 and plotted in the figures are for the whole Earth,

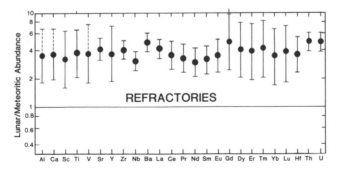

Fig. 3. The refractory lithophile elements in the bulk Moon are essentially unfractionated relative to C1 (~solar) abundance, but evidently are enriched by a factor of 3–6. Upper limits of only three elements (dashed lines) are based on chondritic ratios.

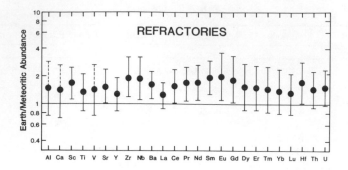

Fig. 4. *The refractory lithophile elements in the bulk Earth are essentially unfractionated relative to C1 (~solar) abundances, but are slightly enriched by a factor of ≈ 1.5. This is consistent with either average ordinary chondrite or devolatilized C1 composition. The values plotted are for bulk Earth, including core. For comparative purposes, if just the terrestrial mantle values are required, all the points are raised by a factor of ≈1.4.*

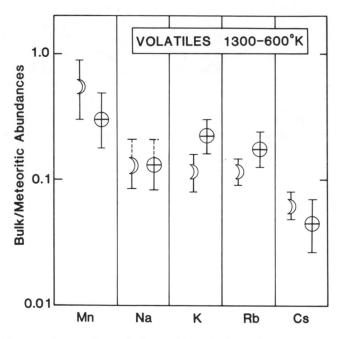

Fig. 5. *Elements of intermediate volatility are depleted in both the Earth and Moon relative to C1 (~solar) abundances. Terrestrial points are for the bulk Earth; for just the silicate fraction the terrestrial points fall higher by a factor of ≈1.4.*

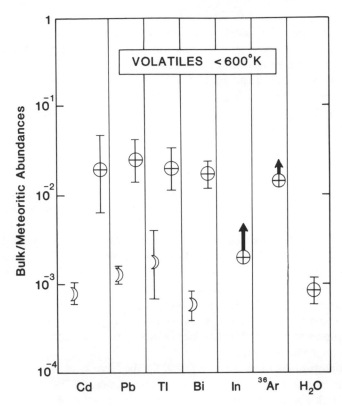

Fig. 6. Highly volatile elements are strongly depleted in both the Earth and Moon, but the depletion factors for several elements, with quite different volatilities, are nearly constant; the lunar values are less than terrestrial values by a factor of ≈30 for Cd, Pb, Tl, and Bi.

including the core. In order to compare the compositions of the Earth's mantle and the Moon, to test the fission hypothesis, each terrestrial value (except for Fe) can simply be divided by 0.7 (or multiplied by 1.43), which is the mass fraction of the Earth's mantle. When this is done, the estimated composition of the terrestrial mantle is consistent with those estimated elsewhere in these proceedings (Drake, 1986; Taylor, 1986).

Uncertainties

It is impossible to attach quantitative values to the extent to which these estimates of bulk composition are uncertain. For each element, the center of the range is much more probable than the upper or lower limits, which are based on extremely improbable element distributions in the bodies. The limits therefore become the greatest potential source of error. These limits are based largely on the crustal concentrations of the elements and the mass of the crust.

At present, the lower limits are of more interest than the upper limits. Current controversy is focused on the question of whether or not the Moon is enriched in refractory elements, relative to either the Earth or solar abundances. There is general agreement that both the Earth and Moon contain at least their solar allotment of these elements, but if the Moon is more enriched than the Earth, then the fission model becomes less attractive. In this study, the minimal extent of enrichment is set by the lower limits on the estimated bulk concentrations. They can be reduced further only if the lunar crust is less massive than assumed or is chemically inhomogeneous, with the crustal concentrations of the elements decreasing substantially with depth.

Here I have used Taylor's estimates of the elemental concentrations in the crust, which he has justified both succinctly (1986) and at length (1982). These estimates are based on the enormous compilation of analytical data on lunar rocks as well as the orbital x-ray and γ-ray data, which constrain the relative proportions of the various rock types on the Moon's surface. The orbital data are sometimes criticized for being incomplete and limited to the equatorial regions. However, there is no objective reason to believe that they grossly misrepresent the actual distribution of rock types on the lunar surface.

The geophysical data indicate a thickness for the lunar crust of between 60 and 80 km; the best average estimates are 70–75 km (Kaula, 1975). The lower limits on the bulk elemental concentrations in Table 2 were based on a crust/global mass ratio of 0.1, corresponding to an average thickness of 60 km. This lower limit of 60 km for the crustal thickness, corresponding to 10–25% less mass than the probable thickness of 70–75 km, was purposely chosen to generate the lowest plausible bulk concentrations of the incompatible, refractory elements. If the crust is actually thinner, then all the lower limits conceivably could be even lower, but they would all be lower by some constant factor.

The question of chemical inhomogeneity can be considered in terms of plagioclase, the dominant mineral in the dominant rock type in the lunar crust. Plagioclase has a comparatively low density and high Al, Ca, Eu, and Sr contents, making it important physically and chemically. The orbital x-ray data provide a clear picture of a plagioclase-rich surface. Variations of plagioclase content with depth are more difficult to ascertain; but the available evidence indicates a fairly uniform distribution. The relatively low density of plagioclase is the key factor because the density of the crust or crustal units is a measure of the plagioclase content. Seismic velocity profiles, the most direct measure of depth-density variability, indicate a uniform density, modified by compaction throughout the crust (Toksöz et al., 1974). Moreover, the Moon's moment of inertia coefficient is explicable in terms of a volumetrically large, low density crust. The center of mass/figure offset also is consistent with a massive, low density crust that is thicker on the far side than on the near side (Kaula, 1975). These geophysical data are discussed elsewhere in these proceedings and have been nicely summarized by Taylor (1982).

In addition, a recent study of large-basin excavation dynamics, in which the crater depths and the distribution of impact debris were considered, suggests a thick crust with a uniform plagioclase content (Spudis and Davis, 1985). Spudis (1985) has also presented persuasive arguments, based on cratering depths and amounts of impact debris generated, that the megaregolith extends to 25–30 km, implying large scale homogenization to at least this depth.

These geochemical and geophysical data all point to a thick (70–75 km), plagioclase-rich ($Al_2O_3 = 25\%$) crust. To establish minimum limits on the mass of each trace element in the lunar crust, I have used the lowest estimates of thickness and mass. It is possible, obviously, that some future data or interpretation will indicate a lunar crust less massive than the adopted estimate. Such a development would then require revisions in Table 2 and Fig. 3, the most important revision being that the lower limits for all incompatible elements would shift to some even lower value. However, the lower limits will all decrease by the same factor, and all the elements will retain their relative positions in the new version of Fig. 3.

4. Discussion and Conclusions

Until now, most estimates of the refractory element content of the Earth and Moon have been based on the best estimates of one or two elements, usually U, and the assumption that all other refractory elements are present in their solar proportions. This assumption is no longer necessary; the limits imposed here constrain the relative proportions of these elements to within a factor of 2 or 3 (Figs. 3 and 4), similar to the uncertainties in the recent comparison of the sun and meteorites (Anders and Ebihara, 1982). The Moon displays an overall enrichment in refractory elements by a factor of 3–6 relative to Type I chondrite abundances. The Earth may be slightly enriched: relative to Type I chondrites, the average enrichment factor for the Earth is 1.5–2, similar to the refractory element content of devolatilized Type I material or ordinary chondrites (H-group in Fig. 2). The refractory elements do not define a perfectly flat line relative to meteoritic abundances, but this is almost certainly an artifact of the calculations, as discussed earlier, and not the result of fractionation.

The data available on the bulk contents of other groups of elements are more limited. The data on the alkali metals plus Mn (Fig. 5) and on the volatile elements (Fig. 6) suggest that within each group the elements are present in nearly solar proportions, yet both groups of elements are strongly depleted in both bodies. The alkali metals and Mn display the greatest relative fractionation, reminiscent of their erratic behavior in chondritic material. Evidently these elements are carried in the chondrule fraction of planetary material and are commonly only partially outgassed or condensed during chondrule formation (Larimer and Anders, 1967).

The bulk metallic element contents cannot be constrained by the methods discussed here; these elements are too highly depleted in crustal rocks, having been sequestered into the core. Their bulk concentrations must be estimated in other, less direct ways. A promising approach is to consider the least noble, most readily oxidized metallic elements: Mo and W. During core formation these elements are less strongly partitioned into the metal phase; measurable quantities remain in the silicate phases, where they behave like typical incompatible elements (Newsom and Palme, 1984; Newsom, 1986). The data appear to be consistent with a chondritic composition for the Earth and a lunar metal content of at least 2%.

Previous estimates of lunar and terrestrial bulk compositions compare favorably with the present estimates (Taylor, 1982; Ganapathy and Anders, 1974; Anders, 1977). Our estimates are very similar to Taylor's, a not entirely fortuitous coincidence; Taylor reasoned that the U content of the crust, together with the heat-flow data, constrained the bulk U content to virtually the same value estimated here. He then *assumed* that all other refractory elements were present in their solar proportions. We have avoided this assumption, or rendered it unnecessary, by demonstrating that the elements are indeed present in their solar proportions. The earlier, multicomponent models also adopted a U content and assumed solar proportions for the remaining refractory elements; most differences between those estimates and the more recent ones can be attributed to the decrease in the estimated lunar U content by nearly a factor of two.

The refractory element patterns are consistent with the view that the Earth and Moon contain a refractory component nearly identical to the one found in chondrites. The data on the other element groups are less complete but consistent with the multicomponent model. One way to test the hypothesis that the Earth and Moon contain the same components as chondrites, albeit mixed in different proportions, and to quantify the amount of each component present, is to compare selected element ratios (Fig. 7). The principles and rationale for this approach have been discussed and its usefulness demonstrated in interpreting chondritic meteorites (Kerridge, 1979; Larimer, 1979).

This method is commonly used to interpret isotopic ratios and the principles are the same used to interpret any substance that is a mixture of several components. When two elements are suspected to have been carried into a mixture in two or more components and a third element is present for normalization, then aliquots of the mixture are drawn and element ratios are determined. The results are analyzed by plotting the two ratios on an x-y diagram. If the ratios define a linear array, then a mixture of two components is inferred, and the composition of the components must fall somewhere on the line. However, if the data array is nonlinear, then at least three components must be present. In mixtures suspected to contain more than two components, additional information on the composition of the components is required to keep the problem tractable.

Here, elements from the various groups are normalized to Al, the most abundant refractory element. The refractory element pattern, illustrated in the top diagram of Fig. 7, is easiest to understand. The average composition of the CaAl-rich inclusions (CAI) in carbonaceous chondrites is used as the composition of the refractory component. The enrichment or depletion of a refractory element E and a more volatile element Mg in a body derived from a system with Type I chondrite (C1), or solar composition, can be inferred by plotting the composition of the body on the diagram. Bodies that are enriched in the refractory component, like the Moon, plot between the CAI and C1 compositions. On the other hand, if a body is depleted in the refractory component, as are some chondrite classes, it would plot along an extension of the line on the opposite side of the C1 point.

Elements more volatile than Mg, including Na, K, Rb, Cs, and Mn, are considered in the middle diagram of Fig. 7. These elements are thought to characterize the chondrule (melted silicate) fraction of planetary matter. All 5 elements are depleted in the Earth by a factor of 5–10. This suggests that the total silicate fraction (melting and unmelted) of the Earth consists of about 80–90 wt % chondrules (i.e., melted silicates) and 10–20 wt % matrix (i.e., unmelted silicates). The Moon evidently contains the same two silicate components in the same proportions, since it plots on the line joining the CAI and the terrestrial compositions. If the silicate fraction of the Moon differed in some way, perhaps containing more or fewer chondrules, then the Moon would plot on some other line.

The Fe contents are more complex. A large fraction of the Earth's Fe inventory is in the core and comprises about 25 wt % of the Earth. The existence of a lunar core is still disputed; if a core exists, it cannot comprise more than about 4 wt % of the Moon (Hood, 1986). Newsom (1986) argues that the Moon must contain at least 2% Fe to account for the excess depletion of several metallic trace elements. Relative to the Earth, the Moon evidently is deficient in metallic Fe by a factor of about 10. Both bodies also contain appreciable amounts of oxidized Fe. The bottom diagram in Fig. 7 compares both the total Fe and FeO contents. The Earth is slightly enriched in total Fe relative to solar abundances or the more common classes of chondrites. At the same time the Earth contains less FeO than common chondrites, and appears more reduced. The FeO is presumed to be incorporated into the silicates. If the same batch of silicates incorporated into the Earth were also incorporated into the Moon, as suggested by the alkali element data, then the Moon should fall on the mixing line, which it does within the uncertainty limits. This diagram does not properly display what appears to be a slight difference in FeO/MgO ratio between the Earth and Moon, which suggests that the Earth's silicates are slightly more reduced (Taylor, 1982). However, elsewhere in these proceedings Warren (1986) argues that there is no difference, that both bodies have the same FeO/MgO ratio. This would agree with the inferences drawn here.

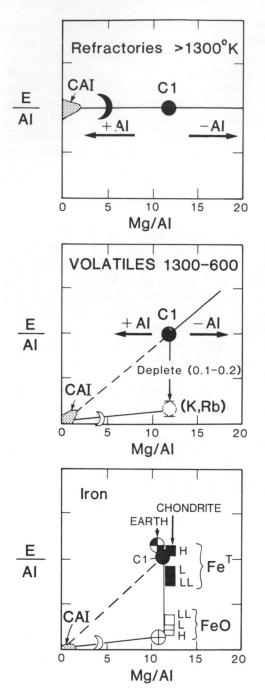

Fig. 7. (Top) Refractory elements (E) are enriched in the Moon, implying an enrichment in the refractory component (CAI). (Middle) Elements more volatile than Mg (Na, K, Rb, etc.) are depleted in the Earth's silicate component (× 0.1–0.2). If the Moon contains the same component, its bulk composition should fall on the mixing line between the depleted Earth point and the CAI point, which it does. (Bottom) The total Fe budget of the Earth slightly exceeds chondritic values, while the FeO content is slightly lower. The Moon contains even less FeO, placing it on the mixing line, implying that the silicate components have the same FeO content.

The most volatile elements are depleted in the Moon by a nearly constant factor of 30 relative to the Earth, or by a factor of 1000 relative to Type I chondrites (Fig. 6). Additional data and the implications of this pattern have been discussed previously (Wolf and Anders, 1980). As has been argued for many years, constant depletion factors imply the existence of a component. The reason is that while these elements are lumped together and referred to as volatiles, they actually condense or vaporize over a temperature range of several hundred degrees. At any specified temperature, one element may be totally vaporized, another partially vaporized, and a third may not yet have begun to evaporate. Models that explain the loss of volatiles by outgassing, whether induced by impact or internal heating, predict differential depletion and therefore are inconsistent with the data.

The chemical data are consistent with a model in which the Earth and Moon are assembled out of a limited number of nebular components. Most other models suffer in comparison, either because they do not offer specific predictions or because their predictions are contradicted by the data. If the Moon originated from the Earth's mantle by fission or impact, its composition should be almost the same as the Earth's. However, there are important differences. The Moon appears to be more enriched in refractory elements and much more depleted in volatile elements. While it is true that the lower concentration limits for refractory elements in the Moon slightly overlap the upper limits for the Earth (Figs. 3 and 4), these limits represent the extreme possibilities. If the lunar crust is as massive and rich in plagioclase as it appears to be, then it contains relatively more Al, Ca, Eu, and Sr than the Earth. Moreover, the main advantage of the fission model, accounting for the depletion of Fe by prior extraction into the Earth's core, has been compromised. Newsom (1986) points out that the lunar mantle and crust are more depleted in metallic elements than the Earth's mantle, implying that fission must be followed by a second extraction that occurs on the Moon. This in turn requires that the Earth's mantle was not entirely devoid of metal at the time of fission. These complexities may not absolutely rule out the fission model but they seriously threaten its viability. In these proceedings, Cameron (1986) suggests that a Moon formed during a giant impact event would be composed largely of debris from the impactor, rather than the Earth. While this model may have some advantages, providing specific chemical predictions is not its strong point.

If the Earth and Moon contain the same components that are found in chondrites, simply mixed together in different proportions, then the problem is to explain how and why the components were accreted in the indicated proportions. Two possibilities might be considered. First, all components could have been present during accretion and the accretion process was discriminatory, with some components given preference over others. The difficulty with this idea is that while metal and silicate possess fundamentally different physical properties that may cause them to behave differently as condensates, there are no obvious physical differences between the refractory and silicate components that might lead to an efficient separation. The second possibility is that only the material present as condensates was accreted and not all the components were equally available throughout accretion. A natural explanation, given the idea of sequential condensation (Fig. 1), is that accretion took place during condensation. This idea was quite popular not too long ago, but has recently fallen into disrepute, as the astrophysical models of the nebula have gradually reduced the peak temperature to the point where hardly any dust could have evaporated. Most students of meteorites have not yet completely abandoned higher peak temperatures, however.

Since the Moon has only 1/80 of the Earth's mass, it was clearly discriminated against in some way. In fact, if we consider the Moon's composition from a different perspective, an interesting relationship emerges. Suppose the Moon is not overendowed in refractory elements, but instead is deficient in all other components. From this perspective, relative to the Earth, the Moon is depleted in Mg silicates by a factor of 2–3, in metal by a factor of 10, and in the highly volatile elements by a factor of 30. This depletion pattern parallels the condensation sequence. It also suggests competition between the Earth and Moon, perhaps implying lunar accretion in orbit around the accreting Earth (Anders, 1977). The accretion process might then be described by modifying the old cliche to read "the rich get their share while the poor get poorer." The model is attractive because it could be developed to offer both chemical and physical predictions.

Acknowledgments. This work was supported in part by NASA grant 9-79.

5. References

Anders E. (1977) Chemical compositions of the Moon, Earth and eucrite parent body. *Phil. Trans. Roy. Soc. London, 285,* 23–40.

Anders E. and Ebihara M. (1982) Solar system abundances of the elements. *Geochim. Cosmochim. Acta, 46,* 2363–2380.

Cameron A. G. W. (1986) The impact theory for origin of the Moon, this volume.

Clayton R. N., Grossman L., and Mayeda T. K. (1973) A component of primitive nuclear composition in carbonaceous meteorites. *Science, 182,* 485–488.

Clayton R. N. and Mayeda T. K. (1984) The oxygen isotope record in Murchison and other carbonaceous chondrites. *Earth Planet. Sci. Lett., 67,* 151–161.

Drake M. J. (1986) Is lunar bulk material similar to Earth's mantle?, this volume.

Ganapathy R. and Anders E. (1974) Bulk compositions of the moon and earth. *Proc. Lunar Sci. Conf. 5th,* pp. 1181–1206.

Grossman L. and Larimer S. W. (1974) Early chemical history of the solar system. *Rev. Geophys. Space Phys., 12,* 71–101.

Hood L. L. (1986) Geophysical constraints on the lunar interior, this volume.

Kaula W. M. (1975) The seven ages of a planet. *Icarus, 26,* 1–15.

Kerridge J. F. (1979) Fractionation of refractory elements in chondritic meteorites (abstract). In *Lunar and Planetary Science X,* pp. 655–657. Lunar and Planetary Institute, Houston.

Langseth M. G., Keihm S. J., and Peters K. (1976) Revised lunar heat flow values. *Proc. Lunar Sci. Conf. 3rd,* pp. 1181–1200.

Larimer J. W. (1971) Composition of the earth. Chondritic or achondritic? *Geochim. Cosmochim. Acta,* *35,* 769–786.

Larimer J. W. (1979) The condensation and fractionation of refractory lithophile elements. *Icarus, 40,* 446–454.

Larimer J. W. and Anders E. (1967) Chemical fractionations in meteorites—II. Abundance patterns and their interpretation. *Geochim. Cosmochim. Acta, 31,* 1239–1270.

Larimer J. W. and Bartholomay M. (1979) The role of carbon and oxygen in cosmic gases: some applications to the chemistry and mineralogy of enstatite chondrites. *Geochim. Cosmochim. Acta, 43,* 1455–1466.

Morgan J. W. and Anders E. (1979) Chemical composition of Mars. *Geochim. Cosmochim. Acta, 43,* 1601–1610.

Morgan J. W. and Anders E. (1980) Chemical compositions of Earth, Venus, and Mercury. *Proc. Natl. Acad. Sci., 77,* 6973–6977.

Newsom H. E. (1986) Constraints on the origin of the Moon from molybdenum and other siderophile elements, this volume.

Newsom H. E. and Palme H. (1984) The depletion of siderophile elements in the Earth's mantle: new evidence from molybdenum and tungsten. *Earth Planet. Sci. Lett., 69,* 354–364.

Palme H. and Nickel K. G. (1985) Ca/Al ratio and the composition of the Earth's upper mantle. *Geochim. Cosmochim. Acta, 49,* 2123–2132.

Rasmussen K. L. and Warren P. H. (1984) Megaregolith thickness, heat flow, and the bulk composition of the moon. *Nature, 313,* 121–124.

Spudis P. D. (1985) Apollo 16 site geology and impact melts: Implications for the geologic history of the lunar highlands. *Proc. Lunar Planet. Sci. Conf. 15th,* in *J. Geophys. Res., 89,* C95–C107.

Spudis P. D. and Davis P. A. (1985) How much anorthosite in the lunar crust?: Implications for lunar crustal origin (abstract). In *Lunar and Planetary Science XVI,* pp. 807–808. Lunar and Planetary Institute, Houston.

Taylor S. R. (1982) *Planetary Science: A Lunar Perspective.* Lunar and Planetary Institute, Houston. 481 pp.

Taylor S. R. (1986) The origin of the Moon: Geochemical considerations, this volume.

Toksöz M. N., Dainty A., Solomon S., and Anderson K. (1974) Structure of the moon. *Rev. Geophys. Space Phys., 12,* 539–567.

Warren P. H. (1986) The bulk Moon MgO/FeO ratio: A highlands perspective, this volume.

Wolf R. and Anders E. (1980) Moon and earth: compositional differences inferred from siderophiles, volatiles and alkalis in basalts. *Geochim. Cosmochim. Acta, 44,* 2111–2124.

Wood J. A. (1984) On the formation of meteoritic chondrules by aerodynamic drag heating in the solar nebula. *Earth Planet. Sci. Lett., 70,* 11–26.

Petrologic Constraints on the Origin of the Moon

JOHN W. SHERVAIS[1,2] AND LAWRENCE A. TAYLOR[1]

[1]*Department of Geological Sciences, University of Tennessee, Knoxville, TN 37996*
[2]*Department of Geology, University of South Carolina, Columbia, SC 29208*

Recent data from the Apollo 14 site show that the lunar crust and mantle are laterally heterogeneous on a regional scale. This heterogeneity extends to the source region of mare basalts and must have formed prior to ≈4.3–4.4 aeons. Lateral heterogeneity in the lunar crust is shown by variations in the petrology and geochemistry of highland rock suites from Apollos 14, 15, 16, and 17. In particular, the Apollo 14 highland suite contains alkali and incompatible element-enriched rocks not found at other sites, and lacks the ferroan anorthosites that characterize the Apollo 16 highlands crust. Later (and vertical) variations in the lunar mantle are shown by contrasts between the >3.8-b.y.-old Apollo 14 mare basalt suite and the younger (<3.9 b.y.) mare basalts from most other Apollo sites. Compositional variations in the mare basalts reflect similar variations in their mantle source regions and subsequent modification by fractional crystallization and assimilation. Apollo 14 basalts show variable enrichments in incompatible elements but are generally rich in alkalis relative to other mare basalts. These differences cannot be explained by fractionation of a single, well-mixed magma ocean. Possible explanations include: (1) asymmetrical fractionation of the lunar magma ocean, as suggested by Wasson and Warren (1980); (2) heterogeneous accretion of two compositionally distinct submoons; and (3) the impact of large, basin-forming projectiles into the primitive lunar crust. We suggest that the third alternative—late accretion by giant impacts—was an important factor in early lunar history. Regardless of whether these mechanisms are responsible or not, the observed large-scale, lateral variations in the petrology and geochemistry of the lunar crust and mantle reflect the processes by which the Moon formed and evolved. These processes are not independent; the general trends in lunar evolution were to a large degree preordained by the circumstances of its origin. By studying one, we may elucidate the other.

Introduction

Any theory for the origin of the Moon must ultimately explain the diversity of lunar rock types and their distribution across the lunar surface. The diversity and distribution of these rocks are fundamental characteristics that reflect the processes involved in lunar origin, including accretionary mechanisms, primordial differentiation, and subsequent melting events.

The lunar crust consists of two main components: ancient (>4.0 aeons) felsic rocks of the densely cratered lunar highlands, and the younger (>3.9 aeons) mare basalts. The felsic highland rocks were originally grouped together as the "ANT" (Anorthosite-Norite-Troctolite) suite, but subsequent studies showed that two distinct suites are present: anorthosites containing Fe-rich mafic minerals (the so-called "ferroan anorthosite" or "FAN" suite), and norites and troctolites containing Mg-rich pyroxene and olivine (the "Mg-rich" suite) (Warren and Wasson, 1977). The Mg-rich rocks are generally considered to represent plutons that intruded a primordial crust of ferroan anorthosite (e.g., James, 1980; Warren and Wasson, 1980b). Mare basalts, which fill impact basins and other low-lying areas, are similar to terrestrial flood basalts. Mare basalts are distinguished by their low alkali and Al contents, high Fe/Mg ratios, and variable Ti concentrations. Three groups are recognized on the basis of TiO_2 = high-Ti, low-Ti, and very low Ti (Papike and Vaniman, 1978). In general, REE and other incompatible elements increase with increasing TiO_2, from very low-Ti to high-Ti basalts.

The Fra Mauro breccias at Apollo 14 contain distinctive suites of mare basalts and highland crustal rocks that contrast significantly with equivalent rocks from other Apollo sites. For the highland suite, major differences include: (1) a lack of ferroan anorthosite (FAN) at Apollo 14, (2) higher REE concentrations in many Apollo 14 Mg-suite rocks, relative to "eastern" rocks of similar mineralogy, (3) alkali-rich lithologies that are scarce at other Apollo sites but are common at Apollo 14 (alkali anorthosite, granophyre), and (4) the common occurrence of evolved ilmenite gabbronorites that are rare at more eastern sites (Warren and Wasson, 1980a; Warren et al., 1981, 1983a,b; James and Flohr, 1983; Hunter and Taylor, 1982; Shervais et al., 1983, 1984a; Lindstrom, 1984a; Lindstrom et al., 1984). For the mare basalts, major differences include: (1) the predominance of low-Ti, high-Al basalts that are rich in alkalis despite their high mg#'s [100 × Mg/(Mg + Fe) molar], (2) REE concentrations that span the entire range observed between very low Ti basalts and high-Ti basalts from other Apollo sites, and (3) compatible element abundances that are intermediate between "normal" low-Ti and high-Ti mare basalt suites (Warner et al., 1980; Shervais et al., 1983, 1984b, 1985a,b; Dickinson et al., 1985). These contrasts, which are discussed in more detail below, imply lateral heterogeneity in the lunar crust and mantle on a regional scale, as originally proposed by Warren and Wasson (1980a). The compatible element data on the Apollo 14 mare basalts further imply that this heterogeneity is at least as old as the source region of mare basalts ($\approx$ 4.4 aeons). The causes of this heterogeneity must be sought in the earliest stages of lunar accretion and differentiation.

In the following review, we place special emphasis on recent data from the Apollo 14 site. The principal reason for this emphasis is the unique nature of the Apollo 14 sample suite, which is petrologically and geochemically distinct from highland

and mare samples at other sites. In fact, it is the uniqueness of this sample suite, corroborated by orbital geochemical data, that establishes the existence of lateral heterogeneities in the lunar crust and mantle. An additional reason is that the unique nature of the Apollo 14 crust and mantle has not been fully appreciated until relatively recently (e.g., Warren *et al.*, 1983a,b; Shervais *et al.*, 1984a, 1985a,b). As a result, this data has not been fully integrated into older syntheses of lunar petrology and geochemistry. We present below brief reviews of previous data on highland and mare petrology, followed by more detailed summaries of the Apollo 14 data.

We conclude by evaluating the implications of large-scale lateral heterogeneities for models of lunar origin and evolution. Whether or not these heterogeneities formed during lunar accretion, or arose later during primordial fractionation, or both, is a fundamental question that must be answered before more detailed models of lunar origin can be formulated.

The Highland Crust

Our petrological and geochemical understanding of the lunar "terrae" or highlands has advanced tremendously since the introduction of the pristine rock concept by Warren and Wasson (1977). This concept has focused attention on the primary igneous components of the highland crust and the relationships between these components and the early magmatic evolution of the crust. In particular, this concept has led to the recognition of three fundamentally different suites of highland crustal rocks: the ferroan anorthosite suite (FAN), the magnesium-rich suite, and the alkali anorthosite suite (Warner *et al.*, 1976; Warren and Wasson, 1980a; James, 1980). Continuing studies by Warren and his coworkers (Warren and Wasson, 1978, 1979a,b, 1980a; Warren *et al.*, 1981, 1983a,b) and by other investigators (e.g., James, 1980, 1981, 1983; James and Flohr, 1983; Lindstrom, 1984a,b; Lindstrom *et al.*, 1984; Hunter and Taylor, 1983; Shervais *et al.*, 1983, 1984a) have detailed variations in the petrology and geochemistry of the highland crust both within and between the Apollo landing sites. Warren and Wasson (1980a) were first to point out a correlation between the petrology/geochemistry of crustal rocks from the Apollo sites and their latitude. They distinguished two regions: the "eastern" sites (Apollos 11, 15, 16, 17) and the "western" sites (Apollos 12, 14). In the following discussion we summarize the petrological and geochemical characteristics of these two regions, with emphasis upon those sites where highland rock-types compose a major fraction of the returned samples: Apollo 14 in the west, and Apollos 16 and 17 in the east. Particular emphasis is placed on recent data from the Apollo 14 site and its constrast to more "easterly" sites.

The eastern crust

Petrologic studies of highland rocks from the eastern Apollo sites established the essential dichotomy of lunar crust: the ferroan anorthosite suite vs. the magnesium-rich suite of troctolite and norite. Additional rock-types that form small but important components of the eastern crust are Mg-rich gabbronorites, alkali gabbronorite, and highly evolved felsic lithologies. Highland crustal rocks were sampled extensively at the Apollo 16 and 17 sites, and the data from these two sites have dominated most models of lunar crustal evolution.

The ferroan anorthosite suite. Ferroan anorthosites are the dominant rock type at the Apollo 16 site and occur sporadically at other Apollo sites. They are characterized by extremely calcic plagioclase (An94–An97) and a wide range in the mg# of the coexisting mafic phase (Fig. 1). As their name implies, ferroan anorthosites are dominantly anorthositic, with more than 85% modal plagioclase (Warren and Wasson, 1980a). Mafic phases include relatively Fe-rich olivine, low-Ca pyroxene, and augite; olivine is the dominant mafic phase. The ferroan anorthosite define a steep trend on plots of An in plagioclase vs. mg# in mafics, with mafic mg#'s ranging from 71 to 50 (Fig. 1). This trend is extended to even lower mg#'s (olivine Fo42) by hyperferroan clasts in lunar meteorite ALHA 81005 (Goodrich *et al.*, 1984).

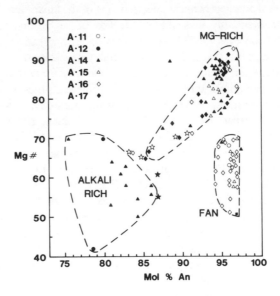

Fig. 1.   Pristine highland rocks are divided into three suites, based on the mol % An in plagio-clase (= 100× Ca/[Ca + Na]) vs. mg# [= 100× Mg/(Mg + Fe)] and mol % MgO/(MgO + FeO) (= mg#) in coexisting mafic minerals. The distribution of these three suites (ferroan anorthosites, Mg-suite, alkali anorthosites) varies between the Apollo sites. Ferroan anortho-sites (FAN) are dominant at Apollo 16 (and 11) but rare at Apollos 14, 15, and 17. Mg-suite rocks are dominant in Apollos 14, 15, and 17, but alkali anorthosites are found only at Apol-los 12 and 14. Mg-gabbronorites are shown with an open star adjacent to site symbol. Alkali gabbronorite from 67965 is shown by solid stars (upper = augite, lower = low-Ca pyroxene).

The ferroan anorthosites have low incompatible element concentrations that are controlled primarily by plagioclase-liquid partitioning (Fig. 2). Rare-earth element (REE) concentrations are generally less than 0.5 to 0.1 times chondrite, with negative slopes on chondrite-normalized REE plots. An exception is divalent Eu, which is strongly enriched in plagioclase, leading to large positive Eu anomalies (Fig. 2). The ferroan anorthosites also have Ti/Sm and Sc/Sm ratios that are near chondritic in rocks with the highest mg#'s (Norman and Ryder, 1980). The decrease in the Sc/Sm ratio with decreasing mg# suggests extensive pyroxene fractionation during crystallization (James, 1980, 1983), but most incompatible elements are relatively unfractionated. Calculation of REE concentrations in the ferroan anorthosite parent magma using revised plagioclase-liquid partition coefficients (McKay, 1982) confirms this observation.

Ferroan anorthosites are the only pristine rock suite whose geochemical and petrological characteristics are compatible with the interpretation that they formed as flotation cumulates in a Moon-wide magma ocean of chondritic relative composition (e.g., Warren and Wasson, 1980b). However, more complex scenarios involving serial magmatism cannot be ruled out (Wetherill, 1975; Walker, 1983; Longhi and Ashwal, 1985).

The Mg-rich suite. Rocks of the Mg-suite are the dominant rock-types at the Apollo 17 site and occur commonly at the Apollo 15 site as well; they are relatively uncommon at Apollo 16.

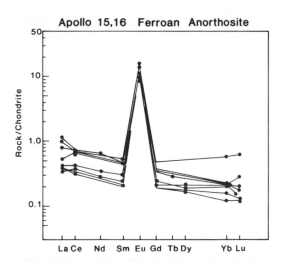

Fig. 2.   *Chondrite-normalized REE concentrations in ferroan anorthosites from Apollos 15 and 16. Note the strong positive Eu anomaly and low overall REE concentrations (e.g., La < 1× chondritic). Circles = Apollo 16, square = Apollo 15. Data from Taylor (1982) and Warren and Wasson (1979, 1980).*

Mg-suite rocks define a shallow trend on plots of An in plagioclase vs. mg# in mafic minerals, plagioclase (An93–An97) and mafic minerals (mg#'s 65–92) varying sympathetically with a slope of about 1:2 (Fig. 1). This suite contains a wide variety of rock-types that vary both in modal mineralogy and in mineral compositions. The most common rock-types are troctolite, spinel troctolite, and norite; less common varieties include gabbro, gabbronorite, feldspathic lherzolite, and dunite (James, 1980, 1983). The modal feldspar content of these rocks ranges from 5–85% plagioclase (Warren and Wasson, 1980a). The dominant mafic phase is Mg-rich olivine in the troctolites or low-Ca pyroxene in the norites. High-Ca is important in the gabbros and gabbronorites (discussed below).

Textural relations suggest that the common Mg-suite lithologies are olivine-plagioclase or pyroxene-plagioclase cumulates that contained minor amounts of trapped interstitial magma prior to final crystallization (Dymek et $al.$, 1975; James and McGee, 1979; James, 1980, 1983). In contrast to the shallow (low-pressure) origin inferred for most ferroan anorthosites, Mg-suite troctolites and spinel troctolites have textural, mineral chemical, and mineral assemblage relationships that require relatively high-pressure equilibration, e.g., 10–30 km depth (Gooley et $al.$, 1974; Herzberg, 1978; Herzberg and Baker, 1980). These relationships include coarse exsolution lamellae, the development of pyroxene-plagioclase symplectites, and the stable coexistence of high-Al pyroxene and spinel. These data suggest that the Mg-suite troctolites and norites represent layered mafic plutons that intrude the preexisting ferroan anorthosite crust (e.g., James, 1980).

The Mg-rich suite has higher concentrations of incompatible elements than the ferroan anorthosites and Ti/Sm ratios that are subchondritic (Norman and Ryder, 1980; Warren et $al.$, 1981). These differences between the Mg-suite and "FAN" have been interpreted to result from the mixing of Mg-suite parent magmas with "urKREEP," the highly fractionated residue of magma ocean crystallization (Norman and Ryder, 1980; Warren et $al.$, 1981). The high Mg/Fe ratios of Mg-suite troctolites rule out any direct connection with the ferroan anorthosite suite, but are a consistent partial melting of either early magma ocean cumulates (if bulk Moon mg# $\approx$ 80) or a primitive lunar interior (if bulk Moon mg# $>$ 86) (Shervais and Taylor, 1983; Warren, 1982; James, 1980).

The Mg-rich gabbronorite association. The "gabbronorite" suite of James and Flohr (1983) consists of six samples, ranging in composition from feldspathic lherzolite 67667 (Warren and Wasson, 1979a) to sodic ferrogabbro 67915 (Taylor et $al.$, 1980). These rocks differ from the more common Mg-norites by having higher modal augite concentrations and high augite/total pyroxene ratios. Plagioclase (An 54–93) and the mafic minerals (mg#'s 35–75) vary sympathetically, but because the plagioclase is more sodic than in comparable Mg-norites, their An-mg# trend is flatter, with a slope near 1:1 (Fig. 1).

Ilmenite is the most common accessory phase in the Mg-gabbronorites; Zr-and Nb-rich phases are absent. Incompatible trace elements are lower than in comparable

Mg-norites, but compatible elements such as Ti and Sc are higher. As a result, the Mg-gabbronorites have higher Ti/Sm and Sc/Sm ratios than Mg-norites with the same mg# (James and Flohr, 1983; James, 1983). These data clearly require a separate parent magma for the Mg-gabbronorite suite (James, 1983). The trace element systematics of the Mg-gabbronorites further suggest that their parent magma did *not* mix with "urKREEP" (James and Flohr, 1983).

The alkali gabbronorites. This rare rock-type has been found in one Apollo 16 breccia ejected from North Ray crater (Lindstrom, 1984b). The alkali gabbronorites are characterized by abundant modal pyroxene, high augite/total pyroxene ratios, and accessory phases that include ilmenite, chromite, K-feldspar, apatite, whitlockite, and zircon (Lindstrom, 1984b). Silicate compositions are more evolved than most Mg-gabbronorites (plagioclase An 84–89, pyroxene mg#'s 55–63) and are most similar to the western alkali anorthosite suite described below.

Trace element concentrations vary widely in the alkali gabbronorites, largely in response to modal variations of accessory phases. They are all much richer in incompatible elements than any other pristine rocks from Apollo 16 and have compositional affinities to KREEP (Lindstrom, 1984b). Their origin remains an enigma at this time.

Evolved lithologies. Highly evolved lithologies are comparatively rare at the eastern sites. Sodic ferrogabbro 67915 is the most evolved member of the Mg-gabbronorite association with 24% modal silica (Taylor *et al.*, 1980). Other evolved samples found in the east include quartz monzodiorite 15405 (Ryder, 1976; Taylor *et al.*, 1980) and small clasts of granophyre from Apollo 17 breccias (James and Hammarstrom, 1977). Quartz and Ba, K-feldspar are the essential minerals and commonly occur in graphic intergrowths. Sodic plagioclase, iron-rich pyroxenes and/or olivine, and ilmenite are common varietal minerals; accessory phases include zircon, whitlockite, and apatite. The petrologic affinities of these rocks are uncertain.

The western crust

Recent petrologic studies of pristine nonmare samples from the Apollo 12 and 14 sites have demonstrated the unique character of the western highlands crust. The task of unravelling the history of this crust is exacerbated by the fact that these samples occur only as small (≤ 3 cm diameter) clasts in the Fra Mauro breccias. Despite this difficulty, at this time some 30 clasts have been characterized geochemically, and dozens more texturally pristine clasts have been characterized petrologically (Warren and Wasson, 1980a; Warren *et al.*, 1981, 1983a,b; Hunter and Taylor, 1983; Shervais and Taylor, 1983; Shervais *et al.*, 1983, 1984a; Knapp *et al.*, 1984; Lindstrom, 1984a; Lindstrom *et al.*, 1984). Four main rock groups have been distinguished (in order of decreasing abundance): (1) the Mg-rich troctolite association, (2) the alkali anorthosite suite, (3) the Mg-rich ilmenite gabbro/norite association, and (4) the alkali-rich granite/granophyre association. Ferroan anorthosites are rare—only one has been characterized geochemically (Warren *et al.*, 1983a) and two others petrographically (Hunter and Taylor, 1983; Shervais *et al.*, 1983).

The Mg-troctolite association. The Apollo 14 Mg-rich troctolite association comprises troctolite, dunite, olivine pyroxenite, and magnesian anorthosite (Lindstrom

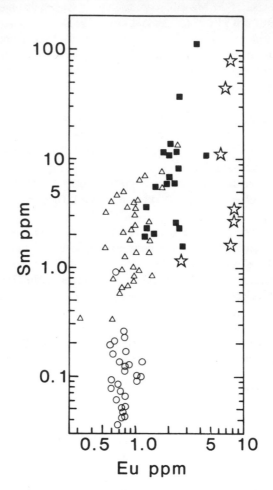

Fig. 3. Samarium vs. Eu (in ppm) in pristine highland rocks, after Warren and Wasson (1980), with additional data from Warren et al. (1981, 1983a,b), Shervais et al. (1983, 1984a), and Lindstrom et al. (1984). FAN = open circles; eastern Mg-suite = open triangles; western Mg-suite = closed squares; alkali anorthosites = stars. Although eastern and western Mg-suite rocks overlap at intermediate Sm and Eu concentrations, western (Apollo 14) Mg-suite troctolites and anorthosites are generally enriched in both elements relative to eastern rocks. Alkali anorthosites show further enrichment, especially in Eu, suggesting a more evolved source.

et al., 1984; Shervais *et al.*, 1984a). Common varietal minerals in the troctolites include spinel, enstatite, and diopside (Warren *et al.*, 1983b; Shervais *et al.*, 1983, 1984a; Knapp *et al.*, 1984). The Apollo 14 troctolites are characterized by calcic plagioclase (An94–96) and a range in olivine compositions (Fo75–Fo90) (Fig. 1). Pyroxene in the relatively pyroxene-rich varieties is magnesian (mg# 90–91), suggesting that the parent magmas became saturated in pyroxene early in their crystallization history (Warren *et al.*, 1983b; Shervais *et al.*, 1983). The Mg-troctolites span a range of incompatible element abundances, with La concentrations ranging from ≈15× to ≈80× chondritic, and Eu anomalies that range from strongly positive to negative as overall REE concentrations increase (Lindstrom *et al.*, 1984). The Apollo 14 troctolites are generally richer in REE overall when compared to eastern troctolites, resulting in a distinction between many "eastern" and "western" troctolites in terms of Ti/Sm, Sc/Sm, and Sm vs. Eu (Warren and Wasson, 1980; Warren *et al.*, 1981, 1983a). However, as more data accrues, these trends are becoming blurred and the distinction between "eastern" and "western" troctolites is less clear (Fig. 3). The differences observed between sites and between samples at a single site may reflect regional trends rather than longitudinal ones (Shervais *et al.*, 1983, 1984a; Lindstrom *et al.*, 1984).

Anorthosites associated with these troctolites are also quite calcic (An95–An97) and magnesian (Fo84–Fo90) (Fig. 1). One magnesian anorthosite contains 1–2% of a REE-rich phosphate, originally thought to be apatite (Shervais *et al.*, 1984a), that has since been identified as whitlockite (Lindstrom *et al.*, 1984, 1985). The

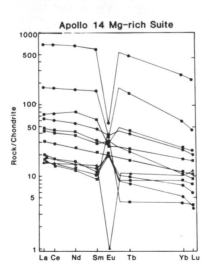

Fig. 4. Chondrite-normalized REE concentrations in Apollo 14 Mg-suite anorthosites (circles), troctolites (squares), and dunite (triangle). Note the wide range of concentrations both in the troctolites and the anorthosites (after Lindstrom et al., 1984).

high modal abundance of whitlockite in this anorthosite has resulted in whole rock REE concentrations over 700× chondritic (Fig. 4) (Lindstrom *et al.*, 1984). As discussed by these authors, the occurrence of whitlockite in a refractory-rich anorthosite is difficult to explain by normal magmatic processes. Other processes, such as metasomatic enrichment, may be responsible (Lindstrom *et al.*, 1984).

Two ultramafic samples have been found in the west—one is a dunite consisting entirely of Fo 88–89 olivine (Lindstrom *et al.*, 1984), the other is an olivine orthopyroxenite consisting of about 90% enstatite (Wo 0.25 En91) and 10% olivine (Fo 89.6) (Shervais *et al.*, 1984a). Both are extremely rich in incompatible elements despite their ultramafic modes and magnesian mineral compositions (Lindstrom *et al.*, 1984; Shervais *et al.*, 1984a). Their unusual trace element chemistry cannot be due to mixing with KREEP because both are fractionated relative to KREEP. Small amounts of a REE-rich phosphate could account for their concentration patterns, however (Lindstrom *et al.*, 1984).

The alkali anorthosite suite. Alkali anorthosites are the most common type of anorthosite found in the west, and they are found *only* in the west, at Apollo 12 and 14 (Warren *et al.*, 1983a; Shervais *et al.*, 1984a), although compositionally similar alkali gabbronorites form a minor component at Apollo 16 (Lindstrom, 1984b). Modes range from nearly pure anorthosite to anorthositic norites with ≈84% plagioclase and 16% pigeonite. Mineral compositions are distinct from both the ferroan anorthosites and the Mg-rich suite: plagioclase ranges from An75–An87, pyroxenes from mg# 40–70 (Fig. 1). Accessory minerals include augite, K-feldspar, ilmenite, whitlockite, a silica polymorph, and Fe-Ni metal (Warren and Wasson, 1980; Warren *et al.*, 1981, 1983a,b; Hunter and Taylor, 1983; Shervais *et al.*, 1983, 1984a).

Most alkali anorthosites are monomict breccias whose primary igneous textures are obscured. The few texturally pristine samples that are available suggest that the alkali anorthosites are plagioclase flotation cumulates (Hunter and Taylor, 1983; Shervais *et al.*, 1983, 1984a). Large, unzoned plagioclase primocrysts are surrounded by postcumulus pyroxene and ilmenite. Accessory phases occur either interstitially or as inclusions within plagioclase primocrysts (Warren *et al.*, 1983b; Shervais *et al.*, 1984).

Whitlockite occurs in alkali anorthosites in amounts ranging from trace to about 9%, although these modes are subject to large uncertainty because of the small sample sizes (Warren *et al.*, 1983b; Shervais *et al.*, 1984a). The whitlockite is generally rich in REE, with concentrations that are 10,000 to 30,000× chondritic (Warren *et al.*, 1983b; Shervais *et al.*, 1984a; Lindstrom *et al.*, 1985). Whole rock REE contents are a function of whitlockite abundance and range from ≈30× to ≈600× chondritic for La (Warren *et al.*, 1981, 1983a,b; Shervais *et al.*, 1984). This is the same range in REE concentrations displayed by the magnesian anorthosites (Figs. 4, 5).

The absence of well-developed exsolution features in pyroxene requires a relatively shallow origin for the alkali anorthosites (Hunter and Taylor, 1983; Shervais *et*

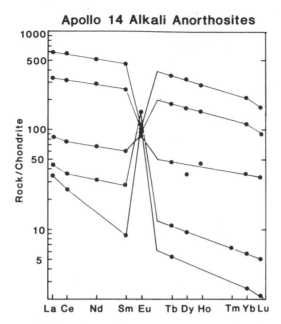

Apollo 14 Alkali Anorthosites

Fig. 5. Chondrite-normalized REE concentrations in Apollo 14 alkali anorthosites. The range in concentrations is the same found in magnesian anorthosites (after Lindstrom et al., 1984).

al., 1983). This rules out accumulation from an urKREEP parent magma at the base of the older ferroan anorthosite crust, unless this older crust had been thinned or removed *before* the alkali anorthosites crystallized.

The Mg-rich ilmenite gabbronorite association. Rocks of this association have been referred to variously as ilmenite gabbros, ilmenite norites, or gabbronorites (Hunter and Taylor, 1983; Warren *et al.*, 1983b; Shervais *et al.*, 1983, 1984a; Knapp *et al.*, 1984). Mineral compositions place these rocks in the Mg-rich suite, with relatively sodic plagioclase that trends into the alkali anorthosite field. The ilmenite gabbronorites consist of cumulus plagioclase and pigeonite, with postcumulus augite, ilmenite, and Fe-Ni metal. The characteristic that tends to unite this rather diverse suite is the modal abundance of ilmenite (0.5–3% by volume).

The ilmenite gabbronorites have petrographic affinities to the gabbronorite suite of James and Flohr (1983) and to cumulates derived from mare basalts. These petrographic similarities include high augite/total pyroxene ratios, common modal ilmenite, relatively sodic plagioclase, and scarce trace element-rich accessory phases (e.g., zircon). However, the two clasts that have been analyzed for trace elements have low Ti/Sm and Sc/Sm ratios that are similar to normal Mg-suite norites and are much lower than ratios seen in either mare basalts or the Apollo 16 Mg-gabbronorites (Warren *et al.*, 1983b; Lindstrom, 1984).

Alkali granites/granophyres. Lunar "granites" comprise a diverse assemblage of silica-saturated rocks that range in composition from quartz monzodiorite to alkali feldspar-quartz granophyres (Ryder, 1976; Quick *et al.*, 1981; Warren *et al.*, 1983c; Shervais and Taylor, 1983; Lindstrom, 1984; Knapp *et al.*, 1984; Salpas *et al.*, 1985). Although granites occur at several Apollo sites, they are most common at Apollo 14.

The most common texture in these granites is that of a quartz + K-feldspar granophyre, either alone or with plagioclase (An60–80) or quartz + ternary feldspar granophyre. The ternary feldspars have compositions of about Or45Ab20 that plot in the "forbidden region" of the feldspar ternary (Shervais and Taylor, 1983). Accessory minerals include pigeonite, augite, ferroaugite, fayalite, ilmenite, zircon, and phosphates (Warren *et al.*, 1983c; Shervais and Taylor, 1983; Knapp *et al.*, 1984). Variations in the assemblages and in mineral compositions (e.g., BaO in K-feldspar, mg# in mafics) imply that at least three or four separate parent magmas are required to account for the Apollo 14 granites.

The abundance of granite at the Apollo 14 site is difficult to estimate. Although crystalline granites are rare, K- and Si-rich "rhyolite" glass is an ubiquitous component of the Apollo 14 soils and breccias. This glass was emplaced hot and probably formed by impact melting of a granite pluton (Shervais and Taylor, 1983). Based on the Apollo 14 soil data of Simon *et al.* (1982), the granite component in the Fra Mauro region can be estimated at a minimum of about 0.5% by volume. The local abundance of granite plutons in the crust is also supported by the chemistry of VHK basalts, which appear to have formed by the assimilation of granite into a basaltic magma (Shih *et al.*, 1984; Shervais *et al.*, 1985b).

An important feature of lunar granites is their fractionated K/La ratios (Warren *et al.*, 1983c; Lindstrom, 1984). Granites are strongly enriched in the incompatible elements K, Th, U, Zr, etc., but less enriched in the REE, suggesting formation by silicate liquid immiscibility (Taylor *et al.*, 1980). However, their V-shaped REE patterns imply phosphate fractionation, and their commonly high Fe contents and the high Mg/Fe ratios of their accessory mafic minerals are not consistent with liquid immiscibility(Salpas *et al.*, 1985). Modelling of granite trace elements shows that neither of the most commonly suggested parent magmas (KREEP and mare basalt) are viable. An origin related to Mg-suite rocks or the alkali anorthosites seems more likely (Salpas *et al.*, 1985).

Discussion of the highlands

Three major rock suites characterize the lunar highland crust: the ferroan anorthosite suite, the Mg-rich suite, and the alkali anorthosite suite. The distribution of these suites at the Apollo sites is summarized in Fig. 1. Ferroan anorthosites, which are characteristically highly depleted in incompatible elements, are the dominant highland rock-type at the Apollo 16 and Apollo 11 sites. The rare Mg-rich rocks at Apollo

16 are members of the Mg-gabbronorite association. The Mg-rich suite is the only major highland suite found at Apollo 17 and also dominates the highland samples from Apollos 14 and 15. There are major petrologic and geochemical differences within this suite, however, that represent either longitudinal (Warren and Wasson, 1980) or regional variations (Shervais *et al.*, 1983, 1984a; Lindstrom *et al.*, 1984). The "western" Mg-suite rocks from Apollo 14 are commonly (but not always) enriched in incompatible elements relative to "eastern" troctolites from Apollo 17 (Fig. 3) (Warren and Wasson, 1980; Warren *et al.*, 1981, 1983a,b; Lindstrom *et al.*, 1984).

Alkali anorthosites are found only at the two westernmost Apollo sites (12 and 14) and constitute the clearest prima facie evidence for lateral heterogeneity in the lunar crust. Lunar granite, which is also enriched in alkalis, is common at the western Apollo sites, is less common at the Apollo 17 site, and is rarely found at the other "eastern" sites.

The ferroan anorthosite suite is the only suite that could have formed directly from the magma ocean prior to the evolution of KREEP. The origin of the other suites is enigmatic. The enrichment of primitive Mg-suite rocks in incompatible elements suggests mixing with small amounts of KREEP. If so, then the alkali anorthosites must represent either a separate parent magma that also mixed with similar proportions of KREEP, or cumulates that formed directly from the primary "KREEP" magma. In any case, the near absence of ferroan anorthosite at several nearside Apollo sites (Apollos 12, 14, 17) suggests two possibilities: (1) A thick crust of ferroan anorthosite may never have formed on the western lunar nearside. Instead, a thin protocrust may have been succeeded directly by later intrusive suites; or (2) Large basin-forming impacts could have removed much of whatever FAN crust did exist.

The distribution of lunar highland rock-types on a broader, Moon-wide scale can be inferred from orbital geochemical data (e.g., Adler and Trombka, 1977). These data are derived primarily from the orbital X-ray fluorescence and gamma-ray experiments (Adler and Trombka, 1977). They show that the lunar farside and most of the lunar nearside highlands are dominantly ferroan anorthositic gabbro in composition: Al/Si ratios are high, while Mg/Si ratios are low. These areas are also low in K, U, and Th (Adler and Trombka, 1977). High K, U, and Th concentrations (=KREEP) are confined virtually to the Mare Imbrium–Oceanus Procellarum region on the western lunar nearside—the site of the so-called "Gargantuan" or "Procellarum" basin (Cadogan, 1974; Wilhelms, 1983).

Cadogan (1974) has suggested that rather than reflecting longitudinal variations in the lunar crust and mantle, the geochemical and petrological differences between the Apollo sites reflect concentric variations radial to the "Gargantuan" impact basin. In Cadogan's scenario, the "Gargantuan" impact event removed most of the existing lunar crust, possibly exposing mantle material. The resulting basin was later filled by KREEP basalt flows (Cadogan, 1974). This model was proposed before it was recognized that the lunar crust consists of at least three major suites of rock, and the model must be revised to account for this additional complexity. In this revised scenario, the upper ferroan anorthosite crust was eroded ballistically, exposing deeper, Mg-troctolite crust on the basin margins, and KREEP-rich rocks toward its center. This model is consistent with the orbital geochemistry, the observed distribution of rock-types and geophysical inferences on the thickness of highlands crust; we will consider it further below.

Mare Basalt Volcanism

Despite their volumetric insignificance in the lunar crust (Hörz, 1978), mare basalts constitute our primary source of information on the Moon's upper mantle. Compositional variations between mare basalt suites reflect variations in the mineralogical and geochemical make-up of the lunar mantle that formed during the earliest stages of lunar evolution. Mare basalt model ages date the early lunar differentiation, and their crystallization ages constrain the Moon's thermal history.

Radiometric ages of mare basalts range from about 4.0 to 3.2 aeons (Papike *et al.*, 1976). Crater density surveys, however, indicate that mare volcanism continued until at least 2.5 aeons (Head, 1976). The onset of significant mare volcanism traditionally has been fixed at about 3.85 aeons—the age of the oldest Apollo 11 basalts (Taylor, 1982). This coincides roughly with the Imbrium event and the termination of the catastrophic impacts that characterized early lunar history.

Recent data, however, demonstrates a much earlier age for the onset of mare volcanism; an Apollo 14 mare basalt cumulate has been dated at 4.2 aeons (Taylor *et al.*, 1983). This is the same age as many pristine highland rocks, and implies that mare volcanism may have played an important role in early crustal genesis on the Moon. In the following sections we present a brief overview of mare volcanism, and then contrast this with the ancient basalt suite from the Apollo 14 site.

The petrology and geochemistry of mare basalts has been reviewed by Papike *et al.* (1976) and Papike and Vaniman (1978). They recognized three broad compositional groups of mare basalt based on Ti-concentrations: high-Ti basalts (TiO_2 = 8–14 wt %), low-Ti basalts (TiO_2 = 1.5–5 wt %), and very low Ti (VLT) basalts (TiO_2 < 1 wt %). The major element compositions of these basalts are summarized in Fig. 6a–d (modified after Papike and Vaniman, 1978). Low-Ti and high-Ti basalts have similar low Al_2O_3 (6–11 wt %) and K_2O (0.03–0.1 wt %), except for the Apollo 11 high-K suite ($K_2O \approx$ 0.1–0.4 wt %). VLT basalts have high Al_2O_3 (10–14 wt %) but even lower K_2O ($\leq$ 0.02 wt %). Sodium contents vary directly with TiO_2 and inversely with mg# (Fig. 6c). Sodium contents are highest in the evolved high-Ti basalts (Na_2O = 0.35 to 0.55 wt %), lower in the low-Ti basalts (Na_2O = 0.15 to 0.40 wt %), and lowest in the VLT basalts (Na_2O = 0.04 to 0.30).

Both Na and Al increase with decreasing mg# within each group, but Ti shows contrasting behavior between groups (Fig. 6). TiO_2 increases with decreasing mg#

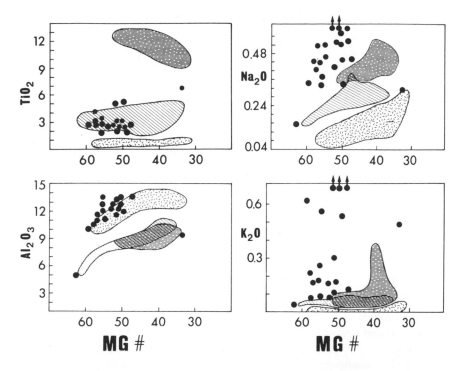

Fig. 6. Mare basalt mg#'s vs. (a) TiO₂, (b) Al₂O₃, (c) Na₂O, and (d) K₂O. Normal mare basalts shown by stipples (VLT basalt), diagonal rule (low-Ti basalts), and horizontal rule (high-Ti basalts). Apollo 14 basalts (= filled circles) are generally low in Ti₂ and high in Al₂O₃, Na₂O, and K₂O. Data from Papike and Vaniman (1978), Shervais et al. (1984b, 1985a,b), Warner et al. (1980), and Hubbard and Gast (1971).

in the low-Ti and VLT basalts, consistent with the conclusion of Papike *et al.* (1976) that fractionation in these basalts is controlled by olivine and magnesian pigeonite. In contrast, TiO_2 *decreases* with decreasing mg# in the high-Ti basalts, reflecting the subtraction of liquidus or near-liquidus Fe-Ti oxides (Papike *et al.*, 1976; Papike and Vaniman, 1978).

Two groups of basalt are characterized by high-Al concentrations (Al_2O_3 = 10–14 wt %). All of the Al-rich basalts represented by the hatchured pattern in Fig. 6 are very low-Ti (VLT) basalts. These basalts were first described at the Apollo 17 site and are common at the Luna 24 site (Papike and Vaniman, 1978). The other group of high-Al basalts have low or intermediate Ti concentrations and are similar petrographically to the low-Ti basalt group (Fig. 6). These high-Al or "feldspathic" basalts are the most common basalt type at the Apollo 14 and Luna 16 sites, and they also occur in the Apollo 12 collection. These basalts differ significantly

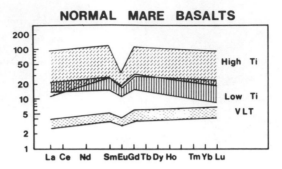

NORMAL MARE BASALTS

Fig. 7. Chondrite-normalized REE concentrations in normal mare basalts. There is a positive correlation between TiO_2 content and REE concentrations: high-Ti basalts have the highest REE, VLT basalts the lowest REE concentrations. The wide range in REE contents requires different source compositions for each group. (Source: Basaltic Volcanism Study Project, 1981).

from both the low-Al, low-Ti basalts and the high-Al VLT basalts in minor and trace element geochemistry. The Apollo 14 high-Al basalt suite will be discussed in the next section.

Trace element abundances of the high-Ti, low-Ti, and VLT basalts also correlate with TiO_2 (Fig. 7). VLT basalts have low concentrations of incompatible elements (REE < 10× chondritic) and LREE depleted patterns. Low-Ti basalts have higher REE concentrations ($\approx$ 10×–20× chondritic) and relatively flat patterns. High-Ti basalts have the highest concentrations ($\approx$ 15×–100× chondritic) and "hump-shaped" patterns that are depleted in the lightest and heaviest rare earths relative to the middle rare earths (Fig. 7). Compatible trace elements (Co, Ni, V, Cr) generally diminish with increasing TiO_2; however, Sc is highest in the high-Ti basalts, reflecting the pyroxene-rich source composition (Fig. 8).

Apollo 14 mare basalts

Twelve compositionally distinct varieties of mare basalt have been documented at the Apollo 14 site. Eight of these are low-Ti, high-Al mare basalts; the others include intermediate-Ti basalt, low-Al ferrobasalt, low-Ti olivine basalt, and very low Ti (VLT) glass spheres (Table 1). High-Ti basalts are not found. The VLT glasses are similar to VLT basalts and glasses from Apollo 17 and Luna 24 and are found only in regolith breccias (Chen *et al.*, 1982). They are not as old as the Fra Mauro breccia clasts and may not be indigenous to the Apollo 14 site. The VLT glasses will *not* be considered in the following discussion.

Major element compositions of the Apollo 14 mare baslts are compared to other mare basalt suites in Fig. 6. The Apollo 14 low-Ti mare basalts are all enriched

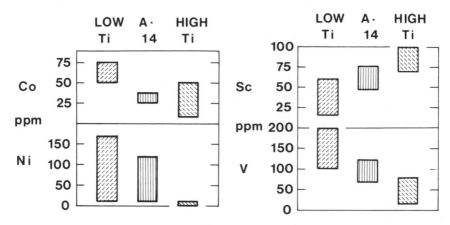

Fig. 8. Compatible element concentrations (Co, Ni, Sc, V) in low-Ti basalts, Apollo 14 basalts, and high-Ti basalts. The compatible element concentrations of the Apollo 14 basalts require a source intermediate in composition to the low-Ti and high-Ti basalt sources. This source cannot be more evolved than the high-Ti basalt source (after Shervais et al., 1985a).

TABLE 1. *Apollo 14 Mare Basalts.*

Basalt Type*	Parent Sample	References
Tridymite ferrobasalt (low-Al)	14321	Shervais *et al.* (1985a)
Ilmenite ferrobasalt (high-Al)	14305	Shervais *et al.* (1985a,b)
14321-type basalt, vitrophyre (high-Al)	14321	Grieve *et al.* (1975); Shervais *et al.* (1985a); Dickinson *et al.* (1985)
14160-type basalt (high-Al)	14160	Warner *et al.* (1980)
14053-type basalt (high-Al)	14053; 14321	Hubbard and Gast (1971); Shervais *et al.* (1985a); Dickinson *et al.* (1985)
Intermediate-Ti basalt (high-Al)	14063; 14004, 10	Ridley (1975); Warner *et al.* (1980)
Low-Ti olivine basalt (low-Al)	14305	Hunter and Taylor (1982); Taylor *et al.* (1983)
14072-type basalt (high-Al)	14072; 14321	Hubbard and Gast (1971); Dickinson *et al.* (1985)
Very High Potassium (VHK) basalt (high-Al)	14305; 14168, 33	Warner (1980); Shervais *et al.* (1984b, 1985b)
14256-type basalt (high-Al)	14256	Warner *et al.* (1980)
LREE-depleted basalt (high-Al)	14321	Shervais *et al.* (1985a); Dickinson *et al.* (1985)
VLT glass spheres	14047; 14049; 14307; 14313 (Soil breccias)	Chen *et al.* (1982)

*Arranged in order of decreasing REE content.

in alkalis relative to other low-Ti mare basalts, with 2× to 14× more Na_2O and 1.5× to 20× more K_2O. Their alkali contents are similar to or higher than high-Ti mare basalts but occur at higher mg#'s than the high-Ti basalts (mg#'s 50–60 vs. mg#'s < 52).

The Apollo 14 basalts have concentrations of compatible trace elements (Co, Ni, Sc, V, Cr) that are intermediate between the concentrations observed in low-Ti basalts and high-Ti basalts (Fig. 8). Incompatible trace element concentrations vary widely and show *no* correlation with TiO_2 content similar to that observed in other mare basalt suites. Rare earth element concentrations vary by more than an order of magnitude in the low- and intermediate-Ti basalts, with La ranging from 8× to 100× chondritic (Fig. 9). This is the same range observed between VLT, low-Ti, and high-Ti mare basalts from other Apollo and Luna sites.

One aspect of the Apollo 14 mare basalt suite that differs from other mare basalts is the evidence that assimilation of crustal materials played an important role in the genesis of *some* Apollo 14 basalts—an idea proposed for mare basalts in general by Binder and Lange (1978). While most Apollo 14 basalts do *not* support the KREEP assimilation model of Binder and Lange (1978), there are two basalt groups that require some form of crustal assimilation: the 14321-type basalts and the VHK basalts. The 14321-type basalts have chondrite-normalized HREE patterns with a

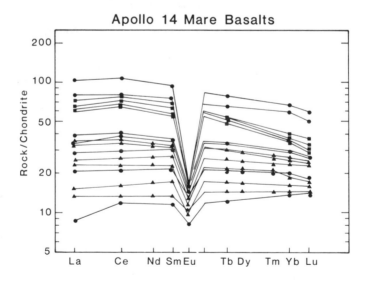

Fig. 9.   *Chondrite-normalized REE concentrations in Apollo 14 mare basalts. VHK basalts = triangles, 14321-type basalts = squares, all other Apollo 14 basalts = dots. The wide range in concentrations requires variations in the source region composition. This variation covers the same range observed between normal VLT, low-Ti, and high-Ti basalts. Data from Shervais et al. (1984b, 1985a,b), Warner et al. (1980), and Hubbard and Gast (1971).*

steep negative slope that is best explained by assimilation of a pristine KREEP component similar to "intermediate-K Fra Mauro" basalt 15386 (Shervais *et al.*, 1985a; Dickinson *et al.*, 1985). The high K/La ratio of VHK basalt can only be explained by the partial assimilation of lunar granite (Shih *et al.*, 1984; Shervais *et al.*, 1985b). The importance of crustal assimilation in the petrogenesis of Apollo 14 basalts is probably related to their eruptive setting in the lunar highlands. Mare basalts erupted in highland settings require more hydraulic head and are more likely to spend time in crustal magma chambers and conduits.

Discussion of the mare basalts

The mare basalt source region has geochemical characteristics that are complementary to the highland crust, and it is generally thought to comprise mafic cumulates from the magma ocean (e.g., Taylor, 1982). The progressive enrichment of mare basalts, in Fe/Mg, alkalis, and incompatible trace elements in the sequence VLT basalt → low-Ti basalt → high-Ti basalt is explained by remelting of cumulates formed at progressively shallower depths in the evolving magma ocean (Fig. 10;

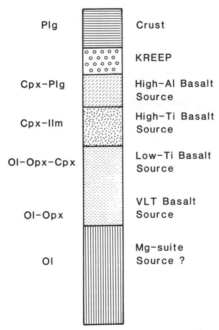

Fig. 10. The generally accepted model for the derivation of mare basalts from mafic cumulates of the magma ocean (Taylor, 1982; Basaltic Volcanism Study Project, 1981). The source region for low-Ti, high-Al basalts is considered to overlie the source region of high-Ti basalts. This conflicts with compatible element data for a less evolved high-Al basalt source, and with Ti/Sm, Sc/Sm, Hf/Lu, and Zr/Sm ratios that show no evidence for the previous fractionation of the cpx-ilm source of high-Ti basalts.

Taylor, 1982; Basaltic Volcanism Study Project, 1981). This is also consistent with the observed decrease in compatible element concentrations and the progressive increase in the negative Eu anomalies. Because high-Al basalts such as those at Apollo 14 are commonly enriched in alkalis and many incompatible trace elements relative to high-Ti basalts, it has been postulated that high-Al mare basalts formed by remelting of a shallow cumulate source that was more evolved than the high-Ti basalt source (Taylor, 1982; Basaltic Volcanism Study Project, 1981). According to this hypothesis, the evolved cumulate source of high-Al basalts *must overlie* the high-Ti basalt source and consist largely of clinopyroxene and plagioclase. This model is presented schematically in Fig. 10.

The data on Apollo 14 mare basalts reviewed earlier reveals that this hypothesis is *incorrect*. This is shown most clearly by the compatible trace elements Ni, Co, Cr, and V. The concentration of these elements in a partial melt do not change appreciably with variations in the degree of melting. Large changes do occur in magmas undergoing fractional crystallization, a fact that is reflected in their crystal cumulates. The decrease in compatible element concentrations in the sequence VLT basalt → low-Ti basalt → high-Ti basalt is consistent with the increase in incompatible trace elements in this same sequence and may be explained by the same process: remelting of cumulates formed at progressively shallower depths in the evolving magma ocean. However, the compatible element concentrations in the Apollo 14 basalts are clearly intermediate between concentrations found in "normal" low-Ti and high-Ti basalts (Fig. 8). This requires that the Apollo 14 basalt source region be *more primitive* than that of high-Ti mare basalts. In addition, major and trace element ratios such as Mg/Fe, Ti/Sm, and Sc/Sm in the Apollo 14 basalts are all *too high* to have formed from a cumulate whose parent liquid had previously fractionated cpx-ilm cumulates.

Although compatible trace element concentrations imply a source region for the Apollo 14 basalts that is more evolved than that of most low-Ti basalts and more primitive than the high-Ti basalt source region, the range of incompatible trace element concentrations requires a more complicated arrangement. These data suggest an interfingering or mixing of the low-Ti and high-Ti basalt source regions, or alternatively, that these distinct source regions never developed in the lunar mantle beneath Apollo 14. The general enrichment of the Apollo 14 basalts in alkalis also suggests that, like the crust, the mantle under this part of the Moon is enriched in alkalis relative to normal lunar values.

The enrichment of most Apollo 14 mare basalts in alkalis and other incompatible trace elements *cannot* be due to the assimilation of KREEP. As shown by Shervais *et al.* (1985a), KREEP assimilation by mare basalts has recognizable geochemical consequences that are observed *only* in the 14321-type basalts and not in the other 11 basalt types from this site. These consequences include lower than normal Ti/Sm, Sc/Sm, and Lu/Hf ratios and steep, negative slopes for chondrite-normalized HREE concentrations (Shervais *et al.*, 1985a). Because these characteristics are only observed in the 14321-type basalts, mixing with KREEP is ruled out as an explanation for the high alkali contents in the other Apollo 14 mare basalts. Enrichment of the mantle source region for these basalts is more likely.

Discussion

The petrological and geochemical data on highland crustal rocks and mare basalts reviewed above support the concept of large-scale lateral heterogeneities in the lunar crust and mantle. These data show that the highland crust in the west is characterized by alkali-rich lithologies (alkali anorthosite, granophyre) and incompatible element enrichment in lithologies of the Mg-rich suite, relative to highland crust in the east. The data also show that old (>3.8 aeons) mare basalts in the west are alkali- and alumina-rich compared to younger mare basalts, have the same range of incompatible trace element concentrations as all other mare basalts combined, and have compatible trace element concentrations intermediate between low-Ti and high-Ti mare basalts. Conclusions based on the distribution of felsic and basaltic rock-types are supported by orbital geochemistry, which shows high concentrations of K, Th, and U in the western highlands (Adler and Trombka, 1977).

Origin of lateral heterogeneities

How can these lateral heterogeneities be explained in terms of plausible mechanisms of lunar origin and evolution? Wasson and Warren (1980) have proposed that this heterogeneity is caused by asymmetrical fractionation of the lunar magma ocean. According to this hypothesis, preferential accumulation of Fe-metal onto the floor of the magma ocean in the nearside hemisphere caused the magma ocean to be some 200 km deeper on the nearside than on the farside. This resulted in a thicker sequence of mafic cumulates on the nearside, less rapid shoaling against the base of the felsic crust, and thus a thicker accumulation of "urKREEP" trapped between the base of the crust and the top of the mafic cumulates. Subsequent magmas rising through this zone would mix with larger fractions of "urKREEP" than equivalent magmas penetrating the crust at more distant, "eastern" locations (Wasson and Warren, 1980).

The asymmetrical fractionation hypothesis was proposed to explain the 2-km offset between the Moon's center of figure (CF) and center of mass (CM), the CM being closer to Earth (Wasson and Warren, 1980). This hypothesis may also be sufficient to explain the distribution of highland rock-types in the crust. The weakest aspect of the asymmetrical fractionation hypothesis is its inability to explain the unique geochemical character of the ancient Apollo 14 mare basalts. As discussed earlier, simple assimilation of variable proportions of KREEP by low-Ti basalt parent magmas *cannot* explain the geochemical variations observed in most Apollo 14 basalts (Shervais *et al.*, 1985a). This is because KREEP assimilation has easily recognizable geochemical consequences that are observed *only* in the 14321-type basalts and *not* in the other 11 varieties of mare basalts that occur in the Fra Mauro breccias (Shervais *et al.*, 1985a). The major and trace element variations observed in these Apollo 14 basalts require that the lateral heterogeneities observed in the lunar crust must exist in the lunar mantle as well. The geochemical characteristics of the Apollo 14 mare basalts are intermediate to VLT, low-Ti, and high-Ti mare basalts, suggesting a source region for the Apollo 14 basalts that is a "mixture" of the VLT, low-Ti, and high-Ti basalt source regions. How does this "mixing" occur? We will address two mechanisms: heterogeneous accretion and late accretion.

Heterogeneous accretion of two submoons

A major problem with the magma ocean hypothesis has always been the nature of the heat source for melting. Accretion of the Moon from a circumterrestrial debris swarm should take about as long as the Earth, some 10^7-10^8 years (Safranov, 1969). This is too long for either short-lived radionuclides (e.g., ^{26}Al) or accretional energy to be effective (Wood, 1983). Ruskol (1973; Ruskol *et al.*, 1975) and Wood (1983) have shown that accretional energy would be sufficient to melt much of the Moon if it formed by the accretion of two submoons, each composing about 0.5 lunar mass. They propose that these two moons form from the circumterrestrial swarm at different times. As the first submoon receded due to tidal friction, the second would begin to form in the repopulated, near-Earth region of the swarm. The second submoon would also recede by tidal friction and gradually overtake the first submoon (Ruskol, 1973; Ruskol *et al.*, 1975; Wood, 1983). The accretion of two bodies with $\approx$0.5 lunar mass each would release the maximum energy in the shortest time of any accretionary scenario. Because the second submoon would form later than the first, from a "new" population of circumterrestrial debris, it is possible that these two submoons could be compositionally distinct.

While this model provides an elegant solution to the heat source problem, it does not appear capable of explaining the compositional variations observed in the lunar crust and mantle as a function of heterogeneous accretion. There are two main problems. First, although the two submoons may be compositionally distinct prior to accretion, it is unlikely that these chemical distinctions would survive a melting event on the scale of a Moon-wide magma ocean. Second, the most likely type of chemical fractionation to result from accretionary processes is between volatile and nonvolatile (refractory) elements. Taylor (1982) has shown, however, that lunar rocks have remarkable constant volatile/refractory elemental ratios, regardless of absolute abundances. The elemental fractionations that are observed are best explained by magmatic processes (Taylor, 1982). This requires that the Moon either accreted from homogenous material, or was homogenized after accretion.

Giant impact(s) during late accretion

Giant impacts provide a mechanism that may create lateral heterogeneities in an already accreted Moon, both directly through ballistic transport and indirectly by altering the thermal budget (Cadogan, 1974; Arkani-Hamed, 1974; Hubbard and Andre, 1983; Wilhelms, 1983). The late accretion of material to the lunar crust is amply demonstrated by the widespread distribution and abundance of multiringed basins in the lunar highlands (Stuart-Alexander and Howard, 1970; Hartmann and Wood, 1971; Howard *et al.*, 1974). The youngest of these basins is about 3.8 aeons (Orientale); traces of the oldest basins may have been obliterated by younger impact events.

The most direct effect of a giant impact would be removal of the preexisting ferroan anorthosite crust—an effect that is observed in the Apollo 12, 14, and 17 sample suites. Following a modified version of Cadogan's (1974) proposal, we suggest that the absence of ferroan anorthosites on the western lunar nearside, the abundance of KREEP, and much of the observed CM/CF offset could be due to ballistic erosion of preexisting ferroan anorthosite crust during formation of the Procellarum ("Gargantuan") basin, and redeposition of the eroded material on the lunar farside. In this scenario, the Mg-rich troctolite suite would represent lower crustal lithologies exposed by the basin-forming event and subsequent rebound of the crust.

Alteration of the lunar thermal budget by giant impacts has been addressed by Arkani-Hamed (1974) and Hubbard and Andre (1983). The direct release of thermal energy during impact events occurs mainly within a few kilometers or tens of kilometers of the surface (Kaula, 1979). This could cause direct impact melting of some lower crustal rocks (not to be confused with the much younger mare basalts) after substantial ballistic erosion of the upper crust, but would not cause remelting of the deeper magma ocean cumulates. Because of the scaling factors involved in impact dynamics (e.g., Kaula, 1979), increasing the size of the projectile does not increase the energy released per unit mass, but only increases the volume of material affected.

Hubbard and Andre (1983) have calculated that impact melting may occur at depths below the transient cavity only if the projectile is sufficiently large and if the materials to be melted are already hot (preferably near their melting temperature). They suggest that deep impact melting may occur in events larger than the Imbrium basin event. The Procellarum basin, with a radius twice that of the Imbrium basin (1200-km radius vs. 650-km radius), is the largest known lunar basin and may have had sufficient energy to cause direct melting of the uppermost magma ocean cumulates (Cadogan, 1974; Hubbard and Andre, 1983).

Probably more efficient for our purpose, which is to locally rehomogenize the mafic magma ocean cumulates and enrich them in incompatible elements, would be impact-triggered sinking of the mafic cumulates (Taylor *et al.*, 1983). The mafic magma ocean cumulates are gravitationally unstable because the shallow, Fe-rich cumulates are denser than the deeper, Mg-rich cumulates. Sinking of the Fe-rich cumulates and diapiric rise of the more Mg-rich cumulates would cause partial melting of both and could rehomogenize the mare basalt source region locally. Extensive remelting would create a magma sea that would mix with residual urKREEP liquids. Alkali anorthosites may represent the second generation highland crust formed from this KREEP-enriched magma sea. Crystallization of this magma sea would entail a second "distillation" of the magma ocean and could create the ultra-KREEPy rocks that characterize the Apollo 14 site. Impact-triggered disruption of magma ocean mafic cumulates could be effected by either a single super-giant impact, or by a large number of smaller basin-forming impacts (Fig. 11).

Giant impacts may be even more effective if they occur prior to final crystallization of the magma ocean. Variations on this theme have been considered by Hartmann

MAGMA SEA

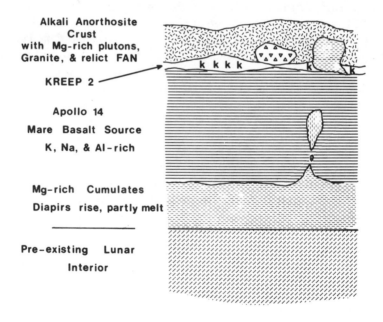

Alkali Anorthosite
Crust
with Mg-rich plutons,
Granite, & relict FAN

KREEP 2

Apollo 14
Mare Basalt Source
K, Na, & Al-rich

Mg-rich Cumulates
Diapirs rise, partly melt

Pre-existing Lunar
Interior

Fig. 11. A cartoon of "magma sea" cumulates formed by rehomogenization of late magma ocean cumulates after a giant impact (Procellarum?) removes the preexisting FAN crust. The new crust consists of alkali anorthosites that form as flotation cumulates on the magma sea and Mg-suite plutons derived by partial melting of Mg-rich diapirs during decompression. The mare basalt source region above the new "seabottom" is homogenous with respect to major elements but is enriched upward in incompatible elements. KREEP 2 is a second generation urKREEP formed by redistilling the magma sea.

(1980). He showed that intense bombardment of the lunar surface during protracted cooling of the magma ocean would affect both the cooling rate of the magma ocean and its chemical composition by puncturing or pulverizing the solid surface layer, and by mixing this material back into the magma ocean. We add that as the magma ocean shoaled, it may have divided into "seas" that were partly interconnected but were separated by surface irregularities in both its floor and ceiling that formed as a result of the bombardment. This hypothesis is not dependent on a single super-giant impact but is favored by a large number of very large impact events.

Conclusions

Lateral heterogeneity of the lunar crust and mantle on a regional scale is shown by (1) variations in the highland rock suites present at each of the Apollo sites, (2) chemical variations within the Mg-suite between "eastern" and "western" sites, (3) geochemical differences between the ancient Apollo 14 mare basalts and younger basalts from the other Apollo and Luna sites, (4) orbital geochemistry measurements of K, U, and Th in the lunar regolith, and (5) the 2-km offset between the Moon's center of figure and center of mass. We emphasize that the origin of these lateral heterogeneities is *not* well understood. However, regardless of whether or not the hypotheses outlined below are responsible, any model for lunar origin ought to explain how these lateral variations might form. Asymmetrical fractionation of the lunar magma ocean is not sufficient to explain all of these anomalies, although it can explain many of them. We suggest that further explanation of these lateral heterogeneities may be found in the dynamics of lunar accretion.

Plausible dynamic processes for creating these heterogeneities are (1) heterogeneous accretion of the Moon from two chemically distinct submoons, and (2) impact of a giant, basin-forming projectile(s) in the later stages of accretion. The former hypothesis provides a mechanism for the rapid release of kinetic energy during accretion, but there are difficulties in trying to preserve preaccretion heterogeneities during Moon-wide fractionation. Further, the constancy of lunar volatile/refractory elemental ratios implies that either the protolunar material was homogenous, or that it was homogenized after accretion. The latter hypothesis is favored by direct evidence for giant basin-forming impacts. The Procellarum-Imbrium-Nubium region of the western lunar nearside may be either a single, highly-eroded ancient basin (the "Gargantuan" basin of Cadogan, 1974), or a compound structure formed by the coalescence of several smaller but equally ancient basins (P. H. Warren, personal communication, 1985). This region correlates directly with orbital mapping of geochemical anomalies and nonferroan crust. Hubbard and Andre (1983) have calculated that single impacts may release enough kinetic energy to cause partial melting deep within the lithosphere if they are larger than the Imbrium event. The combined effect of multiple impacts of this size in the same area should be roughly cumulative because the brecciated basin fill is a good thermal insulator that will decrease heat losses between basin-forming events.

Our knowledge of lunar petrology continues to expand through progressively more detailed studies of existing lunar samples. These studies demonstrate the complex nature of the lunar crust and mantle—a complexity that was not obvious even a few years ago. Models of lunar origin and evolution that can account for this newly recognized complexity are still in the formative stages, as can be seen in the preceding discussion.

Where do we go from here? Clearly, the nature and distribution of petrological and geochemical variations in the lunar crust and mantle must be more tightly constrained before further progress can be made. A first-order priority should be detailed mapping of the lunar surface by a Lunar Polar Orbiter (LPO) that includes radar altimeter, imaging, a magnetometer, and optical, gamma-ray, and X-ray spectrometer. At present, only 20% of the lunar surface has been covered by remote geochemical sensing. Subsequent to detailed mapping by the LPO, further sample return missions are needed to provide ground truth for the sensing experiments and to provide additional samples for detailed petrological and geochemical study. These studies formed the core of the Apollo program, and continue to reveal new insights into the Moon's origin and evolution 16 years after the first sample return. Similar major advances in our understanding of lunar history can be expected from future sample return missions. These advances will help answer the most basic question of lunar history—how did the Moon form?

References

Adler I. and Trombka J. I. (1977) Orbital geochemistry: lunar surface analysis from X-ray and gamma-ray remote sensing experiments. *Phys. Chem. Earth, 10,* 17.

Arkani-Hamed J. (1974) Effect of a giant impact on the thermal evolution of the moon. *The Moon, 9,* 183–209.

Basaltic Volcanism Study Project (1981) *Basaltic Volcanism on the Terrestrial Planets.* Pergamon, New York. 1286 pp.

Binder A. B. and Lange M. A. (1978) On the mare basalt magma source region. *Proc. Lunar Planet. Sci. Conf. 9th,* pp. 337–358.

Cadogan P. (1974) Oldest and largest lunar basin? *Nature, 250,* 315–316.

Chen H. K., Delano J. W., and Lindsley D. H. (1982) Chemistry and phase relations of VLT volcanic glasses from Apollo 14 and Apollo 17. *Proc. Lunar Planet. Sci. Conf. 13th,* in *J. Geophys. Res., 87,* A171–A181.

Dickinson T., Taylor G. J., Keil K., Schmitt R., Hughes S. S., and Smith M. R. (1985) Apollo 14 aluminous mare basalts and their possible relationship to KREEP. *Proc. Lunar Planet. Sci. Conf. 15th,* in *J. Geophys. Res., 90,* C365–C374.

Dymek R. F., Albee A. L., and Chodos A. A. (1975) Comparative petrology of lunar cumulate rocks of possible primary origin: Dunite 72415, troctolite 76535, norite 78235, and anorthosite 62237. *Proc. Lunar Sci. Conf. 6th,* pp. 301–341.

Goodrich C. A., Taylor G. J., Keil K., Boynton W. V., and Hill D. H. (1984) Petrology and chemistry of hyperferroan anorthosites and other clasts from lunar meteorite ALHA81005. *Proc. Lunar Planet. Sci. Conf. 15th,* in *J. Geophys. Res., 89,* C87–C94.

Gooley R., Brett R., Warner J., and Smyth J. R. (1974) A lunar rock of deep crustal origin: Sample 76535. *Geochim. Cosmochim. Acta, 38,* 1329–1340.

Grieve R. A. F., McKay G. A., Smith H. D., and Weill D. F. (1975) Lunar polymict breccia 14321: a petrographic study. *Geochim. Cosmochim. Acta, 39,* 229–245.

Hartmann W. K. (1980) Dropping stones in magma oceans: Effects of early lunar cratering. In *Proceedings of the Conference on the Lunar Highlands Crust* (J. J. Papike and R. B. Merrill, eds.), pp. 155–171. Pergamon, New York.

Hartmann W. K. and Wood C. A. (1971) Moon: Origin and evolution of multi-ring basins. *The Moon, 3,* 3–78.

Head J. W. (1976) Lunar volcanism in space and time. *Rev. Geophys. Space Phys., 14,* 265–300.

Herzberg C. T. (1978) The bearing of spinel cataclasites on the crust-mantle structure of the moon. *Proc. Lunar Planet. Sci. Conf. 9th,* pp. 319–336.

Herzberg C. T. and Baker M. B. (1980) The cordierite- to spinel-cataclasite transition: Structure of the lunar crust. In *Proceedings of the Conference on the Lunar Highlands Crust* (J. J. Papike and R. B. Merrill, eds.), pp. 113–132. Pergamon, New York.

Hörz F. (1978) How thick are lunar mare basalts? *Proc. Lunar Planet. Sci. Conf. 9th,* pp. 3311–3331.

Howard K. A., Wilhelms D. E., and Scott D. H. (1974) Lunar basin formation and highland stratigraphy. *Rev. Geophys. Space Phys., 12,* 309–327.

Hubbard N. and Andre C. G. (1983) Magma genesis in a battered moon: effects of basin-forming impacts. *Moon and Planets, 29,* 15–37.

Hubbard N. and Gast P. (1971) Chemical composition and origin of nonmare lunar basalts. *Proc. Lunar Sci. Conf. 2nd,* pp. 999–1020.

Hunter R. H. and Taylor L. A. (1983) The magma ocean as viewed from the Fra Mauro shoreline: An overview of the Apollo 14 crust. *Proc. Lunar Planet. Sci. Conf. 13th,* in *J. Geophys. Res., 88,* A591–A602.

James O. B. (1980) Rocks of the early lunar crust. *Proc. Lunar Planet. Sci. Conf. 11th,* pp. 365–393.

James O. B. (1981) Petrologic and age relations of the Apollo 16 rocks: Implications for subsurface geology and the age of the Nectaris Basin. *Proc. Lunar Planet. Sci. 12B,* pp. 209–233.

James O. B. (1983) Mineralogy and petrology of the pristine rocks (abstract). In *Workshop on Pristine Highlands Rocks and the Early History of the Moon* (J. Longhi and G. Ryder, eds.), pp. 44–51. LPI Tech. Rpt. 83–02, Lunar and Planetary Institute, Houston.

James O. B. and Flohr M. K. (1983) Subdivision of the Mg-suite noritic rocks into Mg-gabbronorites and Mg-norites. *Proc. Lunar Planet. Sci. Conf. 13th,* in *J. Geophys. Res., 88,* A603–A614.

James O. B. and Hammarstrom J. G. (1977) Petrology of four clasts from consortium breccia 73215. *Proc. Lunar Sci. Conf. 8th,* pp. 2459–2494.

James O. B. and McGee J. J. (1979) Consortium breccia 73255: Genesis and history of two coarse-grained "norite" clasts. *Proc. Lunar Planet. Sci. Conf. 10th,* pp. 713–743.

Kaula W. M. (1979) Thermal evolution of Earth and Moon growing by planetesimal impacts. *J. Geophys. Res., 84,* 999–1008.

Knapp S. A., Shervais J. W., and Taylor L. A. (1984) Consortium breccia 14321: Petrology of the pristine highland clasts (abstract). In *Lunar and Planetary Science XV,* pp. 431–432. Lunar and Planetary Institute, Houston.

Lindstrom M. M. (1984a) Magnesian anorthosites and other clasts from complex breccia 14321 (abstract). In *Lunar and Planetary Science XV,* pp. 481–482. Lunar and Planetary Institute, Houston.

Lindstrom M. M. (1984b) Unique clasts of alkali gabbronorite and ultra-KREEPy melt rock in NRC feldspathic fragmental breccia 67975. *Proc. Lunar Planet. Sci. Conf. 15th,* in *J. Geophys. Res., 89,* C50–C62.

Lindstrom M. M., Knapp S. A., Shervais J. W., and Taylor L. A. (1984) Magnesian anorthosite and associated troctolite and dunite in Apollo 14 breccias. *Proc. Lunar Planet. Sci. Conf. 15th,* in *J. Geophys. Res., 89,* C41–C49.

Lindstrom M. M., Crozaz G., and Zinner E. (1985) REE in phosphates from lunar highlands cumulates: An ion probe study (abstract). In *Lunar and Planetary Science XVI*, pp. 483–494. Lunar and Planetary Institute, Houston.

Longhi J. and Ashwal L. D. (1985) Two-stage models for lunar and terrestrial anorthosites: Petrogenesis without a magma ocean. *Proc. Lunar Planet. Sci. Conf. 15th*, in *J. Geophys. Res., 90*, C571–C584.

McKay G. A. (1982) Partitioning of REE between olivine, plagioclase, and synthetic basaltic melts: Implications for the origin of lunar anorthosites (abstract). In *Lunar and Planetary Science XIII*, pp. 493–494. Lunar and Planetary Institute, Houston.

Norman M. D. and Ryder G. (1980) Geochemical constraints on the igneous evolution of the lunar crust. *Proc. Lunar Planet. Sci. Conf. 11th*, pp. 317–331.

Papike J. J. and Vaniman D. T. (1978) Luna 24 ferrobasalts and the mare basalt suite: Comparative chemistry, mineralogy, and petrology. In *Mare Crisium: The View from Luna 24* (R. B. Merrill and J. J. Papike, eds.), pp. 371–401. Pergamon, New York.

Papike J. J., Hodges F. N., Bence A. E., Cameron M., and Rhodes J. M. (1976) Mare basalts: Crystal chemistry, mineralogy, and petrology. *Rev. Geophys. Space Phys., 14*, 475–540.

Quick J. E., James O. B., and Albee A. L. (1981) Petrology and petrogenesis of lunar breccia 12013. *Proc. Lunar Planet. Sci. 12B*, pp. 117–172.

Ridley W. I. (1975) On high-alumina mare basalts. *Proc. Lunar Sci. Conf. 6th*, pp. 131–146.

Ryder G. (1976) Lunar sample 15405: Remnant of a KREEP basalt-granite differentiated pluton. *Earth Planet. Sci. Lett., 29*, 255–268.

Ruskol E. L. (1973) On the model of the accumulation of the moon compatible with the data on the composition and age of lunar rocks. *The Moon, 6*, 190–201.

Ruskol E. L., Nikolajeva E. V., and Syzdykov A. S. (1975) Dynamical history of coplanar two-satellite systems. *The Moon, 12*, 11–18.

Safranov V. S. (1969) Evolution of the protoplanetary cloud and formation of the Earth and planets, Nauka, Moscow (Translation). *NASA TTF-667*. 206 pp.

Salpas P., Shervais J. W., and Taylor L. A. (1985) Petrogenesis of lunar granites: The result of apatite fractionation (abstract). In *Lunar and Planetary Science XVI*, pp. 726–727. Lunar and Planetary Institute, Houston.

Shervais J. W. and Taylor L. A. (1983) Micrographic granite: More from Apollo 14 (abstract). In *Lunar and Planetary Science XIV*, pp. 696–697. Lunar and Planetary Institute, Houston.

Shervais J. W., Taylor L. A., and Laul J. C. (1983) Ancient crustal components in the Fra Mauro breccias. *Proc. Lunar Planet. Sci. Conf. 14th*, in *J. Geophys. Res., 88*, B177–B192.

Shervais J. W., Taylor L. A., Laul J. C., and Smith M. R. (1984a) Pristine highland clasts in consortium breccia 14305: Petrology and geochemistry. *Proc. Lunar Planet. Sci. Conf. 15th*, in *J. Geophys. Res., 89*, C25–C40.

Shervais J. W., Taylor L. A., and Laul J. C. (1984b) Very High Potassium (VHK) basalt: A new type of aluminous mare basalt from Apollo 14 (abstract). In *Lunar and Planetary Science XV*, pp. 768–769. Lunar and Planetary Institute, Houston.

Shervais J. W., Taylor L. A., and Lindstrom M. M. (1985a) Apollo 14 mare basalts: Petrology and geochemistry of clasts from consortium breccia 14321. *Proc. Lunar Planet. Sci. Conf. 15th*, in *J. Geophys. Res., 90*, C375–C395.

Shervais J. W., Taylor L. A., Laul J. C., Shih C.-Y., and Nyquist L. E. (1985b) Mare basalt petrogenesis: An important link in very high potassium (VHK) basalt. *Proc. Lunar Planet. Sci. Conf. 16th*, in press.

Shih C.-Y., Bansal B. M., Weismann H., and Nyquist L. E. (1984) Rb-Sr chronology and petrogenesis of VHK basalts (abstract). In *Lunar and Planetary Science XV*, pp. 774–775. Lunar and Planetary Institute, Houston.

Simon S. B., Papike J. J., and Laul J. C. (1982) Petrology and geochemistry of Apollo 14 soils. *Proc. Lunar Planet. Sci. Conf. 13th*, in *J. Geophys. Res., 87*, A232–A246.

Stuart-Alexander D. E. and Howard K. A. (1970) Lunar Maria and circular basins—a review. *Icarus*, *12*, 440–456.

Taylor G. J., Warner R. D., Keil K., Ma M. S., and Schmitt R. A. (1980) Silicate liquid immiscibility, evolved lunar rocks and the formation of KREEP. In *Proceedings of the Conference on the Lunar Highlands Crust* (J. J. Papike and R. B. Merrill, eds.), pp. 339–352. Pergamon, New York.

Taylor L. A., Shervais J. W., Hunter R. H., Shih C. Y., Nyquist L., Bansal B. M., Wooden J., and Laul J. C. (1983) Pre-4.2 AE mare-basalt volcanism in the lunar highlands. *Earth Planet. Sci. Lett.*, *66*, 33–47.

Taylor S. R. (1982) *Planetary Science: A Lunar Perspective*. Lunar and Planetary Institute, Houston. 481 pp.

Walker D. A. (1983) Lunar and terrestrial crust formation. *Proc. Lunar Planet. Sci. Conf. 14th*, in *J. Geophys. Res.*, *88*, B17–B25.

Warren P. H. and Wasson J. T. (1977) Pristine nonmare rocks and the nature of the lunar crust. *Proc. Lunar Sci. Conf. 8th*, pp. 2215–2235.

Warren P. H. and Wasson J. T. (1978) Compositional-petrographic investigation of pristine nonmare rocks. *Proc. Lunar Planet. Sci. Conf. 9th*, pp. 185–217.

Warren P. H. and Wasson J. T. (1979a) The compositional-petrographic search for pristine non-mare rocks: Third foray. *Proc. Lunar Planet. Sci. Conf. 10th*, pp. 583–610.

Warren P. H. and Wasson J. T. (1979b) The origin of KREEP. *Rev. Geophys. Space Phys.*, *17*, 73–88.

Warren P. H. and Wasson J. T. (1980a) Further foraging for pristine nonmare rocks: Correlations between geochemistry and longitude. *Proc. Lunar Planet. Sci. Conf. 11th*, pp. 431–470.

Warren P. H. and Wasson J. T. (1980b) Early lunar petrogenesis, oceanic and extraoceanic. In *Proceedings of the Conference on the Lunar Highlands Crust* (J. J. Papike and R. B. Merrill, eds.), pp. 81–99. Pergamon, New York.

Warren P. H., Taylor G. J., Keil K., Marshall C., Wasson J. T. (1981) Foraging westward for pristine nonmare rocks: Complications for petrogenetic models. *Proc. Lunar Planet. Sci. 12B*, pp. 21–40.

Warren P. H., Taylor G. J., Keil K., Kallemeyn G. W., Rosener P. S., and Wasson J. T. (1983a) Sixth foray for pristine nonmare rocks and an assessment of the diversity of lunar anorthosite. *Proc. Lunar Planet. Sci. Conf. 13th*, in *J. Geophys. Res.*, *88*, A615–A630.

Warren P. H., Taylor G. J., Keil K., Kallemeyn G. W., Shirley D. N., and Wasson J. T. (1983b) Seventh foray: Whitlockite-rich lithologies, a diopside-bearing troctolitic anorthosite, ferroan anorthosites and KREEP. *Proc. Lunar Planet. Sci. Conf. 14th*, in *J. Geophys. Res.*, *88*, B151–B164.

Warren P. H., Taylor G. J., Keil K., Shirley D. N., and Wasson J. T. (1983c) Petrology and geochemistry of two large granite clasts from the moon. *Earth Planet. Sci. Lett.*, *64*, 175–185.

Warner J. L., Simonds C. H., and Phinney W. C. (1976) Genetic distinction between anorthosites and Mg-rich plutonic rocks: New data from 76255 (abstract). In *Lunar Science VII*, pp. 915–917. The Lunar Science Institute, Houston.

Warner R. D., Taylor G. J., Keil K., Ma M. S., and Schmitt R. A. (1980) Aluminous mare basalts: New data from Apollo 14 coarse-fines. *Proc. Lunar Planet. Sci. Conf. 11th*, pp. 87–104.

Wasson J. T. and Warren P. H. (1980) Contribution of the mantle to the lunar asymmetry. *Icarus*, *44*, 752–771.

Wetherill G. W. (1975) Late heavy bombardment of the Moon and terrestrial planets. *Proc. Lunar Sci. Conf. 6th*, pp. 1539–1561.

Wetherill G. W. (1976) The role of large bodies in the formation of the Earth and Moon. *Proc. Lunar Sci. Conf. 7th*, pp. 3245–3257.

Wilhelms D. E. (1983) Effects of the Procellarum basin on lunar geology, petrology, and tectonism (abstract). In *Lunar and Planetary Science XIV*, pp. 845–846. Lunar and Planetary Institute, Houston.

Wood J. A. (1983) The lunar magma ocean, thirteen years after Apollo 11 (abstract). In *Workshop on Pristine Highlands Rocks and the Early History of the Moon* (J. Longhi and G. Ryder, eds.), pp. 87–89. LPI Tech Rpt. 83-02, Lunar and Planetary Institute, Houston.

Constraints on the Origin of the Moon from the Abundance of Molybdenum and Other Siderophile Elements

HORTON E. NEWSOM

Institute of Meteoritics and Department of Geology, University of New Mexico, Albuquerque, NM 87131

The depletion of Mo in the Moon has been determined from new analyses of lunar samples. The new data along with literature data indicate that the refractory element Mo is depleted by a factor of 1200 ± 800 in the lunar silicates relative to chondritic abundances normalized to refractory element abundances, and by a factor of 27 ± 20 relative to the Earth's mantle. The new data for Mo confirms that the silicate portion of the Moon is depleted relative to the Earth for many siderophile elements. Assuming that the Moon formed independent of the Earth with CI carbonaceous chondrite ratios of siderophile and refractory elements, the depletion pattern for the siderophile elements W, P, Co, Ni, Mo, Re, and Ir in the Moon is consistent with segregation at low degrees of partial melting (5% to 9%) of an Fe-rich metal core (4.5 wt % to 5.5 wt %) in the Moon. If the Moon formed out of material from the Earth's mantle, the pattern of siderophile element depletion relative to the Earth's mantle is consistent with segregation of a very small amount of Fe-rich metal (0.1 wt % to 0.4 wt %) or a small amount of sulfur-rich metallic liquid (0.2 wt % to 1.2 wt %), at about 20% partial melting. The siderophile element data do not constrain the final amount of the Moon that was melted subsequent to the metal segregation event. Because geophysical data, including paleomagnetic data, are consistent with a metal core as large as 5%, an independent origin is possible. Current theories for the origin of the Moon all fall somewhere between a strictly independent origin and a terrestrial origin. A greater or lesser contribution to the Moon from terrestrial impact ejecta is expected for all models. In addition, the current models allow for the loss of metal from the precursor material out of which the Moon was formed. Therefore, the Moon probably contains somewhat less than the actual amount of the metal responsible for the depletions of the siderophile elements relative to chondritic abundances.

Introduction

The abundance of siderophile (metal-loving) elements in the Moon compared to the Earth has been of central importance to the question of the origin of the

Moon since the return of lunar samples to the Earth. The siderophile elements (for example, W, Co, Ni, and Mo) are of special importance because they provide a record of the processes of accretion and core formation in differentiated planets. The initial abundances of siderophiles in material from which the planets were built are assumed to be the same as in chondrite meteorites. Under reducing conditions siderophiles are preferentially partitioned from silicates into metal (defined here as Fe with <10% Ni and no S). The subsequent segregation of the metal into a planetary core leaves behind a silicate mantle depleted in the siderophile elements relative to the initial chondritic abundances, with the most siderophile elements having the greatest depletion. Very distinct patterns of siderophile element depletion can be established in planetary mantles due to variations in the total amount of metal in each planet, differences in conditions during metal-silicate equilibrium such as temperature, pressure, and oxygen partial pressure, and heterogeneities in the amount of accreting metal at each stage in the accretion of a planet.

The similarities and differences among the patterns of siderophile element concentrations observed in the Moon and in the Earth have been a source of great controversy. This controversy reached a peak with the debate between Anders (1978) and Delano and Ringwood (1978). At that time Ringwood and coworkers suggested that the Earth's mantle and lunar silicates had similar abundances of siderophiles, within a factor of two, suggesting that the Moon may have formed out of the Earth's mantle after core formation. Rammensee and Wänke (1977) also pointed out the similarity in abundance of tungsten (W) in lunar and terrestrial rocks and concluded, based on experimental metal-silicate partitioning experiments, that the depletion of W observed in lunar rocks could not be achieved by segregation of a small metal core in the Moon. Recent work by Newsom and Palme (1984a) and Newsom et al. (1985) has confirmed the similarity of W concentrations in the Earth to those in lunar rocks, when normalized to U to correct for the igneous fractionation of W (Fig. 1). Newsom and Drake (1982a), however, showed that a small lunar metal core (2%–5%) could explain the W depletion, assuming chondritic initial concentrations of W, except for the special case of a totally molten Moon. The important factor is the effect of the degree of partial melting on the metal-silicate partitioning behavior of siderophiles.

In the last few years a much better understanding of the geochemical behavior of siderophile elements has been obtained, especially from the large amount of new experimental work on metal-silicate partition coefficients. Recent work on the abundances of important siderophiles in the Earth and Moon has also been of key importance and has led to a new understanding and a consensus on the magnitude and significance of the siderophile element depletions in the silicate portion of the Moon (Drake, 1983; Newsom, 1984). There has also been a large amount of progress regarding the geophysical evidence for a lunar metal core. A metal content of 5% is consistent with many lines of geophysical evidence, including magnetic moment

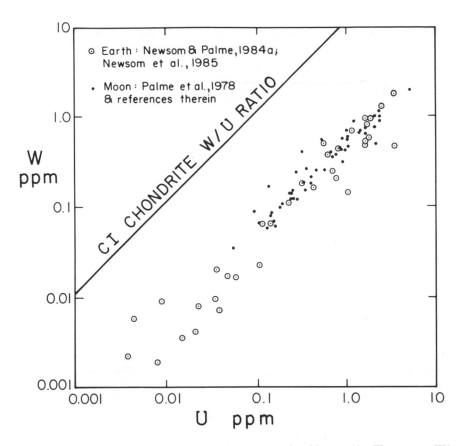

Fig. 1. Tungsten and uranium concentrations in terrestrial and lunar rocks. The constant W/U ratios in the terrestrial and lunar samples are primarily due to the similar incompatible nature of W and U during partial melting. The W/U ratios for the terrestrial and lunar samples are remarkably similar. The W/U ratios of the terrestrial mantle and the silicate portion of the Moon seem to be depleted by the same amount relative to the chondritic W/U ratio. Some of the lunar samples plotted are highlands crustal samples with variable amounts of the KREEP (incompatible K, rare earths, and P) component, but the KREEP component has the same W/U ratio as the mare basalts, indicating a uniform depletion of W in the Moon. The terrestrial samples in the lower left-hand corner are primarily ultramafic nodules for which comparable lunar samples have not yet been found.

(Russell et al., 1981; Russell, 1984), free librations (Yoder, 1981, 1984), and moment of inertia (Hood and Jones, 1985; Hood, 1986). A relatively large metal core could explain the lunar paleomagnetic record (Cisowski et al., 1983; Cisowski and Fuller, 1986; Runcorn, 1983; Dolginov, 1985). The entire lunar metal content also need not have completely accumulated into a core in order to produce the siderophile depletions, although this possibility seems physically unlikely.

In this article, new data will be discussed establishing the abundance and depletion of Mo in the Moon. The Mo data has been previously published in a figure in Newsom (1984, Fig. 2). The overall pattern of siderophile depletion in the Moon, together with the latest information on siderophile depletion factors and partition coefficients, will then be considered in light of possible lunar origins, such as formation from the Earth's mantle or formation by coaccretion. The geochemical constraints on the size of a lunar core are also discussed in relation to various lunar origin models.

Experimental Technique

A new analytical method for determining Mo by neutron activation analysis has been developed by Newsom and Palme (1984b) using a modification of a metal-silicate extraction technique developed by Rammensee and Palme (1982). Samples weighing 0.1 to 0.2 grams were equilibrated with an equal weight of iron metal powder at 1300°C for 45 minutes under reducing conditions near the iron-wüstite buffer in a high-temperature furnace. The temperature was raised to 1600°C for 10 minutes to allow the metal and silicate to segregate in the crucible. The iron bead was then irradiated in a TRIGA Mark 2 reactor for 6 hours with a thermal neutron flux of 7×10^{11} n cm^{-2} s^{-1} (run 1), or at the Kernforschungsanlage Jülich for 21.5 hours with a flux of 1.5×10^{14} n cm^{-2} s^{-1} (runs 2 and 3). The quantitative extraction of Mo into the metal phase before irradiation prevented the production of radioactive Mo by induced fission of ^{238}U. Four analyses of the ultrapure iron powder yielded an average Mo content of 51.1 ± 2 ppb.

The metal bead was dissolved in hot HC1 acid containing a Mo carrier solution. The Mo was then precipitated as a sulfide with H$_2$S gas, leaving Fe in solution in order to remove an interfering ^{59}Fe peak. The precipitate was dissolved in aqua regia and the strongest Mo γ-ray line (actually from ^{99}Tc due to decay of ^{99}Mo) at 140.5 kev was counted on a Ge(Li)-detector. The Mo yields of 60% to 95% were obtained by irradiation of the Mo solution in the TRIGA reactor in Mainz. Corrections were made for self-absorption of neutrons during irradiation (<3%) and, for the Mo added via the Fe metal, as much as 70% for the sample with the lowest Mo concentration. The uncertainty in the standards adds less than 3% error to the accuracy of the Mo determinations (Palme and Rammensee, 1981b). The data together with the total one standard deviation uncertainty are listed in Table 1.

Several factors support the quality of our analyses, although blanks were not run with the samples. Duplicate analyses of the iron metal and the eucrite Stannern (Newsom, 1985) provided consistent results. The same analytical technique was also applied to terrestrial samples and the resulting Mo concentrations are consistent

TABLE 1. *Molybdenum and Neodymium Concentrations in Lunar Samples.*

Sample	Type	Mo(ppb)	Nd(ppm)	Run
14259	fines	125 ± 16	120	1
10057	basalt	129 ± 9	60	2
14305	microbreccia	81 ± 10	140	2
75035	basalt	39 ± 6	36.5	2
15495	basalt	21 ± 3	12.2	2
12002	basalt	50	14	T
12052	basalt	30	17	T
12063	basalt	40	20	T
12038	basalt	50	26	T
12070	fines	30	54	T

Neodymium data for runs one and two are from Wänke et al. (1970, 1972, 1975). Data labeled run T are from Taylor et al. (1971). The uncertainties are the total one standard deviation uncertainty including counting statistics, standard error, and the error from the Mo added during the metal segregation step.

with earlier literature data (Newsom and Palme, 1984a). The Mo concentrations determined in this study are also consistent with the spark source mass spectrometer results of Taylor *et al.* (1971).

Results

The results of the Mo determinations are listed in Table 1. The five samples were analyzed as part of two separate furnace separations and reactor irradiations. The data are plotted in Fig. 2, together with data from Taylor *et al.* (1971). The Mo concentrations for three samples (two fines and one microbreccia) have been corrected in Fig. 2 for a small meteoritic component based on the iridium (Ir) content of these samples as measured in Mainz (references in Table 1), assuming a CI chondritic Mo/Ir ratio. The samples, Ir contents, assumed meteoritic Mo, and corrected Mo concentrations are as follows: 14259, 16 ppm Ir, 30.6 ppb Mo meteoritic, 94.7 ppb corrected Mo concentration; 14305, 10 ppm Ir, 19.2 ppb Mo meteoritic, 62.2 corrected Mo concentration; 12070, 7.5 ppm Ir, 14.4 ppb Mo meteoritic, and 15.6 corrected Mo concentration. The correction based only on Ir is uncertain because there is evidence that the meteoritic component in lunar highlands samples has variable relative concentrations of siderophile elements (Janssens *et al.*, 1978).

Depletion of Molybdenum in the Moon

With the new Mo determinations and the literature data we can evaluate the depletion of Mo in the silicate portion of the Moon compared to the Earth's mantle.

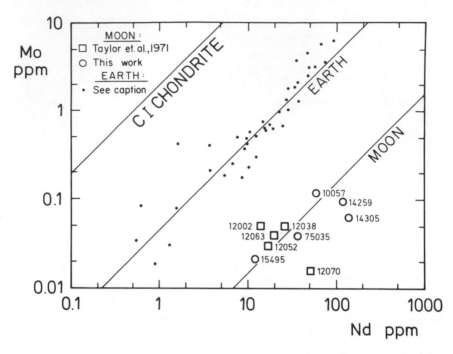

Fig. 2. Molybdenum and neodymium concentrations in terrestrial samples (dot symbols, New-som and Palme, 1984a; Newsom et al., 1985) and in lunar samples (open symbols). The new analyses of Mo in lunar samples confirm the low Mo concentrations reported by Taylor et al. (1971). The lines labeled "CI Chondrite," "Earth," and "Moon" are the average Mo/Nd ratios of the CI carbonaceous chondrites and the silicate portions of the Earth and Moon. The low Mo/Nd ratio of lunar rocks indicates that the silicate portion of the Moon is significantly more depleted in Mo than the Earth's mantle.

A direct comparison between the terrestrial and lunar data can be made on Fig. 2. The Mo/Nd ratios for the lunar samples are much lower than the Mo/Nd ratio for the terrestrial samples.

Plotting the concentration of Mo against the nonsiderophile element Nd (or W against U) serves two purposes. The first is to correct for the igneous fractionation of Mo. The absolute concentrations of Mo in the Earth and Moon are quite variable, but the Mo/Nd ratio is relatively constant. We assume that Mo behaves on the Moon almost identically to Nd during igneous events in the absence of metal, as it does on the Earth (Newsom and Palme, 1984a). The second purpose is to normalize the Mo concentrations to a refractory element, in order to allow the true depletion of the siderophile element to be determined. The Earth and Moon may have slightly different amounts of refractory elements, but the initial relative abundances of the refractory elements, such as W, Mo, U, and the rare earth elements (REE), should be nearly identical in the Earth, Moon, and chondrites. The depletion factors reported in Table 2 are normalized to refractory elements in the same way.

The average normalized depletion for the lunar samples does not depend strongly on the REE used for normalization. For example, the mean depletion for the Mo/Nd ratios is a factor of 1200, for the Mo/Sm ratios a factor of 1300, and for the Mo/Dy ratios a factor of 1100 relative to CI carbonaceous chondrites (Palme et al., 1981). We can therefore adopt a figure of about 1200 ± 800 for the depletion of Mo in the Moon relative to CI chondrites, assuming a Mo/Nd ratio of 0.0016 $\pm$ 0.001 in the Moon. The large uncertainty is due to the statistics of small numbers and the large depletions of the three samples that were corrected for a meteoritic component. Molybdenum is also depleted by a factor of 27 ± 20 relative to the Earth.

Compared to the terrestrial data, the lunar data are somewhat more scattered (Fig. 2). A significant portion of the scatter is probably due to the analytical uncertainty for these difficult Mo determinations. As an example, two of the samples analyzed for Mo by Taylor et al. (1971) were also analyzed by Kharkar and Turekian (1971). Kharkar and Turekian (1971) analyzed seven Apollo 12 samples for Mo by radiochemical neutron activation analysis. Most of their samples have reasonable Mo concentrations in comparison to our data and the data of Taylor et al. (1971). However, one of their samples (12021), which had a very high Mo content of 490 ppb, had high Au, Ag, and Mo concentrations, leading them to suggest that some contamination was present for these elements. The determination of Mo in sample 12070 by Taylor et al. (16 ppb, corrected for meteorite contamination) is the lowest Mo concentration observed in lunar samples in this or other studies. However, the corrected Mo concentration (116 ppb) obtained by Kharkar and Turekian (1971) for 12070 results in a depletion close to the average for all the samples. The analysis of sample 12063 by Kharkar and Turekian gave a result

TABLE 2. *Siderophile Element Depletion Factors Normalized to Refractory Element Contents and CI Abundances (Palme et al., 1981).*

	W	P	Mo	Co	Ni	Re	Ir
Earth	27 ± 10	43 ± 10	44 ± 15	12 ± 0.6	12.5 ± 0.6	420	420
	(b,f)	(a,k)	(b,f)	(g)	(g)	(d)	(d)
Moon	22 ± 7	115 ± 25	1200 ± 800	12 ± 0.4	50 ± 6	3×10^4	9×10^4
	(c)	(a,k)	(l)	(i)	(h,i)	(e)	(e)
Moon Earth	0.8 ± 0.6	3 ± 1	27 ± 20	1.05 ± 0.1	4 ± 0.5	70	200

Sources: (a) Newsom and Drake (1983); (b) Newsom and Palme (1984b); (c) Palme and Rammensee (1981); (d) Chou et al. (1983); (e) Wolf et al. (1979); (f) Newsom et al. (1985); (g) Jagoutz et al. (1979); (h) Delano (1985); (i) Delano (1986); (j) Palme et al. (1981); (k) Weckwerth et al. (1983); and (l) this work.

of 80 ppb Mo, a factor of two greater than the 40 ppb determined by Taylor *et al.* (1971).

Unfortunately, the REE results of Kharkar and Turekian are very erratic, and therefore the data have not been plotted in Fig. 2. Molybdenum data reported by Bouchet *et al.* (1971) and Morrison *et al.* (1970, 1971) are much higher than reported by Taylor *et al.* (1971) or this work, and are probably incorrect.

Another explanation for the scatter in the Mo concentrations are real variations in the siderophile element concentrations in different parts of the Moon (Dickinson and Newsom, 1985). The siderophile concentrations of low-Ti mare basalts and pristine highland rocks are much more variable than in terrestrial rocks (Wolf *et al.*, 1979). These variations could be due to different amounts of metal segregation or addition of primitive material from the interior of the Moon that has not experienced a core formation event.

Discussion

The two major questions to be addressed with the new information on Mo and the other siderophile elements are the existence of a lunar core and the origin of the Moon. The depletion and partition coefficient data for the siderophiles plotted in Figs. 3–8 are discussed first, followed by the method for calculating the theoretical depletion patterns illustrated in each figure. The calculations are finally discussed in terms of two end member models: an origin of the Moon from the Earth's mantle or an essentially independent origin of the Moon from chondritic material. Variations on the endmember models are also considered.

Siderophile element depletions

a. Siderophile depletions relative to chondritic abundances. The conventions used in describing and normalizing siderophile element depletions can be quite confusing. There are two basic methods, however. The first (not used in this paper) is to simply divide the concentration measured in a sample or a whole planet by the concentration in CI carbonaceous chondrites. The disadvantage of this method is that the refractory elements in the Earth and probably the Moon are enriched relative to CI abundances (perhaps by different amounts), such that a concentration ratio of one actually implies a depletion of that element in the sample, relative to the enriched initial abundance in the planet. The second method (used in this paper) is to normalize the concentrations to the refractory lithophile element contents. For example, the Mo depletion is obtained by dividing the Mo/Nd ratio in the Moon by the Mo/Nd ratio in CI chondrites. The "depletion factor" (Table 2) is simply the inverse of the depletion. This normalization with Nd allows us to determine the depletion relative to the actual initial concentration of the element (e.g., Mo) in the planet. Elements commonly used for normalization include Si, U, and the

REE (rare earth elements), with the assumption that the element whose depletion is being determined and the element used for normalization were initially present in the planet with the same relative abundances as in CI chondrites.

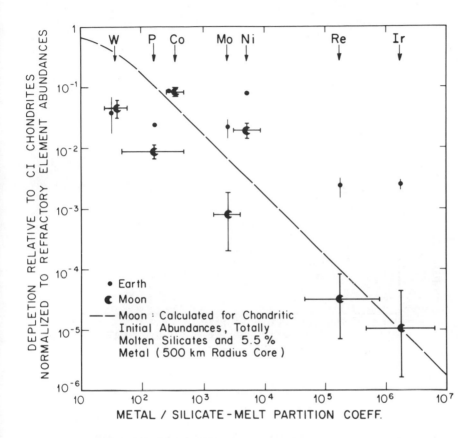

Fig. 3.   *The depletion relative to chondritic abundances of siderophile elements W, P, Co, Ni, Mo, Re, and Ir in the silicate portion of the Moon, and for comparison in the Earth's mantle (Table 2), plotted as a function of the metal/silicate-melt partition coefficient for each element. The uncertainties for the partition coefficients of the terrestrial data (not plotted) are about the same as for the lunar data. The Moon is more depleted than the Earth for every element except W. Also shown is the calculated depletion as a function of metal/silicate-melt partition coefficient, assuming that the Moon started with chondritic relative abundances of the siderophile and refractory elements (e.g., for an independent origin). The calculation assumes a metal core of 500 km radius corresponding to about 5.5 wt % metal and assumes that metal segregation occurred in a totally molten Moon. The partition coefficients used are listed in Table 3. The lunar siderophile depletions are clearly not consistent with the calculated depletions, assuming complete melting of the silicates.*

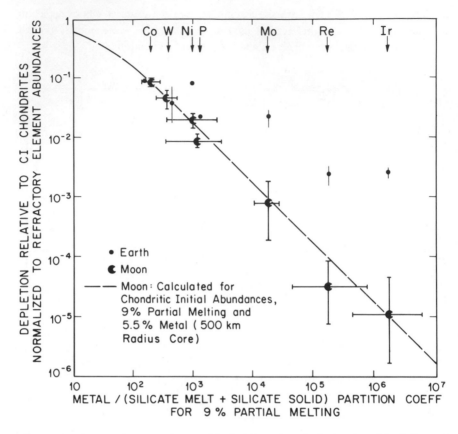

Fig. 4. The lunar and terrestrial siderophile depletions plotted as a function of the bulk metal to total silicate partition coefficient, assuming metal-silicate equilibrium at 9% partial melting. The partition coefficients used in the calculations are listed in Table 3. The data are surprisingly consistent with the illustrated calculation, which represents segregation of 5.5% metal at 9% partial melting, assuming chondritic initial abundances (independent origin).

Figures 3, 4, and 5 illustrate the siderophile depletions in the Moon relative to the chondritic abundances of siderophiles, normalized to the abundance of refractory lithophile elements in the Moon. These figures are most useful for discussing an origin of the Moon independent of the Earth. The depletions for the Earth's mantle are also shown for reference.

b. Siderophile depletion relative to the Earth's mantle. The siderophile element depletions for the Moon shown in Figs. 6–8 are normalized to the siderophile element abundances in the Earth's mantle (Table 2). These figures are especially useful for describing theories for an origin of the Moon out of the Earth's mantle.

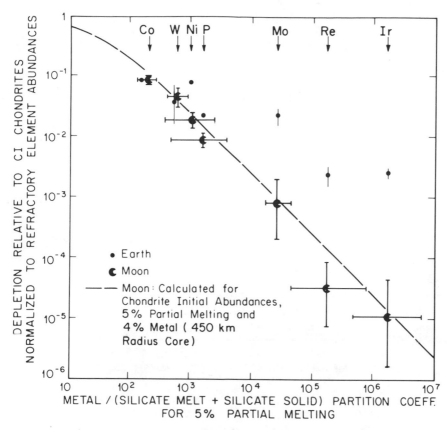

Fig. 5. *This plot is similar to Fig. 4 except for the assumption of 5% instead of 9% partial melting. The calculated depletion for an independent origin assumes 4% metal, corresponding to a smaller core radius of 450 km.*

Until recently the depletion of siderophiles in the Moon relative to the Earth's mantle was highly debated. Rammensee and Wänke (1977) showed that the depletion of W was similar (Figs. 1,3) while Ringwood (1979) argued that P was depleted to a similar extent in lunar silicates and in the Earth's mantle within a factor of two. For other highly siderophile elements, Delano and Ringwood (1978) suggested that the large scatter in the data were due to secondary effects, such as metal segregation in impact melts, and did not reflect the general depletion in the Moon. Ringwood *et al.* (1981) suggested that the large depletion of Re in lunar rocks was possibly due to the volatility of Re, noting that volatile elements such as Na are depleted in lunar rocks relative to the Earth's mantle.

The new data on Mo help settle the question raised by the possible volatility of other elements, since Mo is a refractory element with a very high nebular

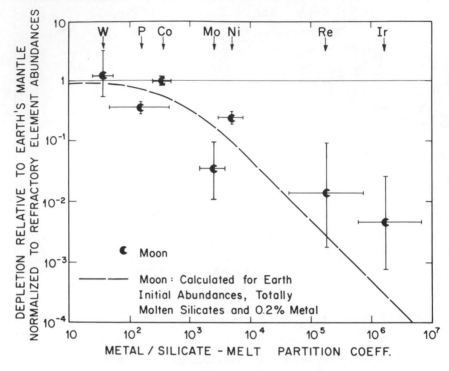

Fig. 6. The depletion of siderophile elements in the silicate portion of the Moon, relative to the abundance of siderophile elements in the Earth's mantle, are plotted against the metal/ silicate-melt partition coefficients. The figure illustrates the case of metal segregation in a totally molten Moon starting with terrestrial siderophile abundances (e.g., origin from Earth mantle material). The calculated depletion line assumes segregation of 0.2% metal from totally molten silicates. The partition coefficients used in the calculation are listed in Table 3. The lunar data are clearly not very consistent with metal segregation from molten silicates.

condensation temperature (1675 K at 0.001 bars; Fegley and Palme, 1985), just less than Ir. The abundance of Mo in carbonaceous chondrites is also constant relative to refractory elements (Palme and Rammensee, 1981b). If the Moon formed out of the Earth's mantle, then the large depletion of Mo in the lunar silicates compared to the Earth, a factor of 27, could not be caused by volatility. The question of the volatility of Re has also been settled recently by Chou *et al.* (1983), who point out that the Moon has a Re/Ir ratio that is a factor of two greater than the Re/Ir ratio in CI chondrites. The lunar interior, therefore, is not depleted in Re relative to the highly refractory element Ir, strongly suggesting that Re was not depleted in the Moon because of its volatility.

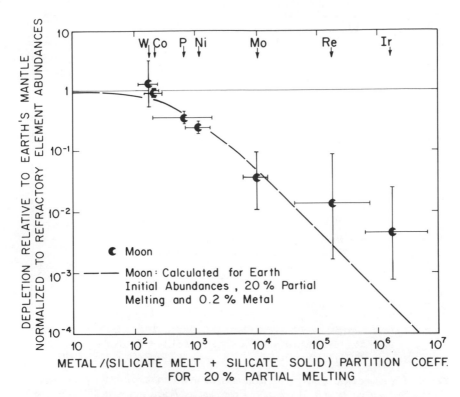

Fig. 7. The depletion of siderophile elements in the Moon relative to the Earth's mantle are plotted against the metal/total-silicate partition coefficient, assuming 20% partial melting. There is a relatively good fit of the calculated depletion (Earth mantle origin) to the lunar data, assuming segregation of 0.2.% metal at 20% partial melting.

The data now clearly indicate that the silicate portion of the Moon is significantly depleted in siderophile elements relative to the Earth's mantle (Figs. 6,7). The least siderophile element considered, W, is depleted to the same extent in lunar samples and in the Earth's mantle, although the uncertainties are still quite large (Fig. 1). The other siderophiles are more depleted relative to the Earth's mantle as their siderophile nature increases.

Siderophile element partition coefficients

a. Metal/silicate-melt. The depletions for each element in Figs. 3–8 are plotted vs. the simple metal/silicate-melt partition coefficients (Figs. 3,6) or vs. the metal/

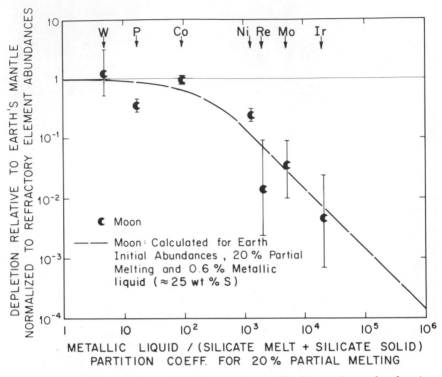

Fig. 8.   The abundances of siderophile elements relative to Earth's mantle are plotted against the metallic liquid (≈ 25 wt % S)/total-silicate partition coefficient, assuming 20% partial melting. The calculated depletion indicates that the lunar siderophile depletions relative to the Earth's mantle can be achieved by segregation of about 0.6% metallic liquid. The metallic-liquid/silicate-melt partition coefficients used in the calculation are listed in Table 3. The uncertainties for the partition coefficients are not shown on the figure, but may be large. The bulk mineral/silicate-melt partition coefficients used are listed in Table 3.

total-silicate partition coefficients (defined below) that represent metal segregation at low degrees of partial melting (Figs. 4, 5, 7, and 8). The partition coefficients used are listed in Table 3. The metal phase can be liquid or solid. The metal/silicate-melt partition coefficient of 5000 for Ni is the average of 3000 (Schmitt, 1984) and 7000 (Jones and Drake, 1985). The metal/silicate-melt partition coefficients for W and Re by Jones and Drake (1985) are consistent with the results of Newsom and Drake (1982b) and Kimura *et al.* (1974), respectively.

There is a large uncertainty in the metal/silicate-melt partition coefficient for P. Of the elements discussed, only P has a large liquid-metal/solid-metal partition coefficient (≈10, Jones and Drake, 1983). The proportion of Fe-Ni metal that is

	W	Co	Ni	P	Mo	Re	Ir
Metal	36	350	5000	160	2500	1.6×10^5	1.7×10^6
silicate melt	(a)	(a)	(a,c)	(d)	(f)	(a)	(a)
S-rich metallic liquid	1	152	5555	4.2	1250	2×10^3	2×10^4
silicate melt	(a)	(a)	(a)	(a)	(a)	(a)	(a)
Bulk silicate mineral	0.01	1.8	5	0.05	0.05	1	1
silicate melt	(e)	(g)	(g)	(g)	(g)	(h)	(h)

Data from: (a) Jones and Drake (1985); (b) Jones and Drake (1983); (c) Schmitt (1984); (d) Newsom and Drake (1983); (e) Newsom and Palme (1984a); (f) Rammensee (1978); (g) Irving (1978); and (h) Chou et al. (1983).

liquid during metal-silicate fractionation is difficult to estimate because the melting point of metal depends strongly on the amount of P and S available.

Several factors influence the metal-silicate partition coefficients. The partition coefficients listed in Table 3 are based on the assumption of reducing conditions, consistent with approximately 13% FeO in equilibrium with Fe metal containing a relatively low amount of Ni (<10%). The FeO content of the Moon is generally considered to be about 13%, although values as low as the Earth's mantle (≈ 8%) cannot be excluded (Drake, 1986; Warren, 1986). If the FeO content of the Moon were as low as 8%, with correspondingly higher partition coefficients, the amount of metal required to achieve the observed siderophile depletion would be a factor of two or three less than for the calculations in Figs. 3–8 (Newsom and Drake, 1982a). Because the dependence of partition coefficients on fO₂ is not the same for all elements, some changes in the relative partition coefficients would also occur (Schmitt, 1984), although, in general, the pattern of siderophile depletion vs. partition coefficients will remain the same.

   b. Silicate-mineral/silicate-melt. The silicate-mineral/silicate-melt partition coefficients are important when metal segregation is assumed to occur at low degrees of partial melting (e.g., Figs. 4, 5, 7, and 8). We are actually concerned with the bulk mineral-melt partition coefficients (Table 3), which are controlled by the actual mineralogical composition of the lunar mantle. The variation expected for different lunar bulk compositions is small, however (Newsom and Drake, 1982a). More work is needed to firmly establish the mineral-melt partitioning behavior of the siderophile elements under lunar conditions.

Siderophile depletion calculations

   The depletion of siderophiles can be calculated for a given metal/silicate partition coefficient and metal content. The basic equation for the weight fraction of metal

(X) required to achieve a certain depletion factor (a) in a single partial melting event has been derived by Rammensee and Wänke (1977), assuming equilibrium

$$X = \frac{a-1}{D^{M/S} + a - 1} \tag{1}$$

where $D^{M/S}$ is the bulk metal/total-silicate partition coefficient, the concentration of a trace element in the metal divided by the concentration in the silicate phase. The depletion factor (a) is the chondritic abundance of the siderophile element normalized to the refractory content of the Moon divided by the abundance of the siderophile in the silicate portion of the Moon.

If silicates are partly molten, the metal-silicate partitioning behavior of the siderophile elements depends strongly on the nature of the partitioning behavior between silicate melt and silicate minerals. The incompatible (magmaphile) siderophiles are excluded from the solid silicate minerals, increasing the concentration of the siderophiles in the remaining phases, including the metal. Thus, for incompatible siderophiles the metal/total-silicate partition coefficient is effectively increased at low degrees of partial melting in the presence of a large fraction of silicate minerals. The incompatible siderophiles include W, P, and Mo. The compatible siderophile elements such as Ni and Co, however, are concentrated in silicate minerals, reducing the concentrations in the silicate melt and the metal. The higher concentration of the siderophiles in the silicates causes the metal/total-silicate partition coefficient for the compatible siderophiles to decrease. Therefore, the presence or absence of silicate minerals during metal-silicate partitioning has a drastic and opposite effect on compatible vs. incompatible siderophile elements.

A few elements, such as germanium (Capobianco and Watson, 1982), are not strongly fractionated between silicate melt and silicate minerals. Although Re is known to be somewhat incompatible and Ir compatible during igneous fractionation on the Earth, Chou et al. (1983) show that there is no evidence of fractionation of Re and Ir relative to each other in the Moon, implying that the mineral/melt partition coefficients are approximately one.

The metal/total-silicate partition coefficent, $D^{M/S}$, can be calculated from the relationship:

$$D^{M/S} = \frac{D(\text{element})}{F_{liq} + (C^{sol}/C^{liq}) \times (1 - F_{liq})} \tag{2}$$

where D(element) is the metal/silicate-melt partition coefficient, F_{liq} is the fraction of silicate melt divided by the total fraction silicates, and C^{sol}/C^{liq} is the solid-silicate/silicate-melt partition coefficient.

a. Origin independent of Earth's mantle: Endmember model. The endmember independent origin model assumes that the Moon was formed from material with chondritic initial siderophile abundances. The dashed line in Fig. 3 represents the calculated depletion as a function of the metal/silicate partition coefficient for the segregation of 5.5 wt % metal in the Moon, assuming the segregation occurred when the silicate portion of the Moon was totally molten. For this case the $D^{M/S} = D(element)$, where D(element) is the metal/silicate-melt partition coefficient. A metal content of 5.5% is equivalent to a core of 500-km radius, about the maximum radius consistent with the geophysical evidence (Hood, 1986). The observed siderophile depletions are clearly not consistent with metal segregation in a totally molten Moon. In particular, the depletion of W, P, and Mo could not be achieved with a reasonable metal content, as pointed out for W by Rammensee and Wänke (1977).

An alternative is metal segregation at a relatively low degree of partial melting. Because of the large density difference between metal and silicate, metal segregation probably occurs at relatively low degrees of partial melting. In Figs. 4 and 5 the siderophile depletions are plotted against the calculated metal/total-silicate partition coefficients, assuming 9% or 5% partial melting and 5.5% or 4% metal, respectively. Figures 4 and 5 show a good correlation between the calculated and observed depletions. Supporting a partial melting model is a new determination of the depletions of Ni and Co in lunar glasses by Delano (1986), which shows that Ni is significantly more depleted in the Moon than Co; this is in contrast to the Earth's mantle, where Co and Ni have identical depletions (Jagoutz *et al.*, 1979). The element Re falls off the calculated line to some extent, but a slight incompatible nature for Re on the Moon would increase the $D^{M/S}$ and move Re closer to the calculated line.

Phosphorus falls on the more depleted side of the calculated line in Figs. 3–5. Some of the extra depletion is probably due to loss of P because of volatility during, or prior to, the formation of the Moon. There is also a large uncertainty in the metal/silicate-melt partition coefficient for P as discussed above.

The metal segregation event probably occurred very early during the accretion of the Moon or in precursor planetesimals before the formation of the Moon. The degree of partial melting during the metal segregation event, therefore, does not constrain the ultimate degree of melting in the Moon. Even a global magma ocean is not ruled out because equilibrium between the metal core and the magma ocean would be very difficult, since the core would be quickly isolated by an olivine blanket.

The calculation in Fig. 5 is for 5% partial melting, which may still allow efficient quantitative segregation of metal into a core. Less metal is needed to explain the observed depletions if a lower degree of partial melting is allowed, but the calculated $D^{M/S}$ for each element becomes extremely sensitive to the assumed mineral/melt partition coefficients used. To allow a metal content as low as 2%, equivalent to a 360-km core, metal segregation would have to occur at about 1% partial melting.

This seems physically unlikely, although essentially no information on this problem exists. Segregation of metal depends on many variables, including the metal and silicate grain size, the density difference, and the time available. The melting behavior of the metal is also a factor—one that depends on the S, P, and Ni content of the metal.

Despite the excellent correspondence between the calculated and observed siderophile depletions in Figs. 4 and 5, a simple chondritic model for the Moon has several difficulties. One problem is the Ni content of metal in the Moon. The calculations described above assume low oxygen fugacities that are consistent with metal low in Ni. Following Prior's Rule for chondrites, however, the low metal content of the Moon would imply metal with a high Ni content, reflecting more oxidized conditions (Morgan et al., 1978). For example, LL chondrites having 5% metal have Ni concentrations of 20% in the metal (Kerridge, 1977). However, there is no a priori reason for expecting the chondritic relationship to hold for the Moon. In fact, Prior's Rule does not hold strictly even for the ordinary chondrites (Wasson, 1974). A reasonable picture can be obtained assuming a CI chondritic Fe/Ni ratio in the Moon, since only 12% Ni in the metal core is required for a 5.5% lunar core (Newsom, 1984). A Ni content of only 12% would be consistent within a factor of two of the expected concentrations of Ni in the most primitive lunar volcanic glasses (Delano, 1986).

If the Ni content of the lunar metal is very high ($\simeq$ 40%), as argued by Ringwood and Seifert (1986), the siderophile depletions could still be achieved for an independent origin if the FeO content was as low as in the Earth's mantle, i.e., for very reducing conditions (Newsom and Drake, 1982a).

b. Variations on an independent lunar origin. Current calculations suggest that capture of a Moon that formed independent of the Earth in heliocentric orbit is very unlikely (Kaula and Harris, 1975), although not impossible (Conway, 1982, 1984). The need to deplete the Moon in volatile elements, siderophile elements, and Fe metal relative to the Earth has led to theories involving formation of the Moon from material that was not processed through the Earth's mantle. One model is a coaccretion model in which the Moon formed from the silicate mantles of differentiated asteroids captured in orbit around the Earth (Weidenschilling et al., 1986; Wasson and Warren, 1984). The depletion of volatiles and perhaps siderophiles is accomplished during the melting of the asteroidal precursors rather than melting of the Moon itself, as in the basaltic eucrite meteorites, which are thought to come from the asteroid Vesta (McCord et al., 1970; Consolmagno and Drake, 1977).

Accretion of the Moon in orbit around the Earth is difficult because of angular momentum problems (Harris, 1977), but capture of material at large geocentric distances together with viscous transport of angular momentum could allow coaccretion to occur (Herbert et al., 1985). To achieve the capture of planetesimals, collision with a circumterrestrial swarm of material may be necessary. Disruption of precursor bodies by a simple passage inside the Earth's Roche limit is not likely (Mizuno and Boss, 1984). With the coaccretion model some unknown contribution to the Moon from the Earth's mantle can be expected, since large impacts were occurring at this time. Material thrown off the Earth's mantle by impacts may even be required to establish the circumterrestrial swarm.

Another feature of the coaccretion model is the possibility that some filtering of the metal cores from the silicate mantles of the disrupted asteroids could occur. This mechanism would allow the Moon to form without a metal core, but still be depleted in siderophile elements (Chapman and Greenberg, 1984). If some metal is not accreted, the depletions could also be caused by an amount of Ni-rich metal much greater than 5.5%.

One of the objections to the coaccretion model is the need to maintain the circumterrestrial swarm for a long time to allow the filtering of metal cores to occur. However, if the metal content of the asteroidal precursors was only 5.5% or less, the filtering mechanism is not even needed. The maximum metal content of the Moon, based on geophysical data ($\simeq$5%; Hood, 1986), is much less than that of the Earth, but evidence exists that late accreting material on the Earth was low in Fe metal. Siderophile elements are not as depleted in the Earth's mantle as expected for equilibrium between the core and mantle. Current models for explaining the Earth's siderophile abundance pattern require a low Fe metal content in late accreting amterial to allow the relatively high siderophile concentrations to build up in the mantle (Wänke, 1981; Newsom and Palme, 1984a; Sun, 1984). Formation of the Moon by coaccretion requires late formation of the Moon after most of the Earth had accreted (Weidenschilling et al., 1986). Lange and Ahrens (1984) have also shown that late accretion of material onto the Earth, containing a full 30% metal, would have prevented the existence of free water at the Earth's surface.

An important difference between the eucrites and the Moon, which may be a difficulty for the coaccretion model, is that the Moon (and the Earth) are significantly depleted in Cr and Mn relative to chondrites, while the eucrites are not (Wänke and Dreibus, 1986). The reason for the depletion of these elements is not known for certain because they are not siderophile, although a link to the size of the depleted planet is suspected (Brey and Wänke, 1983). Because Mn is not chalcophile, having an S-rich metallic-liquid/silicate-liquid partition coefficient of about 0.01 at 1250°C and log fO_2 $\simeq$ –12.5 (Jones and Drake, personal communication, 1985), depletion of Mn by sulfide segregation is unlikely. Cr and Mn could also have been depleted by nebular processes in the material from which the Earth and Moon accreted. For example, the enstatite chondrites are significantly depleted in these elements relative to CI chondrites (Mason, 1979).

Morgan et al. (1978) suggested that a metal-sulfide core in the Moon is compatible with an independent origin for the Moon. The lower partition coefficients for siderophiles between S-rich metallic liquid and silicate melt (Jones and Drake, 1983,

1985) requires a much larger S-rich core of at least 10 wt %, assuming chondritic initial abundances. Therefore, segregation of an S-rich core in the present Moon could not explain the entire siderophile depletion, although segregation of some S-rich metallic liquid in addition to Fe-Ni metal is possible.

c. Origin from the Earth's mantle: Endmember model. Formation of the Moon from the Earth's mantle has been advocated by several people over the years. In 1975, Hartmann and Davis suggested that impact of a small planet with the Earth could have ejected enough material from the Earth's mantle to form the Moon. A possible similarity in the siderophile element abundance pattern for the Earth's mantle and the Moon led Ringwood and coworkers to a terrestrial origin model (Ringwood and Kesson, 1977; Delano and Ringwood, 1978; Ringwood, 1979). A classical fission origin has similar geochemical implications (Binder, 1984), but is considered unlikely from dynamical considerations (Boss and Peale, 1986).

Figure 6 illustrates the possibility of explaining the depletion of siderophile elements with segregation of a small metal core in a totally molten Moon, assuming the Moon starts out with the same ratio of siderophile elements to refractory elements as the Earth. As with chondritic initial abundances (Fig. 3), the observed depletions do not match the calculated depletions for the totally molten model. This probably eliminates the possibility that the depletions were caused by a small amount of metal sinking through a totally molten magma ocean. A much better fit to a calculated depletion trend is seen in Fig. 7, with segregation of 0.2% metal occurring at 20% partial melting. A range of 0.1–0.4% metal would fit the data reasonably well. Rhenium and especially Ir fall off the calculated trend, although these elements have the greatest uncertainties in their depletions and partition coefficients.

Of some concern with this model is the need for quantitative segregation of such a small amount of metal. Even slight variations in the amount of metal segregation would result in large variations in the siderophile concentrations. Perhaps a homogenization event occurred after metal segregation, such as convection in the magma ocean.

Segregation of a sulfur-rich metallic liquid could explain the depletion of siderophiles in the Moon relative to the Earth's mantle (Brett, 1973). A metallic liquid with approximately 25 wt % S is formed near the Fe-FeS eutectic if sufficient S and Fe-metal are present (Jones and Drake, 1983). From the calculation illustrated in Fig. 8, approximately 0.6% metallic liquid is needed to explain the depletions, and a range from 0.2–1.2% metallic liquid might be consistent with the data.

d. Variations on a terrestrial mantle origin. The siderophile element depletions observed in the Moon can be explained by the segregation of a small amount of metal or metallic liquid, if the Moon formed out of the Earth's mantle. Simple mechanisms for forming the Moon from the Earth's mantle, such as fission, are probably not plausible, as shown by recent calculations (Boss and Peale, 1986). The impact of a small planet with the Earth is currently seen as the most likely way to form the Moon with a large component of the Earth's mantle (Hartmann,

1986). The fraction of the Earth's mantle compared to the fraction of the impacting body that finally makes up the Moon, however, is still in doubt. The main problem is the amount of angular momentum transferred from an obliquely impacting body to material from the Earth's mantle (Cameron, 1985). If a large fraction of the material making up the Moon comes from the impacting Mars-sized planet (Cameron, 1985), we would instead be dealing with an independent origin model. In any impact model the depletion of siderophiles could occur in the Earth's mantle, in the impacting body, and in the Moon itself after its formation. This kind of model does have advantages for explaining some of the differences between the Earth and the Moon, such as the higher FeO content of the lunar silicates that could have come from the impacting body (Wänke and Dreibus, 1983). Because of the large number of variables, however, the overall depletion of the siderophile elements in the Moon cannot be used to test this hypothesis and the actual amount of metal in the Moon would be less than the amount of metal responsible for the siderophile element depletion in the Moon.

The question of a core in the Moon is, unfortunately, still in some doubt from the perspective of the siderophile elements. The significant depletion of siderophiles in the lunar silicates, even relative to the Earth's mantle, strongly suggests that at least some metal segregation has occurred in the Moon. The existing models for the origin of the Moon, however, leave open the possibility that some or all of the metal required to explain the siderophile depletions was not incorporated into the Moon. However, some geophysical data, including paleomagnetic data (Cisowski *et al.*, 1983; Runcorn, 1983), strongly favor the existence of a relatively large Fe,Ni metal core. Because an independent origin model requires a large core, and this model actually fits the siderophile data and geophysical data the best, an independent origin may be the closest to being true.

Conclusions

1. New data indicate that the abundance of Mo in lunar samples is much lower than in the Earth's mantle (factor of 27) and much lower than Mo in chondrites (factor of 1200). Because Mo is a refractory element, the depletion of Mo could not be due to the general loss of volatile elements observed in the Moon.

2. The overall depletion pattern for siderophiles in the Moon has now been relatively well established. Weakly siderophile elements such as W are depleted by about the same amount in the Earth's mantle and in lunar silicates, compared to chondritic abundances. Highly siderophile elements, however, are more depleted in lunar silicates compared to the Earth's mantle.

3. For an independent origin of the Moon, the siderophile depletions can be obtained by segregation of approximately 5 wt % metal in the Moon. A simple capture model is unlikely, but coaccretion of the Moon and the Earth from previously

differentiated planetesimals could explain many of the differences between the Earth and the Moon, such as the lower metal and volatile element content of the Moon.

4. For an origin of the Moon primarily out of material from the Earth's mantle, the additional siderophile depletion relative to the Earth's mantle could have occurred in the Moon by segregation of as little as 0.1 to 0.4 wt % metal or as much as 0.4 to 1.2 wt % S-rich metallic liquid.

5. For the impact theory, where the collision of a Mars-sized object with the Earth is required, the Moon could contain large fractions of material from the impacting body and from the Earth's mantle. Some of the lunar siderophile depletion, therefore, may have occurred in the Earth and some in the impacting body.

6. From the depletion of siderophile elements, including Mo, and from the present knowledge of metal-silicate and mineral-melt partition coefficients, the metal segregation almost certainly occurred at low degrees of partial melting of the silicates (5% to 20%), for either an independent or a terrestrial lunar origin. Formation of a large magma ocean subsequent to the metal segregation is not ruled out.

7. In general, there is some consensus about the compositional similarities and differences between the Earth and the Moon, but there is large disagreement about the significance of the chemical relationship. If the Moon formed out of the Earth's mantle, several important mechanisms must have operated to create the differences, such as the siderophile depletions, volatile depletions and enrichments, and other properties (Taylor, 1984; Kreutzberger *et al.*, 1984).

8. Because the currently plausible theories for the origin of the Moon allow for loss of metal from lunar precursor material before the formation of the Moon, the abundances of siderophile elements, while consistent with a large ($\approx$5%) lunar core, do not require that the Moon have a core.

Future Work

What lies ahead for studies of siderophile elements and the lunar core? Additional studies of siderophile depletions in the Moon, as well as studies under lunar conditions of the partition coefficients, will help clarify problems such as the degree of partial melting during metal segregation. The largest advance will come from a definitive geophysical measurement of the size of the lunar core. A very large core ($\approx$5%) would be difficult to reconcile with a terrestrial origin for the Moon, because of the relatively small amount of metal (<1%) needed to explain the differences in the siderophiles between the Earth and Moon. Another question is the possible existence of deep lunar reservoirs that did not experience the same siderophile depletion events. The existence of such reservoirs would point to a cold accretion of the Moon, providing an additional constraint on the origin of the Moon. Clearly a return to the Moon and acquisition of additional lunar samples will be of great importance to furthering these studies.

Acknowledgments. This study could not have been done without the guidance of Dr. H. Palme and the generous support of Prof. H. Wänke. I wish to thank B. Spettel for help with the details of the radiochemistry and W. Rammensee with the metal-silicate extraction technique. The analytical work was done while the author was a postdoctoral fellow at the Max-Planck-Institut für Chemie in Mainz. Significant improvements to the manuscript resulted from careful reviews by M. J. Drake and A. Binder, and comments from E. R. D. Scott and G. J. Taylor. The samples were activated at the TRIGA reactor of the Institut für Anorganische Chemie and Kernchemie der Universität Mainz and at the Kernforschungsanlage Jülich. We wish to thank the reactors staffs. Support for writing this article was provided by NASA grant NAG-9-30 (Klaus Keil, principal investigator).

References

Anders E. (1977) Chemical compositions of the moon, earth and eucrite parent body. *Phil. Trans. Roy. Soc. London, A285*, 23–40.

Anders E. (1978) Procrustean science: indigenous siderophiles in the lunar highlands, according to Delano and Ringwood. *Proc. Lunar Planet. Sci. Conf. 9th*, pp. 161–184.

Binder A. B. (1984) On the origin of the Moon by rotational fission (abstract). In *Papers Presented to the Conference on the Origin of the Moon*, p. 47. Lunar and Planetary Institute, Houston.

Boss A. P. and Peale S. J. (1986) Dynamical constraints on the origin of the Moon, this volume.

Bouchet M., Kaplan G., Voudon A., and Bertoletti M. J. (1971) Spark mass spectrometric analysis of major and minor elements in six lunar samples. *Proc. Lunar Sci. Conf. 2nd*, pp. 1247–1252.

Brett R. (1973) A lunar core of Fe-Ni-S. *Geochim. Cosmochim. Acta, 37*, 165–170.

Brey G. and Wänke H. (1983) Partitioning of Cr, Mn, V and Ni between Fe melt, magnesiowüstite and olivine at high pressures and temperatures (abstract). In *Lunar and Planetary Science XIV*, pp. 71–72. Lunar and Planetary Institute, Houston.

Cameron A. G. W. (1985) Formation of the prelunar accretion disk. *Icarus, 62*, 319–327.

Chapman C. R. and Greenberg R. (1984) A circumterrestrial compositional filter (abstract). In *Papers Presented to the Conference on the Origin of the Moon*, p. 56. Lunar and Planetary Institute, Houston.

Chou C. -L., Shaw D. M., and Crocket J. H. (1983) Siderophile trace elements in the earth's oceanic crust and upper mantle. *Proc. Lunar Planet. Sci. Conf. 13th*, in *J. Geophys. Res., 88*, A507–A518.

Cisowski S. M. and Fuller M. (1986) Lunar paleointensities via the IRMs normalization method and the early history of the moon, this volume.

Cisowski S. M., Collinson D. W., Runcorn S. K., Stephenson A., and Fuller M. (1983) A review of lunar paleointensity data and implications for the origin of lunar magnetism. *Proc. Lunar Planet. Sci. Conf. 13th*, in *J. Geophys. Res., 88*, A691–A704.

Consolmagno G. J. and Drake M. J. (1977) Composition and evolution of the eucrite parent body: evidence from rare earth elements. *Geochim. Cosmochim. Acta, 41*, 1271–1282.

Conway B. A. (1982) On the history of the lunar orbit. *Icarus, 51*, 610–622.

Conway B. A. (1984) The moon's orbit history and inferences on its origin (abstract). In *Papers Presented to the Conference on the Origin of the Moon*, p. 33. Lunar and Planetary Institute, Houston.

Capobianco C. J. and Watson E. B. (1982) Olivine/silicate melt partitioning of germanium: an example of a nearly constant partition coefficient. *Geochim. Cosmochim. Acta, 46*, 235–240.

Delano J. W. (1985) Mare volcanic glasses II: Abundances or trace Ni and the composition of the moon (abstract). In *Lunar and Planetary Science XVI*, pp. 179–180. Lunar and Planetary Institute, Houston.

Delano J. W. (1986) Abundances of cobalt, nickel, and volatiles in the silicate portion of the Moon, this volume.

Delano J. W. and Ringwood A. E. (1978) Siderophile elements in the lunar highlands: Nature of the indigenous component and implications for the origin of the moon. *Proc. Lunar Planet. Sci. Conf. 9th*, pp. 111–159.

Dickinson T. and Newsom H. E. (1985) A possible test of the impact theory for the origin of the moon (abstract). In *Lunar and Planetary Science XVI*, pp. 183–184. Lunar and Planetary Institute, Houston.

Dolginov Sh. Sh. (1985) On the problem of lunar paleomagnetism (abstract). In *Lunar and Planetary Science XVI*, pp. 191–192. Lunar and Planetary Institute, Houston.

Drake M. J. (1983) Geochemical constraints on the origin of the moon. *Geochim. Cosmochim. Acta, 47*, 1759–1767.

Drake M. J. (1986) Is lunar bulk material similar to Earth's mantle?, this volume.

Fegley B., Jr. and Palme H. (1985) Evidence for oxidizing conditions in the solar nebula from Mo and W depletions in refractory inclusions in carbonaceous chondrites. *Earth Planet. Sci. Lett., 72*, 311–326.

Harris A. W. (1977) An analytical theory of planetary rotation rates. *Icarus, 31*, 168–174.

Hartmann W. K. (1986) Moon origin: The impact trigger hypothesis, this volume.

Hartmann W. K. and Davis D. R. (1975) Satellite-sized planetesimals and lunar origin. *Icarus, 24*, 504–515.

Herbert F., Davis D. R., and Weidenschilling S. J. (1985) On forming the moon in geocentric orbit; dynamical evolution of a circumterrestrial swarm (abstract). In *Lunar and Planetary Science XVI*, pp. 341–342. Lunar and Planetary Institute, Houston.

Hood L. L. (1986) Geophysical constraints on the lunar interior, this volume.

Hood L. L. and Jones J. H. (1985) Lunar density models consistent with mantle seismic velocities and other geophysical constraints (abstract). In *Lunar and Planetary Science XVI*, pp. 360–361. Lunar and Planetary Institute, Houston.

Irving A. J. (1978) A review of experimental studies of crystal/liquid trace element partitioning. *Geochim. Cosmochim. Acta, 42*, 743–770.

Jagoutz E., Palme H., Baddenhausen K., Blum K., Cendales M., Dreibus G., Spettel B., Lorenz V., and Wänke H. (1979) The abundances of major, minor and trace elements in the earth's mantle as derived from primitive ultramafic nodules. *Proc. Lunar Planet. Sci. Conf. 10th*, pp. 2013–2050.

Janssens M. J., Palme H., Hertogen J., Anderson A. T., and Anders E. (1978) Meteoritic material in lunar highlands samples from the Apollo 11 and 12 sites. *Proc. Lunar Planet. Sci. Conf. 9th*, pp. 1537–1550.

Jones J. H. and Drake M. J. (1983) Experimental investigations of trace element fractionation in iron meteorites, II: The influence of sulfur. *Geochim. Cosmochim. Acta, 47*, 1199–1209.

Jones J. H. and Drake M. J. (1985) Experiments bearing on the formation and primordial differentiation of the earth (abstract). In *Lunar and Planetary Science XVI*, pp. 412–413. Lunar and Planetary Institute, Houston.

Kaula W. M. and Harris A. W. (1975) Dynamics of lunar origin and orbital evolution. *Rev. Geophys. Space Phys., 13*, 363–371.

Kerridge J. F. (1977) Iron, whence it came, where it went. *Space Sci. Rev., 20*, 3–68.

Kharkar D. P. and Turekian K. K. (1971) Analyses of Apollo 11 and Apollo 12 rocks and soils by neutron activation. *Proc. Lunar Sci. Conf. 2nd*, pp. 1301–1305.

Kimura K., Lewis R. S., and Anders E. (1974) Distribution of gold and rhenium between nickel-iron and silicate melts: Implications for the abundance of siderophile elements on the earth and moon. *Geochim. Cosmochim. Acta, 38*, 683–701.

Kreutzberger E., Drake M. J., and Jones J. H. (1984) Origin of the moon: constraints from volatile elements (abstract). In *Papers Presented to the Conference on the Origin of the Moon*, p. 22. Lunar and Planetary Institute, Houston.

Lange M. A. and Ahrens T. J. (1984) FeO and H_2O and the homogeneous accretion of the earth. *Earth Planet Sci. Lett., 71,* 111–119.

Mason B. (1979) Data of Geochemistry, Sixth edition, Part 1. Meteorites. *U.S. Geol. Surv. Prof. Paper 440-B-1.*

McCord T. B., Adams J. B., and Johnson T. V. (1970) Asteroid Vesta: spectral reflectivity and compositional implications. *Science, 168,* 1445–1447.

Mizuno H. and Boss A. P. (1984) Tidal disruption and the origin of the moon (abstract). In *Papers Presented to the Conference on the Origin of the Moon,* p. 37. Lunar and Planetary Institute, Houston.

Morgan J. W., Hertogen J., and Anders E. (1978) The moon: composition determined by nebular processes. *Moon and Planets, 18,* 465–478.

Morrison G. H., Gerard J. T., Kashuba A. T., Gangadharam E. V., Rothenberg A. M., Potter N. M., and Miller G. B. (1970) Elemental abundances of lunar soil and rocks. *Proc. Apollo 11 Lunar Sci. Conf.,* pp. 1383–1392.

Morrison G. H., Gerard J. T., Potter N. M., Gangadharam E. V., Rothenberg A. M., and Burdo R. A. (1971) Elemental abundance of lunar soil and rocks from Apollo 12. *Proc. Lunar Sci. Conf. 2nd,* pp. 1169–1185.

Newsom H. E. (1984) The lunar core and the origin of the Moon. *EOS (Trans. Am. Geophys. Union), 65,* 369–370.

Newsom H. E. (1985) Molybdenum in eucrites: evidence for a metal core in the eucrite parent body. *Proc. Lunar Planet. Sci. Conf. 15th,* in *J. Geophys. Res., 90,* C613–C617.

Newsom H. E. and Drake M. J. (1982a) Constraints on the moon's origin from the partitioning behavior of tungsten. *Nature, 297,* 210–212.

Newsom H. E. and Drake M. J. (1982b) The metal content of the eucrite parent body: constraints from the paritioning behavior of tungsten. *Geochim. Cosmochim. Acta, 46,* 2483–2489.

Newsom H. E. and Drake M. J. (1983) Experimental investigation of the partitioning of phosphorus between metal and silicate phases: Implications for the Earth, Moon and eucrite parent body. *Geochim. Cosmochim. Acta, 47,* 93–100.

Newsom H. E. and Palme H. (1984a) The depletion of siderophile elements in the Earth's mantle: new evidence from molybdenum and tungsten. *Earth Planet. Sci. Lett., 69,* 354–364.

Newsom H. E. and Palme H. (1984b) The determination of molybdenum in geological samples by neutron activation analysis. *J. Radioanal. Nucl. Chem. Lett., 87,* 273–282.

Newsom H. E., White W. M., and Jochum K. P. (1985) Did the earth's core grow through geological time? (abstract). In *Lunar and Planetary Science XVI,* pp. 616–617. Lunar and Planetary Institute, Houston.

Palme H. and Rammensee W. (1981a) The significance of W in planetary differentiation processes: Evidence from new data on eucrites. *Proc. Lunar Planet. Sci. 12B,* pp. 949–964.

Palme H. and Rammensee W. (1981b) The cosmic abundance of molybdenum. *Earth Planet. Sci. Lett., 55,* 356–362.

Palme H., Baddenhausen H., Blum K., Cendales M., Dreibus G., Hofmeister H., Kruse H., Palme C., Spettel B., Vilcsek E., and Wänke H. (1978) New data on lunar samples and achondrites and a comparison of the least fractionated samples from the earth, moon and the eucrite parent body. *Proc. Lunar Planet. Sci. Conf. 9th,* pp. 25–57.

Palme H., Suess H. E., and Zeh H. D. (1981) Abundances of the elements in the solar system. In *Landolt-Bornstein, VI, 2, Pt. A,* pp. 257–272. Springer-Verlag, New York.

Rammensee W. (1978) Verteilungsgleichgewichte von Spurenelementen zwischen Metallen und Silikaten. Ph.D. thesis, Mainz University, F.R. Germany. 160 pp.

Rammensee W. and Palme H. (1982) Metal-silicate extraction technique for the analysis of geological and meteoritic samples. *J. Radioanal. Chem., 71,* 401–418.

Rammensee W. and Wänke H. (1977) On the partition coefficient of tungsten between metal and silicate and its bearing on the origin of the Moon. *Proc. Lunar Sci. Conf. 8th,* pp. 399–409.

Ringwood A. E. (1979) *Origin of the Earth and Moon.* Springer-Verlag, New York. 295 pp.

Ringwood A. E. and Kesson S. E. (1977) Basaltic magmatism and the bulk composition of the moon, 2. Siderophile and volatile elements in moon, earth and chondrites: Implications for lunar origin. *The Moon, 16,* 425–464.

Ringwood A. E. and Seifert S. (1986) Nickel-cobalt abundance systematics and their bearing on lunar origin, this volume.

Ringwood A. E., Kesson S. E., and Hibberson W. (1981) Rhenium depletion in mare basalts and redox state of the lunar interior (abstract). In *Lunar and Planetary Science XII,* pp. 891–893. Lunar and Planetary Institute, Houston.

Runcorn S. K. (1983) Lunar paleomagnetism, polar displacements and primeval lunar satellites in the earth-moon system. *Nature, 304,* 589–596.

Russell C. T. (1984) On the Apollo subsatellite evidence for a lunar core (abstract). In *Papers Presented to the Conference on the Origin of the Moon,* p. 7. Lunar and Planetary Institute, Houston.

Russell C. T., Coleman P. J., Jr., and Goldstein B. E. (1981) Measurements of the lunar induced magnetic moment in the geomagnetic tail: evidence for a lunar core? *Proc. Lunar Planet. Sci. 12B,* pp. 831–836.

Schmitt W. (1984) Experimentelle Bestimmung von Metall/Sulfid/Silikat-Verteilungskoeffizienten Geochemisch Relevanter Spurenelemente. Ph.D. thesis, Mainz University, F.R. Germany. 102 pp.

Sun S. -S. (1984) Geochemical characteristics of archean ultramafic and mafic volcanic rocks: implications for mantle composition and evolution. In *Archean Geochemistry* (A. Kroner, G. N. Hanson, and A. M. Goodwin, eds.), pp. 25–46. Springer-Verlag, Berlin.

Taylor S. R. (1984) Tests of the lunar fission hypothesis (abstract). In *Papers Presented to the Conference on the Origin of the Moon,* p. 25. Lunar and Planetary Institute, Houston.

Taylor S. R., Rudowski R., Muir R., Graham A., and Kaye M. (1971) Trace element chemistry of lunar samples from the ocean of storms. *Proc. Lunar Sci. Conf. 2nd,* pp. 1083–1099.

Wänke H. (1981) Constitution of terrestrial planets. *Phil. Trans. Roy. Soc. London, A303,* 287–302.

Wänke H. and Dreibus G. (1983) The origin of the moon (abstract). *Fortschr. Mineral., 61,* 215–216.

Wänke H. and Dreibus G. (1986) Geochemical evidence for the formation of the Moon by impact induced fission of the proto-Earth, this volume.

Wänke H., Baddenhausen H., Balacescu A., Teschke F., Spettel B., Dreibus G., Palme H., Quijano-Rico M., Kruse H., Wlotzka F., and Begemann F. (1972) Multielement analyses of lunar samples and some implications of the results. *Proc. Lunar Sci. Conf. 3rd,* pp. 1251–1268.

Wänke H., Palme H., Baddenhausen H., Dreibus G., Jagoutz E., Kruse H., Palme C., Spettel B., Teschke F., and Thacker R. (1975) New data on the chemistry of lunar samples: Primary matter in the lunar highlands and the bulk composition of the moon. *Proc. Lunar Sci. Conf. 6th,* pp. 1313–1340.

Wänke H., Rieder R., Baddenhausen H., Spettel B., Teschke F., Quijano-Rico M., and Balacescu A. (1970) Major and trace elements in lunar material. *Proc. Apollo 11 Lunar Sci. Conf.,* pp. 1719–1727.

Warren P. H. (1986) The bulk moon MgO/FeO ratio: A highlands perspective, this volume.

Wasson J. T. (1974) *Meteorites.* Springer-Verlag, Berlin.

Wasson J. T. and Warren P. H. (1984) The origin of the moon (abstract). In *Papers Presented to the Conference on the Origin of the Moon,* p. 57. Lunar and Planetary Institute, Houston.

Weckwerth G., Spettel B., and Wänke H. (1983) Phosphorus in the mantle of planetary bodies (abstract). *Terra Cognita, 3,* 79–80.

Weidenschilling S. J., Greenberg R., Chapman C. R., Herbert F., Davis D. R., Drake M. J., Jones J., and Hartmann W. K. (1986) Origin of the moon from a circumterrestrial disk, this volume.

Wolf R., Woodrow A., and Anders E. (1979) Lunar basalts and pristine highlands rocks: Comparison of siderophile and volatile elements. *Proc. Lunar Planet. Sci. Conf. 10th,* pp. 2107–2130.

Yoder C. F. (1981) The free librations of a dissipative moon. *Phil. Trans. Roy. Soc. London, A303*, 327–338.

Yoder C. F. (1984) The size of the lunar core (abstract). In *Papers Presented to the Conference on the Origin of the Moon*, p. 6. Lunar and Planetary Institute, Houston.

Abundances of Cobalt, Nickel, and Volatiles in the Silicate Portion of the Moon

JOHN W. DELANO

Department of Geological Sciences, State University of New York, Albany, NY 12222

The trace abundances of Ni have been determined in 20 high-Mg magmas produced by partial melting of the lunar mantle. The silicate portion of the Moon is found to contain 470 ± 50 ppm Ni. Relative to magnesium, nickel is depleted by a factor of 50 ± 6 compared to CI chondrites and by 4.0 ± 0.5 compared to the present Earth's upper mantle. The abundance of Co in the silicate portion of the Moon is estimated to be 90 ± 5 ppm. This yields a depletion factor for Co, relative to magnesium, of 12 ± 1 compared to CI chondrites and 1.05 ± 0.10 compared to the present Earth's upper mantle. The nonchondritic Ni/Co ratio of 5 ± 1 in the Moon's silicate portion was generated by segregation of a small metal/sulfide core. Associated with the eruption of the high-Mg magmas was the release of indigenous volatile elements. Data suggest that these volatiles were derived from primordial blocks of debris from the outer solar system that became entrained in the circumterrestrial accretion disk. The survival of these reservoirs within the lunar interior places constraints on the Moon's thermal history.

1. Introduction

In this study, lunar glasses of volcanic origin (henceforth the designated "pristine" glasses) have been used to estimate the abundances of cobalt and nickel in the silicate portion of the Moon. These pristine glasses are quenched samples of high-Mg magmas of mare affinity that were erupted onto the lunar surface in fire fountains (e.g., Heiken *et al.*, 1974). The source regions of these high-Mg magmas were mafic cumulates in the lunar mantle, which had formed by crystal/liquid fractionation of the magma ocean and were chemically complementary to the highlands crust (e.g., Wood *et al.*, 1970; Brett, 1973a).

Contained within the collection of 25 known varieties of pristine glass (Table 1) are some of the most chemically primitive, magmatic compositions returned from

TABLE 1. Compilation of 25 Varieties of Mare Volcanic Glass Arranged According to Increasing Abundance of TiO_2 (wt %).

	(1)	(2)	(3)	(4)	(5)	(6)	(7)	(8)	(9)	(10)	(11)	(12)	(13)
SiO_2	48.0	45.5	43.9	46.0	45.1	45.2	44.8	46.0	43.7	45.3	44.3	44.1	42.9
TiO_2	0.26	0.38	0.39	0.40	0.41	0.43	0.45	0.55	0.57	0.66	0.91	0.97	3.48
Al_2O_3	7.74	7.75	7.83	7.92	7.43	7.44	7.14	9.30	7.96	9.60	6.89	6.71	8.30
Cr_2O_3	0.57	0.56	0.39	0.55	0.55	0.54	0.54	0.58	0.46	0.40	n.a.	0.56	0.59
FeO	16.5	19.7	21.9	19.1	20.3	19.8	19.8	18.2	21.5	19.6	20.2	23.1	22.1
MnO	0.19	0.22	0.24	n.a.	0.22	0.22	0.24	0.21	n.a.	0.26	0.23	0.28	0.27
MgO	18.2	17.2	16.9	17.2	17.6	18.3	19.1	15.9	17.0	15.0	19.5	16.6	13.5
CaO	8.57	8.65	8.44	8.75	8.43	8.15	8.03	9.24	8.44	9.40	7.40	7.94	8.50
Na_2O	n.d.	n.d.	n.d.	n.d.	n.d.	n.d.	0.06	0.11	n.d.	0.27	0.10	n.d.	0.45
K_2O	n.d.	n.d.	n.d.	n.d.	n.d.	n.d.	0.03	0.07	n.d.	0.04	n.d.	n.d.	n.d.

	(14)	(15)	(16)	(17)	(18)	(19)	(20)	(21)	(22)	(23)	(24)	(25)
SiO_2	40.8	40.5	39.4	38.5	37.9	38.8	37.3	39.2	35.6	35.6	34.0	33.4
TiO_2	4.58	6.90	8.63	9.12	9.12	9.30	10.0	12.5	13.8	15.3	16.4	16.4
Al_2O_3	6.16	8.05	6.21	5.79	5.63	7.62	5.68	5.69	7.15	4.81	4.6	4.6
Cr_2O_3	0.41	0.63	0.67	0.69	0.65	0.66	0.63	0.86	0.77	n.a.	0.92	0.84
FeO	24.7	22.3	22.2	22.9	23.7	22.9	23.7	22.2	21.9	23.7	24.5	23.9
MnO	0.30	0.25	0.28	n.a.	n.a.	0.29	n.a.	0.31	0.25	n.a.	0.31	0.30
MgO	14.8	12.6	14.7	14.9	14.9	11.6	14.3	14.5	12.1	13.0	13.3	13.0
CaO	7.74	8.64	7.53	7.40	7.41	8.55	7.62	7.04	7.89	6.49	6.9	6.27
Na_2O	0.42	0.39	0.41	0.38	0.36	0.39	0.31	0.28	0.49	0.50	0.23	0.05
K_2O	0.10	n.d.	0.04	n.d.	n.d.	n.d.	n.d.	0.29	0.12	n.d.	0.16	0.12

(1) Apollo 15 green C
(2) Apollo 15 green A
(3) Apollo 16 green
(4) Apollo 15 green B
(5) Apollo 15 green D
(6) Apollo 15 green E
(7) Apollo 14 green B
(8) Apollo 14 VLT
(9) Apollo 11 green
(10) Apollo 17 VLT (Warner et al., 1979)
(11) Apollo 17 green (preliminary)
(12) Apollo 14 green A
(13) Apollo 15 yellow

(14) Apollo 14 yellow
(15) Apollo 17 yellow
(16) Apollo 17 orange
(17) Apollo 17 orange (74220-type)
(18) Apollo 15 orange
(19) Apollo 17 orange
(20) Apollo 11 orange
(21) Apollo 14 orange
(22) Apollo 15 red
(23) Apollo 14 red (preliminary)
(24) Apollo 14 black
(25) Apollo 12 red (Marvin and Walker, 1978)

These are the most primitive compositions within each volcanic group.
n.a. = not analyzed; n.d. = not detected.

the Moon. Among these pristine glasses, those with $TiO_2 \leq 1.0$ wt % are the most primitive because they exhibit the following characteristics: (1) the highest Mg/(Mg + Fe) ratios (Fig. 1); (2) the lowest abundances of and least fractionated ratios among the nonvolatile, incompatible elements (Fig. 2); and (3) the highest abundances

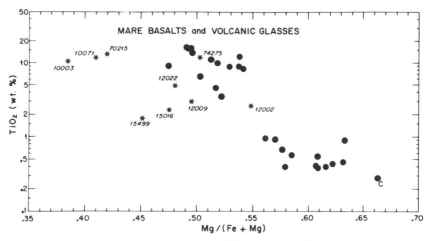

Fig. 1. The Mg/(Mg + Fe) molar ratios in mare basalts (star symbols) and pristine lunar glasses (solid circles, Table 1) have been plotted against TiO₂ (wt %). Each symbol represents a distinct magmatic composition. Note that for any given abundance of Ti the pristine glasses commonly have higher Mg/(Mg + Fe) ratios. Those pristine glasses with TiO₂ ≤ 1.0 wt % are chemically the most primitive (e.g., Delano, 1985) and are the source of constraints discussed in this article. The pristine glass labelled "C" (Apollo 15 green glass, group C) in the lower right corner of this figure is a uniquely peculiar composition (e.g., Delano, 1979) that has been excluded from consideration.

of the compatible elements such as cobalt and nickel. Note also that compared to the fine-grained (i.e., noncumulate) mare basalts the pristine glasses commonly have (1) higher Mg/(Mg + Fe) ratios at any given abundance of Ti (Fig. 1) and (2) higher abundances of nickel (Fig. 3). These chemically primitive aspects exhibited by the pristine lunar glasses demonstrate that these high-Mg magmas underwent little or no crystal/liquid fractionation during ascent from their source regions and therefore appear to be "primary." As a result of this simple emplacement history, the chemistries of these magmas furnish important information about the lunar mantle. The data from the pristine lunar glasses (Delano, 1985) reinforce and extend the constraints obtained from detailed analyses of the crystalline mare basalts. It will be shown in this paper that the low-Ti pristine glasses can be used to constrain the abundances of cobalt and nickel in the silicate portion of the Moon.

2. Nickel

2.1. Geochemical behavior

Nickel exhibits both lithophile and siderophile characteristics. When present in a planet as NiO, its behavior is similar to that of MgO. Demonstration of this

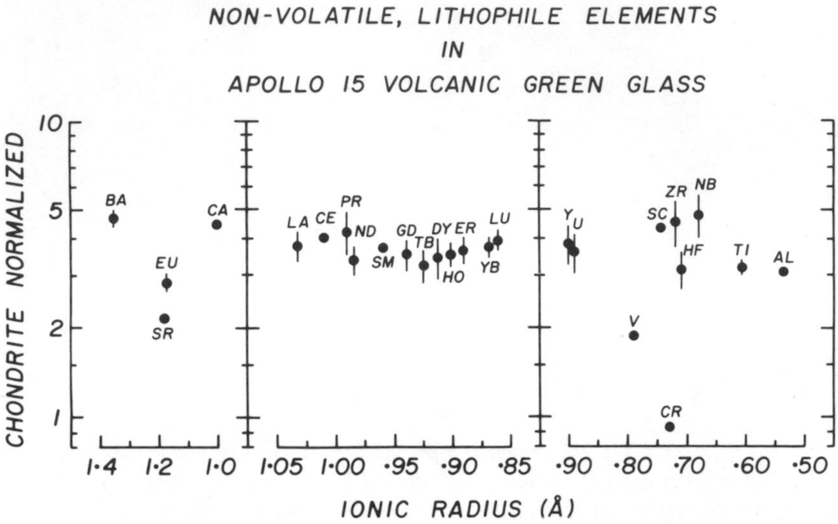

NON-VOLATILE, LITHOPHILE ELEMENTS
IN
APOLLO 15 VOLCANIC GREEN GLASS

Fig. 2. The chondrite-normalized abundances of the nonvolatile, lithophile elements in Apollo 15 green pristine glass have been plotted against ionic radius. In this primitive lunar magma, most of these elements occur in nearly primordial ratios at abundances of $4 \pm 1 \times CI$ chondrites. The negative anomalies evident in Eu + Sr and in V + Cr are indicative of plagioclase (e.g., Taylor et al., 1973) and spinel (e.g., Delano, 1979) fractionation, respectively, during formation of the cumulate source region. Data are from Ma et al. (1981), Taylor et al. (1973), and Wiesmann and Hubbard (1975).

fact is illustrated in Fig. 4 for terrestrial samples having a range in MgO abundances from ~4 wt % to ~50 wt %. Although NiO and MgO are correlated, the correlation is not linear. This nonlinearity is caused by the temperature- and composition-dependence of the nickel partition coefficient between olivine and liquid (e.g., Arndt, 1977; Hart and Davis, 1978; Irving, 1978; Nabelek, 1980). Specifically, the olivine/liquid partition coefficient for Ni increases with declining temperature and with declining MgO in the liquid.

In previous discussions on the abundance of Ni in the Moon's silicate portion, some authors have argued that nickel is unpredictable because it does not correlate with incompatible lithophile elements, such as La (Drake, 1983). However, inspection of the terrestrial data shown in Fig. 4 demonstrates that Ni behaves in a predictable manner when plotted against an element having similar geochemical characteristics (e.g., Mg). An obvious axiom would be the following: Elements having similar bulk partition coefficients in a geochemical system should be correlated (e.g., K and La; Fe and Mn; Ni and Mg).

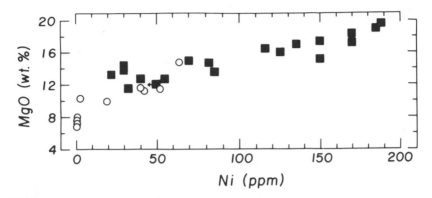

Fig. 3. The abundances of Ni and Mg are directly correlated in the mare basalts (open circles) and pristine lunar glasses (solid squares; Tables 1, 2). Note that the low-Ti, pristine glasses with Ni > 100 ppm have higher abundances of Mg and Ni than any mare basalt of noncumulate origin. This fact further emphasizes, in conjunction with the aspects illustrated in Figs. 1 and 2, the comparatively primitive nature of the low-Ti, pristine glasses. As mentioned in the caption to Fig. 1, the uniquely peculiar green glass from Apollo 15 (group C) has been deliberately omitted (see Delano, 1979). Analyses of noncumulate mare basalts (open circles) are from Compston et al. (1971), Duncan et al. (1974, 1976), and Willis et al. (1971).

2.2. Pristine lunar glasses

The abundances of trace Ni have been determined by electron microprobe in 20 varieties of pristine lunar glass (Table 2). A detection limit of about 20 ppm Ni was achieved using the procedures described in Chen *et al.* (1982), Delano and Livi (1981), and Delano (1979).

Figure 3 shows that Ni and Mg are strongly correlated among the pristine glasses and the mare basalts. This correlation demonstrates that Ni must have been present in these magmas principally as NiO in order for it to behave in a lithophile manner. This similar behavior for Ni and Mg among basaltic samples from the Earth and Moon permits a direct comparison to be made for the purpose of ascertaining the relative abundances of Ni in their respective mantles.

2.3. Abundance of nickel in the Moon

The terrestrial and lunar data are plotted in Fig. 4. The complete absence of any overlap between the two sets of data implies either (1) relative enrichment of Mg in the Moon or (2) relative depletion of Ni in the Moon. Since, as shown in Table 3, the estimated abundance of MgO in the Moon (33.5 ± 1.6 wt %;

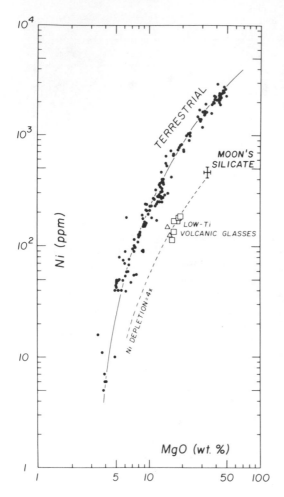

Fig. 4. Nickel and magnesium are observed to be directly correlated in terrestrial mid-ocean ridge basalts (MORB's), high-Mg basalts, basaltic komatiites, komatiites, lherzolites, and peridotites. The low-Ti, pristine lunar glasses (Table 1, 2) are constantly displaced from the terrestrial data, indicating a depletion of Ni in the Moon's silicate portion by a factor of 4.0 ± 0.5 compared to the Earth's upper mantle (refer to the discussion in Section 2 of this article). The terrestrial data are from the following sources: Arndt et al. (1977), Bougault and Hekinian (1974), Clague and Frey (1982), Dickey et al. (1977), Dietrich et al. (1978, 1981), Flanagan et al. (1976), Frey and Green (1974), Gunn (1971), Gunn et al. (1971), Jagoutz et al. (1979), Langmuir et al. (1977), Leeman et al. (1976), Nesbitt and Sun (1976), Stosch and Seck (1980), and Whitford and Arndt (1978).

1σ) is *less* than the value of 38 wt % observed in the Earth's upper mantle (e.g., Jagoutz *et al.*, 1979; Ringwood, 1977; Sun, 1982), option (1) is not applicable. The Moon's silicate portion must therefore be depleted in Ni relative to the terrestrial upper mantle. The dashed line through the cluster of lunar data marks a depletion factor (relative to Earth's upper mantle) of 4. The data shown in Table 4 indicate that the Moon's silicate portion is depleted in Ni by a factor of 4.0 ± 0.5 compared to the terrestrial upper mantle (at *constant* Mg). Since the Moon also appears to be slightly depleted in Mg (Table 3), the silicate portion of the Moon is inferred to contain 470 ± 50 ppm Ni. Relative to Mg, this represents a depletion of Ni by a factor of 50 ± 6 relative to CI chondrites (Table 4). These results are in agreement with the conclusions of Ringwood and Seifert (1986), Wänke and Dreibus

TABLE 2. *Abundances of Ni (ppm) and MgO (wt %) in Pristine Lunar Glasses Demarcated According to their TiO_2 Abundances.* *

Volcanic glass	Ni (ppm)	MgO (wt %)	Ni/Mg ($\times 10^3$)
($TiO_2 \leq 1.0$ wt %)			
Apollo 15 green A	170	17.2	1.64
Apollo 15 green B	150	17.2	1.45
Apollo 15 green E	170	18.3	1.54
Apollo 14 green B	185	19.1	1.61
Apollo 14 VLT	125	15.9	1.30
Apollo 11 green	135	17.0	1.32
Apollo 17 VLT	150	15.0	1.66
Apollo 17 green	188	19.5	1.60
Apollo 14 green A	115	16.6	1.15
Apollo 15 green C	90	18.2	0.82
($TiO_2 \geq 3.5$ wt %)†			
Apollo 15 yellow	85	13.5	1.04
Apollo 14 yellow	82	14.8	0.92
Apollo 17 yellow	55	12.6	0.72
Apollo 17 orange	46	14.7	0.52
Apollo 17 orange (74220)	70	14.9	0.78
Apollo 17 orange	33	11.6	0.47
Apollo 11 orange	30	14.3	0.35
Apollo 14 orange	30	14.5	0.34
Apollo 15 red	<50	12.1	<0.69
Apollo 14 black	≤22	13.3	≤0.27

*Precision for Ni is typically ± 10 ppm with a detection limit of about 20 ppm.

†No volcanic glasses have yet been found having TiO_2 abundances between 1.0 wt % and 3.5 wt %.

(1986), and Wolf and Anders (1980). This depletion of Ni in the silicate portion of the Moon is explicable by formation of a small metal/sulfide (e.g., Newsom, 1986).

3. Cobalt

Wänke *et al.* (1977, 1983) have demonstrated that Co correlates with FeO + MgO in terrestrial basalts/ultramafics and in lunar mare basalts. The Co/(FeO +

TABLE 3.   *Estimated Abundance of MgO (wt %) in the Moon.*

34.8%	(Binder, 1983)
32.7%	(Ringwood, 1979)
32.0%	(Taylor, 1982)
32.4%	(Wänke *et al.*, 1977)
35.5%	(Wänke and Dreibus, 1982)

TABLE 4. Summary of Ni and Mg Systematics in the Earth's Upper Mantle, Moon's Silicate Portion, and CI Chondrites.

	(Ni/Mg)	(Ni/Mg) depletion relative to CI chondrites
Earth's upper mantle[1]	$(9.1 \pm 0.5) \times 10^{-3}$	12.5 ± 0.6
Moon's silicate portion[2]	$(2.3 \pm 0.3) \times 10^{-3}$	50 ± 6
CI chondrites[3]	1.15×10^{-1}	1

	Ni (ppm)	Ni depletion relative to CI chondrites
Earth's upper mantle[1]	2100 ± 100	5.2
Moon's silicate portion[2]	470 ± 50	23 ± 3
CI chondrites[3]	11,000	1

Sources of data:
[1] *Jagoutz et al. (1979).*
[2] *This study.*
[3] *Anders and Ebihara (1982).*

MgO) inferred for the lunar mantle from these correlations is about 1.8×10^{-4}, compared to a value of 2.28×10^{-4} observed in the terrestrial upper mantle (e.g., Jagoutz *et al.*, 1979). This lower ratio in the Moon has been interpreted to indicate that the Moon's silicate portion is depleted in Co relative to the Earth's upper mantle by a factor of 1.25 (Wänke *et al.*, 1977, 1983). While this is a small depletion, it is well to remember that the Moon's silicate portion is generally believed to contain a factor of 2 enrichment in FeO compared to the terrestrial upper mantle. Specifically, the abundance of FeO in the silicate portion of the Moon is thought to be about 13–16 wt % (Binder, 1983; Morgan *et al.*, 1978; Ringwood, 1979; Taylor, 1982; Wänke *et al.*, 1977). For an estimated abundance of MgO in the Moon's silicate portion of 33.5 ± 1.6 wt % (Table 3) and a value for the Co/(FeO + MgO) ratio of about 1.8×10^{-4} from low-Ti mare basalts, the estimated abundance of Co is ~85 ppm. This would correspond to a depletion factor relative to the terrestrial upper mantle of ~1.25 (i.e., indistinguishable from the value concluded by Wänke *et al.*, 1977, 1983).

At present, the only variety of low-Ti, pristine lunar glass to have been analyzed for Co is the Apollo 15 green glass (Ma *et al.*, 1981; Taylor *et al.*, 1973). This primitive magmatic composition has a Co/(FeO + MgO) ratio of $(1.95 \pm 0.05) \times 10^{-4}$, implying an abundance of Co in the Moon's silicate portion of ~92 ppm.

From these data, it is possible to estimate the abundance of Co in the silicate portion of the Moon as being 90 ± 5 ppm. Relative to Mg, the depletion factor for Co in the Moon's silicate portion is only about 1.05 ± 0.10 compared to the terrestrial upper mantle (Table 5). These results are in agreement with those of

Ringwood and Seifert (1986) and represent a minor revision to the original conclusions of Wänke *et al* (1977, 1983) and Wänke and Dreibus (1982, 1986).

An important result of this study is that, contrary to the Earth's upper mantle, the Ni/Co ratio in the Moon's silicate portion has a nonchondritic value of 5 ± 1. This observation is compatible with the existence of a metallic core within the Moon (Newsom, 1984, 1986; Brett, 1973b; Ringwood and Seifert, 1986; Wänke and Dreibus, 1982, 1986).

4. Highly Volatile Elements

4.1. Observations

The pristine lunar glasses were produced by explosive fire-fountaining (e.g., Heiken *et al*, 1974). Associated with these magmatic eruptions was the release of indigenous gas from the Moon's interior (Table 6). The least volatile components occurring within that gas condensed onto the exterior surfaces of the pristine glasses often in the form of chlorides, fluorides, and sulfides (e.g., Clanton *et al*, 1978; Butler, 1978; Meyer *et al*, 1975; Wasson *et al*, 1976), whereas the highly volatile constituents (e.g., N_2, CO, Ar, Kr, Xe) were trapped within vesicular, pristine glasses (Heiken *et al*, 1974; Delano, 1984; Delano and Lindsley, 1983). Isotopic analyses of the highly volatile elements enclosed within the vesicles have recently begun (Barraclough and Marti, 1985).

Isotopic analyses of the Pb (^{204}Pb, ^{206}Pb, ^{207}Pb, ^{208}Pb) that condensed from the fumarolic gas onto the surfaces of the Apollo 15 green and Apollo 17 orange pristine

TABLE 5.   *Summary of Co and Mg Systematics in the Earth's Upper Mantle, Moon's Silicate Portion, and CI Chondrites.*

	(Co/Mg)	(Co/Mg) depletion relative to CI chondrites
Earth's upper mantle[1]	$(4.55 \pm 0.20) \times 10^{-4}$	11.7 ± 0.5
Moon's silicate portion[2]	$(4.45 \pm 0.40) \times 10^{-4}$	12.0 ± 1.0
CI chondrites[3]	5.33×10^{-3}	1

	Co (ppm)	Co depletion relative to CI chondrites
Earth's upper mantle[1]	105 ± 5	4.8 ± 0.3
Moon's silicate portion[2]	90 ± 5	5.7 ± 0.3
CI chondrites[3]	509	1

Sources of data:
[1]*Jagoutz et al (1979).*
[2]*This study.*
[3]*Anders and Ebihara (1982).*

TABLE 6. *Elements Known from Analysis of Pristine Lunar Glasses to be Contained in Indigenous Gas Associated with Magmatic Fire Fountains on the Moon.*

B (1)	Br (3,12,16)
C (20)	Ag (6,12,19)
N (23)	Cd (6–8,12,14,18,19)
F (2–5)	In (6,7,14,19)
Na (6–8,18,21)	Sb (6,19)
S (2,8–10,18,21)	Te (14,19)
Cl (2,3,18,21)	I (16)
Ar (11,23)	Xe (16)
Cu (2,5)	Au (6,7,14,19)
Zn (2,5–9,12–14,18,19,21)	Hg (14,17)
Ga (2,5,7)	Tl (2,12,13,19,22)
Ge (6,7,14,19)	Pb (2,8,13,15,22)
Se (19)	Bi (6)

References:
(1) *Meyer and Schonfeld (1977).*
(2) *Meyer et al. (1975).*
(3) *Jovanovic and Reed (1974).*
(4) *Goldberg et al. (1975, 1976).*
(5) *Wänke et al. (1973).*
(6) *Chou et al. (1975).*
(7) *Wasson et al. (1976).*
(8) *Cirlin et al. (1978).*
(9) *Butler and Meyer (1976); Butler (1978).*
(10) *Grant et al. (1974); Thode and Rees (1976); Gibson and Andrawes (1978).*
(11) *Alexander et al. (1980); Chou et al. (1973, 1974); Eberhardt et al. (1973); Eugster et al. (1980); Huneke (1978); Podosek and Huneke (1973); Schaeffer and Husain (1973); Lakatos et al. (1973).*
(12) *Morgan et al. (1974).*
(13) *Allan et al. (1975).*
(14) *Krähenbühl (1980).*
(15) *Nunes et al. (1974); Silver (1974a,b); Tatsumoto et al. (1973); Tera and Wasserburg (1976).*
(16) *Eugster et al. (1980).*
(17) *Jovanovic and Reed (1979).*
(18) *Cirlin and Housley (1979).*
(19) *Morgan and Wandless (1979, 1984).*
(20) *Epstein and Taylor (1973); Gibson and Moore (1973); Sato (1979); Wszolek et al. (1973).*
(21) *Clanton et al. (1978).*
(22) *Reed et al. (1977).*
(23) *Barraclough and Marti (1985).*

glasses (Meyer *et al.*, 1975; Nunes *et al.*, 1974; Silver, 1974a,b; Tatsumoto *et al.*, 1973; Tera and Wasserburg, 1976) have repeatedly shown it to be enriched in ^{204}Pb, which is nonradiogenic, compared to the Pb-isotopic compositions observed in mare basalts and pristine highlands rocks. Those same authors have concluded that this special, surface-correlated Pb was derived from a source within the Moon characterized by (1) (^{238}U/^{204}Pb) < 35 (i.e., volatile-enriched) and (2) having a U-Pb model age of ~4.55 aeons. This model age is significantly different from the values of 4.3–4.4 aeons (e.g., Oberli *et al.*, 1978) observed in other lunar samples that record the isotopic closure age of differentiated reservoirs produced by crystal/liquid fractionation of the global magma ocean. Finally, Morgan and Wandless (1984) and Delano (1980) have noted that the surface-correlated Ge, Sb, Au, Se, and Te associated with the pristine lunar glasses occur in approximately CI-chondritic ratios.

4.2. Synthesis

Prior to eruption, the indigenous gas associated with the pristine lunar glasses may have resided in primordial, volatile-rich reservoirs within the Moon (Morgan and Wandless, 1984; Wolf *et al.*, 1979; Delano, 1980; Delano and Livi, 1981). The Pb-isotopic data require that this volatile-rich component persisted in the Moon as chemically/isotopically discrete reservoirs at least until the eruptions of pristine volcanic glasses at $\sim 3.5 \times 10^9$ years. It would therefore appear that the early processes of global differentiation associated with the magma ocean and with core formation did *not* affect the entire volume of the Moon (Brett, 1977; Dickinson and Newsom, 1985; Delano, 1980; Delano and Livi, 1981). Since the gravitational potential energy released by core formation would have increased the mean temperature of the Moon by only $\leq 12°$ C (Solomon, 1979), it may not be particularly surprising that primordial matter survived the Moon's global differentiation. Furthermore, the survival of undifferentiated reservoirs would have been even more likely if core formation was initially asymmetric (Stevenson, 1980).

The isotopic compositions of Ne, Ar, Kr, and Xe derived from these primordial reservoirs and trapped in vesicular, pristine lunar glasses should possess important constraints on the exact nature of these volatile-rich reservoirs (e.g., carbonaceous chondrite?). This component may occur as discrete chunks of volatile-rich debris from the outer solar system that became entrained in the circumterrestrial accretion disk. If this view is correct, then this component would be expected to have had a different history/origin from the bulk of the matter comprising the Moon (e.g., Earth-fissioned material).

5. Conclusions

The pristine lunar glasses are samples of primary magmas derived by partial melting of differentiated, cumulate source regions in the Moon's mantle. The low-Ti magmas, which are the most chemically primitive, have been used to constrain the abundances of Ni and Co in the silicate portion of the Moon. New data on the abundances of Ni in the pristine lunar glasses have been presented.

The specific conclusions that emerge from this study are listed below:

1. The abundance of Ni in the Moon's silicate portion is estimated to be 470 ± 50 ppm. Relative to magnesium, Ni is depleted by a factor of 50 ± 6 compared to CI chondrites and by a factor of 4.0 ± 0.5 compared to the present composition of the Earth's upper mantle. A small metallic core within the Moon is required (Drake, 1983; Newsom, 1984, 1986; Brett, 1973b; Ringwood, 1979; Ringwood and Seifert, 1986; Wänke et al., 1977; Wänke and Dreibus, 1982, 1986).

2. The abundance of Co in the Moon's silicate portion is estimated to be 90 ± 5 ppm. Relative to magnesium, Co is depleted in the Moon's silicate portion by a factor of 12 ± 1 compared to CI chondrites and by a factor of only 1.05 ± 0.10 compared to the Earth's upper mantle.

3. The value of the Ni/Co ratio in the silicate portion of the Moon is 5 ± 1. This is clearly distinct from the chondritic value of 20 observed in the Earth's upper mantle. This fractionated value in the Moon's silicate portion is related to core formation.

4. The presence of primordial, volatile-rich reservoirs within the Moon places important constraints on its postaccretion, thermal history (Brett, 1977; Dickinson and Newsom, 1985; Delano, 1980). This component may exist as discrete blocks of chondritic matter that became entrained within the circumterrestrial accretion disk (Wolf et al., 1979; Delano, 1980). Further constraints on the Moon's origin may be furnished through isotopic analyses of the noble gases that occur within vesicular, pristine lunar glasses (Barraclough and Marti, 1985).

Acknowledgments. The author gratefully acknowledges the helpful comments of Robin Brett, Horton Newsom, Ted Ringwood, Graham Ryder, Jeff Taylor, Heinrich Wänke, and Paul Warren on an earlier version of this manuscript. This work was supported by NASA grant NAG 978.

6. References

Alexander E. C., Jr., Coscio M. R., Jr., Dragon J. C., and Saito K. (1980) K/Ar dating of lunar soils IV: Orange glass from 74220 and agglutination from 14259 and 14163. *Proc. Lunar Planet. Sci. Conf. 11th*, pp. 1663–1677.

Allan R. O., Jovanovic S., and Reed G. W., Jr. (1975) Agglutinates: Role in element and isotope chemistry and inferences regarding volatile-rich rock 66095 and glass 64220. *Proc. Lunar Sci. Conf. 6th*, pp. 2271–2279.

Anders E. and Ebihara M. (1982) Solar-system abundances of the elements. *Geochim. Cosmochim. Acta, 46*, pp. 2363–2380.

Arndt N. T. (1977) Partitioning of nickel between olivine and ultrabasic and basic komatiite liquids. *Annual Report of the Director, Geophysical Laboratory Year Book, 76*, pp. 553–557.

Arndt N. T., Naldrett A. J., and Pyke D. R. (1977) Komatiitic and iron-rich tholeiitic lavas of Munro Township, northeast Ontario. *J. Petrol., 18*, 319–369.

Barraclough B. L. and Marti K. (1985) In search of the Moon's indigenous volatiles: Noble gases and nitrogen in vesicular lunar glasses (abstract). In *Lunar and Planetary Science XVI*, pp. 31–32. Lunar and Planetary Institute, Houston.

Binder A. B. (1983) An estimate of the bulk, major oxide composition of the Moon (abstract). In *Workshop on Pristine Highlands Rocks and the Early History of the Moon* (J. Longhi and G. Ryder, eds.), pp. 17–19. LPI Tech. Rpt. 83–02, Lunar and Planetary Institute, Houston.

Bougault H. and Hekinian R. (1974) Rift valley in the Atlantic Ocean near 36°50'N: petrology and geochemistry of basaltic rocks. *Earth Planet. Sci. Lett., 24,* 249–261.

Brett R. (1973a) The lunar crust: a product of heterogeneous accretion or differentiation of a homogeneous Moon? *Geochim. Cosmochim. Acta, 37,* 2697–2703.

Brett R. (1973b) A lunar core of Fe-Ni-S. *Geochim. Cosmochim. Acta, 37,* 165–170.

Brett R. (1977) The case against early melting of the bulk of the Moon. *Geochim. Cosmochim. Acta, 41,* 443–445.

Butler P., Jr. (1978) Recognition of lunar glass droplets produced directly from endogenous liquids: The evidence from S-Zn coatings. *Proc. Lunar Planet. Sci. Conf. 9th,* pp. 1459–1471.

Butler P., Jr. and Meyer C., Jr. (1976) Sulfur prevails in coatings on glass droplets: Apollo 15 green and brown glasses and Apollo 17 orange and black (devitrified) glasses. *Proc. Lunar Sci. Conf. 7th,* pp. 1561–1581.

Chen H.-K., Delano J. W., and Lindsley D. H. (1982) Chemistry and liquidus phase relations of VLT volcanic glasses from Apollo 14 and Apollo 17. *Proc. Lunar Planet. Sci. Conf. 13th,* in *J. Geophys. Res., 87,* pp. A171–A181.

Chou C.-L., Baedecker P. A., and Wasson J. L. (1973) Atmophile elements in lunar soils. *Proc. Lunar Sci. Conf. 4th,* pp. 1523–1533.

Chou C.-L., Baedecker P. A., Bild R. W., and Wasson J. T. (1974) Volatile-element systematics and green glass in Apollo 15 lunar soils. *Proc. Lunar Sci. Conf. 5th,* pp. 1645–1657.

Chou C.-L., Boynton W. V., Sundberg L. L., and Wasson J. T. (1975) Volatiles on the surface of Apollo 15 green glass and trace-element distributions among Apollo 15 soils. *Proc. Lunar Sci. Conf. 6th,* pp. 1701–1727.

Cirlin E. H. and Housley R. M. (1979) Scanning Auger microprobe and atomic absorption studies of lunar volcanic volatiles. *Proc. Lunar Planet. Sci. Conf. 10th,* pp. 341–354.

Cirlin E. H., Housley R. M., and Grant R. W. (1978) Studies of volatiles in Apollo 17 samples and their implications to vapor transport processes. *Proc. Lunar Planet. Sci. Conf. 9th,* pp. 2049–2063.

Clague D. A. and Frey F. A. (1982) Petrology and trace element geochemistry of the Honolulu Volcanic Series, Oahu: Implications for the oceanic mantle beneath Hawaii. *J. Petrol, 23,* 447–504.

Clanton U. S., McKay D. S., Waits G., and Fuhrman R. (1978) Sublimate morphology on 74001 and 74002 orange and black glassy droplets. *Proc. Lunar Planet. Sci. Conf. 9th,* pp. 1945–1957.

Compston W., Berry H., Vernon M. J., Chappell B. W., and Kaye M. J. (1971) Rubidium-strontium chronology and chemistry of lunar material from the Ocean of Storms. *Proc. Lunar Sci. Conf. 2nd,* pp. 1471–1485.

Delano J. W. (1979) Apollo 15 green glass: Chemistry and possible origin. *Proc. Lunar Planet. Sci. Conf. 10th,* pp. 275–300.

Delano J. W. (1980) Chemistry and liquidus phase relations of Apollo 15 red glass: Implications for the deep lunar interior. *Proc. Lunar Planet. Sci. Conf. 11th,* pp. 251–288.

Delano J. W. (1984) Vesicles in four varieties of Apollo 15 volcanic glass (abstract). In *Lunar and Planetary Science XV,* pp. 218–219. Lunar and Planetary Institute, Houston.

Delano J. W. (1985) Pristine lunar glasses: Criteria, data, and implications. *Proc. Lunar Planet. Sci. Conf. 16th,* in *J. Geophys. Res., 90,* in press.

Delano J. W. and Lindsley D. H. (1983) Mare glasses from Apollo 17: Constraints on the Moon's bulk composition. *Proc. Lunar Planet. Sci. Conf. 14th,* in *J. Geophys. Res., 88,* pp. B3–B16.

Delano J. W. and Livi K. (1981) Lunar volcanic glasses and their constraints on mare petrogenesis. *Geochim. Cosmochim. Acta, 45,* 2137–2149.

Dickey J. S., Jr., Frey F. A., Hart S. R., Watson E. B., and Thompson G. (1977) Geochemistry and petrology of dredged basalts from the Bouvet triple junction, South Atlantic. *Geochim. Cosmochim. Acta, 41,* 1105–1118.

Dickinson T. and Newsom H. E. (1985) A possible test of the impact theory for the origin of the Moon (abstract). In *Lunar and Planetary Science XVI,* pp. 183–184. Lunar and Planetary Institute, Houston.

Dietrich V., Emmermann R., Oberhansli R., and Puchelt H. (1978) Geochemistry of basaltic and gabbroic rocks from the West Mariana Basin and the Mariana Trench. *Earth Planet. Sci. Lett., 39,* 127–144.

Dietrich V. J., Gansser A., Sommerauer J., and Cameron W. E. (1981) Paleogene komatiites from Gorgona Island, East Pacific—A primary magma for ocean floor basalts? *Geochem. J., 15,* 141–161.

Drake M. J. (1983) Geochemical constraints on the origin of the moon. *Geochim. Cosmochim. Acta, 47,* 1759–1767.

Duncan A. R., Erlank A. J., Willis J. P., Sher M. K., and Ahrens L. H. (1974) Trace element evidence for a two-stage origin of some titaniferous mare basalts. *Proc. Lunar Sci. Conf. 5th,* pp. 1147–1157.

Duncan A. R., Erlank A. J., Sher M. K., Abraham Y. C., Willis J. P., and Ahrens L. H. (1976) Some trace element constraints on lunar basalt genesis. *Proc. Lunar Sci. Conf. 7th,* pp. 1659–1671.

Eberhardt P., Geiss J., Grögler N., Mauer P., and Stettler A. (1973) ^{39}Ar-^{40}Ar ages of lunar material (abstract). In *Meteoritics, 8,* 360–361.

Epstein S. and Taylor H. P., Jr. (1973) The isotopic composition and concentration of water, hydrogen, and carbon in some Apollo 15 and 16 soils and in the Apollo 17 orange soil. *Proc. Lunar Sci. Conf. 4th,* pp. 1559–1575.

Eugster O., Grögler N., Eberhardt P., and Geiss J. (1980) Double drive tube 74001/2: Composition of noble gases trapped 3.7 AE ago. *Proc. Lunar Planet. Sci. Conf. 11th,* pp. 1565–1592.

Flanagan F. J., Wright T. L., Taylor S. R., Annell C. S., Christain R. C., and Dinnin J. I. (1976) Basalt, BHVO-1, from Kilauea crater, Hawaii. *USGS Prof. Paper 840,* pp. 33–39.

Frey F. A. and Green D. H. (1974) The mineralogy, geochemistry and origin of lherzolite inclusions in Victorian basanites. *Geochim. Cosmochim. Acta, 38,* 1023–1059.

Gibson E. K., Jr. and Andrawes F. F. (1978) Sulfur abundances in the 74001/74002 drive tube core from Shorty Crater, Apollo 17. *Proc. Lunar Planet. Sci. Conf. 9th,* pp. 2011–2017.

Gibson E. K., Jr. and Moore G. W. (1973) Variable carbon contents of lunar soil 74220. *Earth Planet. Sci. Lett., 20,* 404–408.

Goldberg R. H., Burnett D. S., and Tombrello T. A. (1975) Fluorine surface films on lunar samples: Evidence for both lunar and terrestrial origins. *Proc. Lunar Sci. Conf. 6th,* pp. 2189–2200.

Goldberg R. H., Tombrello T. A., and Burnett D. S. (1976) Fluorine as a constituent in lunar magmatic gases. *Proc. Lunar Sci. Conf. 7th,* pp. 1597–1613.

Grant R. W., Housley R. M., Szalkowski F. J., and Marcus H. L. (1974) Auger electron microscopy of lunar samples. *Proc. Lunar Sci. Conf. 5th,* pp. 2423–2439.

Gunn B. M. (1971) Trace element partition during olivine fractionation of Hawaiian basalt. *Chem. Geol., 8,* 1–13.

Gunn B. M., Abranson C. E., Nougier J., Watkins N. D., and Hajash A. (1971) Amsterdam Island, an isolated volcano in the southern Indian Ocean. *Contrib. Mineral. Petrol, 32,* 79–92.

Hart S. R. and Davis K. E. (1978) Nickel partitioning between olivine and silicate melt. *Earth Planet. Sci. Lett., 40,* 203–219.

Heiken G. H., McKay D. S., and Brown R. W. (1974) Lunar deposits of possible pyroclastic origin. *Geochim. Cosmochim. Acta, 38*, 1703–1718.

Huneke J. C. (1978) ^{40}Ar-^{39}Ar microanalysis of single 74220 glass balls and 72435 breccia clasts. *Proc. Lunar Planet. Sci. Conf. 9th*, pp. 2345–2362.

Irving A. J. (1978) A review of experimental studies of crystal/liquid trace element partitioning. *Geochim. Cosmochim. Acta, 42*, 743–770.

Jagoutz E., Palme H., Baddenhausen H., Blum K., Cendales M., Dreibus G., Spettel B., Lorenz V., and Wänke H. (1979) The abundances of major, minor, and trace elements in the earth's mantle as derived from primitive ultramafic nodules. *Proc. Lunar Planet. Sci. Conf. 10th*, pp. 2031–2050.

Jovanovic S. and Reed G. W., Jr. (1974) Labile and nonlabile element relationships among Apollo 17 samples. *Proc. Lunar Sci. Conf. 5th*, pp. 1685–1701.

Jovanovic S. and Reed G. W., Jr. (1979) Regolith layering processes based on studies of low-temperature volatile elements in Apollo core samples. *Proc. Lunar Planet. Sci. Conf. 10th*, pp. 1425–1435.

Krähenbühl U. (1980) Distribution of volatile and nonvolatile elements in grain-size fractions of Apollo 17 drive tube 74001/2. *Proc. Lunar Planet. Sci. Conf. 11th*, pp. 1551–1564.

Lakatos S., Heymann D., and Yaniv A. (1973) Green spherules from Apollo 15: Inferences about their origin from inert gas measurements. *The Moon, 7*, 132–148.

Langmuir C. H., Bender J. F., Bence A. E., Hanson G. H., and Taylor S. R. (1977) Petrogenesis of basalts from the FAMOUS area: Mid-Atlantic Ridge. *Earth Planet. Sci. Lett., 36*, 133–156.

Leeman W. P., Vitaliano C. J., and Prinz M. (1976) Evolved lavas from the Snake River Plain: Craters of the Moon National Monument, Idaho. *Contrib. Mineral. Petrol., 56*, 35–60.

Ma M.-S., Liu Y.-G., and Schmitt R. A. (1981) A chemical study of individual green glasses and brown glasses from 15426: Implications for their petrogenesis. *Proc. Lunar Planet. Sci. 12B*, pp. 915–933.

Marvin U. B. and Walker D. (1978) Implications of a titanium-rich glass clod at Oceanus Procellarum. *Am. Mineral., 63*, 924–929.

Meyer C., Jr. and Schonfeld E. (1977) Ion microprobe study of glass particles from lunar sample 15101 (abstract). In *Lunar Science VIII*, pp. 661–663. The Lunar Science Institute, Houston.

Meyer C., Jr., McKay D. S., Anderson D. H., and Butler P., Jr. (1975) The source of sublimates on the Apollo 15 green and Apollo 17 orange glass samples. *Proc. Lunar Sci. Conf. 6th*, pp. 1673–1699.

Morgan J. W. and Wandless G. A. (1979) 74001 drive tube: Siderophile elements match IIB iron meteorite pattern. *Proc. Lunar Planet. Sci. Conf. 10th*, pp. 327–340.

Morgan J. W. and Wandless G. A. (1984) Surface-correlated trace elements in 15426 lunar glass (abstract). In *Lunar and Planetary Science XV*, pp. 562–563. Lunar and Planetary Institute, Houston.

Morgan J. W., Ganapathy R., Higuchi H., Krähenbühl U., and Anders E. (1974) Lunar basins: Tentative characterization of projectiles, from meteoritic elements in Apollo 17 boulders. *Proc. Lunar Sci. Conf. 5th*, pp. 1703–1736.

Morgan J. W., Hertogen J., and Anders E. (1978) The Moon: Composition determined by nebular processes. *Moon and Planets, 18*, 465–478.

Nabelek P. (1980) Nickel partitioning between olivine and liquid in natural basalts: Henry's Law behavior. *Earth Planet. Sci. Lett., 48*, 293–302.

Nesbitt R. W. and Sun S.-S. (1976) Geochemistry of Archean spinifex-textured peridotites and magnesian and low-magnesian tholeiites. *Earth Planet. Sci. Lett., 31*, 433–453.

Newsom H. E. (1984) The lunar core and the origin of the Moon. *EOS (Trans. Am. Geophys. Union), 65*, pp. 369–370.

Newsom H. E. (1986) Constraints on the origin of the moon from the abundance of molybdenum and other siderophile elements, this volume.

Nunes P. D., Tatsumoto M., and Unruh D. M. (1974) U-Th-Pb systematics of some Apollo 17 lunar samples and implications for a lunar basin excavation chronology. *Proc. Lunar Sci. Conf. 5th*, pp. 1487–1514.

Oberli F., McCulloch M. T., Tera F., Papanastassiou D. A., and Wasserburg G. J. (1978) Early lunar differentiation constraints from U-Th-Pb, Sm-Nd and Rb-Sr model ages (abstract). In *Lunar and Planetary Science IX*, pp. 832–834. Lunar and Planetary Institute, Houston.

Podosek F. A. and Huneke J. C. (1973) Argon in Apollo 15 green glass spherules (15426): [40]Ar-[39]Ar age and trapped argon. *Earth Planet. Sci. Lett., 19*, 413–421.

Reed G. W., Jr., Allen R. O., Jr., and Jovanovic S. (1977) Volatile metal deposits on lunar soils—relation to volcanism. *Proc. Lunar Sci. Conf. 8th*, pp. 3917–3930.

Ringwood A. E. (1977) Basaltic magmatism and the bulk composition of the Moon. I. Major and heat-producing elements. *The Moon, 16*, 389–423.

Ringwood A. E. (1979) *Origin of the Earth and Moon*. Springer-Verlag, New York. 295 pp.

Ringwood A. E. and Seifert S. (1986) Nickel-cobalt abundance systematics and their bearing on lunar origin, this volume.

Sato M. (1979) The driving mechanism of lunar pyroclastic eruptions inferred from the oxygen fugacity behavior of Apollo 17 orange glass. *Proc. Lunar Sci. Conf. 10th*, pp. 311–325.

Schaeffer O. A. and Husain L. (1973) Isotopic ages of Apollo 17 lunar material. *EOS (Trans. Am. Geophys. Union), 54*, p. 614.

Silver L. T. (1974a) Patterns of U-Th-Pb distributions and isotope relations in Apollo 17 soils (abstract). In *Lunar Science V*, pp. 706–708. The Lunar Science Institute, Houston.

Silver L. T. (1974b) Implications of volatile leads in orange, grey, and green lunar soils for an Earth-like Moon (abstract). *EOS (Trans. Am. Geophys. Union), 55*, p. 681.

Solomon S. C. (1979) Formation, history and energetics of cores in the terrestrial planets. *Phys. Earth Planet. Inter., 19*, 168–182.

Stevenson D. J. (1980) Lunar asymmetry and palaeomagnetism. *Nature, 287*, 520–521.

Stosch H.-G. and Seck H. A. (1980) Geochemistry and mineralogy of two spinel peridotite suites from Dreiser Weiher, West Germany. *Geochim. Cosmochim. Acta, 44*, 457–470.

Sun S.-S. (1982) Chemical composition and origin of the earth's primitive mantle. *Geochim. Cosmochim. Acta, 46*, 179–192.

Tatsumoto M., Nunes P. D., Knight R. J., Hedge C. E., and Unruh D. N. (1973) U-Th-Pb, Rb-Sr, and K measurements of two Apollo 17 samples. *EOS (Trans. Am. Geophys. Union), 54*, pp. 614–615.

Taylor S. R., Gorton M. P., Muir P., Nance W., Rudowski R., and Ware N. (1973) Lunar highlands composition: Apennine Front. *Proc. Lunar Sci. Conf. 4th*, pp. 1445–1459.

Taylor S. R. (1982) *Planetary Science: A Lunar Perspective*. Lunar and Planetary Institute, Houston. 481 pp.

Tera F. and Wasserburg G. J. (1976) Lunar ball games and other sports (abstract). In *Lunar Science VII*, pp. 858–860. The Lunar Science Institute, Houston.

Thode H. G. and Rees C. E. (1976) Sulphur isotopes in grain size fractions of lunar soils. *Proc. Lunar Sci. Conf. 7th*, pp. 459–468.

Wänke H. and Dreibus G. (1982) Chemical and isotopic evidence for the early history of the Earth-Moon system. In *Tidal Friction and the Earth's Rotation II* (P. Brosche and J. Sundermann, eds.), pp. 322–344. Springer-Verlag, Heidelberg.

Wänke H. and Dreibus G. (1986) Geochemical evidence for the formation of the Moon by impact induced fission of the proto-Earth, this volume.

Wänke H., Baddenhausen H., Dreibus G., Jagoutz E., Kruse H., Palme H., Spettel B., and Teschke F. (1973) Multielement analyses of Apollo 15, 16 and 17 samples and the bulk composition of the Moon. *Proc. Lunar Sci. Conf. 4th*, pp. 1461–1481.

Wänke H., Baddenhausen H., Blum K., Cendales M., Dreibus G., Hofmeister H., Krause H., Jagoutz E., Palme C., Spettel B., Thacker R., and Vilcsek E. (1977) On the chemistry of lunar samples and achondrites. Primary matter in the lunar highlands: A re-evaluation. *Proc. Lunar Sci. Conf. 8th*, pp. 2191–2213.

Wänke H., Dreibus G., Palme H., Rammensee W., and Weckwerth G. (1983) Geochemical evidence for the formation of the Moon from material of the earth's mantle (abstract). In *Lunar and Planetary Science XIV*, pp. 818–819. Lunar and Planetary Institute, Houston.

Warner R. D., Taylor G. J., Wentworth S. J., Huss G. R., Mansker W. L., Planner H. N., Sayeed U. A., and Keil K. (1979) Electron microprobe analyses of glasses form Apollo 17 rake sample breccias and Apollo 17 drill core. *Inst. of Meteoritics Spec. Publ. No. 20*, Univ. of New Mexico, Albuquerque.

Wasson J. T., Boynton W. V., Kallemeyn G. W., Sundberg L. L., and Wai C. M. (1976) Volatile compounds released during lunar lava fountaining. *Proc. Lunar Sci. Conf. 7th*, pp. 1583–1595.

Whitford D. J. and Arndt N. T. (1978) Rare earth element abundances in a thick, layered komatiite lava flow from Ontario, Canada. *Earth Planet. Sci. Lett., 41*, 188–196.

Wiesmann H. and Hubbard N. J. (1975) A compilation of lunar sample data generated by the Gast, Nyquist, and Hubbard lunar sample PI-ships. JSC preprint, Johnson Space Center, Houston. 50 pp.

Willis J. P., Ahrens L. H., Danchin R. V., Erlank A. J., Gurney J. J., Hofmeyr P. K., McCarthy T. S., and Orren M. J. (1971) Some interelement relationships between lunar rocks and fines, and stony meteorites. *Proc. Lunar Sci. Conf. 2nd*, pp. 1123–1138.

Wolf R. and Anders E. (1980) Moon and Earth: compositional differences inferred from siderophiles, volatiles, and alkalis in basalts. *Geochim. Cosmochim. Acta, 44*, 2111–2124.

Wolf R., Woodrow A., and Anders E. (1979) Lunar basalts and pristine highland rocks: Comparison of siderophile and volatile elements. *Proc. Lunar Planet. Sci. Conf. 10th*, pp. 2107–2130.

Wood J. A., Dickey J. S., Jr., Marvin U. B., and Powell B. N. (1970) Lunar anorthosites and a geophysical model of the moon. *Proc. Apollo 11 Lunar Sci. Conf.*, pp. 965–988.

Wszolek P. C., Simoneit B. R., and Burlingame A. L. (1973) Studies of magnetic fines and volatile-rich soils: Possible meteoritic and volcanic contributions to lunar carbon and light element chemistry. *Proc. Lunar Sci. Conf. 4th*, pp. 1693–1706.

Nickel-Cobalt Abundance Systematics and their Bearing on Lunar Origin

A. E. RINGWOOD AND STEFAN SEIFERT

Research School of Earth Sciences, Australian National University, Canberra, A.C.T. 2601, Australia

The high abundances of nickel and cobalt in the Earth's mantle were established uniquely by the complex processes associated with core-formation. Cobalt abundances are very similar in terrestrial ocean-floor tholeiites and lunar low-Ti mare basalts. This observation, combined with experimental measurements of Co partitions between olivines and primitive lunar volcanic glasses, demonstrates that the cobalt contents of the lunar and terrestrial mantles are also very similar. Theoretical, experimental, and geochemical studies imply that the Ni/Co ratio of the bulk Moon is very close to the chondritic ratio. Moreover, the Ni/Co ratio of the Earth's mantle is known to be very similar to the chondritic ratio. Since the bulk cobalt abundance of the Moon is demonstrably similar to that of the Earth's mantle, it follows that the bulk nickel abundance in the Moon is likewise similar to that of the Earth's mantle. However, investigations of the abundance of nickel in the lunar mantle based on comparison of Ni abundances in terrestrial basalts, lunar mare basalts, and lunar volcanic glasses, combined with experimental measurements of partitions of nickel between olivine and primitive lunar volcanic glasses, show that nickel is depleted in the lunar mantle by about a factor of three, compared to the terrestrial mantle. These observations and inferences can be reconciled if the Moon contains a small, nickel-rich metallic core. Experimental measurements of relevant metal-silicate partition coefficients show that the core would compose 0.4% of the lunar mass and would contain about 40% Ni and 0.7% Co. The presence of a small metallic core is supported by observations on the abundance patterns of other involatile siderophile elements in the Moon. The close similarities between the bulk Co and Ni abundances in the Moon and the Co and Ni abundances in the Earth's mantle strongly support the hypothesis that the Moon was derived from the terrestrial mantle after the Earth's core had formed. The high cobalt and nickel contents of the lunar mantle and the high nickel content of the lunar core are difficult or impossible to explain by current hypotheses of lunar origin, according to which the Moon was formed as a binary or independent planet.

1. Introduction

The abundances of siderophile elements in the Earth's upper mantle display some interesting features that have recently been reviewed by Ringwood (1984). For example, despite the very different siderophilic natures of nickel and cobalt, they are present in the upper mantle approximately in chondritic relative abundances. Moreover, their absolute abundances are much higher than would be expected if the upper mantle had equilibrated with an iron-rich metal phase during core formation at moderately high temperatures (e.g., $1200°-2000°C$). In contrast, the nickel/cobalt ratios of the silicate phases from differentiated meteorites, such as eucrites and pallasitic olivines, depart grossly from the chondritic ratio and the absolute abundances of nickel and cobalt in these meteoritic silicate phases are lower by factors of 5 to 100 than in corresponding terrestrial basaltic and ultramafic rocks from the Earth's mantle (Section 6, Table 1). The distribution of nickel and cobalt in differentiated

TABLE 1. Abundances (ppm) of Cobalt and Nickel in Silicates from the Terrestrial Upper Mantle, Differentiated Meteorites and from a Primitive Lunar Volcanic Glass.

Silicate	Ni	Co	Ni/Co	Ref.
Olivine from Earth's upper mantle	2900	136	21	1,8
Olivine from pallasite	29	9	3.2	2
Olivine from lunar mantle	890	156	5.7	2
Primitive terrestrial oceanic tholeiite	200	41	4.9	3
Average eucrite	3.5	8	0.4	4
Primitive lunar volcanic glass	185	82	2.3	5,6
Average low-Ti mare basalt	59	51	1.2	3,4,9
Average terrestrial tholeiite	75	43	1.7	3,4,9
C1 chondrites	11,100	509	22	7

References:
1. Archbald (1979).
2. Seifert et al. (1985).
3. Ringwood and Kesson (1977).
4. Delano and Ringwood (1978a,b).
5. Delano and Lindsley (1983).
6. Chou et al. (1975).
7. Anders and Ebihara (1982).
8. Stosch (1981).
9. Fig. 3, this paper.

meteorites can readily be explained in terms of the achievement of equilibrium partitioning between metal and silicate phases during differentiation and cooling of their parent bodies at temperatures less than about 1200°C (Stolper, 1977).

The contrast in these abundance patterns is believed to reflect the very different processes involved in core formation in the Earth and in the meteoritic parent body. Core formation in the latter occurred in a relatively low pressure-temperature environment within one or more small parent bodies possessing radii of, at most, a few hundred kilometers. Core formation in the Earth involved metal segregation within a body of planetary size in which the central pressure exceeds 3 Mbar and the central temperature exceeds 4000°C. Distribution of nickel and cobalt between metal and silicate phases would have been substantially influenced by this high P,T environment. Moreover, the Earth's core is known to contain about 10% light elements. Ringwood (1984) showed that the principal light element in the core is probably oxygen and pointed out that its presence would have a significant effect on the distribution of siderophiles between metal and silicate phases. A further process that is believed to have had an important influence on the siderophile abundance patterns of the Earth's mantle was the physical mixing of an oxidized nebula condensate into the mantle at an advanced stage of accretion under conditions that did not permit it to equilibrate with metal phase (e.g., Newsom and Palme, 1984).

In summary, the abundances of nickel and cobalt in the Earth's mantle reflect the very complex processes (as yet incompletely understood) that occurred during accretion and core formation within a body of planetary dimensions. As such, these abundances are likely to be unique to the Earth (or to a similar planet such as Venus). In the light of these observations, it is interesting to study the abundances of nickel and cobalt in the Moon. Before proceeding with this topic, however, it is desirable to discuss the petrological basis that underlies our estimates of the global lunar nickel and cobalt abundances.

2. Petrological Provinces in the Moon

It is widely believed that a large region of the Moon extending to a depth of 400 km or more experienced extensive degrees of partial melting and differentiation during, or soon after, accretion (e.g., Taylor, 1982). One of the products was the plagioclase-rich lunar crust with a mean thickness of about 70 km, which is believed to have crystallized from a deep magma ocean during this early episode of melting and differentiation. The upper mantle of the Moon is believed to be comprised of pyroxene and olivine cumulates from the magma ocean, together with refractory peridotite, depleted in fusible components by the partial melting process that produced the magma ocean (e.g., Ringwood, 1979).

Mare basalts

Mare basalts compose only a small proportion of the volume of the lunar crust. Nevertheless, they are of profound petrologic and geochemical significance because they are believed to have been formed by subsequent partial melting of the ferromagnesian cumulates and peridotitic residua beneath the crust at depths of 100–400 km. Thus, they should be capable of providing key information on the nickel and cobalt contents of an extensive region of the lunar mantle.

Ringwood and Kesson (1976) showed that the low-titanium mare basalts have had a simpler petrogenetic evolution than the high-titanium basalts and retain a more direct "memory" of their mantle source regions. Accordingly, it is preferable to use geochemical data derived from low-titanium basalts in order to constrain the composition of the more primitive regions of the lunar mantle. Most mare basalts have experienced varying degrees of fractionation (mainly of olivine and pyroxene) after leaving their source region. Since nickel and cobalt are preferentially partitioned into these phases, use of the abundances of these elements in mare basalts to derive the Ni and Co contents of their source regions will usually result in underestimates. It is therefore essential to select data from mare basalts that can be shown by standard petrologic criteria such as Mg-numbers* to have experienced the least fractionation after leaving their source regions.

The recognition by Delano and Livi (1981), Delano and Lindsley (1983), and Delano (1986) of some 24 discrete families of volcanic glasses in lunar soils has added a new dimension to lunar basalt petrogenesis. The glassy spherules apparently formed by volcanic fire-fountaining. Their high liquidus temperatures combined with melting relationships at elevated pressures imply an origin at considerable depths, probably 300–400 km. Delano and colleagues have demonstrated that the least fractionated members of each of these families closely approximate primary magmas that have ascended from their source regions without appreciable fractionation of olivine (or pyroxene).

Delano and Livi (1981) and Delano and Lindsley (1983) showed that the source regions from which these primary magmas were derived appear to be comprised of mixtures of two distinct components, Ti-rich and Ti-poor, corresponding to different types of cumulates. Their data provide strong support for a petrogenetic model proposed earlier by Ringwood and Kesson (1976) and depicted in Fig. 1. During crystallization of the magma ocean, intermediate and late ferromagnesian cumulates would have become gravitationally unstable because of their high iron (and titanium) contents.

---

*"Mg-number" is the mol. fraction 100 MgO/(MgO + FeO). "Fe-number" is the mol. fraction 100 FeO/(MgO + FeO).

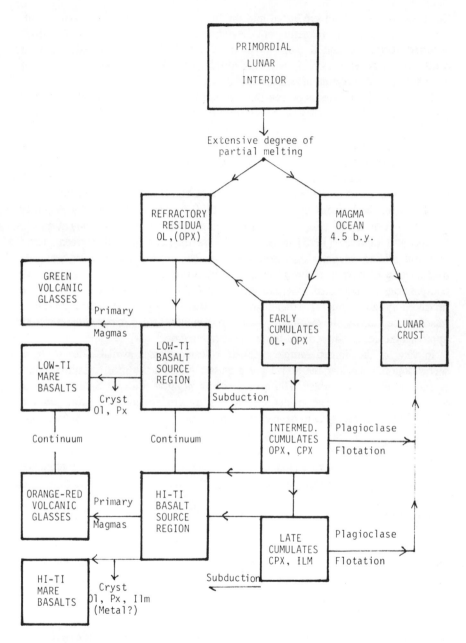

Fig. 1. Petrogenetic relationships between lunar volcanic glasses, mare basalts and their source regions according to Ringwood and Kesson (1976).

Consequently, they were subducted into the lunar interior where hybridization with less dense layers of early, refractory olivine and orthopyroxene cumulates and residua* occurred. Ringwood and Kesson (1976) showed that many characteristics of mare basalts could be explained if they had formed by subsequent partial melting of these hybrid source regions at considerable depth.

The green volcanic glasses possess the highest Mg numbers and Ni contents among primary lunar magmas. They were evidently derived from source regions in which Mg-rich olivine residua and/or early cumulates played a major role. It is from these residua and cumulates that the best estimates of the original nickel and cobalt abundances of the primordial lunar mantle (prior to differentiation of the lunar crust) can be obtained. The most primitive green glass composition contains 185 ppm Ni (Delano and Livi, 1981) and 82 ppm Co (Chou *et al.*, 1975). The composition of olivine on the liquidus of green glass at 1420°C was determined by Seifert *et al.* (1985). The olivine (Fo_{84}) contains 890 ppm of nickel and 156 ppm of cobalt. Delano and Lindsley (1983) demonstrated that the most primitive green glasses represent primary magmas that ascended from their source regions with negligible fractionation of olivine. From a detailed investigation of major and trace element compositions of individual spherules, Ma *et al.* (1981) showed that they had been produced by rather small ($\leq 10\%$) degrees of partial melting of their cumulate source regions. Accordingly, the measured Ni and Co contents of the liquidus olivines closely approach the composition of olivine in the source region.

In view of the limited sample of lunar volcanic glasses available for study, it seems doubtful whether the above green glass represents the most primitive liquid erupted at the lunar surface. The occurrence of negative Eu and Sr anomalies in green glass provides clear evidence that this material contains a more evolved component, which is likely to have experienced depletion in Ni and Co. Accordingly, the experimentally derived compositions of olivine on the liquidus of green glass may tend to underestimate the Ni and Co contents of olivines present in the earliest cumulates from the magma ocean, or from the refractory residua that remained in the lunar mantle after extraction of this magma ocean.

The lunar crust

This region consists dominantly of breccias possessing compositions similar to anorthositic gabbros. Attempts to estimate the abundances of indigenous siderophile elements in the crust have been complicated by the large amount of meteoritic

---

*Ringwood and Kesson's (1976) model described hybridization of late cumulates with the primordial lunar interior. However, they recognized (p. 1700) hybridization with refractory residua and early cumulates as a viable alternative model. Subsequent evidence shows that this latter scenario is to be preferred (e.g., Delano and Livi, 1981).

contamination. Nevertheless, significant quantities of indigenous siderophiles are undoubtedly present. The strong correlations of phosphorus and tungsten with lanthanum in highland rocks (Rammensee and Wänke, 1977; Dreibus et al., 1977; Delano and Ringwood, 1978a,b) and the similarity of W/La and P/La ratios in highland rocks and mare basalts demonstrate that most of the tungsten and phosphorus in highland rocks is indigenous. The good correlation of Co with (MgO + FeO) in highland rocks, mare basalts, and "pristine" lunar rocks (Section 3) likewise shows that most of the cobalt in lunar highland breccias is indigenous (Wänke et al., 1979). Curiously, although there is unequivocal evidence that most of the P, W, and Co* in highland rocks is indigenous, Anders (1975) has claimed that only a very small proportion of the nickel is also indigenous. However, the evidence that follows clearly contradicts this view and demonstrates that a substantial proportion of the nickel in highland breccias is genuinely indigenous.

Palme (1980) showed that in the terrestrial impact melt sheets at Lappajarvi and East Clearwater, variable amounts of contamination by the meteoritic projectile occurred. From an investigation of Ni-Ir and Ni-Co correlations within groups of samples from both localities, he was able to estimate the indigenous Ni and Co contents of the terrestrial target rocks prior to meteoritic contamination. These estimates were found to agree well with observed Ni and Co contents of uncontaminated terrestrial rocks in the vicinity, demonstrating ground-truth for this method. Palme then applied the same technique to two suites of samples from the Apollo 16 and 17 impact melt sheets. Indigenous components of 350 ppm Ni and 25 ppm Co were obtained at the Apollo 16 site, as compared to 40 ppm Ni and 22 ppm Co at the Apollo 17 site.

A second argument demonstrating the presence of substantial quantities of indigenous nickel in lunar highlands breccias was made by Wänke et al. (1979). These authors noted that the meteoritic projectiles that brecciated and contaminated the early lunar crust possessed chondritic Ni/Co ratios, as shown by Delano and Ringwood (1978a, Fig. 2), Wänke et al. (1978), and in Section 3 of this paper. If it is further assumed that *all* of the nickel in Apollo 16 breccias was of meteoritic origin, then the cobalt that would have been supplied by the meteoritic component can be obtained utilizing the chondritic Ni/Co ratio. After subtracting the meteoritic cobalt component from the total cobalt content of Apollo 16 breccias, Wänke et al. found that the ratios of residual cobalt (Co_r) to (Mg + Fe) in the highlands breccias were much smaller than the Co/(Mg + Fe) ratios displayed by both pristine lunar rocks and mare basalts. This is an unacceptable result, since as shown by Wänke et al., and elsewhere in this paper, the Co/(Mg + Fe) ratios of pristine

---

*On the average, about one third of the cobalt in highlands breccias is of meteoritic origin (Wänke et al., 1978).

rocks and mare basalts represent *minimum values* for the Moon because of the effects of metal fractionation in the former and olivine fractionation in the latter.

Clearly, the assumption that all of the nickel in the Apollo 16 breccias has been derived from projectiles of meteoritic origin is untenable. Wänke *et al.* demonstrated that if the indigenous Co/(Mg + Fe) ratios in Apollo 16 breccias were in fact similar to the minimum lunar values for this ratio displayed by pristine rocks and mare basalts, the Apollo 16 breccias must contain from 50–400 ppm of indigenous nickel. Wänke *et al.* (1979) also showed that the Co/Ir ratio of the projectiles at the Apollo 16 site was close to the chondritic ratio. Since, as discussed above, the Ni/Co ratios of the projectiles were also chondritic, it follows that the Ni/Ir ratios of the projectiles were chondritic.

Delano and Ringwood (1978a,b) studied the Ni and Co abundance systematics in lunar highland breccias from a different perspective. They made the most plausible and economical assumption that the Ni/Co and Ni/Ir ratios of the meteoritic projectiles that impacted the highlands were chondritic [this assumption was subsequently verified by studies of Ni, Co, and Ir abundance systematics in lunar breccias by Wänke *et al.* (1979) as discussed above]. After subtracting meteoritic Ni and Co contamination, assuming that all of the iridium in the highlands was of meteoritic origin, and normalizing the results to a constant Al_2O_3 content, the Ni and Co residuals were plotted in Ni/Co vs. Ni diagrams (Figs. 2 and 3).

The residual Ni/Co ratios are seen to define a linear trend with increasing Ni that falls close to the corresponding trend displayed by mare basalts (which are undoubtedly indigenous). These systematic relationships provide strong evidence that

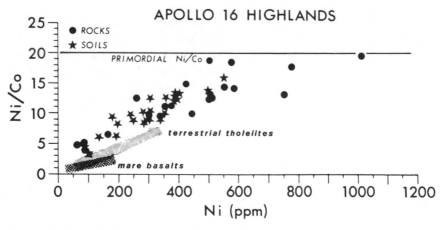

Fig. 2. Ni/Co vs. Ni diagram for Apollo 16 highland breccias after removal of meteoritic contamination and standardisation to a fixed Al_2O_3 content (to correct for plagioclase dilution). Corresponding fields for terrestrial basalts and lunar mare basalts are also shown (from Delano and Ringwood, 1978a).

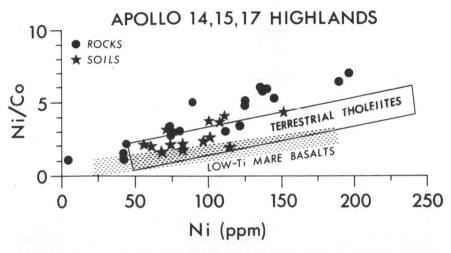

Fig. 3. *Ni/Co vs. Ni diagram for Apollo 17 highland breccias after removal of meteoritic contamination and standardisation to a fixed Al_2O_3 content (to correct for plagioclase dilution). Corresponding fields for terrestrial basalts and lunar mare basalts are also shown (from Delano and Ringwood, 1978a).*

residual Ni and Co abundances in the highlands obtained by the above procedure are also genuinely indigenous. Figures 2 and 3 also show that the Ni/Co vs. Co trends for the highlands are similar to those displayed by terrestrial basaltic rocks.

Two explanations of the trends shown in Figs. 2 and 3 have been proposed. Noting that the analogous terrestrial trend was caused primarily by olivine fractionation, Delano and Ringwood (1978a,b) suggested that the lunar trend was likewise caused by the crystallization of olivine within the magmatic system from which the lunar highlands had formed. Wänke *et al.* (1974, 1978) emphasized the significance of the high Mg-number (70) and high Cr content of the lunar crust and suggested that it contains a "primary component" derived from Earth-orbiting planetesimals that were accreted by the Moon after the anorthositic crust had formed. A related suggestion was made by Taylor and Bence (1975). The linear Ni/Co vs. Ni trends in Figs. 2 and 3 were interpreted by Wänke *et al.* in terms of a mixing relationship between this "primary matter" and indigenous differentiated mafic material from the magma ocean from which the crustal anorthosites crystallized. The planetesimals that provided the "primary matter" in the model of Wänke *et al.* are believed to be representative of the material from which the Moon accreted. Accordingly, the composition of the primary component is essentially that of the bulk Moon. Dreibus *et al.* (1977) estimated this bulk composition by deconvoluting mixing relationships observed in lunar breccias. This estimate was similar to that obtained by other workers using entirely different methods (Ringwood, 1979; Warren and Wasson, 1979; Delano and Lindsley, 1983; Delano, 1986). It was also similar to that of the Earth's mantle. The significance of these observations will be considered subsequently.

Pristine rocks from the lunar highlands

The vast bulk of the rocks from the lunar highlands are comprised of polymict breccias with complex structures. Apparently, they represent mixtures of several petrologic components, one of which arises from meteoritic contamination. In contrast, a small group of rocks has been identified that consists of simpler monomict mineralogical assemblages, sometimes possessing evidence of former igneous or metamorphic textures and free of meteoritic contamination by siderophile elements. These have been called pristine rocks. There are two principal petrologic groups of pristine rocks. The first is comprised of ferroan anorthosites that consist predominantly of anorthite with small amounts of olivine and pyroxene (Mg-number: 40–70). The second "high-Mg group" consists mainly of troctolites and norites with Mg-numbers mostly in the range 70–90.

It was originally believed that pristine rocks had crystallized from the magma ocean that produced the primordial lunar crust and had somehow escaped the intense impact-brecciation and mixing process to which the crust was subsequently exposed. However, it is now recognised that this interpretation represents an oversimplification. Many workers still consider that the ferroan anorthosites may have crystallized from the magma ocean that produced the original crust. However, geochemical and petrologic studies show that the high-Mg group have experienced a complex evolution and are not directly related either to the ferroan anorthosites or to the magma ocean that produced the lunar crust. It is possible that some of these represent cumulates formed in impact melts as they crystallized (e.g., Wood, 1975; Delano and Ringwood, 1978a,b). Others are believed to be derived from differentiated intrusions that were emplaced into the primitive crust (e.g., James, 1980).

Anders (1978) has argued that the indigenous siderophile abundances of the lunar crust can be obtained from averages of chemical analyses of pristine rocks. This view has some serious weaknesses as a working hypothesis:

1. Low abundance of siderophile elements is the property used most widely to recognize and define pristine rocks. It is a circular argument to conclude that because pristine rocks have very low contents of siderophiles, the indigenous abundances of siderophiles in the bulk lunar crust are also low.

2. The fact that the mineralogy and composition of the crust cannot readily be matched by arbitrary mixing of observed pristine rocks (Palme, 1980; Korotev, 1982) implies that the crust contains an important component (probably pyroxene-rich) that has not been sampled adequately by the pristine rock collection. This component could well be the source of the indigenous siderophiles of the crust as discussed earlier.

3. The high Fe-number of the ferroan anorthosites demonstrates that their parent magma had experienced extensive prior separation of olivines and pyroxenes that would be accompanied by strong depletion of nickel. Thus, the nickel present in the ferromagnesian minerals of ferroan anorthosites would not be representative of the parental magma.

4. Pristine rocks are believed to represent cumulates from magmas that experienced extensive differentiation during slow cooling, mostly at high levels in the lunar crust. Under the low prevailing oxygen fugacity, a small amount of metal phase is likely to have separated from the parent magmas, thereby causing drastic depletions in abundances of highly siderophile elements (Ryder, 1983). Further light on this process is shed by low-Ti mare basalts that contain quite high abundances of indigenous nickel (Section 4). It can be demonstrated that their source regions did not contain a metallic phase (e.g., Delano and Ringwood, 1978a,b). However, after eruption and during crystallization near the surface, a small amount of metal phase formed by auto-reduction. If these magmas had crystallized slowly in shallow magma-chambers, the metal would have segregated, thereby removing most of the highly siderophile elements such as Ni and Ir. All memory of the high indigenous siderophile abundances of the parent magma would have been lost. It is only because mare basalts crystallized quite rapidly, thereby preventing metal segregation, that the original nickel contents, prior to eruption, were preserved.

In light of these considerations, it would appear unwise to utilize pristine rocks to estimate the original abundance of a highly siderophile element, such as nickel, in the petrologic system from which the lunar highland breccias were derived. It is far more reasonable to utilize the siderophile abundance systematics displayed by the breccias themselves, as discussed in Sections 3 and 4, to obtain these abundances.

3. Lunar Geochemistry of Cobalt

Wänke et al. (1979) have emphasized the important role played by cobalt as a geochemical tracer in the Moon. Cobalt is only moderately siderophile and is not strongly fractionated by the separation of small amounts of metal phase. Moreover, cobalt is a compatible element that is not strongly fractionated during moderate degrees of crystallization of basaltic magmas. Accordingly, the content of cobalt in the source region of basaltic magmas can be estimated with some confidence.

The abundances of cobalt in low-Ti mare basalts and terrestrial oceanic tholeiites are shown in Fig. 4. The mean Co abundances and dispersions are very similar in both rock-types. Ringwood and Kesson (1977) drew attention to the significance of this relationship and concluded that the cobalt content of the lunar mantle is very similar to that of the terrestrial mantle.

Wänke et al. (1979) compared the Co/(Mg + Fe) ratios of pristine lunar rocks and mare basalts with the corresponding ratios in terrestrial basalts and peridotites.

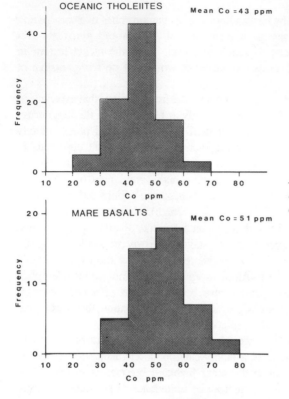

Fig. 4.   Cobalt abundances in low-titanium mare basalts and MORB oceanic tholeiites.

They concluded that the abundance of cobalt in the Moon was about 40% smaller than in the Earth's mantle. Although this difference is quite small, it has nevertheless been exaggerated. The data set used by Wänke *et al.* yield minimum lunar cobalt values because of the prior effects of metal separation from the pristine rocks and crystal fractionation in the mare basalts (Fig. 5). Moreover, a substantial part of the inferred depletion is an artifact arising from the circumstance that the lunar mantle contains substantially more FeO than the terrestrial mantle (Section 5). Delano (1986) compared the Co/Mg ratios of the least fractionated lunar volcanic glasses and mare basalts with those of terrestrial basalts and concluded that the difference did not exceed about 10%.

Similarity in the cobalt contents of the lunar and terrestrial mantles is also demonstrated by recent experimental studies (Seifert *et al.*, 1985). These show that olivine crystallizing near the liquidus of a primitive volcanic green glass (82 ppm Co; Chou *et al.*, 1975) contains 156 ppm Co. This may be compared with the mean value of 136 ppm found in a collection of olivines from peridotite xenoliths derived from the terrestrial mantle (Stosch, 1981).

Wänke *et al.* (1983) noted that the fractionation behaviour of cobalt is intermediate between those of iron and magnesium; hence, it is sometimes convenient to describe the behaviour of cobalt in terms of plots of Co vs. (Fe + Mg) or (FeO + MgO). A plot of Co/(Mg + Fe) ratios for Apollo 16 highland breccias, low-Ti mare basalts, and primitive lunar volcanic glasses is shown in Fig. 5. As discussed previously, the indigenous cobalt contents of the lunar highlands were obtained by subtracting meteoritic cobalt contamination using the chondritic Co/Ir ratio and the assumption that all of the iridium in the highlands was derived from meteoritic contamination.

Figure 5 shows that there is a clear correlation between residual Co and (Fe + Mg) for Apollo 16 highland breccias. Moreover, the line of best fit extrapolates directly to the field of the least fractionated volcanic glasses, the cobalt contents of which are unquestionably indigenous. This provides convincing evidence that the cobalt residuals in the highland breccias are also genuinely indigenous and demonstrates the very strong similarity of Co-Mg-Fe abundance systematics both in the lunar highlands and in primitive lunar basalts. Most of the points for mare basalts fall significantly below the correlation line, owing to the effect of crystal fractionation subsequent to derivation from their source regions (Delano and Ringwood, 1978a,b; Delano, 1986).

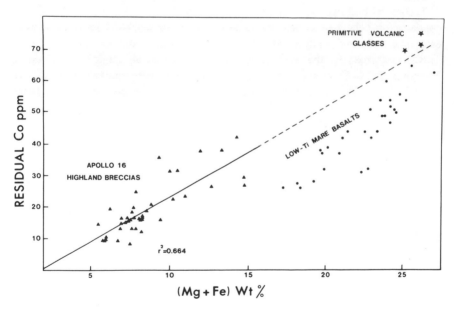

Fig. 5. Cobalt in primitive lunar volcanic glasses, low-Ti mare basalts, and Apollo 16 highland breccias (corrected for meteoritic contamination) vs. (Mg + Fe). Note that the solid line representing the best fit to the highland breccia data projects directly into the field of primitive volcanic glasses.

In Fig. 6, the cobalt residuals from Apollo 16 highland breccias are plotted against corresponding magnesium contents. A distinct correlation is evidence, though it is less strong than that in Fig. 5. This may be explained partly by differing partition coefficients of Co and Mg in olivines and pyroxenes. The slope of the Co-Mg correlation in Fig. 6 is of greater importance. It implies a mean Mg/Co ratio of 2300 in the lunar crust, which is virtually identical to the Mg/Co ratio of 2090 that is characteristic of the Earth's upper mantle (Sun, 1982).

This confirms a conclusion reached by Wänke *et al.* (1978) using rather different methods. These authors concluded that the lunar highlands contain a primitive magnesium-rich component with which the bulk of the indigenous chromium, cobalt, and nickel is associated and showed that the Mg/Co ratio of this component was similar to that of the terrestrial upper mantle.

Implications

The evidence discussed above demonstrates that the absolute abundance of cobalt in the lunar interior (crust + mantle) is very similar to its abundance in the Earth's mantle. This conclusion has direct implications for the corresponding nickel abundances.

Theoretical and experimental studies show that cobalt and nickel display similar condensation behaviour in the solar nebula and can be markedly fractionated from each other only under rather restrictive conditions (Fraser and Rammensee, 1982). This is confirmed by the similarity of Ni/Co ratios in all groups of chondrites. Moreover, the Ni/Co ratio shows only small variations in meteoritic irons and pallasites (Fig. 7) despite the fact that the corresponding Ni/Ir ratios vary by about a factor

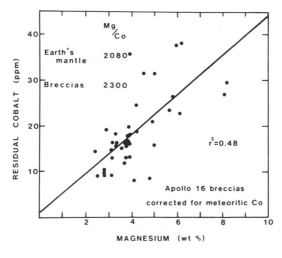

Fig. 6. Plot of cobalt (corrected for meteoritic contamination) vs. magnesium for Apollo 16 highland breccias.

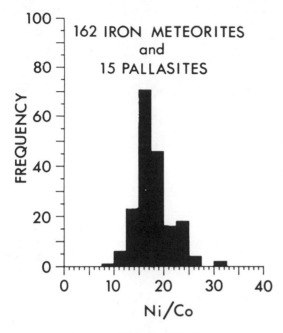

Fig. 7. Histogram of Ni/Co ratios in 177 pallasitic and iron meteorites. Mean value is 17.7 ± 3.6 (1σ). From Delano and Ringwood (1978b).

of 10^4 (Fig. 8), demonstrating that the irons have formed in a very wide range of cosmochemical and geochemical environments.

These observations strongly imply that if the Moon had formed as an independent body in the solar nebula, for example, according to the cosmogonic model of Anders (1977), it should also possess an overall chondritic Ni/Co ratio of 21. Other hypotheses lead to a similar result. Current versions of the binary planet or independent planet models of lunar origin invoke a physical fractionation process that removes metallic iron from the protolunar material prior to accretion (e.g., Orowan, 1969; Ruskol, 1977; Wood and Mitler, 1974). Because of the very high metal/silicate partition coefficients of Ni and Co (e.g., Seifert et al., 1985), the Ni/Co ratio of the metal phase remains essentially identical to the chondritic ratio. It requires only a very small amount of metal in the Moon for this phase to dominate its bulk Ni/Co ratio. If the Moon contains as little as 0.2% metal phase and had been formed according to any of the above hypotheses, its bulk Ni/Co ratio would deviate from the chondritic ratio by less than a factor of two. Lunar models containing 0.5% metal phase closely approach the chondritic Ni/Co ratio.

It is extremely difficult to understand how the binary or independent planet hypotheses could produce a Moon containing such small quantities of metal phase. This would require an extraordinarily efficient mechanism for fractionating metal from silicates. None of the mechanisms yet suggested appear capable of providing this near-perfect separation of phases.

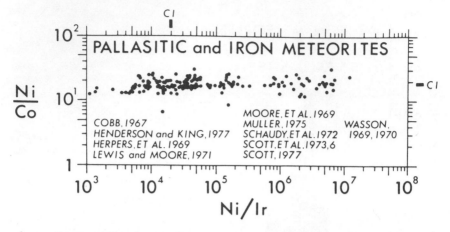

Fig. 8. Ni/Co vs. Ni/Ir in 177 pallasitic and iron meteorites. Diagram shows small (~50%) variation in Ni/Co during fractionation of Ni and Ir by a factor of 10,000. From Delano and Ringwood (1978b).

Widespread agreement now exists among geochemists that the Moon contains a small metallic core (Newsom, 1984). If the abundances of other siderophile elements in the Moon are to be explained on the basis of binary or independent planet hypotheses, the preferred size of the core is about 5% of the lunar mass and its minimum size is about 2% (Newsom, 1984). All of these models would possess chondritic Ni/Co ratios.

In light of these considerations, it seems reasonable to conclude that the Ni/Co ratio of the bulk Moon is close to the chondritic ratio of 21. It so happens that the nickel/cobalt ratio (19) of the Earth's upper mantle is also very close to the chondritic ratio (Sun, 1982). Previously we concluded that the bulk lunar cobalt abundance was similar to that of the Earth's mantle. *Similarity of the terrestrial and lunar Ni/Co ratios therefore implies that the total abundance of nickel in the bulk Moon is also similar to its abundance in the Earth's mantle.* Accordingly, it is of interest to examine the geochemical evidence relating to the occurrence of nickel in the Moon.

4. Lunar Geochemistry of Nickel

The lunar crust

Wänke *et al.* (1976, 1978) employed several recognised geochemical criteria to select a subset of the "least fractionated" Apollo 16 breccias. The criteria were directed towards avoiding samples containing high KREEP contents, abnormally high Fe/

Mg ratios, and correcting for meteoritic contamination via iridium as discussed earlier. Within this subset, Wänke *et al.* observed a strong correlation between Mg and residual nickel. Since magnesium is accepted as being indigenous and has similar crystal-chemical properties to nickel, this relationship provided strong evidence that the nickel residuals were also indigenous. Wänke *et al.* inferred from this and other data that the highlands at the Apollo 16 site contained a primary magnesium-rich component, with which substantial amounts of Cr and Ni are also associated. Morevoer, they showed that the primary component possessed a Mg/Ni ratio similar to that of the Earth's mantle. Wänke *et al.* (1978) interpreted this component to represent planetesimals of the parental lunar swarm that were accreted at a late stage, subsequent to the formation of the lunar crust and separation of the lunar core. The composition of this component would accordingly correspond to that of the bulk moon.

The genetic significance of this Mg-Ni correlation has not been adequately appreciated by the lunar science community. Actually, it can be demonstrated by a simpler procedure than that used by Wänke *et al.* In Fig. 9, the residual nickel values (corrected for meteoritic contamination via Ir) are plotted against magnesium for the entire set of Apollo 16 breccias previously studied by Delano and Ringwood (1982a,b). A significant positive correlation is clearly present. The mean Mg/Ni ratio of 120 thereby obtained is essentially identical to the Mg/Ni ratio of 115 characteristic of the Earth's mantle (Sun, 1982).

Mare basalt source regions

The relationships between nickel contents of terrestrial ocean floor tholeiites and low-Ti mare basalts and their corresponding $MgO/(MgO + FeO)$ ratios are shown

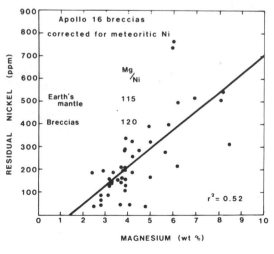

Fig. 9. Plot of nickel (corrected for meteoritic contamination) vs. magnesium for Apollo 16 highland breccias.

in Fig. 10. The solid lines delineate the field occupied by oceanic tholeiites. The observed relationships are caused by fractional crystallization of the respective magmas, accompanied by incorporation of nickel in olivines and pyroxenes. It is seen from Fig. 10 that the corresponding points for Apollo 12 basalts and also for Delano and Livi's (1981) volcanic glasses fall squarely in the middle of the terrestrial field and display a similar overall trend for nickel to fall as the Mg-number decreases. The correspondence in nickel fractionation behaviour in terrestrial and lunar basalt systems (Fig. 10) suggests that both possess generally similar amounts of nickel at equivalent stages of differentiation. This was the basis of the conclusion by Ringwood and Kesson (1977) and Delano and Ringwood (1978a,b) that the nickel abundances in the source regions of terrestrial oceanic tholeiites and low-Ti mare basalts are similar, "within a factor of about two" (Ringwood, 1979). Subsequently, it has become evident that the situation is not quite as straightforward as is suggested by Fig. 10, because the absolute concentration of FeO in the lunar mantle is substantially higher than in the terrestrial mantle, as discussed in Section 5.

Seifert *et al.* (1985) have experimentally determined the nickel content of olivine crystallizing near the liquidus (1420°C) of a primitive lunar volcanic green glass (185 ppm Ni). Delano and Lindsley (1983) demonstrated that the compositions of the green glasses closely approached those of primary magmas that had not experienced any significant crystal fractionation en route to the surface. Accordingly, the nickel content of the liquidus olivine should be close to that of the olivines in the source region, as discussed in Section 2. Seifert *et al.* (1985) obtained a nickel content of 890 ppm in the lunar olivine, which is a factor of 3.3 smaller than average terrestrial mantle olivine that contains 2900 ppm Ni (Archbald, 1979).

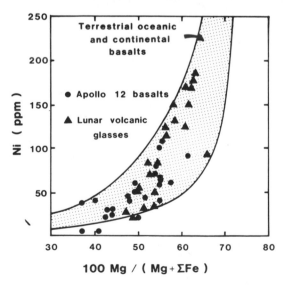

Fig. 10. Nickel contents vs. MgO/ (MgO + FeO) ratios in Apollo 12 basalts and primitive lunar volcanic glasses compared with corresponding field observed for terrestrial basalts. The boundaries of the latter are shown by the stippled region inside the solid lines. Based on Delano and Ringwood (1979b) and Delano and Livi (1981).

This implies that the mare basalt source regions are depleted in nickel by a factor of 3.3 compared to the Earth's mantle.

A referee has suggested caution in applying excessive significance to the composition of green glass. It should be remembered, however, that green glass represents the most primitive endmember of a continuum of volcanic glass and mare basalt compositions that display a wide range of fractionation histories and ages (Delano and Lindsley, 1983). We are selecting and utilizing this primitive composition to estimate its source composition in exactly the same way as the compositions of the most primitive terrestrial basalts are selected and used to estimate that of the Earth's mantle. It should be noted that our estimate of the Ni depletion in the lunar mantle agrees closely with estimates made by Delano (1986) and by Wänke and Dreibus (1986). These authors made their estimates on the basis of Ni vs. Mg correlations for a broad family of lunar basalts, volcanic glasses and terrestrial basalts, and komatiites. A depletion factor of 4.0 ± 0.5 was found for Ni in the lunar mantle relative to the terrestrial mantle.

It was previously concluded that the total nickel concentration of the bulk Moon is very similar to that of the Earth's upper mantle and that the lunar highlands contain a component of "primary matter" possessing a Mg/Ni ratio that also closely resembles that of the terrestrial mantle. On the other hand, the source region of low-Ti mare basalts representing an extensive region of the lunar mantle is substantially depleted in nickel by a factor between 3 and 4 as compared to the terrestrial mantle. These conclusions can be reconciled only if the nickel deficit in the lunar mantle is compensated by the presence of a lunar metallic core containing a significant proportion of nickel.

5. The Lunar Core

Geophysical and geochemical evidence

Geophysical evidence bearing on the existence of a metallic core in the Moon is rather equivocal. Nakamura *et al.* (1974) tentatively proposed the existence of a low-velocity core based upon the interpretation of a single and poorly resolved seismic arrival. Nakamura *et al.* (1976) further state that "The radius of a low-velocity (molten) core cannot be greater than about 350 km because normal transmission of P waves to that depth is observed." This implies that the core does not exceed more than 2% of the mass of the Moon. The observed lunar moment of inertia coefficient of $1/MR^2 = 0.391 \pm 0.002$ (Blackshear and Gapcynski, 1977) has been used to infer the presence of a dense metallic core, but the observations can also be satisfied by accumulation of subducted late iron-rich cumulates (e.g., Ringwood and Kesson, 1976) deep in the lunar interior as discussed by Herbert (1980). Some electromagnetic measurements suggest the existence of an electrically conductive "core" with a radius of 400–500 km (Russell, 1984). However, they are not capable of determining whether this structure is metallic. Silicate rocks at temperatures close to their solidus could yield the inferred electrical conductivity of the "core" (Russell, 1984).

Lunar models (e.g., Buck and Toksöz, 1984) based on attempts to match the observed density, moment of inertia, and seismic velocity distributions are consistent with the existence of metallic cores amounting to 1–2% of the lunar mass. However, a metallic core is not required by the data and could be replaced by a central accumulation of dense, subducted, iron-rich cumulates (Herbert, 1980).

Seismic P- and S-wave velocities, together with petrologic data, strongly suggest that the lunar mantle contains about 12–16% of FeO (Ringwood, 1977, 1979; Wänke et al., 1977; Goins et al., 1979; Buck and Toksöz, 1980; Nakamura et al., 1982; Delano, 1986). This is substantially higher than the terrestrial mantle, which contains about 8% FeO (e.g., Ringwood, 1979), and implies that the lunar mantle is appreciably denser than the terrestrial mantle. In conjunction with the observed mean density of the Moon (3.344 g/cm^3), this places limits on the maximum size of a metallic lunar core. Figure 33 in Toksöz et al. (1974) shows that the core would not exceed about two percent of the mass of the Moon, assuming a plausible range of MgO, SiO$_2$, CaO, and Al$_2$O$_3$ contents for the lunar mantle.

New evidence relating to the presence of a lunar core has recently been presented by Yoder (1984). From a study of the phase shift in the forced precession of the lunar figure, he inferred the presence of a liquid core with a radius of about 330 km (equivalent to ~1.5% of the lunar mass for an Fe-Ni composition).

Recent geochemical evidence also suggests that the Moon may possess a small metallic core. Newsom (1984) has pointed out that there is a close relationship between the depletion of involatile elements in the Moon and the siderophilic nature of the element. Thus, the sequence W, Co, P, Ni, Mo, and Re corresponds to increasing siderophilic nature. Whereas the abundances of W and Co are similar in the Moon and Earth's mantle, P, Ni, Mo, and Re are depleted by factors of 2, 3, 25, and 80, respectively. This pattern can be readily explained if small amounts of a metal phase had segregated from the source regions of mare basalts at an early stage, prior to the partial melting event that generated the basalts. This may well have occurred during the major differentiation that formed the lunar crust and the complementary system of underlying mafic and ultramafic cumulates and residua that compose the lunar upper mantle.

It was concluded in Section 3 that the mean concentrations of Ni (and Co) in the bulk Moon are very similar to those in the Earth's upper mantle. However, olivines present in the lunar mantle contain only a third of the nickel (890 ppm) present in olivines from the terrestrial mantle (2900 ppm). The comparative depletion of nickel in the lunar mantle is readily explained if about two thirds of the total nickel in the Moon is present in a metallic core.

Seifert et al. (1985) have carried out an extensive experimental investigation of the partitions of nickel and cobalt between metal phases, olivines, and a primitive lunar basaltic composition as represented by green glass containing 185 ppm Ni (Delano and Lindsley, 1983). Olivine (Fo_{84}) crystallizing near the liquidus of the green glass at 1420°C contained 890 ppm Ni and this composition is believed to be representative of the parental lunar mantle. The metal phase in equilibrium with this olivine possessed the composition $Fe_{62}Ni_{38}$ (wt %). At 1200°C, which probably represents a lower temperature limit for effective metal segregation during the early lunar differentiation, the composition of metal in equilibrium with lunar olivines containing 890 ppm Ni was $Fe_{43}Ni_{57}$. It is evident, therefore, as previously pointed out by Ringwood (1979), that if a metallic lunar core was formed during early differentiation, it would have been nickel-rich.

This conclusion is also supported by the compositions of the earliest metal phases crystallizing from relatively primitive low-Ti mare basalts, which contain from 30% to 50% nickel (Brett et al., 1981; Hewins and Goldstein, 1974). Because of earlier fractionation experienced by the parental magmas of mare basalts, these Ni-contents are likely to underestimate the Ni contents of metal that would be in equilibrium with their source regions.

A lunar core containing about 40% Ni and amounting to about 0.4% of the mass of the Moon would reconcile our estimates of the present nickel content of the lunar mantle and the bulk lunar Ni abundance, based on that of the Earth's mantle. Such a core would possess a radius of about 200 km.

The behaviour of cobalt provides an important check on the composition and size of the lunar core. It was shown in Section 3 that the cobalt content of the lunar mantle is very similar to that of the Earth's mantle. The olivine in the lunar mantle was estimated to contain 156 ppm Co as compared to 136 ppm Co in olivine from the terrestrial mantle. Seifert et al. (1985) studied the partition of cobalt between lunar olivine and metal phases. The composition of metal phase in equilibrium with an olivine of lunar composition at 1420°C was $Fe_{63}Ni_{36}Co_{0.7}$.

Separation of a core containing 0.7% Co and amounting to 0.4% of a lunar mass would remove about 28 ppm Co from the olivine of the lunar mantle. If lunar olivine were corrected for this loss, it would contain 184 ppm Co as compared to 136 ppm Co for terrestrial olivines. When allowances are made for the experimental and observational uncertainties in estimating the cobalt content of lunar olivine, the agreement can be considered to be satisfactory. Thus, the cobalt partition relationships are reasonably consistent with the model that was derived on the basis of nickel distribution between lunar core and mantle. However, if the core were appreciably larger than considered in the present model, a depletion of cobalt in the lunar mantle would become readily resolvable, contrary to observation. The nickel content of the metal phase also has a significant effect on its cobalt content and tends to lower it. For example, in the absence of nickel, metal in equilibrium with lunar olivine (82 ppm Co) at 1420°C has the composition $Fe_{98.3}Co_{1.7}$ (Seifert et al., 1985). This represents an increase in cobalt content by a factor of 2.4 compared to the Ni-Fe-Co alloy. Thus, a low-nickel core would be substantially more effective in removing cobalt from the lunar mantle. These considerations demonstrate that the cobalt content of the lunar mantle provides an important constraint on the size and composition of the lunar core.

Core models of Newsom and Drake

Newsom and Drake (1982, 1983) and Newsom (1984) have discussed the distributions of Co, Ni, W, and P within the lunar interior on the basis of the "independent planet" hypothesis of lunar origin. They maintain that similarities in abundances of these elements in the Earth's mantle and Moon are coincidental and that the lunar abundances can be explained if the Moon accreted as an independent planet in the solar nebula. After accretion, a "geophysically plausible" lunar core amounting to two percent of the lunar mass separated, causing these elements to be depleted in the lunar mantle, thereby explaining their observed abundances.

In order to obtain the required large depletions of tungsten and phosphorus in the lunar mantle by separation of a small amount of metal phase, the partition coefficients for these elements between metal and silicate phases must be maximized. This requires a highly restrictive and contrived set of relationships between FeO content of the silicate phase, Ni content of metal, temperature and degree of partial melting, all of which exert a strong influence upon the distribution of tungsten and phosphorus between metal and silicate phases. In order to explain the tungsten content of the lunar mantle, Newsom and Drake require separation of metal (6% Ni) after only 1% partial melting of the silicate phase, assuming that the lunar mantle contains about 13% FeO as discussed earlier. [In their discussion, Newsom and Drake (1983) explicitly accept the above evidence that the FeO content of the lunar mantle is substantially higher than that of the terrestrial mantle.] In this paper, however, we have shown that the lunar core contains about 40% Ni, which would reduce the partition coefficient of tungsten between metal and silicate by about a factor of 5 (Rammensee and Wänke, 1977). This would require metal segregation after only 0.2% of partial melting of the silicates. The physical plausibility of separating metal phase at such low degrees of partial melting in the relatively small gravitational field of the Moon is open to serious question.

Newsom and Drake (1983) attempted to explain the phosphorus content of the Moon on the basis of separation of a small core (6% Ni) at 2% partial melting of the lunar mantle containing about 13% FeO (as implied by seismic and petrologic data and accepted by Newsom and Drake). They conceded, however, that their model would not suffice if the lunar core possessed a high nickel content, as shown in this paper. In a subsequent paper, Newsom (1984) reinvestigated the problem and acknowledged the difficulties of explaining the phosphorus abundances in the lunar mantle.

These are by no means the only problems encountered by Newsom and Drake's model. The postulated temperature of metal segregation (1300°C) is not easily reconciled with the very small degrees of partial melting that are required. Newsom (1984) also examined the abundances of Ni and Co in the Moon on the basis of the independent planet hypothesis and concluded that a core amounting to at least 5.5% of the lunar mass would be required to explain the abundances of these elements in the lunar mantle. This is considerably larger than the "geophysically plausible lunar core" referred to in earlier papers by Newsom and Drake. As discussed earlier in this section, a core as large as this is not permitted if the lunar mantle contains ~12–15% FeO, as implied by seismic and petrologic evidence (Ringwood, 1977, 1979; Wänke et al., 1977; Buck and Toksöz, 1980; Nakamura et al., 1982, Delano, 1986). It is also in conflict with the conclusion by Nakamura et al. (1976, 1982) that normal P-wave transmission occurs to a depth of 1400 km in the Moon.

It is evident from the above considerations that Newsom and Drake's interpretation of siderophile abundances in the Moon on the basis of the independent planet hypothesis cannot be sustained.

The role of sulphur

Our previous discussion of the lunar core has neglected the possibility that it may contain sulphur as a major component, as suggested, for example, by Brett (1973). It is well known that the Moon is highly depleted in volatile elements and it seems likely that this general pattern also extends to sulphur, which is an extremely volatile element in the solar nebula. This factor is likely to limit its abundance in the Moon.

If sulphur were an important component of the lunar core, the partition coefficients of cobalt, nickel, tungsten, and phosphorus between metallic and silicate phases would be considerably depressed in comparison with the corresponding partition coefficients between sulfur-free metal and silicates (Ringwood, 1979; Naldrett and Duke, 1980; Jones and Drake, 1982; Newsom and Drake, 1983). It would then be quite impossible to explain the abundances of these siderophile elements in the lunar mantle on the basis of the independent planet hypothesis of lunar origin. This fact was obviously realized by Newsom and Drake (1983) and Newsom (1984) who assumed in their models that the lunar core did not contain significant amounts of sulphur.

If, on the other hand, the Moon had been derived from the Earth's mantle, then the content of sulphur in the Moon would not exceed that of the terrestrial mantle, which is estimated to contain about 320 ppm S (Ringwood and Kesson, 1977). If all of this sulphur had entered a metallic core amounting to 0.4% of the lunar mass, the core would contain about 8% sulphur. This is an upper limit since the lunar mantle has retained a significant amount of sulphur as shown by the presence of 1150 ppm S in low-Ti mare basalts (Ringwood and Kesson, 1977). It seems possible that if the Moon had been derived from the Earth's mantle, the lunar core would contain a few percent of sulphur. Sulphur concentrations as low as this would not substantially affect the conclusions reached earlier in this paper.

6. Origin of the Moon

In previous sections, we have demonstrated that the abundances of cobalt and nickel in the bulk Moon are similar to their abundances in the Earth's upper mantle. Moreover, the Ni/Mg and Co/Mg ratios of an important component of the lunar crust are similar to those in the Earth's upper mantle.

These similarities are of profound genetic significance. It was pointed out in Section 1 that the present Co and Ni abundances in the Earth's mantle were established by several complex processes connected with segregation of the core. It is probable that they are unique to the Earth, or to a body of planetary size that has experienced processes of accretion and core formation very similar to those of the Earth. The simplest explanation of the above observations is that the material that now composes the Moon was derived from the Earth's upper mantle after the Earth's core had segregated.

As an alternative hypothesis it has frequently been suggested that the Moon accreted as an independent planet in the same general region of the parental solar nebula as the Earth and from a common reservoir of metal and silicate particles. According to another version of this hypothesis, the metal-silicate reservoir was first established within the interiors of a preexisting population of chemically differentiated planetesimals. In these independent planet (or binary planet) models, it is assumed that an appropriate physical fractionation mechanism caused almost complete removal of metal from the local region in which the Moon was accreting. This is necessary in order to explain the small size of the lunar core in comparison to the large metallic core in the Earth and the related observation that the Moon is depleted in iron compared to primordial abundances, whereas the iron abundance in the Earth is similar to the primordial or solar value, appropriately normalized.

The mechanisms that have been proposed as causes of this iron fractionation have been rather vague and speculative. Orowan (1969) suggested that the differential ductility and strength of iron compared to silicates would result in iron grains welding together to form relatively large planetesimals during collisions within a geocentric swarm, while silicates would be fragmented into fine dust that might be preferentially enriched in the circumterrestrial swarm from which the Moon ultimately formed. Several authors (e.g., Ruskol, 1977) have suggested variations on this theme. It seems very doubtful, however, whether these processes could operate with the extremely high efficiencies necessary to explain the paucity of metal in the Moon.

Harris and Tozer (1967) invoked ferromagnetic properties of metal grains to explain their preferential aggregation and accretion by the Earth and corresponding depletion in the Moon. However, Banerjee (1967) demonstrated that the mechanism was inadequate (by a factor of 10^4) to cause the desired effects.

Another attempt to explain the depletion of iron in the Moon has been based on models of disintegrative capture (e.g., Wood and Mitler, 1974). It is assumed that a large proportion of planetesimals within the inner solar system melted and differentiated to form metallic cores overlain by silicate mantles. The differentiation is believed to be analogous to that experienced by the parent bodies of eucrites, pallasites, and iron meteorites. At an advanced stage of accretion of the Earth, some of these differentiated planetesimals passed within Roche's limit and were tidally disrupted. It is assumed that the silicate mantles of these objects were captured into geocentric orbits while the iron cores accreted directly onto the Earth. The Moon then formed by coagulation of the Earth-orbiting silicate fragments. This hypothesis encounters severe dynamical difficulties that greatly decrease its plausibility (e.g., Kaula and Harris, 1975). Also, recent studies show that it is improbable that the Earth-approaching planetesimals would spend enough time within Roche's limit to permit tidal disruption (Mizuno and Boss, 1984).

In addition to the difficulties mentioned above, those hypotheses that rely on the physical fractionation of iron from silicates to explain the Moon's low density encounter a further very formidable problem. They imply that the chemical compositions of metal phases accreted by the Earth and Moon should be similar. Accordingly, the compositions of the respective metallic cores should be generally similar, except where they have been modified by internal processes (e.g., incorporation of light element(s) in the Earth's core).

There are strong grounds for believing that the Earth accreted from material possessing approximately chondritic relative abundances of Mg, Fe, Ni, and Co (e.g., Ringwood, 1975, 1979). Accordingly, it can be shown that the Earth's core contains about 6% nickel and 0.3% of cobalt. In contrast, the evidence discussed in this paper implies that the lunar core contains about 40% nickel and 0.7% cobalt. If the nickel content were smaller than 40%, the cobalt content would be correspondingly higher, in the range 1.7–5.0% Co for a nickel-free core separating at 1420°–1200°C (Seifert et al., 1985).

These drastic differences between the compositions of the metallic phases present in the interiors of the Earth and Moon are not explained by any of the current versions of the independent planet hypothesis.

Attempts to relate the composition of the lunar mantle to the composition of the mantles of differentiated meteoritic parent bodies are contradicted by the observed compositions of these bodies. Differentiation in the small meteoritic parent bodies seems to have been a comparatively simple process in which equilibrium was attained between metal and silicate phases (e.g., Stolper, 1977). In consequence, the nickel and cobalt contents of eucrites and of the olivines in pallasites are very low and stand in sharp contrast to the "Earthlike" values characteristic of the lunar interior (Table 1). Moreover, the vast majority of meteoritic irons and all pallasites are much poorer in nickel than the lunar core.

A different version of the binary or independent planet hypothesis of lunar origin has been preferred by Anders (1977). According to his cosmogony (which is based on an interpretation of the chemical fractionations observed in chondritic meteorites), the terrestrial planets were formed from mixtures of several components in the solar nebula, including high-temperature condensates (Ca, Al oxides), intermediate-temperature condensates (Mg silicates), and Fe-Ni metal phase and low-temperature condensate (volatile-rich and oxidised). It is postulated that these components were fractionated from each other by undefined physical processes, after which they accreted into planets. In the case of the Moon, it is assumed that accretion occurred from an appropriate mixture of silicate and oxide condensates but that the metal phase was severely depleted, presumably by an extension of the same process that caused the less marked depletions of metal in the L and LL groups of chondrites. Despite the numerous degrees of freedom of this hypothesis, it retains some testable consequences by virtue of its reliance on chondritic fractionation processes for empirical justification. Chondritic meteorites display a specific relationship between the FeO/ (FeO + MgO) ratio of their silicate phases and the Ni/Fe ratio of coexisting metal phase. This relationship, expressed as Prior's second "law," is maintained irrespective of the degree of physical separation of metal phase, as in L and LL chondrites. (It is a general relationship, arising simply from the fact that nickel is much more readily reduced to the metallic state than iron.) Thus, if planets and the Moon have formed by the processes advocated by Anders, they should also conform to Prior's rule. This would require that if the lunar core contained about 40% Ni, the silicates of the lunar mantle would possess Mg-numbers between 60 and 70, depending on the sulphur content of the core. It can readily be demonstrated that Mg-numbers in this range would lead to densities for the lunar mantle that are substantially higher than observed. Moreover, they are in conflict with the Mg-numbers in the lunar mantle implied by the compositions of primitive mare basalts (Ringwood, 1979; Delano and Lindsley, 1983).

7. Conclusions

1. The mean abundances of cobalt and nickel in the bulk Moon are similar to those in the Earth's mantle. Moreover, the lunar crust appears to contain a primary component possessing similar Mg/Ni and Mg/Co ratios to the Earth's mantle.

2. The cobalt content of the lunar mantle is similar to that of the Earth's mantle, whereas the nickel content of the lunar mantle is depleted by a factor of about three.

3. The above conclusions can be reconciled by the segregation of a small core (0.4% by mass) within the Moon. Equilibrium considerations dictate that the core contains about 40% Ni and 0.7% Co. The existence of a small metallic core in the Moon is supported by previous studies on the P, Mo, and Re abundances in the lunar mantle (Wänke et al., 1983; Newsom, 1984).

4. The high Ni and Co abundances of the Earth's mantle result from the operation of several complex processes associated with the formation of a large metallic core in a planetary sized body. These processes could not have operated in the same manner in the parent bodies of differentiated meteorites or within the Moon.

5. The similarity in Ni and Co contents of the bulk Moon and the Earth's mantle strongly suggests that the material now in the Moon was derived from the Earth's mantle subsequent to separation of the Earth's core.

6. Current hypotheses according to which the Moon was formed as a binary or independent planet fail to explain:

a. The high indigenous concentrations of Ni and Co in the lunar mantle and crust.

b. The high Ni and Co contents of the lunar core.

c. The similarity in abundances of Ni and Co between the bulk Moon and the Earth's mantle.

d. The evidence that the composition of the lunar core differs grossly from that of the Earth's core.

Acknowledgments. The authors are indebted to Drs. S. Sun and S. Kesson for helpful comments on the manuscript.

8. References

Anders E. (1977) Chemical compositions of the Moon, Earth and eucrite parent body. *Phil. Trans. R. Soc. London, A295*, 23–40.

Anders E. (1978) Procrustean science: indigenous siderophiles in the lunar highlands according to Delano and Ringwood. *Proc. Lunar Planet. Sci. Conf. 9th*, pp. 161–184.

Anders E. and Ebihara M. (1982) Solar-system abundances of the elements. *Geochim. Cosmochim. Acta, 46*, 2363–2380.

Archbald P. M. (1979) Abundances and dispersions of some compatible volatile and siderophile elements in the mantle. MSc. thesis, Australian National University. 181 pp.

Banerjee S. (1967) Fractionation of iron in the solar system. *Nature, 216*, 718.

Blackshear W. and Gapcynski J. (1977) An improved value of the lunar moment of inertia. *J. Geophys. Res., 82*, 1699–1701.

Brett R. (1973) A lunar core of Fe-Ni-S. *Geochim. Cosmochim. Acta, 37*, 165–170.

Brett R., Butler P., Meyer C., Reid A., Takeda H., and Williams R. (1981) Apollo 12 igneous rocks 12004, 12008, 12009, 12022: A mineralogical and petrological study. *Proc. Lunar Sci. Conf. 2nd*, pp. 301–317.

Buck W. and Toksöz M. N. (1980) The bulk composition of the Moon based on geophysical constraints. *Proc. Lunar Planet. Sci. Conf. 11th*, pp. 2043–2058.

Chou C. L., Boynton W. V., Sundberg L., and Wasson J. (1975) Volatiles on the surface of Apollo 15 green glass and trace element distributions among Apollo 15 soils. *Proc. Lunar Sci. Conf. 6th*, pp. 1701–1727.

Delano J. W. (1986) Abundances of cobalt, nickel, and volatiles in the silicate portion of the Moon, this volume.

Delano J. W. and Lindsley D. H. (1983) Mare basalts from Apollo 17: Constraints on the Moon's bulk composition. *Proc. Lunar Planet. Sci. Conf. 14th*, in *J. Geophys. Res., 88*, B3–B16.

Delano J. W. and Livi K. (1981) Lunar volcanic glasses and their constraints on mare petrogenesis. *Geochim. Cosmochim. Acta, 45*, 2137–2149.

Delano J. W. and Ringwood A. E. (1978a) Siderophile elements in the lunar highlands: nature of the indigenous component and implications for origin of the Moon. *Proc. Lunar Planet. Sci. Conf. 9th*, pp. 111–159.

Delano J. W. and Ringwood A. E. (1978b) Indigenous abundances of siderophile elements in the lunar highlands: implications for origin of the Moon. *Moon and Planets, 18*, 385–425.

Dreibus G., Spettel B., and Wänke H. (1977) The bulk composition of the Moon and eucrite parent body. *Proc. Lunar Sci. Conf. 8th*, 211–227.

Fraser D. G. and Rammensee W. (1982) Activity measurements by Knudson cell mass spectrometry—the system Fe-Co-Ni and implications for condensation processes in the solar nebula. *Geochim. Cosmochim. Acta, 46*, 549–556.

Goins N., Toksöz M. N., and Dainty A. (1979) The lunar interior, a summary report. *Proc. Lunar Planet. Sci. Conf. 10th*, pp. 2421–2439.

Harris P. and Tozer D. (1967) Fractionation of iron in the solar system. *Nature, 215*, 1449–1451.

Herbert F. (1980) Time dependent lunar density models. *Proc. Lunar Planet. Sci. Conf. 11th*, pp. 2015–2030.

Hewins R. H. and Goldstein J. (1974) Metal-olivine associations and Ni-Co contents in two Apollo 12 mare basalts. *Earth Planet. Sci. Lett., 24*, 59–70.

James O. B. (1980) Rocks of the early lunar crust. *Proc. Lunar Planet. Sci. Conf. 11th*, pp. 365–393.

Jones J. and Drake M. (1982) An experimental geochemical approach to early planetary differentiation, (abstract). In *Lunar and Planetary Science XIII*, pp. 369–370. Lunar and Planetary Institute, Houston.

Kaula W. and Harris A. (1975) Dynamics of lunar origin and orbital evolution. *Rev. Geophys. Space Phys., 13*, 363–371.

Korotev R. L. (1982) Compositional relationship of the pristine non-mare rocks to the highland soils and breccias. In *Workshop on Pristine Highlands Rocks* (J. Longhi and G. Ryder, eds.), pp. 52–55. LPI Tech. Rept. 83–02. Lunar and Planetary Institute, Houston.

Ma M. S., Liu Y., and Schmitt R. (1981) A chemical study of individual green glasses and brown glasses from 15426: Implications for their petrogenesis. *Proc. Lunar Planet. Sci. 12B*, pp. 915–933.

Mizuno H. and Boss A. (1984) Tidal disruption and the origin of the Moon (abstract). In *Papers Presented to the Conference on the Origin of the Moon*, p. 37. Lunar and Planetary Institute, Houston.

Nakamura Y., Latham G., Lammlein D., Ewing M., Dunnebier F., and Dorman J. (1974) Deep lunar interior inferred from recent seismic data. *Geophys. Res. Lett., 1*, 137–140.

Nakamura Y., Latham G., Lammlein D., Ewing M., Dunnebier F., and Dorman J. (1976) Seismic structure of the Moon: A summary of current status. *Proc. Lunar Sci. Conf. 7th*, pp. 3113–3121.

Nakamura Y., Latham G., and Dorman H. J. (1982) Apollo lunar seismic experiment—final summary. *Proc. Lunar Planet. Sci. Conf. 13th*, in *J. Geophys. Res, 87*, A117–A123.

Naldrett A. and Duke J. (1980) Platinum metals in sulphide ores. *Science, 208*, 1417–1424.

Newsom H. (1984) The lunar core and the origin of the Moon. *EOS (Trans. Am. Geophys. Union), 65*, 369–370.

Newsom H. and Drake M. (1982) Constraints on the Moon's origin from the partitioning behaviour of tungsten. *Nature, 297*, 210–212.

Newsom H. and Drake M. (1983) Experimental investigation of the partitioning of phosphorus between metal and silicate phases: Implications for the Earth, Moon and eucrite parent body. *Geochim. Cosmochim. Acta, 47*, 93–100.

Newsom H. E. and Palme H. (1984) The depletion of siderophile elements in the Earth's mantle: New evidence from molybdenum and tungsten. *Earth Planet. Sci. Lett., 69*, 354–364.

Orowan E. (1969) Density of the Moon and nucleation of the planets. *Nature, 222*, 867.

Palme H. (1980) The meteoritic contamination of terrestrial and lunar impact melts and the problem of indigenous siderophiles in the lunar highlands. *Proc. Lunar Planet. Sci. Conf. 11th*, pp. 481–506.

Rammensee W. and Wänke H. (1977) On the partition coefficient of tungsten between metal and silicate and its bearing on the origin of the Moon. *Proc. Lunar Sci. Conf. 8th*, pp. 399–409.

Ringwood A. E. (1975) *Composition and Petrology of the Earth's Mantle*. McGraw Hill, New York. 618 pp.

Ringwood A. E. (1977) Basaltic magmatism and the composition of the Moon I: major and heat-producing elements. *Moon, 16*, 389–423.

Ringwood A. E. (1979) *Origin of the Earth and Moon*. Springer-Verlag, New York. 295 pp.

Ringwood A. E. (1984) The Earth's core: its composition, formation and bearing upon the origin of the Earth. *Proc. R. Soc. London, A397*, 1–46.

Ringwood A. E. and Kesson S. E. (1976) A dynamic model for mare basalt petrogenesis. *Proc. Lunar Sci. Conf. 7th*, pp. 1697–1722.

Ringwood A. E. and Kesson S. E. (1977) Basaltic magmatism and the bulk composition of the Moon II. Siderophile and volatile elements in Moon, Earth and chondrites: Implications for lunar origin. *The Moon, 16*, 425–464.

Ruskol Y. L. (1977) The origin of the Moon. In *The Soviet-American Conference on the Cosmochemistry of the Moon and Planets* (J. H. Pomeroy and N. J. Hubbard, eds.), pp. 815–822. NASA, Washington, DC.

Russell C. (1984) On the Apollo subsatellite evidence for a lunar core (abstract). In *Papers Presented to the Conference on the Origin of the Moon*, p. 7. Lunar and Planetary Institute, Houston.

Ryder G. (1983) Nickel in olivines and parent magmas of lunar pristine rocks. In *Workshop on Pristine Highlands Rocks and the Early History of the Moon* (J. Longhi and G. Ryder, eds.), p. 66–68. LPI Tech. Rpt. 83-02, Lunar and Planetary Institute, Houston.

Seifert S., O'Neill H., and Brey G. (1985) Experimental investigation of the distributions of nickel and cobalt between metal phase, a primitive lunar basaltic magma and its liquidus olivine, unpublished manuscript.

Stolper E. (1977) Experimental petrology of eucritic meteorites. *Geochim. Cosmochim. Acta, 41*, 587–611.

Stosch H. (1981) Sc, Cr, Co and Ni partitioning between minerals from spinel peridotite xenoliths. *Contrib. Mineral. Petrol., 78*, 166–174.

Sun S. (1982) Chemical composition and origin of the Earth's primitive mantle. *Geochim. Cosmochim. Acta, 46*, 179–192.

Taylor S. R. (1982) *Planetary Science: A Lunar Perspective*. Lunar and Planetary Institute, Houston. 481 pp.

Taylor S. R. and Bence A. E. (1975) Evolution of the highland crust. *Proc. Lunar Sci. Conf. 6th*, pp. 1112–1141.

Toksöz M. N., Dainty A., Solomon S., and Anderson K. (1974) Structure of the Moon. *Rev. Geophys. Space Phys., 12*, 539–567.

Wänke H. and Driebus G. (1986) Geochemical evidence for the formation of the Moon by impact-induced fission of the proto-Earth, this volume.

Wänke H., Palme H., Baddenhausen H., Dreibus G., Jagoutz E., Kruse H., Spettel B., Teschke F., and Thacker R. (1974) Chemistry of Apollo 16 and 17 samples: bulk composition, late stage accumulation and early differentiation of the Moon. *Proc. Lunar Sci. Conf. 5th*, pp. 1307–1355.

Wänke H., Palme H., Kruse H., Baddenhausen H., Cendales M., Dreibus G., Hofmeister H., Jagoutz E., Palme F., Spettel B., and Thacker R. (1976) Chemistry of lunar highland rocks: A refined evaluation of the composition of primary matter. *Proc. Lunar Sci. Conf. 7th*, pp. 3479–3499.

Wänke H., Baddenhausen H., Blum K., Cendales M., Dreibus G., Hofmeister H., Kruse H., Jagoutz E., Palme C., Spettel B., Thacker R., and Vilsek E. (1977) On the chemistry of lunar samples and achondrites. Primary matter in the lunar highlands: A re-evaluation. *Proc. Lunar Sci. Conf. 8th*, pp. 2191–2213.

Wänke H., Dreibus G., and Palme H. (1978) Primary matter in the lunar highlands: The case of the siderophile elements. *Proc. Lunar Planet. Sci. Conf. 8th*, pp. 83–110.

Wänke H., Dreibus G., and Palme H. (1979) Non-meteoritic siderophile elements in lunar highland rocks: evidence from pristine rocks. *Proc. Lunar Planet. Sci. Conf. 10th*, pp. 611–626.

Wänke H., Dreibus G., Palme H., Rammensee W., and Weckwerth G. (1983) Geochemical evidence for the formation of the Moon from the Earth's mantle (abstract). In *Lunar and Planetary Science XIV*, pp. 818–819. Lunar and Planetary Institute, Houston.

Warren P. H. and Wasson J. T. (1979) Effects of pressure on the crystallization of a "chondritic" magma ocean and implications for the bulk composition of the Moon. *Proc. Lunar Planet. Sci. Conf. 10th*, pp. 2051–2083.

Wood J. A. (1975) Lunar petrogenesis in a well-stirred magma ocean. *Proc. Lunar Sci. Conf. 6th*, pp. 1087–1102.

Wood J. A. and Mitler H. (1974) Origin of the Moon by a modified capture mechanism (abstract). In *Lunar Science V*, pp. 851–853. The Lunar Science Institute, Houston.

Yoder C. F. (1984) The size of the lunar core (abstract). In *Papers Presented to the Origin of the Moon*, p. 6. Lunar and Planetary Institute, Houston.

The Bulk-Moon MgO/FeO Ratio: A Highlands Perspective

PAUL H. WARREN

Institute of Geophysics and Planetary Physics, Department of Earth and Space Sciences, University of California, Los Angeles, CA 90024

Among the few aspects of the bulk composition of the Moon that can be constrained well enough to have genetic significance is the *mg*, or molar MgO/(MgO + FeO), ratio. This ratio can be assessed from nonmare (highlands) samples, either by considering means of producing olivines with *mg* = 0.92 (and more abundant olivines with *mg* ≥ 0.90) or by considering means of producing compositions that represent aggregates of large volumes of the lunar crust (such as soils, with *mg* typically = 0.69). Both these approaches indicate that the bulk-Moon *mg* ratio is ≥0.85, and most likely ≥0.87. Based on the second approach, the bulk-Moon *mg* ratio is probably <0.91, and almost certainly <0.93. Most of the uncertainty in these calculations stems from poor constraints on the degree of melting during production of melts parental to the rocks of the nonmare crust. The bulk-Moon *mg* ratio is essentially the same, within a few percent uncertainty, as that of the Earth (about 0.895). The recently popular Earth-impact model, wherein the Moon forms out of vaporized ejecta from a collision between the Earth and one of its largest protoplanets, may have difficulty accounting for the similarity in *mg* ratio between the Moon and the Earth. Supporters of the Earth-impact model claim that it would help account for the Moon's volatile-trace element depletions, and even putative refractory-major element enrichments, due to fractional condensation of the vaporized ejecta. However, fractional condensation could easily lead to a different (most likely higher) *mg* ratio for the Moon compared to the Earth. Closely similar *mg* ratios are consistent with models that form the Moon by accretion from a circumterrestrial swarm of fragments of previously differentiated asteroid-sized bodies, provided the mechanism(s) for depleting the Moon in FeNi did not simultaneously enrich it in materials from the shallowest parts of the previously differentiated bodies.

1. Introduction

The main observational constraints on the origin of the Moon involve its orbit (and tidal dissipation parameters), its thermal history, and its bulk composition. Several aspects of the bulk composition can be constrained well enough to have genetic significance. One of these aspects is the ratio MgO/FeO, commonly expressed as *mg* [= molar MgO/(MgO + FeO)]. This ratio evolves during igneous fractionation in ways that are well understood.

The simple fact of the Moon's proximity to the Earth suggests similar provenance of parental materials. Similar provenance is also the simplest explanation for oxygen isotopic similarities (Clayton, 1977) of Earth and Moon materials. However, its low bulk density shows that the Moon is depleted in metallic Fe relative to the Earth (Urey, 1962). Depletions of minor-trace volatile elements are also obvious from geochemical data (e.g., Wasson, 1971; Wänke *et al.*, 1973; Ganapathy and Anders, 1974). The simplest assumption would be that in terms of nonsiderophile, nonvolatile species, including MgO and FeO, the Moon is compositionally similar to the silicate portion of the Earth. Based on such evidence as compositions of mafic volcanic rocks and their xenoliths, the *mg* ratio of the Earth's upper mantle (treating all Fe as FeO) is >0.88 according to most estimates, and almost certainly >0.85 [see Smith, 1977 and Basaltic Volcanism Study Project (BVSP), 1981 for reviews]. The most widely quoted value is 0.895 (Ringwood, 1975). Most estimates of the *mg* ratio of the bulk Moon are substantially lower, <0.82 (Table 1). If real, a disparity of *mg* ratio between the Moon and the Earth would imply either a disparity in provenance of constituent materials, or else a fractionation of the *mg* ratio between the Moon and the Earth as they formed.

In estimating the bulk-Moon *mg* ratio, little attention has been paid to the rocks from the ancient nonmare (highlands) crust. Warren and Wasson (1979b) estimated that the bulk-Moon *mg* ratio is similar to that of the Earth's upper mantle, based on the *mg* ratios of the ferroan anorthosite group of pristine nonmare rocks, assuming that this group formed by flotation of plagioclase over a primordial magma ocean (or magmasphere). But most other estimates place great emphasis on the high FeO contents of mare basalts (e.g., Morgan *et al.*, 1978; Ringwood, 1979; Delano, 1984),

TABLE 1. *Recent Estimates of the Bulk-Moon* **mg** *Ratio.*

Reference	MgO (wt %)	FeO (wt %)	MgO/(MgO + FeO)
Smith (1976)	43.9	6.0	0.929
Warren and Wasson (1979b)	33.5	8.4	0.88
Longhi and Boudreau (1979)	33.6	9.4	0.864
Taylor (1984)	–	–	0.84
Wänke *et al.* (1977)	32.3	13.0	0.816
Ringwood (1979)	32.7	13.9	0.807
Morgan *et al.* (1978)	29.09	12.96	0.800
Buck and Toksöz (1980)	29.02	12.90	0.800
Delano (1984)	34.3	16.5	0.79
Binder (1983)	34.8 ± 1.1	18.0 ± 0.7	0.775

Data for MgO and FeO refer to the silicate fraction of the Moon only. Numerous prior estimates by some of the same authors are not included.

or else on seismic velocity profiles of the lunar mantle (Buck and Toksöz, 1980; Binder, 1983). In this paper the bulk-Moon *mg* ratio will be estimated primarily through consideration of the more magnesian components of the nonmare crust.

2. Critique of Estimates Based on Mare Basalts

One simple model to estimate the bulk-Moon FeO content (G. Dreibus, H. Wänke, and H. Palme, personal communication, 1985) is to compare FeO contents of mare basalts and Earth basalts. Taken at face value, this comparison implies a much higher FeO for the Moon than for the Earth. But this comparison is oversimplified. Due to convective stirring, including subduction of lithosphere by plate tectonics, Earth basalts come from a mantle that is, in terms of major elements like Mg and Fe, almost homogeneous (Ringwood, 1975). In contrast, the most widely accepted genetic model for mare basalts (e.g., Taylor, 1982) holds that the source regions were cumulates from fractional crystallization of a magmasphere. Cumulate sequences (layered intrusions) are notoriously heterogeneous: the FeO content of the melt may increase to several times its initial value as successive layers are deposited (Wager and Brown, 1967). There are many additional complications. For example, most mare basalts, especially the highest-*mg* varieties, are greatly depleted in plagioclase (Al) compared to Earth basalts and, more generally, compared to the plagioclase-saturated liquids that would form by moderate-degree partial melting of an undifferentiated source (BVSP, 1981). If this "missing" plagioclase were restored to the mare basalts, their FeO contents would be diluted to values closer to the FeO contents of Earth basalts.

Typical mare basalts have *mg* ratios in the range 0.32–0.54 (Taylor, 1982). Clearly, their source regions had *mg* ratios far lower than that of the Earth's mantle. But most, if not all, of these sources had been previously differentiated: As reviewed by Taylor (1975), mare basalts have negative Eu anomalies (Eu is less abundant, relative to chondrites, than rare-earth elements of similar atomic number, Sm and Gd), a trait that implies fractionation (loss) of plagioclase, yet experimental data imply that the source region residual phases generally did not include plagioclase. Thus, plagioclase must have been lost during prior magmatism. During (and probably even before) this episode of plagioclase removal, a comparable fraction of the ultimate source material must have been removed as mafic silicates, a process that inevitably reduced the *mg* ratio of the eventual source region.

A few mare basalts and basaltic glasses have flat REE patterns, without large Eu anomalies. Ringwood (1979) estimated the composition of the lower half of the mantle as an average of source regions for two such samples: the Apollo 15 green volcanic glass and Apollo 12 basalt 12002 (Ringwood believes that the mare basalt source regions were at depths >400 km). He assumed that the green glass formed by 50% partial melting (leaving an olivine-orthopyroxene residuum) of a bulk lower mantle composition, and that 12002 formed by 10% partial melting (leaving a clinopyroxene-orthopyroxene-olivine residuum) of a bulk lower mantle composition, plus subsequent crystallization of 10% olivine. This approach led Ringwood (1979) to estimate an *mg* ratio of 0.75 for the bulk lower mantle. (Note: Ringwood's estimate for the bulk composition of the *outer* 400 km of the Moon, derived on the basis of *non*mare rocks, is much higher: 0.86.) In similar fashion, Delano (1984) derived a bulk-Moon composition with *mg* = 0.79 by extrapolating geochemical trends among volcanic glasses to chondritic ratios of refractory lithophile elements (Ca, Al, and Ti). [Curiously, Delano and Lindsley (1983) interpreted data for mare basalts to imply a bulk-Moon *mg* ratio of 0.87.]

To be sure, the bulk-Moon composition must be closely related to the source region compositions inferred by Ringwood (1979) for the green glass, and by Delano (1984) for his hypothetical "genesis glass." But these models are questionable insofar as they suppose, just because elements like REE, Ca, Al, and Ti are unfractionated, that Mg and Fe are also completely unfractionated. The volcanic glasses are closely related to a host of other mare volcanic compositions, for which there is compelling evidence of prior magmatism. If the bulk-Moon content of Ca is roughly chondritic, the most refractory phases deep within the Moon are probably orthopyroxene (Warren and Wasson, 1979b) and olivine, both of which incorporate only minor REE, Ca, Al, and Ti. A major fraction of the original material could have been removed (either by being left behind in a partial melt source region, or by crystallization from a melt) as olivine and/or orthopyroxene, without any manifest effect on ratios among REE, Ca, Al, and Ti. Any loss of olivine or orthopyroxene would have reduced the *mg* ratio of the eventual source region. The most primitive mare compositions are certainly useful for constraining the bulk composition of the Moon, but their source region *mg* ratios should be interpreted as conservative lower limits for the *mg* ratio of the bulk Moon.

During discussion at Kona, it was suggested that highly magnesian rocks might not be representative of an appreciable volume fraction of the lunar crust. If this argument is relevant, it should also be addressed to models based on mare basalts. Depths of the mare basalt source regions are generally estimated at 250 ± 150 km (Taylor, 1982). The uncertainties in these estimates (typically quoted as "100–400 km") do not imply that the source regions were necessarily representative of one-half of the lunar mantle. Head (1975) gave a very rough estimate that mare basalt amounts to "much less than 1%" of the total volume of the crust, which he assumed to be 10 vol % of the whole Moon. Hörz (1978) gave a carefully considered estimate that the "total volume of mare basalt produced throughout lunar history may be as small as $1-2 \times 10^6$ km^3," i.e., just 0.005–0.009 vol % of the Moon. Even with an extremely conservative estimate for the average degree of melting (*f*) during partial melting, say *f* = 0.03, Hörz's (1978) estimate implies that only 0.2% of the total mantle volume was directly involved in mare basalt genesis.

3. MgO/FeO Data for Nonmare Soils and Rocks

3.1. Soils

The overall composition of the ancient nonmare crust is probably best estimated on the basis of data for soils, which form by thorough, perfectly random (impact) mixing of materials from the outer few kilometers of the Moon. Most available soil samples contain large fractions of mare basalt. But soils from three sites are at least nearly free of mare basalt. The *mg* ratios of these soils are: 0.66 for an average Apollo 16 soil (Korotev, 1981); 0.685 for the most Al-rich (i.e., mare basalt-poor) Apollo 17 soil (73141, from Station 2a) (Miller *et al.*, 1974; Rhodes *et al.*, 1974; Rose *et al.*, 1974; Wänke *et al.*, 1974); and 0.69 for an average of Luna 20 soil analyses (Kallemeyn and Warren, 1983). In addition, shock-lithified "fossil" highlands soils are available as meteorites ALHA81005, Y-791197, and Y-82192. The mass-weighted mean of literature analyses of the *mg* ratio of ALHA81005 is 0.727 (Warren and Kallemeyn, 1985). The Y-791197 meteorite appears to be heterogeneous: *mg* analyses range from 0.59 (Fukuoka *et al.*, 1985) to 0.67 (Warren and Kallemeyn, 1985). A single analysis of Y-82192, which is probably paired with Y-791197, indicates an *mg* ratio of 0.67 (K. Yanai, personal communication, 1985). [One further complication: It seems possible, if not probable, that all three lunar meteorites were propelled off the Moon by a single impact (e.g., Warren and Kallemeyn, 1985).]

Thanks to impact mixing, each of these soils is representative of a large volume of the nonmare crust. The simple mean of their five *mg* ratios (counting Y-791197 and Y-82192 as a single sample), 0.68, is an estimate for the *mg* ratio of the nonmare crust as a whole. Without the benefit of the meteorites, Taylor (1975) derived a similar result: 0.70. Likewise, Korotev's (1981) "HON" composition, "the composition of the lunar crust after removal of meteoritic material, mare basalt, KREEP, and cumulate anorthosite" (the latter component contributes negligible Mg and Fe) has *mg* = 0.695.

3.2. Rocks

Soils are mixtures of powdered rocks. The opposite of a soil is a "pristine" rock: a fragment that happened to escape fine-scale mixing with other materials during meteoritic impact, i.e., a piece with its endogenic igneous (or, occasionally, metamorphic) composition intact. Soils inform us about the mean composition of large volumes of the crust, but pristine rocks are essential for assessing compositional diversity in the crust.

There are at least three distinct groups of pristine nonmare rocks (Warren, 1985): ferroan anorthosites, Mg-rich rocks, and KREEP rocks. Mg-rich rocks are usually

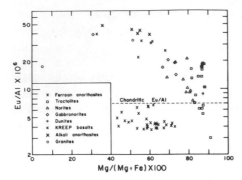

Fig. 1. A plot of bulk-rock **mg** vs. Eu/Al for pristine nonmare rocks (after Warren and Kallemeyn, 1984) illustrates the distinction between ferroan anorthosites and Mg-rich rocks (essentially all other pristine nonmare rocks except for KREEP basalts). Note extremely high **mg** ratios of many Mg-rich troctolites.

feldspathic cumulates, and KREEP (potassium, rare earth element, and phosphorus rich) rocks are usually basaltic in texture and major-element composition. The ferroan anorthosite group is generally interpreted as cumulates from the primordial differentiation of the outer few hundred kilometers of the Moon (the magmasphere), but the other pristine nonmare rocks are generally interpreted as products of more localized melting events (e.g., James and Flohr, 1983; Shirley, 1983; Warren, 1985).

High-*mg* lithologies are a major fraction of the total population of pristine nonmare samples, and are found from every site that provided at least a few kilograms of nonmare rocks. Roughly 100 pristine nonmare lithologies have been described to date. Ryder and Norman (1979) compiled nearly all of the geochemical data available for these lithologies before 1979. Most of the more recent data appeared in Warren and Kallemeyn (1984) and earlier papers of the same "foray" series. Figure 1 shows bulk-rock *mg* ratios for pristine nonmare rocks plotted vs. Eu/Al to illustrate the distinction of ferroan anorthosites from all other lunar rocks (regarding Eu/Al, see Warren and Kallemeyn, 1984). Note that eight of the rocks plotted are troctolites or dunites with bulk-rock *mg* ratios greater than 0.870. Several pristine troctolites have olivine with *mg* = 0.92: Apollo 15 breccia 15445 clast "A" (not plotted because no bulk-rock Fe datum is available) and clasts "ST1" and "ST2" from Apollo 16 breccia 67435 (Ryder and Norman, 1979; Ma *et al.*, 1981). Additional pristine troctolite clasts that contain olivine with *mg* = 0.90 but cannot be shown in Fig. 1 (because matrix-free bulk rock analyses are not available) include one from Apollo 16 breccia 60035 (Warner *et al.*, 1980), one from Apollo 17 breccia 76255 (Ryder and Norman, 1979), one from Y-791197 (K. Yanai, personal communication, 1985), and two probably pristine clasts from Apollo 14 breccia 14305 (Shervais *et al.*, 1984).

From some sites, only few grams of nonmare material were obtained, and consequently few pristine lithologies were recognized. Nonpristine samples can be used to infer original lithic compositions, albeit it must be recognized that a certain fraction of such material is not even indigenous to the Moon. Prinz *et al* (1973) provided a mass of data for soil particles from the Luna 20 probe. Seven of the 330 olivines they analyzed had *mg* >0.920, and five had *mg* >0.930. An olivine with *mg* = 0.92 is among the few that have been analyzed from lunar meteorite Y-791197 (Yanai and Kojima, 1984), and an olivine with *mg* = 0.93 is among the few analyzed from ALHA81005 (Ryder and Ostertag, 1983). Apparently high-*mg* olivines are just as common in the Luna 20 region, and in the region(s) that spawned Y-791197 and ALHA81005, as they are among pristine lithologies from Apollo sites.

4. Partial Melting Models

4.1. Theory

Lunar rocks and soils consist of minerals that crystallized from igneous melts. Those melts presumably formed by partial melting of the lunar interior. Behavior of MgO and FeO during partial melting and crystallization is constrained by a mass of experimental data. In this section an equation will be derived to constrain the *mg* ratio of an equilibrium partial melt for a given *mg* ratio of the source rock.

One of the most important factors governing the *mg* ratio of a melt is the exchange reaction distribution coefficient, K_D, defined (Roeder and Emslie, 1970) as

$$K_D = ([FeO]_{xtl}/[FeO]_{liq})/([MgO]_{xtl}/[MgO]_{liq}) \tag{1}$$

where the square brackets denote molar concentrations, and the subscripts are xtl = crystal, and liq = liquid. Equation (1) implies

$$mg_{xtl} = 1/(1 + K_D/mg_{liq} - K_D) \tag{2}$$

and

$$mg_{liq} = K_D/(K_D + 1/mg_{xtl} - 1) \tag{3}$$

Literature data indicate that at atmospheric pressure K_D for olivine in equilibrium with basaltic melt is 0.30 ± about 0.02 (Table 2). Numerous experimental studies indicate that K_D for olivine increases with increasing pressure, with an apparent slope dK_D/dP (kbar^{-1}) = roughly 0.0007–0.0030 (Grover *et al.*, 1980; Delano, 1980; Longhi *et al.*, 1978); the two studies using the largest arrays of data (Takahashi and Kushiro, 1983; Takahashi, 1984) found about 0.0020 and 0.0015, respectively. But this effect may be offset, because increasing pressure in the 10–40 kbar range (the pressure at the center of the Moon is about 47 kbar) stabilizes pyroxene at

TABLE 2. Experimental Measurements of $K_D = (Fe/Mg)_{xtl}/(Fe/Mg)_{liq}$.

Mineral	System composition	K_D	Pressure	Reference
Olivine	Lunar Al,Mg-rich basalts	0.28–0.31	1 atm	1
"	Terrestrial basalts	0.26–0.33	1 atm	2
"	Miscellaneous terrestrial	0.27–0.38	1 atm	3
"	Peridotitic komatiites	0.314 ± 0.012	1 atm	4
"	Synthetic chondrules	0.33	1 atm	5
"	Peridotitic komatiites	0.34–0.36	0–40 kbar	6
"	Lunar low-Ti mare basalts	0.30–0.36*	0–12.5 kbar	7
"	Lunar magnesian basalt	0.32–0.37*	8–25 kbar	8
"	Part. melts terres. perid.§	0.28–0.37*	5–35 kbar	9
"	Part. melts terres. perid.‡	0.28–0.46*	5–75 kbar	10
Orthopyroxene	Lunar Al,Mg-rich basalts	0.24–0.26	1 atm	1
"	Lunar, Mg-rich	0.30	1 atm	11
"	Part. melts terres. perid.§	0.28–0.32	10–30 kbar	9
"	Peridotitic komatiites	0.33–0.35	15–40 kbar	6
Pigeonite	Lunar low-Ti mare basalts	0.30	1 atm	12
Pigeonite	Lunar, Mg-rich	0.30	1 atm	11
Pigeonite	Eucritic basalts	0.30	1 atm	13
Augite	Lunar low-Ti mare basalts	0.23	1 atm	12
Augite	Lunar, Mg-rich	0.28	1 atm	11
Low-Ca augite	Part. melts terres. perid.§	0.32–0.35	10–20 kbar	9

All system compositions listed feature low oxygen fugacity and TiO_2.
*In these studies, K_D was observed to correlate with pressure (see text).
§Partial melts of terrestrial peridotite HK66.
‡Partial melts of terrestrial peridotite KLB-1.
References: (1) Weill and McKay (1975) and McKay and Weill (1977); (2) Roeder (1974); (3) Takahashi (1978); (4) Bickle (1982) (citing unpublished data by N. T. Arndt); (5) Cirlin et al. (1985); (6) Bickle et al. (1977) and Bickle (1982); (7) Longhi et al. (1978); (8) Grover et al. (1980); (9) Takahashi and Kushiro (1983); (10) Takahashi (1984); (11) Longhi (1980); (12) Grove (1978); (13) Stolper (1977).

the expense of olivine (e.g., Warren, 1985), and pyroxenes tend to have lower K_D than olivine (Table 2). Thus, during high-degree partial melting in the lunar mantle, regardless of depth, the bulk K_D between the melt and residual solids is almost certainly in the range 0.25–0.35, and very probably between 0.30 and 0.35, and K_D between olivine and the same melt, upon emplacement into a crustal (low-pressure) magma body, is almost certainly within a few percent of 0.30.

Another factor governing the mg ratio of a partial melt is the fraction of melting, f. If f is $\ll 1$, the melt will have a lower mg than the initial, system composition. But as f approaches unity, the melt mg approaches the mg of the system. It is generally acknowledged that, so long as the density of the melt is considerably less than the density of the residual crystals, melt fractions during partial melting are unlikely to ever exceed 40% (e.g., Arndt, 1977; Green, 1972; McKenzie, 1984).

Beyond some "critical melt fraction" (van der Molen and Paterson, 1979) the solid matrix of the rock breaks down, allowing the buoyant melt to rapidly rise around and away from the crystals. Based on viscosity experiments, Roscoe (1952), Arzi (1978), and Auer *et al.* (1981) concluded that the "critical" f is about 30%; van der Molen and Paterson (1979) found it to be 30–35%. An absolute upper limit of 55% can be inferred from the results of Arndt (1977), who heated a peridotite enough to make it 55% molten: After two hours, 20% of the liquid separated from the residual crystals and collected at the top of the charge. Jaques and Green (1980, p. 294) allude to similar results with f as low as 40%.

Ultramafic melts (komatiites) were initially interpreted as possible evidence for nearly complete melting of mantle peridotite (Viljoen and Viljoen, 1969). But alternative models have since been proposed for komatiite genesis, such as melting of previously melted source regions (e.g., Arndt, 1977), and/or melting deep in the mantle, where high-degree melts tend to be more Mg-rich because the solid residuum is not pure olivine (e.g., McKenzie, 1984; Warren, 1984). Catastrophic melting occurs in the wake of megaimpacts. Unfortunately, the most detailed models of this process have so far addressed only the Earth (Green, 1972; Grieve and Parmentier, 1984). It is unclear how such models should be "scaled" for the smaller Moon. The Earth model of Grieve and Parmentier (1984) suggests that after a 100-km basin forms the highest-degree melting occurs within about 25 km of the surface, i.e., well within the depth range of the Moon's crust. Melting is not nearly as catastrophic below 50 km, and at 120 km depth the thermal anomaly produced by the impact is never more than about 250 K. If the upper mantle and crust had already differentiated by crystallization of a magmasphere initially several hundred kilometers thick, then the outer 50–100 km of the Moon probably had a lower *mg* ratio than the deeper parts of the former magmasphere (in effect, the "target" material was already a residual liquid). It seems best to assume that despite the potential for occasional deep near-total melting due to megaimpacts, the degree of melting involved in production of lunar partial melts was, as on the Earth, generally less than about 50%.

To calculate mass balance for MgO and FeO between an equilibrium partial melt and residual crystals, we must constrain the ratio a, defined here as

$$a = [\text{MgO} + \text{FeO}]_{xtl}/[\text{MgO} + \text{FeO}]_{liq} \qquad (4)$$

i.e., a equals the mole fraction of (MgO + FeO) in the crystals, divided by the mole fraction of (MgO + FeO) in the melt. Melting experiments using natural Earth peridotites indicate (Fig. 2) that a is consistently close to 2.0, and very seldom <1.6. Calculations by Longhi (1981) of phase compositions during 15–20% partial melting of two suggested lunar bulk compositions gave similar results: $a = 2.05$–2.79.

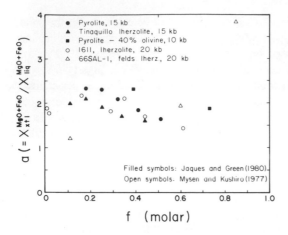

Fig. 2. *Literature data indicate that the ratio "a" (molar concentration of Fe + Mg in the solids)/(molar concentration of Fe + Mg in the melt) is generally close to 2.0 (regardless of degree of melting) during partial melting of peridotites.*

We will now derive an equation to constrain melt mg as a function of system (i.e., bulk source region) mg, and vice versa. First, simple mass balance requires that

$$[MgO]_{sys} = f \cdot [MgO]_{liq} + (1-f) \cdot [MgO]_{xtl} \tag{5}$$

where the subscript sys stands for system, and as usual, the square brackets denote molar concentrations. (Note that f is likewise defined as mole fraction molten, a departure from the standard practice in trace element modeling, where f is defined as weight fraction molten.) Also,

$$[MgO + FeO]_{sys} = f \cdot [MgO + FeO]_{liq} + (1-f) \cdot [MgO + FeO]_{xtl} \tag{6}$$

Since mg is defined as $(MgO)/(MgO + FeO)$, equation (5) is equivalent to

$$[MgO + FeO]_{sys} \cdot mg_{sys} = f \cdot mg_{liq} \cdot [MgO + FeO]_{liq} + (1-f) \cdot mg_{xtl} \cdot [MgO + FeO]_{xtl} \tag{7}$$

Combining equations (6) and (7) yields

$$mg_{sys} = \frac{f \cdot mg_{liq} \cdot [MgO + FeO]_{liq} + (1-f) \cdot mg_{xtl} \cdot [MgO + FeO]_{xtl}}{f \cdot [MgO + FeO]_{liq} + (1-f) \cdot [MgO + FeO]_{xtl}} \tag{8}$$

Combining equations (4) and (8) yields

$$mg_{sys} = \frac{f \cdot mg_{liq} \cdot [MgO + FeO]_{liq} + (1-f) \cdot mg_{xtl} \cdot a \cdot [MgO + FeO]_{liq}}{f \cdot [MgO + FeO]_{liq} + (1-f) \cdot a \cdot [MgO + FeO]_{liq}} \tag{9}$$

which obviously reduces to

$$mg_{sys} = \frac{f \cdot mg_{liq} + (1-f) \cdot mg_{xtl} \cdot a}{f + (1-f) \cdot a} \qquad (10)$$

Combining equations (2) and (10) yields

$$mg_{sys} = \frac{f \cdot mg_{liq} + (1-f) \cdot a / [1 + (K_D / mg_{liq} - K_D)]}{f + (1-f) \cdot a} \qquad (11)$$

Likewise, combining equations (3) and (10) yields

$$mg_{sys} = \frac{f \cdot K_D / (K_D + 1 / mg_{xtl} - 1) + (1-f) \cdot mg_{xtl} \cdot a}{f + (1-f) \cdot a} \qquad (12)$$

Equations (10), (11), and (12) allow us to constrain the value of mg_{sys}, the mg ratio of the bulk source region, during genesis of melts.

4.2. Application

4.2.1. Rocks. Figure 3 and Table 3 illustrate relationships among mg_{sys}, mg_{liq}, mg_{xtl}, f, a, and K_D, based on equations (11) and (12). As discussed above, the relevant K_D was most likely about 0.30, and certainly in the range 0.25–0.35. Figure 3 shows curves (dashed) for $f = 0.60$, but even $f = 0.45$ is probably conservative; the "real" f was probably almost never >0.40. Likewise, although curves are shown for $a = 1.0$, even $a = 1.5$ is probably conservative; $a = 2.0$ is most realistic. Assume that $K_D = 0.30$ and the melt crystallizes olivine with $mg = 0.920$. If $f = 0.30$ and $a = 2.0$, Table 3a implies that the system (initial source region) mg is 0.894. Assume, more conservatively, that $f = 0.30$ and $a = 1.5$; then $mg_{sys} = 0.888$. With greater conservatism, assume $f = 0.45$ and $a = 2.0$; then $mg_{sys} = 0.878$. With still greater conservatism, assume $f = 0.45$ and $a = 1.5$; then $mg_{sys} = 0.869$. Any mg_{sys} significantly less than 0.869 requires highly dubious assumptions about f or a, or else the relevant K_D must be <0.30, in order for olivine with $mg = 0.920$ to crystallize from the melt. Even assuming $K_D = 0.25$, $f = 0.45$, and $a = 1.5$, mg_{sys} must = 0.857, if olivine with $mg = 0.920$ is to crystallize from the melt.

Admittedly, olivines with $mg = 0.92$ are rare. But olivines with $mg = 0.90$ are common. All the curves in Fig. 3 are essentially linear with slopes for mg_{liq} vs. mg_{sys} of about 0.61; and slopes for mg_{xtl} vs. mg_{sys} of about 1.2–1.5. Thus, Table 3a can easily be adapted for any given mg_{xtl}, and Table 3b can easily be adapted for any given mg_{liq}. For example, if the melt is required to produce olivine with

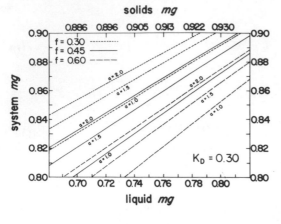

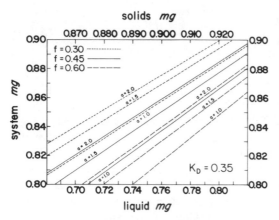

Fig. 3. Equilibrium relationships, based on equations (11) and (12), among mg_{sys}, mg_{liq}, f, and a, assuming (top) that $K_D = 0.30$ and (bottom) that $K_D = 0.35$. Short-dashed lines apply for $f = 0.30$; solid lines for $f = 0.45$; long-dashed lines for $f = 0.60$. As discussed in the text, f is probably seldom $\geqslant 0.3$ and a is probably seldom $\leqslant 2.0$. Note: for any given K_D, a given mg_{liq} implies a specific mg_{xtl}, and vice versa. For any given combination of K_D, a and f with a given mg_{liq} or mg_{xtl}, a specific mg_{sys} (the mg ratio of the initial source region) is implied.

$mg = 0.900$, instead of 0.920, all the mg_{sys} values of Table 3a can be adjusted by subtracting 0.027. If the melt mg is required to be 0.685, instead of 0.695, the mg_{sys} values of Table 3b can be adjusted by subtracting 0.006, and so on.

4.2.2. Soils. For the sake of argument, let us assume that the magnesian olivines are flukes. Mixtures of large volumes of the nonmare crust, such as Korotev's (1981) "HON" or any lunar soil, are essentially random aggregates of numerous solidified liquids produced by partial melting of the lunar interior. Lunar soils contain substantial meteoritic components, and many meteorites have higher $MgO/(MgO + FeO)$ ratios than the bulk soils. However, the vast majority of meteorites are chondrites, which have lower $Mg/(Mg + Fe)$ ratios (including reduced Fe in the denominator) than any lunar soil (Wasson, 1974). Because HON, and all of the averages listed above for mg ratios of nonmare soils, do not distinguish between reduced vs. oxidized Fe, these compositions are unlikely to have their mg ratios raised by addition of

TABLE 3. Relationships among mg_{sys}, mg_{liq}, mg_{xtl}, f, a, and K_D*.

				a. Requirement: $mg_{xtl} = 0.920$				
$K_D=0.25$			$K_D=0.30$			$K_D=0.35$		
			(any given pair of mg_{xtl} and K_D determines the value of mg_{liq})					
$mg_{liq}=0.742$			$mg_{liq}=0.775$			$mg_{liq}=0.800$		
a	f	mg_{sys}	a	f	mg_{sys}	a	f	mg_{sys}
1.0	0.30	0.866	1.0	0.30	0.877	1.0	0.30	0.883
"	0.45	0.840	"	0.45	0.855	"	0.45	0.866
"	0.60	0.813	"	0.60	0.834	"	0.60	0.848
1.5	0.30	0.881	1.5	0.30	0.888	1.5	0.30	0.894
"	0.45	0.857	"	0.45	0.869	"	0.45	0.877
"	0.60	0.831	"	0.60	0.848	"	0.60	0.860
2.0	0.30	0.888	2.0	0.30	0.894	2.0	0.30	0.899
"	0.45	0.868	"	0.45	0.878	"	0.45	0.885
"	0.60	0.844	"	0.60	0.858	"	0.60	0.869
			b. Requirement: $mg_{liq} = 0.695$					
$K_D=0.25$			$K_D=0.30$			$K_D=0.35$		
			(any given pair of mg_{liq} and K_D determines the value of mg_{xtl})					
$mg_{xtl}=0.901$			$mg_{xtl}=0.884$			$mg_{xtl}=0.867$		
a	f	mg_{sys}	a	f	mg_{sys}	a	f	mg_{sys}
1.5	0.30	0.856	1.5	0.30	0.842	1.5	0.30	0.829
"	0.45	0.829	"	0.45	0.817	"	0.45	0.806
2.0	0.30	0.865	2.0	0.30	0.851	2.0	0.30	0.837
"	0.45	0.842	"	0.45	0.829	"	0.45	0.817

*Based on equations (11) and (12).

chondritic matter. Conceivably the meteoritic components of the soils are dominated by nonchondritic materials: The Mainz group (e.g., Wänke et al., 1977) holds that a major fraction of the soils (and polymict breccias) is "primary matter" of bulk-Moon composition, added by tardy accretion of Moon-like material. This model has been sharply criticized by Ryder (1979). Further, Ostertag et al. (1985) cannot fit the composition of lunar meteorite Y-791197 into the primary component model. It seems best to assume that the "primary" meteoritic component of the soils is negligibly small.

Compositions of soils (and HON) may be biased toward lower mg ratios than the initial compositions of the partial melts parental to the nonmare crust: Most nonmare rocks are products of layered mafic intrusions (cumulates). Layering in mafic intrusions is generally controlled by gravity, with the most magnesian cumulates near the bottom, and the least magnesian cumulates near the top (e.g., Wager and

Brown, 1967; Jackson, 1961). Although great impacts have, to a degree, "gardened" the lunar crust, surface materials tend to be of shallow provenance.

Taylor (1975) suggested that an important component of the crust is Mg-rich material from a "frozen crust" over the magmasphere, now thoroughly mixed into highlands soils and polymict breccias. Mass balance calculations indicate that if 1% of frozen crust of Taylor's (1982) estimated bulk-Moon composition were part of an average Apollo 16 soil with $mg = 0.66$, the mg ratio of the balance of the soil would still be 0.65; if the frozen crust component were 2%, the mg ratio of the balance would still be 0.64. But frozen crust components $\geqslant 2\%$ would imply that the mg ratio of the balance of the crust is much lower than the mg ratios of the soils. As discussed by Warren and Wasson (1980), it seems unlikely that a significant thickness of dense, ultramafic crust survived the turmoil of the Moon's first few million years without suffering subduction into the interior, unless the magmasphere to be "frozen" had evolved to the extent that it was crystallizing (saturated with) buoyant plagioclase, by which point its melt fraction had a "ferroan" mg ratio. The magnitude of the frozen crust component in the crust is highly uncertain, but its effect on soil mg ratios is probably more than offset by the effect described in the preceding paragraph.

Assuming that an average ancient crust-forming partial melt had the mg ratio of HON (0.695), Table 3b illustrates the system (average initial source region) mg implied by various combinations of K_D, a, and f. Ultraconservative assumptions like $a = 1.0$ or $f = 0.60$ are not included in Table 3b: such assumptions are scarcely credible in relation to genesis of a small minority of extremely magnesian rocks, and even less credible in relation to genesis of large volumes of the crust. Taken at face value, the $K_D = 0.30$ section of Table 3b implies than an average HON-parental liquid came from a source region with an mg ratio of 0.82 at the very least, and more realistically 0.84–0.85. If the ALHA81005 fossil soil composition ($mg = 0.73$) is treated as an average of partial melts, the average mg_{sys} ratio implied by $K_D = 0.30$ is at least 0.84, and more realistically >0.86.

Aggregates like HON, or soils, can also be used to derive an upper limit for the bulk-Moon mg ratio. However, the likely bias of these aggregates toward lower mg must be compensated, in an unavoidably arbitrary fashion. Because the highest mg for any of these aggregates is 0.73, and the "average" is only about 0.68 (see above), it seems safe to conclude that the crust as a whole has an mg ratio <0.76, and almost certainly <0.80. It may also be assumed that the degree of melting during genesis of the crust averaged at least 10%, because smaller fractions of melt can seldom separate from their source regions (Sleep, 1974). Equally safe assumptions are that the ratio a was no higher than 2.3 (Fig. 2) and that K_D was at least 0.26 (Table 2), as the melts formed. Plugging these values into equation (11) implies that if mg_{liq} is required to be 0.76, mg_{sys} must be <0.917; even under the ultraconservative assumption that mg_{liq} must be 0.80, mg_{sys} must be <0.933.

4.2.3. Possible complications. The mg_{sys} implied by a given mg_{liq} (Table 3b) can be lowered by invoking a K_D greater than 0.30. But then the initial (maximum) mg_{xtl} for olivines crystallized by the melt would be lower, exacerbating the difficulty (Table 3a) of accounting for the numerous olivines with mg in the range 0.90–0.92. Likewise, the mg_{sys} implied by a given mg_{xtl} (Table 3a) can be relaxed by invoking a K_D lower than 0.30. But then, aside from the evidence (Table 2) that K_D for olivine is almost never much less than 0.30, the liquid mg would be lower, exacerbating the difficulty (Table 3b) of accounting for mg ratios of large volumes of the crust.

The mg_{xtl} constraint (Table 3a) might be relaxed without exacerbating the mg_{liq} constraint if the composite K_D of the residual source crystals is greater than the K_D for olivine when the melt is emplaced in the crust. However, as discussed above, the residual crystals are probably olivine plus (particularly if the source is deep in the mantle) pyroxene; and in the pressure range of the lunar mantle composite K_D's for these assemblages are probably very close to the K_D for olivine at crustal pressure. For example, assuming the residuum is predominantly olivine at a pressure of 30 kbar, the bulk K_D for the residuum is most likely about 0.35, whereas the K_D for olivine at surface pressure is about 0.30 (Table 2). If the melt forms with $mg_{sys} = 0.859$, $f = 0.45$, and $a = 2.0$, K_D being 0.35 implies that the melt's mg ratio (mg_{liq}) will be 0.76; and the residuum in equilibrium with this melt will have an mg ratio of 0.900 (Fig. 3). At surface pressure, where K_D for olivine $= 0.30$, a melt with $mg_{liq} = 0.76$ would crystallize olivine with $mg = 0.913$ (Fig. 3). Without the effect of pressure raising the K_D for the residuum, an olivine with $mg = 0.913$ would imply (assuming $K_D = 0.30$, $a = 2.0$, and $f = 0.45$) that mg_{sys} was 0.868. As this example illustrates, the effect of pressure to elevate K_D in the source region does not greatly affect the inferred source region mg ratio.

Another possible complication would be if partial melting were not of the equilibrium (batch) type. In fractional melting, melt is (more or less) continuously removed from the source region as melting proceeds. Ideal (Rayleigh) fractional melting is probably impossible in nature. Melt probably has to accumulate into a finite "batch" before it can segregate from the residual crystals (Sleep, 1974). A more realistic model would be fractional melting with "accumulated fractions" (Shaw, 1977), where the melt is removed in numerous batches, but elsewhere accumulates into a single mass before solidifying. Melting of a previously Fe-depleted, Mg-enriched source region is obviously a third alternative, but unless this Fe-depletion was produced by fractional crystallization, such a model amounts to a type of fractional melting where the batches are not particularly small. It is conceivable that a fraction of the source regions of melts parental to the crust were magnesian cumulates with mg ratios enhanced (beyond the bulk-Moon ratio) by fractional crystallization. But it seems unlikely that primitive, magnesian cumulates would be nearly as prone to remelting as their more "evolved" relations.

Fractional melting might help alleviate the difficulty (Table 3a) of accounting for the numerous olivines with mg in the range 0.90–0.92. Consider a model wherein many small fractions of melt, each amounting to 5% of the initial mass of solid material, are removed in succession; the initial (system) $mg = 0.862$, $a = 2.0$, and $K_D = 0.30$. The first 5% batch is identical to the melt produced by equilibrium melting with $f = 0.05$. Successive batches are increasingly magnesian compared to equilibrium partial melts produced at the same stage of melting of the original material. The mg ratio of the tenth 5% batch, 0.780, is considerably higher than the mg ratio of a 50% equilibrium partial melt, 0.760. The problem with this model is that melts will seldom continue to form after a few tenths of the original material has been removed. The first few batches of melt preferentially remove the "low-melting" major elements (at low pressure = Al, Ca, and to lesser degrees Si and Fe), and almost quantitatively remove the heat-producing elements (K, Th, and U); the source region becomes more and more refractory, and harder and harder to heat, as fractional melting proceeds.

Fractional melting would only exacerbate the difficulty (Table 3b) of accounting for mg ratios of large volumes of the crust. The first (low-mg) batches and the final (high-mg) batches presumably end up at similar depth in the crust. Thanks to meteoritic impacts, these materials are essentially homogenized (along with other partial melts) into soils; so the "accumulated fractions" variant is the only type of fractional melting model germane to the discussion of Table 3b. Consider the example described in the previous paragraph. The mg ratio of the combined first ten 5% fractions of melt, 0.723, is much less than the mg ratio of a 50% equilibrium partial melt, 0.760. For any given source region, an accumulation of fractional melts has a lower mg ratio than an equal mass of equilibrium partial melt.

5. Magmasphere Fractional Crystallization Models

The preceding section addressed nonmare rocks as products of partial melting deep in the interior of the Moon, where f is seldom much greater than 0.4 because buoyant melt tends to separate from residual crystals. Primordial heating is commonly assumed to have produced a magmasphere by total or near-total melting of the outer few hundred kilometers on the Moon. Of course, a total melt of the Moon would begin crystallization with a higher mg ratio than any partial melt. However, except for a single dunite (Dymek et al., 1975), all of the high-mg pristine lunar rocks contain abundant cumulus plagioclase, which can only mean that their parent melts were saturated with plagioclase. Plagioclase saturation in turn implies that the parent melt was far "evolved" from a total melt of the Moon: either the degree of melting was never extremely high, or else sufficient mafic silicate crystallization occurred to cause plagioclase saturation, before these rocks formed.

There is a strong consensus among literature estimates of the Moon's bulk composition (for a review see Warren, 1983) that its (normative) plagioclase content is less than 20 wt %. Warren and Wasson (1979b) hold that 9–10 wt % is more realistic. Plagioclase saturation requires the (normative) plagioclase content of a magnesian melt to be about 55 wt % (Walker *et al.*, 1973). Assuming that Al (which controls normative plagioclase) is perfectly incompatible with mafic silicates, the fraction (M) of the ultimate source material that must be removed as mafic silicates before a lunar melt can be saturated with plagioclase is given by

$$M = 1 - n_b/n_s \qquad (13)$$

where n_b = the normative plagioclase content of the Moon (<20 wt %) and n_s is the normative plagioclase content required for plagioclase saturation (about 55 wt %). Thus, at least 64 wt % "loss" of mafic silicates is a prerequisite for the parent melts of all but one of the pristine nonmare rocks.

Paradoxically, due to this prerequisite, plagioclase-mafic silicate cumulates derived from a total (or near-total) melt of bulk-Moon composition will have lower *mg* ratios than plagioclase-mafic silicate cumulates from moderate-degree equilibrium partial melts. Cumulates form by fractional crystallization. Fractional crystallization of mafic silicates leads to drastic reduction of the *mg* ratio of the residual melt (Fig. 4). For example, if 64 wt % of a melt with *mg* initially = 0.80 is removed as olivine with $K_D = 0.30$, the residual melt has *mg* = only about 0.30 (in equilibrium with *mg*–0.55 olivine). In contrast, if most of the Al fractionation occurs by mafic silicates being left as residual solids from equilibrium partial melting, the *mg* ratio of the melt will depend primarily on the *mg* ratio of the source region, and only secondarily on the degree of melting. For example, assuming the source *mg* (mg_{sys}) = 0.887, $a = 2.0$, and $K_D = 0.30$, inversion of equation (11) implies that the melt *mg* = 0.80 at $f = 0.5$, 0.72 at $f = 0.2$, and 0.71 at $f = 0.05$. For direct comparison with the above fractional crystallization example, assume the partial

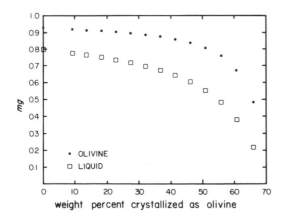

Fig. 4. Melt and olivine mg ratio vs. f during fractional crystallization of olivine by an ultramafic melt that starts out with mg = 0.80; calculated by a simple finite-difference program, assuming $K_D = 0.30$.

melt source $mg = 0.80$; at $f = 0.36$ (i.e., $1-0.64$), the melt mg will be 0.63 (in equilibrium with mg–0.85 olivine).

In short, as far as rocks that contain cumulus plagioclase are concerned, a model with the parent melt being an ultrahigh-degree (beyond exhaustion of plagioclase in the source) partial melt implies a higher mg ratio for the ultimate source material than does a model with the parent melt being a moderate-degree partial melt. The bulk-Moon (normative) plagioclase content is by all accounts <25 wt % (Warren, 1983), and experimental petrology results (Walker et al., 1973) imply that any partial melt up to the point of plagioclase exhaustion in the source will contain about 50 wt % (normative) plagioclase. Therefore the maximum f before plagioclase exhaustion during partial melting of a primitive lunar source region is not more than 50 wt % (and probably far lower, being directly proportional to the estimated bulk-Moon normative plagioclase content). Higher degrees of melting, as in a magmasphere, would only lead to more "ferroan" mg ratios for the plagioclase-mafic silicate cumulates, or, given the moderate mg ratios of lunar plagioclase-mafic silicate cumulates, would only imply a higher mg ratio for the ultimate source material.

Indeed, Warren and Wasson (1979b) concluded from their model of fractional crystallization [expanding on the work of Longhi (1977)] that only the ferroan anorthosites subset of pristine nonmare rocks (Fig. 1) formed directly from the magmasphere. This model, which is too elaborate to be completely reviewed here, indicates that a "pyrolite"-like (Ringwood, 1979) magmasphere with mg initially $= 0.895$ will evolve to $mg = 0.47$ (in equilibrium with mg–0.75 olivine) before reaching plagioclase saturation. According to this model, assuming that the magmasphere was formed by high-degree melting ($f \geq 0.4$) of bulk-Moon material, a bulk-Moon mg ratio of about 0.88 yields the best fit to the mg ratios of ferroan anorthosites (assuming, conservatively, that ferroan anorthosite mg ratios are controlled mainly by cumulus olivine or low-Ca pyroxene, as opposed to "trapped liquid"). The higher mg ratios of most Mg-rich rocks (Fig. 1) would imply commensurably higher bulk-Moon mg ratios ($\geqslant 0.90$), if they were magmasphere products. The mg ratio implied by the ferroan anorthosites would be slightly (0.01) higher if the magmasphere formed as a total melt of bulk-Moon composition, but considerably lower if the magmasphere formed by moderate-degree ($f < 0.4$) partial melting. However, it seems highly unlikely that both the ferroan anorthosites and the Mg-rich rocks formed from melts produced by moderate-degree partial melting, because in such a case there should be a compositional continuum, instead of two clearly distinct clusters, on Fig. 1.

6. Discussion: Other Constraints

6.1. KREEP basalts

In terms of volume, KREEP basalt is at least comparable to mare basalt. Warren and Wasson (1979a) estimated, assuming all KREEP originated as a "high-K" variety (the residuum of the magmasphere), that its total mass corresponded to about 0.17 wt % of the Moon. KREEP is now found mainly in diluted form as a component in impact breccias. "Low-K" KREEP breccia ("LKFM") is one of the most abundant rock types in the crust (e.g., Taylor, 1975).

Despite their extremely high incompatible element contents, KREEP basalts have moderate *mg* ratios, typically 0.5–0.7 (Warren and Wasson, 1979a). According to one model of KREEP genesis, fractional crystallization in the primordial magmasphere generated pockets of KREEPlike residual liquid ("urKREEP") (Warren and Wasson, 1979a), which were probably frequently assimilated into Mg-rich melts (Warren *et al.*, 1981). But it is instructive to consider simpler models, assuming that KREEP basalt forms by partial melting. For example, Weill and McKay (1975) estimated that the source region of pristine KREEP basalt 15386 had *mg* = 0.83. This basalt has light REE enriched to about 200× chondritic, so its source obviously was greatly differentiated relative to the bulk-Moon composition, and such a source should, according to any simple model, have an *mg* ratio commensurably low. This discrepancy probably says more about simple partial melting models for genesis of KREEP basalt than it does about the bulk composition of the Moon. But it also suggests that caution should be exercised in interpreting other lunar basalts (mare basalts) as gauges of the lunar *mg* ratio.

6.2. Geophysical constraints

6.2.1. Density, moment of inertia, and induced magnetic field data.
The mean density of the Moon is 3.3437 ± 0.0016 g cm^{-3}, and its mean moment of inertia parameter is 0.3904 ± 0.0023 (Ferrari *et al.*, 1980). Its density proves that the Moon is depleted in total iron, compared to the Earth or to chondritic meteorites. On the other hand, depending upon the size of any possible high-density core, the bulk density and moment of inertia factor could be interpreted to suggest that the Moon is enriched in FeO compared to the Earth's mantle (e.g., Buck and Toksöz, 1980).

According to Ferrari *et al.* (1980), their precisely determined moment of inertia factor suggests, although it does not prove, that the Moon possesses a small core; the most likely radius, assuming the core is mainly metallic FeNi, is 300 km, but it might conceivably be as large as 550 km, if the composition is mainly FeS. Levin (1979) interprets the moment of inertia factor similarly. Herbert (1980) suggested that the moment of inertia constraint could be met without a core, but his models assumed bulk-Moon FeO contents of 24–25 wt % (highly implausible, as shown in the next paragraph) and *mg* ratios of 0.65–0.74. The induced magnetic field has been studied using various approaches, which indicate that the possible radius of lunar core is: <435 km (Hobbs *et al.*, 1983); slightly > 400 km (Russell *et al.*, 1981; Russell, 1984); or 0–5.8 wt % of the Moon (Daily and Dyal, 1979) (a radius of 400 km would correspond to 1.2 vol % of the Moon). Seismic data give an exceedingly tentative suggestion of a small core (radius of the order 300 km) (Nakamura *et al.*, 1976). Runcorn (1979) invokes a core to account for lunar paleomagnetism. Stevenson (1980) and Stevenson and Yoder (1981) argue that side-effects of a core might explain the offset between the Moon's center of figure and its center of mass, as well as its tidal dissipation. In summary, the Moon probably has a significant core, but the core is probably not much more than 1.2 vol %, or 3 wt %, of the Moon.

Parkin *et al.* (1974) interpreted the induced magnetic field data not only in terms of reduced Fe, but also in terms of total Fe (reduced Fe + FeO). They used two simple models: one in which the only magnetic components in the Moon are metallic FeNi + olivine, and one in which the only magnetic components are metallic FeNi + orthopyroxene (the paramagnetic susceptibility of olivine is greater than that of orthopyroxene by a factor of roughly 1.6). In order to fit the Apollo magnetometry data, they concluded that, assuming the content of reduced Fe = 2.0 wt %, the FeO content must be about 5.0 wt % for the olivine model, or 14.2 wt % for the orthopyroxene model, or assuming reduced Fe = 1.0 wt %, the FeO content must be about 7.1 wt % for the olivine model, or 16.0 wt % for the orthopyroxene model. This model is probably oversimplified. Other phases such as high-Ca pyroxene, garnet, and spinel might be significant carriers of FeO. For any given *mg* ratio, all of these phases have lower paramagnetic susceptibility (by factors of roughly 0.48, 0.86, and 0.80, respectively) than orthopyroxene (Nagata, 1961, pp. 76–123, and references therein). But most bulk-Moon compositional models imply that olivine is far more abundant than any other mafic silicate [the only exceptions are the models of Warren and Wasson (1979b) and Buck and Toksöz (1980)]. If half of the mafic silicates are olivine, the reduced Fe = 1.0 wt % model of Parkin *et al.* (1974) implies that FeO = about 11.5 wt %; the reduced Fe = 2.0 wt % model implies FeO = only about 9.6 wt %.

Despite huge uncertainties in the reduced Fe content and the olivine/pyroxene ratio, the work of Parkin *et al.* (1974) indicates that the bulk-Moon FeO content is probably less than 12 wt %, and 14 wt % is a safe upper limit. With one exception (Buck and Toksöz, 1980), all of the models listed in Table 1 with *mg* <0.84 also have FeO ≥ 12.9 wt %. These models probably underestimate the bulk-Moon *mg* ratio.

6.2.2. Seismic velocities. Buck and Toksöz (1980) interpreted their seismic velocity profiles for the lunar interior in terms of *mg* ratio for the constituent mafic silicates. They concluded that the bulk-Moon *mg* ratio must be close to 0.80. Binder (1983) factored their conclusion into his model.

There are several problems with this interpretation. First, the seismic data pertain mainly to the Moon's central near side, which may not be representative in terms of *mg* ratio. A 2.0 km offset between the Moon's center of figure and its center of mass (Bills and Ferrari, 1977) indicates that the nearside hemisphere is considerably denser than the farside hemisphere. One possible explanation for this asymmetry is that the low-density anorthositic crust is much thicker on the far side than on the near side. However, the density asymmetry may be at least partially a result of the nearside mantle having an *mg* ratio several percent lower than that of the farside mantle (Wasson and Warren, 1980).

Second, Buck and Toksöz (1980) probably underestimated the uncertainty in their seimic velocity profiles. They claim that the brackets on their seismic velocity curves are "maximum possible variations in velocity for a given depth and are not error bars." But Nakamura *et al.* (1982), using a data base they claimed is more complete than that of Buck and Toksöz (1980), derived very different profiles. Buck and Toksöz (1980) show S-wave velocity for depths >500 km bracketed between 4.08 and 4.33 km/sec, but Nakamura *et al.* (1982) found S-wave velocity for depths >500 km to be 4.65 ± 0.16 km/sec. If the "correct" velocity is 4.65 km/sec and the olivine/pyroxene ratio is roughly unity, Fig. 2 of Buck and Toksöz (1980) would imply that the mean *mg* ratio of the lower mantle is roughly 0.95; as drawn, the figure implies *mg* is about 0.7, and definitely <0.78. As this example illustrates, seismic data do not constrain the mantle's *mg* ratio quite so precisely as implied by Buck and Toksöz (1980).

7. Implications for Lunar Origin

Surprisingly little has been written about implications of the bulk-Moon *mg* ratio for the origin of the Moon. The foregoing analysis implies that the bulk-Moon *mg* ratio is probably within 2–3% of the Earth's upper mantle *mg* ratio (about 0.895). The closeness of this similarity can be appreciated by considering the range in *mg* ratio as measured among chondritic meteorites—0.55–0.99 (Wasson, 1974)— or the ranges of estimates of bulk *mg* ratios for other planetary bodies. As reviewed by BVSP (1981, Fig. 4.5.22) ranges of estimates of bulk *mg* ratios are 0.65–0.80 for the parent asteroid of the eucrite meteorites, 0.67–0.77 for Mars, and 0.94–1.00 for Mercury.

In the solar nebula, the bulk of the Fe and Mg left the gas in favor of metallic FeNi and olivine, respectively, when the temperature fell to about 1400 K; *later* the metallic Fe tended to be gradually oxidized into FeO when the temperature fell below 1000 K (BVSP, 1981, p. 644). The MgO/(MgO + FeO) ratio of a given chondrite is essentially a function of the lowest temperature at which its constituent materials ceased equilibrating with the nebula; the lower the temperature was, the higher is the FeNi content, and the lower is the *mg* ratio, of the chondrite. If the equilibration temperature was low enough, essentially all of the metallic Fe was oxidized into mafic silicates. In this respect, the enstatite chondrites (which have *mg* ratios >0.98) are the highest-temperature chondrites, and are inferred to have originated close to the sun; the carbonaceous chondrites are the lowest temperature chondrites, and are inferred to have originated >5 AU from the sun (Wasson, 1977). This reasoning has even been extended (e.g., Goettel and Barshay, 1978) to relate "formation temperatures" of planets to their FeO contents.

Assuming that its core is mainly FeNi, the Earth is about 30 wt % FeNi. Even enstatite chondrites only contain roughly 25 wt % FeNi, so it is not surprising that the Earth's upper mantle has a higher *mg* than most chondrites (the highest *mg* ratio among "ordinary" chondrites is about 0.84) (Wasson, 1974). If the modes of formation of the Moon and the Earth were otherwise analogous, the FeNi-poor Moon might be expected to have an *mg* ratio much lower than that of the Earth. The paradox that the Moon has an *mg* ratio at least nearly as high as that of the Earth suggests that the Moon's protoplanet(s) originally had similar high content(s) of FeNi, but somehow most of the FeNi was separated from the silicates, as or before the Moon assembled. Many mechanisms have been proposed to account for this efficient metal-silicate fractionation. But are these models consistent with an Earth-silicate-like composition, and in particular an Earthlike *mg* ratio, for the bulk Moon?

7.1. Capture models

As reviewed by Wood (1977), a popular pre-Apollo model of lunar origin holds that the Moon was captured essentially intact from heliocentric orbit into orbit about the Earth. Although this hypothesis seems dynamically improbable (Kaula and Harris, 1975; Wood, 1977), it is not easily tested using geochemical criteria. An exotic, captured Moon would probably have an *mg* ratio unlike that of the Earth. But the two bodies might just happen to have very similar *mg* ratios, particularly if the Moon's orbit prior to capture had a semimajor axis close to 1 AU (in which case its "feeder zone" would have largely overlapped that of the Earth—but then the Moon would not be expected to have an FeNi content greatly different from that of the Earth).

7.2. Fission and Earth-impact models

Many models seek to explain the low metallic FeNi content by producing the Moon out of the Earth's mantle. Classical, "pure" fission models have lost favor due to dynamical constraints (e.g., Kaula and Harris, 1975; McKinnon and Mueller, 1984). More recent variants of the fission model invoke formation of the Moon out of a vaporized mass of the Earth's mantle. Ringwood (1966, 1975) suggested that the outer Earth was volatilized by primordial heat (accretion, core formation, etc.), and the Moon formed out of condensates from a disk of this vaporized material, spun off of the Earth. Hartmann and Davis (1975) proposed impact-induced fission:

Mantle material was ejected from the Earth due to a relatively late impact by the second largest protoplanet in the Earth's neighborhood. Ringwood (1979) advocated a hybrid between his earlier model and the Hartmann-Davis model, suggesting that the Moon was fissioned from the Earth as a result of several impacts onto an outer Earth that was already largely molten. Cameron and Ward (1976), Kaula (1979), Cameron (1984), Kaula and Beachey (1984), Stevenson (1984), Melosh (1985), and (to a degree) Wetherill (1985) also advocate Earth-impact models of lunar origin. This general class of models was perceived (Kerr, 1984) to be very popular at the Conference on the Origin of the Moon, from which this volume resulted.

Hartmann and Davis (1975) noted that their model implies that the Moon would be enriched in refractory elements and depleted in volatile elements relative to the Earth. Likewise, Cameron and Ward (1976) suggested that refractory elements, including Ca and Al, were enriched in the Moon because during rapid expansion and cooling of vaporized impact ejecta, more volatile elements condensed later, and hence into finer grains, and "most of this fine dust probably escapes from the system or is otherwise segregated from the refractory materials by magnetic effects." In Stevenson's (1984) impact model, the Moon forms out of condensates from vapor that "bleeds out" of the Earth's atmosphere at a temperature of the order 10^4 K, and the cooling time of the protolunar disk is of the order 10^2 years. In all of these Earth-impact models, the protolunar material cools rapidly in a disk around the Earth, and only a fraction of the material initially in the disk ends up accreting to the Moon. This rapid cooling, in the presence of mechanical forces such as tides, magnetic forces, the Poynting-Robertson effect, etc., can be expected (as proponents of these models acknowledge) to effect a fractional condensation of the initial material. How would this fractional condensation have affected the *mg* ratio of the condensate?

Ringwood (1966, 1970) employed thermodynamics to estimate volatilization/condensation temperatures for major elements in an extended atmosphere of the primordial Earth. His results suggest that Si and Mg would be considerably more volatile than Al, Ca, and *reduced* Fe. But because he did not calculate the temperature at which the Fe would oxidize into FeO, these calculations have only slight relevance to the problem of the bulk *mg* ratio of condensates from a protolunar disk. In addition to temperature, total pressure, and total Fe content, important variables for stability of FeO at the expense of reduced Fe include: the amount of H_2O, CO, and other phases in competition with FeO for oxygen; the amount of CH_4, NH_3, and other phases in competition with H_2O for hydrogen; the amount of FeS in competition with FeO for iron; etc. Few or none of these parameters can be precisely constrained for the type of scenario in question. In any case, the main solids available to FeO would presumably be mafic silicate solid solutions (olivine and pyroxene), among which FeO competes, at a disadvantage, for space with MgO. It seems likely that the aggregate *mg* ratio of the solids would always be higher than the *mg* ratio of the gas during fractional condensation of these phases, just as the aggregate *mg* ratio of the solids is always higher than the *mg* ratio of the melt during fractional crystallization of these phases from basaltic magmas.

Ringwood (1979, pp. 251–252) noted a factor that may have tended to give an impact-generated Moon a lower *mg* ratio in comparison to the Earth's mantle: If the outer Earth was largely molten, it would probably be differentiatied into a low-*mg*, largely molten uppermost layer, atop a high-*mg*, mostly solid lower layer, and the upper layer would be the main contributor to the Moon. Conceivably this effect was exactly balanced by the effect of fractional condensation of the impact-generated protolunar disk, allowing the Moon to form with an *mg* ratio close to that of the Earth.

7.3. Circumterrestrial accretion models

Another class of models form the Moon by accretion of small bodies in geocentric orbit, without invoking terrestrial derivation for those bodies. At first glance, such models seem ill-suited to explain the Moon's low FeNi content, unless the Moon is assumed to have a low (by Earth standards) "formation temperature," which would lead to a low *mg* ratio. One sure means of separating metallic FeNi from silicates is by core formation. In fission models, this process is assumed to occur within the Earth, shortly before the fission event. An alternative is to assume that core formation occurred on numerous asteroid-sized bodies that were broken into smaller fragments, some of which accreted in a geocentric orbit to form the Moon. Several mechanisms have been proposed whereby a Moon that was formed from a circumterrestrial swarm of such fragments (during the later stages of accretion of the Earth) would preferentially accrete more silicate fragments than FeNi fragments.

The "modified capture" model of Wood and Mitler (1974) suggested that many asteroid-sized objects were broken apart by passage through the Earth's Roche limit, and fragments from the deeper (and, hence, more FeNi-rich) layers of these objects were less likely to be captured into geocentric orbit than were fragments from shallower (silicate-rich) layers. However, Wood (1977) acknowledged that this sorting mechanism would be "very inefficient." Smith (1974) proposed a "disintegrative capture" model that considered numerous other mechanisms, as well as tidal disruption, for breaking up the differentiated protoplanets and sorting the resultant fragments.

More recent models have emphasized collisions as a means of breaking the postulated differentiated asteroid-sized bodies apart. Kaula (1977) suggested that when collisions broke the bodies apart, their shallow FeNi-poor parts tended to acquire higher than average energy/mass relative to Earth, and hence tended to take up orbits farther from the Earth. Wasson and Warren (1979, 1985) also emphasized collisions as a means of break-up, but suggested that the Moon's low FeNi content resulted from a tendency for the collisions to yield larger FeNi fragments than silicate fragments, due to the greater mechanical strength of FeNi (cf. Orowan, 1969; Ruskol, 1977).

Because larger fragments have greater momentum, and hence greater probability of passing through a circumterrestrial swarm of planetesimals and striking the Earth, the swarm was depleted in FeNi and enriched in silicates, compared to the Earth. The "Tucson Lunar Origin Consortium" (Greenberg et al., 1984; Chapman and Greenberg, 1984) advocates a similar model.

An Earthlike *mg* ratio is consistent with some of these models, but not all. The postulated asteroid-sized precursor bodies, presumably from an Earthlike "feeding zone" near 1 AU, probably had bulk *mg* ratios similar to that of the Earth. But the modified capture model of Wood and Mitler (1974) implies that the Moon's *mg* ratio should be substantially lower than that of an average precursor body, because fragments from the disrupted bodies' crusts are more likely to be captured than fragments from their lower mantles. Differentiation in the precursor bodies presumably concentrated low-*mg* silicates in the crusts, and high-*mg* silicates in deep layers near the cores. Likewise, Kaula's (1977) model predicts that the Moon will be enriched in "outer" materials; he specifies plagioclase, but by implication low-*mg* mafic silicates would be enriched as well. However, in circumterrestrial accretion models such as those of Wasson and Warren (1979, 1985) and Greenberg et al. (1984), where the compositional sorting is based mainly on whether the previously differentiated fragments are ductile metal or brittle silicates, the Moon is predicted to have an *mg* ratio similar to that of the Earth.

8. Conclusions

1. The bulk-Moon *mg* ratio is ≥ 0.85, and most likely ≥ 0.87. This conclusion can be derived either by considering means of producing olivines with *mg* = 0.92 (and more abundant olivines with *mg* $\geq$ 0.90) or by considering means of producing compositions that represent aggregates of large volumes of the lunar crust (such as soils, with *mg* typically = 0.69). By the latter method, the bulk-Moon *mg* ratio is probably <0.91, and almost certainly <0.93.

2. Conversely, if the bulk-Moon *mg* is <0.85 (as often alleged), then: (a) during production of melts parental to Mg-rich troctolites the degree of melting was probably often >50%; and (b) in order to have these melts be saturated with plagioclase, the mantle source regions, presumably similar in composition to the bulk Moon, probably comprised >25 wt % plagioclase (i.e., >8 wt % Al_2O_3).

3. The bulk-Moon *mg* ratio is essentially the same, within a few percent uncertainty, as that of the Earth (about 0.895). This similarity is an important constraint on models of lunar origin:

(a) Closely similar *mg* ratios may be difficult to accommodate with Earth-impact models, holding that the Moon formed from vaporized ejecta after a collision between the Earth and one of its largest protoplanets. Condensation under such a scenario would probably be fractional, and therefore might greatly change the *mg* ratio of the Moon relative to that of the Earth.

(b) Closely similar *mg* ratios are consistent with models that form the Moon by accretion from a circumterrestrial swarm of fragments of previously differentiated asteroid-sized bodies, *if* the mechanism(s) for depleting the Moon in FeNi does not simultaneously enrich it in materials from the shallowest parts of the previously differentiated bodies.

Acknowledgments. I thank J. T. Wasson for stimulating discussions, A. G. W. Cameron, J. W. Delano, and M. J. Drake for constructive criticism of an early draft, and J. Longhi and H. Palme for helpful reviews. This research was supported by NASA grants NAG 9-87 and NAG 9-96 (both to J. T. Wasson).

9. References

Arndt N. T. (1977) Ultrabasic magmas and high-degree melting of the mantle. *Contrib. Mineral. Petrol., 64*, 205–221.

Arzi A. A. (1978) Critical phenomena in the rheology of partially melted rocks. *Tectonophysics, 44*, 173–184.

Auer F., Berckhemer H., and Oehlschlegel G. (1981) Steady state creep of fine grain granite at partial melting. *J. Geophys., 49*, 89–92.

Basaltic Volcanism Study Project (BVSP) (1981) *Basaltic Volcanism on the Terrestrial Planets.* Pergamon, New York. 1286 pp.

Bickle M. J. (1982) The magnesium contents of komatiitic liquids. In *Komatiites* (N. T. Arndt and E. G. Nisbet, eds.), pp. 479–494. Allen and Unwin, London.

Bickle M. J., Ford C. E., and Nisbet E. G. (1977) The petrogenesis of peridotitic komatiites: evidence from high-pressure melting experiments. *Earth Planet. Sci. Lett., 37*, 97–106.

Bills B. G. and Ferrari A. J. (1977) A lunar density model consistent with topographic, gravitational, librational, and seismic data. *J. Geophys. Res., 82*, 1306–1314.

Binder A. B. (1983) An estimate of the bulk, major oxide composition of the Moon (abstract). In *Workshop on Pristine Lunar Highlands Rocks and the Early History of the Moon* (J. Longhi and G. Ryder, eds.), pp. 17–19. LPI Tech. Rpt. 83-02, Lunar and Planetary Institute, Houston.

Buck W. R. and Toksöz M. N. (1980) The bulk composition of the Moon based on geophysical constraints. *Proc. Lunar Planet. Sci. Conf. 11th*, pp. 2032–2058.

Cameron A. G. W. (1984) Formation of the prelunar accretion disk (abstract). In *Papers Presented to the Conference on the Origin of the Moon*, p. 58. Lunar and Planetary Institute, Houston.

Cameron A. G. W. and Ward W. R. (1976) The origin of the Moon (abstract). In *Lunar Science VII*, pp. 120–122. The Lunar Science Institute, Houston.

Chapman C. R. and Greenberg R. (1984) A circumterrestrial compositional filter (abstract). In *Papers Presented to the Conference on the Origin of the Moon*, p. 56. Lunar and Planetary Institute, Houston.

Cirlin E.-H., Taylor L. A., and Lofgren G. E. (1985) Fe/Mg K for olivine/liquid in chondrules: Effects of cooling rate (abstract). In *Lunar and Planetary Science XVI*, pp. 133–134. Lunar and Planetary Institute, Houston.

Clayton R. N. (1977) Genetic relationships among meteorites and planets. In *Comets, Asteroids and Meteorites: Interrelations, Evolution and Origins* (A. H. Delsemme, ed.), pp. 545–550. Univ. of Toledo, Ohio.

Daily W. D. and Dyal P. (1979) Magnetometer data errors and lunar induction studies. *J. Geophys. Res., 84*, 3313–3326.

Delano J. W. (1980) Chemistry and liquidus phase relations of Apollo 15 red glass: Implications for the deep lunar interior. *Proc. Lunar Planet. Sci. Conf. 11th*, pp. 251–288.

Delano J. W. (1984) Abundances of Ni, Cr, Co, and major elements in the silicate portion of the Moon: Constraints from primary lunar magmas (abstract). In *Papers Presented to the Conference on the Origin of the Moon*, p. 15. Lunar and Planetary Institute, Houston.

Delano J. W. and Lindsley D. H. (1983) Mare glasses from Apollo 17: Constraints on the Moon's bulk composition. *Proc. Lunar Planet. Sci. Conf. 14th*, in *J. Geophys. Res., 88*, B3–B16.

Dymek R. F., Albee A. L., and Chodos A. A. (1975) Comparative petrology of lunar cumulate rocks of possible primary origin: Dunite 72415, troctolite 76535, norite 78235, and anorthosite 62237. *Proc. Lunar Sci. Conf. 6th*, pp. 301–341.

Ferrari A. J., Sinclair W. S., Sjogren W. L., Williams J. G., and Yoder C. F. (1980) Geophysical parameters of the Earth-Moon system. *J. Geophys. Res., 85*, 3939–3951.

Fukuoka T., Laul J. C., Smith M. R., Hughes S. S., and Schmitt R. A. (1985) Chemistry of Yamato-791197 meteorite: evidence for lunar highland origin (abstract). In *Papers Presented to the 10th Symposium on Antarctic Meteorites*, pp. 41.1–41.2. National Inst. Polar Research, Tokyo.

Ganapathy R. and Anders E. (1974) Bulk compositions of the Moon and Earth, estimated from meteorites. *Proc. Lunar Sci. Conf. 5th*, pp. 1181–1206.

Goettel K. A. and Barshay S. S. (1978) The chemical equilibrium model for condensation in the solar nebula: Assumptions, implications, and limitations. In *The Origin of the Solar System* (S. F. Dermott, ed.), pp. 611–627. Wiley, New York.

Green D. H. (1972) Archean greenstone belts may include terrestrial equivalents of lunar maria? *Earth Planet. Sci. Lett., 15*, 263–270.

Greenberg R., Chapman C. R., Davis D. R., Drake M. J., Hartmann W. K., Herbert F. L., Jones J., and Weidenschilling S. J. (1984) An integrated dynamical and geochemical approach to lunar origin modelling (abstract). In *Papers Presented to the Conference on the Origin of the Moon*, p. 51. Lunar and Planetary Institute, Houston.

Grieve R. A. F. and Parmentier E. M. (1984) Considerations of large scale impact and the early Earth (abstract). In *Lunar and Planetary Science XV*, pp. 326–327. Lunar and Planetary Institute, Houston.

Grove T. L. (1978) Experimentally determined FeO-MgO-CaO partitioning between pyroxene and liquid in lunar basalts (abstract). *EOS (Trans. Amer. Geophys. Union), 59*, 401.

Grover J. E., Lindsley D. H., and Bence A. E. (1980) Experimental phase relations of olivine vitrophyres from breccia 14321: The temperature- and pressure-dependence of Fe-Mg partitioning for olivine and liquid in a highlands melt-rock. *Proc. Lunar Planet. Sci. Conf. 11th*, pp. 179–196.

Hartmann W. K. and Davis D. R. (1975) Satellite-sized planetesimals and lunar origin. *Icarus, 24*, 504–511.

Head J. W. (1975) Lunar mare deposits: Areas, volumes, sequence, and implication for melting in source areas (abstract). In *Papers Presented to the Conference on Origins of Mare Basalts and their Implications for Lunar Evolution*, pp. 66–69. The Lunar Science Institute, Houston.

Herbert F. (1980) Time-dependent lunar density models. *Proc. Lunar Planet. Sci. Conf. 11th*, pp. 2015–2030.

Hobbs B. A., Hood L. L., Herbert F., and Sonett C. P. (193) An upper bound on the radius of a highly electrically conducting lunar core. *Proc. Lunar Planet. Sci. Conf. 14th*, in *J. Geophys. Res., 88*, B97–B102.

Hörz F. (1978) How thick are lunar mare basalts? *Proc. Lunar Planet. Sci. Conf. 9th*, pp. 3311–3331.

Jackson E. D. (1961) Primary textures and mineral associations in the ultramafic zone of the Stillwater Complex, Montana. *U.S. Geol. Surv. Prof. Paper 358*. 106 pp.

Jaques A. L. and Green D. H. (1980) Anhydrous melting of peridotite at 0–15 Kb pressure and the genesis of tholeiitic basalts. *Contrib. Mineral. Petrol., 73*, 287–310.

James O. B. and Flohr M. K. (1983) Subdivision of the Mg-suite noritic rocks into Mg-gabbronorites and Mg-norites. *Proc. Lunar Planet. Sci. Conf. 13th,* in *J. Geophys. Res., 87,* A603–A614.

Kallemeyn G. W. and Warren P. H. (1983) Compositional implications regarding the lunar origin of the ALHA81005 meteorite. *Geophys. Res. Lett., 10,* 833–836.

Kaula W. M. (1977) Mechanical processes affecting differentiation of protolunar material. In *The Soviet-American Conference on the Cosmochemistry of the Moon and Planets* (J. H. Pomeroy and N. J. Hubbard, eds.), pp. 805–813. NASA, Washington, D.C.

Kaula W. M. (1979) Thermal evolution of Earth and Moon growing by planetesimal impacts. *J. Geophys. Res., 84,* 999–1008.

Kaula W. M. and Beachey A. E. (1984) Mechanical models of close approaches and collisions of large protoplanets (abstract). In *Papers Presented to the Conference on the Origin of the Moon,* p. 59. Lunar and Planetary Institute, Houston.

Kaula W. M. and Harris A. W. (1975) Dynamics of lunar origin and orbital evolution. *Rev. Geophys. Space Phys., 13,* 363–371.

Kerr R. A. (1984) Making the Moon from a big splash. *Science, 226,* 1060–1061.

Korotev R. (1981) Compositional trends in Apollo 16 soils. *Proc. Lunar Planet. Sci. 12B,* pp. 577–605.

Levin B. J. (1979) On the core of the Moon. *Proc. Lunar Planet. Sci. Conf. 10th,* pp. 2321–2323.

Longhi J. (1977) Magma oceanography 2: Chemical evolution and crustal formation. *Proc. Lunar Sci. Conf. 8th,* pp. 601–621.

Longhi J. (1980) A model of early lunar differentiation. *Proc. Lunar Planet. Sci. Conf. 11th,* pp. 289–315.

Longhi J. (1981) Preliminary modeling of high pressure partial melting: Implications for early lunar differentiation. *Proc. Lunar Planet. Sci. 12B,* pp. 1001–1018.

Longhi J. and Boudreau A. E. (1979) Complex igneous processes and the formation of the primitive lunar crustal rocks. *Proc. Lunar Planet. Sci. Conf. 10th,* pp. 2085–2105.

Longhi J., Walker D., and Hays J. F. (1978) The distribution of Fe and Mg between olivine and lunar basaltic liquids. *Geochim. Cosmochim. Acta, 42,* 1545–1558.

Ma M.-S., Schmitt R. A., Taylor G. J., Warner R. D., and Keil K. (1981) Chemical and petrographic study of spinel troctolite in 67435: Implications for the origin of Mg-rich plutonic rocks (abstract). In *Lunar and Planetary Science XII,* pp. 640–642. Lunar and Planetary Institute, Houston.

McKay G. A. and Weill D. F. (1977) KREEP petrogenesis revisited. *Proc. Lunar Sci. Conf. 8th,* pp. 2339–2355.

McKenzie D. (1984) The generation and compaction of partially molten rock. *J. Petrol., 25,* 713–765.

McKinnon W. B. and Mueller S. W. (1984) A reappraisal of Darwin's fission hypothesis and a possible limit to the primordial angular momentum of the Earth (abstract). In *Papers Presented to the Conference on the Origin of the Moon,* p. 34. Lunar and Planetary Institute, Houston.

Melosh H. J. (1985) When worlds collide: Jetted vapor plumes and the Moon's origin (abstract). In *Lunar and Planetary Science XVI,* pp. 552–553. Lunar and Planetary Institute, Houston.

Miller M. D., Pacer R. A., Ma M.-S., Hawke B. R., Lookhart G. L., and Ehmann W. D. (1974) Compositional studies of the lunar regolith at the Apollo 17 site. *Proc. Lunar Sci. Conf. 5th,* pp. 1079–1086.

Morgan J. W., Hertogen J., and Anders E. (1978) The Moon: Composition determined by nebular processes. *Moon and Planets, 18,* 465–478.

Mysen B. O. and Kushiro I. (1977) Compositional variations among coexisting phases with degree of melting of peridotite in the upper mantle. *Amer. Mineral., 62,* 843–865.

Nagata T. (1961) *Rock Magnetism.* Maruzen, Tokyo. 350 pp.

Nakamura Y., Latham G. V., Dorman H. J., and Duennebier F. K. (1976) Seismic structure of the Moon: A summary of current status. *Proc. Lunar Sci. Conf. 7th,* pp. 3113–3121.

Nakamura Y., Latham G., and Dorman H. J. (1982) Apollo lunar seismic experiment—final summary. *J. Geophys. Res., 87*, 117–123.

Orowan E. (1969) Density of the Moon and nucleation of planets. *Nature, 222*, 867.

Ostertag R., Stöffler D., Palme H., Spettel B., Weckwerth G., and Wänke H. (1985) Lunar meteorite Yamato 791197: A weakly shocked regolith breccia from the far side of the Moon (abstract). In *Lunar and Planetary Science XVI*, pp. 635–636. Lunar and Planetary Institute, Houston.

Parkin C. W., Daily W. D., and Dyal P. (1974) Iron abundance and magnetic permeability of the Moon. *Proc. Lunar Sci. Conf. 5th*, pp. 2761–2778.

Prinz M., Dowty E., Keil K., and Bunch T. E. (1973) Mineralogy, petrology and chemistry of lithic fragments from Luna 20 fines: origin of the cumulate ANT-suite and its relationship to high-alumina and mare basalts. *Geochim. Cosmochim. Acta, 37*, 979–1006.

Rhodes J. M., Rodgers K. V., Shih C., Bansal B. M., Nyquist L. E., Wiesmann H., and Hubbard N. J. (1974) The relationships between geology and soil chemistry at the Apollo 17 landing site. *Proc. Lunar Sci. Conf. 5th*, pp. 1097–1117.

Ringwood A. E. (1966) Chemical evolution of the terrestrial planets. *Geochim. Cosmochim. Acta, 30*, 41–104.

Ringwood A. E. (1970) Origin of the Moon: The precipitation hypothesis. *Earth Planet. Sci. Lett., 8*, 131–140.

Ringwood A. E. (1975) *Composition and Petrology of the Earth's Mantle*. McGraw-Hill, New York. 618 pp.

Ringwood A. E. (1979) *Origin of the Earth and Moon*. Springer-Verlag, New York. 295 pp.

Roeder P. L. (1974) Activity of iron and olivine solubility in basaltic liquids. *Earth Planet. Sci. Lett., 23*, 397–410.

Roeder P. L. and Emslie R. F. (1970) Olivine-liquid equilibrium. *Contrib. Mineral. Petrol., 29*, 275–289.

Roscoe R. (1952) The viscosity of suspension of rigid spheres. *Brit. J. Appl. Phys., 3*, 267–269.

Rose H. J., Jr., Cuttitta F., Berman S., Brown F. W., Carron M. K., Christian R. P., Dwornik E. J., and Greenland L. P. (1974) Chemical composition of rocks and soils at Taurus-Littrow. *Proc. Lunar Sci. Conf. 5th*, pp. 1119–1133.

Runcorn S. K. (1979) An iron core in the Moon generating an early magnetic field? *Proc. Lunar Planet. Sci. Conf. 10th*, pp. 2325–2333.

Ruskol Ye. L. (1977) The origin of the Moon. In *The Soviet-American Conference on the Cosmochemistry of the Moon and Planets* (J. H. Pomeroy and N. J. Hubbard, eds.), pp. 815–822. NASA, Washington, D.C.

Russell C. T. (1984) On the Apollo subsatellite evidence for a lunar core (abstract). In *Papers Presented to the Conference on the Origin of the Moon*, p. 7. Lunar and Planetary Institute, Houston.

Russell C. T., Coleman P. J., Jr., and Goldstein B. E. (1981) Measurements of the lunar induced magnetic moment in the geomagnetic tail: Evidence for a lunar core? *Proc. Lunar Planet. Sci. 12B*, pp. 831–836.

Ryder G. (1979) The chemical components of highlands breccias. *Proc. Lunar Planet. Sci. Conf. 10th*, pp. 561–581.

Ryder G. and Norman M. (1979) *Catalog of Pristine Non-mare Materials, Part 1: Non-anorthosites (Revised)*. Publication 14565, NASA Johnson Space Center Curatorial Facility, Houston.

Ryder G. and Ostertag R. (1983) ALHA 81005: Moon, Mars, petrography, and Giordano Bruno. *Geophys. Res. Lett., 10*, 791–794.

Shaw D. M. (1977) Trace element behavior during anatexis. In *Magma Genesis* (H. J. B. Dick, ed.), pp. 189–213. State of Oregon Dept. of Geology and Mineral Industries Bulletin 96.

Shervais J. W., Taylor L. A., Laul J. C., and Smith M. R. (1984) Pristine highland clasts in consortium breccia 14305: petrology and geochemistry. *Proc. Lunar Planet. Sci. Conf. 15th*, in *J. Geophys. Res., 89*, C25–C40.

Shirley D. N. (1983) A partially molten magma ocean model. *Proc. Lunar Planet. Sci. Conf. 13th,* in *J. Geophys. Res., 88,* A519–A527.

Sleep N. H. (1974) Segregation of magma from a mostly crystalline mush. *Geol. Soc. Amer. Bull., 85,* 1225–1232.

Smith J. V. (1974) Origin of the Moon by disintegrative capture with chemical differentiation followed by sequential accretion (abstract). In *Lunar Science V,* pp. 718–720. The Lunar Science Institute, Houston.

Smith J. V. (1976) Development of the Earth-Moon system with implications for the geology of the early Earth. In *The Early History of the Earth* (B. F. Windley, ed.), pp. 3–18. Wiley, New York.

Smith J. V. (1977) Possible controls on the bulk composition of the Earth: Implications for the origin of the Earth and the Moon. *Proc. Lunar Sci. Conf. 8th,* pp. 333–369.

Stevenson D. J. (1980) Lunar asymmetry and paleomagnetism. *Nature, 287,* 520–521.

Stevenson D. J. (1984) Lunar origin from impact on the Earth: Is it possible? (abstract). In *Papers Presented to the Conference on the Origin of the Moon,* p. 60. Lunar and Planetary Institute, Houston.

Stevenson D. J. and Yoder C. (1981) A fluid outer core for the Moon and its implications for lunar dissipation, free librations and magnetism (abstract). In *Lunar and Planetary Science XII,* pp. 1043–1045. Lunar and Planetary Institute, Houston.

Stolper E. M. (1977) Experimental petrology of eucritic meteorites. *Geochim. Cosmochim. Acta, 41,* 587–611.

Takahashi E. (1978) Partitioning of Ni^{+2}, Co^{+2}, Fe^{+2}, Mn^{+2} and Mg^{+2} between olivine and silicate melts: compositional dependence of partition coefficient. *Geochim. Cosmochim. Acta, 42,* 1829–1844.

Takahashi E. (1984) Melting study of peridotites and the Earth's partial melt zone (abstract). *U.S.–Japan Seminar on Partial Melting Phenomena in the Earth and Planetary Evolution.* Univ. of Oregon, Eugene.

Takahashi E. and Kushiro I. (1983) Melting of a dry peridotite at high pressures and basalt magma genesis. *Amer. Mineral., 68,* 859–879.

Taylor S. R. (1975) *Lunar Science: A Post-Apollo View.* Pergamon, New York. 372 pp.

Taylor S. R. (1982) *Planetary Science: A Lunar Perspective.* Lunar and Planetary Institute, Houston. 481 pp.

Taylor S. R. (1984) Tests of the lunar fission hypothesis (abstract). In *Papers Presented to the Conference on the Origin of the Moon,* p. 25. Lunar and Planetary Institute, Houston.

Urey H. C. (1962) The origin of the Moon and its relationship to the origin of the solar system. In *IAU Symposium No. 14: The Moon* (Z. Kopal and Z. K. Mikailov, eds.), pp. 133–148. Academic, New York.

van der Molen I. and Paterson M. S. (1979) Experimental deformation of partially-melted granite. *Contrib. Mineral. Petrol., 70,* 299–318.

Viljoen M. J. and Viljoen R. P. (1969) Evidence for the existence of a mobile extrusive peridotitic magma from the Komati Formation of the Onverwacht Group. *Spec. Publ. Geol. Soc. South Africa, 2,* 87–112.

Wager L. R. and Brown G. M. (1967) *Layered Igneous Rocks.* Freeman, San Francisco. 588 pp.

Walker D., Grove T. L., Longhi J., Stolper E. M., and Hays J. F. (1973) Origin of lunar feldspathic rocks. *Earth Planet. Sci. Lett., 20,* 325–336.

Wänke H., Baddenhausen H., Dreibus G., Jagoutz E., Kruse H., Palme H., Spettel B., and Teschke F. (1973) Multielement analyses of Apollo 15, 16 and 17 samples and the bulk composition of the Moon. *Proc. Lunar Sci. Conf. 4th,* pp. 1461–1481.

Wänke H., Baddenhausen H., Blum K., Cendales M., Dreibus G., Hofmeister H., Kruse H., Jagoutz E., Palme C., Spettel B., Thacker R., and Vilcsek E. (1977) On the chemistry of lunar samples and achondrites. Primary matter in the lunar highlands: A re-evaluation. *Proc. Lunar Sci. Conf. 8th,* pp. 2191–2213.

Wänke H., Palme H., Baddenhausen H., Dreibus G., Jagoutz E., Kruse H., Spettel B., Teschke F., and Thacker R. (1974) Chemistry of Apollo 16 and 17 samples: Bulk composition, late stage accumulation and early differentiation of the Moon. *Proc. Lunar Sci. Conf. 5th*, pp. 1307–1335.

Warner R. D., Taylor G. J., and Keil K. (1980) Petrology of 60035: Evolution of a polymict ANT breccia. In *Proceedings of the Conference on the Lunar Highlands Crust* (J. J. Papike and R. B. Merrill, eds.), pp. 377–394. Pergamon, New York.

Warren P. H. (1983) Models of bulk Moon composition: A review. In *Workshop on Pristine Lunar Highlands Rocks and the Early History of the Moon* (J. Longhi and G. Ryder, eds.), pp. 75–79. LPI Tech Rpt. 83-02, Lunar and Planetary Institute, Houston.

Warren P. H. (1984) Primordial degassing, lithosphere thickness, and the origin of komatiites. *Geology, 12*, 335–338.

Warren P. H. (1985) The magma ocean concept and lunar evolution. *Ann. Rev. Earth Planet. Sci., 13*, 201–240.

Warren P. H. and Kallemeyn G. W. (1984) Pristine rocks (8th foray): "Plagiophile" element ratios, crustal genesis, and the bulk composition of the Moon. *Proc. Lunar Planet. Sci. Conf. 15th*, in *J. Geophys. Res., 89*, C16–C24.

Warren P. H. and Kallemeyn G. W. (1985) Geochemistry of lunar meteorites Yamato-791197 and ALHA81005 (abstract). In *Papers Presented to the 10th Symposium on Antarctic Meteorites*, pp. 40.1–40.2. National Inst. Polar Research, Tokyo.

Warren P. H. and Wasson J. T. (1979a) The origin of KREEP. *Rev. Geophys. Space Phys., 17*, 73–88.

Warren P. H. and Wasson J. T. (1979b) Effects of pressure on the crystallization of a "chondritic" magma ocean and implications for the bulk composition of the Moon. *Proc. Lunar Planet. Sci. Conf. 10th*, pp. 2051–2083.

Warren P. H. and Wasson J. T. (1980) Early lunar petrogenesis, oceanic and extraoceanic. In *Proceedings of the Conference on the Lunar Highlands Crust* (J. J. Papike and R. B. Merrill, eds.), pp. 81–99. Pergamon, New York.

Warren P. H., Taylor G. J., Keil K., Marshall C., and Wasson J. T. (1981) Foraging westward for pristine nonmare rocks: Complications for petrogenetic models. *Proc. Lunar Planet. Sci. 12B*, pp. 21–40.

Wasson J. T. (1971) Volatile elements on the Earth and the Moon. *Earth Planet. Sci. Lett., 11*, 219–225.

Wasson J. T. (1974) *Meteorites—Classification and Properties*. Springer, New York. 316 pp.

Wasson J. T. (1977) Relationship between the composition of solid solar-system matter and distance from the Sun. In *Comets, Asteroids and Meteorites: Interrelations, Evolution and Origins* (A. H. Delsemme, ed.), pp. 551–559. Univ. of Toledo, Ohio.

Wasson J. T. and Warren P. H. (1979) Formation of the Moon from differentiated planetesimals of chondritic composition (abstract). In *Lunar and Planetary Science X*, pp. 1310–1312. Lunar and Planetary Institute, Houston.

Wasson J. T. and Warren P. H. (1980) Contribution of the mantle to the lunar asymmetry. *Icarus, 44*, 752–771.

Wasson J. T. and Warren P. H. (1985) The origin of the Moon (abstract). In *Papers Presented to the Conference on the Origin of the Moon*, p. 57. Lunar and Planetary Institute, Houston.

Weill D. F. and McKay G. A. (1975) The partitioning of Mg, Fe, Sr, Ce, Sm, Eu, and Yb in lunar igneous systems and a possible origin of KREEP by equilibrium partial melting. *Proc. Lunar Sci. Conf. 6th*, pp. 1143–1158.

Wetherill G. W. (1985) Giant impacts and the formation of the Moon (abstract). In *Lunar and Planetary Science XVI*, p. 901. Lunar and Planetary Institute, Houston.

Wood J. A. (1977) Origin of the Earth's Moon. In *Planetary Satellites* (J. A. Burns, ed.), pp. 513–529. Univ. of Arizona, Tucson.

Wood J. A. and Mitler H. E. (1974) Origin of the Moon by a modified capture mechanism, *or* half a loaf is better than a whole one (abstract). In *Lunar Science V*, pp. 851–853. The Lunar Science Institute, Houston.

Yanai K. and Kojima H. (1984) Yamato-791197: A lunar meteorite in the Japanese collection of Antarctic meteorites. *Proc. Ninth Symp. Antarc. Meteorites*, 18–34.

Implications of Isotope Data for the Origin of the Moon

D. L. TURCOTTE AND L. H. KELLOGG

Department of Geological Sciences, Cornell University, Ithaca, NY 14853

Measurements of isotope ratios and the associated concentration ratios have provided a wealth of data on the age of lunar rocks as well as constraints on the origin of the Moon. On the Earth the uniformity of isotope and concentration ratios for midocean ridge basalts has been taken as strong evidence for vigorous mantle convection. This convection mixes and homogenizes the upper mantle beneath the lithospheric plates. Isotope and concentration ratios for lunar mare basalts show much more variability. This has been taken as evidence for distinct source regions. Heterogeneity of the source regions implics little or no mixing and, therefore, little or no mantle convection. Currently, it is popular to associate the origin of the Moon with a massive collision between the Earth and a large planetesimal. Such a hypothesis implies a hot early Moon. If the early Moon was hot, strong mantle convection would have been expected to stir and homogenize the lunar mantle, and distinct source regions would not be available for the subsequent mare volcanism. An alternative hypothesis for the origin of the Moon is accretion from relatively cool material. In this case the deep interior of the Moon would be cool but the near surface rocks would have been heated and melted by the gravitational energy associated with accretion. This would have resulted in a stable density distribution that would have inhibited mantle convection. The heating of an initially cool interior has been hypothesized to explain the mare volcanism. The distinct source regions implied by the isotope data supports a relatively cool origin for the Moon.

Introduction

Isotope systematics provide important constraints on the evolution of planetary bodies. Isotope data are available for at least four bodies: the Earth, the Moon, the eucrite parent body, and the shergottite parent body. In addition, isotope data for chondritic meteorites provides important reference information. Isotope data has been used extensively to date the crystallization ages of igneous rocks. In addition, isotope data has been used to obtain information on global evolution.

Of particular importance is the concept of global reservoirs. For the Earth, the most important reservoirs are a depleted mantle reservoir and the complementary enriched continental crustal reservoir. Other reservoirs include the core and a possible undepleted mantle reservoir. Systematic reservoir studies of the samarium-neodymium and rubidium-strontium systems have been carried out by many authors (Allègre et al., 1983; Anderson, 1983; Armstrong, 1981; Chase, 1981; DePaolo, 1983, 1979, 1980; Hofmann and White, 1982; Jacobsen and Wasserburg, 1980a; Turcotte and Kellogg, 1985). One conclusion is that the mean age of the continents is 2.1 ± 0.2 Ga. The data favor a depleted mantle reservoir that is confined to the upper mantle. The data also favor significant amounts of crustal recycling. Thus isotope data place important constraints on major evolutionary processes in the Earth.

A major question associated with midocean ridge basalts (MORB) is the extent of mantle heterogeneities. At a midocean ridge the oceanic crust is enriched in incompatible elements while the upper mantle lithosphere beneath is depleted. The continental crust is also enriched when it forms, most likely at oceanic trench systems. In order for the entire upper mantle to have a depleted isotope signature, the depleted portion of it that is complementary to the enriched continental crust must be mixed throughout the upper mantle. This mixing can be accomplished by mantle convection (Richter and Ribe, 1979; Olsen et al., 1984a,b). The efficiency of this mixing process can be tested by isotope measurements. It has been suggested that there are large-scale global anomalies (Dupré and Allègre, 1980; Hart, 1984). There are also small-scale anomalies.

With regard to the Moon, an important question is whether the source region for mare basalts was uniform. If the lunar mantle was nearly uniform with small heterogeneities, that would be strong evidence for active mantle convection and a hot lunar interior. However, if there were distinct source regions for the lunar mare basalts this would imply an absence of vigorous mantle convection.

If the Moon accreted as a hot body, vigorous solid-state mantle convection would almost certainly have occurred (Turcotte et al., 1972). However, if the accreting material was relatively cool, the interior of the Moon would initially have been cool. As the Moon grew in size, the gravitational energy of accretion would cause the temperature to increase. With a conversion efficiency of about 35%, sufficient melting would have occurred to form a magma ocean (Turcotte and Pflugrath, 1985). With a cold interior and a hot exterior, the density distribution would be stable and mantle convection would be suppressed. Thus heterogeneities associated with either accretion or the surface melting and differentiation would be preserved.

A number of authors have reviewed the implications of lunar isotope data (Basaltic Volcanism Study Project, 1981; Carlson and Lugmair, 1979; Nyquist, 1977; Wasserburg et al., 1977; Jacobsen and Wasserburg, 1980b; Unruh et al., 1984). Most conclude that the source region for mare basalts was made up of distinct source regions. In this paper we review the data in terms of lunar reservoir models.

Lunar Reservoir Models

A model for the evolution of the Moon into a number of reservoirs is illustrated in Fig. 1. The early Moon was certainly strongly depleted in volatiles. Very early

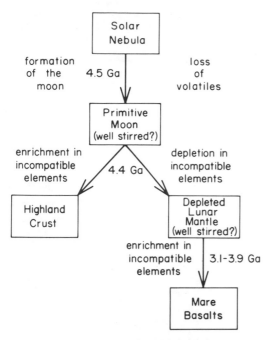

Fig. 1. A reservoir model for the evolution of the Moon. The primitive Moon formed from the solar nebula either directly or by ejection from the Earth. This primitive Moon was fractionated to form an enriched highland crustal reservoir and the complementary depleted mantle reservoir. The mare basalts were subsequently produced by partial melting of the lunar mantle.

in the evolution of the Moon a differentiation event led to the formation of the lunar crust. This is evidence that the outer portion of the Moon was hot but does not constrain the temperature of the deep interior.

If the Moon accreted as an independent body and if the accreting material was cool, then it follows that the deep interior of the Moon would initially have been cool. As discussed above, a lunar temperature that increases with radius would suppress solid state convection. Thus heterogeneities could have been preserved. The heating of a Moon with a cool interior has been used to explain the delayed mare volcanism (Solomon and Toksöz, 1973). However, it is now popular to associate the formation of the Moon with a cataclysmic collision between a large planetesimal and the proto-Earth. Such a collision would have produced a circumferential cloud of vaporized material and debris. Such a scenario for the origin of the Moon would imply a hot initial state for the lunar interior. Both secular cooling of the lunar interior and the heat generation by radioactive isotopes would require vigorous convection within the lunar mantle (Turcotte et al., 1979). For the Earth, vigorous mantle convection results in a nearly homogeneous mantle, and mixing due to mantle convection on the Moon would also be expected to homogenize the lunar mantle beneath the lunar lithosphere.

Isotope Systems

We will consider reservoir models for the evolution of the Moon. We will first study how the samarium-neodymium and rubidium-strontium isotopic systems evolve in terms of a two-reservoir model. The radioactive parent isotopes are ^{147}Sm and ^{87}Rb with a mole density (moles per unit mass) j, the radiogenic daughter isotopes are ^{143}Nd and ^{87}Sr with a mole density i*, and the nonradiogenic reference isotopes are ^{144}Nd and ^{86}Sr with a mole density i.

$$j \qquad i^* \qquad i$$

$$^{147}Sm \rightarrow {}^{143}Nd({}^{144}Nd)$$

$$^{87}Rb \rightarrow {}^{87}Sr \ ({}^{86}Sr)$$

We will use bulk silicate Earth values as a reference. For the samarium-neodymium system these values are the same as chondritic values, but they differ somewhat for the rubidium-strontium systems, as we will show. The reference system evolves according to

$$j_s = j_{so} \, e^{-\lambda t} = j_{so} \, (1 - \lambda t) \tag{1}$$

$$i_s^* = i_{so}^* + j_{so}(1 - e^{-\lambda t}) = i_{so}^* + j_{so}\lambda t \tag{2}$$

where λ is the decay constant ($\lambda_{Sm} = 6.54 \times 10^{-3}$ Ga^{-1}, $\lambda_{Rb} = 1.42 \times 10^{-2}$ Ga^{-1}) and t is time measured forward from the creation of the Earth. The subscript "0" refers to initial values at t = 0. For both systems studied here $\lambda \tau_e \ll 1$ (where $\tau_e = 4.55$ Ga, the age of the solar system) so that the linear approximation is appropriate.

Measurements are usually expressed in terms of composition ratios $\mu \equiv j/i$ and isotope ratios $\alpha \equiv i^*/i$. Following DePaolo and Wasserburg (1976) we find it convenient to express our results in terms of fractionation factors defined by

$$f \equiv \frac{\mu}{\mu_s} - 1 = \frac{(j/i)}{(j_s/i_s)} - 1 \tag{3}$$

and isotope ratios defined by

$$\epsilon \equiv \left[\frac{\alpha}{\alpha_s} - 1 \right] \times 10^4 = \left[\frac{(i^*/i)}{(i_s^*/i_s)} - 1 \right] \times 10^4 \tag{4}$$

where the subscript s refers to the bulk silicate Earth. Our approach follows that of DePaolo (1979, 1980) and Jacobsen and Wasserburg (1979, 1980a).

We assume that a crust of mass M_c was instantaneously separated from a mantle reservoir of mass M_m at a time τ_c before the present. We further assume that vigorous convection in the lunar mantle homogenized the mantle reservoir. We introduce

---

mean enrichment factors for the concentrations of species i and j in the crust relative to the undifferentiated mantle reservoir at the time of separation

$$D_{si} = \frac{i_c}{i_s} = \frac{i_c^*}{i_s^*} \tag{5}$$

$$D_{sj} = \frac{j_c}{j_s} \tag{6}$$

It is assumed that the various isotopes of an element have the same enrichment factors.

Introducing conservation equations for the species

$$i_s (M_c + M_m) = i_c M_c + i_m M_m \tag{7}$$

$$j_s (M_c + M_m) = j_c M_c + j_m M_m \tag{8}$$

Combining equations (5) through (8) gives

$$\frac{i_m}{i_s} = \frac{i_m^*}{i_s^*} = 1 - \frac{M_c}{M_m}(D_{si} - 1) \tag{9}$$

$$\frac{j_m}{j_s} = 1 - \frac{M_c}{M_m} (D_{sj} - 1) \tag{10}$$

These equations are valid at the time τ_c when separation takes place. Utilizing equations (1), (2), (5), (6), (9), and (10), we find that the mean concentrations of the species for the crust and mantle are given by

$$j_c = j_{so}D_{sj} [1 - \lambda(\tau_e - \tau)] \tag{11}$$

$$i_c^* = D_{si} [i_{so}^* + j_{so}\lambda(\tau_e - \tau_c)] + D_{sj} j_{so}\lambda(\tau_c - \tau) \tag{12}$$

$$j_m = j_{so}\left[1 - \frac{M_c}{M_m}(D_{sj} - 1)\right][1 - \lambda(\tau_e - \tau)] \tag{13}$$

$$i_m^* = [i_{so}^* + j_{so}\lambda(\tau_c - \tau)]\left[1 - \frac{M_c}{M_m}(D_{si} - 1)\right] + j_{so}\lambda(\tau_c - \tau)\left[1 - \frac{M_c}{M_m}(D_{sj} - 1)\right] \tag{14}$$

where τ is time measured back from the present ($\tau < \tau_c$) and quadratic terms in $\lambda\tau$ have been neglected. Substitution of equations (5), (9), (11), (12), (13), and (14) into (3) and (4) gives

$$f_c = \frac{D_{sj}}{D_{si}} - 1 \tag{15}$$

$$f_m = \frac{1 - \dfrac{M_c}{M_m}\,(D_{sj} - 1)}{1 - \dfrac{M_c}{M_m}\,(D_{si} - 1)} \tag{16}$$

$$\epsilon_c = Q\left(\frac{D_{sj}}{D_{si}} - 1\right)(\tau_c - \tau) = Qf_c(\tau_c - \tau) \tag{17}$$

$$\epsilon_m = Q\left[\frac{1 - \dfrac{M_c}{M_m}\,(D_{sj} - 1)}{1 - \dfrac{M_c}{M_m}\,(D_{si} - 1)} - 1\right](\tau_c - \tau) = Qf_m(\tau_c - \tau) \tag{18}$$

where

$$Q = 10^4\,\frac{j_{so}}{i^*_{so}}\,\lambda \tag{19}$$

We neglect terms of order $\lambda\tau_c$ in writing equations (15) and (16) since we will keep only the leading terms. This is equivalent to neglecting the loss of the parent due to isotopic decay. We assume that $j_{so}\lambda\tau_c/i^*_{so} \ll 1$ in writing equations (17) and (18); however, we keep the terms of this order since it is the leading term in writing ϵ.

We first consider the neodymium-samarium system. The relevant data are summarized in Table 1. The values of initial ϵ_i are given as a function of age τ in Fig. 2. The values of f are given as a function of age τ in Fig. 3. These values are referenced to a chondritic or bulk silicate Earth value $\mu_s = 0.1967$ (Jacobsen and Wasserburg, 1980b; Allègre *et al.*, 1983). Also included in Fig. 2 is the predicted evolution of a reservoir from equation (18) taking $Q = 25.3$, $\tau_c = 4.4$ Ga, and various values for f.

First it is necessary to make several assumptions. It is assumed that the f and ϵ values of lunar mare basalts are equal to the values in the depleted source region. This is a good approximation for ϵ since isotope fractionation is insignificant, but

TABLE 1. Sm and Nd Data for Lunar Samples.

Sample-Type	μ	α	$T_{isochron}$	ϵ'_i	f**	Source
10062 M	0.2157	0.512524 ± 19	3.88 ± 0.06	2.4	0.09659	(2)
10072 M	0.2061	0.512238 ± 17	3.57 ± 0.03	1.9	0.04779	(2)
12008,54 M	0.2261	0.513033 ± 16	(3.25)*	9.9 ± 0.4	0.1495	(1)
12014,18 M	0.2042	0.512312 ± 17	3.29 ± 0.11†	4.8 ± 0.5	0.03813	(1)
12031,25 M	0.2022	0.512199 ± 23	3.23 ± 0.11†	3.6 ± 0.5	0.02796	(1)
12034 K						
MG K	0.1680	0.511905 ± 18	(3.9)*	−1.4 ± 0.4	−0.1459	
CL8 K	0.1669	0.511859 ± 21	(3.9)*	−1.7 ± 0.5	−0.1515	
12038 M	0.1847	0.511699 ± 19	3.28 ± 0.23	1.5 ± 0.6	−0.0610	(8)
	0.1866	0.511714 ± 40			−0.0513	
12038,222 M	0.1864	0.511669 ± 19	3.28 ± 0.21†	−0.3 ± 0.5	−0.0524	(1)
12039,19 M	0.2090	0.512351 ± 18	3.20 ± 0.05	3.9 ± 0.5	0.06253	(1)
PL M	0.1727	0.511575 ± 71			−0.1220	
PX M	0.2434	0.513071 ± 71			0.2374	
12051,135 M	0.2183	0.512825 ± 19	3.16 ± 0.04	9.6 ± 0.4	0.1098	(1)
12056,13 M	0.2261	0.513096 ± 36	3.20 ± 0.14	11.4 ± 0.9	0.1495	(1)
PX M	0.3085	0.514838 ± 45			0.5684	
12063,287 M	0.2250	0.513094 ± 38	3.30 ± 0.13†	11.4 ± 0.8	0.1439	(1)
12076,18 M	0.2055	0.512301 ± 21	(3.29)*	4.1 ± 0.5	0.0447	(1)
14307 K						(5)(3)
MTX K	0.1690	0.511931 ± 23	(3.9)*	−1.4 ± 0.5	−0.1408	
CLAST K	0.1674	0.511876 ± 22	(3.9)*	−1.6 ± 0.4	−0.1490	
15382 K						(5)(3)
A K	0.1679	0.511889 ± 21	(3.9)*	−1.6 ± 0.4	−0.1464	
B K	0.1683	0.511894 ± 16	(3.9)*	−1.7 ± 0.3	−0.1444	
15426 O						(3)
EG1 O	0.2016	0.512842 ± 34	3.79 ± 0.08	0.0 ± 0.4	0.0249	
EG2 O	0.2015	0.512831 ± 18	3.38 ± 0.06	0.4 ± 0.4	0.0244	
YG1 O	0.2011	0.512823 ± 24			0.0224	
15555 M			3.32 ± 0.04†	2.1 ± 0.3		(3)
65015 K	0.1673	0.511883 ± 19	(3.9)*	−1.5 ± 0.4	−0.1495	(5)
70017 M			3.68 ± 0.18†	6.2 ± 0.3		(3)
70135,27 M	0.2473	0.513500 ± 16	3.77 ± 0.06	6.6 ± 0.6	0.2572	(1)
PLAG M	0.1988	0.512309 ± 18			0.0107	
PX M	0.4721	0.514464 ± 15			1.400	
75035 M			3.81 ± 0.14†	6.2 ± 0.5		(3)
75055 M			3.78 ± 0.04†	6.0 ± 0.5		(3)
75075,58,1 M	0.2538	0.513666 ± 18	3.70 ± 0.07	7.2 ± 0.6	0.2903	(1)
2 M	0.2536	0.513648 ± 15			0.2893	
76535 T	0.1556	0.511556 ± 14	4.26 ± 0.06	−0.1 ± 0.3	−0.2089	(4)
77075 D	0.1739	0.512050 ± 19	4.13 ± 0.82*	−0.7	−0.1159	(7)
77215,37 B	0.1780	0.51200 ± 7	4.37 ± 0.7*	−3.8, −4.5	−0.0951	(6)

*Not an isochron age.

**Calculated with respect to modern $\mu_s= 0.1967$ (Allegre et al., 1983).

†Rb-Sr isochron age is used.

M = mare basalt, K = KREEP, O = other (glass), T = troctolite, D = olivine "dikelets", and B = microbreccia.

Sources of data: (1) Nyquist et al. (1979); (2) Papanastassiou et al. (1977); (3) Lugmair and Marti (1978); (4) Lugmair et al. (1976); (5) Lugmair and Carlson (1978); (6) Nakamura et al. (1976); (7) Nakamura and Tatsumoto (1977); and (8) Nyquist et al. (1981) and Lugmair et al. (1975). Note: references (6) and (7) are subject to significant laboratory comparison uncertainties.

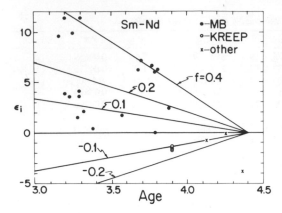

Fig. 2. The measured initial samarium-neodymium isotope ratios ϵ_i for a variety of lunar rocks are given as a function of their age. The data are tabulated in Table 1. Also included is the predicted evolution of a mantle reservoir from equation (18) taking $Q = 25.3$, $\tau_c = 4.4$, and various values for the fractionation factor f.

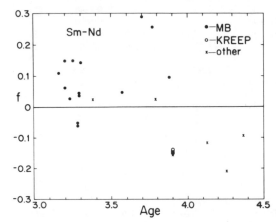

Fig. 3. The measured samarium-neodymium fractionation factors f for a variety of lunar rocks are given as a function of their age. The values are referenced to a chondritic or bulk silicate Earth value $\mu_s = 0.1967$. The data are tabulated in Table 1.

may be a bad approximation for f if the degree of partial melting is low. The ratio of the composition ratio of the melt μ_m to the composition ratio of the unmelted parental material μ_f is plotted against the degree of partial melting χ in Fig. 4 (Jacobsen and Wasserburg, 1979; Depaolo, 1979). With a substantial degree of partial melting, essentially all the incompatible elements go into the melt. It is also necessary to assume that the mare basalts were not contaminated when they passed through the preexisting lunar crust.

It is seen in Fig. 2 that there is considerable scatter in the data. Discrete groups appear to correlate with variations in major element chemistry. Several authors (Lugmair and Marti, 1978; Papanastassiou *et al.*, 1977; Nyquist *et al.*, 1979) argue that this variation in ϵ_i values for the lunar basalts indicates early lunar differentiation into multiple distinct reservoirs. Lugmair and Marti (1978) identify at least three reservoirs, the first of which is the source of KREEPy basalts whose initial ϵ is

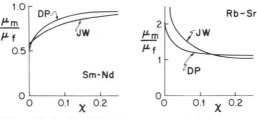

Fig. 4. The ratio of the composition ratio of the melt μ_m to the composition ratio of the unmelted parental material μ_f as a function of the degree of partial melting χ for the samarium-neodymium and rubidium-strontium systems. Results given by Jacobsen and Wasserburg (1979) (JW) and DePaolo (1979) (DP) are included.

negative. The remaining samples have positive initial ϵ. Lugmair and Marti (1978) distinguish between green glass (nearly chondritic) and mare basalts with high initial ϵ. Papanastassiou et al. (1977) present a model of early differentiation of the upper Moon (crust formation about 4.4 aeons ago). They propose that the lunar mantle may have undergone further differentiation at later ages, and that the varying ϵ's of the mare basalts reflect different source layers. They also observe that the Sm/Nd ratios of lunar rocks differ widely from chondritic values. The lunar ratios contrast with those of terrestrial basalts, whose Sm/Nd ratios are very nearly chondritic. Papanastassiou et al. (1977) point out that this variation makes estimation of the bulk Sm/Nd ratios of the Moon extremely difficult, and argue that to assume a chondritic bulk Moon may be wrong. Nyquist et al. (1979) argue that a plot of initial Sr ratios against initial Nd ratios for lunar rocks confirms the layered nature of the lunar mantle. The plot shows a lack of correlation between initial Nd and Sr ratios in the rocks, indicating that the rocks came from distinct sources. A correlation would have indicated that the variations in ϵ could be an artifact of mixing of the magma sources of the rocks.

It is seen from Fig. 3 that the measured values of fractionation factors are not consistent with the values inferred from the isotope evaluation diagram (Fig. 2). For a model of instantaneous formation of the lunar crust, a number of data points in Fig. 2 require f > 0.3. However, no measured values for f are this high. Accepting the validity of the measured values, this difference can be attributed to a low degree of partial melting that leads to a fractionation of samarium and neodymium or to contamination during eruption. For comparison the preferred value for the neodymium-samarium fractionation in the MORB source is f = 0.19 (Turcotte and Kellogg, 1985).

We next consider the rubidium-strontium system. The relevant data are summarized in Table 2. The values of initial ϵ are given as a function of age τ in Fig. 5. The values of initial f are given as a function of age τ in Fig. 6. These values are referenced to a mean silicate Earth value $\mu_s = 0.0892$. This value was given by Allègre et al. (1983) based on a cross correlation of the neodymium-samarium and rubidium-strontium systems. Also included in Fig. 5 is the predicted evolution

TABLE 2. Rb and Sr Data for Lunar Samples.

Sample-Type		μ	α	τ_{isochron}	ϵ'_i	f**	Source
10024,24	M	0.1062	0.70480 ± 19	3.61 ± 0.07	−10.3	0.1906	(7)
10050,24	M	0.00164	0.69914 ± 4	(3.65)*	−13.7	−0.9816	(7)
10062	M	0.0122	0.69970 ± 6	4.01 ± 0.11	−7.5	−0.8632	(12)
10072	M	0.1032	0.70463 ± 5	3.64 ± 0.05	−10.7	0.1570	(12)
12002	M			3.36 ± 0.10	−13.0		(7)
12004	M	0.03183	0.70094 ± 5	3.29 ± 0.07	−14.5	−0.6432	(6)
12008	M	0.0154	0.70001 ± 4	(3.25)*	−18.0	−0.8274	(11)
12009,1	M	0.03095	0.70096 ± 5			−0.6530	(7)
2	M	0.03116	0.70098 ± 6			−0.6507	(7)
12011	M	0.0313	0.70099 ± 9			−0.6591	(11)
12014,18	M	0.0296	0.70096 ± 7	3.29 ± 0.11	−12.2	−0.6682	(11)
12015,18	M	0.0294	0.70097 ± 8			−0.6704	(11)
12021	M			3.33 ± 0.06	−14.7		(7)
12031,25	M	0.0151	0.70020 ± 6	3.23 ± 0.11	−14.3	−0.8307	(11)
12035,18	M			3.20	−16.0		(7)
12038	M	0.00887	0.69964 ± 4	3.35 ± 0.09	−17.1	−0.9006	(10)
12039,1	M	0.0270	0.70091 ± 6	3.19 ± 0.06	−13.9	−0.6973	(2)
2	M	0.0266	0.70083 ± 7			−0.7018	(2)
12040	M	0.02147	0.70041 ± 8	3.30 ± 0.04	−16.1	−0.7593	(7)
12051	M			3.16 ± 0.04	−18.4		(2)(7)
12052	M			(3.28)*	−11.1		(2)
12055,28	M	0.0274	0.70087 ± 5	3.19 ± 0.06	−14.4	−0.6928	(2)
12056,13	M	0.0141	0.69997 ± 4	(3.25)*	−17.7	−0.8419	(2)
12063	M	0.0141		3.30 ± 0.13	−18.6	−0.8419	(7)
12064,26	M	0.01766	0.70025 ± 6	3.18 ± 0.09	−17.3	−0.8020	(7)
12065	M	0.03115	0.70094 ± 6	3.16 ± 0.09	−15.6	−0.6508	(7)
12076,18	M	0.0316	0.70112 ± 4	(3.29)*	−12.6	−0.6457	(2)
14001,7,1	K	0.3325	0.71948 ± 6			2.728	(6)
7,3	K	0.2384	0.71360 ± 4	3.89 ± 0.03	9.5	1.673	(6)
14053	O	0.05724	0.70276 ± 7	3.96 ± 0.04	−1.7	−0.3583	(6)
14072	O			3.99 ± 0.14	−5.6		(8)
14073	K	0.2107	0.71200 ± 5	3.88 ± 0.04	9.04	1.3621	(6)
14305,122	M	0.0448	0.70184 ± 4	4.23 ± 0.05	−1.83	−0.4978	(9)
14310	K	0.1824	0.71041 ± 6	3.87 ± 0.04	9.13	1.0448	(6)
14321	O	0.0964	0.70484 ± 6	3.95 ± 0.04	−2.8	0.0807	(8)(6)
15058,85	M	0.02262	0.70040 ± 6	3.46 ± 0.04	−14.1	−0.7464	(1)
15065	M			3.28 ± 0.04	−16.2		(5)
15076	M			3.33 ± 0.08	−16.7		(5)
15085	M			3.40 ± 0.04	−16.0		(5)
15117	M			3.35 ± 0.04	−16.2		(5)
15379	M	0.0249	0.70048 ± 5			−0.7209	(5)
15386,15	K	0.285	0.71640 ± 7	3.94 ± 0.04	10.8	2.195	(15)
15434	K			3.91 ± 0.04	12.2		(8)
15555,147	M	0.0262	0.70046 ± 15	3.34 ± 0.09	−16.5	−0.7063	(1)
15682	M	0.02576	0.70048 ± 5	3.44 ± 0.07	−14.5	−0.7112	(5)
67075	A			3.66 ± 0.63	−12.7		(3)
17		0.0092	0.69958 ± 3			−0.8969	
53		0.0118	0.69984 ± 7			−0.8677	

TABLE 2. (Continued)

Sample-Type		μ	α	T_{isochron}	ϵ_i'	f**	Source
70017,35	M	0.00565	0.69945 ± 9	3.68 ± 0.18	–11.1	–0.9367	(15)
70135,27	M	0.0127	0.69995 ± 5	3.75 ± 0.09	–9.2	–0.8576	(11)
70215	M	0.0085	0.69965 ± 7	(3.8)*	–9.21	–0.9047	(3)
72275	O			4.01 ± 0.04	–2.6		(8)
74255,25	M	0.0217	0.70045 ± 6	3.83 ± 0.06	–7.6	–0.7567	(3)
74275,56	M	0.0226	0.70042 ± 5	3.83 ± 0.06	–7.6	–0.7466	(3)
75035	M	0.1	0.69977 ± 20	3.81 ± 0.14	–8.8	–0.89	(16)
75075,58	M	0.0081	0.69968 ± 4	3.84 ± 0.12	–8.0	–0.9092	(1)(15)
76535	T			4.61 ± 0.07	3.93		(4)
77075	D	0.1060	0.70554 ± 4	4.09 ± 0.08	–3.5	0.1883	(13)
77215	M			4.42 ± 0.04	0.4		(14)
37		0.2733	0.71641 ± 12			2.064	
145		0.0777	0.70397 ± 3			–0.1289	

*Not an isochron age.

**Calculated with respect to a mean silicate Earth value $\mu_s = 0.0892$ (Allègre et al., 1983).

M = mare basalt, K = KREEP, O = nonmare "mare-like" basalt, T = troctolite, D = olivine "dikelets", A = anorthosite, and B = microbreccia.

Sources of data: (1) Birck et al. (1975); (2) Nyquist et al. (1977); (3) Nyquist et al. (1976); (4) Papanastassiou and Wasserburg (1976); (5) Papanastassiou and Wasserburg (1973); (6) Papanastassiou and Wasserburg (1971a); (7) Papanastassiou and Wasserburg (1971b); (8) Ryder and Spudis (1980); (9) Taylor et al. (1983); (10) Nyquist et al. (1981); (11) Nyquist et al. (1979); (12) Papanastassiou et al. (1977); (13) Nakamura and Tatsumoto (1977); (14) Nakamura et al. (1976); (15) Nyquist et al. (1975); and (16) Murthy and Coscio (1976).

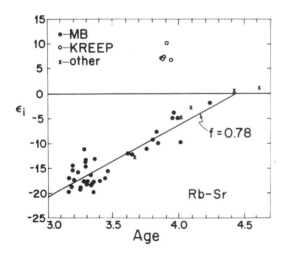

Fig. 5. The measured initial rubidium-strontium isotope ratios ϵ_i for a variety of lunar rocks are given as a function of their age. The data are tabulated in Table 2. Also included is the predicted evolution of a mantle reservoir from equation (18) taking $Q = 17.7$, $\tau_c = 4.4$, and $f = -0.78$.

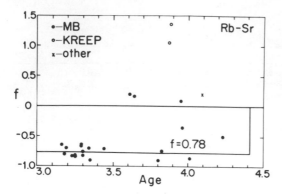

Fig. 6. The measured rubidium-strontium fractionation factors for a variety of lunar rocks are given as a function of their age. These values are referenced to a mean silicate Earth value $\mu_s = 0.0892$. The data are tabulated in Table 1.

of a reservoir from equation (18) taking $Q = 17.7$, $\tau_c = 4.4$ Ga, and $f = -0.78$. The values of f measured directly are in quite good agreement with the values inferred from isotope evolution. However, a word of caution is in order. The maximum negative value of f is -1, this corresponding to the complete removal of rubidium. Nevertheless, the behavior of the rubidium-strontium system seems to be more systematic than the behavior of the neodymium-samarium system.

The results given in Figs. 5 and 6 show a strong depletion in rubidium relative to strontium. This could be associated with either the formation of the Moon or the differentiation of the lunar crust. Clearly the KREEP basalts are enriched in rubidium and may represent a complementary enriched reservoir. However, the Moon is generally depleted in volatiles and this general depletion during formation would be expected to lead to a relative depletion in rubidium.

Data on the composition ratio μ for the rubidium-strontium system is summarized in Fig. 7. Our best fit value for the lunar mare basalts, $f = -0.78$, is identical to the best fit value for MORB, $f = -0.78$, given by Allègre et al. (1983). Since the processes leading to the depletion of MORB must have been different from the conditions leading to the depletion of the lunar mantle, it is difficult to provide

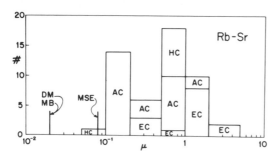

Fig. 7. A histogram of measured rubidium-strontium composition ratios μ for enstatite (EC), hypersthene (HC), and amphoterite (AC) chondrites (Gopalan and Wetherill, 1968, 1969, 1970). Also included are the mean values for the bulk silicate Earth (MSE), the Earth's depleted mantle (DM), and the lunar mare basalts (MB) corresponding to the correlation given in Fig. 5.

a chemical basis for this similarity. Also included in Fig. 7 is the mean silicate Earth value given by Allègre *et al.* (1983) and measurements from a variety of chondritic meteorites (Gopalan and Wetherill, 1968, 1969, 1970). The depletion of the Earth relative to the chondritic meteorites is clearly illustrated.

We will next consider the uranium-thorium-lead system. The radioactive parent isotopes are ^{238}U, ^{235}U, and ^{232}Th with mole densities j, j', and j". The radioactive daughter isotopes are ^{206}Pb, ^{207}Pb, and ^{208}Pb with mole densities i*, i*', and i*". The nonradiogenic reference isotope is ^{204}Pb with a mole density i.

$$
\begin{array}{ccc}
j & i* & i
\end{array}
$$

$$^{238}U \rightarrow {}^{206}Pb\,({}^{204}Pb)$$

$$^{235}U \rightarrow {}^{207}Pb\,({}^{204}Pb)$$

$$^{232}Th \rightarrow {}^{208}Pb\,({}^{204}Pb)$$

It is common practice to introduce the composition ratios $\mu = j/i$, $\nu = j'/j$, and $\kappa = j''/j$ and the isotope ratios $\alpha = i*/i$, $\beta = i*'/i$, $\gamma = i*''/i$. The time evolution of a closed system is given by

$$\alpha = \mu_0(1 - e^{-\lambda t}) + \alpha_0 \tag{20}$$

$$\beta = \nu_0\mu_0(1 - e^{-\lambda' t}) + \beta_0 \tag{21}$$

$$\gamma = \kappa_0\mu_0(1 - e^{-\lambda'' t}) + \gamma_0 \tag{22}$$

$$\mu = \mu_0 e^{-\lambda t} \tag{23}$$

$$\nu = \nu_0 e^{(\lambda - \lambda')t} \tag{24}$$

$$\kappa = \kappa_0 e^{(\lambda - \lambda'')t} \tag{25}$$

The decay constants are $\lambda = 1.551 \times 10^{-10}$ a^{-1}, $\lambda' = 9.848 \times 10^{-10}$ a^{-1}, and $\lambda''= 4.948 \times 10^{-11}$ a^{-1}. Initial values for the lead isotope ratios as obtained from studies of meteorites (Tatsumoto *et al.*, 1973) are $\alpha_0 = 9.307$, $\beta_0 = 10.294$, and $\gamma = 29.476$; initial values of the composition ratios were υ_0, ν_0, and κ_0.

We hypothesize that a homogeneous lunar mantle reservoir was formed early in the evolution of the moon ($\tau_c \approx \tau_e = 4.5$ Gyr). The initial composition ratios in this mantle reservoir were μ_0 and ν_0. The present reference composition values are

$$\mu_r = \mu_0 e^{-\lambda \tau_c} \tag{26}$$

$$\nu_r = \nu_0 e^{(\lambda - \lambda')\tau_c} = \frac{1}{137.8} \tag{27}$$

The mantle reservoir evolved until the mare basalts were extracted at a time τ.

From equations (20), (21), (23), (24), (26), and (27), this reservoir evolved according to

$$\alpha_i = \mu_r(e^{\lambda \tau_c} - e^{\lambda \tau}) + 9.307 \tag{28}$$

$$\beta_i = \frac{\mu_r}{137.8}(e^{\lambda' \tau_c} - e^{\lambda' \tau}) + 10.294 \tag{29}$$

where α_i and β_i were the isotope ratios at the time the mare basalts erupted. The assumed isotopic evolution of the lunar mantle reservoir is given by the solid lines in Fig. 8. The corresponding ages τ for mare basalt eruption are shown by the dashed lines. This is known as the Holmes-Houtermans model.

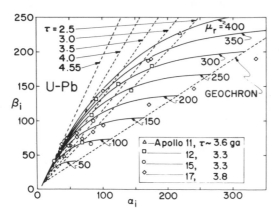

Fig. 8. The measured initial uranium-lead isotope ratios β_i for a variety of mare basalts are given as a function of the initial uranium-lead isotope ratios α_i. The data are tabulated in Table 3. The solid lines are evolution curves obtained from equations (28) and (29) for various values of the reference composition ratio μ_r. The dashed lines give the expected initial values (β_i and α_i) corresponding to the time τ when the mare basalts were erupted.

Isotope data for the uranium-thorium-lead system is summarized in Table 3. Initial values were obtained using equations (20) and (21) and ages obtained independently (usually using rubidium-strontium data). The initial values for a variety of mare basalts are plotted in Fig. 8. If the lunar mantle reservoirs were uniform, all the data would lie on a $\mu = $ constant curve at the intersection of the time line corresponding to the age of the basalt.

TABLE 3. Lead Data, Mare Basalts.

Sample	α	β	γ	μ	κ	τ	α_i	β_i	Source
10017	410.0	191.9	435.0	492.8	4.07	3.575	44.7	74.55	(2)
10003	423.9	198.0	448.9	491.3	3.96	3.76	34.9	56.9	(2)
10057	1241.5	590.1	1281.3	1392.0	4.08	3.55	218.6	226.9	(2)
10071	199.7	95.8	206.7	222.5	4.06	3.47	41.0	48.2	(2)
12064	449.6	234.9	444.6	510.6	3.64	3.18	124.0	153.7	(4)
12021	672.2	270.3	671.9	871.9	3.69	3.33	82.6	108.6	(4)
12038	391.2	158.8	428.5	473.0	4.05	3.35	68.9	69.24	(4)
12063,49 a	720.2	299.9	710.5	863.3	3.67	3.3	143.1	144.6	(4)
12063,49 b	739.9	306.9	724.8	972.3	3.45	3.3	89.5	131.9	(4)
12063,50	186.0	75.0	156.0	–	–	3.3			(5)
12052,66	796.7	331.3	789.2	985.6	(3.65)	3.15	175.7	179.32	(4)
15555	623.0	249.0	–	792.0	–	3.34	85.3	100.6	(6)
15085 1 P	158.0	77.8	182.0	–	–	3.40			(8)
15085 1 C	183.0	86.8	–	227.0	4.09	3.40	25.3	41.6	(8)
15085 2 C	232.0	110.0	–	217.0	4.10	3.40	81.3	66.8	(8)
75055 C	260.1	150.0	–	250.0	3.40	3.78	37.0	65.9	(1)
75055 P	236.4	139.2	229.0	250.0	3.40	3.78	60.7	76.7	(1)
75035 P	579.2	341.0	520.5	507.0	3.35	3.81	170.6	187.9	(1)
75035 C	582.2	342.9	–	507.0	3.35	3.81	173.6	189.8	(1)
75035 C	509.1	296.3	–	507.0	3.35	3.81	100.5	143.2	(1)
74275 P	449.9	226.6	4|8.3	430.0	3.54	3.83	101.0	94.1	(1)
74275 C	519.9	256.3	–	430.0	3.54	3.83	171.0	123.8	(1)
74275 C	397.4	196.9	–	430.0	3.54	3.83	48.5	64.4	(1)
74255 P	680.7	321.5	621.1	427.0	3.48	3.83	334.2	189.9	(1)
74255 C	586.7	278.8	–	427.0	3.48	3.83	240.2	147.2	(1)
75075,1	490.2	247.0	446.0	528.0	3.78	3.84	60.27	82.84	(3)
75075,2	389.1	200.0	339.6	373.0	3.13	3.84	85.4	83.9	(3)
70017 P	248.9	135.5	221.7	234.0	3.12	3.68	68.77	73.5	(8)
70017 C	230.1	125.8	–	234.0	3.12	3.68	49.97	63.82	(8)

P = composition data and C = concentration data.
Sources of data: (1) Nunes et al. (1974); (2) Tatsumoto and Rosholt (1970); (3) Chen et al. (1978); (4) Tatsumoto et al. (1971); (5) Cliff et al. (1971); (6) Tera and Wasserburg (1974); (7) Mattison et al. (1977); and (8) Unruh and Tatsumoto (1977).

The large values of α and β indicate a strong depletion of naturally occurring lead relative to uranium. This leads to severe measurement problems concerning the reference lead isotope ^{204}Pb. Some of the scatter in the data can be attributed to this difficulty. Apollo 17 data have a particularly large scatter and had a particularly low μ value. There appears to be a weak correlation of the data with $\mu = 250$. This compares with $\mu = 9$ for the Earth's mantle. Clearly the lunar mantle was strongly depleted in lead probably prior to its formation. It is difficult to envision a lunar reservoir for the missing lunar lead.

Conclusions

Isotope data on the lunar rocks for the rubidium-strontium system show the least variability. Both isotope and concentration data show a strong depletion of rubidium relative to terrestrial values in the source region for mare basalts. The KREEP basalts show enrichment. It is interesting to note that the mean depletion in mare basalts is equal to the mean depletion in midocean ridge basalts. However, the significance of this equality, if any, is not clear. The depletion of rubidium in the source region for mare basalts is probably due to fractionation during the formation of the Moon. The systematic depletion of rubidium in the Moon relative to terrestrial values would seem to favor a hot origin for the Moon; however, this does not seem to be consistent with the concept of multiple source regions as discussed above.

The isotope and concentration data for the samarium-neodymium system show considerable variability. This variability has led a number of authors to postulate distinct source regions for the mare basalts. As in the case of the source region for MORB, neodymium is depleted relative to samarium and the magnitude of the depletion is about the same as on the Earth.

The uranium-lead system also shows considerable scatter. However, it is clear that the source region for mare basalts is strongly depleted in lead relative to the Earth's mantle; an estimated $\mu \sim 250$ vs. $\mu = 9$. It is difficult to envision a reservoir on the Moon for this missing lead, and thus the fractionation must have occurred during the formation of the Moon. The low concentrations of the reference stable lead isotope ^{204}Pb cause serious experimental difficulties. Thus the scatter in the lead data cannot be taken as strong evidence for distinct source regions for the mare basalts but is certainly consistent with this hypothesis.

The isotope data from lunar rocks show much more variability than the similar data from terrestrial rocks. In particular, the midocean ridge basalts yield consistent isotope signatures on a worldwide basis. This is taken as strong evidence that vigorous mantle convection has mixed and homogenized the upper mantle beneath the lithosphere plates. The variability of the lunar data is taken as evidence for distinct source regions. It is also evidence that mantle convection did not mix and homogenize the lunar mantle prior to the eruption of the mare basalts.

If the lunar interior was hot at the time of accretion, then strong mantle convection would be expected. If the Moon accreted from relatively cool material, then the deep interior of the Moon would be cool but the near surface rocks would have been heated and melted by the gravitational energy associated with accretion. This would have resulted in a stable density distribution that would have inhibited mantle convection. The heating of an initially cool lunar interior has been hypothesized to explain the late episodes of mare volcanism. Thus the distinct source regions implied by the isotope data supports a relatively cool origin for the Moon.

Acknowledgments. This research has been supported by grant NGR-33-010-108 from the National Aeronautics and Space Administration. The authors would like to thank Robert Kay and Claude Allègre for many enlightening discussions. The authors would also like to thank Larry Nyquist and D. A. Papanastassiou for comprehensive and informative reviews. This paper is contribution 821 of the Department of Geological Sciences at Cornell University.

References

Allègre C. J., Hart S. R., and Minster J. F. (1983) Chemical structure and evolution of the mantle and continents determined by inversion of Nd and Sr isotope data, II. Numerical experiments and discussion. *Earth Planet. Sci. Lett.*, *66*, 191–213.

Anderson D. L. (1983) Chemical composition of the mantle. *Proc. Lunar Planet. Sci. Conf. 14th*, in *J. Geophys. Res.*, *88*, B41–B52.

Armstrong R. L. (1981) Radiogenic isotopes: The case for crustal recycling on a near-steady-rate no-continental-growth Earth. *Phil. Trans. Roy. Soc. London*, *A301*, 443–472.

Basaltic Volcanism Study Project (1981) *Basaltic Volcanism on the Terrestrial Planets*. Pergamon, New York. 1286 pp.

Birck J. L., Fourcade S., and Allègre C. J. (1975) ^{87}Rb-^{86}Sr age of rocks from the Apollo 15 landing site and significance of internal isochrons. *Earth Planet. Sci. Lett.*, *26*, 29–35.

Carlson R. W. and Lugmair G. W. (1979) Sm-Nd constraints on early lunar differentiation and the evolution of KREEP. *Earth Planet. Sci. Lett.*, *45*, 123–132.

Chase C. G. (1981) Oceanic island Pb: Two-state histories and mantle evolution. *Earth Planet. Sci. Lett.*, *52*, 277–284.

Chen J. H., Tilton G. R., Mattinson J. M., and Vidal Ph. (1978) Lead isotope systematics of mare basalt 75075. *Proc. Lunar Planet. Sci. Conf. 9th*, pp. 509–521.

Cliff R. A., Lee-Hu C., and Wetherill G. W. (1971) Rb-Sr and U, Th-Pb measurements on Apollo 12 material. *Proc. Lunar Sci. Conf. 2nd*, pp. 1493–1502.

DePaolo D. J. (1979) Implications of correlated Nd and Sr isotopic variations for the chemical evolution of the crust and mantle. *Earth Planet. Sci. Lett.*, *43*, 201–211.

DePaolo D. J. (1980) Crustal growth and mantle evolution: Inferences from models of element transport and Nd and Sr isotopes. *Geochim. Cosmochim. Acta*, *44*, 1185–1196.

DePaolo D. J. (1983) The mean life of continents: Estimates of continent recycling rates from Nd and Hf isotopic data and implications for mantle structure. *Geophys. Res. Lett.*, *10*, 705–708.

DePaolo D. J. and Wasserburg G. J. (1976) Nd isotopic variations and petrogenetic models. *Geophys. Res. Lett.*, *3*, 249–252.

Dupré B. and Allègre C. J. (1980) Pb-Sr-Nd isotopic correlations and the chemistry of the North Atlantic mantle. *Nature*, *286*, 17–22.

Gopalan K. and Wetherill G. W. (1968) Rubidium-strontium age of hypersthene (L) chondrites. *J. Geophys. Res.*, *73*, 7133–7136.

Gopalan K. and Wetherill G. W. (1969) Rubidium-strontium age of amphoterite (LL) chondrites. *J. Geophys. Res.*, *74*, 4349–4358.

Gopalan K. and Wetherill G. W. (1970) Rubidium-strontium studies on enstatite chondrites: Whole meteorite and mineral isochrons. *J. Geophys. Res.*, *75*, 3457–3467.

Hart S. R. (1984) A large-scale isotope anomaly in the Southern Hemisphere mantle. *Nature*, *309*, 753–757.

Hofmann A. W. and White W. M. (1982) Mantle plumes from ancient oceanic crust. *Earth Planet. Sci. Lett.*, *57*, 421–436.

Jacobsen S. B. and Wasserburg G. J. (1979) The mean age of mantle and crustal reservoirs. *J. Geophys. Res.*, *84*, 7411–7427.

Jacobsen S. B. and Wasserburg G. J. (1980a) A two-reservoir recycling model for mantle-crust evolution. *Proc. Natl. Acad. Sci.*, *77*, 6298–6302.

Jacobsen S. B. and Wasserburg G. J. (1980b) Sm-Nd isotope evolution of chondrites. *Earth Planet. Sci. Lett.*, *50*, 139–155.

Lugmair G. W. and Carlson R. W. (1978) The Sm-Nd history of KREEP. *Proc. Lunar Planet. Sci. Conf. 9th*, pp. 689–704.

Lugmair G. W. and Marti K. (1978) Lunar initial $^{143}Nd/^{144}Nd$: Differential evolution of the lunar crust and mantle. *Earth Planet. Sci. Lett., 39,* 349–357.

Lugmair G. W., Marti J. P., Kurtz P., and Scheinin N. B. (1976) History and genesis of lunar troctolite 76535 or: How old is it? *Proc. Lunar Sci. Conf. 7th,* pp. 2009–2033.

Lugmair G. W., Scheinin B., and Marti K. (1975) Sm-Nd age and history of Apollo 17 basalt 75075: Evidence for early differentiation of the lunar exterior. *Proc. Lunar Sci. Conf. 6th,* pp. 1419–1429.

Mattinson J. M., Tilton G. R., Todt W., and Chen J. H. (1977) Lead isotope studies of mare basalt 70017. *Proc. Lunar Sci. Conf. 8th,* pp. 1473–1487.

Murthy V. R. and Coscio M. R., Jr. (1976) Rb-Sr ages and isotopic systematics of some Serenitatis mare basalts. *Proc. Lunar Sci. Conf. 7th,* pp. 1529–1544.

Nakamura N. and Tatsumoto M. (1977) The history of the Apollo 17 station 7 boulder. *Proc. Lunar Sci. Conf. 8th,* pp. 2301–2314.

Nakamura N., Tatsumoto M., Nunes P. D., Unruh D. M., Schwab A. P., and Wildeman T. R. (1976) 4.4 b.y.-old clast in Boulder 7, Apollo 17: A comprehensive chronological study by U-Pb, Rb-Sr and Sm-Nd methods. *Proc. Lunar Sci. Conf. 7th,* pp. 2309–2333.

Nunes P. D., Tatsumoto M., and Unruh D. M. (1974) U-Th-Pb systematics of some Apollo 17 lunar samples and implications for a lunar basin excavation chronology. *Proc. Lunar Sci. Conf. 5th,* pp. 1487–1514.

Nyquist L. E. (1977) Lunar Rb-Sr chronology. *Phys. Chem. Earth, 10,* 103–142.

Nyquist L. E., Bansal M., and Wiesmann H. (1975) Rb-Sr ages and initial $^{87}Sr/^{86}Sr$ for Apollo 17 basalts and KREEP basalts 15386. *Proc. Lunar Sci. Conf. 6th,* pp. 1445–1465.

Nyquist L. E., Bansal B. M., and Wiesmann H. (1976) Sr isotope constraints on the petrogenesis of Apollo 17 mare basalts. *Proc. Lunar Sci. Conf. 7th,* pp. 1507–1528.

Nyquist L. E., Bansal B. M., Wooden J. L., and Wiesmann H. (1977) Sr isotopic constraints on the petrogenesis of Apollo 12 mare basalts. *Proc. Lunar Sci. Conf. 8th,* 1383–1415.

Nyquist L. E., Shih C.-Y., Wooden J. L., Bansal B. M., and Wiesmann H. (1979) The Sr and Nd isotopic record of Apollo 12 basalts: Implications for lunar geochemical evolution. *Proc. Lunar Planet. Sci. Conf. 10th,* pp. 77–114.

Nyquist L. E., Wooden J. L., Shih C.-Y., Wiesmann H., and Bansal B. M. (1981) Isotopic and REE studies of lunar basalt 12038: Implications for petrogenesis of aluminous mare basalts. *Earth Planet. Sci. Lett., 55,* 335–355.

Olson P., Yuen D. A., and Balsiger D. (1984a) Mixing of passive heterogeneities by mantle convection. *J. Geophys. Res., 89,* 425–436.

Olson P., Yuen D. A., and Balsiger D. (1984b) Convective mixing and the fine structure of mantle heterogeneity. *Phys. Earth. Planet. Inter., 36,* 291–304.

Papanastassiou D. A. and Wasserburg G. J. (1971a) Rb-Sr ages of igneous rocks from the Apollo 14 mission and the age of the Fra Mauro formation. *Earth Planet. Sci. Lett., 12,* 36–48.

Papanastassiou D. A. and Wasserburg G. J. (1971b) Lunar chronology and evolution from Rb-Sr studies of Apollo 11 and 12 samples. *Earth. Planet. Sci. Lett., 11,* 37–62.

Papanastassiou D. A. and Wasserburg G. J. (1973) Rb-Sr ages and initial strontium in basalts from Apollo 15. *Earth Planet. Sci. Lett., 17,* 324–337.

Papanastassiou D. A. and Wasserburg G. J. (1976) Rb-Sr ages of troctolite 76535. *Proc. Lunar Sci. Conf. 7th,* pp. 2035–2054.

Papanastassiou D. A., DePaolo D. J., and Wasserburg G. J. (1977) Rb-Sr and Sm-Nd chronology and geneaology of mare basalts from the Sea of Tranquillity. *Proc. Lunar Sci. Conf. 8th,* pp. 1639–1672.

Richter F. M. and Ribe N. M. (1979) On the importance of advection in determining the local isotopic composition of the mantle. *Earth Planet. Sci. Lett., 43,* 212–222.

Ryder G. and Spudis P. (1980) Volcanic rocks in the lunar highlands. *Proceedings of the Conference on the Lunar Highlands Crust* (J. J. Papike and R. B. Merrill, eds.), pp. 353–375. Pergamon, New York.

Solomon S. C. and Toksoz M. N. (1973) Internal constitution and evolution of the Moon. *Phys. Earth Planet. Inter.*, *7*, 15–38.

Taylor L. A., Shervais J. W., Hunter R. H., Shih C.-Y., Bansal B. M., Wooden J. L., Nyquist L. E., and Laul J. C. (1983) Pre-4.2 AE mare basalt volcanism in the lunar highlands. *Earth Planet. Sci. Lett.*, *66*, 33–47.

Tatsumoto M. and Rosholt J. N. (1970) Age of the Moon; an isotopic study of U-Th-Pb systematics of lunar samples. *Science, 167*, 461.

Tatsumoto M., Knight R. J., and Allègre C. J. (1973) Time differences in the formation of meteorites as determined from the ratio of lead-207 to lead-206. *Science, 180*, 1279–1283.

Tatsumoto M., Knight R. J., and Doe B. R. (1971) U-Th-Pb systematics of Apollo 12 lunar samples. *Proc. Lunar Sci. Conf. 2nd*, pp. 1521–1546.

Tera F. and Wasserburg G. J. (1974) U-Th-Pb systematics of lunar rocks and inferences about lunar evolution and the age of the Moon. *Proc. Lunar Sci. Conf. 5th*, pp. 1571–1599.

Turcotte D. L. and Kellogg L. H. (1985) Isotopic modelling of the evolution of the mantle and crust. *J. Geophys. Res.*, in press.

Turcotte D. L. and Pflugrath J. C. (1985) Thermal structure of the accreting Earth. *Proc. Lunar Planet. Sci. Conf. 15th*, in *J. Geophys. Res.*, *90*, C541–C544.

Turcotte D. L., Cook F. A., and Willeman R. J. (1979) Parameterized convection within the Moon and the terrestrial planets. *Proc. Lunar Planet. Sci. Conf. 10th*, pp. 2375–2392.

Turcotte D. L., Hsui A. T., Torrance K. E., and Oxburgh E. R. (1972) Thermal structure of the Moon. *J. Geophys. Res., 71*, 6931–6939.

Unruh D. M. and Tatsumoto M. (1977) Evolution of mare basalts: The complexity of the U-Th-Pb system. *Proc. Lunar Sci. Conf. 8th*, pp. 1673–1696.

Unruh D. M., Stille P., Patchett P. J., and Tatsumoto M. (1984) Lu-Hf and Sm-Nd evolution in lunar mare basalts. *Proc. Lunar Planet. Sci. Conf. 14th*, in *J. Geophys. Res., 89*, B459–B477.

Wasserburg G. J., Papanastassiou D. A., Tera F., and Huneke J. C. (1977) Outline of a lunar chronology. *Phil. Trans. Roy. Soc. London, A285*, 7–22.

I-Pu-Xe Dating and the Relative Ages of the Earth and Moon

T. D. SWINDLE, M. W. CAFFEE, C. M. HOHENBERG

McDonnell Center for the Space Sciences and Department of Physics, Washington University, St. Louis, MO 63130

S. R. TAYLOR

Research School of Earth Sciences, Australian National University, Canberra, Australia

Using I-Pu-Xe dating, we estimate that retention of excess fission xenon in lunar samples began no more than 63 ± 42 m.y. after t_o, the time of formation of primitive meteorites. This result is consistent with the oldest crystallization ages observed for lunar rocks, several of which fall within 50 m.y. of t_o, but all of which have uncertainties of 50–100 m.y. Dynamical studies of planetary accretion also suggest timescales on the order of 10 to 100 m.y. I-Pu-Xe ages of the Earth's atmosphere and mantle suggest formation 75–100 m.y. after t_o. These ages indicate that some lunar rocks underwent isotopic closure (and thus that the Moon must have already formed) between 80 m.y. before and 30 m.y. after formation of the Earth. The lunar ages should be equal to or later than those of the Earth if the Moon was derived from the Earth by fission or impact. Better understanding of the excess fission xenon effect could lead to a much stricter younger limit of the formation age of the Moon than is currently available. For lunar samples, I-Pu-Xe ages are determined from the $^{129}Xe/^{136}Xe$ ratio in excess fission xenon, which varies within samples in a systematic fashion that is consistent with chronological interpretations. Several aspects of the excess fission xenon effect are not well understood, but the only aspect that causes a large uncertainty in I-Pu-Xe ages is the question of whether the source of the xenon is degassing of the bulk Moon or degassing of a more local area. In either case, the I/U ratio of the source region (which is used to establish the initial $^{129}I/^{244}Pu$ ratio) is apparently about the same, but it is more difficult to constrain the I/U ratio of a local source region.

Introduction

How old is the Moon?

That is a fundamental question whose answer has to be a strong constraint on any theory of the origin of the Moon. Since, in many theories, the early history of the Moon is closely related to the early history of the Earth, what we really need to know is how the age of the Moon compares with that of the Earth. Finding the relative ages of the Earth and Moon is a difficult proposition, requiring a clear understanding of just what is being dated and high resolution of events occurring long ago. The Earth has seen extensive geological and biological processing that has wiped out almost all traces of its first billion years. The Moon, meanwhile, has undergone extensive bombardment and some differentiation, leaving few records of its earliest history. There are, however, a few clues remaining to the early history of both bodies.

Previous attempts to estimate formation ages of either the Earth or the Moon have generally relied on U-Th-Pb or Rb-Sr (initial $^{87}Sr/^{86}Sr$) systematics to establish the times of early differentiation events, or (for the Moon) the crystallization ages of the oldest rocks. As we will discuss, these studies have shown that both the Earth and the Moon are quite old (certainly more than 4400 m.y. old), but they have not succeeded in constraining the relative ages of the two bodies to any significant degree.

A technique that shows promise for determining the relative ages of the Earth and the Moon with considerably more precision is I-Pu-Xe dating. In this system, calculation of the age of the Earth is based on the addition of gas from the decay of the extinct radionuclides ^{244}Pu and ^{129}I to xenon reservoirs of either the terrestrial atmosphere or the mantle (Wetherill, 1975; Pepin and Phinney, 1976; Bernatowicz and Podosek, 1978; Staudacher and Allégre, 1982). For the Moon, the I-Pu-Xe age is based on the varying ratio of iodine-derived to plutonium-derived xenon in "excess fission xenon" (xenon presently residing on grain surfaces in gas-rich highland breccias, but produced by the decay of ^{244}Pu and ^{129}I elsewhere in the Moon).

In this paper we will discuss the ages of the Earth and Moon as determined by various chronometric systems, but our primary focus will be on developing an I-Pu-Xe chronometer. Toward that end, we will review the data on excess fission xenon, with particular emphasis on the strengths and weaknesses of the assumptions required for lunar I-Pu-Xe chronometry. The oldest lunar ages we obtain by using the I-Pu-Xe chronometer are similar to those obtained using other well-established chronometers. The uncertainties, too, are currently comparable, but I-Pu-Xe has the potential for great improvement. Most of the uncertainties in I-Pu-Xe dating relate to weak constraints on lunar chemical abundances and a lack of a detailed understanding of the excess fission xenon effect, while the uncertainties in other chronometers relate to analytical precision and to the difficulty in obtaining information about the early history of the Earth.

Decay parameters

Except for the I-Pu-Xe system, we will not discuss details of the dating schemes but will simply point out that the "clock" in each technique is the decay of a naturally occurring radioactive "parent" to a stable "daughter" isotope of another element. Wetherill (1975) has reviewed the basic premises of radiometric dating techniques, and other reviews of specific techniques contain more detailed discussions of those particular systems. In Table 1, we summarize the characteristics of the decays involved and list the values adopted for certain crucial parameters, such as half-lives. For most of the techniques, we will be comparing results obtained over the last 15 or more years by a variety of investigators. It is an unfortunate result of progress that the accepted values for many of these parameters have evolved during this period. To make comparisons more meaningful, we have recalculated all results to conform to the common set of parameters listed in Table 1.

TABLE 1. *Chronometric Parameters.*

		Half-lives		
Parent Isotope	Daughter Isotope(s)	Type of Decay	Half-life (Total)	Ref.
^{87}Rb	^{87}Sr	Beta	48813	(1)
^{238}U	^{206}Pb	Alpha chain	4468	(1)
^{235}U	^{207}Pb	Alpha chain	704	(1)
	^{208}Pb	Alpha chain	14010	(1)
^{40}K	^{40}Ar	E. capture	1250	(1)
^{147}Sm	^{143}Nd	Alpha	106000	(2)
^{129}I	^{129}Xe	Beta	17	(3)
^{244}Pu	^{131}Xe-^{136}Xe	Fission	82	(4)

	Isotopic Ratios at Time t_o	
Ratio	Value	Ref.
^{129}I/^{127}I	1.095×10^{-4}	(5)
^{244}Pu/^{238}U	6.47×10^{-3}	(6)

All half-lives in 10^6 years (m.y.)
References: (1) Steiger and Jäger (1977); (2) Nakamura et al. (1976); (3) Katcoff et al. (1951); (4) Fields et al. (1966); (5) Hohenberg and Kennedy, (1981); (6) Hudson, personal communication (1985), recalculation of data in Hudson et al. (1982).

Established Chronometers

The oldest rocks

A lower limit to the age of a body is determined by the crystallization age of its oldest rocks. The utility of this approach depends on how much processing the body has undergone since formation. Consequently, this approach works well for meteorites, but it does not set very strict limits on the age of the Earth because of its continued igneous activity. For the Moon, a few lunar rocks (or fragments of rocks) give ages that approach those expected for the body as a whole.

Meteorites. The most precise meteoritic ages are the U-Th-Pb system ages on the carbonaceous chondrite Allende and the achondrite Angra dos Reis. As shown in Table 2, studies of Allende in three different laboratories all give Pb-Pb ages of about 4560 m.y., although the stated uncertainties ($2\sigma = 4$ m.y. in each case) are less than the 10 m.y. range in determined ages.

Samarium-neodymium ages of most meteorites are also consistent with formation about 4550 m.y. ago. The most comprehensive Rb-Sr and ^{40}Ar-^{39}Ar studies on meteorites give ages of 4498 ± 15 and 4480 ± 30 m.y., respectively. The difference

TABLE 2.   *The Oldest Rocks.*

Sample	Age (m.y.)	Method	Reference
Meteorites			
Allende	4565 ± 4	U-Th-Pb	Chen and Tilton, 1976
	4553 ± 4	U-Th-Pb	Tatsumoto et al., 1976
	4567 ± 4	U-Th-Pb	Arden and Cressey, 1984
Chondrites[a]	4498 ± 15	Rb-Sr	Minster et al., 1982
	4480 ± 30	^{40}Ar-^{39}Ar	Turner et al., 1978
Angra dos Reis	4550 ± 40	Sm-Nd	Lugmair and Marti, 1977
	4544 ± 1	U-Th-Pb	Wasserburg et al., 1977b
Earth			
Mt. Narryer zircons	4100–4200	U-Th-Pb	Froude et al., 1983
Moon			
65015 plagioclase[b]	4470	^{40}Ar-^{39}Ar	Jessberger et al., 1974
72417 dunite clast	4470 ± 100	Rb-Sr	Papanastassiou and Wasserburg, 1975
73217 zircons	4356(+23,−14)	U-Th-Pb	Compston and Williams, 1983
74235[c]	4510–4550	U-Th-Pb	Nunes et al., 1974
76535	4530 ± 70	Rb-Sr	Papanastassiou and Wasserburg, 1976
76535 olivines	(50 ± 80)[d]	Pu-REE-Xe	Caffee et al., 1981
77215	4370 ± 70	Sm-Nd	Nakamura et al., 1976
	4350 ± 40	Rb-Sr	Nakamura et al., 1976

[a]*Rb-Sr age is whole-rock isochron, all samples. Whole-rock Rb-Sr isochrons for H, E, and LL clasts are indistinguishable.* ^{40}*Ar-*^{39}*Ar age is average age of 16 unshocked chondrites. Individual* ^{40}*Ar-*^{39}*Ar ages ranged from 4520 ± 30 m.y. to 4420 ± 30 m.y.*

[b]*Extractions containing last 2% of gas released.*

[c]*But see alternate interpretation given by Tera and Wasserburg (1974).*

[d]*Formation time relative to primitive meteorites.*

between these ages and the U-Th-Pb ages may indicate an incorrect choice of half-lives (Minster *et al.*, 1982; Turner *et al.*, 1978), although the difference between U-Th-Pb and ^{40}Ar-^{39}Ar ages could also be explained by ongoing metamorphism, which results in isotopic closure for argon occurring somewhat later than for less mobile elements (Turner *et al.*, 1978). This overall consistency, as well as the narrow spread of I-Xe ages (Hohenberg *et al.*, 1967), suggests rapid formation of most meteorites 4550 m.y. ago. The time of formation of primitive meteorites is thus considered a reference point and is used as t_0 for relative chronometers.

Earth and Moon. The oldest terrestrial rocks analyzed are four zircons from Mt. Narryer, Western Australia (Froude *et al.*, 1983), which give nearly concordant U-Th-Pb ages of 4100–4200 m.y., far younger than the apparent age of the Earth (see below). For the Moon, however, there are a few samples with crystallization ages of about 4500 m.y. In Table 2, we list the oldest lunar samples, as determined by several dating techniques.

One lunar sample in which two different dating techniques give very old ages is troctolite 76535. Papanastassiou and Wasserburg (1976) found a Rb-Sr internal isochron of 4530 ± 70 m.y. The isochron was defined by 10 points, but the points with the highest Rb/Sr, which were most important in determining the slope, were from olivine mineral separates. Caffee *et al.* (1981) detected xenon from the in situ decay of ^{244}Pu in an olivine separate of 76535 and determined a Pu-REE formation age of 50 m.y. after primitive meteorites, which is consistent with the Rb-Sr age. Samarium-neodymium (Lugmair *et al.*, 1976) and ^{40}Ar-^{39}Ar (Husain and Schaeffer, 1975; Huneke and Wasserburg, 1975; Bogard *et al.*, 1975) studies give significantly younger ages of about 4250 m.y. Since olivine is the highest-temperature mineral analyzed, it was suggested that either the rock cooled slowly (Lugmair *et al.*, 1976), or that a thermal event at 4250 m.y. reset some isotopic systems, but not those protected by high-temperature minerals such as olivine (Papanastassiou and Wasserburg, 1976; Caffee *et al.*, 1981). Both the Rb-Sr and the Pu-REE-Xe systems show that some grains or inclusions within 76535 formed within about 50 m.y. of the meteorites.

Other lunar rocks give ages that are old, but they are either not as old or not as definitive as those for 76535. In addition to the samples listed in Table 2, for which internal isochrons (or other multiple consistent measurements) are available, many other samples give model Rb-Sr ages of about 4500 m.y. (cf. reviews by Wasserburg *et al.*, 1977a; Tatsumoto *et al.*, 1977; Nyquist, 1977).

Uranium-thorium-lead

The uranium-thorium-lead system can, in principle, be used to date not only rocks but also differentiation events. Extensive data exist for lead isotopes in both lunar and terrestrial samples, but multiple interpretations are possible for both bodies. In any of these interpretations, a major event that differentiated lead from uranium occurred no more than about 150 m.y. after the formation of primitive meteorites. The U-Th-Pb systems for both bodies have been interpreted as indicating formation contemporaneous with the primitive meteorites.

The U-Th-Pb system of the Moon is reviewed in Wasserburg et al. (1977a) and Tatsumoto et al. (1977). For the Moon, many samples fall on or near a single line on a "concordia" plot (for example, a plot of $^{207}Pb/^{206}Pb$ vs. $^{238}U/^{206}Pb$). One intersection of this line with the concordia curve falls at about 3900 m.y. and is interpreted as major lunar impact metamorphism. Wasserburg et al. (1977a) argue that the other intersection of the line and the concordia curve, at 4420 m.y., dates the formation of the lunar crust. Tera and Wasserburg (1976) also analyzed glass balls from 74420 that define a line intersecting concordia at about 4590 m.y. They suggested that the simplest interpretation for these results would be that these samples came from a depth in the Moon that was not differentiated at 4420 m.y. Tatsumoto et al. (1977) argued that some Apollo 17 rocks were indeed concordant at ages somewhat older than 4420 m.y. Their data for 74235, in particular, give nearly concordant ages of 4510–4550 m.y. (Nunes et al., 1974). Data from both groups, therefore, suggest that the Moon is nearly contemporaneous with the primitive meteorites.

For the Earth, the simplest assumption, namely, that the isotopic ratios of modern lead evolved with a single U/Pb ratio, gives an age of about 4430 m.y. (Patterson, 1956; recalculated by Bugnon et al., 1979). Several authors have found the apparent initial lead isotopic ratios in 2700-m.y.-old granites from various parts of the world and calculated the age of the source, again assuming single-stage evolution. Results range from 4470 m.y. to 4560 m.y. (Gancarz and Wasserburg, 1977; Tilton and Steiger, 1965; Bugnon et al., 1979). However, the assumption of single-stage evolution may not be correct. Tera (1981) has argued that the evidence indicates a multistage evolution, with the first differentiation occurring at the same time as the formation of primitive meteorites. Tera (1981) also points out that this could be the time of formation of the planetesimals from which the Earth formed, rather than the time of formation of the Earth as a planet.

Initial $^{87}Sr/^{86}Sr$

An additional possibility of placing constraints on the time of origin of the Moon comes from the great lunar depletion in volatile elements. For example, the lunar K/U ratio is 2500 compared to 10,000 for the Earth and 60,000 for CI chondrites. For the volatile/refractory element pair Rb-Sr, the lunar elemental ratio is 0.009 compared to 0.031 for the Earth and 0.30 for CI chondrites (Taylor, 1982). The mechanisms for these extensive depletions in both Earth and Moon remain conjectural, but they impose constraints on all theories of lunar and planetary origin. The time of these depletions can, in principle, be determined from initial $^{87}Sr/^{86}Sr$ ratios. However, lack of knowledge of the Earth's initial strontium isotopic composition is a major stumbling block.

Scenarios for the accretion of the terrestrial planets involve sweep-up times for planetesimals, in a gas-free environment, of 50–100 m.y. (Wetherill, 1980). The planetesimals that accreted to form the Earth were already depleted in volatile elements. Fractionated meteorites (e.g., eucrites) record a loss of Rb relative to Sr at times not clearly distinguishable from 4560 m.y. and are derived by partial melting of an asteroidal mantle already depleted in siderophile and volatile elements (e.g., Stolper, 1977). The concentrations for these elements are strikingly similar to those in lunar mare basalts, so that the conditions leading to lunar chemistry are not unique and occurred close to t_0.

The population of fractionated precursor planetesimals that existed at about 4500 m.y. had low initial $^{87}Sr^{86}Sr$. These ratios include that of basaltic achondrites (BABI; 0.69898 ± 3) and a uniquely low value for Angra dos Reis (ADOR; 0.69883 ± 2; Papanastassiou and Wasserburg, 1969). In addition, a lower ratio exists for the Allende white inclusions (ALL; 0.69877 ± 2; Gray et al., 1973).

The initial $^{87}Sr/^{86}Sr$ for the Earth is not known. However, in the 100 m.y. required for accretion from planetesimals in Wetherill-Safronov models, $^{87}Sr/^{86}Sr$ will increase, for a bulk Earth Rb/Sr of 0.031, by 0.00011. Thus, if the Earth accreted from planetesimals with initial $^{87}Sr/^{86}Sr$ equal to BABI, the terrestrial ratio would be 0.69909 by the time that accretion was complete, core formation had occurred, and lunar fission could occur.

The lunar initial $^{87}Sr/^{86}Sr$ (LUNI) is well constrained by measurement on ancient low Rb/Sr anorthosites, such as 60025 (with a measured $^{87}Sr/^{86}Sr$ of 0.69895 ± 3). LUNI must be equal to or lower than this value, which is itself slightly lower than BABI (Nyquist, 1977).

What if the Earth accreted from unfractionated planetsimals? H, L, and LL chondrites possess an initial $^{87}Sr/^{86}Sr$ ratio of 0.69885 ± 10, which is close to that of ADOR (Minster et al., 1982). Their Rb/Sr ratio of about 0.3 (Mason, 1979), however, would lead to large increases in $^{87}Sr/^{86}Sr$ (about 0.001) in 100 m.y. An Earth accreted from such material would have $^{87}Sr/^{86}Sr$ of about 0.70 at 4450 m.y. in addition to an unacceptably high Rb/Sr ratio for the bulk Earth.

These data place constraints on models for lunar origins (or on the Earth's initial strontium composition). If the Moon was derived in some fashion from the terrestrial mantle following core formation (fission), then either the Earth's initial $^{87}Sr/^{86}Sr$ must be much lower than BABI and close to ADOR, or all events occurred much closer to t_0 than predicted by dynamical models.

Iodine-Plutonium-Xenon

A promising technique for establishing limits on the relative ages of the Earth and Moon is I-Pu-Xe dating. However, one must carefully delineate the specific events dated in each case. For the Earth, the present application uses xenon isotopic ratios in the atmosphere or mantle and thus addresses the specific formation times of these. For the Moon, iodine-derived and plutonium-derived xenon are observed as a redistributed component that appears on the surfaces of grains of all gas-rich highland breccias. This component, which has most commonly been identified as "excess fission xenon," is not well understood, but it undoubtedly contains chronometric information. The difficulty lies in knowing how to interpret that information.

The I-Pu-Xe method is based on the simultaneous decay, early in the history of the solar system, for two now-extinct radionuclides, ^{129}I and ^{244}Pu (see Table 1). At some time t, the ratio of the production rate of ^{129}Xe (from beta decay of ^{129}I) to that of ^{136}Xe (from spontaneous fission of ^{244}Pu) is given by:

$$\frac{d[^{129}Xe]}{d[^{136}Xe]}(t) = \left[\frac{^{129}I}{^{244}Pu} \right]_o \times \left[\frac{\lambda^{129}}{\lambda^{244}} \right] \times \frac{1}{By^{136}} \times \exp[\lambda(t-t_o)] \tag{1}$$

The ratio of ^{129}Xe to ^{136}Xe in the decay products accumulated since time t is as follows.

$$\left[\frac{^{129}Xe}{^{136}Xe} \right](t) = \left[\frac{^{129}I}{^{244}Pu} \right]_o \times \frac{1}{By^{136}} \times \exp[\lambda(t-t_o)] \tag{2}$$

Here y^{136} is the yield of ^{136}Xe in ^{244}Pu fission, B is the branching ratio for spontaneous fission of ^{244}Pu, and superscripted λ's are the decay constants of the respective radionuclides. The effective decay constant for the $^{129}I/^{244}Pu$ ratio, denoted here simply by λ, is equal to $\lambda^{129}-\lambda^{244}$. Note that we are using a time scale where t is defined to be 0 at present and positive in the past (e.g., $t_o = 4550$ m.y.) The $^{129}I/^{244}Pu$ ratio at time t_o can be obtained from the expression

$$\left[\frac{^{129}I}{^{244}Pu} \right]_o = \left[\frac{^{129}I}{^{127}I} \right]_o \times \left[\frac{^{238}U}{^{244}Pu} \right]_o \times \left[\frac{^{127}I}{^{238}U} \right]_{now} \times \exp[-\lambda^{238}t_o] \tag{3}$$

where the initial ratios refer to t_o, the time of formation of meteorites. We will discuss elemental abundances as weight ratios, but the I/U weight ratio should be multiplied by 238/127 to obtain the molar ratio required in this equation. Values for the ratios $(^{129}I/^{127}I)_o$ and $(^{244}Pu/^{238}U)_o$ are given in Table 1. In using the $(^{244}Pu/^{238}U)_o$ value from Table 1, we implicitly assume that Pu and U were not significantly fractionated on global scales.

A pure I-Xe age, which does not involve the amount of fission ^{136}Xe, can be calculated if the total amount of radiogenic ^{129}Xe is known. The amount of radiogenic ^{129}Xe produced since time t is equal to the amount of ^{129}I that was present at that time, or

$$[^{129}Xe]_{rad} = \left[\frac{^{129}I}{^{127}I} \right]_o \times [^{127}I] \times \exp\left[\lambda^{129}(t-t_o)\right] \tag{4}$$

However, knowledge of both the absolute I content of the source and the extent of source degassing is required.

I-Pu-Xe age of the Earth

For the Earth, it is possible to calculate I-Pu-Xe ages of either the atmosphere or the mantle, using equation (2). First, of course, it is necessary to determine the ratio of iodine-derived ^{129}Xe to plutonium-derived ^{136}Xe. Since these represent relatively small perturbations, this requires reasonably detailed knowledge of the composition of the xenon to which those components are added.

In the terrestrial atmosphere, the isotopic structure of the light, nonradiogenic isotopes of xenon (^{124}Xe, ^{126}Xe, ^{128}Xe, and ^{130}Xe) is similar to that of trapped xenon in meteorites, except for a general trend, suggestive of mass-dependent fractionation. If one assumes that the $^{129}Xe/^{130}Xe$ ratio of the Earth's primordial xenon is similarly related, the $^{129}Xe/^{130}Xe$ ratio in meteorites with little radiogenic xenon (such as Novo Urei) provides a measure for the nonradiogenic terrestrial value. On this basis, 6.7% of the Earth's ^{129}Xe must be iodine-derived. The fraction of the ^{136}Xe that comes from ^{244}Pu fission (4.65%) is determined in a similar fashion (Pepin and Phinney, 1976), although meteoritic trapped xenon is somewhat more complicated at the heavy isotopes. Perhaps the most elusive factor in I-Pu-Xe dating is the initial $^{129}I/^{244}Pu$ ratio, which, as shown in equation (3), requires an estimate of the I/U ratio of the source region.

The I-Pu-Xe ages calculated by different authors are in reasonable agreement (see Table 3) at 85–100 m.y. after xenon closure in primitive meteorites. However, the uncertainty in the results is larger than the spread in calculated ages might suggest. Wetherill (1975) and Pepin and Phinney (1976) have pointed out that it is possible to adjust the parameters (particularly the I/U ratio of the source region) in a reasonable fashion and obtain a difference in formation age of the Earth and meteorites of zero.

Alternatively, an I-Xe age of the Earth's atmosphere can be calculated [using equation (4)] without having to rely on the amount of fission ^{136}Xe in the atmosphere. However, this I-Xe age requires knowledge of the iodine content of the source, the extent that the source region has been degassed, and the total amount of xenon in the Earth's atmosphere. These are all difficult to determine, but the best estimates lead to I-Xe ages that are comparable to I-Pu-Xe ages (see Table 3).

To understand the implications of the I-Pu-Xe age of the Earth's atmosphere, we must first understand what it is that is being dated. Xenon retention in the Earth's atmosphere did not necessarily begin at an early stage of accretion. Even though xenon is gravitationally bound, it is possible that the atmosphere undergoes "erosion" when it is bombarded by meteoroids (Cameron, 1983). The tenuous

TABLE 3. I-Pu-Xe Ages of the Earth.

Reference	Type	Age
Wetherill, 1975	I-Xe (Atmos.)	113
	I-Pu-Xe (Mantle)	128
Pepin and Phinney, 1976	I-Xe (Atmos.)	63—120
	I-Pu-Xe (Atmos.)	91 ± 20
	I-Pu-Xe (Mantle)	0–160
Bernatowicz and Podošek, 1978	I-Pu-Xe (Atmos.)	84
Staudacher and Allégre, 1982	I-Xe (Atmos.)	75–100
	I-Pu-Xe (Atmos.)	100
	I-Pu-Xe (Mantle)	65–85

Note: All numbers are 24 m.y. larger than given in the original references, to adjust the calculations to the present best estimate of the initial $^{244}Pu/^{238}U$ ratio (see Table 1).

atmosphere of Mars suggests that this or similar mechanisms might have been operative in the solar system. If degassing were very rapid early in the Earth's history, then the I-Pu-Xe age represents the time of the last quantitative loss of atmosphere. On the other hand, if the Earth (or the portion that degassed to form the atmosphere) was not completely degassed at the time of such losses (Bernatowicz and Podosek, 1978), then the xenon that the atmosphere now contains represents a time earlier than the last major atmospheric loss. Thus, this age probably represents some time between the time of accretion of most of the Earth and the time when the atmosphere was no longer suffering significant losses. However, this time interval is probably no more than a few tens of millions of years.

Another method of calculating an age for the Earth from xenon isotopic structure [also using equation (2)] involves fission and iodine contributions to mantle xenon extracted from well gases (Wetherill, 1975) or basalts (Staudacher and Allégre, 1982). Again, this technique (denoted "mantle" in Table 3) gives ages of about 100 m.y. after the formation of primitive meteorites. Unlike the other techniques, which calculate the age of the atmosphere, this method calculates the age of the mantle. This technique has the advantage that the mantle is less susceptible to loss of xenon than the atmosphere. The disadvantage is that whereas the xenon isotopic composition of the atmosphere is well known, that of the mantle is less certain. In particular, a component with the isotopic spectrum of ^{244}Pu fission has been observed in the atmosphere (Pepin and Phinney, 1976), but the plutonium spectrum has never been confirmed in fission-like excesses in any terrestrial mantle samples. Staudacher and Allégre (1982), for example, assumed that enhancements in the $^{134}Xe/^{130}Xe$ ratios of mantle samples they studied were due to plutonium fission, but their measurements were not sufficiently precise to determine the isotopic spectrum of the fission component. However, the agreement of their mantle age with atmospheric ages suggests that this assumption is reasonable.

Despite the uncertainty in the various terrestrial ages, the collective evidence tends to support its overall validity. These ages provide potentially useful constraints on the origin of the Moon, especially when compared to lunar I-Pu-Xe ages. Such comparisons can give a significant amount of the information about models in which the material that formed the Moon was once a part of the Earth's mantle. In these models, whether the material was ejected from the Earth by fission or by impact(s), it seems likely that the I-Pu-Xe clock in the Earth's atmosphere would be reset by such disruptive events. Since the Earth's accretion and differentiation must have been largely complete by that time, further loss of terrestrial atmosphere is less likely. Therefore, if the Moon derived from the Earth by fission or spallation, the I–Pu–Xe age of the terrestrial atmosphere or mantle should be older than the Moon or any materials on the Moon.

General properties of parentless lunar xenon

We now wish to turn to the I-Pu-Xe age of the Moon, for which we will need to determine the ratio of iodine-derived to plutonium-derived xenon and an appropriate I/U elemental ratio for the source region. The only lunar samples in which both iodine- to plutonium-derived xenon are observed are those that contain "excess fission xenon."

"Excess fission xenon" was first noted in lunar breccia 14301 (Crozaz *et al.*, 1972; Drozd *et al.*, 1972). In a stepwise heating experiment, the xenon released at the lowest temperatures was found to be enriched in the heavy isotopes with a component whose isotopic spectrum matched that of the spontaneous fission yield of ^{244}Pu (with the addition of some ^{129}Xe). This enrichment was surprising: Plutonium and uranium are refractory elements, and their fission products usually remain in high-temperature sites. Further studies have revealed the following general properties of excess fission xenon.

1. Parentless xenon is a widespread lunar effect. As can be seen in Table 4, parentless xenon has been observed in gas-rich highland breccias from both the Apollo 14 and Apollo 16 sites. In fact, the effect has been observed in all highland breccias that are rich in solar wind gases, suggesting that the effect is probably global in scale. It has also been observed in 67915,67, a troctolitic anorthosite clast, and in 76535, a troctolitic cumulate.

2. The excesses at the heavy isotopes come from spontaneous fission of ^{244}Pu. In every case in which the isotopic spectrum of the excess component has been identified, whether by analysis of stepwise heating results or by analysis of size separates, it has matched that of ^{244}Pu (Drozd *et al.*, 1972; Behrmann *et al.*, 1973; Reynolds *et al.*, 1974; Bernatowicz *et al.*, 1978, 1979, 1980; Swindle *et al.*, 1985). Excesses at ^{129}Xe (presumably from ^{129}I decay) sometimes accompany the ^{244}Pu fission in varying proportions. Thus, the component is not purely "fission," but rather comes from the decay of two extinct radionuclides, ^{129}I and ^{244}Pu. This not withstanding,

TABLE 4. Excess Fission Xenon.

Sample	$^{136}Xe_{ex}$	$^{130}Xe_{SW}$	$\left[\dfrac{^{129}Xe}{^{136}Xe}\right]_{Low-T}$	$\left[\dfrac{^{129}Xe}{^{136}Xe}\right]_{Bulk}$	Ref.
Breccias	1.8	48.4			
14047					
14055	2.1	55.1			
14301	1.4	8.6	0.26 ± 0.03	1.21 ± 0.61	2,1
	0.7		0.24 ± 0.06	1.32 ± 0.14	3
14307	2.6–4.0	136.0		3.5 ± 2.5[a]	4,5
14313	4.8–6.2	51.6			6,5
14318	2.2–2.3	5.4	0.05 ± 0.07	0.23 ± 0.21[a]	6
	2.3		0.03 ± 0.04	0.25 ± 0.23[a]	7
			0.15 ±0.15	0.56 ± 0.04	8
60019	1.57 ± 0.16	5.3	0.64 ± 0.09[d,e]	0.79 ± 0.11	5
60275	0.79 ± 0.08	3.51	0.00 ± 0.01[d]	0.86 ± 0.19	5
67455	0.09 ± 0.01	0.8		0.66 ± 0.43'	5
Soil	0.6	9.8		0.0024	1
14149					
Rocks	0.0001[b]	0.007	4.7 ± 1.7[b]		9
67915,67					
76535	0.005–0.0067[c]	0.1	0.01 ± 0.04		10

All amounts in units of 10^{-10} $cm^3 STP/gm$. Where two references are given, data are given in the first reference, but some of the calculations are in the second.

[a] *Excess ^{129}Xe abundance calculated assuming mixture of spallation xenon ($^{126}Xe/^{129}Xe/^{130}Xe = 1.0 \pm 0.2/1.75 \pm 0.5/1$), solar wind xenon (SUCOR; Podosek et al., 1971) ratios with uncertainties large enough to overlap the values given for BEOC 10084 (Eberhardt et al., 1970) and BEOC 12001 (Eberhardt et al., 1972), and excess ^{129}Xe.*

[b] *Based on amount of fission ^{134}Xe, assuming all fission ^{134}Xe in $750°C$ extraction (which contains excess ^{129}Xe) is surface-correlated. Data for ^{136}Xe agree within errors but have larger uncertainties.*

[c] *Assuming $4 \times 10^{-12} cm^3 STP/gm$ in situ fission from modal abundances (Gooley et al., 1974) and amounts of in situ fission in mineral separates (Caffee et al., 1981)).*

[d] *From unpublished calculations of the authors.*

[e] *$^{129}Xe/^{136}Xe$ ratios in $700°$, $800°$, and $900°C$ steps are, in order, 0.60 ± 0.10, 0.58 ± 0.07, and 0.71 ± 0.10.*

References: (1) Drozd et al. (1975); (2) Drozd et al. (1972); (3) Bernatowicz et al. (1979); (4) Bernatowicz et al. (1977); (5) Bernatowicz et al. (1978); (6) Behrmann et al. (1973); (7) Reynolds et al. (1974); (8) Swindle et al. (1985); (9) Aeschlimann et al. (1983); (10) Hohenberg et al. (1980).

the name "excess fission xenon" has been established in the literature and we will continue to use it as an operational description.

3. This xenon is "excess": it does not come from in situ fission. In many cases, the amount is far in excess of what could have been produced in the sample, given the actinide content and age (determined by other radiometric dating techniques)

of the rock. Furthermore, some of the "excess" xenon is released at low temperatures in stepwise heating experiments, which is not consistent with in situ fission. Plutonium, like uranium and the rare earths, is refractory, so xenon from in situ fission of ^{244}Pu will have a release pattern similar to that of xenon derived from these other sources. However, the release of excess fission xenon does not correlate with the release of either xenon from neutron-induced fission of ^{235}U (Reynolds et al., 1974) or xenon from spallation of rare earth elements.

4. Excess fission xenon is surface-correlated. The low-temperature release pattern suggested grain-surface siting, and studies of grain-size separates from a soil (Basford et al., 1973) and a disaggregated breccia (Bernatowicz et al., 1979) confirmed this.

5. The ratio of excess ^{129}Xe (from ^{129}I) to excess ^{136}Xe (from ^{244}Pu) varies from sample to sample (Table 4) and within a given sample. In some cases, no ^{129}Xe accompanies the excess fission. It is the variability of this ratio that first led to the suggestion that I-Pu-Xe dating might be possible (Behrmann et al., 1973), and it is this application that we will focus on.

In the cases of 14301 and 14318, where excess fission xenon has been most extensively studied, the $^{129}Xe/^{136}Xe$ ratio observed in the low-temperature component is reproducible from one stepwise heating experiment to another (Table 4). However, the surface-correlated component obtained through analysis of size separates has a higher $^{129}Xe/^{136}Xe$ ratio. This variation of the $^{129}Xe/^{136}Xe$ ratio within a single sample seems puzzling at first, but actually may be providing clues for understanding the process and at the same time giving sensitive chronometric information (Bernatowicz et al., 1979; Swindle et al., 1985). If acquisition of excess fission xenon was an ongoing process in which initially lightly bound surface components became progressively more tightly bound by chemical or physical surface alteration, the order of acquisition might be preserved in the thermal release pattern. When we analyze such a sample today, long after compaction of the breccia, the gas least tightly bound (which would be released first in a stepwise heating experiment) would be the last gas acquired. If the acquisition process continued for tens of millions of years, we would expect the ratio of iodine-derived ^{129}Xe to fission ^{136}Xe to be decreasing with time, since the ratio of ^{129}I to ^{244}Pu decreases exponentially with a 20 m.y. half-life. Thus, we would expect the least tightly bound component (the low-temperature component) to have a lower $^{129}Xe/^{136}Xe$ ratio than the excess as a whole. This is, in fact, what is observed in both 14301 and 14318, the only samples for which precise information is available.

There is one other set of samples where such a comparison can be made, but with much less precision. In analyzing gas-rich Apollo 16 breccias, Bernatowicz et al. (1978) calculated the apparent isotopic spectrum of the excess xenon component for each temperature extraction, in part to separate terrestrial contamination from excess fission xenon. The spectra show that terrestrial contamination is present in most of the low-temperature (T < 1000°C) extractions (Bernatowicz, personal communication, 1984). But there are a few instances where the ratios of the heavy isotopes are indicative of fission, not terrestrial contamination. The ^{129}Xe/^{136}Xe ratios for these extractions are slightly less than those of the bulk samples. In 60275 the 600°C extraction, containing about 0.1% of the excess fission ^{136}Xe, has no detectable ^{129}Xe excess, while the bulk sample has a ^{129}Xe/^{136}Xe ratio of 0.88 ± 0.09 in the excess component (Table 4). In 60019 the 700°, 800°, and 900°C extractions, with a total of about 1% of the excess fission ^{136}Xe, have ^{129}Xe/^{136}Xe ratios in the excess component ranging from 0.58 ± 0.07 to 0.71 ± 0.10, compared to a ratio for the entire excess in the sample of 0.82 ± 0.09.

Evidence presented by the Apollo 16 samples is, of course, not conclusive in itself, but it adds strength to the case presented by 14301 and 14318. Acquisition of excess fission xenon seems to have been an ongoing process, with the ^{129}Xe/^{136}Xe ratio in the gas available for acquisition controlled by the (exponentially decreasing) ^{129}I/^{244}Pu ratio. All four samples demonstrate that excess fission xenon acquired early (and released last in stepwise heating) has a higher ^{129}Xe/^{136}Xe ratio than that acquired later (and released in the lower-temperature extractions).

Interpretations

The above evidence suggests that excess fission xenon could be used as a chronometer. However, not unlike the Earth, consideration must be given to how to interpret the I-Pu-Xe clock. In particular, we must consider exactly what it is that is being dated. To answer this question, we must consider the history of excess fission xenon from production to incorporation.

The problem of the acquisition of excess fission xenon can, in principle, be separated into three parts (Drozd *et al.*, 1976): the source of the xenon (where the iodine and plutonium were located when they decayed), the transportation (including perhaps, interim storage) of the xenon between decay and incorporation, and the incorporation of the xenon onto grain surfaces. Regrettably, none of the three is completely understood, so we must evaluate how much our lack of knowledge in each of these areas hinders our use of the I-Pu-Xe chronometer.

Method of incorporation. The part of the excess fission xenon problem that has received the most attention recently is the question of how xenon is incorporated onto grain surfaces. There are two types of models that are generally considered. One is electromagnetic implantation, in which xenon atoms, exhaled into the transient lunar atmosphere, are ionized and then accelerated onto the lunar surface by the electrical and magnetic fields of the solar wind (Manka and Michel, 1970). The other model involves adsorption of xenon onto grain surfaces, either from the transient atmosphere or directly from exhaled gases percolating up through the regolith (Drozd *et al.*, 1976). While adsorption seems to require more additional fixing (perhaps from shock due to meteoroid impact) to drive the atoms into more retentive sites, the observed ordering of the ^{129}Xe/^{136}Xe ratio of the surface component requires postincorporation modification in either case. There are several observations that are pertinent to the question of method of incorporation.

1. Electromagnetic implantation is widely accepted as a mechanism for implanting noble gases into grains residing on the lunar surface. The occurrence of excess surface-correlated ^{40}Ar in modern lunar soils (Heymann and Yaniv, 1970) can be explained by electromagnetic implantation (Manka and Michel, 1970), as can certain observed variations in the surface-correlated ^{40}Ar/^{36}Ar ratios (Manka and Michel, 1973).

2. Laboratory experiments indicate that conventional adsorption seems to be quantitatively unable to account for the amount of excess fission xenon observed (Podosek et al., 1981; Bernatowicz et al., 1982). However, Caffee et al. (1981) suggested that trapping sites on the surfaces of lunar grains may be altered by exposure to more active gases, such as those in the terrestrial atmosphere. Thus, the soil sample used by Podosek et al. (1981) might not have the same sorptive properties it had just after new active surface traps were created. As evidence, they pointed out that analyses of 76535 showed progressively less excess fission xenon as time since collection increased, since the trapping sites may have been preempted by the more active gases in the terrestrial atmosphere. Initial analyses of 76535 obtained an order of magnitude more excess fission xenon than analyses less than a decade later. The sample of Bernatowicz et al. (1982), which was crushed under vacuum, did adsorb more efficiently than the soil sample of Podosek et al. (1981), but its adsorption characteristics are still far short of those apparently needed to account for excess fission xenon. Furthermore, the adsorption parameters obtained by Bernatowicz et al. (1982) give regolith transit times comparable to those calculated by Hodges (1977) on the basis of observed abundances of ^{40}Ar and ^{222}Rn in the lunar atmosphere.

3. Two samples (76535 and 67915,67) that are not rich in solar wind, and therefore probably have not been exposed to the lunar atmosphere for any significant length of time, do contain excess fission xenon. Podosek et al. (1981) acknowledged that 76535 could not have acquired its excess fission xenon through electromagnetic implantation but argued, on the basis of their adsorption experiment, that electromagnetic implantation is probably the dominant process. But Caffee et al. (1981) noted that the surface concentration of excess fission xenon in 76535 is comparable to that found in regolith breccia 14301, and they therefore argued that 76535 should not be considered an exceptional case, and hence electromagnetic implantation is not the dominant process.

4. Another serious problem for any model that would have the xenon spend time in the lunar atmosphere (electromagnetic implantation or adsorption from the atmosphere) is the lack of a quantitative correlation between excess fission xenon and solar wind gas. Whereas the amount of solar wind contained in a sample correlates well with other measures of soil maturity, the amount of fission ^{136}Xe available for adsorption should have been decreasing with time, just as the ^{129}Xe/^{136}Xe ratio decreases. Thus, if the solar wind flux is constant, the ratio of excess ^{129}Xe to excess ^{136}Xe might be expected to correlate with the ratio of fission xenon to solar wind. This is not the case (see Table 4).

If acquisition of excess fission xenon does not require exposure to the lunar atmosphere (e.g., if it depends on adsorption within the regolith), then the amount of solar wind xenon would not be expected to correlate with the properties of excess fission xenon. However, one might not expect perfect correlation between solar wind and excess fission xenon even if it is acquired through the lunar atmosphere. For instance, the fixing process (shock?) is almost certainly stochastic. Hodges (1977) has suggested that the amount of ^{40}Ar in the lunar atmosphere varies with seismicity, so a similar effect for xenon might also be expected. Therefore, whereas the association with the solar wind is certainly expected, a one-to-one correspondence might not be required.

Transport. What is the time scale from production to incorporation? It is apparently short, at least when compared to the time scales of interest here. ^{222}Rn, a very short-lived (3.8 day half-life) member of the ^{238}U decay chain, was observed at the lunar surface by the alpha particle spectrometer on Apollo 15 (Bjorkholm *et al.*, 1973). This indicates that some noble gases can escape quickly from the interior. Hodges (1977) calculates that the mean time between production and release into the atmosphere for radon is about 70 days, assuming common source regions for the radon and ^{40}Ar. Using their adsorption results, Podosek *et al.* (1981) calculate a mean regolith residence time for xenon of 22 days. If atmospheric residence is involved, the time scale for this must also be short, because the lunar atmosphere is quickly lost to either reimplantation or expulsion by the solar wind. Hodges (1977) has calculated a mean atmospheric residence time for ^{40}Ar of about 100 days (80% of which is spent adsorbed on cold surfaces), and xenon would presumably have a mean residence time of the same order of magnitude. Perhaps some xenon could be stored for long periods of time in a cold region of the Moon before being released to move up to the surface (Drozd *et al.*, 1976). If so, then the ^{129}Xe/^{136}Xe ratio of the excess would reflect an average over the length of time of storage, and any age determined using this ratio would be an average and, thus, a younger limit for the beginning of xenon retention.

Source. The identification of the source is crucial to any chronometric interpretation, since we need to know the initial ^{129}I/^{244}Pu ratio. Reynolds *et al.* (1974) reviewed the possible source regions that had been proposed and concluded that all the models had some difficulties. We now believe the source to be the lunar interior, although we still do not know whether to consider the entire Moon as the source, or whether the xenon observed in a single sample only comes from more localized regions.

The simplest assumption (for purposes of computing ages) is that the xenon comes from degassing of a substantial fraction of the Moon and that the appropriate initial ^{129}I/^{244}Pu ratio is that computed from the iodine/uranium ratio of the bulk Moon. In support of this view, Hodges (1977) has calculated that the current rate at which ^{40}Ar is being supplied to the lunar atmosphere is about six percent of the rate at which it is being produced in the interior of the Moon, indicating that a substantial fraction of the Moon is currently being degassed. Also, excess fission xenon has been observed at three different sites, indicating that it is not a local phenomenon. However, discrete local source regions (pockets of material enriched in actinides or iodine) cannot be ruled out, so it possible that grain surfaces at different locations may sample xenon from localized (rather than global) sources. The possibility of local source domination is increased if the xenon is incorporated as it percolates up through the regolith. For local production, the iodine/uranium ratio we should address is that of the material in the first few kilometers beneath each site, rather than that of the bulk Moon. On the other hand, if the xenon spends time in a transient atmosphere before incorporation (which is true for electromagnetic reimplantation and for some adsorption models), then global mixing should occur.

Another source for excess fission xenon that has been considered, but that we feel is unlikely, is meteoritic infall. There is certainly xenon from plutonium and iodine in meteorites, and the ratio of iodine-derived ^{129}Xe to ^{136}Xe varies. However, the ratio of excess ^{129}Xe to ^{136}Xe in chondrites is typically 10 or more, far higher than any ratios seen in lunar excess fission xenon. Also, if the source were infalling meteorites, variations in the ^{129}Xe/^{136}Xe ratio within a single breccia would be expected to be random, rather than the regular, apparently time-ordered, pattern of increasing ratio with increasing temperature of extraction that is observed. Previous authors have suggested other reasons why meteoritic infall seems unlikely. Reynolds *et al.* (1974), for example, pointed out that the surface-correlated components in 14301 and 14318 have similar ^{40}Ar/^{36}Ar ratios, despite having different ^{129}Xe/^{136}Xe ratios, which would be difficult to explain if most of the xenon (or even most the ^{129}Xe) were from meteoritic infall. Bernatowicz *et al.* (1978) argued that mass balance considerations make it unlikely that the ^{136}Xe comes from infall, since fission ^{136}Xe is typically much more abundant in gas-rich breccias than in chondrites.

How do these uncertainties in the history of excess fission xenon affect our ability to use the I-Pu-Xe chronometer to find the age of the Moon? The uncertainty in time of transport is not important, since the I-Pu-Xe method can only establish a minimum lunar age. The uncertainty in method of incorporation does not directly affect the age we would compute (however, if we knew the method of incorporation, it might help us determine which particular source region to consider). The uncertainty in source region is important, since we need to know the present-day I/U ratio of the source. We will, therefore, now consider the I/U ratio of the Moon as a whole (and how well we know that ratio) and then consider how much the I/U ratio of more localized source regions might differ.

The iodine/uranium ratio of the source

If we are to obtain I-Pu-Xe ages, we need to know the initial ^{129}I/^{244}Pu ratio of the source. As we have seen, this can be calculated from equation (3) if we know the present ^{127}I/^{238}U ratio of the appropriate source region. If we assume that the xenon comes from nearly continuous degassing of a large portion of the Moon, then the quantity we desire is the present iodine to uranium ratio of the bulk Moon. Otherwise, the appropriate ratio is that of a more local source region.

Lunar I/U ratio. Various estimates of the lunar uranium abundance are available. We will adopt 33 ppb (Taylor, 1982) as the bulk lunar uranium abundance, although higher (46 ppb, Langseth *et al.*, 1976) and lower (19 ppb, Rasmussen and Warren, 1985) estimates are also available.

The Moon is highly depleted in volatile and very volatile elements. Iodine is present in surface samples at levels of a few ppb (cf. Wänke *et al.*, 1974; Dreibus *et al.*, 1977; Jovanovic and Reed, 1973, 1978; Reed *et al.*, 1972). The variation in the measurements makes any direct estimates of bulk lunar iodine values uncertain, and indirect methods must be used. Here we use two approaches to the problem, noting that the volatility and mobility of the halogens make all direct estimates hazardous. The abundances of the halogens are reasonably well established in CI meteorites, although we note that they have been exposed to aqueous phases. Nevertheless, we assume that the relative abundances of Cl, Br, and I remain fixed in their solar system ratios: Cl = 698 ppm; Br = 3.56; I = 430 ppb (Anders and Ebihara, 1982).

In the first approach we use the abundances of elements of similar volatility, whose concentrations can be estimated with more reliability than those of the halogens, to calculate the depletion factors between the Earth and CI chondrites and between the Moon and Earth. From these depletion factors, we can place upper limits on the lunar iodine abundance.

Bismuth, cadmium, and thallium are elements classified, along with the halogens, as "very volatile" under solar nebula conditions (Anders and Ebihara, 1982). The abundances in CI meteorites (Anders and Ebihara, 1982) are given in Table 5a. Terrestrial abundances from Taylor and McLennan (1985) are listed. These are for the primitive mantle (present mantle + crust), assuming these elements do not enter the core. The CI/Earth ratio gives the measure of depletion in the Earth. If we assume the same depletion for iodine, then the bulk terrestrial values for iodine so obtained are listed in Table 5a. The range from 18–38 ppb must represent upper limits for iodine, since it is more volatile and, hence, presumably more depleted than Bi, Cd, or Tl.

An estimate of the bulk lunar iodine value can be obtained in the following manner. Values for Bi, Cd, and Tl in the bulk Moon are not well established, but are reasonably well known in lunar mare basalts and terrestrial oceanic basalts. Table 5b lists the values from Wolf and Anders (1980) for low-Ti lunar basalts and terrestrial oceanic basalts. Both rock types are derived from planetary interiors, albeit with differing differentiation histories, by partial melting. The relative abundances provide a guide to the bulk mantle values, although the hazards of such a comparison are obvious. Nevertheless, they indicate that the Earth is enriched by factors of 35–65 relative to the Moon. Calculation from the terrestrial iodine values in Table

TABLE 5. Iodine Abundances.

a. Estimates of bulk terrestrial iodine abundance				
Element	CI[1]	Earth[2]	CI/Earth	Terr. Iodine
Bi	114	10	11.4	38
Cd	720	40	18	24
Tl	142	6	23.7	18

b. Estimates of bulk lunar iodine abundance				
Element	Terrestrial Oceanic Basalt[3]	Low-Ti Mare Basalt[3]	Terrestrial Enrichment Factor	Bulk Lunar Iodine
Bi	7.0	0.23	30.4	1.25
Cd	129	1.99	64.8	0.37
Tl	11.8	0.34	34.7	0.52

All elemental abundances in ppb.
(1) Anders and Ebihara (1982).
(2) Taylor and McLennan (1985).
(3) Wolf and Anders (1980).

5a for each element yields bulk lunar iodine values ranging from 0.37–1.25 ppb (Table 5b). These values again must represent upper limits, since iodine is more volatile than Bi, Cd, or Tl. On this basis, an absolute upper limit of 1 ppb for the lunar iodine value can be set, with the probable value less than 0.4 ppb.

The second approach is to use the chlorine abundances in lunar rocks (since chlorine abundances are more reliable than those for iodine) ratioed to a suitable nonvolatile element. A La/Cl ratio of about 1.7 appears to be typical of lunar surface rocks. The bulk lunar La value is 0.90 ppm, yielding a chlorine value of 0.53 ppm. The CI chlorine to iodine ratio of 1623 yields a bulk lunar iodine value of 0.33 ppb. This estimate depends on the assumption that the incompatible elements (La and I) retain their bulk lunar values during differentiation. The La/Cl correlation lends support to this assumption, although one would expect that the halogens would be more readily concentrated toward the surface, so that this estimate as well must represent an upper limit and one that is similar to the previous estimate (though somewhat lower). Accordingly, it seems safe to place an upper limit of about 0.3 ppb on the bulk lunar iodine content. This leads to a value of the I/U ratio of less than 0.01. Note from equations (1–3) that an upper limit for this ratio places an upper limit on the age difference between the Moon and the primitive meteorites.

Near-surface I/U ratio. If the source region for excess fission xenon is only the near-surface region of the Moon, then it is the I/U ratio of that region in which we are interested. While this may be even more difficult to calculate, it is true that both iodine and uranium are concentrated near the surface. Iodine and uranium are both incompatible elements and concentrate in residual melts (e.g., KREEP) from the magma ocean. Accordingly, they will be concentrated in the crust along with other KREEP components. Although the crystallization of the magma ocean was complete by about 4300–4400 m.y. ago, the intensive basin-forming bombardment continued down to 3850 m.y. In this interval, at least 80 basins greater than 300 km in diameter and 10,000 craters (30–300 km diameter) formed (Wilhelms, 1985). Accordingly the megaregolith probably extends to depths of 25 km (Spudis, 1984). Thus, both U and I are unlikely to be concentrated in a thin, near-surface zone but are probably distributed throughout the upper 25 km, although the volatile character of iodine may result in a near-surface rise in the I/U ratio.

Measured iodine and uranium contents of lunar samples do indeed lead to I/U ratios near the bulk lunar value we calculated. The uranium contents of Apollo 14 breccias are usually 3 to 4 ppm (Bernatowicz et al., 1978), while the measured iodine contents of Apollo 14 samples range from 2 to 200 ppb (Reed et al., 1972). If these values are representative of the source of the xenon in Apollo 14 samples, then the I/U ratio is within an order of magnitude of 0.01 (note that an order of magnitude uncertainty here corresponds to an uncertainty in time of about 70 m.y.). For Apollo 16 breccias, measured iodine contents range from 0.4 to 70 ppb, while uranium contents range from <1 ppb to about 1 ppm and are typically a few hundred parts per billion (Jovanovic and Reed, 1978; Wänke et al., 1974). Thus the I/U ratio again appears to be within an order of magnitude of 0.01.

I-Pu-Xe ages of lunar samples

In this study, we are interested in the I-Pu-Xe age of the oldest grains we can find. The highest ^{129}Xe/^{136}Xe ratio listed in Table 4 is that of troctolitic clast 67915,67 (4.7 ± 1.7). If we assume that the source of the excess fission xenon is the bulk Moon, and that the Moon has a present-day ^{127}I/^{238}U ratio of 0.01, we use equations (1) and (3) to calculate a time of incorporation for the excess fission xenon in 67915,67 of 25 m.y. after the most primitive meteorites. To be technically accurate, this refers to the time that the isotopically distinct excess xenon component, now observed in this sample, was established.

The ^{129}Xe/^{136}Xe ratio for 67915,67, however, is quite uncertain. The stated statistical uncertainty is 38%. In addition, the 67915,67 value was obtained by isotopic decomposition of a single low-temperature step in which little gas was released. Finally, it is not certain that xenon effects seen in 67915,67 and 76535 are exactly the same as the excess fission effect in gas-rich highland breccias, where we have seen systematic variations of the ^{129}Xe/^{136}Xe ratio with temperature. The highest of the more precisely determined ^{129}Xe/^{136}Xe ratios is that of the surface-correlated component in 14301, 1.32 ± 0.14. Assuming that the source is the bulk Moon, we calculate an age of 63 m.y. after meteorites for incorporation of this component.

What are the uncertainties related to these calculations? By far the most uncertain quantity in the calculation is the I/U ratio of the source. If the source is the bulk

Moon, the I/U ratio should be considered uncertain by at least a factor of two, corresponding to an uncertainty in the final result of at least 20 m.y. A bigger uncertainty is the question of whether we have indeed chosen the most appropriate source region. If the source was a local region of the Moon, the I/U ratio was almost certainly no more than an order of magnitude different, and probably within about a factor of four of the bulk Moon value. We will adopt an uncertainty of 42 m.y. This is large enough to allow a factor of four variation in the I/U ratio (larger than the uncertainty in the bulk Moon value but probably comparable to the variation in large local sources) and/or a $^{129}Xe/^{136}Xe$ ratio as high as the apparent value for 67915,67. Thus, a reasonable estimate of the beginning of retention of excess fission xenon would be 63 ± 42 m.y. after formation of primitive meteorites.

What are the implications of such an old I-Pu-Xe age? One implication is that there were lunar reservoirs capable of maintaining unique xenon isotopic compositions at that time. Either there were grain surfaces capable of retaining xenon or the xenon was stored for some length of time between production and incorporation (which would mean that the Moon itself is older than the calculated age). Therefore, in any case, the oldest age we observe must be considered a younger limit to the age of the Moon.

It is not necessary that all, or even a large fraction, of the grains in any of these breccias have retained xenon since such early times. In fact, if more than a few percent of the grains were that old, we would see a significant amount of in situ fission xenon. We can set an upper limit to the amount of in situ fission ^{136}Xe by considering the volume-correlated components calculated for 14301 (Bernatowicz et al., 1979) and 14318 (Swindle et al., 1985). All the isotopic ratios calculated for the volume-correlated component depend on the amount of in situ fission assumed. By demanding that the $^{126}Xe/^{130}Xe$ ratio of the volume-correlated component fall with the range of values observed for spallation (Hohenberg et al., 1978), we calculate the maximum amount of in situ fission and find that at most 15% of the grains can be 4500 m.y. old. In addition, if a large fraction of the grains were this old, there should be signs of older ages in ^{40}Ar-^{39}Ar studies (Reynolds et al., 1974). Thus, these observations, considered in their entirety, suggest that only a few percent of the grains in these breccias (certainly less than 15%) are very old. Whether or not these old grains represent a morphological subset (perhaps the submicron fraction) is not clear. In the only study that attempted to find the specific grains containing excess fission xenon, the excess component was lost in the disaggregation procedure, apparently being replaced in tenuous surface sites by more active atmospheric species (Caffee et al., 1981).

With the assumptions we have made, the oldest ages we obtain for lunar samples containing excess fission xenon are very similar to the ages of other ancient lunar samples, as determined by other dating techniques. This agreement could simply be fortuitous, but it does tend to confirm the assumptions of I-Pu-Xe dating. Unfortunately, since the ancient component in highland breccias is apparently only a small fraction of the breccia, it will be difficult to confirm these ages using any other radiometric technique unless these grains can be identified by properties other than their xenon systematics.

Summary

While there is chronometric information contained in the $^{129}Xe/^{136}Xe$ ratio of excess fission xenon, interpreting this requires assumptions about its acquisition. In every case observed (two samples with well-defined isotopic spectra and two whose spectra can be identified only with marginal precision), this ratio varies in the manner we would expect if the variations are due to the differing decay rates of ^{129}I and ^{244}Pu, with acquisition closely coupled with production.

We can provide accurate chronometry if we know the initial $^{129}I/^{244}Pu$ ratio of the source, which can be determined from the present I/U ratio of the source region. There are, however, two difficulties. One is that we cannot say with certainty whether the source is a local region of the Moon or whether it is global, with the xenon representing production from the bulk Moon. Even if we know the source region, we still don't know the present I/U ratio to any better than a factor of two or three, primarily because iodine values are so uncertain. The best present estimate is that this ratio is 0.01 (perhaps somewhat lower for Apollo 14 samples if the sources are local). Working in our favor, however, are the short half-lives involved: Each factor of two uncertainty in the I/U ratio translates into only about 20 m.y. uncertainty in time.

We calculate that the oldest identifiable xenon component in 14301 was acquired 63 ± 42 m.y. after t_0 (the time of formation of primitive meteorites), where the uncertainty allows for either a local or global source region. The ages of the oldest lunar samples (in particular, 76535, 74235, 72417, and 65015), as determined by other radiometric techniques, also fall within 50 m.y. of t_0. However, the uncertainties in all these determinations are large enough to allow a formation time as late as 100 m.y. after primitive meteorites. The ages of the oldest rocks, as determined by other dating techniques, are typically uncertain by 50–100 m.y., being primarily limited by analytical precision. Ages determined by the I-Pu-Xe system currently have comparable uncertainties, but these uncertainties come from a lack of understanding of the details of the excess fission xenon effect (in particular, a lack of knowledge of the source). Thus, improved understanding of the excess fission xenon effect can, in principle, lead to the most precise ages yet obtained for very old lunar samples.

The I-Pu-Xe ages of the Earth's atmosphere and mantle, as calculated by several authors using slightly different assumptions, are about 75–100 m.y. after t_0 (between 80 m.y. after and 30 m.y. before the oldest lunar I-Pu-Xe ages). This age, like the I-Pu-Xe ages of lunar samples, is sensitive to the choice of the I/U ratio of the source, which leads to comparable difficulties. If the Moon were derived from the Earth by fission or by spalling, this age should predate the age of any lunar samples. Under the current best estimates, it does not. While the present uncertainties in the ages of both bodies are still too large to say definitively that the Earth is younger than the Moon, the data tend to be inconsistent with fission or impact models.

Arguments based on initial $^{87}Sr/^{86}Sr$ ratios are limited by lack of information about the Earth's initial strontium isotopic composition. However, the lunar value of 0.69895 ± 3 or less is quite primitive. In fact, if the Moon formed from the Earth, the initial lunar value is low enough to demand that either the Earth's initial $^{87}Sr/^{86}Sr$ ratio was lower than BABI and close to ADOR, or that the Earth accreted, differentiated, and spawned the Moon much closer to t_0 than is indicated by dynamical studies or by the I-Pu-Xe ages of the atmosphere and mantle. For any model in which all or much of the Moon's material does not come from the Earth, such as a double-planet model (cf. Ruskol, 1977) or a Mars-sized impactor model (cf. Cameron, 1986), the constraints of the initial strontium ratios are relaxed (Taylor, 1986).

Acknowledgments. We are deeply indebted to T. J. Bernatowicz, who patiently shared and discussed his own experimental results. We also appreciate helpful discussions with F. A. Podosek, G. Dreibus, and K. D. McKeegan. This work was supported in part by NASA grant NAG 9-7.

References

Aeschlimann U., Eberhardt P., Geiss J., Groegler N., Jost D., Ma M. -S., Marti K., Schmitt R., and Taylor J. (1983) Consortium 67915: Ages and composition of troctolitic anorthosites and sodic ferrogabbro clasts (abstract). In *Lunar and Planetary Science XIV*, pp. 1–2, Lunar and Planetary Institute, Houston.

Anders E. and Ebihara M. (1982) Solar system abundances of the elements. *Geochim. Cosmochim. Acta, 46*, 2363–2380.

Arden J. W. and Cressey G. (1984) Thallium and lead in the Allende C3V carbonaceous chondrite. A study of the matrix phase. *Geochim. Cosmochim. Acta, 48*, 1899–1912.

Basford J. F., Dragon J. C., Pepin R. O., Coscio M. R., Jr., and Murthy V. R. (1973) Krypton and xenon in lunar fines. *Proc. Lunar Sci. Conf. 4th*, pp. 1915–1955.

Behrmann C. J., Drozd R. J., and Hohenberg C. M. (1973) Extinct lunar radioactivities: xenon from ^{244}Pu and ^{129}Xe in Apollo 14 breccias. *Earth Planet. Sci. Lett., 17*, 446–455.

Bernatowicz T. J. and Podosek F. A. (1978) Nuclear components in the atmosphere. In *Terrestrial Rare Gases* (E. C. Alexander, Jr. and M. Ozima, eds.), pp. 99–135. Japan Science Societies Press, Tokyo.

Bernatowicz T., Drozd R. J., Hohenberg C. M., Lugmair G., Morgan C. J., and Podosek F. A. (1977) The regolith history of 14307. *Proc. Lunar Sci. Conf. 8th*, pp. 2763–2783.

Bernatowicz T. J., Hohenberg C. M., Hudson B., Kennedy B. M., and Podosek F. A. (1978) Excess fission xenon at Apollo 16. *Proc. Lunar Planet. Sci. Conf. 9th*, pp. 1571–1597.

Bernatowicz T. J., Hohenberg C. M., Hudson B., Kennedy B. M., Laul J. C., and Podosek F. A. (1980) Noble gas component organization in 14301. *Proc. Lunar Planet. Sci. Conf. 11th*, pp. 629–668.

Bernatowicz T. J., Hohenberg C. M., and Podosek F. A. (1979) Xenon component organization in 14301. *Proc. Lunar Planet. Sci. Conf. 10th*, pp. 1587–1616.

Bernatowicz T. J., Kramer F. E., Podosek F. A., and Honda M. (1982) Adsorption and excess fission Xe: Adsorption of Xe on vacuum crushed minerals. *Proc. Lunar Planet. Sci. Conf. 13th*, in *J. Geophys. Res., 87*, A465–A476.

Bjorkholm P. J., Golub L., and Gorenstein P. (1973) Distribution of ^{222}Rn and ^{210}Po on the lunar surface as observed by the alpha particle spectrometer. *Proc. Lunar Sci. Conf. 4th*, pp. 2793–2802.

Bogard D. D., Nyquist L. E., Bansal B. M., Wiesmann H., and Shih C. Y. (1975) 76535: an old lunar rock. *Earth Planet. Sci. Lett., 26*, 69–80.

Bugnon M.-F., Tera F., and Brown L. (1979) Are ancient lead deposits chronometers of the early history of Earth? *Annual Report of the Director, Department of Terrestrial Magnetism, Carnegie Institution, 1978–1979*, pp. 346–352. Washington, D. C.

Caffee M., Hohenberg C. M., and Hudson B. (1981) Troctolite 76535: a study in the preservation of early isotopic records. *Proc Lunar Planet Sci. 12B*, pp. 99–115.

Cameron A. G. W. (1983) Origin of the atmosphere of the terrestrial planets. *Icarus, 56*, 195–201.

Cameron A. G. W. (1986) Formation of the prelunar accretion disk, this volume.

Chen J. H. and G. R. Tilton (1976) Isotopic lead investigations on the Allende carbonaceous chondrite. *Geochim. Cosmochim. Acta, 40*, 635–643.

Compston W. and Williams I. S. (1984) U-Pb geochronology of zircons from lunar breccia 73217 using a Sensitive High-mass Resolution Ion Microprobe. *Proc. Lunar Planet. Sci. Conf. 14th*, in *J. Geophys. Res., 89*, B525–B534.

Crozaz G., Drozd R., Graf H., Hohenberg C. M., Monnin M., Ragan D., Ralston C., Seitz M., Shirck J., and R. M. Walker (1972) Uranium and extinct ^{244}Pu effects in Apollo 14 materials. *Proc. Lunar Sci. Conf. 3rd*, pp. 1623–1636.

Dreibus G., Spettel B., and Wänke H. (1977) Lithium and halogens in lunar samples. *Philos. Trans. R. Soc. London, A285*, 49–54.

Drozd R., Hohenberg C., and Morgan C. (1975) Krypton and xenon in Apollo 14 samples: fission and neutron capture effects in gas-rich samples. *Proc. Lunar Sci. Conf. 6th*, pp. 1857–1877.

Drozd R., Hohenberg C. M., and Ragan D. (1972) Fission xenon from extinct ^{244}Pu in 14301. *Earth Planet. Sci. Lett., 15*, 338–346.

Drozd R. J., Kennedy B. M., Morgan C. J., Podosek F. A., and Taylor G. J. (1976) The excess fission xenon problem in lunar samples. *Proc. Lunar Sci. Conf. 7th*, pp. 599–623.

Eberhardt P., Geiss J., Graf H., Grögler N., Krähenbühl U., Schwaller H., Schwartzmuller H., and Stettler A. (1970) Trapped solar wind noble gases, exposure age and K/Ar-age in Apollo 11 lunar fine material. *Proc. Apollo 11 Lunar Sci. Conf.*, pp. 1037–1070.

Eberhardt P., Geiss J., Graf H., Grögler N., Mendia M. D., Morgeli M., Schwaller H., Stettler A., Krähenbühl U., and von Gunten H. R. (1972) Trapped solar wind noble gases in Apollo 12 lunar fines 12001 and Apollo 11 breccia 10046. *Proc Lunar Sci. Conf. 3rd*, pp. 1821–1856.

Fields P. R., Friedman A. M., Milsted J., Lerner J., Stevens C. M., Metta D., and Sabine W. K. (1966) Decay properties of plutonium-244, and comments on its existence in nature. *Nature, 212*, 131–134.

Froude D. O., Ireland T. R., Kinny P. D., Williams I. S., Compston W., Williams I. R., and Myers J. S. (1983) Ion microprobe identification of 4100 to 4200 Ma-old terrestrial zircons. *Nature, 304*, 616–618.

Gancarz A. J. and Wasserburg G. J. (1977) Initial Pb of Amitsoz gneiss, West Greenland, and implications for the age of the Earth. *Geochim. Cosmochim. Acta, 41*, 1283–1301.

Gooley R., Brett R., Warner J., and Smyth J. R. (1974) A lunar rock of deep crustal origin: sample 76535. *Geochim. Cosmochim. Acta, 38*, 1329–1339.

Gray C. M., Papanastassiou D. A., and Wasserburg G. J. (1973) The identification of early condensates from the solar nebula. *Icarus, 20*, 213–239.

Heymann D. and Yaniv A. (1970) Ar^{40} anomaly in lunar samples from Apollo 11. *Proc. Apollo 11 Lunar Sci. Conf.*, pp. 1261–1267.

Hodges R. R., Jr. (1977) Release of radiogenic gases from the Moon. *Phys. Earth Planet. Inter., 14*, 282–288.

Hohenberg C. M. and Kennedy B. M. (1981) I-Xe dating: intercomparisons of neutron irradiations and reproducibility of the Bjurbole standard. *Geochim. Cosmochim Acta, 45*, 251–256.

Hohenberg C. M., Hudson B., Kennedy B. M., and Podosek F. A. (1980) Fission xenon in troctolite 76535. In *Proc. Conf. Lunar Highlands Crust* (J. J. Papike and R. B. Merrill, eds.), pp. 419–439. Pergamon, New York.

Hohenberg C. M., Marti K., Podosek F. A., Reedy R. C., and Shirck J. R. (1978) Comparisons between observed and predicted cosmogenic noble gases in lunar samples. *Proc. Lunar Planet. Sci. Conf. 9th*, pp. 2311–2344.

Hohenberg C. M., Podosek F. A., and Reynolds J. H. (1967) Xenon-iodine dating: sharp isochronism in chondrites. *Science, 156*, 233–236.

Hudson G. B., Hohenberg C. M., Kennedy B. M. and Podosek F. A. (1982) ^{244}Pu in the early solar system (abstract). In *Lunar and Planetary Science XIII*, pp. 346–347. Lunar and Planetary Institute, Houston.

Huneke J. C. and Wasserburg G. J. (1975) Trapped ^{40}Ar in troctolite 76535 and evidence for enhanced ^{40}Ar-^{39}Ar age plateaus (abstract). In *Lunar Science VI*, pp. 417–419. The Lunar Science Institute, Houston.

Husain L. and Schaeffer O. A. (1975) Lunar evolution: the first 600 million years. *Geophys. Res. Lett., 2*, 29–32.

Jessberger E. K., Huneke J. C., Podosek F. A., and Wasserburg G. J. (1974) High-resolution argon analysis of neutron-irradiated Apollo 16 rocks and separated minerals. *Proc. Lunar Sci. Conf. 5th*, pp. 1419–1449.

Jovanovic S. and Reed G. W., Jr. (1973) Volatile trace elements and the characterization of the Cayley Formation and the primitive lunar crust. *Proc. Lunar Sci. Conf. 4th*, pp. 1313–1324.

Jovanovic S. and Reed G. W., Jr. (1978) Trace element evidence for a laterally inhomogeneous moon. *Proc. Lunar Planet. Sci. Conf. 9th*, pp. 59–80.

Katcoff S., Schaeffer O. A., and Hastings J. M. (1951) Half-life of I^{129} and the age of the elements. *Phys. Rev., 82*, 688–690.

Langseth M. G, Keihm S. J., and Peters K. (1976) Revised lunar heat-flow values. *Proc. Lunar Sci. Conf. 7th*, pp. 3143–3171.

Lugmair G. W. and Marti K. (1977) Sm-Nd-Pu timepieces in the Angra dos Reis meteorite. *Earth Planet. Sci. Lett, 35*, 273–284.

Lugmair G. W., Marti K., Kurtz J. P., and Scheinin N. B. (1976) History and genesis of lunar troctolite 76535 or: How old is old? *Proc. Lunar Sci. Conf. 7th*, pp. 2009–2033.

Manka R. H. and Michel F. C. (1970) Lunar atmosphere as a source of argon-40 and other lunar surface elements. *Science, 169*, 278–280.

Manka R. H. and Michel F. C. (1973) Lunar ion energy spectra and surface potential. *Proc. Lunar Sci. Conf. 4th*, pp. 2897–2908.

Mason B. H. (1979) Data of geochemistry. *U.S. Geol. Surv. Prof. Pap. 440–B–1*, 132 pp.

Minster J.-F., Birck J. L., and Allégre C. J. (1982) Absolute age of formation of chondrites studied by the ^{87}Rb-^{87}Sr method. *Nature, 300*, 414–419.

Nakamura N., Tatsumoto M., Nunes P. D., Unruh D. M., Schwab A. P., and Wildeman T. R. (1976) 4.4 b.y.-old clast in Boulder 7, Apollo 17: A comprehensive chronological study by U-Pb, Rb-Sr and Sm-Nd methods. *Proc. Lunar Sci. Conf. 7th*, pp. 2309–2333.

Nunes P. D., Tatsumoto M., and Unruh D. M. (1974) U-Th-Pb systematics of some Apollo 17 lunar samples and implications for a lunar basin excavation chronology. *Proc. Lunar Sci. Conf. 5th*, pp. 1487–1514.

Nyquist L. E. (1977) Lunar Rb-Sr chronology. *Phys. Chem. Earth, 10*, 103–142.

Papanastassiou D. A. and Wasserburg G. J. (1969) Initial strontium isotopic abundances and the resolution of small time differences in the formation of planetary objects. *Earth Planet. Sci. Lett.*, 5, 361–376.

Papanastassiou D. A. and Wasserburg G. J. (1975) Rb-Sr study of a lunar dunite and evidence for early lunar differentiates. *Proc. Lunar Sci. Conf. 6th*, pp. 1467–1489.

Papanastassiou D. A. and Wasserburg G. J. (1976) Rb-Sr age of troctolite 76535. *Proc. Lunar Sci. Conf. 7th*, pp. 2035–2054.

Patterson C. (1956) Age of meteorites and the earth. *Geochim. Cosmochim. Acta*, 10, 230–237.

Pepin R. O. and Phinney D. (1976) The formation interval of the earth (abstract). In *Lunar Science VII*, pp. 682–684. The Lunar Science Institute, Houston.

Podosek F. A., Bernatowicz T. J., and Kramer F. E. (1981) Adsorption and excess fission xenon. *Proc. Lunar Planet. Sci. 12B*, pp. 891–901.

Podosek F. A., Huneke J. C., Burnett D. S., and Wasserburg G. J. (1971) Isotopic composition of xenon and krypton in the lunar soil and in the solar wind. *Earth Planet. Sci. Lett.*, 10, 199–216.

Rasmussen K. L. and Warren P. H. (1985) Megaregolith thickness, heat flow, and the bulk composition of the Moon. *Nature*, 313, 121–124.

Reed G. W., Jr., Jovanovic S., and Fuchs L. (1972) Trace element relations between Apollo 14 and 15 and other lunar samples, and the implications of a moon-wide Cl-KREEP coherence and Pt-metal noncoherence. *Proc. Lunar Sci. Conf. 3rd*, pp. 1989–2001.

Reynolds J. H., Alexander E. C., Jr., Davis P. K., and Srinivasan B. (1974) Studies of K-Ar dating and xenon from extinct radioactivities in breccia 14318; implications for early lunar history. *Geochim. Cosmochim. Acta*, 38, 401–417.

Ruskol E. Y. (1977) The origin of the Moon. *NASA SP-370*, 85–822.

Spudis P. D. (1984) Apollo 16 site geology and impact melts: implications for the geologic history of the lunar highlands. *Proc. Lunar Planet. Sci. Conf. 15th*, in *J. Geophys. Res.*, 89, C95–C107.

Staudacher T. and Allégre C. J. (1982) Terrestrial xenology. *Earth Planet. Sci. Lett.*, 60, 389–406.

Steiger R. H. and Jäger E. (1977) Subcommission on geochronology: convention on the use of decay constants in geo- and cosmochronology. *Earth Planet. Sci. Lett.*, 36, 359–362.

Stolper E. (1977) Experimental petrology of eucritic meteorites. *Geochim. Cosmochim. Acta*, 41, 587–611.

Swindle T. D., Caffee M. W., Hohenberg C. M., Hudson G. B., Laul J. C., Simon S. B., and Papike J. J. (1985) Noble gas component organization in Apollo 14 breccia 14318; ^{129}I and ^{244}Pu regolith chronology. *Proc. Lunar Planet. Sci. Conf. 15th*, in *J. Geophys. Res.*, 90, C517–C539.

Tatsumoto M., Nunes P. D., and Unruh D. M. (1977) Early history of the Moon: Implications of U-Th-Pb and Rb-Sr systematics. In *The Soviet-American Conference on the Cosmochemistry of the Moon and Planets* (J. H. Pomeroy and N. J. Hubbard, eds.), pp. 507–523. NASA SP-370. Science and Technical Information Office, NASA, Washington.

Tatsumoto M., Unruh D. M., and Desbrough G. A. (1976) U-Th-Pb and Rb-Sr systematics of Allende and U-Th-Pb systematics of Orgueil. *Geochim. Cosmochim. Acta*, 40, 617–634.

Taylor S. R. (1982) *Planetary Science: A Lunar Perspective*. Lunar and Planetary Institute, Houston. 481 pp.

Taylor S. R. (1986) The origin of the moon: Geochemical constraints, this volume.

Taylor S. R. and McLennan S. M. (1985) *The Continental Crust: Its Composition and Evolution*. Blackwell, Oxford. In press.

Tera F. (1981) Aspects of isochronism in Pb isotope systematics—application to planetary evolution. *Geochim. Cosmochim. Acta*, 45, 1439–1448.

Tera F. and Wasserburg G. J. (1974) U-Th-Pb systematics of lunar rocks and inferences about lunar evolution and the age of the Moon. *Proc. Lunar Sci. Conf. 5th*, pp. 1571–1599.

Tera F. and Wasserburg G. J. (1976) Lunar ball games and other sports (abstract). In *Lunar Science VII*, pp. 858–860. The Lunar Science Institute, Houston.

Tilton G. R. and Steiger R. H. (1965) Lead isotopes and the age of the earth. *Science, 150,* 1805–1807.

Turner G., Enright M. C., and Cadogan P. H. (1978) The early history of chondrite parent bodies inferred from ^{40}Ar-^{39}Ar ages. *Proc. Lunar Planet. Sci. Conf. 9th,* pp. 989–1025.

Wänke H., Palme H., Baddenhausen H., Dreibus G., Jagoutz E., Kruse H., Spettel B., Teschke F., and Thacker R. (1974) Chemistry of Apollo 16 and 17 samples: Bulk composition, late state accumulation and early differentiation of the moon. *Proc. Lunar Sci. Conf. 5th,* pp. 1307–1335.

Wasserburg G. J., Papanastassiou D. A., Tera F., and Huneke J. C. (1977a) Outline of a lunar chronology. *Phil. Trans. Roy. Soc. London,* A285, 7–22.

Wasserburg G. J., Tera F., Papanastassiou D. A., and Huneke J. C. (1977b) Isotopic and chemical investigations on Angra dos Reis. *Earth Planet. Sci. Lett., 35,* 294–316.

Wetherill G. W. (1975) Radiometric chronology of the early solar system. *Ann. Rev. Astron. Astrophys., 18,* 77–113.

Wetherill G. W. (1980) Formation of the terrestrial planets.*Ann. Rev. Astron. Astrophys, 18,* 77–113.

Wilhelms D. E. (1985) Lunar impact rates reconsidered (abstract). In *Lunar and Planetary Science XVI,* pp. 904–905. Lunar and Planetary Institute, Houston.

Wolf R. and Anders E. (1980) Moon and Earth: compositional differences inferred from siderophiles, volatiles, and alkalies in basalts. *Geochim. Cosmochim. Acta, 44,* 2111–2124.

IV. Geophysical Constraints

Geophysical Constraints on the Lunar Interior

L. L. HOOD

Lunar and Planetary Laboratory, University of Arizona, Tucson, AZ 85721

Geophysical evidence relating to the structure, composition, and thermal state of the lunar interior is reviewed with emphasis on aspects that are of the most interest in evaluating lunar origin models. Probable limits on crustal composition and mean thickness indicate an average crustal Al_2O_3 content of 17–25 wt % and a crustal contribution to the lunar bulk abundance of about 1.8–2.6 wt %. If whole-Moon differentiation is allowed, then bulk Al_2O_3 contents comparable to chondritic and terrestrial abundances may not be excluded. If differentiation to shallower depths is assumed, then an enrichment as great as two times chondritic may be estimated. The sparsity of surface heat flow determinations and differences in their interpretation allow inferred bulk U abundances that are either consistent with or larger than chondritic and terrestrial values. Heat flow estimates derived from electrical conductivity data are equally uncertain. Thus geophysical evidence for a lunar refractory lithophile element enrichment is tenuous. Seismic velocities in the upper mantle are consistent with a predominantly olivine/pyroxene composition with Mg/(Mg + Fe) ratios in the range 0.70–0.85. Seismic velocities at depths > 500 km are more uncertain but the most probable available model is characterized by increased velocities, implying either a large garnet fraction or an increase in Mg/(Mg + Fe) ratio for mafic silicates. Extrapolations of laboratory elastic constant data for normative mineral assemblages to the pressures and temperatures of the lunar middle mantle indicate that the model seismic velocity increase can be attributed to the presence of garnet alone only if inordinately large Al_2O_3 abundances (> 7–10 wt %) are assumed at these depths. Alternatively, a transition to more MgO-rich mafic silicates in addition to garnet would be sufficient to explain the velocity increase. However, in the latter case the corresponding density models are characterized by density decreases below 500 km depth that would have been unstable to convective overturn during early lunar history. It is therefore concluded that current lunar seismic velocity models and/or their interpretations are not yet adequate to strongly constrain changes in composition and density in the middle and lower mantle. Consequently, observational limits on the lunar moment of inertia do not yet strongly constrain the existence and radius of a possible metallic core. Certain independent observations including those derived from paleomagnetic data and seismic data for a single farside impact also do not provide definitive evidence for or against the presence of a metallic core. Other observations including the measured negative lunar induced magnetic dipole moment in the geomagnetic tail and the 0.2" advance of the lunar spin axis from the Cassini alignment suggest but do not prove the presence of a small metallic core with radius ~330–460 km. Iron cores of this size represent ~2–4% of the lunar mass and are larger than would be expected if the Moon formed entirely from terrestrial mantle material and if segregation of this amount of metal caused the depletions of lunar siderophiles. However, additions of metal to a proto-Moon composed largely of terrestrial mantle material by differentiated circumterrestrial planetesimals or by a large terrestrial impactor would circumvent this difficulty. Alternatively, models such as the binary accretion hypothesis would be capable of producing a bulk Moon in agreement with available geophysical constraints.

1. Introduction

The purpose of this paper is to review available geophysical constraints relating to the structure, composition, and thermal state of the Moon. Of particular interest in the context of this book are constraints on the bulk composition of the crust and mantle, on the existence and size of a possible metallic core, and on the depth of initial melting and differentiation, issues that are fundamental to the problem of determining the most probable mode of lunar origin and early internal evolution. Relevant data sets include those derived from seismic, gravity, topography, magnetic, laser ranging, and surface heat flow measurements. None of these data sets is individually definitive in the sense of yielding unique structural or compositional information. However, syntheses of available constraints and comparisons with independent geochemical and petrological inferences are instructive and are emphasized.

Aspects of crust and mantle bulk composition that are both geophysically addressable and of fundamental interest in evaluating lunar origin models include the proposed lunar refractory element enrichment and the $Mg/(Mg + Fe)$ ratio for the bulk Moon. Geophysical evidence for a lunar enrichment of refractory elements comes primarily from interpretations of surface heat flow data yielding model-derived limits on the bulk U abundance (Keihm and Langseth, 1977) and from estimates for the bulk lunar Al_2O_3 abundance based in part on crustal composition and thickness models derived from seismic, gravity, and topography data (Taylor and Bence, 1975; Kaula, 1977). Evidence that the bulk lunar mantle $Mg/(Mg + Fe)$ atomic ratio is significantly less than that of the terrestrial mantle (0.75–0.81 as opposed to 0.89) is based largely on petrological inferences (Ringwood and Essene, 1970; Ringwood, 1979) and on interpretations of earlier seismic velocity models (Buck and Toksöz, 1980; Binder, 1980). Both the refractory element enrichment and FeO enrichment have been cited as basic tests of the hypothesis that the Moon originated primarily from material of the Earth's mantle (e.g., Kaula, 1977; Taylor, 1982). Acceptance of the FeO enrichment has required advocates of the latter hypothesis to postulate either a gradual transfer of FeO from the terrestrial mantle to the core over geologic time (Jagoutz and Wänke, 1982) or the addition of Earth-bound planetesimals to a proto-Moon composed largely of recondensed terrestrial mantle material (Ringwood, 1984). The necessity of proposed refractory lithophile and FeO enrichments has been questioned most notably by Warren and Wasson (1979) (see also Warren, 1982, 1984).

From the observed mean density and absence of volatiles, it has long been realized that the Moon must be severely depleted in iron in comparison to other terrestrial bodies so that a metallic core must be small or nonexistent. More recently, the observed depletion of lunar siderophiles relative to CI chondrites and the terrestrial mantle has been interpreted in terms of an indigenous lunar metal-silicate fractionation event, suggesting the existence of at least a small metallic core (Newsom and Drake, 1982, 1983; Drake, 1983). Although the alternate hypothesis of a distribution of metal pods cannot be observationally eliminated, gravitational segregation of the metal fraction through a mantle sufficiently plastic to allow core formation would be a plausible end result. Geochemical data cannot be applied to derive the segregated metal fraction and hence the core size since this depends on unknown initial siderophile abundances and on the assumed depth of differentiation. Hence Newsom (1984) has reversed the argument and suggests that a geophysical determination of the core size would help to distinguish among lunar origin models. If the Moon formed entirely from terrestrial mantle material, which is already depleted in siderophiles relative to chondrites, then segregation of only 0.1–1% metal would be required to produce the lunar siderophile abundances; the radius of a predominantly iron core would be < 285 km. If the Moon formed entirely from undifferentiated solar nebula material, then Newsom estimates that segregation of at least 2% metal would be required for the lowest plausible degree of silicate partial melting and the expected iron core radius would be > 360 km if whole-Moon differentiation is assumed. In reality, it is possible that only part of the Moon differentiated, resulting in a smaller core size than the predicted limits. Alternatively, although the Moon may have formed primarily from material derived from the Earth's mantle, metal may have been added to the proto-Moon from Earth-orbiting planetesimals or from the postulated large impactor of Hartmann and Davis (1975) and Cameron and Ward (1976). Thus an exact evaluation of the lunar metallic core radius and mass may not unequivocally distinguish the mode of lunar origin. We should not lose sight of the fundamental observation of an iron-depleted Moon indicating, for instance, selective accretion of iron-poor differentiated planetesimals under the influence of a nearby Earth or formation from terrestrial mantle material. However, the siderophile data do allow quantitative tests of particular origin models under different assumptions if the core mass is known, so a determination of the core radius and density continues to be of basic interest (Newsom, 1984).

Finally, a long-standing question relevant to models of lunar origin and internal evolution concerns the depth of the initial lunar melting/differentiation event that yielded the highland crust. Although Wood et al. (1970) and Smith et al. (1970) originally suggested whole-Moon partial melting and differentiation, many models of crustal and mare basalt petrogenesis have adopted an outer differentiated zone 300–500 km thick overlying a primordial undifferentiated interior (e.g., Ringwood and Kesson, 1977; Herbert *et al.*, 1977; Taylor, 1978). Such models were considered as more plausible on the basis of (1) a consideration of available heat sources for melting the deep interior, (2) global thermal expansion and contraction estimates derived from an apparent lack of surface compressive tectonic features (Solomon and Chaiken, 1976), and (3) interpretations of early seismic velocity models (e.g., Dainty *et al.*, 1976; Taylor, 1978). The requirement of differentiating only a fraction of the lunar interior now seems less stringent for several reasons. First, more recent seismic velocity models (Nakamura, 1983) indicate that earlier models suggesting a more FeO-rich primordial interior below 400–500 km depth may have been less reliable. Second, the surface tectonic constraint on the depth of melting has been questioned on at least two grounds. Herbert *et al.* (1977) consider that present-day internal temperatures may be higher than assumed by Solomon and Chaiken so that a greater depth of melting would be allowed by a radius change of ~1 km. Binder and Lange (1980) suggest the existence of sufficient surface faulting and subsurface solid state creep early in lunar history to relieve the compressional stress resulting from a larger (~5 km) radius decrease required for initial whole-Moon partial melting and differentiation. Finally, although the heat source for producing an initially molten or partially molten Moon continues to be obscure, a similar problem exists for the parent bodies of certain meteorites and remains generally unresolved. Thus for the purposes of this paper, it appears that restricted depths of early lunar melting and differentiation are no more certain than whole-Moon differentiation; consequently, no such restrictions will be imposed.

The remainder of this review is divided into six main sections. Sections 2 and 3 summarize independent quantitative constraints derivable from seismic, gravity, and topography data on the structure and composition of the crust and mantle. Section 4 discusses constraints on the present-day temperature profile from seismic, electromagnetic sounding, surface heat flow, and gravity data. In Section 5, earlier efforts to construct lunar bulk composition and density models consistent with a composite of available geophysical constraints as well as geochemical and petrologic constraints are critically reviewed. An initial effort is made to apply similar methods using the later mantle seismic velocity model of Nakamura (1983). Section 6 summarizes independent geophysical constraints on the existence and radius of a possible metallic core obtained from interpretations of seismic, electromagnetic sounding, laser ranging (physical librations), and paleomagnetic data.

2. Crustal Structure and Composition

2.1. Seismic constraints

General detailed reviews of lunar seismicity as observed from the four stations of the Apollo seismic network have been given by Lammlein *et al.* (1974) and by Toksöz *et al.* (1974). A concise summary of most of the final conclusions has been given by Nakamura *et al.* (1982). Sources of lunar seismic signals include artificial and natural impacts, weak repetitive deep-focus moonquakes triggered by tidal stresses (Toksöz *et al.*, 1977; Nakamura, 1978; Koyama and Nakamura, 1980), and more energetic but rare shallow moonquakes that are most probably tectonic in origin (Nakamura *et al.*, 1979). The total lunar seismic energy release is small (less than 10^{18} ergs annually compared to the terrestrial output of about 10^{25} ergs); recorded moonquake magnitudes range up to 3.3 on the Richter scale for shallow events and up to ~1 for deep events (Goins *et al.*, 1981a). In addition to the relatively weak signal amplitudes, intense scattering occurs in a near-surface brecciated zone, reducing the signal-to-noise ratio and complicating the interpretation of lunar seismograms.

In the case of near-surface and crustal seismic velocity models, the inversion problem was simplified through the use of astronaut-activated seismic energy sources and nearby artificial impacts of the Lunar Module ascent stage and upper stage of the Saturn rocket whose event times, energies, and positions were accurately known. A crustal P-wave velocity model, valid for the Mare Cognitum region near the Apollo 12 and 14 landing sites, satisfying all available travel time data and producing theoretical seismograms closely matching the amplitudes and wave character of the

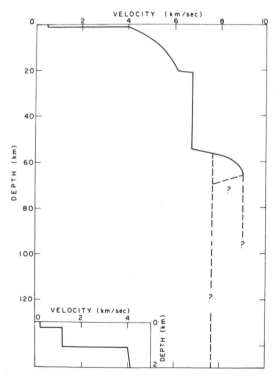

Fig. 1.   Crustal P-wave velocity model for the Mare Cognitum region beneath the Apollo 12 and 14 landing sites (Toksöz et al., 1974).

observed signals, was calculated by Toksöz *et al* (1974) and is shown in Fig. 1. Velocity increases within the uppermost 1–2 km are ascribable to the presence of rock with different microcrack properties (Simmons *et al*, 1980) although a change from mare basalt composition to highland anorthosite may also contribute (Cooper *et al*, 1974). An early hypothesis that the 20-km depth discontinuity marked the base of mare basalt flows (Toksöz *et al*, 1972) was shown to be implausible on the basis of photogeologic data (e.g., Head, 1979, 1982) and the probable observation of a similar discontinuity beneath the Apollo 16 highlands site (Goins *et al*, 1981b). The continuous increase in compressional velocity between 1.4 and 20 km depth is believed to be primarily due to the effect of crack closure with increasing pressure, while the discontinuous increase near 20 km is considered to be indicative of a composition change (Todd *et al*, 1973; Toksöz *et al*, 1974). The nearly constant inferred velocity between 20 and 55 km and gradual increase to a distinctly higher value near 60 km is consistent with a compositionally homogeneous and competent lower crust underlain by a transition zone approximately 3–5 km thick to the mantle. Conservatively, the inferred crustal thickness of 55 km may be accurate only to within ±10 km as values ranging from 45 km (Koyama and Nakamura, 1979) to 65 km (Toksöz *et al*, 1972) have been reported. For comparison, Nakamura (1983) has derived an estimate of 58 ± 8 km for the same region near the Apollo 12 and 14 stations based on a reappraisal of the same data set. Beneath the Apollo 16 Descartes highlands site, a somewhat greater crustal thickness of 75 ± 5 km has been inferred through the identification of apparent reflected phases from the transition zone (Goins *et al*, 1981b). Goins *et al* note that the increased thickness relative to that in the Mare Cognitum region would be sufficient to isostatically compensate the increased elevation of 1.5 km at the Apollo 16 site if laterally constant crustal and mantle densities of 3.0 and 3.3 g cm^{-3} are assumed.

With respect to crustal compositional inferences, the velocity model of Fig. 1 in the 20–55 km depth range was shown to be generally consistent with laboratory data for lunar anorthositic gabbros (i.e., representative of surface highland samples) after correction for microfracturing effects (Toksöz *et al*, 1972, 1974). However, it was emphasized that other compositions could not be ruled out and a later independent evaluation concluded that it is impossible to distinguish between anorthositic gabbro, gabbro, or anorthosite compositions on the basis of seismic data alone (Liebermann and Ringwood, 1976). The latter authors suggest that a differentiated crustal model characterized by anorthosite enrichment in the upper crust and a corresponding depletion in the lower crust is more plausible and would significantly reduce the inferred crustal Al_2O_3 content (from 25% Al for a mean composition of anorthositic gabbro to 17% Al for a mean composition of gabbro).

In support of a differentiated crustal model, both sample studies and spectral reflectance data later indicated the occurrence of low-K Fra Mauro basalt components in areas where deep crustal material sampled by large basin impacts would be preserved, e.g., near the rims of the Imbrium and Serenitatis basins (Ryder and Wood, 1977;

Charette *et al.*, 1977). On this basis, two- or three-layer models were proposed in which the most accessible anorthositic gabbro layer comprises less than one-third of the crustal column and is underlain by a 15–20-km-thick low-K Fra Mauro basalt ("noritic") layer and possibly a third mafic anorthosite layer (Fig. 2). The mean Al_2O_3 content of such a stratigraphic column would be 18–20 wt % and the mean density would be 3.00–3.05 g cm^{-3}. A correlation between estimated transient cavity size (and hence depth of excavation) and the total fraction of "noritic" material identified spectrally within the ejecta of 11 lunar basins was also demonstrated (Spudis *et al.*, 1984), further supporting the concept of a grossly layered crust with anorthositic gabbro components concentrated mainly in an uppermost 15–20 km thick layer.

2.2. Gravity/topography constraints

Precise measurements of the lunar gravitational field using observed accelerations of artificial satellites were limited during the Apollo era by lack of direct line-of-sight Doppler tracking data on the far side and by a less-than-complete set of satellite

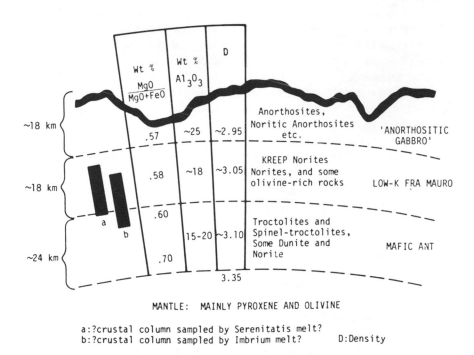

Fig. 2. Proposed structural and compositional model for the crust consistent with geophysical constraints and with the observation that impact melts near the rims of the Serenitatis and Imbrium basins have the low-K Fra Mauro composition (Ryder and Wood, 1977).

orbital parameters (e.g., Michael and Blackshear, 1972). Although harmonic analysis techniques were developed to deduce approximate long-wavelength gravity fields over the entire Moon from nearside tracking data (Ferrari and Ananda, 1977; Bills and Ferrari, 1980), high-resolution coverage needed for detailed modeling studies of individual crustal structures was restricted to low-latitude portions of the near side. Similarly, precise lunar topographic data are absent for large regions of the far side and polar zones so that global maps are presently limited to spherical harmonic models with wavelengths representative of basin-sized or larger-scale features (Bills and Ferrari, 1977a).

The observed crustal gravity field is generally mild reflecting near isostatic compensation and is dominated by the mascon anomalies associated with major Imbrium-aged impact basins (Muller and Sjogren, 1968). Analyses of Bouguer gravity anomalies indicate that isostatic compensation is nearly complete for pre-Imbrian highland topography but is incomplete for certain Imbrian and post-Imbrian structures such as the Apennine mountains (Ferrari et al., 1978) and the mascons, indicating global cooling and thickening of the elastic lithosphere since about 4 aeons. From detailed studies of several basins including Grimaldi and Serenitatis over which high-resolution gravity and topography measurements were obtained, it has been provisionally concluded that the mascons are due to a combination of rapid mantle rebound at the times of basin-forming impacts followed by an additional mass excess resulting from later mare basalt filling [see Phillips and Lambeck (1980) for a more complete discussion].

Assuming that lateral density variations are entirely the result of thickness variations of a nearly constant density crust (Airy isostatic compensation), Bills and Ferrari (1977b) have appled gravity and topography spherical harmonic models together with nearside seismic thickness estimates to show that (1) the average farside crustal thickness would be greater than that of the near side (see also Wood, 1973), and (2) crustal thickness would range from 30–35 km beneath the mascons to 90–110 km beneath the highlands with a mean of approximately 70 km. It is also possible in principle that a significant Pratt compensation component (lateral density variations in a constant thickness crust) is present, implying a somewhat different mean crustal thickness. An apparent correlation of surface major element chemistry with topographic elevation in the highlands was proposed as suggestive of a significant Pratt component (Solomon, 1978). However, later analyses of the orbital geochemical data found little evidence for lateral crustal density variations over most of the Moon (Haines and Metzger, 1979). On the extreme assumption that the crustal composition is vertically uniform so that surface compositions can be applied to calculate lateral density differences, Haines and Metzger (1980) concluded that a compensation model that is about 70% Airy still provides the best fit to the gamma-ray Fe data. If vertical inhomogeneity is allowed as suggested above, then Airy isostasy may prevail to a greater extent. These results combined with the greater crustal thickness beneath the Apollo 16 highlands site derived from seismic data indicates that a dominantly

Airy compensation model is more probable. However, additional seismic thickness determinations at other locations around the Moon would be needed to confirm this inference.

The observed center-of-figure to center-of-mass offset of 2.0–2.5 km (Kaula *et al.*, 1972, 1974) can in principle be explained by either a less dense or a thicker farside crust. Assuming the dominance of Airy isostasy, the relation between the observed offset and crustal thickness differences can be illustrated by a simple model in which the part of the Moon below the crust is approximated as a uniform-density sphere displaced from the center of figure (Kaula *et al.*, 1972). In this case, the amount of the offset ΔC corresponding to a difference in thickness ΔT between the nearside and farside crust is given by

$$\Delta C = \frac{\Delta T \, r_m \, (\rho_m - \rho_c)}{2\left[r_m^3 \, (\rho_m - \rho_c) + r_c^3 \rho_c \right]} \tag{1}$$

where r_c is the mean lunar radius (1738 km), r_m is the radius of the subcrustal sphere, ρ_c is the mean crustal density, and ρ_m is the mean density of the subcrustal sphere. For example, choosing minimum nearside and maximum farside thicknesses of 60 km and 100 km (cf. Bills and Ferrari, 1977b) so that $\Delta T = 40$ km and $r_m = 1658$ km, and taking $\rho_c = 3.0$ g cm^{-3} and $\rho_m = 3.4$ g cm^{-3} yields $\Delta C = 2.1$ km. Considering the simplicity of the model and uncertainties in the offset determination, the agreement is reasonable. Although alternate hypotheses involving, for example, differing nearside and farside mantle compositions (Wasson and Warren, 1980) cannot be eliminated, the crustal explanation is consistent with available seismic, gravity, and topography data. A nearside/farside crustal thickness asymmetry further helps to explain the predominance of the nearside maria by preferential flooding of thin-crusted, low-lying regions and may have originated from uneven bombardment of the proto-Moon under the influence of a nearby Earth (Wood, 1973; but see also Kobrick, 1976).

3. Seismic Constraints on Mantle Structure and Composition

Derivation of accurate limits on lunar mantle seismic velocities has been hindered by the small number and areal distribution of stations, by the finite available data set, and by a low signal-to-noise ratio as previously noted. Nevertheless, any such limits would represent a primary existing means of establishing the structure of the mantle, including providing evidence for possible changes in composition, mineralogical phase, and density. Determination of crustal and upper mantle velocities is largely controlled by data from near-surface sources including artificial and natural impacts and shallow moonquakes while determination of velocities at greater depths depends mainly on data from deep moonquake sources. The focal depths of the latter range from 800 to 1100 km and impose a basic limit on the maximum depth to which velocities are usefully determinable.

Early attempts to invert the limited arrival time data then available for the purpose of constructing seismic velocity profiles for the mantle showed that compressional (P) and shear (S) wave velocities at the top of the mantle were near 8 and 4.5 km s^{-1}, respectively, with evidence for a gradual decrease at greater depths (Lammlein et al., 1974; Toksöz et al., 1974; Nakamura et al., 1974). Nakamura et al. further demonstrated that these velocities are compatible with an upper mantle composition consisting largely of olivine and pyroxene with Mg/(Mg + Fe) ratios near 0.70–0.85, an inference consistent with petrologic data indicating ratios near 0.75–0.80 for mare basalt source regions at comparable depths (Ringwood and Essene, 1970; Morgan et al., 1978; Ringwood, 1979).

The seismic attenuation structure of the lunar mantle was established at an early stage although important refinements have occurred more recently. The inverse dissipation factor Q is in the range 4000–7000 for both P- and S-waves in the upper mantle, decreasing to values near 1500 in the middle mantle (500–1000 km depth), and decreasing more markedly at greater depths to values of the order of 100 or less for S-waves (Latham et al., 1970; Lammlein et al., 1974; Toksöz et al., 1974; Nakamura et al., 1976; Dainty et al., 1976; Nakamura and Koyama, 1982). The extremely large Q-values for the upper mantle approach those of high-quality ceramics and are indicative of temperatures well below the solidus, absence of volatiles, and possibly the importance of additional contributions to Q by processes such as stress corrosion cracking (Schreiber, 1977; Tittman et al., 1980; Spetzler et al., 1980). In general, the decrease in Q in the middle mantle may reflect increasing volatile content and/or temperatures while the major increase in S-wave attenuation below depths of 1000–1100 km may imply a closer approach of the selenotherm to the solidus and the occurrence of partial melting (Nakamura et al., 1973; Dainty et al., 1976; Goins et al., 1981c).

Predictably, establishment of the variation of seismic velocities with depth in the mantle has proved to be the most challenging problem of lunar seismic data investigations. As discussed by Cleary (1982), details of the shear wave velocity profile in the terrestrial mantle are still debatable so it would be surprising if lunar mantle velocities have been accurately determined, particularly when the small number of lunar surface stations and the low signal-to-noise ratio of lunar seismic records are considered. Based on analyses of a portion of the final arrival time data set, numerous suggestions of discontinuous S-wave velocity decreases at depths ranging from 300 to 500 km and possible identifications of reflected phases from these velocity discontinuities were reported (Nakamura et al., 1974, 1976; Dainty et al., 1976; Goins, 1978; Goins et al., 1981c). However, the identification of reflected phases remains somewhat subjective despite the application of polarization filtering signal processing techniques (see Fig. 3 of Goins et al., 1981c). In addition, the significance of derived S-wave velocity decreases based on the limited data sets that were analyzed is questionable (Cleary, 1982). Nevertheless, the Goins (1978) model, shown in Fig. 3a, represents a possible mantle velocity profile based on

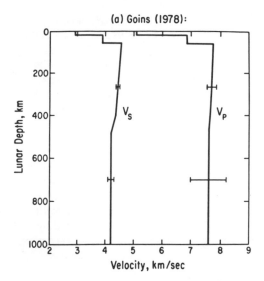

(a) Goins (1978):

Fig. 3a. Mantle P- and S-wave velocity model for the lunar mantle derived by Goins (1978) from a subset of the final Apollo seismic network arrival time data set. Brackets are taken from Buck and Toksöz (1980), who suggest that these limits represent maximum possible variations in velocity at a given depth for the Goins model. The S-wave velocity discontinuities at 400 and 480 km depth depend on identifications of reflected phases in polarization-filtered seismograms.

a subset of the final data and has been applied on several occasions to constrain models of lunar bulk composition and differentiation (Basaltic Volcanism Study Project, 1981; Buck and Toksöz, 1980; Taylor, 1978).

The most recent mantle seismic velocity model is based on the complete five-year data set acquired when the four Apollo seismometers were simultaneously operative (Nakamura, 1983). Signals were analyzed from 41 independent deep moonquake sources, as compared to 24 deep moonquake sources used by Goins (1978) and Goins *et al.* (1981c) and 11 sources used by Nakamura *et al.* (1976). To minimize subjectivity in selection of arrival times, averages of times read by up to four different analysts were used and a linearized least-squares matrix inversion procedure was applied in steps to obtain the most probable velocity structure. It was concluded that no more than a three-layer mantle velocity model is justified by data resolution. The result is shown in Fig. 3b with one standard deviation formal error limits resulting mainly from the spread in arrival time readings. As emphasized by Nakamura, the constant-velocity zones are approximations used for computational convenience and the discontinuities occurring at 270 and 500 km are not necessarily real. In particular, a gradual velocity decrease in the upper mantle is more probable with estimated gradients of -0.0013 ± 0.0013 s^{-1} for P-waves and -0.0011 ± 0.0005 s^{-1} for S-waves. Possible systematic errors due for example to lateral heterogeneities between the nearby Apollo 12 and 14 stations have also been discussed by Nakamura, who stresses that these uncertainties must be considered in interpreting the results.

In comparing Figs. 3a and 3b, it can be seen that both P- and S-wave velocities in the upper mantle are in substantial agreement, although the negative upper mantle

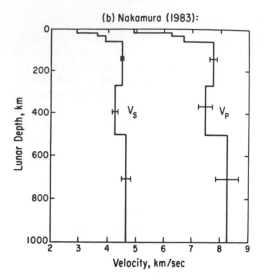

(b) Nakamura (1983):

Fig. 3b. P- and S-wave velocity model for the lunar mantle derived by Nakamura (1983) using a nearly complete arrival time data set, including significantly more deep moonquake signals than were available to Goins (1978). Constant-velocity zones were employed for computational convenience in the inversion procedure; actual velocities may gradually change within zones and discontinuities in velocity are not necessarily real. One standard deviation error limits resulting mainly from the spread in arrival time readings are indicated.

velocity gradients are larger in Nakamura's model than in Goins' model. Apparently because of the expanded deep moonquake data set, the major difference between the two models occurs in the middle mantle (500–1000 km depth), where Nakamura's model is characterized by velocity increases while Goins' model is characterized by velocity decreases. The velocity increases of Nakamura's model are significant at the one standard deviation level (formal errors only) meaning that the statistical probability that velocities are within the derived range is about 65%. As concluded by Nakamura (1983), the derived velocity increases would imply either an increase in Mg content below 500 km or a phase change, presumably involving the transition from spinel- to garnet-bearing assemblages. Because of the more complete arrival time data set, the Nakamura limits represent the most probable available model of lunar mantle seismic velocity structure, although significant uncertainties in this model still exist.

In order to examine in more detail the interpretation of the velocity models of Figs. 3a and 3b, available laboratory measurements for relevant minerals must be extrapolated to the approximate temperature and pressure regime of the lunar mantle. Assuming that at least the upper mantle ($\leq$ 400–500 km depth) was differentiated to produce the highland crust, aluminum-bearing phases such as plagioclase and spinel will be minor and assemblages in this zone will consist dominantly of olivines and pyroxenes (e.g., Basaltic Volcanism Study Project, 1981). Considering the range of allowed mantle temperatures (Fig. 7) and the stability fields of appropriate assemblages (Green and Ringwood, 1967), the spinel-to-garnet phase transition may occur at depths as shallow as 400 km (~20 kbar) or as deep as 550 km (~25 kbar). For our purposes, we assume that the middle and lower mantle lie entirely within the garnet stability field. If this zone is fully differentiated, then the Al_2O_3 content may be substantially less than that of the bulk Moon. However, if this zone is primordial, the composition will be essentially the same as the bulk composition so that larger Al and garnet abundances would be allowed.

A compilation of relevant laboratory measurements for olivine, orthopyroxene, and garnet is given in Table 1. For simplicity, only shear wave velocities and their temperature and pressure derivatives are considered because the error limits of Figs.

TABLE 1. *Shear Wave Velocities, Pressure, and Temperature Derivatives of Velocities for Olivine, Pyroxene, and Garnet at Standard Temperature and Pressure.*

	Mg/(Mg + Fe) mol %	V_s km s^{-1}	$\partial V_s/\partial P$ km s^{-1} kb^{-1}	$\partial V_s/\partial T$ km s^{-1} °C^{-1}
Olivine	100	4.95[a]		
	90	4.80		
	80	4.64	3.60×10^{-3}[b]	-3.40×10^{-4}[b]
	70	4.47		
Pyroxene	100	4.93[c]		
	90	4.82		
	80	4.70	4.33×10^{-3}[d]	-3.39×10^{-4}[e]
	70	4.59		
Garnet	100	(5.4)[f]		
	66	(5.0)	2.20×10^{-3}[f,g]	-2.17×10^{-4}[g]
	22	4.76		
	0	(4.6)		

(a) Chung (1971); (b) Kumazawa and Anderson (1969); (c) Chung, unpublished; Nakamura et al. (1974); Kumazawa (1969); (d) Birch (1969); (e) Birch (1943); (f) Wang and Simmons (1974); (g) Simmons and Wang (1971).
Values in parentheses are estimated values.

3a and 3b are generally smaller for S-waves than for P-waves. In the case of olivines and pyroxenes, velocities are listed for compositions ranging from Fo_{100} to Fo_{70} and from En_{100} to En_{70} (see Buck and Toksöz, 1980, for a similar tabulation). In the case of garnet, only velocities for the pyrope $(Mg_3Al_2(SiO_4)_3)$ and almandine $(Fe_3Al_2(SiO_4)_3)$ end members and combinations thereof are listed. Extrapolations of the values of Table 1 to the lunar interior are obtained using a constant-density pressure model and a nominal thermal history model temperature profile (Toksöz et al., 1978) to allow one-to-one comparisons with earlier work (Basaltic Volcanism Study Project, 1981; Buck and Toksöz, 1980). Results are shown graphically with the S-wave velocity model of Goins (1978) in Fig. 4a and with the S-wave velocity limits of Nakamura (1983) in Fig. 4b.

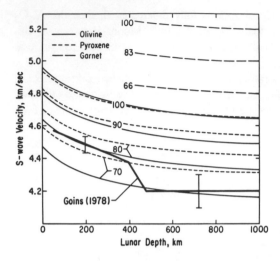

Fig. 4a. Superposition of the S-wave velocity limits of Fig. 3a onto theoretical curves representing extrapolations of the laboratory data of Table 1 to the pressures and nominal temperatures of the lunar interior.

In the case of the Goins model (Fig. 4a), it is evident that an olivine/pyroxene composition with Mg/(Mg + Fe) ratios in the range 0.70–0.80 is preferred in the upper mantle. The velocity decrease below 400–480 km depth can in principle be explained as due to an increase in FeO content of mafic silicates, possibly suggesting a transition to a primitive undifferentiated interior (Dainty *et al.*, 1976; Taylor, 1978). An alternate possibility suggested by Nakamura *et al.* (1976) and Buck and Toksöz (1980) is the presence of free Fe-FeS metal, which would also tend to lower the seismic velocities. However, it is notable that the velocity limits shown for the upper and lower mantle are nearly consistent with a constant-composition model, especially when uncertainties in the mantle temperature profile (and hence in the theoretical velocity curves) are considered. Thus the primary evidence for a change in mantle material properties and/or composition rests on the interpretation of a velocity transition zone, which comes from the identifications of mantle reflected phases. Uncertainties in these identifications have been mentioned above.

In the case of the Nakamura model (Fig. 4b), upper mantle compositions consisting of olivines and pyroxenes with Mg/(Mg + Fe) ratios in the range 0.65–0.80 are indicated. The velocity decrease in the two-layer upper mantle velocity model may indicate increasing FeO content with depth, although the effect of increasing temperature again cannot be eliminated because of uncertainties in the upper mantle temperature gradient. In agreement with the conclusion of Nakamura (1983), the inferred velocity increase in the middle mantle (500–1000 km depth) can be most directly explained as a consequence of increased MgO abundance and/or the presence of garnet. According to Fig. 4b, if the former is true, then the preferred Mg/(Mg + Fe) ratio for the middle mantle is in the range 0.85–1.0, implying a bulk mantle Mg/(Mg + Fe) ratio significantly larger than that of the lunar upper mantle.

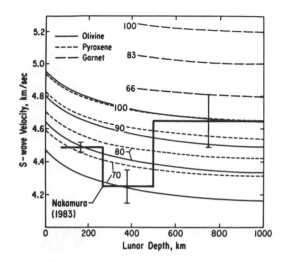

Fig. 4b. Same format as Fig. 4a with the S-wave velocity model of Fig. 3b superposed.

In order to estimate the volume fraction of garnet required to match the middle mantle S-wave velocity limits of Nakamura, the Mg/(Mg + Fe) ratio of pyrope-almandine garnet in equilibrium with olivine and pyroxene with given Mg/(Mg + Fe) ratios at appropriate temperatures and pressures must first be specified. From the data and model of Oka (1978; see Fig. 28 of Deer *et al.*, 1982), the Fe-Mg partition coefficient between olivine and garnet is in the range 1.5 to 2 at pressures of 25–35 kbar and temperatures of 1000–1200°C, i.e., approximately equivalent to the regime of the lunar middle mantle. If a bulk Moon Mg/(Mg + Fe) ratio of 0.80 is assumed to be appropriate for the middle mantle (assumed to be undifferentiated) and if olivine is the major phase, then the Mg/(Mg + Fe) ratio of pyrope-almandine garnet in equilibrium with such an olivine would be in the range 0.67–0.73. From Fig. 4b, the S-wave velocity of pyrope-almandine in these proportions would be 4.82–4.88 km s^{-1} while that of Fo$_{80}$ is near 4.35 km s^{-1} at 750 km depth. Thus a minimum of ~26 vol % or 28 wt % garnet would be required to match Nakamura's one standard deviation lower limit of 4.49 km s^{-1} for this depth range. (Simple weighted averaging is used to calculate the net S-wave velocity.) Calculations of normative mineral assemblages using measured Al$_2$O$_3$ solubilities in orthopyroxene coexisting with garnet (Akella, 1976) for a range of possible compositions (e.g., Table 4.5.7 of Basaltic Volcanism Study Project, 1981) indicate that approximately 9–10 wt % Al$_2$O$_3$ would be needed to supply 28 wt % pyrope-almandine garnet in the lunar middle mantle. If a significant fraction of En$_{80}$ is also present, the garnet abundance required to match the seismic velocity lower limit is still near 22 wt %, corresponding to an Al abundance of 7–8 wt %.

A range of 7–10 wt % for the Al_2O_3 abundance below 500 km depth is higher than most recent estimated limits on the bulk lunar Al_2O_3 abundance. If the crust contains 17 to 25 wt % Al_2O_3, corresponding to mean compositions ranging from gabbro to anorthositic gabbro, and the crustal mean density and thickness are 3.0 g cm^{-3} and 70 km, respectively (Section 2), then the crust contributes between 1.8 and 2.6 wt % Al_2O_3 to the bulk Moon. Allowing for some inefficiency in the removal of aluminous phases from the differentiated zone, the mean Al_2O_3 content at depths between the base of the crust and the top of the primordial interior may be estimated as ~1 wt % (Taylor and Bence, 1975; Kaula, 1977). For whole-Moon differentiation, the bulk lunar Al_2O_3 content would then be between 2.7 and 3.5 wt %. For comparison, the CI chondritic abundance is 2.4 wt % Al_2O_3, while the primitive terrestrial mantle abundance has been estimated as 3.3 wt % Al_2O_3 (e.g., Anders, 1976). If initial melting and differentiation extended only to 500 km depth and homogeneous accretion is assumed, then the bulk Al_2O_3 content would be between 3.8 and 5.2 wt %. The latter range compares to estimates of 5.0 wt % Al_2O_3 by Kaula (1977), who assumed initial differentiation to 200–300 km depth, and ≤6.0 wt % Al_2O_3 by Taylor and Bence (1975), who considered a more aluminous crust and differentiation to 1000 km depth. However, current geochemical arguments based in part on the volume of parental material needed to produce the observed concentration of elements in the crust are more consistent with whole-Moon differentiation (Taylor, 1982). On this basis, it is tentatively concluded that an explanation for the Nakamura (1983) S-wave velocity increase below 500 km solely in terms of a transition to garnet-bearing assemblages would require an Al_2O_3 abundance that is larger than predicted by current bulk composition models. A transition to more MgO-rich mafic silicates in addition to (or instead of) garnet therefore appears to be required.

4. Thermal State

4.1. Surface heat flow data

Heat flow probe measurements were successfully obtained at only two of the Apollo landing sites, Apollo 15 and 17, yielding final estimates of 21 and 16 mW/m^2, respectively (Langseth et al., 1976). Derivation of globally representative averages from these isolated measurements is necessarily difficult and values ranging from 18 mW/m^2 (Langseth et al., 1976) to 11 mW/m^2 (Rasmussen and Warren, 1985) have been proposed. The former authors extrapolated the observed values to compute global averages using orbital measurements of surface thorium abundances and inferred crustal thicknesses. The latter authors point out that the Apollo 15 and 17 sites were near the edges of maria where megaregolith thickness is likely to be unusually shallow, leading to anomalously high thermal conductivities and heat flow values (see also Warren, 1982; Conel and Morton, 1975). Correcting for the effect of locally thin megaregolith, they derived substantially lower global heat flow averages.

Assuming nominal radioactive element ratios and a steady-state balance between heat production and loss, mean surface rates of 11 and 18 mW/m^2 would imply bulk Moon uranium abundances of 29 and 46 ppb, respectively, compared to 14 ppb for Cl chondrites and 18 ppb for the bulk Earth. However, the concept of a steady-state balance between heat production and loss may not be appropriate for planetary-sized objects in which the effect of secular cooling can be important (Schubert et al., 1979), so that the above global heat flow estimates may be consistent with a bulk uranium abundance of as little as 20 and 35 ppb, respectively (Rasmussen and Warren, 1985). The larger of these two bulk Moon uranium abundances is significantly larger than terrestrial and chondritic values, supporting the view that the Moon is generally enriched in refractory elements (Kaula, 1977; Taylor, 1982, 1984). However, the lower of these abundances would not imply a significant lunar enrichment. Thus establishment of the true globally averaged lunar heat flow value and further theoretical investigation of the relationship between surface heat flux and heat source content of bodies with thick lithospheres is needed before final conclusions may be drawn. Although Keihm and Langseth (1977) estimated temperatures near 300 km depth in the range 800–1100°C using the steady-state balance assumption and a global heat flow of 18 mW/m^2, actual uncertainties are larger than this for the reasons discussed above.

4.2. Seismic/gravity data

A qualitative constraint on the selenotherm in the upper mantle is the high average seismic Q-values (4000–7000; Nakamura and Koyama, 1982), implying temperatures well below the solidus at these depths. In addition, the apparent maintenance of mascon anisostasy over 3–4 aeons is consistent with relatively cool upper mantle selenotherms (but see also Solomon and Head, 1979). An attempt to quantify the mascon maintenance constraint was made by Pullan and Lambeck (1980), who applied laboratory-determined creep laws and calculated stress differences for the interior beneath the mascons to estimate an upper limit of about 800°C on the selenotherm at 300 km depth. We note that this limit is only marginally consistent with the range estimated by Keihm and Langseth.

In the middle mantle, Q-values > 1000 together with the locations of deep moonquake foci between 800 and 1100 km (implying the accumulation of tidal stresses and the existence of large-scale heterogeneities) are again most consistent with temperatures well below the solidus, although the reduced Q may indicate increased temperatures. Finally, the possible sharp increase in shear wave attenuation below ~1100 km where deep moonquake sources either cease to exist or are undetectable would be consistent with temperatures closely approaching or exceeding the solidus in the deep interior. However, it should be noted that current views on the origin of the reduced Q in the terrestrial low-velocity zone (Anderson, 1981) indicate that the latter implication is not necessarily required.

4.3. Electromagnetic sounding data

In principle, inferred bounds on the mantle electrical conductivity profile combined with laboratory measurements of conductivity vs. temperature for relevant mineral assemblages provide an alternate means of deducing limits on the selenotherm as a function of depth. Current limitations of this approach include (1) uncertainties in lunar mantle composition, (2) lack of sufficiently complete laboratory measurements of electrical conductivity for relevant mineral compositions, (3) uncertainties in how physical properties (e.g., fabric, grain boundaries, intracrystalline defects) affect the conductivity at depth in the Moon; and (4) uncertainties and nonuniqueness in the determination of the mantle electrical conductivity profile.

Determination of electrical conductivity as a function of depth in the Moon requires simultaneous measurements of external magnetic fields far from the Moon (forcing fields) and magnetic fields at or near the lunar surface (forcing fields plus induced fields) as a function of time (or frequency). In addition, the plasma environment experienced by the Moon during the period of observation must be carefully considered in deriving an appropriate theoretical model for interpretation of the measurements [see Sonett (1982) for a general review]. A summary of estimates and limits on electrical conductivity as a function of lunar depth is presented in Fig. 5. The dashed lines show limits estimated by Dyal *et al.* (1976) using classical vacuum induction theory from an analysis of a large six-hour geomagnetic tail field transient event. Possible error sources include the occurrence of finite plasma densities in the tail lobes, although these densities appear to be of minimal significance for this event (Hood and Schubert, 1978), and questions of nonuniqueness. The remaining estimates and limits shown were obtained from analyses of more lengthy (30–70 hours) records

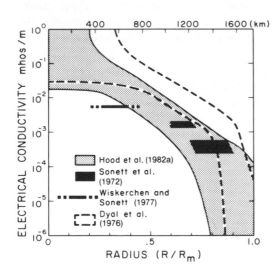

Fig. 5. Summary of estimated limits on electrical conductivity as a function of depth in the lunar interior derived using simultaneous surface and orbital magnetometer data.

when the Moon was immersed in the supermagnetosonic solar wind or terrestrial magnetosheath plasma so that induced magnetic fields could be assumed to be confined to the lunar interior and the downstream cavity. In practice, a spherically symmetric theory in which plasma confinement is assumed to be both perfect and to cover the entire Moon, i.e., no downstream cavity is present, has been used to interpret the measurements. The latter is justified for measurements obtained on the sunward hemisphere at sufficiently low frequencies (see Fig. 10 of Schubert et al., 1973). Possible remaining error sources include possible incompleteness of external plasma confinement of induced fields and questions of nonuniqueness. For example, in the work of Hood et al. (1982a,b), forward modeling and Monte Carlo calculations were used to define the conductivity bounds under the assumption that the conductivity increases continuously or discontinuously with increasing depth. Hence conductivity decreases and highly conducting layers such as that originally proposed by Sonett et al. (1971) were not considered. As can be seen in the figure, general bounds on lunar mantle electrical conductivity derived under the stated assumptions via each major electromagnetic sounding approach are in reasonable (order of magnitude) agreement and show that the conductivity rises from 10^{-4}–10^{-3} S m^{-1} at depths of a few hundred km to 10^{-2}–10^{-1} S m^{-1} at depths of 1000–1200 km. In the following, the conductivity limits of Hood et al. (1982a), which were derived from more lengthy data records than those of Dyal et al. (1976), will be applied to estimate temperature limits. However, it is emphasized that the reason for the amplitude difference between the electrical conductivity limits of Dyal et al. and those of Hood et al. has not yet been quantitatively resolved.

A more detailed review of laboratory mineral electrical conductivity measurements and their application to limit lunar internal temperatures can be found in Sonett (1982). Briefly, semiconduction in minerals is exponentially dependent on reciprocal temperature and is approximately expressible in terms of an activation energy E_n as

$$\sigma = \sum_{n=1}^{3} \sigma_n \exp(-E_n/kT) \tag{2}$$

where the three terms, n = 1,2,3, represent electronic, ionic, and impurity conduction, respectively, k is Boltzmann's constant, T is temperature in Kelvin, and σ_n is a nearly constant, slowly varying function of temperature. The work of Huebner et al. (1978, 1979) and Duba et al. (1979) indicates the importance of the impurity or trace element term. Specifically, the presence of small concentrations of trivalent cations, particularly Al^{+3}, can significantly increase the orthopyroxene conductivity at a given temperature (Fig. 6). Also, the conductivity of olivine is a function of Mg/(Mg + Fe) ratio and varies by at least five orders of magnitude between forsterite and fayalite at 400°C (Tolland and Strens, 1972; Bradley et al., 1964). Thus the bulk composition parameters to which the interpretation of electrical conductivity in terms of temperature is most sensitive, namely Mg/(Mg + Fe) ratio and Al_2O_3

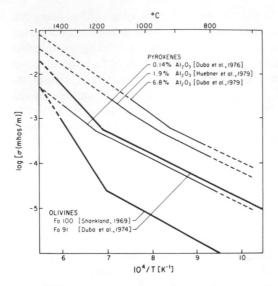

Fig. 6. Schematic summary of some representative laboratory measurements of electrical conductivity vs. temperature for orthopyroxenes and olivines. The continuous lines represent smoothed means of direct measurements and the dashed extensions are linear extrapolations to higher and lower temperatures. The pyroxene conductivity at a given temperature increases with Al_2O_3 content while that of olivine increases with FeO content (Hood et al., 1982b).

abundance, are the same parameters that are most critical in the interpretation of mantle seismic velocities.

Three examples of the translation of lunar electrical conductivity profiles into temperature profiles that illustrate current uncertainties arising from composition models alone are shown in Fig. 7. The electrical conductivity profiles were calculated to be in agreement with the inductive response data of Hood *et al.* (1982a, dark shaded area in Fig. 5) and were converted into temperature profiles using the laboratory data of Fig. 6. Although only one solidus (that of Ringwood and Essene) is plotted,

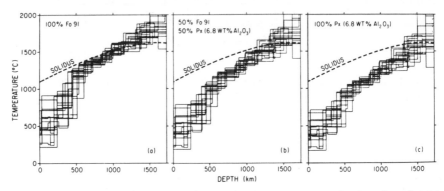

Fig. 7. Possible lunar temperature profiles (selenotherms) appropriate for the indicated composition calculated by Hood et al. (1982b) using electrical conductivity profiles that are consistent with the inductive response data of Hood et al. (1982a) and with the mineral conductivity vs. temperature relations of Fig. 6.

other solidus profiles appropriate for different model compositions are also possible (e.g., Hodges and Kushiro, 1974; Green and Ringwood, 1967), so that more realistic uncertainties in the solidus at a given depth are of the order of 100°C. It can be seen from Fig. 7 that the addition of aluminous orthopyroxenes to a given model composition results in a significant lowering of inferred temperatures.

Prior to the availability of the aluminous orthopyroxene data, several studies derived relatively high temperature estimates for the upper mantle based on electrical conductivity determinations for nonaluminous orthopyroxenes and/or olivine (Dyal et al., 1974; Sonett and Duba, 1975; Duba et al., 1976). These estimates were within several hundred degrees Celsius of the anhydrous basalt solidus of Ringwood and Essene (1970) (e.g., Fig. 7a) and were not easily reconciled with independent constraints on the shallow selenotherm, including maintenance of mascon anisostasy for the past 3–4 aeons (Sonett and Duba, 1975; Pullan and Lambeck, 1980) and the high inferred seismic Q-values for this depth range. Later studies showed that consideration of laboratory data for more aluminous orthopyroxenes allowed lower upper mantle temperatures that were more in accord with independent constraints (Huebner et al., 1979; Hood and Sonett, 1982; Hood et al., 1982b). However, the interpretations of the Nakamura seismic velocity model discussed in Section 3 indicate that more probable compositional models for the mantle below 500 km depth contain Mg-rich olivine and pyroxene rather than aluminous orthopyroxene. On the other hand, relatively small concentrations of trace oxides (1–2% Al_2O_3, Cr_2O_3) have been shown to result in major increases in conductivity of pyroxenes (En_{86}) at temperatures representative of the lunar mantle (Huebner et al., 1979). Consequently, more accurate bulk conductivity models based on more detailed compositional models and a more complete set of laboratory measurements of electrical conductivity vs. temperature may be required before lunar electrical conductivity data can be effectively applied to limit the form of the selenotherm. Although some attempts have been made to compare available limits with the predictions of thermal history models (e.g., Hood et al., 1982b), these comparisons are somewhat premature and final conclusions will not be possible until compositional uncertainties and their effect on mineral electrical conductivity have been minimized.

5. Density Models

5.1. Moment of inertia

The lunar mean density has been known with reasonable precision for many decades (e.g., Kopal, 1967), but accurate determination of the moment of inertia required a relatively precise evaluation of low-degree gravity coefficients that did not occur until the middle and late 1970's. Calculation of the moment of inertia requires independent estimates for at least three of the following parameters:

$$\beta = (C - A)/B \qquad (3)$$

$$\gamma = (B - A)/C \tag{4}$$

$$C_{20} = [(A + B)/2 - C]/MR^2 \tag{5}$$

$$C_{22} = (B - A)/4MR^2 \tag{6}$$

where β and γ are physical libration parameters, C_{20} and C_{22} are second-degree lunar gravity coefficients, A, B, and C are principal lunar moments of inertia in order of increasing amplitude, and M and R are the lunar mass and mean radius. Approximate values for β and γ were known prior to the Apollo missions from astronomical data but were refined, especially in the case of γ, to high accuracy from early analyses of laser ranging data (e.g., Williams *et al.*, 1974). Initial evaluations of C_{20} and C_{22} using relatively low-altitude orbiter tracking data were of reduced accuracy because of limited data coverage and effects of gravitational potential series truncation (Gapcynski *et al.*, 1975). More accurate values for these coefficients were obtained from analyses of long-term variations of periapsis and node for the high orbiting Explorer 35 and 49 spacecraft yielding a value for C/MR^2 of 0.391 ± 0.002 (Blackshear and Gapcynski, 1977). A later combined analysis of laser ranging data and Doppler tracking data from the high orbiting Lunar Orbiter 4 yielded a value of 0.3905 ± 0.0023 as well as an improved value for the mean density of 3.344 ± 0.003 (Ferrari *et al.*, 1980). The congruency of these results using independent, optimally selected data sets increases the likelihood that the true moment of inertia value lies within the error limits of these two determinations.

5.2. Model calculations

Values of the dimensionless moment of inertia between 0.388 and 0.393 require some form of density increase with depth in the Moon, but the form of the density increase is not uniquely determinable. In particular, in the absence of other information, models with and without dense metallic cores are equally plausible. To illustrate this point, it is instructive to consider a Moon model with N constant density layers so that mean density and moment of inertia may be written (e.g., Bills and Ferrari, 1977b)

$$\bar{\rho} = \sum_{k=1}^{N} \rho_k (x_k^3 - x_{k+1}^3) \tag{7}$$

$$I/MR^2 = \frac{0.4}{\bar{\rho}} \sum_{k=1}^{N} \rho_k(x_k^5 - x_{k+1}^5) \tag{8}$$

where ρ_k is the density of the kth layer and x_k is the ratio of the radius of the kth layer to the mean lunar radius R, beginning with the outermost layer

$(x_1 \equiv 1; x_N \equiv 0)$. For the special case of a three-layer model, the unknowns are ρ_1, ρ_2, ρ_3, x_2, and x_3. If it is assumed that the density ρ_1 and thickness $1-x_2$ of the outer layer (i.e., the mean crustal density and thickness) are known, and ρ_2 is eliminated between equations (7) and (8), one obtains a single equation relating the remaining unknowns ρ_3 and x_3:

$$[(\bar{\rho} \, I/0.4MR^2) - \rho_1 (1 - x_2^5) - \rho_3 x_3^5] (x_2^3 - x_3^3) - [\bar{\rho} - \rho_1 (1 - x_2^3) - \rho_3 x_3^3] (x_2^5 - x_3^5) = 0 \quad (9)$$

Thus even if mean crustal thickness and density are given and density changes with depth in individual layers are neglected, the radius and density of any central dense core cannot both be evaluated. For example, as shown in Fig. 8, a model with about a 10% density increase in the mantle can satisfy equation (9) as well as models with dense Fe or FeS cores.

It is evident therefore that application of the moment of inertia constraint to impose meaningful limits on the existence and mass of a possible metallic core requires limits on the variation of density in the mantle. Radial density variations in the mantle will in general occur from the effects of hydrostatic compression, thermal expansion, and changes in composition or mineralogic phase. However, in the Moon, the compression effect is approximately offset by thermal expansion so that the most important cause of radial density variations is composition/phase changes. In the upper mantle (≤ 500 km depth), composition changes can be partly constrained by petrologic studies of mare basalts, which may have originated from depths as great as 400–500 km. However, at greater depths, changes in composition or phase and therefore density are constrained almost entirely by seismic velocity limits and their interpretation.

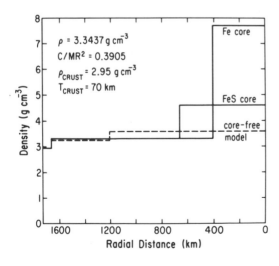

Fig. 8. Three-layer lunar density models calculated from equation (9) for the indicated parameters. Models with small (~10%) mantle density increases can satisfy the moment of inertia constraint as well as models with dense metallic cores.

Earlier efforts to construct lunar density models consistent with available geophysical and petrologic data were made using the Goins (1978) seismic velocity model (Buck and Toksöz, 1980; Binder, 1980). Binder used the Goins model as a qualitative guide for limiting mantle composition changes and considered only stepwise density increases of ≤ 0.2 g cm^{-3} at 850 km depth. Under this restriction and using a nominal crustal and upper mantle (≤ 250 km depth) density model, Fe cores with radii between 250 and 400 km were needed to produce moment of inertia factors between 0.390 and 0.393. However, the search for allowed mantle composition models was not exhaustive and it is possible that models exist for which no metallic core is required. For example, a larger density increase of 0.4 g cm^{-3} at 850 km depth negates the requirement for a core and the possibility of density increases at other depths was not considered. In the work of Buck and Toksöz (1980; see also Basaltic Volcanism Study Project, 1981), mantle composition models consistent with the Goins seismic velocity model were more explicitly calculated. Model compositions were converted to mineral abundances using a temperature- and pressure-dependent normative scheme and seismic velocities were calculated from laboratory data such as that of Table 1 for comparison to the Goins model. Compositions were then iterated upon to obtain models in agreement with all geophysical constraints, including the seismic velocity limits. Their final preferred density model (Fig. 9) is characterized by a small density increase at 400 km depth and the presence of an assumed FeS core with radius $\lesssim 300$ km. The core amounts to 1.2% of the lunar mass and is therefore poorly determined. Both the mantle density increase and the requirement for a metallic core are dependent on several assumptions driven in part by the seismic velocity model. First, it was assumed that initial melting and differentiation extended only to 400 km depth based on the seismic velocity change at this depth (Fig. 4a) and on the thermal contraction argument of Solomon (1977). Second, because the primitive interior composition was assumed to contain 5 wt % Al$_2$O$_3$, the presence of garnet

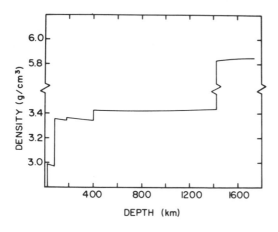

Fig. 9. Density profile calculated for a model composition that matches all geophysical constraints using the seismic velocity model of Goins (1978; Fig. 3a) and assuming initial differentiation of the lunar mantle to a depth of 400 km (Buck and Toksöz, 1980). The model composition includes several weight percent Fe-FeS metal at depths > 400 km in order to lower the seismic velocities to agree with the Goins model. The assumed FeS core represents approximately 1% of the lunar mass.

(~10 wt %) at depths below 400 km produced a small velocity increase in disagreement with the velocity decrease of Fig. 4a. In order to lower the seismic velocities, two methods were considered: (1) lowering the Mg/(Mg + Fe) ratio (Dainty et al., 1976; Goins, 1978), and (2) adding a few weight percent Fe-FeS metal to the primitive interior composition (Nakamura et al., 1976). The latter option was elected and resulted in a modest mantle density increase at 400 km depth. Clearly, the resulting density model and the requirement for a metallic core are strongly dependent on these assumptions.

In order to investigate further the range of density models consistent with the seismic velocity models of Goins (1978) and Nakamura (1983), the following constraints are adopted: (1) mean density and moment of inertia (ρ = 3.344 ± 0.003 g cm^{-3}; C/MR^2 = 0.3905 ± 0.0023; Ferrari et al., 1980), (2) approximate limits on crustal mean thickness and density (t$_{cr}$ = 60–85 km; ρ_{cr} = 2.90–3.05 g cm^{-3}; Section 2); and (3) mantle S-wave velocity limits of Nakamura (1983) (<270 km depth: v$_s$ = 4.49 ± 0.03 km s^{-1}; 270–500 km: v$_s$ = 4.25 ± 0.10 km s^{-1}; 500–1000 km: v$_s$ = 4.65 ± 0.16 km s^{-1}) or Goins (1978) interpolated to match the depth ranges of Nakamura (< 270 km: v$_s$ = 4.50 ± 0.05 km s^{-1}; 270–500 km: v$_s$ = 4.40 ± 0.05 km s^{-1}; 500–1000 km: v$_s$ = 4.20 ± 0.12 km s^{-1}). In the upper mantle (< 500 km depth), only olivine-pyroxene mixtures with arbitrary Mg/(Mg + Fe) ratios are considered, while in the middle and lower mantle the addition of garnet is allowed in amounts up to 20 wt % corresponding to a maximum bulk Al$_2$O$_3$ content of approximately 7 wt % (Section 3). No assumptions are made with respect to the initial depth of differentiation. Although the interpretation of Nakamura's velocity limits given in Section 3 indicates that a more Mg-rich middle mantle may be needed to match these limits, the possibility is retained here of a primitive interior at depths > 500 km, including a substantial garnet fraction in combination with increased Mg/(Mg + Fe) ratios for olivines and pyroxenes.

Following Buck and Toksöz (1980), density was calculated as a function of pressure and temperature at each integration step by solving the Murnaghan-Birch equation of state

$$P(\rho,T) = \frac{3K(T)}{2} \left[\left(\frac{\rho}{\rho_0(T)}\right)^{7/3} - \left(\frac{\rho}{\rho_0(T)}\right)^{5/3} \right] \tag{10}$$

where K(T) is temperature-dependent bulk modulus and ρ_0(T) is temperature-corrected zero-pressure density. Pressure was determined by integrating the hydrostatic equation and temperature was estimated using the thermal history model result of Toksöz et al. (1978). STP densities for olivine, orthopyroxene, and garnet were taken from Table 4.4.1 of Basaltic Volcanism Study Project (1981) while bulk modulus and thermal expansion coefficient values and their temperature derivatives were taken from Table 4.4.2 of the same reference. Because seismic velocity limits are determined only to ~1000 km depth, density profiles can be integrated to greater depths only via additional assumptions. Again following Buck and Toksöz (1980), the approach is taken here of (1) assuming no change in bulk composition between the middle and lower mantle, and (2) allowing for the possible presence of an Fe-rich core to meet the mean density and moment of inertia constraints. Results have been reported by Hood and Jones (1985) and are shown in Fig. 10 for a series of compositional models that match the seismic velocity limits of Fig. 4. The mantle density discontinuities are not necessarily real and may be replaced by more gradual density changes; they result from the assumed three-layer composition model, which originates from the three-layer velocity model of Nakamura.

In the case of the Goins (1978) S-wave velocity limits, reductions in the middle mantle Mg/(Mg + Fe) ratio combined with the allowed presence of up to 20 wt % garnet result in significant density increases at 500 km depth. These increases are sufficient in some cases to negate the requirement for a metallic core of either Fe or FeS composition, although Fe cores with radii ≤ 400 km are allowed. In the case of the Nakamura (1983) S-wave velocity limits, increases in the Mg/(Mg + Fe) ratio in the middle mantle needed to match the S-wave velocity increase

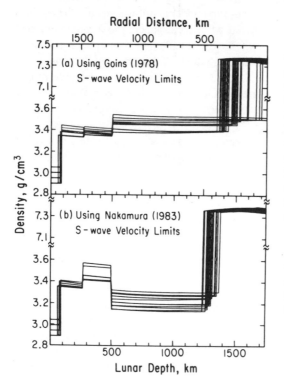

Fig. 10a. Density models for simplified mantle compositions consistent with the mantle S-wave velocity model of Fig. 3a. In the upper mantle (< 500 km depth), only olivine/pyroxene mixtures with arbitrary Mg/(Mg + Fe) ratios were considered while at greater depths the presence of garnet in amounts up to 20 wt % was allowed. No assumptions about the depth of initial differentiation were made.

Fig. 10b. Same format as Fig. 10a using the S-wave velocity model of Fig. 3b. See text for discussion.

invariably result in density decreases despite the allowed presence of ≤ 20% garnet. Such density decreases necessitate the existence of substantial Fe cores with radii between approximately 360 and 500 km to match the mean density and moment of inertia constraints.

In general, these results demonstrate the sensitivity of lunar bulk composition and density models to mantle seismic velocities and their interpretation. As discussed in Section 3, because of the more complete arrival time data set and method of analysis, the Nakamura (1983) seismic velocity model is presently the most probable available model. However, error limits are estimated by Nakamura only at the one standard deviation level (~65% confidence). Therefore, although the density models of Fig. 10b are the most probable set given the available seismic and gravity data, confidence limits remain low and alternate models such as those of Fig. 10a cannot be rigorously excluded. Nevertheless, it is interesting to note that the density inversion with depth of Fig. 10b due to the' presence of more Mg-rich olivine and pyroxene in the middle and lower mantle was predicted by early petrologic models that supposed initial whole-Moon differentiation. Both Wood *et al.* (1970) and Smith *et al.* (1970) emphasized that initially crystallizing Mg-rich olivine crystals will sink gravitationally through a less dense parental magma, leaving a more FeO-rich residuum at higher levels. Solidification of the latter would result in an inverse density stratification.

There is, however, a serious difficulty with the density models of Fig. 10b. At some point in early lunar history when mantle temperatures were higher than at present, any density inversion at depths as large as 500 km, whether abrupt or gradual, should have been relaxed by subsolidus convective overturn. According to Schubert *et al.* (1977), the critical Rayleigh number for the onset of convection in the lunar mantle for the case of a 300-km-thick lithosphere is $\sim 7 \times 10^3$. Adopting nominal values for the coefficient of thermal expansion $\alpha \sim 3 \times 10^{-5}$ K^{-1}, acceleration of gravity g $\sim 1.1 \times 10^2$ cm s^{-2}, thermal diffusivity $\kappa \sim 10^{-2}$ cm^2 s^{-1}, depth of the convecting layer d $\sim 1 \times 10^8$ cm, and temperature gradient $\beta \sim 8 \times 10^{-6}$ K cm^{-1} (Fig. 8), the corresponding minimum value of the kinematic viscosity v for which subsolidus convection is inhibited is found to be of the order of v_{crit} = $(\alpha\beta gd^4/\kappa Ra_{crit}) \sim 4 \times 10^{22}$ cm^2 s^{-1}. As discussed by Schubert *et al.* (1977), viscosities of this order are achieved at temperatures between about 1150° and 1250°C. According to Fig. 8, present-day temperatures at 500 km depth may easily be lower than this range so that subsolidus convection may presently be inhibited at these depths (Hood *et al.*, 1982b). But at earlier stages of lunar history, temperatures were higher and would have approached and exceeded the solidus at the time of lunar formation if initial melting and differentiation extended to 500 km. Therefore, any primordial density inversion would presumably have been relaxed. Consequently, it must be concluded that the density models of Fig. 10b are unlikely to be realized, implying inaccuracies in either the Nakamura seismic velocity limits or in the simplified modeling of those limits as described herein. Further work to more precisely determine the range of mantle composition models that are consistent with the Nakamura velocity limits is warranted.

According to the results reviewed in this section, lunar density models based partly on the moment of inertia constraint currently provide little reliable evidence for or against the existence of a dense metallic core. Although the models of Fig. 10b require such cores, they lead to a contradiction when the condition of stability against subsolidus convection throughout lunar history is imposed. Thus independent evidence for the presence of a metallic core remains highly desirable and would place an important lower boundary condition on lunar density models.

6. Independent Constraints on the Existence of a Metallic Core

6.1. Seismic P-wave arrival delay

Possible seismic evidence for a 170–360 km radius low-velocity core was tentatively suggested by Nakamura et al. (1974) from an early interpretation of a single P-wave arrival from a farside meteoroid impact. The event time and location were estimated from arrivals at the three closest stations so that the arrival time at the fourth station could be predicted. The estimated location was nearly antipodal to the fourth station and the observed arrival time was delayed by about 50 seconds. Velocity limits for the core were placed at 3.7–5.1 km s^{-1} suggesting a metallic composition (for reference, P-wave velocities for metallic Fe and Ni at STP are near 5.9 and 5.8 km s^{-1}, respectively). However, as discussed for example by Goins (1978), the particular event in question (occurring on day 262, 1973) was relatively weak with a correspondingly low signal-to-noise ratio. Independent measurements by Goins and others indicated that true onset times of the emergent P-waves at each of the four stations are ambiguous so that actual uncertainties in P-wave arrival times are sufficient to explain the apparent P-wave arrival delay at the fourth station. Similarly, a somewhat larger core radius of 400–450 km may also be accommodated by these uncertainties. Unfortunately, no additional detectable distant farside impacts occurred during the remainder of the five-year operation period of the seismic network (e.g., Goins et al., 1981c; Nakamura et al., 1982). Accordingly, this observation remains undefinitive. It follows that deductions based on a combination of this observation and the moment of inertia constraint (Levin, 1979) are also undefinitive.

6.2. Electromagnetic sounding data

Limits on electrical conductivity in the Moon derived from electromagnetic sounding data provide an alternate but equally challenging means of constraining the radius of a metallic core. (The electrical conductivity of metallic iron at central lunar temperatures is ~10^5 S m^{-1}.) Two primary techniques were developed and applied using Apollo surface and orbital magnetometer data. The first has been previously discussed in Section 4 and requires at least one orbiting and one surface magnetometer to define external and induced magnetic fields as a function of time. Inversion of the measurements via a suitable theoretical model then yields limits on the electrical conductivity as a function of lunar depth. Conductivity profile bounds derived from both geomagnetic tail transient data (Dyal *et al.*, 1976) and solar wind/magnetosheath field variation data (Sonett *et al.*, 1972; Wiskerchen and Sonett, 1977; Hood *et al.*, 1982a) have been summarized in Fig. 5. No evidence for a highly conducting core (conductivity $\lesssim 10$ S m^{-1}) was obtained in any of these studies. The most restrictive upper limit (360 km) was estimated by Hood *et al.* (1982a) from forward modeling calculations but this was later revised upward to 435 km by Hobbs *et al.* (1983) using inverse theory. Additional uncertainties resulting from accuracy of the theoretical model used to interpret the data have been discussed by Hood *et al.* (1982a). A further source of nonrandom error arises from the possibility of intercalibration or gain differences between the orbiting Explorer 35 (Ames Research Center) and Apollo 12 surface magnetometers. Daily and Dyal (1979) have shown that gains and offsets between the two instruments differed by as much as 1–2% and 0.5 nT, respectively, during the relevant observation period (the first four lunations of operation of the Apollo 12 magnetometer). The gain error is comparable to or less than the one standard deviation error in the induced response measurement (transfer function) at the lowest sampled frequency of 10^{-5} Hz. However, if the gain difference was characterized by a nonzero mean during the observation period, then a bias would be introduced, causing the upper limit on the core radius to be too large or too small. On this basis, considering the latter error source and referring to Fig. 4 of Hobbs *et al.* (1983), a more conservative upper bound for the metallic core radius from time-dependent induction studies is ~500 km.

The second approach toward lunar core detection via induced magnetic fields used a single orbiting magnetometer and was therefore not subject to intercalibration or gain difference errors. However, physical error sources and interpretational questions must still be considered. The objective of the analysis was to identify time periods when the Moon was in a near-vacuum plasma environment as occurs on occasion in the geomagnetic tail lobes [see Schubert and Lichtenstein (1974) for a general review of Moon-plasma interactions]. During such intervals, exposure of the Moon to a spatially uniform and temporally steady ambient magnetic field would result in an induced dipolar field with a moment oriented opposite to the applied field and with an amplitude that decays with time as the ambient field diffuses through the poorly conducting mantle. For the case of a constant conductivity sphere with radius a and conductivity σ, the dipole moment amplitude is

$$m_i/B_0 = -\frac{3a^3}{\pi^2} \sum_{n=1}^{\infty} \left[\exp\left(-n^2\pi^2 t/\beta\right)\right]/n^2 \tag{11}$$

where m_i is the induced moment, B_0 is the amplitude of the applied field, $\beta =$

$4\pi\sigma a^2/c^2$ (c is the speed of light; σ is in Gaussian units), t is time since the sphere was exposed to the ambient field, and the negative sign indicates that the moment vector is oriented opposite to the applied field. For example, if the mantle is approximated as an outer conducting layer with radius 1400 km and conductivity 10^{-3} S m^{-1} superposed on an inner conducting sphere of radius 600 km and conductivity 10^{-1} S m^{-1} (cf. Fig. 5), one finds that the negative induced magnetic field due to mantle conductivity alone would decay to negligibly small values (undetectable at the surface) in four to six hours. However, existence of a smaller core with substantially higher conductivity is allowed by the conductivity limits of Fig. 5. The induced moment due to a molten silicate core of radius 400 km and conductivity ~10 S m^{-1} would decay to less than half of its initial value only after a time of ~30 hours while that of a metallic core of 400 km radius and conductivity ~10^5 S m^{-1} would not decay significantly even after a period of 1000 hours. In the latter case, after all mantle currents have decayed, the negative induced moment would approach an asymptotic value, obtained by setting $\sigma = \infty$ in equation (11),

$$m_i/B_0 = -a^3/2 \tag{12}$$

For example, if a = 400 km, $m_i/B_0 = -3.2 \times 10^{22}$ Gauss-cm^3 per Gauss of applied field, while if a = 500 km, $m_i/B_0 = -6.25 \times 10^{22}$ Gauss-cm^3 per Gauss.

Measurements of the lunar induced magnetic moment in the geomagnetic tail using single orbiting magnetometers on the Apollo 15 and 16 subsatellites have been reported by Russell *et al.* (1974a,b), Goldstein *et al.* (1976), and Russell *et al.* (1981). As discussed by these authors, the major difficulty encountered in measuring induced moments that would be interpretable in terms of the simple theory outlined above is that the Moon is seldom in a quasi-vacuum environment and the geomagnetic tail field is seldom steady and spatially uniform (Fig. 11). Significant ambient plasma densities encountered for instance during passages through the magnetospheric plasma sheet diamagnetically reduce ambient field amplitudes producing an apparent positive lunar magnetic moment. To exclude such intervals, plasma measurements at the subsatellite, at the lunar surface, and inferred from magnetic field measurements in the tail were applied. Final editing eliminated all but 21 orbits (42 hours) of Apollo 15 and 16 subsatellite data from the original 7-month data set. A consistently negative induced moment with a final estimated value of $-4.23 \pm 0.64 \times 10^{22}$ Gauss-cm^3 per Gauss of applied field was determined (Russell *et al.*, 1981), corresponding to a highly conducting core radius of 439 ± 22 km. Since incomplete removal of intervals with significant plasma densities would only result in a less negative moment, it is unlikely that errors resulting from plasma diamagnetism alone can explain this observation.

Other error sources include the possibility of a significantly paramagnetic Moon (mean permeability > 1) and possible incomplete decay of mantle induced fields resulting from short-term changes in ambient field amplitude. The former error source

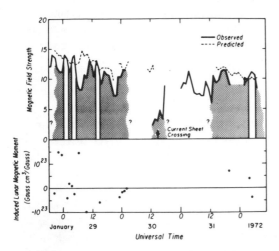

Fig. 11. Sample of lunar induced magnetic moment measurements in the geomagnetic tail obtained with the Apollo 15 subsatellite magnetometer during late January, 1972. The heavy solid line is the orbital average magnetic field strength. The dashed line is the predicted field strength based on hourly averages of the solar wind dynamic pressure. Intervals when the Moon is believed to be outside the plasma sheet are unshaded. Induced moment estimates are shown in the lower panel (Russell et al., 1981).

is unlikely to contribute since the permeability of a silicate lunar mantle would be ~1.001, a value that is consistent with but not strongly constrained by available magnetometer measurements (Daily and Dyal, 1979). Also, a lunar permeability substantially greater than unity would again only reduce the magnitude of the observed negative moment and so cannot explain it. The second error source also appears to be excludable as a cause of the observed mean moment although it is more difficult to eliminate completely. As shown in Fig. 11, there is an average tendency for ambient field amplitudes to increase slightly at times when the Moon is not in or near the plasma sheet (unshaded intervals); such a tendency is understandable because of the reduction of diamagnetic fields when the Moon exits regions with larger plasma densities. These increases are typically $\leqslant 10\%$ of ambient field strengths for the events shown in the figure and occur within approximately two hours of the times when negative induced moments are observed. Considering again a two-layer mantle conductivity model with central conductivity 10^{-1} S m^{-1} and radius 600 km, one finds that expected induced moment amplitudes after a decay interval of one hour would be of the order of $(0.1) \times 3 \times 10^{22}$ Gauss-cm^3 per Gauss of applied field, i.e., less than one standard deviation from the mean of the measured induced moment. These mantle contributions would act in such a direction as to bias upward the magnitude of the negative induced moment and hence the radius of a highly conducting core. However, this sample calculation indicates that the resulting bias may amount to no more than 20 km.

Finally, as emphasized by Russell et al. (1981), the Apollo subsatellite data do not impose strong restrictions on the actual electrical conductivity of a highly conducting core. Only a lower limit of ~10 S m^{-1} may be estimated from equation (11) using the lengths of typical observation intervals (10–30 hours) during which the Moon experienced a relatively steady unidirectional field. Thus, in principle, a molten silicate core can explain the induced moment measurement as well as can a metallic core. Further refinements of the conductivity lower limit as well as the radius estimate should be attempted using an independent data set such as may be acquired by a future polar orbiting satellite.

6.3. Laser ranging data

An additional line of evidence for a fluid central core has been derived from measurements of physical libration parameters, indicating a larger dissipation of rotational energy in the Moon than would be expected in the absence of a fluid core (Yoder, 1981, 1984; Ferrari et al., 1980). The observed parameter is a small (0.2″) advance in the lunar rotation axis from the Cassini alignment caused by internal dissipation. Assuming that solid body friction is responsible, this advance would imply a value for the ratio k_2/Q of $1.08 \pm 0.05 \times 10^{-3}$, where k_2 is the lunar potential Love number, a measure of the extent to which the Moon is deformed by tidal forces, and Q^{-1} is the fraction of tidal energy dissipated per flexing cycle. The lunar k_2 has been weakly determined in the analysis of Ferrari et al. (1980) as $k_2 = 0.022 \pm 0.011$ while theoretical estimates are in the range 0.029–0.034 (Cheng and Toksöz, 1978). Hence, one concludes that the lunar solid Q must be in the range 8–37 with preferred values between 20 and 30. However, these values are much lower than expected for the Moon based on seismic mantle Q-values and comparisons with other bodies. Thus Yoder (1981) has proposed viscous friction at the interface between a fluid metallic core and the solid mantle as an alternate means of explaining the observed rotation axis advance. For particular laminar and turbulent core-mantle coupling models, he concludes that turbulent coupling is preferred and derives a rough estimate of 330 km for the core radius. In support of this proposal, maintenance of the outer part of a metallic Fe-FeS core in a fluid state to the present day is theoretically feasible (Stevenson and Yoder, 1981). However, the core density and hence composition is not determinable from the model; in principle, a molten silicate core may also explain the libration signature. Additionally, the inferred core radius value depends at least partly on the core-mantle coupling model and should not be considered as accurately determined. Nevertheless, the physical libration signature generally provides qualitative evidence for a fluid lunar core.

6.4. Paleomagnetism

Lastly, the pervasive magnetization of lunar surface materials, if ascribable in whole or in part to the existence of a former core dynamo, would imply the presence of a metallic core and may allow further inferences regarding lunar internal and dynamical history (Runcorn et al., 1970; Strangway et al., 1970, 1973a; Fuller, 1974; Hood and Cisowski, 1983; Runcorn, 1983; Collinson, 1984). Relevant observational data include magnetization properties and paleointensity estimates for returned samples, inferences from surface magnetic field measurements, correlations of orbital magnetic anomalies with surface geology, and directional properties of large-scale magnetization inferred from orbital magnetometer measurements.

The main ferromagnetic carriers in lunar materials, metallic iron grains, are produced principally by reduction of preexisting iron silicates via shock and heat during impacts (e.g., Fuller, 1974). Thus mare basalt samples that contain relatively little metallic iron exhibit relatively low natural remanent magnetization (NRM) intensities in comparison to highland breccias and soils. Stable NRM intensities for the latter samples range up to 10^{-4} Gauss-cm^3 g^{-1} while values for mare basalts are typically less than 2×10^{-6} Gauss-cm^3 g^{-1}. Consistently, surface magnetic field strengths and orbital magnetic anomalies are generally weak over the maria but are stronger over the highlands (Strangway et al., 1973b; Dyal et al., 1974). The stable NRM component of mare basalts appears to be mainly thermoremanent in origin; i.e., the primary NRM was most probably acquired by cooling through the Curie point during formation of the basalt (Fuller, 1974; Strangway et al., 1973a), although exceptions may also exist (Fuller et al., 1979). The same is likely to be true for strongly annealed breccias whose primary NRM is generally unstable and less intense because of the predominance of iron grains in the multidomain size range. However, the primary NRM component of unannealed breccias and soils that carry the most intense and stable NRM of any sampled materials is not clearly thermoremanent in origin and shock magnetization effects appear to be significant (Fuller et al., 1974; Cisowski et al., 1974, 1976). The interpretation of the NRM of the latter samples is particularly important for understanding measurements of surface and orbital magnetic fields that will tend to be produced mainly by the most strongly magnetized surface materials.

Reliable estimates for the intensities of lunar paleofields as a function of sample age have been hindered by irreversible chemical changes and magnetic interactions within the constituent ferromagnetic iron grains during laboratory heatings (Pearce et al., 1976). As a result, accurate paleointensity determinations using the preferred Koenigsberger-Thellier-Thellier (KTT) method have been successful for only a small number of samples; these data alone provide little evidence for a time variation of lunar paleofield amplitudes (Sugiura and Strangway, 1980). Alternate techniques that are less accurate but relatively insensitive to modifications resulting from sample heating have therefore been sought to expand the number of samples for which paleointensity values can be estimated. One such technique involves normalizing the stable NRM of a given sample to its saturation isothermal remanent magnetization (IRM$_s$), i.e., normalizing the observed primary magnetization to the total remanence-carrying ability of the sample (Cisowski et al., 1975, 1983). Calibration of the normalized intensities to derive paleointensity estimates is achieved empirically using data for those samples whose paleointensities have been estimated by other methods including the KTT method. The IRM$_s$ normalization method has been criticized for not providing a criterion for identifying the NRM of a sample as thermoremanent magnetization (Sugiura and Strangway, 1980), but is supported as accurate to within half an order of magnitude by tests on terrestrial rocks for which paleointensities are known (Cisowski and Fuller, 1985). Using the IRM$_s$ method, Cisowski *et al.* (1983) obtain evidence for relatively weak magnetization of the oldest available samples (>3.9 aeons) followed by an interval between 3.9 and 3.6 aeons during which paleointensities maximize near 1 Gauss and decrease within a poorly defined time interval by approximately one order of magnitude (Fig. 12). Their results differ from earlier analyses that proposed an approximately constant decline of lunar paleofield amplitudes in the interval 4.0 to 3.2 aeons (Stephenson *et al.*, 1975, 1977).

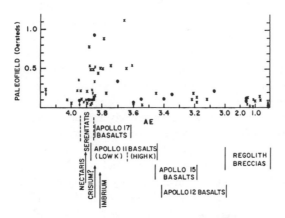

Fig. 12. Plot of sample paleointensity vs. age, where paleointensities are calculated from normalized remanence measurements using an empirical calibration factor to convert to absolute units (Cisowski et al., 1983).

Cisowski *et al.* (1983) conclude that the inferred high-field epoch between 3.9 and 3.6 aeons is most directly interpreted in terms of a temporary core dynamo while weaker paleofields outside of this interval may be ascribable to alternate generation mechanisms associated with local surface processes such as impact processes. In a second review of lunar paleointensity data, Collinson (1984) also favors an early core dynamo and suggests the importance of transient field generation mechanisms associated with impact processes for samples such as regolith breccias and glasses, which acquired their magnetization between ~3.0 aeons ago and the present. A well-studied case in the latter category is 70019, an Apollo 17 impact glass sample dated as <100 m.y. old that yielded a KTT paleointensity of 0.025 G, significantly larger than observed present-day surface fields (Sugiura *et al.*, 1979; see also Pearce and Chou, 1977). It is difficult to conceive of a process other than the impact that formed the glass for providing such a field.

Orbital measurements of lunar paleomagnetic fields have been obtained from the Apollo 15 and 16 subsatellites directly with magnetometers (Coleman *et al.*, 1972; Russell *et al.*, 1975; Hood *et al.*, 1981) and indirectly by measuring the fluxes of electrons reflected from surface field gradients (Howe *et al.*, 1974; Anderson *et al.*, 1975). Early mapping of the magnetometer data considered only measurements obtained within the relatively undisturbed plasma environment of the geomagnetic tail lobes. Consequently, useful (i.e., low-altitude) coverage was limited to portions of the far side, primarily the Van de Graaff-Aitken region where especially strong anomalies were found. Later mapping also considered intervals when the Moon was in the solar wind and the subsatellite was in the lunar wake; additional coverage at altitudes as low as 20 km was obtained within a narrow equatorial band across a portion of the near side. In agreement with expectations from sample studies and surface field measurements (e.g., Strangway *et al.*, 1973b), anomalies were found to be generally weak across the nearside maria but significant anomalies were detected in areas dominated geologically by the Fra Mauro and Cayley formations (Hood *et al.*, 1979a,b; Fig. 13). The latter are generally interpreted as primary and/or secondary basin ejecta, supporting the view that many orbital anomalies are associated with strongly magnetized low-grade breccias and soils produced in impacts. No

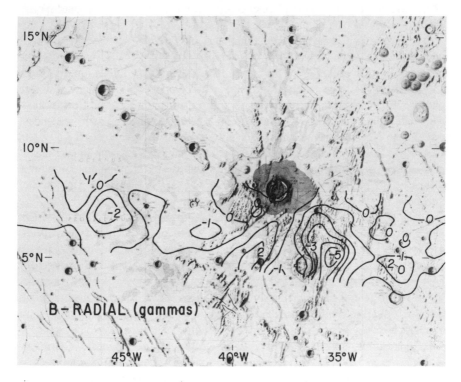

Fig. 13. Superposition of a radial magnetic field anomaly map derived from low-altitude Apollo 16 subsatellite magnetometer measurements onto a geologic map of the Kepler quadrangle. The largest maxima are correlated with exposures of the Fra Mauro formation (roughly textured unit representing primary Imbrium basin ejecta), while the craters Kepler and Encke and areas dominated by mare basalt are magnetically relatively weak (Hood, 1980).

signatures were observed in association with impact craters (e.g., Kepler) or mare edges indicating that mare basalt flows are too weakly magnetized and/or too thin to produce detectable anomalies. The strongest single anomaly within the limited zone covered by low-altitude magnetometer data was correlated with an unusual swirl-like albedo marking called Reiner Gamma on western Oceanus Procellarum (Hood et al., 1979a,b). Later mapping yielded a further strong anomaly centered on the farside crater Gerasimovich where swirl-like albedo markings of the same class are found (Hood, 1981). Similar markings are present within Mare Ingenii adjacent to Van de Graaff and are visible within Van de Graaff itself at high sun angles. Suggestions for the origin of the swirls range from unusually magnetized secondary crater ejecta (Hood et al., 1979b) to residues from cometary impacts (Schultz and Srnka, 1980) to albedo contrasts resulting from deflection of the solar wind ion bombardment by the associated magnetic anomalies (Hood and Schubert, 1980). Evidence for the origin of Reiner Gamma from Earth-based reflectance spectra has been evaluated by Bell and Hawke (1981).

Generalized maps of electron reflection fluxes interpreted as surface magnetic field intensities have been derived from available Apollo 15 and 16 subsatellite measurements (Lin et al., 1976; Frontispiece, *Proceedings of the Eighth Lunar Science Conference*, 1977). In general, these measurements demonstrate the paucity of surface magnetic fields across the nearside maria and the predominance of stronger fields in the highlands, in agreement with the orbital magnetometer data and with independent indirect mapping methods such as the limb compression technique (Lichtenstein et al., 1978). Detectable surface fields are generally uncorrelated with impact craters in the maria, again implying that mare basalt flows are too thin and/or too weakly magnetized to produce detectable edge effects (Lin, 1979). However, a statistical correlation in mare regions between inferred surface field strength and geologic age was reported (Lin, 1979) that has been cited by Cisowski et al. (1983) as consistent with a decrease in sample paleointensities after ~3.6 aeons. Anderson and Wilhelms (1979) have further applied the electron data to identify anomalously magnetized regions on the far side located in areas where basin ejecta units are observed or inferred. These magnetized regions may be analogous to the Fra Mauro anomalies on the near side and further indicate the importance of impact-produced materials as major sources of orbital anomalies. Perhaps the most potentially important single anomaly detected via the electron reflection technique is that appearing to be associated with Rima Sirsalis, an extensional graben-like feature located southeast of Grimaldi (Anderson et al., 1977). The latter authors report the detection of strong electron reflection peaks located within 40 km of Rima Sirsalis and approximately paralleling the trend of the rille (Fig. 14). A continued occurrence of reflection peaks at least 60 km beyond the end of the visible rille over Oceanus Procellarum was also reported, suggesting that the magnetization feature continues on beneath the mare surface. If the magnetization responsible for the anomaly is directly associated with the rille, then a deep-seated source implying slow cooling in the presence of a steady, large-

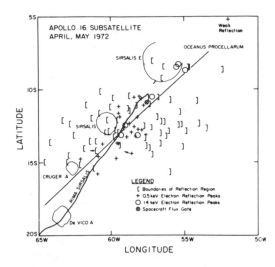

scale magnetic field would be indicated. This would be in contrast with other orbital anomalies that appear to be associated primarily with surficial materials predominantly of impact origin. An investigation of possible source models of the Rima Sirsalis anomaly was conducted by Srnka *et al.* (1979). A model consisting of a gap in a uniformly magnetized crust with the dimensions of the rille was shown to require a vertical magnetization intensity of $2-3 \times 10^{-3}$ Gauss cm^3 g^{-1}. Although comparable to magnetization intensities inferred for the Reiner Gamma formation in Oceanus Procellarum (Hood, 1980), this magnetization exceeds by more than an order of magnitude that of any returned sample. An alternate model consisting of a uniformly magnetized vertical prism at least as wide as the visible rille and extending to depths of at least 25 km yielded more plausible required magnetization levels of 10^{-5}–10^{-4} Gauss cm^3 g^{-1}. Since Rima Sirsalis predates the major episode of mare volcanism that flooded southern Procellarum, the latter model may be consistent with intrusion of the rille by a system of dikes during that episode (Srnka *et al.*, 1979). Future availability of lower-altitude magnetometer data over Rima Sirsalis may significantly improve our understanding of the origin of this anomaly and of the paleomagnetism in general.

In considering the probable origin(s) of lunar paleomagnetic fields, it is first necessary to evaluate the core dynamo hypothesis based on our presently incomplete knowledge of planetary dynamo field generation. Possible energy sources for planetary dynamos include convection driven by thermal or chemical buoyancy and turbulence driven by precessional torques. Assuming a convection-driven dynamo in which the Coriolis force dominates, Busse (1978) has suggested an upper bound on the strength of the generated magnetic field expressible in the form

$$M \leq K\rho_c^{1/2}R_c^4\Omega \tag{13}$$

where M is the planetary magnetic dipole moment, ρ_c and R_c are the core mean density and radius, Ω is the angular frequency of rotation, and K is a function of core properties including thermal diffusivity, magnetic diffusivity, and the effective wavenumber of convective eddies. Jacobs (1979) has proposed a relation similar to equation (13) using a simple dimensional argument. Russell (1979) noted that equation (13) is consistent in order of magnitude with the four observed planetary magnetic moments provided that the function K is approximately constant. Assuming that equation (13) is also applicable to the lunar core during the time when a former magnetic field may have been generated, i.e., assuming a convection-driven dynamo in which the Coriolis force dominates and K is similar to that of present-day planetary cores, Anderson (1983) has argued that a lunar core dynamo would be unlikely to provide surface fields of order 1 Gauss as required by paleointensity data. The essence of Anderson's argument can be seen in Fig. 15, which is a plot of observed planetary dipole moments vs. the quantity $\rho_c^{1/2} R_c^4 \Omega$ using approximate core parameters for Mercury, Earth, Jupiter, and Saturn [see Russell (1979) for a discussion of numerical values]. For the case of a lunar surface field intensity of 1 Gauss (right-hand scale), a significant departure from the trend established by observable planetary field amplitudes would be necessary if this field were to be attributable to a core dynamo, even if an upper limit core radius of 500 km and angular velocity 20 times that of the present Moon are assumed. Although this argument does not eliminate the possibility of a former lunar core dynamo, it does indicate that an inordinately efficient convective dynamo was operative or that an unusual energy source was available for driving dynamo field generation. Possibilities in the latter category include a precessional energy source when the Moon was nearer to the Earth and an addition of gravitational potential energy

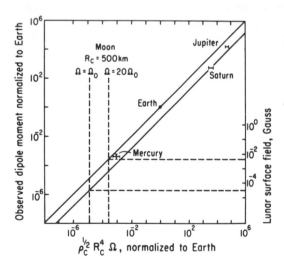

Fig. 15. Log-log plot of planetary magnetic dipole moments vs. the quantity $\rho_c^{1/2} R_c^4 \Omega$, where ρ_c and R_c are the density and radius of the respective planetary cores and Ω is the angular frequency of rotation. Proportionality lines are drawn through the terrestrial and Saturnian data points. Estimates for the lunar surface field assuming that a former lunar core dynamo existed and obeyed a similar amplitude rule are indicated by the dashed lines on the right-hand scale. Ω_0 is the present lunar angular frequency of rotation. (After Anderson, 1983.)

at the time of core formation. The data of Cisowski *et al.* (1983) would constrain either of these energy inputs to occur primarily in the 3.9 to 3.6 aeons time period.

Alternate hypotheses for the origins of lunar paleomagnetic fields have been reviewed by Daily and Dyal (1979). Here we explicitly discuss only two of these, production of transient magnetic fields in impacts and generation of local fields by thermoelectric currents between adjacent surface magma exposures. As previously noted, both Cisowski *et al.* (1983) and Collinson (1984) have suggested the need for impact-generated fields to explain the paleointensities derived for relatively young breccias and glasses such as 70019. Transient field generation in both meteoroid impacts (Hide, 1972; Srnka, 1977) and cometary impacts (Gold and Soter, 1976; Schultz and Srnka, 1980) have been hypothesized. A recent investigation of the meteoroid impact hypothesis has shown that hypervelocity impact-produced plasmas can significantly amplify local magnetic field intensities during times when ejected brecciated materials are likely to have been magnetized (Hood and Vickery, 1984). However, generation of fields as large as ~1 Gauss as required by paleointensity data was not demonstrated. While transient fields in this category may explain part of the magnetization detected from orbit associated with impact-produced materials, they do not readily explain the observed NRM of crystalline basalts and annealed breccias; as discussed above, the latter is apparently thermoremanent in origin, implying slow cooling in the presence of a steady field. If a core dynamo is ultimately shown to be implausible for providing the required paleofields for these samples, then alternate steady-field generation mechanisms must be considered. For example, Dyal *et al.* (1977) propose a model in which cooling mare lava flows generate a thermoelectric current by the Seebeck effect. These flows are considered to be connected electrically below by subsurface magma and above by the solar wind plasma producing an electric current between differentially cooling conductors. Field sources of this type, while not proven, are difficult to eliminate as an alternate explanation for the 3.9–3.6 aeon high-field epoch, which coincides with the onset of widespread mare volcanism.

One empirical means for evaluating the evidence for global scale vs. local surface lunar paleofield sources involves determination of approximate bulk directions of magnetization for major anomaly sources using orbital magnetometer data (Hood *et al.*, 1978; Hood, 1980, 1981; Runcorn, 1982, 1983). If a steady core dynamo was responsible for the magnetization of these anomaly sources, then sources of comparable age should ideally be magnetized along field lines oriented in the shape of a dipole centered in the Moon. Unfortunately, the coverage of orbital magnetometer data is presently limited to narrow equatorial bands on the near side and to a few limited areas on the far side. Consequently, a total of only 28 sources with widely varying probable ages can currently be studied in this manner so that the directional evidence for or against a core dynamo is difficult to evaluate. When all 28 independent sources are considered together, the corresponding distribution of paleomagnetic pole positions is not significantly different from random (see Fig.

9 of Hood, 1981). Also, adjacent or nearby anomaly sources are generally inferred to be magnetized in very different directions, suggesting a local scale size for the magnetizing field (Hood *et al.*, 1978; Hood, 1980). A case in point is the Fra Mauro formation. As shown in Fig. 13, two adjacent Fra Mauro exposures correlate with radial field anomalies that are of opposite sign. The simplest interpretation, supported by detailed modeling (Hood, 1980), is that the bulk directions of magnetization of these exposures are very different. Since these exposures were presumably emplaced contemporaneously, a large-scale uniform magnetizing field is not favored. On the other hand, Runcorn (1982, 1983) finds that three subsets of the 28 anomaly sources do have significantly clustered paleopoles and he proposes that these may represent sources magnetized during three different epochs of lunar history when the rotational pole was greatly displaced from its present position. Although the number of sources in each group is small and their contemporaneous ages are questionable, it is difficult to eliminate the possibility of a former core dynamo on the basis of the presently limited data set.

7. Discussion

The Moon is the only solar system body other than the Earth for which reasonably diverse geophysical measurements are available. However, these measurements are still largely incomplete so that current assessments must be classified as tentative. It is first clear that geophysical evidence for a lunar refractory lithophile element enrichment and a lower bulk $Mg/(Mg + Fe)$ ratio as compared to the Earth's mantle and CI chondrites is extremely tenuous. Surface heat flow data are sufficiently sparse and global extrapolations thereof sufficiently disparate to allow inferences of bulk U abundances that do not differ significantly from terrestrial mantle and chondritic estimates (Section 4; Rasmussen and Warren, 1985). Only lower limits on the bulk lunar Al_2O_3 abundance of ~1.8–2.6 wt % may be deduced from probable limits on crustal composition and mean thickness (Sections 2 and 3). If whole-Moon differentiation is allowed, bulk Al_2O_3 abundances comparable to chondritic and terrestrial values (2.4 and 3.3 wt %, respectively), may not be excluded. Earlier seismic velocity models (e.g., Goins, 1978) that were interpretable in terms of a more FeO-rich interior below 400–500 km depth are now considered to be less reliable based on the analysis by Nakamura (1983) of a more complete arrival time data set, although uncertainties associated with the Nakamura model must also be considered.

Current geophysical efforts to resolve further the gross compositional structure of the lunar mantle center on interpretation and modeling of the Nakamura (1983) seismic velocity model. The velocity increase below ~500 km depth that characterizes this model may in principle be explained either by the presence of a substantial garnet fraction [> 20 wt % if $Mg/(Mg + Fe)$ ratios of ~0.80 for coexisting olivine and pyroxene are assumed] or by an increase in $Mg/(Mg + Fe)$ ratio, or by some combination of the two. Provisional calculations (Section 3) indicate that an explanation in terms of garnet alone requires minimum Al_2O_3 abundances of 7–10%, which would be larger than most recent estimates for the bulk Moon. Thus models with a more MgO-rich deep interior as well as a possible garnet fraction appear to be favored if the Nakamura velocity limits are accepted. However, the corresponding density models (Section 5) are generally characterized by an inverse density contrast near 500 km. Such models would most probably have been unstable to convective overturn during early lunar history when mantle temperatures were substantially higher than at present. Thus seismic velocity limits on middle and lower lunar mantle composition are still uncertain. In particular, bulk lunar Mg/(Mg + Fe) ratios equal to that of the terrestrial upper mantle (0.89) are neither determinable nor excludable on the basis of geophysical (primarily seismic) data alone.

Because of the uncertainties in middle and lower mantle composition and density structure as well as crustal mean thickness and density, observational bounds on the lunar moment of inertia do not yet place strong constraints on the existence or radius of a possible metallic core. Predominantly iron cores with radii up to ~400 km and FeS cores with radii as large as ~600 km are allowed in principle by mantle density profiles that remain constant with radius. However, core-free models in combination with mantle density increases of ~10% can also be accommodated. Independent constraints on the existence of a metallic core, reviewed in Section 6, are consistent with the absence of fluid or highly electrically conducting cores with radii larger than about 450 km. Observations that suggest the existence of either a molten silicate or a metallic core include the measured negative lunar induced magnetic dipole moment in the geomagnetic tail (Russell et al., 1981) and the small (0.2″) advance in the lunar rotation axis from the Cassini state (Yoder, 1981). The former observation is consistent with a highly conducting core radius of ~420–460 km while the latter observation has been interpreted by Yoder as implying a fluid core radius of ~330 km. Refinements of the core-mantle coupling model of Yoder and/or the induced moment estimate using future data may eventually allow a reconciliation of these two results. Paleomagnetic data are not currently sufficiently complete to rigorously confirm or deny the former core dynamo hypothesis, although it is clear that alternate field generation mechanisms are needed to explain part of the observed magnetization.

Implications of present geophysical constraints for lunar origin models are equally tentative. If the Nakamura velocity model is correct to within the stated one standard deviation (~65% confidence) error limits, then the bulk Mg/(Mg + Fe) ratio may be closer to that of the terrestrial upper mantle so that variants of the fission hypothesis may be accommodated. However, estimates for the radius of a possible metallic core derived from magnetic and laser ranging data of 330–460 km imply a metal content of 2–4 wt % of the lunar mass, somewhat larger than would be needed to produce the observed depletions of lunar siderophiles if the Moon formed entirely from terrestrial mantle material (Newsom, 1984). Thus only variants that allow incorporation into the proto-Moon of additional metal, e.g., from a large terrestrial impactor or from circumterrestrial planetesimals, would be acceptable if these core radius estimates are valid. Binary accretion models as advocated for example by Newsom and Drake (1982, 1983) can also be accommodated by available geophysical constraints.

The preferred long-term approach for establishing more definitively the structure, composition, and thermal state of the lunar interior is the seismic method supplemented by electromagnetic sounding and heat flow determinations using a distributed network of surface instruments. However, significant improvements in orbital geophysical measurements will also yield useful refinements and will most probably predate establishment of an improved surface geophysical network by a considerable period. Measurements from a single polar orbiting satellite that would assist in resolving issues discussed in this paper include (1) radio brightness mapping of global surface heat flux; (2) increased coverage of orbital measurements of crustal magnetic fields and independent measurements of the lunar induced magnetic moment in the geomagnetic tail; and (3) increased coverage of high-resolution gravity/topography data allowing more detailed modeling of global crustal thickness and/or density variations.

Acknowledgments. I wish to express my thanks to the original organizing committee of the Conference on the Origin of the Moon for suggesting the problem of the existence of a lunar core from which the present paper evolved. S. C. Solomon is thanked for an especially helpful review. Discussions with colleagues, particularly S. M. Cisowski, M. J. Drake, M. Fuller, F. Herbert, J. Jones, Y. Nakamura, R. J. Phillips, S. K. Runcorn, C. T. Russell, and C. P. Sonett, are appreciated. This work was supported under NASA grant 7020.

8. References

Akella J. (1976) Garnet pyroxene equilibria in the system $CaSiO_3$-$MgSiO_3$-Al_2O_3 and in a natural mineral mixture. *Am. Mineral., 61*, 589–598.

Anders E. (1976) Chemical compositions of the moon, earth and eucrite parent body. *Phil. Trans. Roy. Soc. London, A285*, 23–40.

Anderson K. A. (1983) Magnetic dipole moment estimates for an ancient lunar dynamo. *Proc. Lunar Planet. Sci. Conf. 13th,* in *J. Geophys. Res., 88*, A588–A590.

Anderson K. A. and Wilhelms D. E. (1979) Correlation of lunar far side magnetized regions with ringed impact basins. *Earth Planet. Sci. Lett., 46*, 107–112.

Anderson K. A., Lin R. P., McGuire R. E., and McCoy J. E. (1975) Measurement of lunar and planetary magnetic fields by reflection of low energy electrons. *Space Sci. Instrum., 1*, 439–470.

Anderson K. A., Lin R. P., McGuire R. E., McCoy J. E., Russell C. T., and Coleman P. J., Jr. (1977) Linear magnetization feature associated with Rima Sirsalis. *Earth Planet. Sci. Lett., 34*, 141–151.

Anderson O. L. (1981) A decade of progress in earth's internal properties and processes. *Science, 213*, 82–88.

Basaltic Volcanism Study Project (1981) *Basaltic Volcanism on the Terrestrial Planets.* Pergamon, New York. 1286 pp.

Bell J. F. and Hawke B. R. (1981) The Reiner Gamma Formation: Composition and origin as derived from remote sensing observations. *Proc. Lunar Planet. Sci. 12B,* pp. 679–694.

Bills B. G. and Ferrari A. J. (1977a) A harmonic analysis of lunar topography. *Icarus, 31*, 244–259.

Bills B. G. and Ferrari A. J. (1977b) A lunar density model consistent with topographic, gravitational, librational, and seismic data. *J. Geophys. Res., 82*, 1306–1314.

Bills B. G. and Ferrari A. J. (1980) A harmonic analysis of lunar gravity. *J. Geophys. Res., 85*, 1013–1025.

Binder A. (1980) On the internal structure of a moon of fission origin. *J. Geophys. Res., 85*, 4872–4880.

Binder A. and Lange M. A. (1980) On the thermal history, thermal state, and related tectonism of a moon of fission origin. *J. Geophys. Res., 85*, 3194–3208.

Birch F. (1943) Elasticity of igneous rocks at high temperatures and pressures. *Geol. Soc. Amer. Bull., 54*, 263–286.

Birch F. (1969) Density and composition of the upper mantle; first approximation as an olivine layer. In *The Earth's Crust and Upper Mantle*, pp. 18–36. Amer. Geophys. Union Monograph No. 13.

Blackshear W. T. and Gapcynski J. P. (1977) An improved value for the lunar moment of inertia. *J. Geophys. Res., 82*, 1699–1701.

Bradley R. S., Jamil A. K., and Munro D. C. (1964) The electrical conductivity of olivine at high temperatures and pressures. *Geochim. Cosmochim. Acta, 28*, 1669–1678.

Buck W. R. and Toksöz M. N. (1980) The bulk composition of the moon based on geophysical constraints. *Proc. Lunar Planet. Sci. Conf. 11th*, pp. 2043–2058.

Busse F. H. (1978) Magnetohydrodynamics of the earth's dynamo. *Ann. Rev. Fluid Mech., 10*, 435–462.

Cameron A. G. W. and Ward W. R. (1976) The origin of the moon (abstract). In *Lunar Science VII*, pp. 120–122. The Lunar Science Institute, Houston.

Charette M. P., Taylor S. R., Adams J. B., and McCord T. B. (1977) The detection of soils of Fra Mauro basalt and anorthositic gabbro composition in the lunar highlands by remote spectral reflectance techniques. *Proc. Lunar Sci. Conf. 7th*, pp. 1049–1061.

Cheng C. H. and Toksöz M. N. (1978) Tidal stresses in the moon. *J. Geophys. Res., 83*, 845–853.

Chung D. H. (1971) Elasticity and equations of state of olivines in the Mg_2SiO_4-Fe_2SiO_4 system. *Geophys. J., 25*, 511–538.

Cisowski S. M. and Fuller M. (1985) Lunar sample paleointensity, via the IRM_s normalization method (abstract). In *Lunar and Planetary Science XVI*, pp. 135–136. Lunar and Planetary Institute, Houston.

Cisowski S. M., Collinson D. W., Runcorn S. K., Stephenson A., and Fuller M. (1983) A review of lunar paleointensity data and implications for the origin of lunar magnetism. *Proc. Lunar Planet. Sci. Conf. 13th*, in *J. Geophys. Res., 88*, A691–A704.

Cisowski S. M., Dunn J. R., Fuller M., Rose M. F., and Wasilewski P. J. (1974) Impact processes and lunar magnetism. *Proc. Lunar Sci. Conf. 5th*, pp. 2841–2858.

Cisowski S. M., Fuller M., Wu Y., Rose M. F., and Wasilewski P. J. (1975) Magnetic effects of shock and their implications for magnetism of lunar samples. *Proc. Lunar Sci. Conf. 6th*, pp. 3123–3141.

Cisowski S. M., Dunn J. R., Fuller M., Wu Y., Rose M. F., and Wasilewski P. J. (1976) Magnetic effects of shock and implications for the origin of lunar magnetism. *Proc. Lunar Sci. Conf. 7th*, pp. 3299–3320.

Cleary J. R. (1982) Comment on "Lunar seismology: The internal structure of the moon" by N. R. Goins, A.M. Dainty, and M. N. Toksoz. *J. Geophys. Res., 87*, 5495–5496.

Coleman P. J., Jr., Schubert G., Russell C. T., and Sharp L. R. (1972) Satellite measurements of the moon's magnetic field: A preliminary report. *The Moon, 4*, 419–429.

Collinson D. W. (1984) On the existence of magnetic fields on the moon between 3.6 Ga ago and the present. *Phys. Earth Planet. Inter., 34*, 102–116.

Conel J. E. and Morton J. B. (1975) Interpretation of lunar heat flow data. *The Moon, 14*, 263–289.

Cooper M. R., Kovach R. L., and Watkins J. S. (1974) Lunar near-surface structure. *Rev. Geophys. Space Phys., 12,* 291–308.

Daily W. D. and Dyal P. (1979) Magnetometer data errors and lunar induction studies. *J. Geophys. Res., 84,* 3313–3326.

Dainty A. M., Toksöz M. N., and Stein S. (1976) Seismic investigations of the lunar interior. *Proc. Lunar Sci. Conf. 7th,* pp. 3057–3075.

Deer W. A., Howie R. A., and Zussman J. (1982) *Rock-Forming Minerals,* Vol. 1A. Wiley and Sons, New York. 919 pp.

Drake M. J. (1983) Geochemical constraints on the origin of the moon. *Geochim. Cosmochim. Acta, 47,* 1759–1767.

Duba A., Dennison M., Irving A. J., Thornber C. R., and Huebner J. S. (1979) Electrical conductivity of aluminous orthopyroxene (abstract). In *Lunar and Planetary Science X,* pp. 318–319. Lunar and Planetary Institute, Houston.

Duba A., Heard H. C., and Schock R. N. (1974) Electrical conductivity of olivine at high pressure and under controlled oxygen fugacity. *J. Geophys. Res., 79,* 1667–1673.

Duba A., Heard H. C., and Schock R. N. (1976) Electrical conductivity of orthopyroxene to 1400°C and the resulting selenotherm. *Proc. Lunar Sci. Conf. 7th,* pp. 3173–3181.

Dyal P., Parkin C. W., and Daily W. D. (1974) Magnetism and the interior of the moon. *Rev. Geophys. Space Phys., 12,* 568–591.

Dyal P., Parkin C. W., and Daily W. D. (1976) Structure of the lunar interior from magnetic field measurements. *Proc. Lunar Sci. Conf. 7th,* pp. 3077–3095.

Dyal P., Parkin C. W., and Daily W. D. (1977) Global lunar crust: Electrical conductivity and thermoelectric origin of remanent magnetism. *Proc. Lunar Sci. Conf. 8th,* pp. 767–783.

Ferrari A. J. and Ananda M. P. (1977) Lunar gravity: A long-term Keplerian rate method. *J. Geophys. Res., 82,* 3085–3097.

Ferrari A. J., Nelson D. L., Sjogren W. L., and Phillips R. J. (1978) The isostatic state of the lunar Apennines and regional surroundings. *J. Geophys. Res., 83,* 2863–2871.

Ferrari A. J., Sinclair W. S., Sjogren W. L., Williams J. G., and Yoder C. F. (1980) Geophysical parameters of the earth-moon system. *J. Geophys. Res., 85,* 3939–3951.

Fuller M. (1974) Lunar magnetism. *Rev. Geophys. Space Phys., 12,* 23–70.

Fuller M., Meshkov M. E., Cisowski S. M., and Hale C. J. (1979) On the natural remanent magnetism of certain mare basalts. *Proc. Lunar Planet. Sci. Conf. 10th,* pp. 2211–2233.

Fuller M., Rose F., and Wasilewski P. J. (1974) Preliminary results of an experimental study of the magnetic effects of shocking lunar soil. *The Moon, 9,* 57–61.

Gapcynski J. P., Blackshear W. T., Tolson R. H., and Compton H. R. (1975) A determination of the lunar moment of inertia. *Geophys. Res. Lett., 2,* 353–356.

Goins N. R. (1978) The internal structure of the moon. Ph.D. thesis, Massachusetts Institute of Technology, Cambridge. 666 pp.

Goins N. R., Dainty A. M., and Toksöz M. N. (1981a) Seismic energy release of the moon. *J. Geophys. Res., 86,* 378–388.

Goins N. R., Dainty A. M., and Toksöz M. N. (1981b) Structure of the lunar crust at highland site Apollo 16. *Geophys. Res. Lett., 8,* 29–32.

Goins N. R., Dainty A. M., and Toksöz M. N. (1981c) Lunar seismology: The internal structure of the moon. *J. Geophys. Res., 86,* 5061–5074.

Gold T. and Soter S. (1976) Cometary impact and the magnetization of the moon. *Planet. Space Sci., 24,* 45–54.

Goldstein B. E., Phillips R. J., and Russell C. T. (1976) Magnetic evidence concerning a lunar core. *Proc. Lunar Sci. Conf. 7th,* pp. 3321–3341.

Green D. H. and Ringwood A. E. (1967) The stability field of aluminous pyroxene peridotite and garnet peridotite and their relevance in upper mantle structure. *Earth Planet. Sci. Lett., 3,* 151–160.

Haines E. L. and Metzger A. E. (1979) The variation of iron concentration in the lunar highlands and resultant implications for crustal models (abstract). In *Lunar and Planetary Science X,* pp. 488–490. Lunar and Planetary Institute, Houston.

Haines E. L. and Metzger A. E. (1980) Lunar highland crustal models based on iron concentrations: Isostasy and center-of-mass displacement. *Proc. Lunar Planet. Sci. Conf. 11th,* pp. 689–718.

Hartmann W. K. and Davis D. R. (1975) Satellite-sized planetesimals and lunar origin. *Icarus, 24,* 504–515.

Head J. W. (1979) Lava flooding of earth planetary crusts: Geometry, thickness, and volumes of flooded impact basins (abstract). In *Lunar and Planetary Science X,* pp. 516–518. Lunar and Planetary Institute, Houston.

Head J. W. (1982) Lava flooding of ancient planetary crusts: Geometry, thickness, and volumes of flooded lunar impact basins. *Moon and Planets, 26,* 61–88.

Herbert F., Drake M. J., Sonett C. P., and Wiskerchen M. J. (1977) Some constraints on the thermal history of the lunar magma ocean. *Proc. Lunar Sci. Conf. 8th,* pp. 573–582.

Hide R. (1972) Comments on the moon's magnetism. *The Moon, 4,* 39.

Hobbs B. A., Hood L. L., Herbert F., and Sonett C. P. (1983) An upper bound on the radius of a highly electrically conducting lunar core. *Proc. Lunar Planet. Sci. Conf. 14th,* in *J. Geophys. Res., 88,* B97–B102.

Hodges F. N. and Kushiro I. (1974) Apollo 17 petrology and experimental determination of differentiation sequences in model moon compositions. *Proc. Lunar Sci. Conf. 5th,* pp. 505–520.

Hood L. L. (1980) Bulk magnetization properties of the Fra Mauro and Reiner Gamma Formations. *Proc. Lunar Planet. Sci. Conf. 11th,* pp. 1879–1896.

Hood L. L. (1981) Sources of lunar magnetic anomalies and their bulk directions of magnetization: Additional evidence from Apollo orbital data. *Proc. Lunar Planet. Sci. 12B,* pp. 817–830.

Hood L. L. and Cisowski S. M. (1983) Paleomagnetism of the moon and meteorites. *Rev. Geophys. Space Phys., 21,* 676–684.

Hood L. L. and Jones J. (1985) Lunar density models consistent with mantle seismic velocities and other geophysical constraints (abstract). In *Lunar and Planetary Science XVI,* pp. 360–361. Lunar and Planetary Institute, Houston.

Hood L. L. and Schubert G. (1978) A magnetohydrodynamic theory for the lunar response to time variations in a spatially uniform ambient magnetic field. *Proc. Lunar Planet. Sci. Conf. 9th,* pp. 3125–3135.

Hood L. L. and Schubert G. (1980) Lunar magnetic anomalies and surface optical properties. *Science, 208,* 49–51.

Hood L. L. and Sonett C. P. (1982) Limits on the lunar temperature profile. *Geophys. Res. Lett., 9,* 37–40.

Hood L. L. and Vickery A. (1984) Magnetic field amplification and generation in hypervelocity meteoroid impacts with application to lunar paleomagnetism. *Proc. Lunar Planet. Sci. Conf. 15th,* in *J. Geophys. Res., 89,* C211–C223.

Hood L. L. Coleman P. J., Jr., and Wilhelms D. E. (1979a) The moon: Sources of the crustal magnetic anomalies. *Science, 204,* 53–57.

Hood L. L., Coleman P. J., Jr., and Wilhelms D. E. (1979b) Lunar nearside magnetic anomalies. *Proc. Lunar Planet. Sci. Conf. 10th,* pp. 2235–2257.

Hood L. L., Herbert F., and Sonett C. P. (1982a) The deep lunar electrical conductivity profile: Structural and thermal inferences. *J. Geophys. Res., 87,* 5311–5326.

Hood L. L., Herbert F., and Sonett C. P. (1982b) Further efforts to limit lunar internal temperatures from electrical conductivity determinations. *Proc. Lunar Planet. Sci. Conf. 13th*, in *J. Geophys. Res., 87*, A109–A116.

Hood L. L., Russell C. T., and Coleman P. J., Jr. (1978) The magnetization of the lunar crust as deduced from orbital surveys. *Proc. Lunar Planet. Sci. Conf. 9th*, pp. 3057–3078.

Hood L. L., Russell C. T., and Coleman P. J., Jr. (1981) Contour maps of lunar remanent magnetic fields. *J. Geophys. Res., 86*, 1055–1069.

Howe H. C., Lin R. P., McGuire R. E., and Anderson K. A. (1974) Energetic electron scattering from the lunar remanent magnetic field. *Geophys. Res. Lett., 1*, 101–104.

Huebner J. S., Duba A., and Wiggins L. B. (1979) Electrical conductivity of pyroxene which contains trivalent cations: Laboratory measurements and the lunar temperature profile. *J. Geophys. Res., 84*, 4652–4656.

Huebner J. S., Duba A., Wiggins L. B., and Smith H. E. (1978) Electrical conductivity of orthopyroxene; Measurements and implications (abstract). In *Lunar and Planetary Science IX*, pp. 561–563. Lunar and Planetary Institute, Houston.

Jacobs J. A. (1979) Planetary magnetic fields. *Geophys. Res. Lett., 6*, 213–214.

Jagoutz E. and Wänke H. (1982) Has the earth's core grown over geologic times? (abstract). In *Lunar and Planetary Science XIII*, pp. 358–359. Lunar and Planetary Institute, Houston.

Kaula W. M. (1977) On the origin of the moon, with emphasis on bulk composition. *Proc. Lunar Sci. Conf. 8th*, pp. 321–331.

Kaula W. M., Schubert G., Lingenfelter R. E., Sjogren W. L., and Wollenhaupt W. R. (1972) Analysis and interpretation of lunar laser altimetry. *Proc. Lunar Sci. Conf. 3rd*, pp. 2189–2204.

Kaula W. M., Schubert G., Lingenfelter R. E., Sjogren W. L., and Wollenhaupt W. R. (1974) Apollo laser altimetry and inferences as to lunar structure. *Proc. Lunar Sci. Conf. 5th*, pp. 3049–3058.

Keihm S. J. and Langseth M. G. (1977) Lunar thermal regime to 300 km. *Proc. Lunar Sci. Conf. 8th*, pp. 499–514.

Kobrick M. (1976) Random processes as a cause of the lunar asymmetry. *The Moon, 15*, 83–89.

Kopal Z. (1967) *The Measure of the Moon.* D. Reidel, Dordrecht. 464 pp.

Koyama J. and Nakamura Y. (1979) Re-examination of the lunar seismic velocity structure based on the complete data set (abstract). In *Lunar and Planetary Science X*, pp. 685–687. Lunar and Planetary Institute, Houston.

Koyama J. and Nakamura Y. (1980) Focal mechanism of deep moonquakes. *Proc. Lunar Planet. Sci. Conf. 11th*, pp. 1855–1865.

Kumazawa M. (1969) The elastic constants of single-crystal orthopyroxene. *J. Geophys. Res., 74*, 5961–5972.

Kumazawa M. and Anderson O. L. (1969) Elastic moduli, pressure derivatives, and temperature derivatives of single-crystal olivine and single-crystal forsterite. *J. Geophys. Res., 74*, 5961–5972.

Lammlein D. R., Latham G. V., Dorman J., Nakamura Y., and Ewing M. (1974) Lunar seismicity, structure, and tectonics. *Rev. Geophys. Space Phys., 12*, 1–21.

Langseth M. G., Keihm S. J., and Peters K. (1976) Revised lunar heat-flow values. *Proc. Lunar Sci. Conf. 7th*, pp. 3143–3171.

Latham G., Ewing M., Press F., Sutton G., Dorman J., Nakamura Y., Toksöz M. N., Wiggins R., Derr J., and Duennebier F. (1970) Passive seismic experiment. *Science, 167*, 455–467.

Levin B. J. (1979) On the core of the moon. *Proc. Lunar Planet. Sci. Conf. 10th*, pp. 2321–2323.

Lichtenstein B. R., Coleman P. J., Jr., and Russell C. T. (1978) A comparison of contour maps derived from independent methods of measuring lunar magnetic fields. *Proc. Lunar Planet. Sci. Conf. 9th*, pp. 3079–3092.

Liebermann R. C. and Ringwood A. E. (1976) Elastic properties of anorthite and the nature of the lunar crust. *Earth Planet. Sci. Lett., 31*, 69–74.

Lin R. P. (1979) Constraints on the origins of lunar magnetism from electron reflection measurements of surface magnetic fields. *Phys. Earth Planet. Inter., 20,* 271–280.

Lin R. P., Anderson K. A., Bush R., McGuire R. E., and McCoy J. E. (1976) Lunar surface remanent magnetic fields detected by the electron reflection method. *Proc. Lunar Sci. Conf. 7th,* pp. 2691–2703.

Michael W. H., Jr. and Blackshear W. T. (1972) Recent results on the mass, gravitational field, and moments of inertia of the moon. *The Moon, 3,* 388–402.

Morgan J. W., Hertogen J., and Anders E. (1978) The moon: Composition determined by nebular processes. *Moon and Planets, 18,* 465–478.

Muller P. M. and Sjogren W. L. (1968) Mascons: Lunar mass concentrations. *Science, 161,* 680–684.

Nakamura Y. (1978) A_1 moonquakes: Source distribution and mechanism. *Proc. Lunar Planet. Sci. Conf. 9th,* pp. 3589–3607.

Nakamura Y. (1983) Seismic velocity structure of the lunar mantle. *J. Geophys. Res., 88,* 677–686.

Nakamura Y. and Koyama J. (1982) Seismic Q of the lunar upper mantle. *J. Geophys. Res., 87,* 4855–4861.

Nakamura Y., Duennebier F. K., Latham G. V., and Dorman H. J. (1976) Structure of the lunar mantle. *J. Geophys. Res., 81,* 4818–4824.

Nakamura Y., Lammlein D., Latham G., Ewing M., Dorman J., Press F., and Toksöz M. N. (1973) New seismic data on the state of the deep lunar interior. *Science, 181,* 49–51.

Nakamura Y., Latham D., Lammlein D., Ewing M., Duennebier F., and Dorman J. (1974) Deep lunar interior inferred from recent seismic data. *Geophys. Res. Lett., 1,* 137–140.

Nakamura Y., Latham G. V., Dorman H. J., Ibrahim A. K., Koyama J., and Horvath P. (1979) Shallow moonquakes: Depth, distribution, and implications as to the present state of the lunar interior. *Proc. Lunar Planet. Sci. Conf. 10th,* pp. 2299–2309.

Nakamura Y., Latham G. V., and Dorman H. J. (1982) Apollo lunar seismic experiment—final summary. *Proc. Lunar Planet. Sci. Conf. 13th,* in *J. Geophys. Res., 87,* A117–A123.

Newsom H. E. (1984) The lunar core and the origin of the moon. *Eos (Trans. Amer. Geophys. Union), 65,* 369–370.

Newsom H. E. and Drake M. J. (1982) Constraints on the moon's origin from the partitioning behavior of tungsten. *Nature, 297,* 210–212.

Newsom H. E. and Drake M. J. (1983) Experimental investigation of the partitioning of phosphorus between metal and silicate phases: Implications for the earth, moon and eucrite parent body. *Geochim. Cosmochim. Acta, 47,* 93–100.

Oka Y. (1978) Experimental study on the partitioning of Fe and Mg between garnet and olivine and its applications to kimberlites. *J. Fac. Soc. Hokkaido Univ., Ser. IV, 18* (Yagi volume), 351–376.

Pearce G. W. and Chou C.-L. (1977) On the origin of sample 70019 and its suitability for lunar magnetic field intensity studies. *Proc. Lunar Sci. Conf. 8th,* pp. 669–677.

Pearce G. W., Hoye G. S., Strangway D. W., Walker B. M., and Taylor L. A. (1976) Some complexities in the determination of lunar paleointensities. *Proc. Lunar Sci. Conf. 7th,* pp. 3271–3297.

Phillips R. J. and Lambeck K. (1980) Gravity fields of the terrestrial planets: Long-wavelength anomalies and tectonics. *Rev. Geophys. Space Phys., 18,* 27–76.

Proceedings of the Eighth Lunar Science Conference (1977) Frontispiece. Pergamon, New York.

Pullan S. and Lambeck K. (1980) On constraining lunar mantle temperatures from gravity data. *Proc. Lunar Planet. Sci. Conf. 11th,* pp. 2031–3041.

Rasmussen K. L. and Warren P. H. (1985) Megaregolith thickness, heat flow, and the bulk composition of the moon. *Nature, 313,* 121–124.

Ringwood A. E. (1979) *Origin of the Earth and Moon.* Springer-Verlag, New York. 295 pp.

Ringwood A. E. (1984) Origin of the moon (abstract). In *Papers Presented to the Conference on the Origin of the Moon,* p. 46. Lunar and Planetary Institute, Houston.

Ringwood A. E. and Essene E. (1970) Petrogenesis of Apollo 11 basalts, internal constitution and origin of the moon. *Proc. Apollo 11 Lunar Sci. Conf.*, pp. 769–799.

Ringwood A. E. and Kesson S. E. (1977) Composition and origin of the moon. *Proc. Lunar Sci. Conf. 8th*, pp. 371–398.

Runcorn S. K. (1982) Primeval displacements of the lunar pole. *Phys. Earth Planet. Inter., 29*, 135–147.

Runcorn S. K. (1983) Lunar magnetism, polar displacements and primeval satellites in the earth-moon system. *Nature, 304*, 589–596.

Runcorn S. K., Collinson D. W., O'Reilly W., Battey M. H., Stephenson A., Jones J. M., Manson A. J., and Readman P. W. (1970) Magnetic properties of Apollo 11 lunar samples. *Proc. Apollo 11 Lunar Sci. Conf.*, pp. 2369–2387.

Russell C. T. (1979) Planetary magnetic moments: A scaling law and two predictions. *Nature, 281*, 552–553.

Russell C. T., Coleman P. J., Jr., Fleming B. K., Hilburn L., Ioannidis G., Lichtenstein B. R., and Schubert G. (1975) The fine-scale lunar magnetic field. *Proc. Lunar Sci. Conf. 6th*, pp. 2955–2969.

Russell C. T., Coleman P. J., Jr., and Goldstein B. E. (1981) Measurements of the lunar induced magnetic moment in the geomagnetic tail: Evidence for a lunar core. *Proc. Lunar Planet. Sci. 12B*, pp. 831–836.

Russell C. T., Coleman P. J., Jr., and Schubert G. (1974a) Lunar magnetic field: Permanent and induced dipole moment. *Science, 186*, 825–826.

Russell C. T., Coleman P. J., Jr., Lichtenstein B. R., and Schubert G. (1974b) The permanent and induced dipole moment of the moon. *Proc. Lunar Sci. Conf. 5th*, pp. 2747–2760.

Ryder G. and Wood J. A. (1977) Serenitatis and Imbrium impact melts: Implications for large-scale layering in the lunar crust. *Proc. Lunar Sci. Conf. 8th*, pp. 655–668.

Schreiber E. (1977) The moon and Q. *Proc. Lunar Sci. Conf. 8th*, pp. 1201–1208.

Schubert G. and Lichtenstein B. R. (1974) Observations of moon-plasma interactions by orbital and surface experiments. *Rev. Geophys. Space Phys., 12*, 592–626.

Schubert G., Cassen P., and Young R. E. (1979) Subsolidus convective cooling histories of terrestrial planets. *Icarus, 38*, 191–211.

Schubert G., Sonett C. P., Schwartz K., and Lee H. J. (1973) Induced magnetosphere of the moon, 1. Theory. *J. Geophys. Res., 78*, 2094–2110.

Schubert G., Young R. E., and Cassen P. (1977) Subsolidus convection models of the lunar internal temperature. *Phil. Trans. Roy. Soc. London, A285*, 523–536.

Schultz P. H. and Srnka L. J. (1980) Cometary collisions on the moon and Mercury. *Nature, 284*, 22–26.

Shankland P. J. (1969) Transport properties of olivine. In *The Application of Modern Physics to the Earth and Planetary Interior* (S. K. Runcorn, ed.), pp. 175–190. Wiley-Interscience, New York.

Simmons G., Batzle M. L., and Harlow A. L. (1980) Thermal modification of microcracks in lunar rocks and revised estimates for the elastic properties of the shallow moon (abstract). In *Lunar and Planetary Science XI*, pp. 1030–1032. Lunar and Planetary Institute, Houston.

Simmons G. and Wang H. (171) *Single Crystal Elastic Constants and Calculated Aggregate Properties: A Handbook.* 2nd edition, M.I.T. Press, Cambridge. 370 pp.

Smith J. V., Anderson A. T., Newton R. C., Olsen E. J., and Wylie P. J. (1970) Petrologic history of the moon inferred from petrography, mineralogy, and petrogenesis of Apollo 11 rocks. *Proc. Apollo 11 Lunar Sci. Conf.*, pp. 897–925.

Solomon S. C. (1977) The relationship between crustal tectonics and internal evolution in the moon and Mercury. *Phys. Earth Planet. Inter., 15*, 135–145.

Solomon S. C. (1978) The nature of isostasy on the moon: How big a Pratt-fall for Airy models. *Proc. Lunar Planet. Sci. Conf. 9th*, pp. 3499–3511.

Solomon S. C. and Chaiken J. (1976) Thermal expansion and thermal stress in the moon and terrestrial planets: Clues to early thermal history. *Proc. Lunar Sci. Conf. 7th*, pp. 3229–3243.

Solomon S. C. and Head J. W. (1979) Vertical movement in mare basins: Relation to mare emplacement, basin tectonics, and lunar thermal history. *J. Geophys. Res., 84*, 1667–1682.

Sonett C. P. (1982) Electromagnetic induction in the moon. *Rev. Geophys. Space Phys., 20*, 411–455.

Sonett C. P. and Duba A. (1975) Lunar temperature and global heat flux from laboratory electrical conductivity and lunar magnetometer data. *Nature, 258*, 118–121.

Sonett C. P., Colburn D. S., Dyal P., Parkin C. W., Smith B. F., Schubert G., and Schwartz K. (1971) Lunar electrical conductivity profile. *Nature, 230*, 359–362.

Sonett C. P., Smith B. F., Colburn D. S., Schubert G., and Schwartz K. (1972) The induced magnetic field of the moon: Conductivity profiles and inferred temperature. *Proc. Lunar Sci. Conf. 3rd*, pp. 2309–2336.

Spetzler H. A., Getting I. C., and Swanson P. L. (1980) The contribution of activated processes to Q. *Proc. Lunar Planet. Sci. Conf. 11th*, pp. 1825–1835.

Spudis P. D., Hawke B. R., and Lucey P. (1984) Composition of Orientale basin deposits and implications for the lunar basin-forming process. *Proc. Lunar Planet. Sci. Conf. 15th*, in *J. Geophys. Res., 89*, C197–C210.

Srnka L. J. (1977) Spontaneous magnetic field generation in hypervelocity impacts. *Proc. Lunar Sci. Conf. 8th*, pp. 785–792.

Srnka L. J., Hoyt J. L., Harvey J. V. S., and McCoy J. E. (1979) A study of the Rima Sirsalis lunar magnetic anomaly. *Phys. Earth Planet. Inter., 20*, 281–290.

Stephenson A., Runcorn S. K., and Collinson D. W. (1975) On changes in the intensity of the ancient lunar magnetic field. *Proc. Lunar Sci. Conf. 6th*, pp. 3049–3062.

Stephenson A., Runcorn S. K., and Collinson D. W. (1977) Paleointensity estimates from lunar samples 10017 and 10020. *Proc. Lunar Sci. Conf. 8th*, pp. 679–687.

Stevenson D. J. and Yoder C. F. (1981) A fluid outer core for the moon and its implications for lunar dissipation, free librations, and magnetism (abstract). In *Lunar and Planetary Science XII*, pp. 1043–1044. Lunar and Planetary Institute, Houston.

Strangway D. W., Larson E. E., and Pearce G. W. (1970) Magnetic studies of lunar samples—breccia and fines. *Proc. Apollo 11 Lunar Sci. Conf.*, pp. 2435–2451.

Strangway D. W., Sharpe H., Gose W., and Pearce G. (1973a) Magnetism and the history of the moon. In *Magnetism and Magnetic Materials—1972* (C. D. Graham, Jr. and J. J. Rhyne, eds.), pp. 1178–1187. American Institute of Physics, New York.

Strangway D. W., Gose W., Pearce G., and McConnell R. K. (1973b) Lunar magnetic anomalies and the Cayley Formation. *Nature, 246*, 112–114.

Sugiura N. and Strangway D. W. (1980) Comparisons of magnetic paleointensity methods using a lunar sample. *Proc. Lunar Planet. Sci. Conf. 11th*, pp. 1801–1813.

Sugiura N., Wu Y. M., Strangway D. W., Pearce G. W., and Taylor L. A. (1979) A new magnetic paleointensity value for a young lunar glass. *Proc. Lunar Planet. Sci. Conf. 10th*, pp. 2189–2198.

Taylor S. R. (1978) Geochemical constraints on melting and differentiation in the moon. *Proc. Lunar Planet. Sci. Conf. 9th*, pp. 15–23.

Taylor S. R. (1982) *Planetary Science: A Lunar Perspective*. Lunar and Planetary Institute, Houston. 481 pp.

Taylor S. R. (1984) Tests of the lunar fission hypothesis (abstract). In *Papers Presented to the Conference on the Origin of the Moon*, p. 25. Lunar and Planetary Institute, Houston.

Taylor S. R. and Bence A. E. (1975) Evolution of the lunar highland crust. *Proc. Lunar Sci. Conf. 6th*, pp. 1121–1141.

Tittman B. R., Clark V. A., and Spencer T. W. (1980) Compressive strength, seismic Q, and elastic modulus. *Proc. Lunar Planet. Sci. Conf. 11th*, pp. 1815–1823.

Todd T., Richter D. A., Simmons G., and Wang H. (1973) Unique characterization of lunar samples by physical properties. *Proc. Lunar Sci. Conf. 4th*, pp. 2639–2662.

Toksöz M. N., Dainty A. M., Solomon S. C., and Anderson K. (1974) Structure of the moon. *Rev. Geophys. Space Phys., 12*, 539–567.

Toksöz M. N., Goins N. R., and Cheng C. H. (1977) Moonquakes: Mechanisms and relation to tidal stresses. *Science, 196*, 979–981.

Toksöz M. N., Hsui A. T., and Johnston D. H. (1978) Thermal evolutions of the terrestrial planets. *Moon and Planets, 18*, 281–320.

Toksöz M. N., Press F., Anderson K., Latham G., Ewing M., Dorman J., Lammlein D., Nakamura Y., Sutton G., and Duennebier F. (1972) Velocity structure and properties of the lunar crust. *The Moon, 4*, 490–504.

Tolland H. G. and Strens R. G. J. (1972) Electrical conduction in physical and chemical mixtures: Application to planetary mantles. *Phys. Earth Planet. Inter., 5*, 380–386.

Wang H. and Simmons G. (1974) Elasticity of some mantle crystal structures, 3. Spessartite-almandine garnet. *J. Geophys. Res., 79*, 2607–2613.

Warren P. H. (1982) The moon is not necessarily enriched in refractory elements (abstract). In *Lunar and Planetary Science XI*, pp. 837–838. Lunar and Planetary Institute, Houston.

Warren P. H. (1984) The bulk moon MgO/FeO ratio: A highlands perspective (abstract). In *Papers Presented to the Conference on the Origin of the Moon*, p. 19. Lunar and Planetary Institute, Houston.

Warren P. H. and Wasson J. T. (1979) Effects of pressure on the crystallization of a "chondritic" magma ocean and implications for the bulk composition of the moon. *Proc. Lunar Planet. Sci. Conf. 10th*, pp. 2051–2083.

Wasson J. T. and Warren P. H. (1980) Contribution of the mantle to the lunar asymmetry. *Icarus, 44*, 752–771.

Williams J. G., Sinclair W. S., Slade M. A., Bender P. L., Hauser J. P., Mulholland J. D., and Shelus P. J. (1974) Lunar moment of inertia constraints from lunar laser ranging (abstract). In *Lunar Science V*, pp. 845–847. The Lunar Science Institute, Houston.

Wiskerchen M. J. and Sonett C. P. (1977) A lunar metal core? *Proc. Lunar Sci. Conf. 8th*, pp. 515–535.

Wood J. A. (1973) Bombardment as a cause of the lunar asymmetry. *The Moon, 8*, 73–103.

Wood J. A., Dickey J. S., Jr., Marvin U. B., and Powell B. N. (1970) Lunar anorthosites and a geophysical model of the moon. *Proc. Apollo 11 Lunar Sci. Conf.*, pp. 965–988.

Yoder C. F. (1981) The free librations of a dissipative moon. *Phil. Trans. Roy. Soc. London, A303*, 327–338.

Yoder C. F. (1984) The size of the lunar core (abstract). In *Papers Presented to the Conference on the Origin of the Moon*, p. 6. Lunar and Planetary Institute, Houston.

Lunar Paleointensities via the IRMs Normalization Method and the Early Magnetic History of the Moon

S. M. CISOWSKI AND M. FULLER

Department of Geological Sciences, University of California, Santa Barbara, CA 93106

The determination of a planetary body's magnetic field environment through time represents a fundamental set of information that may shed light on both the internal and orbital evolution of the body. Normalization of the magnetization that a rock has acquired in nature (NRM) to its saturation remanence [IRMs, i.e., remanence acquired after exposure to a strong (saturating) magnetic field] can provide an estimate of the relative strength of the ancient magnetizing field, if the magnetic carriers of all samples involved are of roughly the same grain size and composition, and if the carriers are assumed to have acquired a thermal remanence (acquired by cooling through the Curie temperature). The application of this paleointensity method to terrestrial rocks known to have recorded polarity transitions indicates that the method is well suited to delineating order-of-magnitude changes in magnetizing fields. The NRM/IRMs ratio values tabulated for 67 lunar samples of suitable character and diverse ages suggests that an anomalously high lunar surface field existed from about 3.8–3.6 aeons. The high field era is reflected only in Apollo 17 and Apollo 11 "low K" mare basalts, Apollo 16 North and South Ray crater breccias, and Apollo 17 green-grey breccias. Although the absolute strength of the high field remains uncertain, such a temporal variation in the lunar magnetizing field may reflect important physical and/or chemical events in the early history of the Moon, possibly relating to core formation and/or orbital evolution. Either of these processes, if further delineated by additional independent physical data, would put additional constraints on the primitive composition and origin of the Earth's satellite.

Introduction

Unlike the Earth, the Moon does not presently possess a magnetic field of internal origin. However, surface and subsatellite measurements made during the Apollo missions indicated highly variable surface fields that seem to relate to coherently magnetized geologic units (e.g., ejecta blankets). Initial measurements on lunar samples returned to Earth verified that many, although not all, lunar specimens retained some record of an ancient magnetizing field. The importance of such a field to

---

the problem of the origin of the Moon depends upon the nature of its source. If, for example, future lunar polar orbiters demonstrate the fossil field to be of internal origin, the presence of a once-actively-convecting core would be inferred in order to provide the necessary dynamo site. Such a conclusion would have important implications concerning the composition of the material that formed the Moon, particularly with respect to siderophile elements. This in turn could lead to further constraints on the source of this matter. If, on the other hand, it is demonstrated to be of external origin, then a major new puzzle would arise to account for such a field in the vicinity of the Moon at the required time. This result might then relate to the early orbital evolution of the Moon, as its proximity to the Earth's magnetic field would become of critical importance. Unfortunately, with the present limited directional data, the source of the ancient lunar field remains a matter of speculation and contention.

Since the discovery of lunar sample remanence, much attention has focused on experiments designed to determine the intensity of the lunar paleofield, but, in many cases, with results less than ideal. The chief difficulty is in attempting to simulate the acquisition of thermal magnetic remanence by heating and cooling rocks containing complex, metastable magnetic phases. On the other hand, relatively little attention has been paid to the directional description of the paleofield, due to the lack of oriented bedrock samples. However, directional data have been inferred from subsatellite measurements with varying interpretations of their meaning (Hood, 1981; Runcorn, 1983).

IRMs Normalization Technique

In order to fully utilize magnetic intensity data from all workers in the field, we have devised a relative paleointensity method based on the normalization of natural remanence (NRM) to saturation remanence (IRMs) magnetization, as measured after exposing each sample to a strong (~6000 oersteds) magnetic field (Cisowski et al., 1983). Our approach differed from previous attempts to use IRMs as a normalization parameter in that alternating field (AF) demagnetized IRMs as well as NRM intensities are used, thus totally eliminating from consideration all coarse-grained remanence carriers. Demagnetization to 200 oersteds AF was chosen as the optimum alternating field, as lower AF values might not be sufficient to remove secondary (i.e., nonlunar) components of magnetization, whereas higher AF values might introduce spurious remanence effects associated with the demagnetization process. The assumption is made that in most cases the remanence carried by lunar samples is a thermal remanence, i.e., it was acquired on cooling through the Curie temperature of the component magnetic phases. This seems to be a reasonable assumption for the majority of lunar samples for which magnetic data is available, as they consist primarily of mare and highland basalts, high-grade recrystallized breccias, and glass-bearing, well-lithified regolith breccias. IRMs normalization has the distinct

---

advantage of taking into account the total remanence carrying potential of each sample, although the character of the weak field remanence carriers may be somewhat obscured by this high field technique.

The IRMs normalization technique is based on the observation that the acquisition of thermal remanence is linearly proportional to ambient field strength during blocking (acquisition of remanence) of fine-grained magnetic particles, at least in the low field (<5 oersteds) region. Figure 1 shows the results of a TRM acquisition experiment done on a native iron-bearing synthetic lunar basalt (Cirlin and Housley, 1980), illustrating reasonably linear acquisition of normalized thermal remanence from 17,000–98,000 gammas. Although the rate of remanence acquisition can be expected to change with the composition and size of the magnetic carrier, all lunar samples are dominated magnetically by Fe-Ni alloys, and the use of demagnetized data generally limits consideration to fine-grained remanence carriers. The lack of correlation in the weak field portion of this experiment may result from the presence of iron sulfide compounds in addition to native iron.

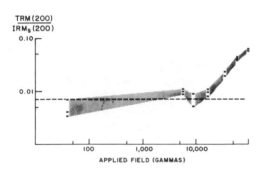

Fig. 1. TRM/IRMs ratio after demagnetization to 200 oersteds vs. applied laboratory field for iron-bearing synthetic lunar sample. Dashed line shows boundary between strongly magnetized (3.6–3.9 aeons) and weakly magnetized lunar samples.

The IRMs normalization method has not been commonly used in determining paleointensities on terrestrial rocks, since it lacks the precision needed to detect small changes in paleointensity. In order to test the sensitivity of the IRMs normalization technique to larger changes in magnetic field strength, we have examined specimens from two terrestrial igneous bodies known to have recorded polarity reversals in the Earth's dipole field (Dodson et al., 1978; Williams and Fuller, 1982). The magnetic properties of both intrusions reflect a mixture of coarse- and fine-grained magnetic carriers. The absence of the dipole field during reversal intervals is known to decrease the intensity of the geomagnetic field by about an order of magnitude (Wilson et al., 1972). Figure 2 shows the sequential variation in normalized intensities, at varying AF demagnetization levels, in going from the reversed polarity zone (R), through intermediate polarities (I), to normal polarity (N) for both igneous bodies. The order-of-magnitude change in field intensity is clearly reflected in the NRM/IRMs ratios, particularly after demagnetization. At the Agno intrusion, sampling was confined to normal and reversed zones immediately adjacent to the transition zone, whereas

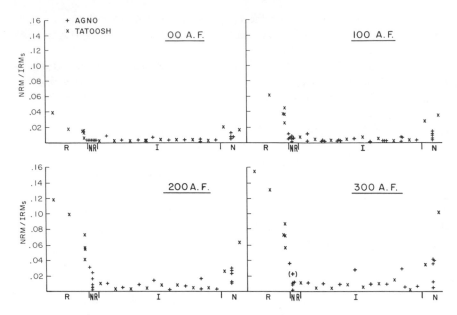

Fig. 2.   NRM/IRMs ratio at varying demagnetization levels for samples from Tatoosh and Agno igneous bodies. Sites are plotted according to relative position with respect to polarity transition interval. R: reversed; NR: normal before demagnetization, reversed after; I: intermediate; N: normal.

the Tatoosh reversed samples were in general taken at considerable distances from the reversal plane. As intensity loss and recovery have been observed to occur over a longer time frame than the observed directional transition during field reversals (Hillhouse and Cox, 1976; Dodson *et al.*, 1978), it is likely that none of the Agno samples reflect the initial field strength before the reversal took place.

Figure 3 shows the results of an experiment in which 41 terrestrial igneous rocks of widely varying compositions (with opaque minerals primarily as iron-titanium oxides) and magnetic grain size were heated to 700°C (above their component magnetic mineral Curie temperatures) and cooled in an ambient field of 0.2 oersteds (20,000 gammas). Eighty percent of the samples display TRM intensities, normalized to postcooling IRMs values after demagnetization, within a factor of 5 range.

These three experiments indicate that the IRMs normalization method can reflect relative magnetizing field intensity well within an order of magnitude for the majority of samples, and can define order-of-magnitude changes in the strength of the geomagnetic field. Because the principal magnetic carrier in lunar samples is Ni-Fe rather than Fe oxide, ratio values between lunar and terrestrial samples may not be directly comparable. However, as the same principal of linear acquisition

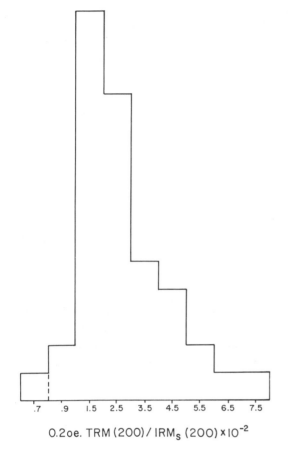

Fig. 3.   Histogram of TRM/IRMs ratio after demagnetization to 200 oersteds for 41 terrestrial igneous samples heated and cooled to 700° C in a field of 0.20 oersteds. Height of small square at bottom left equals one sample.

$$0.2\text{oe. TRM}\,(200)\,/\,\text{IRM}_s\,(200)\times 10^{-2}$$

of TRM with applied field is involved, the method itself should be equally valid to the lunar data set.

IRMs Normalization Results on Lunar Samples

NRM and IRMs AF demagnetization data are now available for over 90 lunar samples for which some age determinations or estimates have been made. However, in the course of compiling and collecting this data it became clear that the IRMs normalization technique is unsuitable for certain types of lunar samples. The first type includes samples dominated by coarse-grained carriers, such that IRMs demagnetized to 200 oersteds is reduced to less than 25% of its initial value. This group of 12 samples consists primarily of mare basalts whose coarse grain textures indicate cooling rates of <1°C/hr (Lofgren and Usselman, 1975; Usselman *et al.*,

1975). These samples are deemed unsuitable in that repeat demagnetizations at the same alternating field, or even at subsequently lower fields, result in large variations in both intensity and magnetic vector direction. Mare basalt sample 15475 (Fig. 4a) is an example of this type of behavior. In this and other similar samples the measured remanence seems to represent a ground state of magnetization below which the sample cannot be demagnetized, rather than a record of the ancient magnetizing field. The existence of such a ground state in iron-bearing samples is well illustrated in Fig. 1, where normalized TRM intensities acquired by the synthetic lunar sample in 40 and 9,000 gamma fields overlap. TRM's acquired by this synthetic sample in fields of less than 17,000 gammas also showed erratic directional behavior, further mimicking lunar samples in this "ground" magnetic state. This behavior is similar to what other workers have described as zig-zag or textural remanence behavior (Hoffman and Bannerjee, 1975; Brecher, 1976). Because of the multidomain nature of these samples, we believe their erratic demagnetization behavior is related to random motion of domain walls between several different pinning sites. An additional fine-grained mare basalt sample (15499) has also been included in this group as it exhibits high directional instability and an extremely low coercive force (7 oersteds; Fuller *et al.*, 1979), and is believed to be dominated by strongly interacting Fe-Ni grains (Chowdhary, personal communication, 1984).

A second smaller group of five samples is deemed unacceptable because individual subsamples from the same rock have given large (order-of-magnitude) variations in demagnetized intensity. Figure 4b (mare basalt 10017) is an example of this phenomenon. This variation is *not* typical of most lunar samples that have undergone

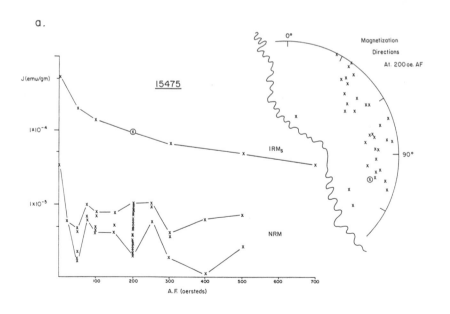

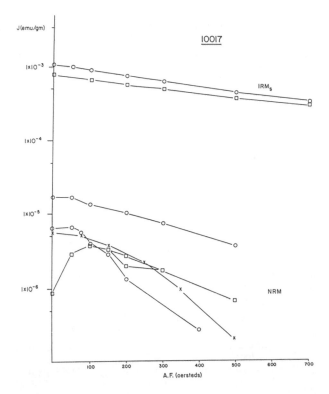

b.

Fig. 4. NRM and IRMs demagnetization curves and vector directions for (a) coarse-grained mare basalt 15475; (b) all measured subsamples of mare basalt 10017 (data from Hoffman et al., 1979; Sugiura et al., 1978; Cisowski et al., 1983).

multiple magnetic measurements, even though individual subsamples were often measured on different magnetometers in different laboratories, sometimes years apart. Although the cause of the variations within these five samples remains enigmatic, some form of postsampling contamination may be involved. As several of the anomalously high subsamples (e.g., 10017,28) have prominent sawblade markings on them, the possibility of contamination during sawing is suspected. Tables 1 and 2 list all samples of these two groups along with the NRM intensity ranges for group 2.

Figure 5 is our latest compilation of normalized intensity data on all samples measured to date (excluding the two groups above) with some age determination or estimate. Among these samples, the unshocked mare basalts and aphanitic highland samples (primarily recrystallized breccias and impact melts) are the most likely to have recorded a thermal remanence (TRM). The most lithified regolith breccias

TABLE 1. Samples with "Soft" IRM Demagnetization Curves (13).

Mare Basalt (Cooling Rates <1°C/Hr Except 15499 and 70135)	Highland Samples
15075	14053
15085	14072
15475	67455
15499#	68415
70017	
70035	
70135#	
75075	
78505	

TABLE 2. Samples with Large Variations in Intensity Between Subsamples (5).

Sample #	NRM Range at 200 Oersteds AF ($\times 10^{-6}$ emu/gm)	
10017	1.3–11.0	
10047	0.3–4.9	Mare Basalts
15597	0.3–40.0	
14267	6.7–150.0	
77215	0.3–4.8	Highland Samples

are also likely to carry a TRM resulting from shock heating. The data continue to suggest a period of relatively high magnetic field intensity for the Moon from 3.6–3.85 aeons (Cisowski *et al.*, 1983), but with highly variable intensities from about 3.85–3.95 aeons. Error bars have been shown only for samples immediately to either side of these time intervals. The small number of dated samples (three) with ages older than 3.95 aeons makes it difficult to state with confidence whether a high field may have existed at any time beyond 3.95 aeons, although the suggestion is that the field was either lower or fluctuated widely before then. Note that one of the oldest samples (14063) is a fragmental matrix breccia that most likely was not heated to above its Curie temperature during assembly.

Considerable overlap in radiometric age error bars exists for samples older than 3.8 aeons. It was thought that part of the wide scatter in normalized intensities

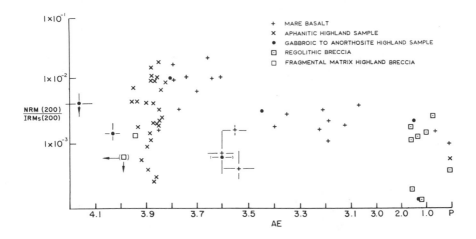

Fig. 5. NRM/IRMs ratio after demagnetization to 200 oersteds vs. age for 67 lunar samples. Radiometric ages are adjusted for the new radiometric decay constants (Steiger and Jager, 1977).

at about 3.85 aeons could be better resolved if a relative age relationship between samples was established. In order to better separate the large number of samples whose radiometric ages cluster around this time into relative age groupings, two alternate chronological approaches have been attempted. The first is to arrange certain groups of samples into relative stratigraphic ages, when known. Thus Fig. 6 displays all measured lunar basalts in their age sequence, extending from the old VHA and KREEP highland basalts and the distinctly oldest mare basalt, 10003, to the young Apollo 12 mare basalts. Likewise, Fig. 7a contrasts the normalized intensities of the Apollo 17 green-grey breccias, which are known to be stratigraphically younger (Schmitt, 1973; Stettler *et al.*, 1978), to the Apollo 17 blue-grey breccias. The second approach is to separate lunar highland breccias into age groupings as defined by their meteoritic trace element abundance patterns. By comparing clast-matrix relations, a relative age scale has been devised for most breccias (Hertogen *et al.*, 1977). Although the assignment of individual groups to specific basin-forming events requires a high degree of interpretation, the separation of the breccias into older and younger groupings is less subjective. Figure 7b presents the distribution of normalized intensities between post-Serenitatis age and older breccias, as defined by meteoritic metal abundance patterns.

Both approaches clearly reinforce the observation that the older lunar samples exhibit a lower range of normalized intensities, although the younger breccia samples show a wide variation in intensities. While still somewhat tentative, the implication is that the strong lunar magnetic field came into existence shortly after Serenitatis

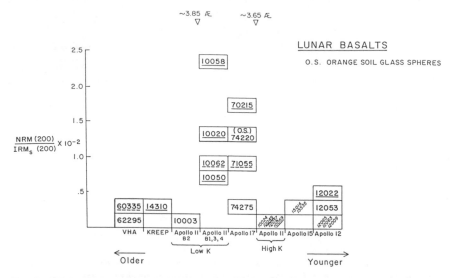

Fig. 6. NRM/IRMs ratio values after demagnetization to 200 oersteds vs. age for lunar basalts. Underlined samples exhibit high directional stability to AF demagnetization. Dashed samples exhibit moderate directional stability.

time, about 3.9 aeons, and that before that time the field was not significantly stronger than it was during the time of extrusion of the younger mare basalts. Note also that the high intensity samples are generally stable in direction with demagnetization (as indicated by underlining in Fig. 6), as would be expected if they carried a strong field thermal remanence.

Conclusions

Figure 8 is a linear representation of the normalized intensity data, calibrated to an absolute paleointensity scale calculated from a correlation plot for samples on which both relative and absolute paleointensity estimates have been made (see Fig. 3 of Cisowski *et al.*, 1983). This calibration would indicate that magnetic fields comparable to the Earth's present field strength existed on the surface of the Moon for several hundred million years. However, paleointensity estimates involving heating of lunar samples may have overstated the intensity of the paleofield due to progressive destruction of fine-grained magnetic carriers (Walton, 1983, p. 90–91). Interestingly, a comparison of the lunar normalized intensity ratios to those of the synthetic lunar sample (Fig. 1) and of terrestrial samples (Figs. 2 and 3) suggests that the lunar high field era may have primarily involved intensities of only thousands to a few tens of thousands of gammas, rather than the up to 1 oersted (100,000 gammas) fields indicated from other lunar paleointensity studies (Collinson *et al.*, 1977).

a.

APOLLO 17 GREEN-GREY BRECCIAS

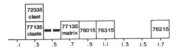

BLUE-GREY & LIGHT GREY BRECCIAS

b.

POST-SERENITATIS BRECCIAS
(Groups 1H&5H)

OLDER BRECCIAS
(Groups 2,3,6,&7)

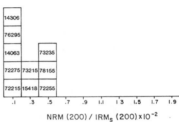

NRM(200) / IRM$_s$ (200) ×10^{-2}

Fig. 7. Histographs of NRM/IRMs ratio values for (a) Apollo 17 green-grey (younger) vs. blue-grey (older) boulder breccias; (b) post-Serenitatis vs. older lunar breccias, as inferred from meteoritic trace metal abundance groupings (see Hertogen et al., 1977).

Whatever the absolute strength and source of any ancient lunar surface field, a clear picture has emerged from this analysis of lunar sample magnetic data. Figure 9a summarizes the normalized intensity data for the four types of lunar samples that fall in age between 3.6 and 3.9 aeons. These samples show a distinct distribution of intensities compared to samples of all other types and ages. AF demagnetization

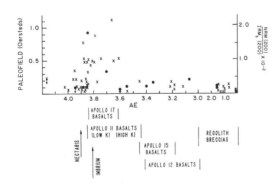

Fig. 8. Absolute (left scale) and relative (right scale) paleointensity estimates on 67 lunar samples vs. age. Absolute intensities are derived from correlation plot based on 12 samples (bullet) with both relative and absolute paleointensity estimates (see Cisowski et al., 1983).

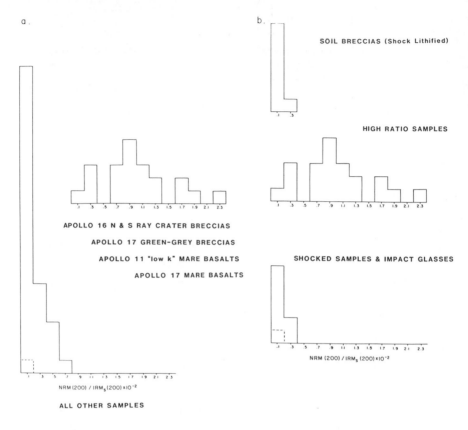

Fig. 9. Histographs of NRM/IRMs ratio values for (a) four groups of samples with ages between 3.6–3.9 aeons vs. all other samples; (b) the same high ratio samples vs. soil breccias, pervasively shocked samples, and soil breccias.

to 300 and 400 oersteds shows a similar distribution of normalized intensity values, with median ratio values of 1.0 and 1.6×10^{-02} for the high intensity (3.6–3.9 aeon) samples, as compared to a median ratio value of 1.0×10^{-02} at 200 oersteds for these same samples. Considering the diverse nature but similar age of the high intensity samples, the simplest explanation is that the lunar surface field was at times significantly stronger between 3.6 and 3.9 aeons. Because normalized intensities for samples outside this time range fall in the range of the ground state of magnetization as illustrated for the synthetic lunar sample of Fig. 1, absolute paleointensities for these samples cannot be obtained using the IRMs normalization method. The most that can be said for these samples is that they acquired their magnetization in fields of less than 17,000 gammas.

Figure 9b, which compares the high ratio samples to shock-lithified soil breccias and impact glasses, and other lunar samples with pervasive mineralogic shock effects, indicates that the high fields implied by these samples are not associated with piezomagnetic or other impact-related phenomena. Possible sources for this high field include a short-lived, core-residing lunar dynamo, close approach to the Earth and its own internal dipole field, or some combination of the above. The first of these potential sources is perhaps the most significant to the problem of the origin of the Moon, as it would imply that the material now composing the lunar crust and mantle underwent some depletion of siderophile elements before 3.9 aeons. Any calculations of predifferentiated lunar compositions based on surface sample abundances would have to be adjusted for this possibility, thus more rigorously defining the source material for the Moon. Recent interpretations of various geophysical data on the internal properties of the Moon seem to leave open the possibility of a small iron core (~400 km radius) (Hood and Jones, 1985), which may or may not be of sufficient size to have facilitated the highly efficient lunar dynamo needed to produce a surface field of even thousands of gammas intensity. The restricted period of time over which the strong lunar magnetic field apparently existed might also be related to such phenomena as the formation and subsequent solidification of a lunar core, or to an enhanced rotational speed of the Moon, resulting from its projected close proximity to the Earth early in its history. In the latter case both an increase in Coriolis force and tidal heating effects would serve to make a self-regenerating core dynamo move tenable, while the proximity to the Earth's magnetic field might allow for an amplification of the ambient field via the lunar core, as envisioned by Levy for the moons of the giant planets (Levy, 1979).

Acknowledgments. This work was supported under NASA grant NSG 7586. The authors would like to thank Robert Housley for his assistance in supplying the synthetic lunar sample and J. R. Dunn for his assistance in performing laboratory work.

References

Brecher A. (1976) Textural remanence: A new model of lunar rock magnetism. *Earth Planet. Sci. Lett.,* 29, 131–145.

Cirlin E. H. and Housley R. M. (1980) Redistribution of volatiles during lunar metamorphism. *Proc. Lunar Planet. Sci. Conf. 11th,* pp. 349–364.

Cisowski S. M., Collinson D. W., Runcorn S. K., Stephenson A., and Fuller M. (1983) A review of lunar paleointensity data and implications for the origin of lunar magnetism. *Proc. Lunar Planet. Sci. Conf. 13th,* in *J. Geophys. Res., 87,* A691–A704.

Collinson D. W., Stephenson A., and Runcorn S. K. (1977) Intensity and origin of the ancient lunar magnetic field. *Phil. Trans. Roy. Soc. London, A285,* 241–247.

Dodson R., Dunn J. R., Fuller M., Williams I., Ito H., Schmidt V. A., and Wu Y.-M. (1978) Paleomagnetic record of a late Tertiary field reversal. *Geophys. J. Roy. Astron. Soc., 53,* 373–412.

Fuller M., Meshkov E., Cisowski S. M., and Hale C. J. (1979) On the natural remanent magnetism of certain mare basalts. *Proc. Lunar Planet. Sci. Conf. 10th,* pp. 2211–2233.

Hertogen J., Janssens H., Takahashi H., Palme H., and Anders E. (1977) Lunar basins and craters: Evidence for systematic compositional changes of bombarding population. *Proc. Lunar Sci. Conf. 8th*, pp. 17–45.

Hillhouse J. and Cox A. (1976) Brunhes-Matuyma polarity transition. *Earth Planet. Sci. Lett., 29*, 51–64.

Hoffman K. A. and Bannerjee S. K. (1975) Magnetic "zig-zag" behavior in lunar rocks. *Earth Planet. Sci. Lett., 25*, 331–337.

Hoffman K. A., Baker J. R., and Bannerjee S. K. (1979) Combining paleointensity methods: a dual-valued determination or. lunar sample 10017,135. *Phys. Earth Planet. Inter., 20*, 317–323.

Hood L. L. (1981) Sources of lunar magnetic anomalies and their bulk directions of magnetization: Additional evidence from Apollo orbital data. *Proc. Lunar Planet. Sci. 12B*, pp. 817–830.

Hood L. L. and Jones J. (1985) Lunar density models consistent with mantle seismic velocities and other geophysical constraints (abstract). In *Lunar and Planetary Science XVI*, pp. 360–361. Lunar and Planetary Institute, Houston.

Levy E. H. (1979) Planetary dynamo amplification of ambient magnetic fields. *Proc. Lunar Planet. Sci. Conf. 10th*, pp. 2335–2342.

Lofgren G. E. and Usselman T. M. (1975) Cooling history of Apollo 15 quartz normative basalts determined from cooling rate experiments (abstract). In *Lunar Science VI*, pp. 516–517. The Lunar Science Institute, Houston.

Runcorn S. K. (1983) Lunar magnetism, displacements of the pole and primeval moons in the Earth-Moon system. *Nature, 304*, 589–596.

Schmitt H. H. (1973) Apollo 17 report on the valley of Taurus-Littrow. *Science, 182*, 681–690.

Steiger R. H. and Jager E. (1977) Subcommission on geochronology: Convention on the use of decay constants in geo- and cosmochronology. *Earth Planet. Sci. Lett., 36*, 359–362.

Stettler A., Eberhardt P., Geiss J., Grogler N., and Guggisberg S. (1978) Chronology of the Apollo 17 Station 7 boulder and the south Serenitatis impact (abstract). In *Lunar and Planetary Science IX*, pp. 1113–1115. Lunar and Planetary Institute, Houston.

Sugiura N., Strangway D. W., and Pearce G. W. (1978) Heating experiments and paleointensity determinations. *Proc. Lunar Planet. Sci. Conf. 9th*, pp. 3151–3163.

Usselman T. M., Lofgren G. E., Donaldson C. H., and Williams R. J. (1975) Experimentally reproduced textures and mineral chemistries of high-titanium mare basalts. *Proc. Lunar Sci. Conf. 6th*, pp. 997–1020.

Walton D. (1983) The reliability of ancient intensities obtained by thermal demagnetization. In *Geomagnetism of Baked Clays and Recent Sediments* (K. M. Creer, P. Tucholka, and C. E. Barton, eds.), pp. 88–97. Elsevier, Amsterdam.

Williams I. and Fuller M. (1982) A Miocene polarity transition (R-N) from the Agno Batholith, Luzon. *J. Geophys. Res., 87*, 9408–9418.

Wilson R. L., Dagley P., and McCormack A. G. (1972) Paleomagnetic evidence about the source of the geomagnetic field. *Geophys. J. Roy. Astron. Soc., 28*, 213–224.

The Initial Thermal State of the Moon

ALAN B. BINDER

Code SN4, NASA Johnson Space Center, Houston, TX 77058

Two independent sets of data that support the possibility that the Moon was initially totally molten are reviewed. First, studies of mare basalts and highland anorthosites indicate that the concentration level of the nonvolatile, incompatible trace elements in the ~200-km-deep residual magma at the time the anorthosites of the primary crust and the mare basalt source region first began to form was 10 to 20 × CI. In order for these results to be consistent with a U and Th content of $\leq 3 \times$ CI for the bulk Moon, as derived from the heat flow data, the incompatible elements must have been concentrated in the residual magma by the earlier fractional crystallization of olivine or orthopyroxene by a factor of ~5. This requires that most or all of the Moon was initially molten. Second, the existence of young (≤ 0.5 to 1×10^9-year-old), 10-km-scale, highland thrust faults and >1-kbar stress drop, shallow moonquakes imply that the current global, thermoelastic stresses in the crust are in the few kilobar range. Such large stresses favor lunar models that call for a totally molten origin.

1. Introduction

The initial thermal state of the Moon is a direct consequence of its mode of formation and to a large extent determined the course of its initial chemical differentiation and subsequent petrological and endogenetic tectonic evolution. For example, the Moon was most probably initially totally molten if it formed by binary fission (e.g., O'Keefe, 1969; Wise, 1969; Binder, 1980) or as a result of a large impact on the proto-Earth (Hartmann and Davis, 1975; Cameron and Ward, 1976). Alternatively, the Moon was probably molten only in its outer few hundred kilometers if it formed by accretion (e.g., Ransford and Kaula, 1979; Wood, 1982). Attempts to define the Moon's initial thermal state on the basis of thermoelastic stress models, petrological models, or geophysical models have been made by a number of investigators.

Solomon and Chaiken (1976) and Solomon and Head (1979) investigated the thermal and thermoelastic stress histories of the Moon assuming that it initially had a shallow magma ocean and a cool deep interior. Solomon and Chaiken (1976) initially argued that since the Moon does not have any large-scale thrust fault scarps like those found on Mercury by Strom *et al.* (1975), the lunar radius could not have thermally contracted by more than 1 km during lunar history. Using this value as a boundary condition, Solomon and Chaiken concluded that the magma ocean was about 200 km deep. Later, Solomon and Head (1979) revised the depth of the magma ocean to about 300 km on the basis of their assumption that the absence of young lunar thrust faults indicates that the crustal thermoelastic stresses currently must be less than 1 kbar.

In contrast to these results, Runcorn (1977) has argued that the Moon was totally molten during its earliest history on the basis of his concept that the Moon had a very early dipole magnetic field and hence had a ~500-km-radius molten iron core. As part of their study of the thermal history of an initially totally molten Moon, Binder and Lange (1980) pointed out that the Strom *et al.* (1975) discussion of the amount of contraction needed to produce the large Mercurian fault scarps was misinterpreted by Solomon and Chaiken (1976). A correct extrapolation of the data derived from the Mercury observations to the lunar case limits its maximum radius change to somewhere between 1.7 and 17 km before large-scale fault scarps should be present. Since the thermal contraction of the model Moon presented by Binder and Lange (1980) is 5.4 km (and the current level of the thermoelastic stresses in its crust is 3.5 kbar), they concluded that the Moon could have been initially totally molten and should have young, modest-scale thrust faults in the highlands. Subsequently, Binder (1982a) and Binder and Gunga (1985) have shown that young, 10-km-scale thrust fault scarps do exist in the lunar highlands and Binder and Oberst (1985) have shown that kilobar-level stress drop crustal moonquakes have been observed in the highlands. These authors argue that their observations support the concept that the Moon was initially totally molten.

There are also a number of relatively unconstrained petrological models (e.g., Taylor, 1975, pp. 252, 318–324; Delano, 1979) that suggest that the Moon was initially molten to depths of 500–1000 km.

Given these varied results, the purpose of this paper is to review two sets of independent data: (1) the trace element distributions within lunar samples, and (2) evidence for the current tectonic regime and to explore their implications for the early thermal state of the Moon.

2. Trace Element Distributions

The trace element distributions in mare basalts and highland rocks can be used to define the trace element concentrations in the magma system from which the mare basalt source region and the primary crust formed. Given this information on the incompatible trace elements and similar data on their concentrations in the bulk Moon, it would be possible, via mass balance considerations, to determine how much of the Moon was initially molten.

The wide range of mare basalt and pyroclastic glass characteristics (major mineralogy, siderophile and incompatible trace element concentrations and distribution patterns, Eu anomalies, and Rb-Sr and Nd-Sm isotopic systematics) are successfully accounted for by a comprehensive model of mare basalt petrogenesis in which the mare basalt source region began to form (1) when the outer ~20% of the Moon crystallized, and (2) when the concentrations of the nonvolatile, incompatible trace elements [as represented by Sr, Ba, and REE (rare earth elements)] in the magma system were 15 to 20 × CI values (Binder, 1982b, 1985). Preliminary tests of this model suggest that it may also be able to account for the incompatible trace element concentrations and distribution patterns of at least the average of the highland ferroan anorthosites and the lunar granites.

Similarly, on the basis of the concentrations of the nonvolatile, incompatible trace elements in the plagioclases of ferroan anorthosites, Palme et al. (1984) have determined that the concentrations of these elements in the parental magma of these rocks were from 6 to 23 × CI, and were on average about 12 × CI. Given that (1) the amount of plagioclase in the crust is about 8% of the lunar mass (Taylor, 1975, pp. 252, 318–324), (2) the amount of plagioclase in the mare basalt source region is about 1% of the mass of the Moon (Binder, 1982b), and (3) the magma system from which the crust and mare basalt source region formed was cotectic (~50% normative plagioclase and ~50% normative pyroxene and quartz) (e.g., Warren and Wasson, 1979; Binder, 1982b), this magma system contained at least 18% of the lunar mass.

Thus the results of Binder (1982b, 1985) and Palme et al. (1984) are consistent with one another and together suggest that the primary feldspathic crust and the complimentary mafic mare basalt source region formed from a magma system that contained about 20% of the lunar mass and had initial concentrations of Sr, Ba, and REE in the range of 10 to 20 × CI. Given these results, there are two limiting possibilities as to how that came about. The first possibility is that the bulk Moon has a 10 to 20 × CI concentration of these elements and that only the outer 20% of the Moon was initially molten, i.e., a magma ocean model. The second possibility is that the entire Moon differentiated, i.e., the Moon was initially totally molten, and the magma system from which the crust and mare basalt source region formed was a residual magma formed after 80% of the Moon crystallized as olivine or orthopyroxene. In this case, the concentrations of Sr, Ba, and REE in the bulk Moon are 2 to 4 × CI. The question is, How can we determine which of these cases, or what intermediate case, is correct?

The answer can be sought using the limited amount of available heat flow data. The heat flow experiments carried out at the Apollo 15 and 17 landing sites yield values of 21 and 14 ergs/cm^2/s, respectively (Langseth et al., 1976). Assuming for the moment that the Moon is in thermal steady state, that nothing else has affected these values, and that they are reasonably representative of the global heat flow, the average of these limited measurements implies that the concentrations of U and Th in the bulk Moon are about 4.5 × CI. However, the thermal history models presented by Binder and Lange (1977) indicate that about 30% of the global heat flow is due to the initial heat of the Moon. Also, Conel and Morton (1975) and Warren and Rasmussen (1984) have done thermal model studies based on the differences in the thickness of the megaregolith and hence the conductivity under the maria and the adjacent basin ejecta blankets. They find that the heat flow at the edges of the maria (i.e., where the Apollo 15 and 17 sites are located) is enhanced by >25% over the global value. However, these investigators used depths of only a few kilometers for the megaregolith. According to the seismic data (Töksoz *et al.*, 1974; Mizutani and Usako, 1974) and thermal conductivity/porosity data (Keihm and Langseth, 1977), the megaregolith could be up to 20 km deep. If so, the enhancement of the heat flow at the edges of the mare due to the megaregolith/conductivity differences is less than these authors have found by some as yet undetermined value. In addition to this effect, the excavation of deep basins almost certainly removed KREEP from the subbasin areas and deposited it as part of the rim ejecta (a suggestion supported by the concentrations of U and Th seen around Mare Imbrium in the orbital gamma ray maps). This effect also leads to an undetermined enhancement of heat flow at the basin rims (Binder, 1975). Thus, the Apollo 15 and 17 values are almost certainly higher than the global value. Taking into consideration these ill-defined basin effects and the corrections on the heat flow values due to the initial heat of the Moon, the U and Th contents of the Moon are probably $\leq 3 \times$ CI.

Uranium, Th, Sr, Ba, and REE are all refractory, incompatible elements that condensed out of the solar nebula at essentially the same temperature and therefore should show no effects of fractionation (see Fig. 2 of Grossman and Larimer, 1974, or Fig. 4 of Anders and Ebihara, 1982). Thus their CI-normalized concentrations in the bulk Moon must all be very nearly the same. Since the heat flow data indicate that the bulk Moon concentrations of U and Th are probably $\leq 3 \times$ CI, then the concentrations of Sr, Ba and REE must be the same, i.e., within the 2 to $4 \times$ CI range derived on the basis of the mare basalt and anorthosite modeling. As discussed above, this constraint implies that the Moon was initially essentially totally molten.

3. Current Tectonic State of the Moon

Thermal history and thermoelastic stress models indicate that if only the outer few hundred kilometers of the Moon were initially molten, and if it had a relatively cool interior, then the global, compressional stresses in the crust would be <1 kbar, and therefore the highlands would be free from young thrust faults and shallow, crustal, tectonic moonquakes (Solomon and Chaiken, 1976; Solomon and Head, 1979). If the Moon was initially totally molten, the global, compressional stresses in the crust would be $\geqslant 1$ kbar, and therefore the highlands should have young (≤ 0.5 to 1×10^9-year-old), 10-km-scale thrust faults and shallow (<6 km deep),

>1-kbar stress drop tectonic moonquakes (Binder and Lange, 1980; Binder, 1982a). Young thrust faults have been found throughout the 4.4% of the highlands that were photographed at low-to-moderate sun angles by the Apollo panoramic cameras. Extrapolation of these data indicate that some 2000 such thrust faults exist in the highlands (Binder, 1982a; Binder and Gunga, 1985). The observed characteristics of these highland thrust faults, as determined by Binder and Gunga (1985), along with their predicted characteristics, are given in Table 1. As can be seen in Table 1, the agreement between the model predictions and the observations is excellent. Also, spectral analyses of the 28 shallow moonquakes observed with the Apollo seismic network show that three of them have stress drops that are >1 kbar (Binder and Oberst, 1985). Together, these results provide evidence that the Moon was initially totally molten.

TABLE 1.   Highland Thrust Fault Data.

	Predicted*	Observed
Number	–	~2000†
Average Length	10-km-scale	9.0 ± 0.4 km
Maximum Depth	<6 km	3.8 km
Maximum Age	≤0.5 to 1×10^9 yr	$0.7 \times 10^{9+(X2)}_{-(X4)}$ yr

*Based on thermoelastic stress models and thrust fault theory.
§Extrapolated from 4.4% of highlands observed using the Apollo panoramic images, to the entire highlands. All values are from Binder and Gunga (1985).

The interpretation of highland thrust faults as being due to kilobar-level global stresses has been questioned. The counterarguments are basically that the definition of the lower limit of the stress needed to cause the lunar thrust faults is incorrect since (1) the coefficient of friction used in the lunar calculations is higher than that defined empirically by Byerlee (1978) for terrestrial materials, and (2) the crushing strength of the material in the outer few kilometers of the Moon is near zero due to brecciation.

Analyses of the stresses associated with failure of an elastic medium during thrust faulting result in the following generalized equation (Anderson, 1951, pp. 7–11, 155–158):

$$\sigma \geq k\rho gz + C \tag{1}$$

where σ is the compressional stress needed to cause faulting, ρ is the density of the rock, g is the gravity, z is the depth of the fault, C is the crushing strength of the rock, and k is a function of the coefficient of friction (f),

$$k = (\sqrt{1 + f^2} + f)/(\sqrt{1 + f^2} - f) \tag{2}$$

The first term in (1) is due to the friction on the fault plane, caused mainly by the lithostatic load, and the second term is due to the crushing strength of the material. Data derived from an analysis of observed arcuate lunar thrust faults indicate that $f = 1.1$ (Binder and Gunga, 1985). Therefore, by (2), the coefficient of the depth-dependent frictional term in (1) is 0.3 kbar/km. The same data indicate that the lunar thrust faults are at least as deep as 4 km (Table 1). Thus the frictional term in (1) alone requires a minimum compressional stress of >1.2 kbar in order for the lunar thrust faults to have formed. However, it has been argued that f, and therefore k, are much lower than the values of 1.1 and 0.3 kbar/km, respectively, as derived from the lunar data, since Byerlee's (1978) empirical data shows that $f = 0.85$ for essentially all terrestrial materials. Using this value, the frictional term in (1) is 0.2 kbar/km, and hence only 0.8 kbar would be required to overcome friction on 4-km-deep lunar faults. Accepting this as a significant difference for the moment, note first that the Byerlee data are for terrestrial rocks, not for lunar rocks. As has been known for over a decade, the physical characteristics of rocks and fines are very different under lunar conditions (vacuum and no water) than under terrestrial conditions. For example, the rupture strength under lunar conditions is double that on Earth (Mizutani et al., 1977), the coefficient of internal friction (Q) is orders of magnitude higher in the lunar case (e.g., Tittmann et al., 1974), and the friction angle of lunar fines is higher (35°–52°; Mitchell et al., 1972) than it is for equivalent terrestrial fines (26°–34°; Lambe and Whitman, 1969, p. 149). These differences are attributed to (1) lubricating effects and (2) the hydraulic effects on the formation and propagation of ruptures in the rocks of water and atmospheric gases (Mizutani et al., 1977). These differences also demonstrate that the physical parameters derived for terrestrial rocks cannot be used when considering lunar problems. Specifically, as discussed by Byerlee (1978), the static friction he measured is caused not only by friction, but also by the rupturing of welded points of contact between the rocks. Since rupture under lunar conditions requires higher stresses than under terrestrial conditions (Mizutani et al., 1977), it follows that f for lunar rocks must be higher than the value of $f = 0.85$ found by Byerlee for terrestrial rocks. This is consistent with the results of Binder and Gunga (1985). Hence, as discussed above, the frictional term in (1) alone requires stresses >1.2 kbar if the highland faults are up to 4 km deep as the analyses of Binder and Gunga (1985) indicate.

Whether or not the crushing strength of the rock (C in equation (1)) is zero in the outer few km to 20 km of the Moon is questionable. Relatively coherent layers of rock were observed by the Apollo 15 astronauts at depths of a few meters in the wall of Hadley Rille, on the side of Silver Spur, and possibly in Mount Hadley (Swann et al., 1972). In the case of Hadley Rille, these layers are clearly mare basalt flows, while those in Silver Spur and Mount Hadley may be due to highland volcanism, impact melt, or sintering of hot ejecta blankets. Thus in the maria and in highland areas where coherent volcanic and impact units are found, the second term in (1) will be $\gg 0$ and might even approach the 4 ± 3 kbar stress needed to crush firm rocks under lunar conditions (Binder, 1982a).

Even if C is close to zero due to brecciation, the brecciated nature of the outer few kilometers of the Moon requires that a considerable amount of crustal shortening takes place before thrust faulting can occur. As shown in laboratory experiments, brecciated material must undergo considerable compaction before thrust faulting can begin (Binder and Gunga, 1985). Since the compressional thermoelastic stresses discussed are caused by the thermal contraction of the lunar radius and hence a shortening of the circumference, a large amount of this shortening is taken up by compaction of the brecciated outer crust before the stresses can build up and cause faulting. The amount of shortening required can be expressed in terms of an equivalent stress (i.e., the stress caused by the equivalent amount of elastic shortening of the crust). Based on model calculations and laboratory experiments, the equivalent stress needed to compact the brecciated parts of the crust is equal to or more than the crushing strength of the rocks (Binder and Gunga, 1985). Thus in order for thrust faulting to occur in the outer few kilometers of the crust, 10 km to a few tens of km of crustal shortening, which would be equivalent to several kilobars of thermoelastic stress, are required before thrust faulting can begin.

Together, the frictional term (which for 4-km-deep faults requires >1.2 kbar of stress) and the crushing term (C), which in the limiting cases is the crushing strength of coherent rocks (4 ± 3 kbar) or the compactional equivalent stress (several kilobars) for brecciated rocks, in (1) indicate that the observed highland thrust faults could have formed only if the global stresses were in the $\gg 1$ kbar range. This is possible only if the Moon was initially essentially totally molten.

In addition to the above arguments, there is the supportive evidence from the spectral analyses of the shallow moonquakes that indicate that 3 (out of the observed 28) of these moonquakes have stress drops >1 kbar (Binder and Oberst, 1985). Thus the seismic results are consistent with the theoretical stresses required to form the faults of the observed scarps.

4. Summary and Conclusions

Two independent data sets, (1) the mare basalt and highland anorthosite data on the incompatible trace element concentrations in the bulk Moon vs. those in the magma from which the crust and mare basalt source region were formed and (2) the data on the young, highland thrust faults and associated high stress, shallow moonquakes, indicate that the Moon was largely or totally molten early in lunar history. As has been shown earlier (Binder, 1974; Warren and Wasson, 1979), the chemical/petrological fractionation of an initially totally molten Moon is consistent with the known properties of the Moon, e.g., the massive feldspathic crust, the mare basalt source region, the Moon's bulk composition, etc. Given these conclusions, any lunar origin model must be able to explain how the Moon became molten in its earliest history if the model is to be acceptable.

Acknowledgments. I thanks Drs. V. L. Sharpton, P. Salpas, and R. Phillips for their most helpful reviews of the paper. This work was done while the author was a NRC Senior Fellow at the NASA/ Johnson Space Center.

5. References

Anders E. and Ebihara M. (1982) Solar-system abundances of the elements. *Geochim. Cosmochim. Acta,* *46*, 2363–2380.

Anderson E. M. (1951) *The Dynamics of Faulting.* Oliver and Boyd, Edinburgh. 208 pp.

Binder A. B. (1974) On the origin of the moon by rotational fission. *The Moon, 11,* 53–76.

Binder A. B. (1975) On the heat flow of a gravitionally differentiated moon of fission origin. *The Moon, 14,* 237–290.

Binder A. B. (1980) The first few hundred years of evolution of a moon of fission origin. *Proc. Lunar Planet. Sci. Conf. 11th,* pp. 1931–1939.

Binder A. B. (1982a) Post-Imbrium global tectonism: Evidence for an initially totally molten moon. *Moon and Planets, 26,* 117–133.

Binder A. B. (1982b) The mare basalt magma source region and mare basalt magma genesis. *Proc. Lunar Planet. Sci. Conf. 13th,* in *J. Geophys. Res., 87,* A37–A53.

Binder A. B. (1985) Mare basalt genesis: Trace elements and isotopic ratios. *Proc. Lunar Planet. Sci. Conf. 16th,* in *J. Geophys. Res., 90,* in press.

Binder A. B. and Gunga H.-C. (1985) Young thrust fault scarps in the highlands: Evidence for an initially totally molten moon. *Icarus,* in press.

Binder A. B. and Lange M. A. (1977) On the thermal history of a moon of fission origin. *The Moon, 17,* 29–45.

Binder A. B. and Lange M. A. (1980) On the thermal history, thermal state, and related tectonism of a moon of fission origin. *J. Geophys. Res., 85,* 3194–3208.

Binder A. B. and Oberst J. (1985) High stress shallow moonquakes: Evidence for an initially totally molten moon. *Earth Planet. Sci. Lett., 74,* 149–154.

Byerlee J. D. (1978) Friction of rocks. *Pure Appl. Geophys., 116,* 615–626.

Cameron A. G. W. and Ward W. A. (1976) The origin of the moon (abstract). In *Lunar Science VII,* pp. 120–122. The Lunar Science Institute, Houston.

Conel J. and Morton J. B. (1975) Interpretation of lunar heat flow data. *The Moon, 14,* 263–290.

Delano J. W. (1979) Apollo 15 green glass: Chemistry and possible origin. *Proc. Lunar Planet. Sci. Conf. 10th,* pp. 275–300.

Grossman L. and Larimer J. (1974) Early chemical history of the solar system. *Rev. Geophys. Space Phys., 12,* 71–101.

Hartmann W. K. and Davis D. (1975) Satellite-sized planetesimals and lunar origin. *Icarus, 24,* 504–515.

Keihm S. J. and Langseth M. G. (1977) Lunar thermal regime to 300 km. *Proc. Lunar Sci. Conf. 8th,* pp. 499–514.

Lambe T. W. and Whitman R. V. (1969) *Soil Mechanics.* Wiley and Sons, New York. 553 pp.

Langseth M. G., Keihm S. J., and Peters K. (1976) Revised lunar heat flow. *Proc. Lunar Sci. Conf. 7th,* pp. 3143–3171.

Mitchell J. K., Houston W. N., Scott R. F., Costes N. C., Carrier W. D., III, and Bromwell L. G. (1972) Mechanical properties of lunar soil: Density, porosity, cohesion, and angle of internal friction. *Proc. Lunar Sci. Conf. 3rd,* pp. 3235–3253.

Mizutani H. and Osako M. (1974) Elastic-wave velocities and thermal diffusivities of Apollo 17 rocks and their geophysical implications. *Proc. Lunar Sci. Conf. 5th*, pp. 2891–2901.

Mizutani H., Spetzler H., Getting I., Martin R. J., III, and Soga N. (1977) The effect of outgassing upon the closure of cracks and the strength of lunar analogues. *Proc. Lunar Sci. Conf. 8th*, pp. 1235–1248.

O'Keefe J. A. (1969) Origin of the moon. *J. Geophys. Res., 75*, 6565–6574.

Palme H., Spettel B., Wänke H., Bischoff A., and Stöffler D. (1984) Early differentiation of the Moon: Evidence from trace elements in plagioclase. *Proc. Lunar Planet. Sci. Conf. 15th*, in *J. Geophys. Res., 89*, C3–C15.

Ransford G. A. and Kaula W. M. (1979) A comparison of accretion models (abstract). In *Lunar and Planetary Science X*, pp. 998–1000. Lunar and Planetary Institute, Houston.

Runcorn S. K. (1977) Early melting of the moon. *Proc. Lunar Sci. Conf. 8th*, pp. 463–469.

Solomon S. C. and Chaiken J. (1976) Thermal expansion and thermal stress in the moon and terrestrial planets: Clues to early thermal history. *Proc. Lunar Sci. Conf. 7th*, pp. 3229–3243.

Solomon S. C. and Head J. W. (1979) Vertical movement in mare basins: Relation to mare implacement, basin tectonics, and lunar thermal history. *J. Geophys. Res., 84*, 1667–1682.

Strom R. G., Trask N. J., and Guest J. E. (1975) Tectonism and volcanism on Mercury. *J. Geophys. Res., 80*, 2478–2507.

Swann G. A., Bailey N. G., Batson R. M., Freeman V. L., Hait M. H., Head J. W., Holt H. E., Howard K. A., Irwin J. B., Larson K. B., Muehlberger W. R., Reed V. S., Rennilson J. J., Schaber G. G., Scott D. R., Silver L. T., Sutton R. L., Ulrich G. E., Wilshire H. G., and Wolfe E. W. (1972) Preliminary geologic investigation of the Apollo 15 landing site. In *Apollo 15 Preliminary Science Report*, pp. 5–1 to 5–112. NASA SP 289, NASA, Washington, DC.

Taylor S. R. (1975) *Lunar Science: A Post-Apollo View*. Pergamon, New York. 372 pp.

Tittmann B. R., Housley R. M., Alders G. A., and Cirlin E. H. (1974) Internal friction in rocks and its relationship to volatiles on the moon. *Proc. Lunar Sci. Conf. 5th*, pp. 2913–2918.

Toksöz M. N., Dainty A. M., Solomon S. C., and Anderson K. R. (1974) Structure of the moon. *Rev. Geophys. Space Phys., 12*, 539–567.

Warren P. H. and Rasmussen K. L. (1984) Megaregolith thickness, heat flow, and the bulk composition of the moon (abstract). In *Papers Presented to the Conference on the Origin of the Moon*, p. 18. Lunar and Planetary Institute, Houston.

Warren P. H. and Wasson J. T. (1979) Effects of pressure on the crystallization of a "chondritic" magma ocean and implications for the bulk composition of the moon. *Proc. Lunar Planet. Sci. Conf. 10th*, pp. 2051–2083.

Wise D. U. (1969) Origin of the moon from the earth: Some new mechanisms and comparisons. *J. Geophys. Res., 74*, 6034–6045.

Wood J. A. (1982) The lunar magma ocean, thirteen years after Apollo 11 (abstract). In *Workshop on Pristine Highlands Rocks and the Early History of the Moon* (J. Longhi and G. Ryder, eds.), pp. 87–89. LPI Tech Rpt. 83–01, Lunar and Planetary Institute, Houston.

On the Early Thermal State of the Moon

SEAN C. SOLOMON

Department of Earth, Atmospheric, and Planetary Sciences, Massachusetts Institute of Technology, Cambridge, MA 02139

New theories for the formation of the Moon from an accretion disk thrown into circumterrestrial orbit after the collision of a planet-size object with the Earth have led to a reexamination of the tectonic consequences of an initially molten Moon. Even the smallest estimates of radial contraction that would accompany cooling of the Moon from an initially molten state predict accumulated near-surface horizontal compressive stresses considerably in excess of the compressive strength of the upper lunar crust, estimated to be 0.5 to 1 kbar on the basis of topographic relief, the stress levels necessary to form mare ridges in mascon mare basins, and measurements of rock friction. Various mechanisms for relieving or modifying such large near-surface stresses are considered, including viscoelastic effects, widespread development of major fault systems, impact gardening, and opposing stresses arising from other global-scale processes. All of these mechanisms face substantial difficulties when tested against geological and mechanical information from the Moon and other terrestrial planets. These considerations pose a serious problem for theories of lunar origin that call for an initially molten state.

Introduction

As a planetary body warms or cools, its volume increases or decreases. The thermal stress associated with significant differential expansion or contraction of the interior can be manifested in the formation of diagnostic tectonic features at the planetary surface (Solomon and Chaiken, 1976). While a number of bodies in the solar system display global patterns of tectonic features suggestive of extended episodes of contraction (Mercury) or expansion (Ganymede and perhaps Mars), it has long been recognized (MacDonald, 1960) that the Moon lacks large-scale global tectonic features indicative of substantial volume changes. This fact may be used to test lunar thermal history models against the predicted changes in lunar radius and the predicted magnitudes of near-surface stresses.

A suite of lunar thermal history models was tested by Solomon and Chaiken (1976) and Solomon (1977) against the constraints that the Moon has changed in radius by no more than about 1 km and has accumulated globally no more than about 1 kbar of near-surface thermal stress since the end of heavy impact bombardment 3.8 b.y. ago. The thermal models shared many simplifying assumptions and adopted physical constants and were parameterized by the initial depth of an early lunar "magma ocean" (Wood et al., 1970) and the initial temperature of the deep interior. Given the adopted constraints and parameterization, an initial magma ocean depth of about 200–300 km was the preferred solution. This conclusion and the constraints on which it was based were later challenged by Binder and Lange (1980) and Binder (1982), who argued that with different assumptions the thermal evolution from an initially molten state could be made to be compatible with the tectonic history of the Moon.

This is both an appropriate time and an appropriate volume in which to reexamine this issue. Recent theoretical work on models for the formation of the Moon from an accretion disk thrown into circumterrestrial orbit after the collision of a single Mars-size object with the Earth (Hartmann and Davis, 1975; Cameron and Ward, 1976) indicates that the newly formed Moon would likely be completely molten (Thompson and Stevenson, 1983; Cameron, 1984; Stevenson, 1984). Examining the tectonic consequences of such an initial thermal state for the Moon thus provides an important test for these new theories of lunar formation. Further, considerable progress has been made in the last few years on understanding the state of stress in planetary lithospheres (e.g., Goetze and Evans, 1979; Brace and Kohlstedt, 1980; Phillips and Lambeck, 1980; Willemann and Turcotte, 1981), and this insight can profitably be applied to the question of permissible levels of thermal stress in the Moon.

In the next section we briefly review the arguments concerning the tectonic constraints on lunar thermal history models, with an emphasis on the magnitude of thermal stress. We then examine critically the limits on stress magnitudes and the routes by which initially molten lunar history models might be made to be consistent with such limits. Finally, we comment on the implications of this examination for models of lunar origin.

Tectonic Constraints on Lunar Thermal History

The first systematic application of lunar tectonics as a constraint on global thermal history was that of Solomon and Chaiken (1976). Following MacDonald (1960), it was argued that the absence of large-scale global tectonic features indicates that the Moon has not experienced significant expansion or contraction, at least since the time of heavy bombardment and impact basin formation prior to 3.8 b.y. ago. To quantify this constraint, it was further suggested that in the last 3.8 b.y. the change in lunar radius was limited to approximately ±1 km and the accumulated horizontal thermal stress at the lunar surface was limited to approximately ±1 kbar.

Several reasons were given for these choices (Solomon and Chaiken, 1976; Solomon, 1977). The limit of ±1 kbar thermal stress was suggested on the basis of the estimated strength of the uppermost lunar crust (see further discussion of this point in the next section). It is straightforward to show (see Appendix) that a limit on accumulated thermal stress σ_t is equivalent to a limit on net radius change ΔR through the relation

$$\frac{\sigma_t}{\Delta R} = \frac{E}{R(1-\nu)} \tag{1}$$

where E and ν are Young's modulus and Poisson's ratio, respectively, of the upper crust and R is the lunar radius. For instance, when $E = 10^{12}$ dyne/cm^2, $\nu = 0.25$, and $R = 1738$ km, (1) gives $\sigma_t/\Delta R = 0.8$ kbar/km, so that a ±1 kbar limit on σ_t is roughly equivalent to a ±1 km limit on ΔR. It should be noted that this value for Young's modulus is representative only of the average properties of the lunar crust and that lower values are more appropriate for the uppermost layers. A P-wave velocity equivalent to a Young's modulus in excess of 0.5×10^{11} dyne/cm^2 has been measured at 1.4 km depth at the Apollo 17 landing site (Cooper et al., 1974), however, and (1) may still be applied with negligible modification at such shallow depths to infer accumulated thermal stress.

An additional rationale for the selection of quantitative constraints on radius change and thermal stress for the Moon is given by a comparison with Mercury. Mercury may be distinguished tectonically from the Moon on the basis of the widespread presence of prominent lobate scarps, interpreted as thrust and reverse faults indicative of horizontal shortening of the lithosphere (Murray et al., 1974; Strom et al., 1975). Because these scarps are seen more or less evenly distributed throughout the area of Mercury imaged by Mariner 10, they have been taken to indicate that Mercury experienced global contraction and associated surface horizontal compression from the time of heavy bombardment until after the emplacement of the smooth plains (Strom et al., 1975; Strom, 1979). Planetary despinning may have been an additional source of stress, but such a possibility would not obviate the need for global contraction to explain the observed distribution of scarps (Melosh and Dzurisin, 1978). Since the Moon lacks features comparable in scale and distribution to Mercury's lobate scarps, a reasonable inference is that the thermal stress generated in any contraction stage of lunar thermal history was less than the compressive stress that led to scarp formation on Mercury (Solomon and Chaiken, 1976).

No direct estimate is available of the magnitude of stress that produced the lobate scarps. Estimates can be made, however, of the change in radius of Mercury associated with displacements on the scarps and, through (1), of the implied level of surface stress. The change in the radius of Mercury during scarp formation is given approximately by (Strom et al., 1975)

$$\frac{\Delta R}{R} = \left[\frac{1}{8\pi R^2} \sum_{i=1}^{N} \frac{L_i h_i}{\tan \theta_i} \right] \qquad (2)$$

where L_i, h_i, and θ_i are the length, average throw, and average angle of dip of the ith scarp, and N is an estimate of the global number of scarps. Strom et al. (1975) estimated that h_i is typically 1 km and θ_i is roughly $35 \pm 10°$, and they measured the total length of scarps over the 24% of the surface mapped to be 1.5×10^4 km. Equation (2) then gives $\Delta R/R = 0.4$ to 0.9×10^{-3} or $\Delta R = 1$–2 km. Strom et al. (1975) cautioned that this estimate had a considerable uncertainty. A more thorough analysis was conducted by Cordell (1977), who measured the dip angle of about 15 scarps that transect impact craters. Cordell obtained an average value of $\bar{\theta} = 41 \pm 25°$, a figure he estimated to be low because several of the values used to estimate the average were only lower limits. Shadow measurements of 40% of the scarps gave an average height or throw of $\bar{h} = 0.61 \pm 0.31$ km, and the average length was found to be $\bar{L} = 106 \pm 80$ km (Cordell, 1977). If these average values are representative, (2) gives $\Delta R = 0.74$ km. The errors in $\bar{\theta}$ alone permit this figure to be larger by a factor of three, or smaller by a factor of two, although the larger values are less favored because of the probable low bias to $\bar{\theta}$. In sum, the best estimate of ΔR for Mercury during the last ~4 b.y. is 1–2 km (Strom et al., 1975; Cordell, 1977; Strom, 1979). By (1), assuming E $= 10^{12}$ dyne/cm^2 and $\nu = 0.25$, a limit of 0.5 to 1 kbar may be estimated for the associated surficial compressive stress.

To estimate the radius change and thermal stress accompanying lunar thermal history, Solomon and Chaiken (1976) made a number of simplifying assumptions to both the temperature and thermal stress models. A suite of thermal history models was considered; the models differed only in the assumed values of two parameters that specified the initial temperature profile: the depth of an initially molten "magma ocean" and the central temperature. The abundances of important radiogenic heat sources (present bulk U concentration of 30 ppb) were set to match the Apollo heat flow values (Langseth et al., 1976), and the heat sources within the initially molten zone were assumed to have differentiated into the crust. The thermal history models included conduction as the only heat transport mechanism in the solid state. Where melting reactions occurred, the calculations included heat of fusion but not the change in specific volume. A uniform volumetric coefficient of thermal expansion $(4 \times 10^{-5}\,°C^{-1})$ was assumed. With these assumptions, the predicted radius change and accumulated thermal stress over the last 3.8 b.y. satisfied $\Delta R < 1$ km and $\sigma_t < 1$ kbar (to within a factor of two) for a magma ocean depth of 200 ± 100 km and an initial central temperature less than $500°C$. In contrast, an initially molten model contracted by 16 km over 4.6 b.y. By (1), with the parameters adopted earlier, this is equivalent to an accumulated thermal stress σ_t of 12 kbar.

These calculations were repeated by Solomon (1977) under a somewhat improved scheme for determining radius change. The thermal expansion coefficent was taken to differ in the crust and mantle and to be temperature dependent (Baldridge and Simmons, 1971; Skinner, 1962). First-order finite-strain theory, subject to conservation of mass, was employed to find the lunar radius at each time increment in the thermal history model. Somewhat greater heat source abundances (40 ppb U) were also assumed. The resulting calculations were in essential agreement with the earlier ones: an initial magma ocean depth of 200–300 km and an initial central temperature less than 500°C best fit the constraint that $\Delta R < 1$ km in the last 3.8 b.y.

Cassen et al. (1979) calculated the predicted change in lunar radius for thermal history models with similar initial states, but they explicitly included the effect of solid-state convection on heat transport in the lunar mantle. Though they did not explore fully the range of permissible initial conditions for their models, they found that a model with initial melting in the outermost 300 km and an initial central temperature of less than about 650°C predicts a radius change of less than 1 km for the last 3.5 b.y. In contrast, their models with initially hotter profiles contracted by more than 1 km over the same time interval.

These results are not surprising. The restriction on ΔR and σ_t since heavy bombardment amounts to the constraint of nearly constant lunar volume (MacDonald, 1960). The outer 300 km of the Moon represents half the lunar volume. Thus the preferred thermal history models, given this constraint and the parameterized initial state, are those with the outer half of the lunar volume initially hot and the inner half initially cold (Fig. 1a). As the Moon evolves in such models, the outer half tends to cool and the inner half generally warms, maintaining the desired nearly constant volume. In contrast, an initially molten Moon model generally cools and contracts throughout its history (Fig. 1b), leading to a potentially substantial accumulation of compressive surface stress. As an aside, it may be noted that Turcotte (1983) has argued that the mean tangential thermal stress in the elastic lithosphere is likely to have been extensional throughout lunar evolution, even for models with hot initial states. While this argument is valid, the local thermal stress at and near the lunar surface is compressive for all such models and it is this stress that is constrained by surficial tectonic features.

The specific limits on change in lunar radius and accumulated thermal stress discussed above were questioned by Binder and Lange (1980) and Binder (1982), who argued that a thermal history model with an initially molten state could be made to be consistent with the absence of large-scale tectonic features on the Moon. The specific model of lunar formation envisioned by these authors was fission from a rapidly rotating Earth (Binder, 1974), but the thermal history calculation was expected to be representative of alternative models of lunar formation with early whole-Moon melting (e.g., Runcorn, 1977). The thermal history model of Binder and Lange (1980) was based on an initial state in which the Moon is everywhere

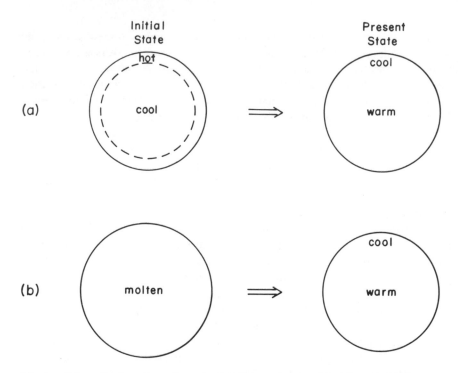

Fig. 1. Schematic view of two alternative initial thermal states of the Moon. In (a) the outer half by volume is initially hot and the inner half initially cool; in (b) the Moon is initially molten. Both evolve to similar present states, but (a) maintains a nearly constant volume over the past 3.8 b.y. while (b) contracts throughout lunar history.

at the solidus temperature; this does not strictly represent a molten Moon since neither the heat of fusion nor the liquidus-solidus temperature interval were included as contributions to cooling and contraction. Solid-state heat transport was assumed to occur by conduction only; a depth-dependent and rather low thermal conductivity was adopted. The present bulk U abundance was taken to be 35 ppb, and the distribution of radiogenic heat sources was prescribed by a model of lunar igneous differentiation; in particular, 27 percent of the radiogenic heat sources were assumed to have been retained in the mantle following whole-Moon differentiation. The thermal expansion coefficient was taken equal to weighted averages of values for feldspar, pyroxene, and olivine, according to a model of lunar composition, and dependent on temperature and pressure. Radius change and thermal stress were calculated from (A1-A3) for a Moon of uniform elastic constants; no account was made of changes in specific volume during melting or freezing. With these assumptions, the radius change predicted from the Binder and Lange (1980) thermal model was 5.4 km and the accumulated surface compressional stress was 3.5 kbar over the last 4.5

b.y. The considerably lower value of ΔR compared to the 16 km figure of Solomon and Chaiken (1976) for an initially molten state was ascribed by Binder and Lange (1980) principally to the different thermal conductivity profiles (and thus characteristic cooling times) assumed in the two sets of calculations, though the different initial states (molten vs. solid and at the solidus temperature) and the different mantle heat source abundances are also major factors.

For several reasons, the 5 km radial contraction and 3.5 kbar accumulated thermal stress determined by Binder and Lange (1980) are likely to be only lower bounds for an initially molten Moon. Their initial temperature profile is cooler than a truly molten state and does not include the heat loss contributed by initial heat of fusion or cooling during solidification. If solid-state convection were important in lunar thermal history, heat loss would be expected to have occurred more rapidly than they assumed. Finally, no account was made in the Binder and Lange (1980)—or any earlier—calculations of the contribution to lunar volume change from changes in specific volume during freezing and remelting. This last omission can be quite important for Moon-size objects (Solomon, 1977; Squyres, 1980).

Even if we accept the results of Binder and Lange (1980) as a possibly valid outcome of an initially molten Moon, the essential question is whether five or more kilometers of radial contraction and three or more kilobars of compressive stress at and near the lunar surface are consistent with other information on the mechanical characteristics and tectonic evolution of the lunar lithosphere. We now attempt to answer this question.

Possible Accommodations of High Thermal Stress

A number of mechanisms can be considered as possible factors that will accommodate the accumulation of 3 kbar or more of compressive thermal stress at the lunar surface within the constraints of the known tectonic history of the Moon. These factors include finite strength, viscoelasticity, brittle failure, impact gardening, and other sources of stress. We consider each of these factors in turn.

Finite strength

If the shallowest portions of the lunar crust have a finite compressive strength in excess of 3 kbar, then support of such a stress level would be possible with little or no tectonic activity. A suggestion to this effect was made by Binder (1982). Several independent lines of evidence, however, suggest that the compressive strength of the uppermost lunar crust is considerably less than 3 kbar.

A simple measure of crustal strength on a planetary body is provided by the topographic relief. Shallow stress differences are on the order of ρgh, where ρ is the density of the upper crustal material, g is the gravitational acceleration, and h is the relief (e.g., Jeffreys, 1976, pp. 263–285). Since the maximum topographic relief on the Moon is about 10 km (e.g., Kaula *et al.*, 1974), the indicated stress differences and the implied strength are in the neighborhood of 500 bars.

A second measure of strength may be obtained from examples of failure in situations where the near-surface horizontal compressive stress may be estimated within reasonable bounds. Such an example is provided by mare ridges in mascon mare bains, features plausibly interpreted as the result of compressive failure (Howard and Muehlberger, 1973; Maxwell *et al.*, 1975; Lucchitta, 1976) in response to lithospheric flexure (Solomon and Head, 1979, 1980). Flexural models have been developed by Solomon and Head (1980) for eight mascon mare basins on the basis of the gravitational excess mass, the distribution of mare basalt units, the locations of tectonic features, and the present topography. The flexural models predict compressive horizontal stresses of 200–400 bars in the central mare regions at the time of formation of ridges postdating the youngest major mare basalts units in Serenitatis, Humorum, and Crisium. Somewhat greater compressive stresses (~700 bars) are predicted earlier in the volcanic evolution of these basins, but these stress levels are proportional to the less well-constrained magnitude of the lithospheric load at such times.

The 200- to 400-bar figure is likely to be an underestimate of the total stress at the time of faulting for several reasons. The lithospheric load in the flexure models is taken from the present excess mass and does not include the negative contribution to excess mass from flexure in the central basin region. The stresses predicted at the time of the youngest mare ridges should be increased by 15 to 20% to account for this effect. Further, additional sources of stress are likely to have added to the compressive stress field in the central basin regions, including local basin thermal stress (Bratt *et al.*, 1985) and topographic stress. Global thermal stress accumulated in the interval between mare emplacement and ridge formation should also be considered. The local thermal stress contributed during the final phases of mare volcanism, 600–800 m.y. after basin formation for the basins in question, was likely to be negligible (Bratt *et al.*, 1985). The topographically induced stress should be proportional to the difference in elevation between the basin floor and the surrounding highlands; by the arguments given earlier the 2–4 km average elevation difference between basin and adjacent highlands for Serenitatis and Crisium would contribute 100–200 bars of additional compressive stress in the basin center.

We must also consider the global thermal stress accumulated between the emplacement of the mare units and the formation of the mare ridges that fault these units. Because the ages of the youngest mare ridges are poorly constrained (except that they postdate the mare basalt unit on which they formed), the argument might be advanced that an arbitrarily large global compressive stress was necessary to initiate mare ridge formation, which would leave unconstrained the magnitude of stress necesary to induce faulting. Such an argument, however, can be challenged on two grounds. First, the close association of most major mare ridges with mare units supports the view that local, rather than global, sources of stress were more important. Second, evidence from both mare basalt flow trajectories (Schaber, 1973; Pieters *et al.*, 1980) and radar sounding (Peeples *et al.*, 1978; Sharpton and Head, 1982) indicates that a number of mare ridges formed prior to emplacement of some of the younger mare basalt units, even if the faults marked by these ridges were later reactivated following cessation of volcanism. Since the formation of such ridges must have followed emplacement of the underlying mare basalt units by no more than 100–300 m.y. in the case of Serenitatis, the accumulated global thermal stress (whether compressive or extensional) over such a time interval would not have been a dominant contributor to failure.

On the basis of all of these considerations we conclude that mare ridges formed in response to horizontal compressive stresses of 500 bars to 1 kbar in the uppermost lunar crust. Such stress levels were primarily the product of volcanic loading of the lunar lithosphere, with additional lesser contributions from topographic relief and perhaps global thermal stress.

A final measure of the strength of the uppermost lunar crust comes from rock mechanics experiments and the distribution of stress and strength vs. depth in the Earth's lithosphere. Available measurements of stress vs. depth in the Earth are consistent with the hypothesis that strength in the upper part of the lithosphere is limited by friction on existing faults (Goetze and Evans, 1979; Brace and Kohlstedt, 1980). The relationship between fault friction and confining pressure is quite simple and is in large measure insensitive to rock type and temperature (Byerlee, 1978). For values of the least compressive principal stress less than 1.1 kbar, this relation for frictional strength—which Brace and Kohlstedt (1980) call Byerlee's law—is given by

$$\sigma_V - \sigma_H \simeq 4 \; \rho g z \qquad (3)$$

under horizontal deviatoric compression, where z is depth and σ_V and σ_H are the vertical and largest horizontal principal stresses, respectively, and by

$$\sigma_V - \sigma_H \simeq -0.8 \; \rho g z \qquad (4)$$

under horizontal deviatoric extension. Extensional stress is positive in both (3) and (4).

Byerlee's law of friction is also likely to be a measure of strength in the upper portion of the lunar lithosphere in both mare and highland regions. This result follows if we make the assumption that failure at shallow depths on the Moon, as on Earth, is limited by friction on preexisting faults. Such an assumption requires the shallow portions of the Moon to be pervasively faulted and cracked at a variety of scales, but this state should be a natural consequence of several ubiquitous processes, including impact, differential cooling, and thermal cycling. The large vertical gradient in seismic velocity in the uppermost 20 km of the lunar crust is evidence for pervasive fracturing

(Toksöz *et al.*, 1974), but the distribution of length scales of such fractures is essentially unconstrained.

The frictional strength of the uppermost 20 km of lunar crust, according to Byerlee's law, is shown in Fig. 2 for both compressive and extensional horizontal deviatoric stress. The magnitude of stress difference that will induce compressive failure averages 500 bars in the uppermost 5 km of lunar crust and 1 kbar in the uppermost 10 km. A compressive strength of 3.5 kbar is not reached until almost 20 km depth. Independent support for at least the extensional portion of Fig. 2 comes from the

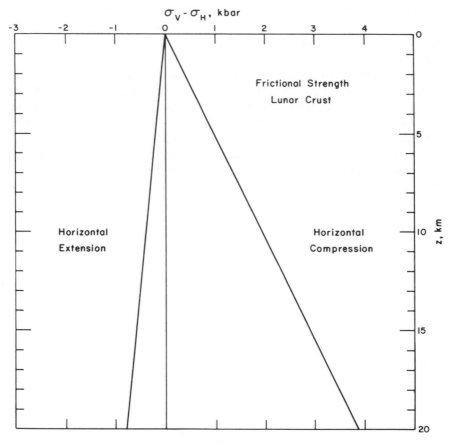

Fig. 2. *Strength of the upper lunar crust under horizontal deviatoric compression and extension. It is assumed that strength in this depth range is controlled by friction (Byerlee, 1978) on preexisting faults and that ρ and g are constant at 3 g/cm³ and 162 cm/s², respectively. The quantities σ_V and σ_H are the vertical stress and the principal horizontal stress of greatest absolute value, respectively; all stresses are positive in extension.*

depth of faulting inferred during the formation of graben circumferential to mascon mare basins (Golombek, 1979). The maximum depth of faulting is in reasonable agreement with the hypothesis that faulting occurs over the depth interval where flexural extensional stress exceeds the extensional strength predicted by Byerlee's law (Solomon, 1985).

Several lines of reasoning thus lead us to the same result. The compressive strength of the uppermost 5–10 km of lunar crust is 0.5 to 1 kbar. Horizontal compressive stresses in excess of this range would induce compressive failure and surface tectonic features such as mare ridges. In particular, near-surface horizontal compressive stresses of 3 kbar or more cannot be supported elastically on the Moon.

Viscoelasticity

If horizontal compressive stresses of 3 kbar or more cannot be sustained in the shallow lunar crust, we should consider possible mechanisms by which global thermal stress might be relaxed or relieved on time scales comparable to its accumulation time. One such mechanism is viscoelasticity. If the rheological behavior of the lunar lithosphere includes a viscous component with a sufficiently low viscosity, then global thermal stress might relax at a rapid enough rate so that the accumulated near-surface stress never exceeds globally the frictional compressive strength (Fig. 2).

There are at least two immediate difficulties with this possibility, however. The first is that the near-surface portions of the lunar crust have remained at low temperature (less than ~100°C) since the end of heavy bombardment in the highlands and since the end of mare volcanism in the maria; viscous effects are not expected to be significant under such conditions. Second, the persistence of the mascons since the time of mare basalt emplacement provides strong evidence that any viscous component to the rheology of the lithosphere must be minor. For 3 kbar or more of potentially accumulating global thermal stress to have been continuously relaxed so as never to have exceeded the near surface strength (0.5 to 1 kbar) would have required relaxation of essentially all the thermal stress contributed in the first 2 b.y. after mare emplacement; such an eventuality could not have occurred without relaxation as well of the lithospheric stress differences supporting the mascon loads over the same time interval. Viscoelastic effects do not appear capable of reconciling the stress levels predicted for an initially molten Moon with the lack of large-scale tectonic features.

Brittle failure

If 3 kbar or more of near-surface compressive stress on the Moon can neither have been supported elastically nor relaxed viscously, an obvious question is whether the accumulation of such stress was episodically relieved by brittle failure. Such a possibility was advanced by Binder and Lange (1980), who suggested that such failure may have been accommodated by largely unobservable slip on faults of the "lunar grid," and by Binder (1982), who argued that highland scarps and at least some of the high-frequency teleseisms (HFTs)—or shallow moonquakes (Nakamura, 1977)—are manifestations of this failure.

Highland scarps are generally regarded as compressive tectonic features analogous to mare ridges, both because of morphological similarities and because of well-documented instances where mare ridges continue across mare-highland contacts as highland scarps (Howard and Muehlberger, 1973; Lucchitta, 1976). The attribution of highland scarp formation in regions distant from mare basins to global thermal stress must be made within a consistent context of stress, strength, and formation time. Binder (1982) has asserted that a number of highland scarps are geologically young (less than 1 b.y.) on the basis of the apparent morphological freshness of craters faulted by the scarps. If this assertion were generally correct, then highland scarp formation by global thermal stress would be inconsistent with both 3 kbar or more of accumulated stress and the compressive strength of the shallow lunar crust (Fig. 2). An initially molten Moon should lead to episodic and widespread compressive failure either more or less evenly distributed in time or perhaps concentrated in the first half of lunar history when global cooling rates might have been comparatively high. Such a situation evidently governed the history of scarp formation on Mercury. In contrast, a Moon with nearly constant volume would not be expected to display evidence of vigorous global-scale tectonic activity over an extended period of lunar history. Modest global thermal stresses of 0.5 to 1 kbar might nonetheless have exceeded the compressive strength in the uppermost few kilometers of crust (Fig. 2) in a number of locations, particularly in settings where topographic or other local stresses added constructively (e.g., within craters or other topographic depressions).

It is difficult to relate HFT moonquakes to any particular source of stress since their focal mechanisms are unknown and even their hypocentral locations are poorly constrained because of their great distance from the Apollo seismic stations. Nakamura *et al.* (1979) have made a convincing case that the amplitude-distance relation for HFT seismograms requires that most, if not all, of these events occurred below the crust in the uppermost lunar mantle. While a small number of crustal HFTs cannot be excluded by this argument, these moonquakes as a class are not relevant to the magnitude or rate of release of tectonic stress in the upper lunar crust.

We conclude that while brittle failure in the uppermost crust may occur at modest levels even for lunar thermal models with nearly constant global volumes, the level of total accumulated thermal stress in models with initially molten states would so often exceed the compressive strength of the near-surface regions of the Moon as to lead to more widespread thrust faulting of a diversity of ages and of greater effective crustal shortening than is observed.

Impact gardening

The possibility should be considered that there are processes other than viscoelasticity that might preferentially relieve the level of stress in the uppermost crust before it has exceeded the failure strength. The accumulated effects of many impacts might be such a process. Certainly the heavy bombardment during the era of major basin formation would likely have overprinted any clear signature of global thermal stress in the first 800 m.y. of lunar and terrestrial planetary history (Solomon and Chaiken, 1976). Could the more modest impact flux subsequent to 3.8 b.y. ago have "gardened" the uppermost crust to a sufficient degree to relieve global thermal stress? This question deserves further quantitative attention. To raise this possibility to a viable level, however, will require the demonstration that impact cratering can generally reduce rather than augment lithospheric stress, that the effect of gardening on stress extends below the rather shallow thickness (~1 km) indicated for the low-velocity regolith and impact-derived breccias by seismic refraction measurements (Cooper *et al.*, 1974; Nakamura *et al.*, 1982), and that impact gardening can relieve accumulated stress at shallow levels on the Moon without also doing so for Mercury.

Other sources of stress

In the 3.8 b.y. of lunar evolution since heavy bombardment, other global sources of stress may have modified the stress field generated by global warming and cooling. The principal such sources are despinning and tidal evolution as the Moon receded from the Earth, and polar wander if the principal axes of the lunar inertia tensor were ever rearranged by large impacts or mare volcanism (Melosh, 1975; Runcorn, 1982). These processes, however, produce stresses with diagnostic patterns of compression and extension at a planetary surface (Melosh, 1977, 1980a,b) and would thus not offset large compressive thermal stresses except in at most specific regions. Considerations of orbital evolution also limit the magnitude of these sources of stress to levels substantially less than those postulated for global thermal stress (Binder, 1982; Grimm and Solomon, 1985).

Concluding Discussion

Thermal history models for the Moon that start from a nearly molten or molten initial state predict that horizontal compressive stresses of 3 kbar or more have accumulated at or near the lunar surface over the past 4 b.y. Such stress levels exceed the compressive strength of the upper lunar crust, indicated by several independent arguments to be in the range 0.5 to 1 kbar. A variety of possible routes by which such high stress levels might be accommodated within the tectonic history of the Moon have been considered, but all of these possible explanations encounter serious difficulties.

This result poses an obstacle to theories of lunar origin that call for a molten initial state, including at least the most straightforward scenario for the formation of the Moon from an accretion disk thrown into circumterrestrial orbit by a giant impact on the Earth (Thompson and Stevenson, 1983; Cameron, 1984; Stevenson, 1984). The obstacle is not necessarily a fatal one, however, particularly inasmsuch as this class of lunar formation theories is only beginning to be explored in a quantitative fashion. An effort should be made to examine the conditions under which a large impact event gives rise to several protomoons, and to the question of the evolution of multiple protomoons formed as a result of more than one large impact. Among a set of such scenarios may be at least one in which the early Moon formed by an aggregation of objects, some of which had cooled while others were still largely molten.

Any specific model for the initial state of the Moon must be tested against the predicted evolution of internal temperature and lithospheric thermal state. The simply parameterized models that formed the basis for this discussion hardly exhaust the range of possible initial states or evolutionary tracks for the Moon. Such thermal stress calculations should account insofar as possible for all sources of heat as well as changes in volume induced by melting reactions and any solid-solid phase changes.

The Moon and the smaller terrestrial planets, with their global lithospheres and large portions of their surface stable since the time of heavy bombardment, have left us a record of their internal evolution for nearly the last 4 billion years. This record for the Moon can be read in the history of tectonic activity and the level of stress differences known to be currently supported. All new hypotheses for the formation and earliest history of the Moon should be pitted against that record as an essential test of their longevity.

Appendix

We derive here a simple relation between the radius change and the surface horizontal stress for a planetary body undergoing nonuniform temperature change and thermal stress. Let the body consist of n concentric spherical shells, each of uniform elastic constants, with n increasing inward; some of the inner shells can behave as fluid rather than elastic material without changing the results here. In each elastic shell, the radial displacement u and the radial and tangential stress are of the form (Timoshenko and Goodier, 1970, pp. 452–453):

$$u(r) = rI(r) + A_i r + B_i R^3/r^2 \qquad (A1)$$

$$\sigma_r(r) = -\frac{2E_i I(r)}{1 + \nu_i} + \frac{E_i A_i}{1-2\nu_i} - \frac{2E_i B_i}{1 + \nu_i}\frac{R^3}{r^3} \qquad (A2)$$

$$\sigma_t(r) = -\frac{E_i}{1 + \nu_i}I(r) + \frac{E_iA_i}{1-2\nu_i} + \frac{E_iB_i}{1 + \nu_i}\frac{R^3}{r^3} - \frac{E_i\alpha(r)\Delta T(r)}{3(1-\nu_i)} \qquad (A3)$$

where r is the radial coordinate, E_i and ν_i are Young's modulus and Poisson's ratio in the ith shell, $\alpha(r)$ is the volumetric coefficient of thermal expansion, $\Delta T(r)$ is the temperature change, R is the outer radius of the spherical body, and

$$I(r) = \left[\frac{1 + \nu}{1-\nu}\right]\frac{1}{3r^3}\int_0^r r^2\alpha(r)\Delta T(r)dr \qquad (A4)$$

The quantities A_i and B_i in (A1-A3) are dimensionless integration constants to be determined after application of the boundary conditions that u and σ_r are continuous across each interface, that $\sigma_r = 0$ at $r = R$, and that all quantities are finite at the origin. The integral in (A4) is seen to be proportional to the volumetrically averaged value of $\alpha\Delta T$. In our application we take $\Delta T(R) = 0$; i.e., the surface temperature is assumed to be constant.

From (A2) and the free-surface boundary condition at $r = R$:

$$\frac{-2E_1I(R)}{1+ \nu_1} + \frac{E_1A_1}{1-2\nu_1} - \frac{2E_1B_1}{1 + \nu_1} + 0 \qquad (A5)$$

$$\text{or } B_1 = \frac{(1 + \nu_1)}{2(1-2\nu_1)}A_1 - I(R)$$

Then from (A1), (A3) and (A5), the radial displacement and horizontal stress at $r = R$ are given by

$$u(R) = \frac{3R(1-\nu_1)}{2(1-2\nu_1)}A_1 \qquad (A6)$$

$$\sigma_t(R) = \frac{3E_1A_1}{2(1-2\nu_1)} \qquad (A7)$$

To solve for A_1 requires the application of all boundary conditions and the substitution or solution for all 2n constants. Since both u(R) and $\sigma_t(R)$ in (A6) and (A7) are proportional to A_1, however, their ratio is simply

$$\frac{\sigma_t}{u} = \frac{E_1}{R(1-\nu_1)} \qquad (A8)$$

Thus given a fractional change in radius u/R experienced by a spherical body of constant surface temperature in response to differential warming or cooling of the interior, the corresponding horizontal stress accumulated at the surface follows from (A8).

Acknowledgments. This paper grew out of a summary talk in the closing session of the Conference on the Origin of the Moon. I thank the conveners Bill Hartmann, Roger Phillips, and Jeff Taylor for inviting me to prepare the summary. I also thank Al Cameron, Jay Melosh, Roger Phillips, and Don Turcotte for comments on the paper and Jan Nattier-Barbaro for assistance in manuscript preparation. My attendance at the conference was supported by the Lunar and Planetary Institute. Additional support for this work has come from the NASA Planetary Geology and Geophysics Program under grants NSG-7081 and NSG-7297.

References

Baldridge W. S. and Simmons G. (1971) Thermal expansion of lunar rocks. *Proc. Lunar Sci. Conf. 2nd*, pp. 2317–2321.

Binder A. B. (1974) On the origin of the Moon by rotational fission. *The Moon, 11*, 53–76.

Binder A. B. (1982) Post-Imbrian global lunar tectonism: Evidence for an initially totally molten Moon. *Moon and Planets, 26*, 117–133.

Binder A. B. and Lange M. A. (1980) On the thermal history, thermal state, and related tectonism of a Moon of fission origin. *J. Geophys. Res., 85*, 3194–3208.

Brace W. F. and Kohlstedt D. L. (1980) Limits on lithospheric stress imposed by laboratory experiments. *J. Geophys. Res., 85*, 6248–6252.

Bratt S. R., Solomon S. C., and Head, J. W. (1985) The evolution of impact basins: Cooling, subsidence and thermal stress. *J. Geophys. Res.*, in press.

Byerlee J. D. (1978) Friction of rocks. *Pure Appl. Geophys., 116*, 615–626.

Cameron A. G. W. (1984) Formation of the prelunar accretion disk (abstract). In *Papers Presented to the Conference on the Origin of the Moon*, p. 58. Lunar and Planetary Institute, Houston.

Cameron A. G. W. and Ward W. R. (1976) The origin of the Moon (abstract). In *Lunar Science VIII*, pp. 120–122. The Lunar Science Institute, Houston.

Cassen P., Reynolds R. T., Graziani F., Summers A., McNellis J., and Blalock L. (1979) Convection and lunar thermal history. *Phys. Earth Planet. Inter., 19*, 183–196.

Cooper M. R., Kovach R. L., and Watkins J. S. (1974) Lunar near-surface structure. *Rev. Geophys. Space Phys., 12*, 291–308.

Cordell B. M. (1977) Tectonism and the interior of Mercury. Ph.D. thesis, Univ. of Arizona, Tucson. 124 pp.

Goetze C. and Evans B. (1979) Stress and temperature in the bending lithosphere as constrained by experimental rock mechanics. *Geophys. J. Roy. Astron. Soc., 59*, 463–478.

Golombek M. P. (1979) Structural analysis of lunar grabens and the shallow crustal structure of the Moon. *J. Geophys. Res., 84*, 4657–4666.

Grimm R. E. and Solomon S. C. (1985) Tectonic tests of proposed polar wander paths for Mars and the Moon (abstract). In *Lunar and Planetary Science XVI*, pp. 298–299. Lunar and Planetary Institute, Houston.

Hartmann W. K. and Davis D. R. (1975) Satellite-sized planetesimals and lunar origin. *Icarus, 24*, 504–515.

Howard K. A. and Muehlberger W. R. (1973) Lunar thrust faults in the Taurus-Littrow region. *Apollo 17 Preliminary Science Report*, NASA SP-330, pp. 31-22 to 31-25.

Jeffreys H. (1976) *The Earth*, 6th ed., Cambridge University Press, 574 pp.

Kaula W. M., Schubert G., Lingenfelter R. E., Sjogren W. L., and Wollenhaupt W. R. (1974) Apollo laser altimetry and inferences as to lunar structure. *Proc. Lunar Sci. Conf. 5th*, pp. 3049–3058.

Langseth M. G., Keihm S. J., and Peters K. (1976) Revised lunar heat-flow values. *Proc. Lunar Sci. Conf. 7th*, pp. 3143–3171.

Lucchitta B. K. (1976) Mare ridges and related highland scarps—result of vertical tectonism? *Proc. Lunar Sci. Conf. 7th*, pp. 2761–2782.

MacDonald G. J. F. (1960) Stress history of the Moon. *Planet. Space Sci.*, 2, 249–255.

Maxwell T. A., El-Baz F., and Ward S. H. (1975) Distribution, morphology, and origin of ridges and arches in Mare Serenitatis. *Geol. Soc. Amer. Bull.*, 86, 1273–1278.

Melosh H. J. (1975) Mascons and the Moon's orientation. *Earth Planet. Sci. Lett.*, 25, 322–326.

Melosh H. J. (1977) Global tectonics of a despun planet. *Icarus*, 31, 221–243.

Melosh H. J. (1980a) Tectonic patterns on a tidally distorted planet. *Icarus*, 43, 334–337.

Melosh H. J. (1980b) Tectonic patterns on a reoriented planet: Mars. *Icarus*, 44, 745–751.

Melosh H. J. and Dzurisin D. (1978) Mercurian global tectonics: A consequence of tidal despinning? *Icarus*, 35, 227–236.

Murray B. C., Belton M. J. S., Danielson G. E., Davies M. E., Gault D. E., Hapke B., O'Leary B., Strom R. G., Suomi V., and Trask N. (1974) Mercury's surface: Preliminary description and interpretation from Mariner 10 pictures. *Science*, 185, 169–179.

Nakamura Y. (1977) HFT events: Shallow moonquakes? *Phys. Earth Planet. Inter.*, 14, 217–223.

Nakamura Y., Latham G. V., Dorman H. J., Ibrahim A.-B. K., Koyama J., and Horvath P. (1979) Shallow moonquakes: Depth, distribution and implications as to the present state of the lunar interior. *Proc. Lunar Planet. Sci. Conf. 10th*, pp. 2299–2309.

Nakamura Y., Latham G. V., and Dorman H. J. (1982) Apollo lunar seismic experiment—final summary. *Proc. Lunar Planet. Sci. Conf. 13th*, in *J. Geophys. Res.*, 87, A117–A123.

Peeples W. J., Sill W. R., May T. W., and Ward S. H. (1978) Orbital radar evidence for lunar subsurface layering in Maria Serenitatis and Crisium. *J. Geophys. Res.*, 83, 3459–3468.

Phillips R. J. and Lambeck K. (1980) Gravity fields of the terrestrial planets—long wavelength anomalies and tectonics. *Rev. Geophys. Space Phys.*, 18, 27–76.

Pieters C., Head J. W., Adams J. B., McCord T. B., Zisk S., and Whitford-Stark J. W. (1980) Late high-titanium basalts of the western maria: Geology of the Flamsteed region of Oceanus Procellarum. *J. Geophys. Res.*, 85, 3913–3938.

Runcorn S. K. (1977) Early melting of the Moon. *Proc. Lunar Sci. Conf. 8th*, pp. 463–469.

Runcorn S. K. (1982) Primeval displacements of the lunar pole. *Phys. Earth Planet. Inter.*, 29, 135–147.

Schaber G. G. (1973) Lava flows in Mare Imbrium: Geologic evaluation from Apollo orbital photography. *Proc. Lunar Sci. Conf. 4th*, pp. 73–92.

Sharpton V. L. and Head J. W. (1982) Stratigraphy and structural evolution of southern Mare Serenitatis: A reinterpretation based on Apollo Lunar Sounder Experiment data. *J. Geophys. Res.*, 87, 10983–10998.

Skinner B. J. (1962) Thermal expansion of ten minerals. *U.S. Geological Survey Prof. Paper 450D*, pp. 109–112.

Solomon S. C. (1977) The relationship between crustal tectonics and internal evolution in the Moon and Mercury. *Phys. Earth Planet. Inter.*, 15, 135–145.

Solomon S. C. (1985) The elastic lithosphere: Some relationships among flexure, depth of faulting, lithospheric thickness, and thermal gradient (abstract). In *Lunar and Planetary Science XVI*, pp. 799–800. Lunar and Planetary Institute, Houston.

Solomon S. C. and Chaiken J. (1976) Thermal expansion and thermal stress in the Moon and terrestrial planets: Clues to early thermal history. *Proc. Lunar Sci. Conf. 7th*, pp. 3229–3243.

Solomon S. C. and Head J. W. (1979) Vertical movement in mare basins: Relation to mare emplacement, basin tectonics, and lunar thermal history. *J. Geophys. Res.*, 84, 1667–1682.

Solomon S. C. and Head J. W. (1980) Lunar mascon basins: Lava filling, tectonics, and evolution of the lithosphere. *Rev. Geophys. Space Phys., 18,* 107–141.

Squyres S. W. (1980) Volume change in Ganymede and Callisto and the origin of grooved terrain. *Geophys. Res. Lett., 7,* 593–596.

Stevenson D. J. (1984) Lunar origin from impact on the Earth: Is it possible (abstract)? In *Papers Presented to the Conference on the Origin of the Moon,* p. 60. Lunar and Planetary Institute, Houston.

Strom R. G. (1979) Mercury: A post-Mariner 10 assessment. *Space Sci. Rev., 24,* 3–70.

Strom R. G., Trask N. J., and Guest J. E. (1975) Tectonism and volcanism on Mercury. *J. Geophys. Res., 80,* 2478–2507.

Thompson A. C. and Stevenson D. J. (1983) Two-phase gravitational instabilities in thin disks with application to the origin of the Moon (abstract). In *Lunar and Planetary Science XIV,* pp. 787–788. Lunar and Planetary Institute, Houston.

Timoshenko S. P. and Goodier J. N. (1970) *Theory of Elasticity,* 3rd ed., McGraw-Hill, New York. 567 pp.

Toksöz M. N., Dainty A. M., Solomon S. C., and Anderson K. R. (1974) Structure of the Moon. *Rev. Geophys. Space Phys., 12,* 539–567.

Turcotte D. L. (1983) Thermal stresses in planetary elastic lithospheres. *Proc. Lunar Planet. Sci. Conf. 13th,* in *J. Geophys. Res., 88,* A585–A587.

Willemann R. J. and Turcotte D. L. (1981) Support of topographic and other loads on the Moon and on the terrestrial planets. *Proc. Lunar Planet. Sci. 12B,* pp. 837–851.

Wood J. A., Dickey J. S., Jr., Marvin U. B., and Powell B. N. (1970) Lunar anorthosites and a geophysical model of the Moon. *Proc. Apollo 11 Lunar Sci. Conf.,* pp. 965–988.

Origin of the Moon and its Early Thermal Evolution

TAKAFUMI MATSUI AND YUTAKA ABE

Geophysical Institute, Faculty of Science, University of Tokyo, Bunkyo-ku, Tokyo 113, Japan

Using a binary accretion model, we have studied the early thermal evolution of a Moon growing by planetesimal impacts. It is shown that due to the blanketing effect of an impact-induced atmosphere, a magma ocean with a depth of several hundred kilometers can be formed during accretion, even when the accretion time is as long as 10^7 years.

Introduction

Major models on the origin of the Moon can be classified into three categories as schematically shown in Fig. 1: (1) fission from the proto-Earth, (2) intact capture, and (3) binary accretion. The binary accretion model for the origin of the Moon may include accretion of lunesimals (planetesimals that form the Moon) in geocentric orbit produced by (4) collisional ejection or (5) disintegrative capture. Among these models only the capture hypothesis assumes an independent origin of the Moon from the Earth. However, the capture model seems to be inconsistent with the oxygen isotope systematics that suggest that formation of the Moon was intimately associated with formation of the Earth (e.g., Goettel, 1984). The fission model for the origin of the Moon also appears implausible because of its dynamic difficulties (e.g., McKinnon and Mueller, 1984). Therefore, the binary accretion model appears to be the most plausible explanation for the origin of the Moon.

It has been widely accepted that a magma ocean was formed when the Moon was formed (e.g., Wood *et al.*, 1970; Warren, 1985), although skepticism seems to be increasing (Walker, 1983). One of the main reasons for skepticism is that it is not clear how sufficient heat could accumulate to form a magma ocean. One of the most important problems, therefore, in considering binary accretion is to provide a theoretical background for the formation of the lunar magma ocean. In this paper, we will study the early thermal evolution of a Moon growing by planetesimal impacts and discuss the formation of a magma ocean.

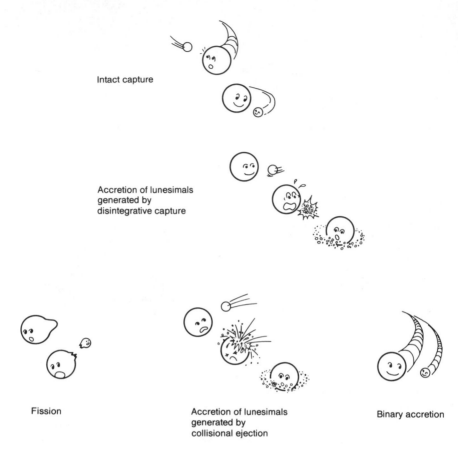

Intact capture

Accretion of lunesimals
generated by
disintegrative capture

Fission

Accretion of lunesimals
generated by
collisional ejection

Binary accretion

Fig. 1. A cartoon of models for the origin of the Moon.

Background

The planets were formed by accretion of planetesimals in heliocentric orbits (Goldreich and Ward, 1973; Safronov, 1972). Two different scenarios on the accretion process of planets have been proposed: one is the gas (solar nebula)-free accretion model (Safronov, 1972), and the other is accretion in a gaseous environment (Hayashi, 1981). Early evolution of the terrestrial planets is completely different for these two mechanisms. The latter scenario results in formation of a solar-nebula-type protoatmosphere (Hayashi *et al.*, 1979). It is, however, hard to accept the existence of such a solar-nebula-type dense atmosphere surrounding the Earth in view of the rare gas data for the present atmosphere (e.g., Fanale, 1971). Therefore, we consider gas-free accretion in the models presented in this paper.

Our Scenario for the Origin of the Moon

The Moon is depleted in volatiles and iron and has a chemical composition similar to the Earth's mantle, although the mean FeO content and depletion of siderophile elements are different between them (Ringwood, 1979). The bulk Earth is enriched in elemental Fe as easily suggested from the comparison of densities between the Earth, Moon, and the mixtures of cosmic levels of Mg, Si, and Fe (Wood, 1986). Ruskol (1977), and recently Chapman and Greenberg (1984), noted the ductile properties of metallic iron to explain a difference in bulk iron content between the Earth and Moon. They suggested a collisional filtering effect due to the difference in mechanical properties between iron and silicate in a circumterrestrial swarm. This seems to be plausible, but, if it were the case, why did such a difference in mechanical properties of planetesimals not play a significant role in accretion of the Earth? We have pointed out that some nucleating agent is required for gaining mass following collisions in the early stage of planetary formation, and have suggested a heterogeneous accretion model for the origin of the terrestrial planets; that is, iron accretes first and silicate second (Matsui and Mizutani, 1977; Matsui, 1978, 1979). Recent laboratory simulations of planetesimal collision show that rocky materials are fragmented completely upon collision once the impact velocity exceeds several tens of meters per second (Matsui et al., 1982, 1984), and the ejecta velocity of the fragments reaches a level almost similar to the impact velocity for the case of rock-rock collision (Waza et al., 1985). On the other hand, iron meteorite and metallic iron can coagulate and gain mass after collision even at relatively high impact velocities (Matsui and Schultz, 1984). These experimental results seem to support the heterogeneous accretion model. The iron depletion in the Moon may be simply interpreted in the context of the heterogeneous accretion model as suggested by Orowan (1969): At first the proto-Earth was formed by selectively sweeping iron planetesimals and the proto-Moon grew around the growing proto-Earth by accumulation of silicate-rich planetesimals in geocentric orbit. These planetesimals are produced by either collisional ejection from the proto-Earth or intact or disintegrative capture of planetesimals in heliocentric orbit. We use the heterogeneous accretion model for the origin of the Moon in the following.

It is, however, unlikely that lunesimals are depleted in volatiles, specifically H_2O, even if they experienced either complete or erosional fragmentation before accretion of the Moon. This is because the impact velocity required for complete fragmentation of silicate planetesimals is several tens of meters per second (e.g., Matsui et al., 1982, 1984), which is much lower than the critical impact velocity required for the initiation of impact dehydration (several kilometers per second according to Lange and Ahrens, 1982). Therefore, we may reasonably assume that the water content of the Moon-forming planetesimals is similar to that of the Earth-forming planetesimals. Even if we consider the accretion of the collisional ejecta from the proto-Earth, we may also assume that some fraction of water is retained in the ejecta. This is because water content in the mantle of the proto-Earth may be almost identical with the initial water content of planetesimals (Abe and Matsui, 1985b; Matsui and Abe, 1986). Judging from the present water budget of the Earth, we consider that the Moon-forming planetesimals contain at least 0.1 wt % water.

Impact velocity of the planetesimals, V_i, increases with growth of a proto-Moon. V_i is given by

$$V_i^2 = (2GM/R) + u^2 \qquad (1)$$

where G, M, R, and u are the gravitational constant, the mass and radius of a growing Moon, and the velocity of planetesimals relative to a growing Moon at very great distance. According to Safronov (1972), $u^2 = GM/\theta R$, where θ is the Safronov number. High-velocity impact of planetesimals into the surface of a growing Moon results in crater formation. The cratering is the process of not only forming a cavity and releasing impact energy but also ejecting dust and evaporating volatile gases such as H_2O and CO_2 retained in the planetesimals and the surface layer. Since very frequent high-velocity collision occurs during accretion, impact-induced volatiles are expected to surround the entire surface of a growing Moon. However, escape of an impact-induced atmosphere and growth of the Moon are considered to be competitive processes. If the characteristic decay time of an impact-induced atmosphere is longer than the incremental accretion time, Δt, an impact-induced atmosphere is expected to play an important role in insulating the thermal radiation from the surface layer and thus in heating the surface layer due to the release of impact energy. Recently, we showed that due to the blanketing effect of an impact-induced atmosphere, the surface of the Earth growing by planetesimal impacts was heated up to the temperature much higher than the melting temperature; thus a magma ocean covering the entire surface was formed (Matsui and Abe, 1984, 1985, 1986; Abe and Matsui, 1985a,b). For the case of the Moon, because of its small size, accumulation of such an impact-induced atmosphere may be only possible during incremental accretion of each radial shell. Therefore, in order to discuss the early thermal state of the Moon, we need to take into account both generation and escape of an impact-induced atmosphere simultaneously. Since H_2O is the most abundant volatile in the Earth and in carbonaceous chondrites, for simplicity we consider an impact-induced atmosphere constituted of only H_2O.

Generation of an H_2O Atmosphere During the Accretion Stage

The formation of an H_2O atmosphere is very much dependent on the surface temperature. When the surface temperature is lower than ~900 K, impact dehydration is possibly one of the most important atmospheric sources. According to Lange and Ahrens (1982), serpentine loses its structural water when the peak impact pressure,

P_{imp}, exceeds the critical impact pressure, P_{cr}, required for the dehydration reaction of serpentine. P_{imp} is given by

$$P_{imp} = \rho \, [C_0 + (K' + 1) \, V_i/8] \, V_i/2 \qquad (2)$$

where ρ, C_0, and K' are density, bulk sound velocity, and partial derivative of incompressibility with respect to pressure, respectively. P_{cr} is a function of the porosity: $P_{cr} = 6 \times 10^{10}$ Pa when the porosities of the surface materials and planetesimals are zero, and P_{cr} reduced to 2.28×10^{10} Pa for the case of porosity of ~17% (Lange and Ahrens, 1982). Once the peak pressure exceeds the critical impact pressure, the dehydration reaction is assumed to proceed completely. However, some of the released water may undergo rehydration reactions, and some may not be able to escape from the interior. Therefore, only some fraction, f, of the released water is assumed to be added to the atmosphere. We can define f as follows: f = (amount of the water added to the atmosphere)/(amount of the water released by shock heating). f is obviously a function of the atmospheric pressure, the surface temperature, and the physical properties of the surface layer. Since the reaction rate of the hydration reaction is fast compared to the geological time scale, the loss of impact-induced H_2O due to the subsequent hydration reaction may be limited by the amount of available reactants. That is, the thickness of the layer wherein the hydration reaction can occur determines the loss of atmospheric H_2O. Shock dehydration occurs even in relatively deep regions, but the hydration reaction occurs only in the uppermost layer, which is in contact with the atmosphere for a long time. Then, qualitatively, the amount of released water is larger than the amount of rehydrated water, and hence f is larger than 0. It is, however, difficult to estimate f. The growth of an impact-induced atmosphere during the Moon's accretion from M to M + ΔM can be expressed by

$$\Delta M_a = \Delta M W_{prj} f \qquad (3)$$

where ΔM_a is the increase of mass of the water-vapor atmosphere and W_{prj} is the water content of planetesimals.

In addition to the rehydration reaction, oxidation of metallic iron is also one of the most dominant sinks of atmospheric water (Abe and Matsui, 1985a). If abundant reactant is available, significant amounts of water may be expended to oxidize Fe, which is not the case in this study, however, since it is assumed that most of the metallic iron was accumulated into the Earth.

When the surface temperature is higher than 900 K and lower than the melting temperature, hydrous minerals become unstable. Then, instead of a hydration reaction, a dehydration reaction occurs, even without impact heating. We consider f = 1 in this case.

Once the surface of the accreting Moon reaches the melting temperature, the amount of H_2O in the atmosphere may decrease. This is because H_2O will dissolve into a silicate melt (Fricker and Reynolds, 1968). The amount, X_w (in wt %), of water dissolved into silicate melt at pressure, P_a (in Pascals), is given by

$$X_w \approx 2.08 \times 10^{-4} P_a^{0.54} \tag{4}$$

where $P_a = M_a g / 4\pi R^2$ (M_a is the total mass of the atmosphere and g is gravitational acceleration). The growth of the atmosphere in this temperature range can be simply calculated by

$$\Delta M_a = \Delta M (W_{prj} - \alpha X_w) \tag{5}$$

where α is the degree of melt determined from the distribution of the latent heat between the solidus and liquidus.

Escape of an Impact-induced H_2O Atmosphere

Since the Moon is much smaller than the Earth, thermal escape may play an important role in evolution of the atmosphere. Using Houghton's formulation of the number of molecules escaping per unit time per unit area (Houghton, 1977), the mass loss rate of the atmosphere is given by

$$M_a = [(\pi m)/(2kT)]^{1/2} GM (1 + B) \exp (-B) n'$$
$$\text{and } B = (GMm)/(RkT) \tag{6}$$

where m is the mass of H_2O molecule, k and G are the Boltzman constant and the gravitational constant, M and R are the mass and radius of a growing Moon, and n' is the surface density of the atmospheric layer from which most of the escape occurs. T is the temperature of the exosphere for $B(T_s) > B_{cr}$ or the surface temperature for $B(T_s) < B_{cr}$, where T_s is the surface temperature. B is called the escape parameter. According to Öpik (1963), B_{cr} is ~1 irrespective of the atmospheric composition. n' is very much dependent on the scale height of the atmosphere. Following Öpik's (1963) suggestion, we assumed $n' = m/s$ (s is the cross section of the H_2O molecule) for the high scale height case, $B(T_s) > B_{cr}$, and $n' = M_a/(4\pi R^2)$ for the low scale height case $B(T_s) < B_{cr}$. n' differs significantly between the high and low scale height cases, as will be discussed in the later section. B also differs between these cases, because we use the surface temperature as T for the low scale height case but the temperature of the exosphere as T for the high scale height case. Therefore, rapid and slow mass losses are expected to occur alternately depending on the surface temperature.

Energy Balance at the Surface Layer

Next, we need to estimate the blanketing effect of the H_2O atmosphere on the surface temperature of a Moon growing by planetesimal impacts. For simplicity, we assume the size distribution of the impacting planetesimals may be described by a δ-function. The surface of a growing Moon is excavated and stirred by planetesimal impacts. Therefore, we define the surface layer as a layer that is homogeneous in physical properties, isothermal and thermally equilibrated with the atmosphere. The thickness of the surface layer, ΔR, is considered to be comparable to the typical crater depth. The subsurface layer is defined by a layer below the surface layer with a thickness corresponding to that of the surface layer. We assume that the released impact energy is imparted to a layer with thickness $2\Delta R$: we call this layer the thermally affected layer. Then, the following equations can be derived for the energy balance at the surface and subsurface layers:

$$(1-h)\Delta M V_i^2/2 = 4\pi R^2 F_{atm}\Delta t + C_p\Delta M(T_s-T_p)$$
$$h\Delta M V_i^2/2 = C_p\Delta M'(T_g-T_s') \tag{7}$$

where h is the fraction of the released impact energy retained in the subsurface layer, ΔM and $\Delta M'$ are the masses of the surface and subsurface layers, Δt is the time interval required for the growth from M to $M + \Delta M$, C_p is the specific heat of the surface layer, T_s and T_g are the temperatures of the surface and subsurface layers, T_p is the temperature of planetesimals before impact, and T_s' is the past surface temperature during the growth from $M - \Delta M'$ to M. We can assume $\Delta M \simeq \Delta M'$ and $T_s \simeq T_s'$ because ΔR is much smaller than R. F_{atm} is the energy flux escaping from the surface into the interplanetary space through the atmosphere. We assume the atmosphere is gray and radiatively equilibrated for the optically thin atmosphere. In this case, F_{atm} is given by

$$F_{atm} = 2\sigma (T_s^4 - T_0^4)/(\tau_s^* + 2) \tag{8}$$

where σ is the Stefan-Boltzmann constant and T_0 is the black body equilibrium temperature of the proto-Moon heated only by solar irradiation. τ_s^* is one of the most important parameters in this study, and corresponds approximately to an optical depth for the optically-thin atmosphere. τ_s^* can be defined by

$$\tau_s^* = (3/2)\int_0^\infty \kappa \, \rho_a dz = 3 \, \kappa \, M_a/8\pi R^2 \tag{9}$$

where κ and ρ_a are the coefficient of absorption and the density of the generated H_2O atmosphere, respectively. For simplicity, to derive the above equation we assumed hydrostatic equilibrium of a plane-parallel atmosphere and that κ is constant. κ

is given by $\kappa = (\kappa_0 g/3P_0)^{1/2}$, where κ_0 is the coefficient of absorption at pressure P_0 and g is the gravity at the surface of the growing Moon (Abe and Matsui, 1985a). The absorption coefficient averaged over a wide range of wavelengths should be used because of the assumption of a gray atmosphere. However, we adopted a value of the smoothed absorption coefficient around the window region (about 1000 cm^{-1}) as κ_0: $\kappa_0 = 0.01$ m^2 kg^{-1} and $P_0 = 101,325$ Pa (Yamamoto, 1952). It should be noted that the κ_0 value thus determined is a minimum.

When the total mass of the atmosphere increases, the lower atmosphere becomes optically thick and is convectively equilibrated. In this case, a straightforward estimate of the surface temperature is:

$$T_s = (P_s/P_{eff})^{(\gamma-1)/\gamma} T_{eff} = (\tau_s^*)^{(\gamma-1)/\gamma} T_{eff} \qquad (10)$$

where T_{eff} is the effective temperature given by $T_{eff} = (F_{atm}/\sigma + T_0^4)^{1/4}$, γ is the ratio between the specific heat at constant pressure and the specific heat at constant volume ($\gamma = 1.33$ for the H$_2$O atmosphere), and P_{eff} and P_s are the atmospheric pressure at the photosphere (at which the optical depth is 2/3) and the bottom of the atmosphere, respectively. In this case, F_{atm} is given by

$$F_{atm} = \sigma (T_s^4/\tau_s^* - T_0^4) \qquad (11)$$

Compared to equation (8) this gives a lower energy flux and hence results in a slightly higher surface temperature. For simplicity of the calculation scheme, we use (8) to calculate the surface temperature even for the optically thick atmosphere.

We need an accretion model to solve the energy balance equations at the surface layers. We use a constant growth rate model: $\dot{R} = R_0/\tau_{acc}$, where R_0 and τ_{acc} are the final radius of the Moon and the accretion time, respectively. This may appear to be oversimplified, but the constant growth rate is a good approximation when enough materials exist to form the Moon.

An internal temperature profile within the accreting Moon can be determined by solving the conventional conduction equation. We used the following melting relation (Kaula, 1979)

$$T_{sol} = 1480 + 2.6 \times 10^{-8} P - 9.2 \times 10^{-20} P^2 \qquad (12)$$

where T_{sol} is the solidus temperature in Kelvin, and P is the pressure in Pa. The liquidus is assumed to be 200 K higher than the solidus. The latent heat (400 kJ kg^{-1} K^{-1}) was assumed to be uniformly distributed between solidus and liquidus phases.

Numerical Results and Discussion

The early thermal evolution of a Moon growing by planetesimal impacts is primarily dependent on the release rate of the impact energy and its loss efficiency. Therefore, the main parameters that affect the numerical results are the accretion time, Safronov number, initial water content, and critical escape parameter. Description of the models and the numerical results are summarized in Table 1. At first, in order to study whether an impact-induced atmosphere plays an important role in insulating the thermal radiation from the surface layer, we show the results of Moon models in which no thermal escape of an impact-induced atmosphere was taken into account (models 1 and 2 in Table 1). The parameters used in this calculation are $\theta = 0.02$, $W_{prj} = 0.1\%$, $f = 0.2$, and $P_{cr} = 2.28 \times 10^{10}$Pa. As shown in Fig. 2, the blanketing effect of an impact-induced atmosphere is significant. When the growth rate is $R_0/10^6$ years, a magma ocean with the depth of ~800 km is formed. The depth of the magma ocean is, however, strongly dependent on the accretion time. Deflection of the curve above 900 K is due to the increase in H_2O by complete dehydration. The surface temperature of Fig. 2a becomes almost constant after ~0.6 R_0. This is because the solubility of water into the silicate melt controls the amount

TABLE 1. Descriptions of the Models and Numerical Results.

Model	Parameters				Numerical results*		
	τ_{acc} (10^6 years)	θ	W_{prj} (%)	B_{cr}	T_s (K)	W_f (10^{18}kg)	D (km)
1	1	0.02	0.1	–	1588	6.9	810
2	10	0.02	0.1	–	1425	47.	0
3	1	0.02	0.1	1	1588	6.9	810
4	1	0.02	0.1	2	1588	6.9	500
5	1	0.02	0.1	3	1588	6.9	200
6	1	0.02	0.1	4	1450	4.8	0
7	1	0.02	0.1	5	1244	2.4	0
8	1	0.1	1.0	1	2206	130	760
9	1	0.1	1.0	2	2176	124	476
10	10	0.02	0.1	1	1425	47.	0
11	10	0.02	1.0	1	1848	140	800
12	10	0.02	10.	1	4054	3400	740
13	10	0.02	1.0	2	1839	134	480
14	0.01	4	10.	–	2906	2020	40
15	0.01	4	0.	–	948	0	0
16	0.001	4	0.	–	1500	0	0

*T_s, W_f and D are the surface temperature at $R = R_0$, final amount of an impact-induced atmosphere, and depth of a magma ocean, respectively. However, T_s and W_f for models 6 and 7 are the maximum surface temperature and the maximum amount of the atmosphere.

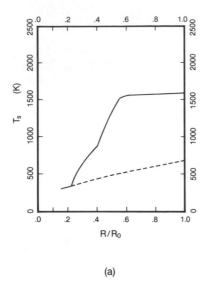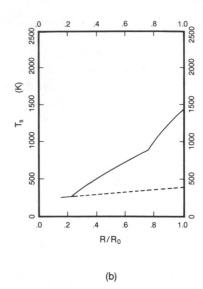

(a)   (b)

Fig. 2.   The traces of the surface temperature (solid curve) for the no-escape models and for the no-atmosphere models (broken curve): (a) model 1 and (b) model 2 in Table 1. The radius is normalized by the final radius, R_0.

of an impact-induced H_2O atmosphere and thus the atmospheric mass becomes almost constant (see Fig. 3).

The evolution of an impact-induced atmosphere for $\tau_{acc} = 10^6$ years (model 1) is shown in Fig. 3. Once the surface temperature exceeds the melting temperature, the amount of the atmosphere becomes almost constant, as mentioned above. It is shown that the final mass of the H_2O atmosphere is about 10^{19} kg when a magma ocean is formed. τ_s^* corresponding to this amount of atmosphere is about 100, and the degree of partial melt in a magma ocean is about 50% in this case. The final amount of an impact-induced H_2O atmosphere for the $\tau_{acc} = 10^7$ years model (model 2) is about 5×10^{19} kg, but no magma ocean is formed in this case. The difference in thermal evolution between these two models indicates that the impact-induced atmosphere is not dense enough to prevent sufficient radiative heat loss from the surface so as to melt the surface layer. A magma ocean with a depth of ~800 km, which is similar to that shown in Fig. 2a, can be formed for a model with $\theta = 0.02$, $W_{prj} = 1\%$, $f = 0.2$, and $\tau_{acc} = 10^7$ years (model 11 in Table 1). In this case, however, the amount of an impact-induced atmosphere (about 10^{20} kg) is higher than that of the previous models (about 10^{19} kg). Since τ_s^* is a linear function of atmospheric mass, τ_s^* of model 11 is about 1000. This suggests that a τ_s^* of 1000 is required for melting the surface layer in lower impact

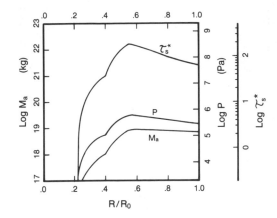

Fig. 3. Evolution of an impact-induced H_2O atmosphere (the mass, M_a, the pressure, P, at the bottom of the atmosphere, and τ_s^* for model 1 in Table 1). The radius is normalized by the final radius, R_0.

energy release rate models ($\tau_{acc} \simeq 10^7$ years), as suggested by Matsui and Abe (1985).

Since the escape parameter is proportional to g/T, we cannot neglect the thermal escape of an impact-induced atmosphere for the Moon. Figure 4 shows the internal temperature profiles of a growing Moon with $\tau_{acc} = 10^6$ years (models 3, 4, 5, and 6 in Table 1) in which the thermal escape of an impact-induced atmosphere was taken into account (other parameters are the same as those of model 1). The internal temperature increases almost monotonically with increase in radius for model 3 ($B_{cr} = 1$), which is very similar to the trace of the surface temperature shown in Fig. 2a. However, for the other models, the internal temperature profile shows oscillations over about half of the final radius. This is because the surface temperature alternately increases and decreases with increasing radius for the models with $B_{cr} > 1$. At first glance, such an oscillating feature of the surface temperature would seem to be the result of a numerical instability. However, this is not the case, because the oscillation period is much longer than the numerical time step. This phenomenon is due to the crude treatment of atmosphere escape in this study. We assumed that the thermal escape occurs from the exosphere for $B > B_{cr}$ (higher scale height case) and from the bottom of the atmosphere for $B < B_{cr}$ (lower scale height case). As seen in equation (6), the escape rate of the atmosphere is proportional to the number density, which differs very much (several orders of magnitude) between these two cases; for example, $n' = m/s = 4.27 \times 10^{-8}$ kgm^{-2} and $n' = M_a/4\pi R^2 = 8 \times 10^5$ kgm^{-2}, where $M_a = 10^{19}$ kg and $R = 10^3$ km. The escape parameter, B, determines whether the lower scale height case or higher scale height case is applied. B is a function of the temperature and mass of a growing Moon. The thermal escape rate is smaller than the production rate of the atmosphere until the surface temperature reaches the critical temperature determined by $T_{cr} = (GMm)/(RkB_{cr})$. When the surface temperature is lower than the critical temperature, an

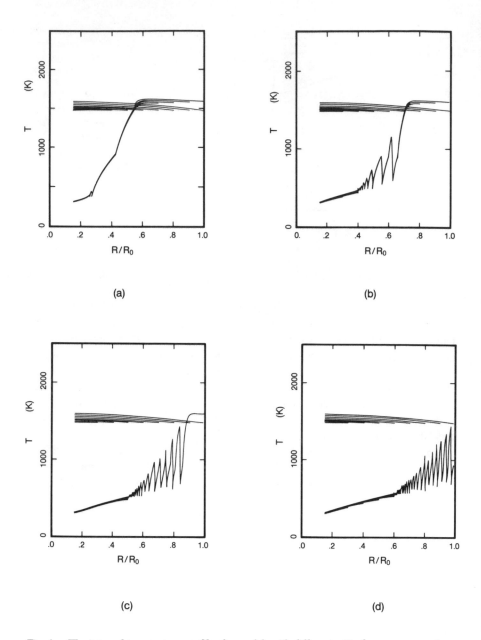

Fig. 4.   *The internal temperature profiles for models with different critical escape parameters [(a) model 3, (b) model 4, (c) model 5, and (d) model 6 in Table 1] for radii of the accreting Moon of 0.2, 0.3, 0.4, 0.5, 0.6, 0.7, 0.8, 0.9, and 1.0 R_0. The upper curves represent the melting temperature distribution of each stage.*

impact-induced atmosphere can be accumulated and the surface temperature can increase due to the increase of the atmospheric mass. However, once the surface temperature exceeds the critical temperature, the thermal escape rate becomes much larger than the production rate of the atmosphere and thus the total amount of the atmosphere turns out to decrease rapidly. With the decrease in atmospheric mass, the surface temperature decreases simultaneously to the equilibrium surface temperature without an atmosphere, and the cycle thus repeats. Since the thermal escape occurs intermittently with the increase in surface temperature, the trace of surface temperature oscillates.

We show the critical temperature and the equilibrium surface temperature without an atmosphere and with $\theta = 0.02$ in Fig. 5. By comparing Figs. 4 and 5, the reader can understand that the surface temperature will oscillate between the critical temperature and the equilibrium temperature. This also supports the view that the oscillating feature is not the result of a numerical instability. The reason why the surface temperature of model 3 ($B_{cr} = 1$) does not show a significant oscillation is that the surface temperature never reaches the critical temperature and thus the production rate of the atmosphere is higher than the escape rate at any time for this model. In the other models ($B_{cr} > 1$), the more plausible temperature distribution seems to be between the two extreme cases of the critical temperature and the equilibrium surface temperature without an atmosphere. It is hard to predict whether the actual surface temperature will be close to the critical temperature profile or

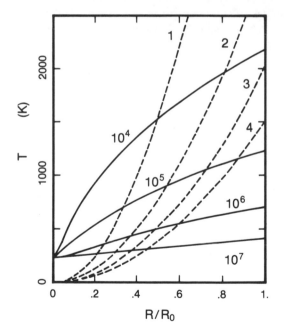

Fig. 5. The critical temperature (broken curve) and equilibrium surface temperature (solid curve) of a growing Moon without an atmosphere and with $\theta = 0.02$ are plotted against the normalized radius. The number attached to each curve represents the critical escape parameter for the broken curve and the accretion time (in years) for the solid curve, respectively.

the equilibrium temperature profile. It will depend on the growth model of each accreting shell layer. Qualitatively, however, if the growth of the Moon is a smooth function of time, the surface temperature profile would be close to the equilibrium temperature profile. On the other hand, if impacts of large planetesimals contribute to the growth of the Moon (that is, the growth is not a smooth function of time), the surface temperature profile would be close to the critical temperature profile.

The depth of a magma ocean is roughly given by the depth at which the critical temperature crosses over the melting temperature. Therefore, it depends on B_{cr}. If we use $B_{cr} = 1$ as an appropriate value (Öpik, 1963), the depth of a magma ocean is about 800 km. In general, the early thermal history of the Moon is shown here to be very much dependent on B_{cr} and the accretion time. The final amount of an impact-induced atmosphere for the models shown in Fig. 4 does not differ between the models and is about 10^{19} kg. This is because the solubility of H_2O into the silicate melt and the degree of partial melt of a magma ocean control the final amount of the H_2O atmosphere. Initial water content does not affect the amount of an impact-induced H_2O atmosphere but controls the internal water budget (Matsui and Abe, 1986). The water content of the interior of the proto-Moon is almost identical with the assumed initial water content except for several dry layers.

It has been widely accepted that the Moon is depleted in volatiles. This seems to be contradictory with the models shown in this study. However, this is not the case for the lower water content (less than 1%) models. According to the petrological point of view, a water content of 0.1% is considered dry (I. Kushiro, personal communication, 1985). For example, calcium-rich plagioclase (anorthosite) would float in a lunar magma with 0.1% water content. As shown in this study, an impact-induced atmosphere can surround the entire surface of a growing Moon during accretion of each radial shell, but it will be dissipated within a few 10^8 years after formation of the Moon because of its low gravity. Water retained in a magma ocean will also be lost simultaneously with solidification of a magma ocean. Therefore, we believe those models with less than 1% water are not inconsistent with a dry Moon concept.

In summary: (1) The blanketing effect of an impact-induced atmosphere may play a key role in controlling the surface temperature of a Moon growing by planetesimal impacts when $\theta < 0.1$ and $\tau_{acc}/(W_{prj}f) < 5 \times 10^8$ years. (2) If the critical escape parameter is around 2, a magma ocean with the depth of several hundred kilometers may be formed. The depth of a magma ocean is, however, strongly dependent on the critical escape parameter. (3) If the thermal escape of an impact-induced H_2O atmosphere occurs, we may need to take into account either the blanketing effect of CO_2 and dust for the early thermal history of the Moon or a rapid accretion.

If the existence of a lunar magma ocean with a depth of several hundred kilometers is required to explain the origin of the lunar anorthositic crust, this study suggests that binary accretion is a favorable hypothesis for the origin of the Moon.

Acknowledgments. This research was partially supported by grants-in-aid for Scientific Research (No. 60540247) sponsored by the Ministry of Education, Science, and Culture of Japan.

References

Abe Y. and Matsui T. (1985a) The formation of an impact-generated H$_2$O atmosphere and its implications for the early thermal history of the earth. *Proc. Lunar Planet. Sci. Conf. 15th*, in *J. Geophys. Res., 90*, C545–C559.

Abe Y. and Matsui T. (1985b) Early evolution of the earth (abstract). In *Lunar and Planetary Science XVI*, pp. 1–2. Lunar and Planetary Institute, Houston.

Chapman C. R. and Greenberg R. (1984) A circumterrestrial compositional filter (abstract). In *Papers Presented to the Conference on the Origin of the Moon*, p. 56. Lunar and Planetary Institute, Houston.

Fanale F. P. (1971) A case for early catastrophic degassing of the earth. *Chem. Geol., 8*, 79–105.

Fricker P. E. and Reynolds R. T. (1968) Development of the atmosphere of Venus. *Icarus, 9*, 221–230.

Goettel K. A. (1984) Bulk composition of the moon in the context of models for condensation in the solar nebula (abstract). In *Papers Presented to the Conference on the Origin of the Moon*, p. 20. Lunar and Planetary Institute, Houston.

Goldreich P. and Ward W. R. (1973) The formation of planetesimals. *Astrophys. J., 183*, 1051–1061.

Hayashi C. (1981) Formation of the planets. In *Fundamental Problems in the Theory of Stellar Evolution* (D. Sugimoto, D. Q. Lamb, and D. N. Schramm, eds.), pp. 113–128. IAU Symposium 93, Reidel, Dordrecht.

Hayashi C., Nakazawa K., and Mizuno H. (1979) Earth's melting due to the blanketing effect of the primordial dense atmosphere. *Earth Planet. Sci. Lett., 43*, 22–28.

Houghton J. T. (1977) *The Physics of Atmosphere*. Cambridge University Press, Cambridge. 203 pp.

Kaula W. M. (1979) Thermal evolution of earth and moon growing by planetesimal impacts. *J. Geophys. Res., 84*, 999–1008.

Lange M. A. and Ahrens T. J. (1982) The evolution of an impact-generated atmosphere. *Icarus, 51*, 96–120.

Matsui T. (1978) Collisional evolution of mass-distribution spectrum of planetesimals. *Proc. Lunar Planet. Sci. Conf. 9th*, pp. 1–13.

Matsui T. (1979) Collisional evolution of mass-distribution spectrum of planetesimals II. *Proc. Lunar Planet. Sci. Conf. 10th*, pp. 1881–1895.

Matsui T. and Abe Y. (1984) The formation of an impact-generated atmosphere and its implications for the early thermal history of the earth (abstract). In *Lunar and Planetary Science XV*, pp. 517–518. Lunar and Planetary Institute, Houston.

Matsui T. and Abe Y. (1985) Formation of a "magma ocean" on the terrestrial planets due to the blanketing effect of an impact-induced atmosphere. *Earth, Moon and Planets*, in press.

Matsui T. and Abe Y. (1986) Evolution of an impact-induced atmosphere and magma ocean of the accreting Earth. *Nature*, in press.

Matsui T. and Mizutani H. (1977) Why is a minor planet minor? *Nature, 270*, 506–507.

Matsui T. and Schultz P. H. (1984) On the brittle-ductile behavior of iron meteorites: New experimental constraints. *Proc. Lunar Planet. Sci. Conf. 15th*, in *J. Geophys. Res., 89*, C323–C328.

Matsui T., Waza T., and Kani K. (1984) Destruction of rocks by low velocity impact and its implications for accretion and fragmentation processes of planetesimals. *Proc. Lunar Planet. Sci. Conf. 14th*, in *J. Geophys. Res., 89*, B700–B706.

Matsui T., Waza T., Kani K., and Suzuki S. (1982) Laboratory simulation of planetesimal collision. *J. Geophys. Res., 87*, 10968–10982.

McKinnon W. B. and Mueller S. W. (1984) A reappraisal of Darwin's fission hypothesis and a possible limit to the primordial angular momentum of the earth (abstract). In *Papers Presented to the Conference on the Origin of the Moon*, pp. 34–35. Lunar and Planetary Institute, Houston.

Öpik E. J. (1963) Selective escape of gases. *Geophys. J.*, 7, 490–509.

Orowan E. (1969) Density of the moon and nucleation of planets. *Nature, 222*, 867.

Ringwood A. E. (1979) *Origin of the Earth and Moon*. Springer-Verlag, New York. 295 pp.

Ruskol Ye. L. (1977) The origin of the moon. In *The Soviet-American Conference on Cosmochemistry of the Moon and Planets* (J. H. Pomeroy and N. J. Hubbard, eds.), pp. 815–822. NASA SP-370, NASA, Washington, DC.

Safronov V. S. (1972) *Evolution of the Protoplanetary Cloud and Formation of the Earth and Planets.* Translated by the Israel Program for Scientific Translation, Jerusalem. 206 pp.

Walker D. (1983) Lunar and terrestrial crust formation. *Proc. Lunar Planet. Sci. Conf. 14th*, in *J. Geophys. Res.*, 88, B17–B25.

Warren P. H. (1985) The magma ocean concept and lunar evolution. *Ann. Rev. Earth Planet. Sci*, 13, 201–240.

Waza T., Matsui T., and Kani K. (1985) Laboratory simulation of planetesimal collision II—Ejecta velocity distribution. *J. Geophys. Res., 90*, 1995–2012.

Wood J. A. (1986) Moon over Mauna Loa: A review of hypotheses of formation of Earth's Moon, this volume.

Wood J. A., Dickey J. S., Marvin U. B., and Powell B. N. (1970) Lunar anorthosites and a geophysical model of the moon. *Proc. Apollo 11 Lunar Sci. Conf.*, pp. 965–988.

Yamamoto G. (1952) On a radiation chart. *Sci. Rept. Tohoku Univ., Ser. 5, Geophysics, 4*, 9–23.

V. Theories and Processes of Origin 1:
Lunar Formation Involving Capture or Fission

Origin of the Moon by Capture

S. FRED SINGER

George Mason University, Fairfax, VA 22030

A coherent account is presented here based on the hypothesis that the Moon formed separately in a heliocentric orbit similar to the Earth's and was later captured by the Earth. The adoption of this hypothesis, together with the observed depletion of iron in the Moon, sets some important constraints on the condensation and agglomeration phenomena in the primeval solar nebula that led to the formation of planetesimals, and ultimately to planets. Capture of the Moon also defines a severe heating event whereby the Earth's kinetic energy of rotation is largely dissipated internally by the mechanism of tidal friction. From this melting event dates the geologic, atmospheric, and oceanic history of the Earth. An attempt is made to account for the unique development of the Earth, especially in relation to Mars and Venus, its neighboring planets. A capture origin of the Moon that employs a "push-pull" tidal theory does not strain the laws of physics, involves a minimum of *ad hoc* assumptions, and has a probability that is commensurate with the evidence of the existence of a unique Moon.

Introduction

This paper is a survey of published work, mostly my own, dealing with a particular capture theory of lunar origin. Nearly 20 years ago I showed, by development of a frequency-dependent ("push-pull") version of tidal theory, that capture need not violate the laws of physics and, in addition, had a respectable probability (Singer, 1968). Some time later I became satisfied that the Moon's low iron content (Singer, 1972) and low volatile content (Singer and Bandermann, 1970) were both in accord with a capture process at an early stage of the development of the planetary system.

I see no need to modify these papers. Therefore, following a brief historical introduction, I will: (1) review "frequency-dependent (push-pull) tidal theory" and explain how it affects the evolution of the Moon's early orbit, so as to make capture acceptable; (2) turn to the probability of capture and discuss several plausible mechanisms to accomplish capture involving a three-body theory; (3) explain why the Moon's chemical differences from Earth (including low iron content, depletion of volatiles, similar oxygen isotope ratios) are entirely consistent with a lunar formation away from the Earth (but in an Earth-like orbit) followed by later capture; (4) point to inevitable physical consequences to the Earth of a capture process and show that it will lead to rapid heating, melting, and core formation; the despinning of the Moon at the beginning of capture will create a magma ocean. (5) I speculate that the early formation of Earth's water oceans, atmosphere, and even life may be due to the capture of the Moon.

Historical Note

Professor E. J. Öpik first called my attention to the work of Gerstenkorn (1955), who had extended earlier calculations (by Darwin and others) to calculate (looking back in time) the evolution of the lunar orbit. Starting with the present orbit, Gerstenkorn based his work on tidal theory; he found that the normally prograde lunar orbit became retrograde and parabolic as its inclination to the Earth's equator increased beyond 90°. This result suggested to him capture of the Moon from an initially retrograde orbit (Gerstenkorn, 1955). However, MacDonald (1964), while confirming these orbital evolution results, pointed out that the required lunar angular momentum change was far too large. It would require a corresponding complementary spin change for the Earth, leading to a dissipation of kinetic energy of rotation large enough to vaporize the Earth! This criticism effectively discredited the capture theory, although Alfven and Arrhenius (1968) tried to circumvent MacDonald's objection by resorting to a "resonance" capture process. [Even earlier, Urey (1952) had suggested lunar capture, but his argument was not based on dynamics: the low lunar density suggested to him a formation well beyond the Earth's orbit.] But MacDonald's result of an initially retrograde lunar orbit was based on a tidal perturbation that did not depend on frequency.[1] A corrected tidal theory leads to a dramatically different result, suggesting lunar capture from a *prograde* orbit, thus overcoming MacDonald's objections to capture (Singer, 1968).

Frequency-Dependent (Push-Pull) Tidal Theory

A conventional way to model the tidal perturbation is by means of a tidal "bulge" on the (solid or ocean-covered) Earth that lags in time, and therefore leads the Earth-Moon line by a phase angle. The angle δ of the bulge depends on the elastic constants of the Earth; its current value is taken to be 2.16° (MacDonald, 1964), corresponding to Earth angular velocity Ω and Moon mean motion n. The angle has been allowed to remain constant in some orbital evolution calculations, in spite of large variations in both Ω and n. More realistically, δ is taken to be proportional to the relative angular velocity $(\Omega-n)$, but such a calculation, often termed "frequency-dependent," also leads to high inclinations of the lunar orbit (Conway, 1982). More complicated formulations of the Darwin model have been given by Kaula (1964) (i.e., phase lag proportional to "frequency" Ω–n is a special case).

We now describe a theoretical extension of tidal theory that departs from the existing theory in one important aspect: the phase angle is made dependent on instantaneous frequency. Thus[2]

$$\delta = b(\Omega\text{–}df/dt) \tag{1}$$

It is important to note that the phase angle may vary, therefore, in magnitude and in *sign* during a *single* orbit of the satellite, so that the perturbing force may both push and pull on the satellite during a single orbit. (For the time being, a simple linear dependence is assumed; b is chosen to give $\delta = 2.16°$ for the current angular velocities of Moon and Earth.)

A simple exposition of the theory follows. The phase angle is zero, and therefore the perturbing torque is zero, when the planet is perfectly elastic and exhibits no dissipation and therefore no time delay in raising a tidal bulge. However, an imperfectly elastic planet may also experience no dissipation provided the frequency of the applied forcing function is zero. This occurs when the satellite is in a synchronous orbit in which it revolves with the same angular velocity as the planet so that $df/dt = n = \Omega$. If the satellite's orbit is elliptic and has a perigee that is within the synchronous orbit, then we obtain the situation shown in Fig. 1. Near apogee the satellite will be moving slower than the planet; therefore, the tidal bulge will lead

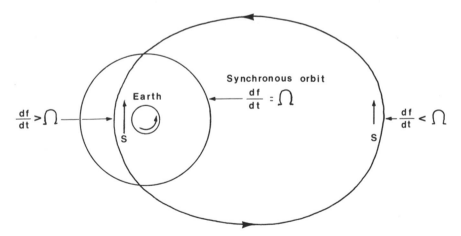

Fig. 1. *Illustrating the tangential perturbing forces, S, acting on an elliptic orbit whose perigee is within the Earth's synchronous orbit limit. Note the reversal in the sign of S as the relative angular velocity of the Earth and Moon reverses sign.*

and produce a perturbation force, S, as shown. Near perigee, however, the satellite will be moving more rapidly than the planet, the tidal bulge will lag, and the perturbation force will be opposite and, of course, much larger because of the strong dependence of tidal perturbation on distance. At perigee, the tidal bulge "pulls back" on the satellite and therefore brings the apogee in; at apogee, the tidal bulge "pulls forward" on the satellite and therefore raises the perigee slightly. The net effect of this push-pull mechanism will be to decrease the semi-major axis and eccentricity. Whereas in MacDonald's calculations it was permissible to perform certain averaging operations over the orbit, one now must calculate the value of (da/dt) at each point of the orbit and then sum these over a complete elliptic orbit. It is necessary to proceed in equal time steps, and therefore to convert at each step from mean anomaly to eccentric and then to true anomaly. In this manner one obtains average values of rates of change of a and e as a function of the orbit's parameters a and e. By taking the ratio, one obtains (da/de) as a function of a and e, with t eliminated.

The major result of the push-pull tidal theory is the fact that under certain conditions an orbit can be "reflected," i.e., approach to a minimum distance and then expand again, even if the inclination of the orbit is zero. This result follows since δ may reverse its sign over a sufficient portion of the orbit to yield an effect due to force S which reverses sign. In frequency-independent theory (MacDonald, 1964) it is necessary for (cos Σ – n/Ω) to reverse sign before the orbit can be "reflected"; as a consequence, large values of inclination angle Σ develop, and large amounts of kinetic energy must be dissipated through tidal friction. The use of push-pull theory thus removes a major objection against the capture origin of the Moon.

Doubts have occasionally been expressed (see, for example, Kaula and Harris, 1975) about the validity of the backward extrapolation of the lunar orbit, in view of (1) higher terms of Legendre polynomials of tidal perturbation (term P_3 involving Love number k_3); (2) higher terms of the general Fourier series (Kaula, 1964) that involve higher-power terms of eccentricity (such as e^2, e^4, . . .); (3) nonlinear viscosity for large distortions; and (4) break-up of the Moon within the Roche limit. While these factors might affect the time scale of orbital evolution, they should not affect the major results of the orbit calculation (Singer, 1968, 1970a), namely: (A) At the instant of capture the Moon's inclination is moderate and certainly not retrograde. (B) The Moon's orbit has never been equatorial while near the Earth. The first result deals with MacDonald's (1964) objection to capture; the latter result, as Goldreich (1966) first pointed out, presents a severe obstacle to all theories of lunar origin (such as fission, precipitation, etc.) that involve an assembly of the Moon in a near-Earth orbit. Result (B) thus supports indirectly a capture origin of the Moon.

The essence of the orbital evolution calculations can be grasped by considering only the principle of conservation of angular momentum.

The Earth's angular momentum, L_E, is due to its rotation and is given by

$$L_E = C\Omega \tag{2}$$

where C is the moment of inertia, currently 8.1×10^{44} g cm^2, and Ω the spin angular velocity, currently 7.29×10^{-5} rad sec^{-1}, corresponding to a 24-hour rotation period.

The Moon's angular momentum, L_M, is due to its orbit around the Earth and is given by

$$L_M = m[GMa(1-e^2)]^{1/2} \qquad (3)$$

with m and M the masses of Moon and Earth, respectively, G the gravitational constant, and a and e the semi-major axis and eccentricity of the lunar orbit.

If in this simple treatment we neglect the inclination of the lunar orbit, then the total angular momentum, K, is given by

$$K = C\Omega + m[GMa(1-e^2)]^{1/2} = \text{constant} \qquad (4)$$

The detailed orbit calculations show that—going backward in time—the Moon approaches closer to the Earth until it reaches the Earth's synchronous orbit, i.e., until its angular velocity around the orbit, n, matches the spin angular velocity of the Earth, Ω. In other words, the length of the month approaches the length of the day.

We therefore set $n = \Omega$, but note also Kepler's third law, i.e.

$$n^2a^3 = G(M + m) \qquad (5)$$

We can now derive the minimum size of the (nearly circular) lunar orbit, a_{min}, by substituting into (4)

$$Ca_{min}^{-3/2} + ma_{min}^{1/2} = K[G(M + m)]^{-1/2} \qquad (6)$$

to obtain $a_{min} = 2.32$ R_E. (If the Earth's core had not yet formed, then C would be 9.5×10^{44} g cm^2 and $a_{min} = 2.6$ R_E.)

It can be shown that the size of the "synchronous" orbit in terms of a planet's radius is given by

$$a_s/R = 1.93 \times 10^{-3} \, (P^2d)^{1/3} \qquad (7)$$

where P is the spin period in seconds and d the density in g cm^{-3}. Currently, $a_s = 6.60$ R_E. Conversely, a synchronous distance of ~2.5 R would correspond to a spin period (length of day) of about 5 hours.

Calculating backward in time beyond the point of minimum orbit size, the Moon's orbit becomes large and highly elliptic and develops a moderate inclination (Singer,

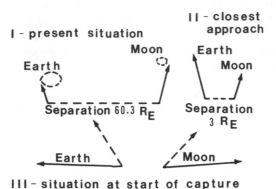

I - present situation

II - closest approach

III - situation at start of capture

Fig. 2. Angular momentum vectors of Earth and Moon shift with respect to ecliptic as distance between the two changes. (I) At top is the present situation. Earth's angular momentum is its spin while Moon's is mainly orbital angular momentum. Both precess about the ecliptic pole (small circles). (II) Center shows the situation near closest Earth-Moon approach a few billion years ago. Earth's spin has decreased from 24 to about 5 hours and spin angular momentum has increased. (III) Bottom shows angular momenta corresponding to situation at start of capture. Solid lines show classical tidal calculation in which the tidal bulge phase angle is frequency-independent and always leading. The Moon has retrograde orbit. Dashed lines show new calculations that assume the tidal bulge is frequency-dependent: it then turns out that the Moon is not captured from a retrograde but from a direct orbit (Singer, 1968).

1968, 1970a) (see also Fig. 2). However, the term a $(1-e^2)$ increases only moderately while the Earth's angular momentum changes only slightly.[3]

Thus the major effect of the Moon on the Earth is the despinning of the Earth as the lunar orbit recedes from the Earth after capture. The Earth's kinetic energy of rotation is dissipated into heat by means of tidal friction. As can be seen from Fig. 3, most of the dissipation takes place before the Moon reaches a distance of about 10 Earth radii, i.e., within a few thousand years after capture. At closest approach, the kinetic energy density is 23.1 times the present value of 3.6×10^8 erg g^{-1}. Thus under the present calculations (frequency-dependent tidal perturbation) the total energy dissipation is $\sim 10^{10}$ erg g^{-1}, just enough to melt silicate rock. Under the earlier calculations (MacDonald, 1964), capture of the Moon would have led to a dissipation of over 10^{11} erg g^{-1}—enough to vaporize the Earth!

The same capture process despins the Moon during its initial close approach to the Earth. We may assume an initial lunar spin period of a few hours. The tidal dissipation of this spin kinetic energy leads to melting of an outer layer of the Moon, in good accord with other evidence for a lunar magma ocean (Wood, 1986). The thickness of the layer D can be derived from the dissipated energy per unit mass of the layer, roughly $(R_L^3 \Omega_L^2 / 15D)$.

The time scale of orbital evolution presents no real problem. MacDonald, using $\delta = 2.16°$, obtains 1.8 b.y. But this result should not be taken literally, since it depends on the assumption that the Earth's elastic parameters have always been the same as the present. Changes in the ocean-continent configuration, such as an

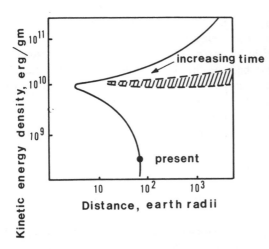

Fig. 3. Kinetic energy density heats Earth. If Moon is captured from retrograde orbit, as in classical theory, the initial spin velocity of Earth increases essentially without limit and initial kinetic-energy density increases in the same manner (solid line). Dashed region shows the kinetic-energy density for frequency-dependent tidal theory. Difference between the initial (~10^{10} erg/gm) and present kinetic-energy density must be dissipated by tidal friction, mainly within Earth's interior, causing melting within a few thousand years.

equatorial ocean belt (before continental drift), might well reduce δ and thereby lengthen the time scale to 4.5 b.y. (Singer, 1968). This is an important point since under my capture process the Moon forms in an Earth-like orbit and cannot be "stored" for any length of time.

The Capture Process

In my original paper (Singer, 1968) I pointed out that capture has a low probability, but I also gave plausible mechanisms for the capture process. To begin with, some dozen moons may have formed, all in essentially Earth's orbit. The most likely outcome of a moon-Earth encounter is nothing but a hyperbolic pass (as seen from a geocentric view point). The next most likely event is probably an impact, with the moon becoming part of the Earth. But there are many moons, and only one needs to be captured.

Capture is quite plausible if viewed within the restricted three-body formalism (sun-Earth-Moon). The orbit of Fig. 14 of Singer (1968) is an example of (temporary) "Jacobi Capture," showing three successive passes near the Earth before the moon again escapes; one pass happens to be extremely close to the Earth's surface.[4] A slight change in the parameters of the problem, and the transient "Jacobi Capture" can turn into permanent capture, as the jaws of the Jacobi (zero-velocity) surfaces close off the moon's escape (Johnson, 1972). Here are four separate plausible mechanisms for such an event:

1. Energy is dissipated by tidal interaction (or aerodynamic drag) as the moon passes close to the Earth's surface one or more times. The tidal energy losses of each encounter are cumulative and can be added, a fact that greatly increases the

probability of capture. Kaula and Harris (1975) argue to the contrary, i.e., that such elastic encounters "will add random energy to the moon exceeding by orders of magnitude the energy that might be lost by tidal interactions." Their argument against capture is incorrect, as can be seen directly by appealing to the constancy of the Jacobi integral of the three-body problem (see, e.g., Wood, 1986). In other words, the zero-velocity surfaces will not expand as a result of elastic "collisions" between Earth and moon; rather, they will contract as a result of the irreversible energy losses (caused by tidal friction) that produce impulsive changes in the Jacobi integral. [Their argument, based on Öpik's (1972) earlier discussion, would hold only if the moon were subject to significant stochastic orbit perturbations between close encounters with the Earth.]

2. On a close pass the tidally distended moon is disrupted, with part escaping but the remainder captured (Öpik, 1972).

3. As pointed out independently by Lyttleton (1967), energy pumping by the Earth's orbit eccentricity could lead to permanent capture. I have viewed this as an annual, quasiadiabatic pulsation of the Jacobi surfaces, with the moon losing energy upon reflection from the surface, depending on the phase of the pulsation.

4. If by chance the Earth should accrete a moon, thus increasing its mass, while another moon is undergoing Jacobi Capture, then the latter moon can now be permanently captured (Clark *et al.*, 1975).

Once captured, i.e., into an elliptical orbit, the orbit will evolve quite rapidly by tidal perturbation, provided the moon has a close perigee and a large mass. An initially retrograde orbit will remain retrograde and shrink into the Earth with the moon disappearing. An initially prograde orbit may evolve into the orbit of the present Moon. I have speculated elsewhere (Singer, 1970b) that retrograde capture of a moon by Venus may have removed its initial angular momentum. The conventionally assumed mechanism, namely solar tidal friction, is not adequate to despin Venus.

Two of the suggested capture mechanisms, (1) and (2), require a close pass; the others do not. Of course, a combination of capture mechanisms can make capture more certain.

Chemical Composition

The gross features of the Moon are its low density, indicating a low iron content, and its low content of volatiles relative to the Earth. These features have been used to argue against capture, but such reasoning may be erroneous. We grant that the Moon must have had an Earth-like orbit in order to have the small geocentric velocity that gives capture an appreciable probability.[5] But to explain the different composition requires an extension of the conventional theory of planet formation from a primordial solar nebula.

As the primordial solar nebula cools, the vapor pressure of its various constituents rapidly decreases and eventually falls below the partial pressure of each constituent. Presumably at this stage the condensation of small particles begins. This process has been recognized for some time (Lord, 1965; Larimer, 1967; Larimer and Anders, 1970; Lewis, 1972). From the present composition of the terrestrial planets, it appears that the primary condensates in the inner solar system are iron, nickel, and iron and magnesium silicates (Larimer, 1967). In the outer nebula, temperatures are lower and the bulk of the condensate is made up of water and water-ammonia ices (Lewis, 1972).

The condensation into grains, the nucleation process, is followed by the growth of grains from the gas in a manner described many years ago by Hoyle (1946). At the same time a dust-disk forms perpendicular to the rotational axis in a manner described by Lyttleton (1972) and independently by Goldreich and Ward (1973). This disk is gravitationally unstable and breaks down into planetesimals in spite of the disruptive influence of the tidal forces of the sun.[6] Goldreich and Ward (1973) give radii of about 100 m and time scales of the order of one year for the first-generation planetesimals. These are grouped into clusters containing approximately 10,000 members. On a time scale of a few thousand years these clusters contract to form second-generation planetesimals having radii of the order of 5 km. Further coalescence then proceeds by direct collisions of the planetesimals and seems capable of producing growth at the rate of 15 cm per year at one astronomical unit.[7]

The final accumulation of the planets then proceeds by the gravitational interactions and actual collisions of the secondary planetesimals in a manner described by Öpik (1972), Arnold (1965), Safronov (1969), and Wetherill (1975, 1985). In particular, Öpik (1972) and Safronov (1969) have pointed out that the observed obliquities of the planets can be explained in terms of final impacts by bodies with masses of the order of a few percent of the planet's mass.

At this stage it becomes necessary to combine the separate discussions of the condensation process and the agglomeration process. One early attempt in that direction was made by Singer (1972) as an extension of ideas by Turekian and Clark (1969). He focused attention on two important time scales, a time scale for cooling and condensation, T_c, and a time scale for agglomeration and accumulation, T_a. It is the ratio of these two time scales that is important. He argued that T_c is greater than T_a within 2 AU, and that T_c is smaller than T_a beyond that. The hypothesis was developed in order to explain why and how the Earth and Moon could have such different gross chemical compositions in spite of having been formed from solar cloud material at 1 AU. He used it also as a means of accounting for the different gross properties of the terrestrial planets and outer planets. Significantly, the asteroid belt occurs in the region where T_c is on the same order as T_a.

The cooling is calculated by considering radiation loss, both from the gas and from the dust particles that have already condensed. The time scale of interest,

T_c, is the interval between the time that the condensation temperature of iron is reached and the time that the condensation temperature of the silicates is reached. It is then argued that if T_c is greater than T_a, the iron grains will rapidly accumulate into large planetesimals before the silicates have a chance to condense.

As a consequence, there is formed at the Earth's orbit, at 1 AU, a single iron body, which is the core of the proto-Earth. Several conventional reasons support the formation of a single iron core (near 1 AU): (1) The core assembles before the silicates condense, during a regime of high gas and dust densities that moderate the orbital inclinations and eccentricities of the iron-rich planetesimals and therefore produce lower relative impact velocities. (2) Iron has greater strength than silicate rock and is therefore less likely to shatter and break up upon impact. (3) The magnetic properties of iron are sometimes put forward to explain a more rapid assembly; there appears to be no strong support for this idea.

While the accumulation of this iron core is progressing, the gas has cooled sufficiently to allow the condensation of the silicates into grains, which then follow a similar sequence. He postulates as a result the formation of some 10–50 lunar-sized objects containing mainly silicates.

The final accumulation of the Earth takes place by gravitational interactions among 10 to 50 bodies, with the iron core being the largest, and therefore gravitationally most important. With these bodies in very similar orbits, they have a chance to approach each other frequently. In most cases, corresponding to distant "impacts," nothing will happen, except for a slight perturbation of the orbit. Occasionally there would be a real impact in which the proto-Earth gradually acquires a silicate mantle.

The explanation of the chemical difference between Earth and Moon described here makes use of the fractionation we know to have occurred in the formation of the solar system. Other theories of lunar origin (such as fission, precipitation, etc.) require a special fractionation of Earth material subsequent to the formation of the Earth.

The depletion of volatiles on the lunar surface can be explained directly by noting that an object of lower mass will accrete less of the condensed volatiles than would the Earth. It has been suggested (Ganapathy et al., 1970) that the observed depletion would follow if the Moon accumulated in a close Earth orbit. But their proposal does not stand up to a detailed analysis (Singer and Bandermann, 1970) that instead supports capture.

Finally, we have the evidence of the similarity of oxygen isotope ratios for Earth and Moon. This similarity supports a nearby origin of lunar material rather than in some remote part of the solar nebula (see, for example, Taylor, 1986). However, Taylor's objections to capture must be carefully qualified: they apply only to certain capture theories and *not* to those in which the Moon is assumed to be assembled in an Earth-like orbit—which is the only dynamically plausible capture theory.

Speculations on the Evolution of the Earth

Why has the Earth evolved so differently from its neighboring planets, Venus and Mars? Presumably, the starting composition could not have been very different: Venus has 97% of the size of the Earth (82% of the mass); Mars, although half the size, has the same spin rate as the Earth. There are, of course, obvious differences in the distance to the sun, but there are especially differences in the number and nature of the satellites, Venus having none and Mars having two tiny satellites (Singer, 1971). Other important differences relate to the thickness and nature of the atmosphere, but these may well be of secondary origin. It is possible also that the difference in the spin rate between Venus and Earth is of secondary origin, and can be explained in terms of satellites (Singer, 1970b). At present, it is speculative to guess which of the many parameters is responsible for the uniqueness of the Earth and its cycle of evolution.

The capture of the Moon represents a very unique event in the history of the solar system. It is tempting, therefore, to link this unique event of capture to the unique development of the Earth's atmosphere and oceans, and to the subsequent evolution of life, which may also be unique in the solar system (Singer, 1977).

The oldest rocks dated so far are about 3.5 b.y. old; all of them represent terminal processes with respect to geological cycles, the beginnings of which are inferred at about 4 b.y. ago (Donn *et al.*, 1965). This suggests that erosion by running water occurred at least 4 b.y. ago and places the origin of the Earth's atmosphere and oceans at a time even earlier than this; in other words, at most a few hundred million years after the formation of the Earth.

If one adopts the view that the Earth was formed cold and without any atmosphere and that the atmosphere and oceans evolved from the degassing of rocks (Rubey, 1951), then it is difficult to explain such an early degassing phase on the basis of radioactive heating alone. It has become evident that a chondritic composition of the Earth is not appropriate and that the concentration of the short-lived K^{40} should be reduced, with important thermal consequences. For example, in studies by Phinney and Anderson (1967), the mantle has been taken as a mixture of peridotite plus tholeiitic basalt; these oceanic tholeiitic basalts have a potassium content of about 20% that of typical basalts. In considering the integrated heat production for various pyrolite models, Phinney and Anderson conclude that a 3:1 mixture will reach melting in 2 b.y. and a 6:1 mixture will barely reach melting in 4.5 b.y. [For comparison, Ringwood (1975) adduces evidence that the segregation of the core could not have occurred later than about 10^8 years after accumulation of the Earth.] The rotational kinetic energy that is dissipated during lunar capture, however, would give more than adequate amounts of heating, leading to at least partial melting. Thus, the energy required to raise one gram of silicates to $1000°K$ and melt it is 1.7×10^{10} erg g^{-1} or 47 times the present rotational kinetic energy of 3.6×10^8 erg g^{-1} (compare also Fig. 3).

This heating gives rise to rapid degassing of rocks and to intense volcanic activity, which releases the gases and creates in a sudden way an atmosphere consisting mainly of water vapor, carbon monoxide and dioxide, and nitrogen; a more reducing atmosphere containing methane, ammonia, and hydrogen has also been proposed (Holland, 1974). The water vapor subsequently condenses out and forms the oceans, while the carbon dioxide gradually diminishes as it is precipitated into limestone. In the absence of oxygen, complex organic molecules can be formed, such as amino acids and even protein-like substances.

At a certain instant of time, perhaps at a number of locations around the Earth, molecules form that are self-replicating, and living molecules evolve into anaerobic organisms. Several examples of Precambrian microflora have now been described (Schopf, 1975). On the basis of such events, Cloud (1968) has placed "biogenesis" at 3.5–3.8 b.y. and the start of oxygen-generating photosynthesis in the hydrosphere at 3.2–3.4 b.y.; the latter is given by the earliest occurrence of banded iron formations consisting of Fe_2O_3 and FeO. As the oxygen content becomes appreciable, organisms adapt themselves to using oxygen in respiration and at the same time create more oxygen by photosynthesis. The subsequent rapid rise in the oxygen concentration may have made possible the evolution of more complicated life forms in a manner well known to biologists (Berkner and Marshall, 1965).

There are additional implications to the particular picture of the early history of the Earth-Moon system just presented. Until the return of the first lunar samples, about 1970, most authors supported a time scale of lunar capture of only 2 b.y. (MacDonald, 1964) to maybe 700 m.y. (Alfven and Arrhenius, 1968). But there had to be some corresponding geological event on the Earth to mark this lunar event. Some geologists (Olson, 1966) argued for the reality of a Precambrian unconformity. Others (Urey and MacDonald, 1971) argued that the Moon was captured at a large enough distance from the Earth (20–40 radii) so that it would have produced no terrestrial effect. However, such a physical capture mechanism is difficult to imagine. Capture of the Moon can best be understood if it occurred shortly after the formation of the Earth—in other words, over 4 b.y. ago. In order to make this time compatible with the presently observed tidal dissipation, one has to assume that this dissipation was less than the present value during much of the Earth's history. Since most of the dissipation occurs in the shallow basins of the ocean, "large variations in the dissipation rate are possible depending on changes in the ocean—continent configuration" (Singer, 1968). Hargraves (1976) has brought forth independent evidence for a Precambrian history during which the ocean extended over the whole equatorial region of the Earth so that tides could propagate without much energy dissipation. The implications for the Precambrian evolution of life in such a stratified global sea are discussed by Chamberlain and Marland (1977).

Conclusion

· Contrary to published reports, capture theory of the Moon is alive and well—but only the version based on push-pull tidal theory that uses the instantaneous angular velocity df/dt of the Moon. The conventional calculations that use the mean motion n always lead to impossibly high inclinations (retrograde orbits) at the time of capture.

Capture from an eccentric heliocentric orbit (à la Urey and MacDonald, 1971) is neither feasible nor necessary; the low iron content and low volatile content can be explained in terms of a Moon formed in an Earth-like orbit.

The Kant-Laplace hypothesis of the origin of the solar system still survives, but in a modern form. With the addition of condensation chemistry and agglomeration mechanics, one can try to understand the process of planet formation. The hypothesis that the Moon was captured sets an important constraint on this process, and at the same time illuminates it greatly. The capture hypothesis can also be used to shed light on the earliest phase of the development of the Earth and may help to explain some of its unique features.

Notes

1. Since the elastic parameters of the Earth do not depend on the frequency of the applied stress forces, MacDonald argued that the tidal phase angle should be independent of frequency. (In his treatment of tidal evolution Jeffreys also uses a constant phase angle of tidal dissipation.) The misunderstanding arises since "frequency" is later used in a different sense, i.e., as the relative angular rates of the Earth and Moon.

2. Here Ω is the Earth's angular velocity, held constant during a single lunar orbit. But instead of the mean motion n we employ df/dt, the instantaneous lunar angular velocity, which varies during the Moon's elliptic orbit; f is the true anomaly, a the semi-major axis, and e the eccentricity of the Moon's orbit. The inclination of the lunar orbit is given by Σ.

3. I have verified by detailed calculation (Singer, 1970a) that the higher-order terms of the tidal potential that become important at close Moon-Earth separation do not affect these results. In particular, the Moon's orbit never becomes equatorial at close distances. Goldreich (1966) first pointed out that this fact puts important constraints on the origin of the Moon. Conway's (1982) calculations support Goldreich's.

4. See also the supporting discussion by Wood (1986), and especially his Fig. 6, showing seven consecutive passes of a moon near the Earth. (I have used the word "moon," with lower-case "m," to describe an object that could later become the Moon; others use the word "luna.")

5. Urey (1952) originally assumed that the low density implied formation beyond the Earth's orbit, followed by later capture. Such an event is dynamically highly improbable, as are other capture theories (Anderson, 1972; Cameron, 1972) that assume an initial lunar orbit very different from that of the Earth (Kaula and Harris, 1975).

6. Safronov (1969) has credited Gurevich and Lebedinskii (1950) with a similar discussion.

7. The gas density drops due to the expulsion of hydrogen and helium from the inner solar system because of a T-Tauri event of the sun (Cameron, 1973), but it is not quite clear at just what stage of the agglomerations this event occurs. In any case, once the bodies become kilometer-sized, their motions

will not be affected by any reasonable gas densities, although the gas may serve to dampen the internal rotational and random energies of the clusters of the primary planetesimals.

References

Alfven H. and Arrhenius G. (1968) Two alternatives for the history of the Moon. *Science, 165*, 11–17.

Anderson D. L. (1972) The origin of the moon. *Nature, 239*, 263–265.

Arnold J. R. (1965) The origin of meteorites as small bodies. *Astrophys. J., 141*, 1548–1556.

Berkner L. V. and Marshall L. C. (1965) On the origin and rise of oxygen concentration in the Earth's atmosphere. *J. Atmos. Sci., 22*, 225–261.

Cameron A. G. W. (1972) Orbital eccentricity of Mercury and the origin of the Moon. *Nature, 240*, 299–300.

Cameron A. G. W. (1973) Accumulation processes in the primitive solar nebula. *Icarus, 18*, 407–450.

Chamberlain W. M. and Marland G. (1977) Precambrian evolution in stratified global sea. *Nature, 265*, 135–136.

Clark S. R., Turcotte D. L., and Nordman J. C. (1975) Accretional capture of the Moon. *Nature, 258*, 219–220.

Cloud P. E. (1968) Atmospheric and hydrospheric evolution on the primitive Earth. *Science, 160*, 729–736.

Conway B. A. (1982) On the history of the lunar orbit. *Icarus, 51*, 610–622.

Donn W. L., Donn B. D., and Valentine W. G. (1965) On the early history of the Earth. *Geol. Soc. Am. Bull., 76*, 287–306.

Ganapathy R., Keays J. C., Laul E., and Anders E. (1970) Trace elements in Apollo-11 lunar rocks: implications for the meteorite influx origin of Moon. *Proc. Apollo 11 Lunar Sci. Conf.*, pp. 1117–1142.

Gerstenkorn H. (1955) Uber Gezeitenreibung beim Zweikorperproblem. *Z. Astrophys., 36*, 245–274.

Goldreich P. (1966) History of the lunar orbit. *Rev. Geophys., 4*, 411–439.

Goldreich P. and Ward W. R. (1973) Formation of planetesimals. *Astrophys. J., 183*, 1051–1061.

Gurevich L. E. and Lebedinskii A. I. (1950) Ob obrazovanii planet. *Izv. Akad. Nauk. SSSR (Ser. Fiz.), 14*, 765–799.

Hargraves R. B. (1976) Precambrian geologic history: continents emerged from beneath the primordial sea. *Science, 193*, 363–365.

Holland H. D. (1974) Aspects of the geologic history of seawater. *Orig. Life, 5*, 87–91.

Hoyle F. (1946) On the condensation of the planets. *Mon. Not. Roy. Astron. Soc., 106*, 406–422.

Johnson F. (1972) Restricted Three Body Problems and Lunar Capture. Thesis, University of Houston, Texas.

Kaula W. M. (1964) Tidal dissipation by solid friction and the resulting orbital evolution. *Rev. Geophys., 2*, 661–685.

Kaula W. M. and Harris A. W. (1975) Dynamics of lunar origin and orbital evolution. *Rev. Geophys. Space Phys., 13*, 363–371.

Larimer J. W. (1967) Chemical fractionations in meteorites—I. Condensation of the elements. *Geochim. Cosmochim. Acta, 31*, 1215–1238.

Larimer J. W. and Anders E. (1970) Chemical fractionations in meteorites—III. Major element fractionations in chondrites. *Geochim. Cosmochim. Acta, 34*, 367–387.

Lewis J. S. (1972) Low temperature condensation from the solar nebula. *Icarus, 16*, 241–252.

Lord H. C. (1965) Molecular equilibria and condensation in a solar nebula and cool stellar atmospheres. *Icarus, 4*, 279–288.

Lyttleton R. A. (1967) Early history of the Moon: Dynamical capture of the Moon by the Earth. *Proc. Roy. Soc., A296*, 285–292.

Lyttleton R. A. (1972) On the formation of planets from a solar nebula. *Mon. Not. Roy. Astron. Soc., 158*, 463–483.

MacDonald G. J. F. (1964) Tidal friction. *Rev. Geophys., 2*, 467–544.

Olson W. S. (1966) Origin of the Cambrian-Precambrian unconformity. *Am. Sci., 54*, 458.

Öpik E. J. (1972) Comments on lunar origin. *Ir. Astron. J., 10*, 190–238.

Phinney R. A. and Anderson D. L. (1967) Present knowledge about the thermal history of the Moon. In *The Physics of the Moon* (S. F. Singer, ed.), pp. 161–179. American Astronautical Society, Washington, D.C.

Ringwood A. E. (1975) *Composition and Petrology of the Earth's Mantle.* McGraw-Hill, NY.

Rubey W. W. (1951) Geologic history of sea water. *Bull. Geol. Soc. Am., 62*, 1111–1147.

Safronov V. S. (1969) Evolution of the protoplanetary cloud and formation of the Earth and the planets. *NASA TTF-677*, NASA, Washington, D.C.

Schopf J. W. (1975) Precambrian paleobiology: problems and perspectives. *Ann. Rev. Earth Planet. Sci., 3*, 213–249.

Singer S. F. (1968) The origin of the Moon and geophysical consequences. *Geophys. J. Roy. Astron. Soc., 15*, 205–226.

Singer S. F. (1970a) Origin of the Moon by capture and its consequences. *EOS (Trans. Amer. Geophys. Union), 51*, 637–641.

Singer S. F. (1970b) How did Venus lose its angular momentum? *Science, 170*, 1196–1197.

Singer S. F. (1971) The Martian satellites. In *Physical Studies of the Minor Planets* (T. Gehrels, ed.), pp. 399–405. NASA SP-267, NASA, Washington, D.C.

Singer S. F. (1972) Lunar composition as a clue to the early history of the solar system. *Proc. 24th Intl. Geol. Congr.*, pp. 11–17.

Singer S. F. (1977) The early history of the Earth-Moon system. *Earth Sci. Rev., 13*, 171–189.

Singer S. F. and Bandermann L. W. (1970) Where was the Moon formed? *Science, 170*, 438–439.

Taylor S. R. (1986) The origin of the Moon: Geochemical considerations, this volume.

Turekian K. and Clark S. P. (1969) Inhomogeneous accumulation of the Earth from the primitive solar nebula. *Earth Planet. Sci. Lett., 6*, 346–348.

Urey H. C. (1952) *The Planets.* Yale Univ. Press, New Haven, CT.

Urey H. C. and MacDonald G. J. F. (1971) Origin and history of the Moon. In *Physics and Astronomy of the Moon* (Z. Kopal, ed.), pp. 481–523. Academic, NY.

Wetherill G. W. (1975) Late heavy bombardment of the Moon and terrestrial planets. *Proc. Lunar Sci. Conf. 6th*, pp. 1539–1559.

Wetherill G. W. (1985) Occurrence of giant impacts during the growth of the terrestrial planets. *Science, 228*, 877–879.

Wood J. A. (1986) Moon over Mauna Loa: A review of hypotheses of formation of Earth's Moon, this volume.

Numerical Simulations of Fission

RICHARD H. DURISEN

*MS 245-3, Space Science Division, NASA-Ames Research Center, Moffett Field, CA 94035,
and Department of Astronomy, Swain West 319, Indiana University, Bloomington, IN 47405*

ROBERT A. GINGOLD

*Mt. Stromlo and Siding Spring Observatories, Research School of Physical Sciences, Australian
National University, Private Bag, Woden P.O., A.C.T. Australia 2606*

In this paper, we use the term "fission" to refer to the breakup of an equilibrium celestial body driven by rapid rotation. Historically, it was conjectured that fission would lead to splitting of a body directly into two or more pieces. Numerical hydrodynamic simulation techniques have now become sufficiently powerful to study the outcome of dynamic fission instabilities. We summarize recent work and present new simulations spanning a range of rotation rates and fluid compressibility. In the best resolved cases, dynamic fission instability always leads to ejection of a ring or disk of debris rather than one or a few discrete bodies. In this case, just as in most other lunar origin theories, a fission-product Moon must accrete out of a geocentric swarm of material. Intrinsic nonaxisymmetry of the remnant Earth after fission would prevent rapid recollapse of the swarm. This revised picture alleviates some of the problems associated with earlier versions of the fission theory. The two most serious remaining objections are that it is difficult to make the proto-Earth rotate fast enough to undergo fission and that the proto-Earth must be largely molten at the time it fissions. To overcome the first objection, it may be necessary to combine fission with the planetesimal impact theory. Some advantages of such a hybrid theory are discussed. The second objection cannot be fully assessed until more is known about the thermal history and accretion of the proto-Earth.

1. Introduction

1.1. Background on fission

In astrophysical fluid dynamics, the term "fission" usually refers to breakup of a rapidly rotating, self-gravitating, equilibrium celestial body caused by growth of a surface distortion. The subject has a long history and a correspondingly extensive literature, in which the origin of the Moon is only one among many possible applications. Detailed reviews of fission and related topics can be found in Tassoul (1978) and in Durisen and Tohline (1985). A brief and simplified summary of what was known about fission prior to this decade will now be provided in order to put our calculations in context. Primary references for these results can be found in the cited reviews.

Historically, our most complete knowledge about fission instabilities comes from work on the spheroidal and ellipsoidal equilibrium states of uniform, incompressible fluids (cf. Chandrasekhar, 1969). The simplest of these are the Maclaurin spheroids, which are rigidly rotating and are symmetrically flattened about their rotation axes. With proper normalization, the Maclaurin spheroids can be completely distinguished by a single parameter. An especially convenient choice of this parameter is $\beta = T/|W|$, where T = total rotational kinetic energy and W = total gravitational energy. For $\beta > \beta_s \approx 0.1375$, new triaxial equilibrium states, called Jacobi ellipsoids, become possible. These are configurations with ellipsoidal (i.e., triaxial) surface figures that rotate about their shortest axes. For $\beta > \beta_s$, both Maclaurin spheroids and Jacobi ellipsoids exist. Because the Jacobi ellipsoid of the same total angular momentum has lower total energy, a viscous Maclaurin spheroid with $\beta > \beta_s$ will evolve toward a Jacobi ellipsoid state on a dissipative time scale. The surface distortions that grow viscously for $\beta > \beta_s$ turn the Maclaurin spheroids slowly into tumbling bars, and so the Maclaurin spheroids are said to be *secularly* unstable to bar distortions in the presence of viscosity for $\beta > \beta_s$. We refer to the bars as "tumbling" rather than "rotating" about the short axis because the pattern speed of the bar does not necessarily match the rotation rate of the fluid. The only bars that are rigid rotators are the Jacobi ellipsoids themselves.

Actual breakup of the configuration probably does not result from purely secular evolution, as was believed decades ago (Chandrasekhar, 1969, Chapter 1), but rather through the onset of dynamic instability at larger values of β. These dynamic instabilities could be reached either directly by accretion of high angular momentum material or after an intermediate phase of secular evolution. For $\beta > \beta_d \approx 0.2738$, the Maclaurin spheroids become *dynamically* unstable to the growth of a barlike surface distortion. Dissipation is not necessary, and the growth time scale is a rotation period. For $\beta > \beta_p \approx 0.1628$, the Jacobi ellipsoids become dynamically unstable to the growth of a tumbling pear-shaped distortion. Until 1977, little was known about the outcome of these instabilities, but most researchers speculated that they would lead to breakup of the configuration into two or more discrete pieces. We will refer to this conjectured outcome specifically as "binary fission." We will use "fission" to refer more generally to any form of breakup.

Equilibrium and stability calculations have also been done numerically for nonuniform, rapidly rotating configurations of compressible fluids. The work relevant to the fission problem has mostly concerned polytropic fluids for which

$$P \sim \rho^{1+1/n} \tag{1}$$

where P is the pressure, ρ is the mass density, and n is called the "polytropic index." When n = 0, the fluid is incompressible. Ordinary stars are approximated by polytropes with n = 3/2 to 3. Terrestrial material in an Earth-sized planet is about as compressible as an n = 1/2 fluid. Remarkably, with few exceptions, studies to date have shown that, for n $\lesssim$ 3, compressible configurations exhibit essentially the same types of equilibrium structures and stability limits as n = 0 fluids. Particularly, the β_s, β_p, and β_d given above seem to generalize with quantitative accuracy (to within ±0.02 or better). In general terms, then, a fluid celestial body will fission if some evolutionary process leads the body up to and beyond the bar or pear-shaped dynamic instability point. The strength of dissipative forces will determine which instability point is reached.

1.2. A fission origin of the Moon

Darwin's (1879) original suggestion for how the Moon may have "fissioned" from the Earth actually invoked a tidal resonance with the sun for a $\beta < \beta_s$ (see also McKinnon and Mueller, 1984). However, more recent versions of the fission origin hypothesis, as well as a later revision by Darwin, have invoked the instabilities described in the preceding section (for instance, Ringwood, 1960; Wise, 1966; O'Keefe, 1966; Binder, 1974; and O'Keefe and Sullivan, 1978). The relative merits of this theory compared with its competitors from a geochemical standpoint are extensively (and hotly) debated elsewhere in this volume. Here, we are primarily concerned with the dynamics of the process. Until recently, proponents of fission have been forced to make only crude estimates about the outcome of fission instabilities. Just as workers in other areas of astrophysics, they usually assumed that binary fission would occur, breaking off the Moon all at once as a discrete body. There are several serious dynamical problems with this picture: (1) the current Earth-Moon system does not have enough angular momentum to make the proto-Earth surpass even the secular stability limit β_s; (2) a much more massive body than the Moon is likely to fission; and (3) the Moon's orbit was significantly inclined to the Earth's equator in the past.

Over the last eight years, several groups have been able to study the outcome of dynamic fission instabilities using numerical hydrodynamic simulation techniques (Lucy, 1977, 1981; Gingold and Monaghan, 1978, 1979; Durisen and Tohline, 1980; and Durisen et al., 1985). The last of these studies is the most reliable, because results of high-resolution calculations with several different codes were compared. For n = 3/2 and β = 0.33, Durisen et al. (1985) found that binary fission does not occur. Instead, the dynamically growing bar distortion develops into a trailing spiral pattern. These spiral arms wrap up into a ring or disk of material that detaches from the remnant central object. The central remnant is a stable tumbling barlike structure. The ejected ring or disk contains 18% of the mass and about half the angular momentum. As already pointed out by Durisen and Scott (1984), this view of the fission process actually has a lot in common with other lunar origin hypotheses. After the fission event, the Moon must accrete from a geocentric swarm of material, just as postulated in the models that invoke capture by tidal disruption, accretion in Earth orbit, or planetesimal impact. Furthermore, if the Moon accreted from only a small portion of the ejecta and the rest is lost, then we can understand the small present Earth-Moon angular momentum.

This paper reevaluates the results of Durisen *et al.* (1985) and the arguments of Durisen and Scott (1984) in light of contributions by other researchers at the Conference on the Origin of the Moon, from which this volume resulted. We have also performed several new fission simulations that are somewhat more directly relevant to terrestrial conditions. Section 2 describes the new results and compares them with Durisen *et al.* (1985). Implications are then discussed in Section 3 with particular emphasis on the most vexing problems for the fission hypothesis—the proto-Earth's angular momentum and the validity of a fluid treatment. Section 4 provides a brief summary of our major conclusions.

2. Numerical Simulations

2.1. Techniques

In Durisen *et al.* (1985), three different 3D hydrodynamics codes were used to treat the same problem. One of these (Gingold and Monaghan, 1977) was a smoothed-particle hydrodynamics code where the fluid was represented by a finite number of sample fluid-element particles that were followed by means of the Lagrangian equations of motion. The other two codes (Tohline, 1980; Boss, 1980) solved the Eulerian hydrodynamics equations in a donor-cell finite-difference form. Axisymmetric equilibrium models for $n = 3/2$ polytropes with $\beta = 0.33$ and 0.38 were calculated by the self-consistent field method (Bodenheimer and Ostriker, 1973). These models were loaded into each code as initial conditions in a manner appropriate to its particle or finite-difference grid structure. In most cases, the initial models were also given an initial nonaxisymmetric perturbation, usually with a barlike character. The final outcomes of evolutions with different initial perturbations were usually indistinguishable.

Because of the multicode, multicollaborator nature of the study, only two β values for one n were considered. One of the important findings, however, was that the codes tended to agree qualitatively about how the dynamic instability for $\beta > \beta_d$ develops and what it ultimately leads to. Quantitatively, the codes agreed with surprising accuracy (within about ±5%) concerning the percentages of mass and angular momentum ejected. In no case did a clear-cut example of classic binary fission occur, not even when the initial perturbations were strongly biased in its favor. Such agreement between radically different numerical schemes is gratifying given the difficulty of obtaining consistent answers in even 1D or 2D calculations. It makes us confident that we are resolving the true physical outcome of fission instabilities and that we can now use any one of these codes to extend our investigations.

Of the three codes, the one using smoothed-particle hydrodynamics produces reliable results with the greatest ease. For this paper, we have again loaded equilibrium polytropic models as initial conditions, but this time for a wider range of n and β. The angular momentum distribution was chosen to be the same as in Durisen et al. (1985). No perturbations were added because the use of a finite number of particles to represent the model provided sufficient initial nonaxisymmetry for the dynamic instability to manifest itself in a few rotations. The evolutions with n = 1/2 are close in degree of compressibility to terrestrial material (see Durisen and Scott, 1984). The n=3/2 cases are more appropriate to gaseous protoplanets or stars. We have done simulations with both 500 and 2000 particles.

2.2. Results

Table 1 summarizes the results of 2000 particle calculations from Durisen et al. (1985) and from our work. Additional evolutions with 500 particles were run for both n = 1/2 and 3/2 with β = 0.26, 0.28, 0.30, and 0.33. Only the 2000 particle evolutions are shown in Table 1 because they are more quantitatively reliable. In all cases, a substantial fraction of the mass and angular momentum was ejected through the development of spiral structure. Within a few rotation periods, the spiral arms wrapped up and detached. The end state always consisted of a central nonaxisymmetric structure, surrounded by a ring or disk of ejected debris. In some evolutions, the central remnant was followed for several tumble periods without changing and, in this sense, seemed to be a stable equilibrium configuration. The central remnants were distinctly barlike for the lowest β values but exhibited somewhat more complex pear or crosslike structures at the highest β's. In no case did binary fission occur, and there was no evidence for condensation of a discrete body or bodies out of the ejecta.

The results confirm the speculation in Durisen and Scott (1984) that dynamic fission instabilities in n = 1/2 polytropes will be qualitatively similar to those for n = 3/2. There were, however, some quantitative differences between n =

TABLE 1. *Results of Fission Simulations.*

n	β	Mass Ejected	Ang. Mom. Ejected	Source
1/2	0.33	10%	30%	This paper
3/2	0.33	18%	51%	Durisen et al. (1985)
	0.38	36%	77%	Durisen et al. (1985)

1/2 and n = 3/2 and other unexpected features. The n = 1/2 polytropes generally ejected less mass and angular momentum by factors of two to four, depending on β. In all evolutions, the ejection occurred via a two-armed spiral pattern, except for n = 1/2, β = 0.33. Two-armed evolutions are illustrated in Durisen *et al.* (1985) and in Durisen and Tohline (1985). The unusual three-armed behavior is shown in Fig. 1. After a few rotations, two of the arms merge to form an asymmetric two-armed pattern. The appearance of unstable higher-order distortions at high β is predicted by linear stability theory. The n = 3/2, β = 0.38 case was also not quite bilaterally symmetric and developed a crosslike rather than barlike central remnant. For given β, the higher-order distortions seem to be more prominent in less compressible fluids. Near marginal instability, two-armed distortions dominate for both n. The 500 particle runs suggest, as previously argued by Durisen and Scott (1984), that even near marginal stability ($\beta \approx \beta_d$) finite amounts of mass are ejected—a few percent for n = 1/2 and closer to 10% for n = 3/2. This will need to be confirmed by evolutions with more particles.

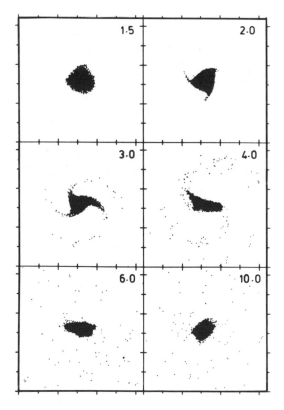

Fig. 1. Dynamic fission in an n = 1/2, β = 0.33 polytrope. These panels show the positions of the particles in the smoothed-particle hydrodynamics code projected onto the equatorial plane. The evolution starts at time zero near equilibrium with the only nonaxisymmetry due to representation of the polytrope by a finite number of particles. Evolutionary times are given in the upper right-hand corner in units of the initial central rotation period of the polytrope. The sense of fluid rotation is counter-clockwise.

The next section discusses the aspects of our results that are directly relevant to lunar origin.

3. Discussion

3.1. Implications of the numerical results

We have now mapped the outcome of dynamic bar instability over a range of β for fluid bodies with compressibilities appropriate to both stars (n = 3/2) and terrestrial planets (n = 1/2). Most of the arguments and speculations in Durisen and Scott (1984) are confirmed. Regardless of compressibility, the dynamic bar instability leads to spiral arm ejection of substantial mass and angular momentum fractions; it does not lead to classic binary fission. A ring or disk of debris around a central, usually barlike remnant results. This resembles the starting point of many other lunar origin theories and represents a natural way to produce a "geocentric swarm" of planetesimal-like bodies. We would like to emphasize that similar results for the dynamic bar mode instability of n = 3/2 polytropes were obtained independently by three different state-of-the-art numerical codes (Durisen et al., 1985). Nevertheless, it is premature to conclude that classic binary fission never occurs, because we have not yet explored the pear-shaped instability of Jacobi ellipsoids. There is only weak evidence so far (see Sections II.A and II.C of Durisen and Tohline, 1985) that the pear-shaped instability will also lead to debris-shedding, not binary fission. For completeness, we should mention that binary fission has been seen in some other simulations of n = 1/2 polytropes (Lucy, 1977; Gingold and Monaghan, 1978, 1979), but the conditions in these simulations were almost certainly inappropriate to lunar origin. The latter results do suggest, however, that binary fission might occur when assumptions are significantly different from our own.

There are important differences between a geocentric swarm produced by fission and those produced in other lunar origin theories. The ejected matter is the initially outermost part of the unstable body and so would be predominantly mantle material if the Earth's core had already segregated. Because the remnant Earth after fission is nonaxisymmetric, the swarm will not collapse rapidly back onto the Earth, as it does in other lunar theories. As illustrated most clearly in Figs. 4 to 7 of Durisen and Tohline (1985), fission-ejected debris is separated from the remnant by a gap. [Smoothed-particle simulations, as in our Fig. 1, do not seem to show the gap for reasons discussed in Durisen et al. (1985).] The gap is maintained by gravitational torques arising from the central body's intrinsic nonaxisymmetry. This process is well known in solar nebula theory and is referred to as "tidal truncation" [see Durisen and Tohline (1985) and Weidenschilling et al. (1986) for references and discussions]. Thus fission produces a swarm that is much more massive than the Moon but that must be mostly lost from the system, not reaccreted by the Earth. The swarm, once in place, can be contaminated by heliocentric material, and so a Moon formed as a product of fission need not resemble Earth mantle material exactly. In addition, the collisional aggregation process would tend to deplete volatile elements. The inclination of the Moon's orbit to the Earth's equator plane in the past poses no particular difficulty for our modified fission theory, because the Moon can form at many Earth radii and will contain only a fraction of the original ejecta. A last few collisions or gravitational scatterings with large geocentric (or heliocentric) planetesimals could have disturbed its inclination.

One aspect of the n = 1/2 results that somewhat weakens the case for fission is that the more Earth-like n = 1/2 fluid bodies do not eject as much angular momentum. Durisen and Scott (1984), knowing only the n = 3/2 results, supposed that more than half the angular momentum could be lost from the Earth-Moon system simply by expelling most of the debris. Now, to meet the angular momentum constraint of the current Earth-Moon system, we must suppose that long-term evolution beyond the end point of our simulations causes substantial additional transfer of angular momentum from the central remnant to the debris. This will happen naturally to some extent in a geocentric swarm as a consequence of the central body's nonaxisymmetry, but, without detailed models, it is not clear whether the process is ultimately efficient enough. The same time-dependent gravitational field of the central body, aided by the nascent Moon itself, could also serve to disperse ejecta eventually into heliocentric orbit. These possibilities suggest directions for further development of a comprehensive fission theory.

It is important to remember that, so far, we have explored only the dynamic distortional instabilities that set in for $\beta > \beta_d$ in axisymmetric initial structures. If the viscous time scale is shorter than the accretion or spin-up time, then the proto-Earth will distort into a bar shape for $\beta > \beta_s$ and encounter the pear-shaped distortional instability at β_p. We still have little trustworthy information about the pear-shaped instability. The configuration starts with less total angular momentum and may conceivably eject a higher fraction of it during fission. We plan to explore this possibility further by adding viscous stresses to our simulations.

3.2. The angular momentum problem

Table 2 compares the current angular momentum of the Earth-Moon system with that required to attain each of the secular and dynamic instability points we have discussed earlier. The angular momentum J is given in the middle column in units of $(G\ M_E{}^3 r_E)^{1/2}$ where M_E and r_E are the current mass and mean radius of the Earth. The critical β and J values for instabilities are those for uniform density incompressible configurations as given in Chandrasekhar (1969). For simplicity, we ignore the fact that compressibility and the inclusion of the ejecta mass will

	β	$J/(GM_E^3 r_E)^{1/2}$	$J/(r_E v_e)$
Current Earth-Moon System	0.026	0.115	0.081 M_E
Secular bar instability	0.138	0.304	0.215 M_E
Dynamic pear instability	0.163	0.390	0.275 M_E
Dynamic bar instability	0.274	0.509	0.359 M_E

cause small differences in these limiting values. In this context, r_E is chosen so that the Earth's volume does not change as its surface figure distorts. It is clear from this table just how far the current Earth-Moon system is from containing enough angular momentum to cause a proto-Earth to fission. The β value for the "current" Earth-Moon system in Table 2 is the β that a uniform density, incompressible Earth would have if it were given all the angular momentum of the present system.

The last column of Table 2 expresses J in terms of an amount of mass moving tangentially at escape velocity v_e at the Earth's surface. This normalization of J emphasizes the difficulty of imparting angular momentum to the proto-Earth by accretion of heliocentric planetesimals. As discussed elsewhere in this volume, it is hard to understand how the Earth-Moon system acquired even its current J value, much less how it acquired values three to five times larger. The problem is that heliocentric planetesimals are almost as often prograde and retrograde as seen from the Earth in an encounter, with only a small prograde bias. To achieve the $J/(M_E r_E v_e)$ values in Table 2, the proto-Earth would probably need to do so stochastically through a relatively small number of accretional collisions with large planetesimals. For illustrative purposes, suppose the Earth accreted from N equal mass bodies. Then, stochastically, we expect $J/(r_E v_e) \lesssim M_E/\sqrt{N}$. If the Earth accreted predominantly from Moon-sized planetesimals, J values near the current one are reasonably likely. To reach the critical values for dynamic pear and bar instabilities requires Mars-sized bodies. Of course, the Earth probably accreted from a spectrum of planetesimal sizes. Still, our simplistic estimate suggests that one or several grazing collisions by Mars-sized or larger bodies are necessary to trigger fission.

If this reasoning holds up, the fission theory becomes embarrassingly similar to the giant planetesimal impact theory. In fact, it may ultimately prove necessary to incorporate elements of both theories in a successful and complete explanation of lunar origin (see Hartmann, 1986). Fission instabilities triggered by a last large impact would guarantee ejection of a geocentric swarm that would not rapidly reaccrete onto the Earth, because the remnant Earth would be nonaxisymmetric and would truncate the inner edge of the swarm through gravitational torques. Fission would also guarantee that a substantial fraction of the swarm's mass would come from the Earth as well as the planetesimal. Proponents of the planetesimal-impact model need to be aware of these possible, if not likely, complications.

3.3. Solid vs. fluid proto-Earth

A serious potential difficulty for the fission theory may be its reliance on the stability properties and behavior of *fluid* systems, whereas the proto-Earth might well have been mostly solid rather than molten. A discussion of whether an accreting proto-Earth would in fact have been molten goes beyond the scope of this paper. However, we should point out that dynamic fission does not occur in recent numerical calculations by Boss and Mizuno (1985) that were designed to simulate a solid in high-density regions and rocky debris in low-density regions. These authors use a modified version of the Boss (1980) 3D hydrodynamics code. Equation (1) was replaced by a Murnaghan equation of state, which is much less compressible even than n = 1/2. The resistance of the solid to any shear is modeled by explicitly damping all nonrotational motions by a large fraction each numerical time step. Boss and Mizuno (1985) find that, for damping rates equivalent to the viscosity of solid mantle materials, the dynamic bar instability is completely suppressed. According to these authors, the dynamic bar instability in the Earth would be inhibited or suppressed for kinematic viscosities greater than about 10^{14} cm^2/s. The "viscosity" of solid mantle material is somewhat model dependent but is probably many orders of magnitude larger. The viscosity of magmas, on the other hand, is many orders of magnitude smaller. Thus material physics and the thermal and accretional history of the proto-Earth, though poorly known, are critical for deciding whether dynamic fission would actually occur, even when β is large enough. It is not yet clear how much of the proto-Earth would need to be molten for a significant amount of mass to be ejected by dynamic fission. Boss and Mizuno (1985) also show by a separate analytic calculation that the secular bar instability is *not* suppressed in viscous solids. This raises the interesting possibility that a more traditional binary mode of breakup might occur by purely secular evolution in solid protoplanets.

4. Conclusions

Numerical studies of dynamic bar instabilities in rapidly rotating, inviscid fluid bodies show that fission leads to ejection of mass and angular momentum in the form of a ring or disk, rather than in one or more discrete bodies, as usually envisioned. Fission theory, modified to account for these new results, thus comes to resemble most other lunar origin theories in that the Moon is formed out of a "geocentric swarm" of debris. However, in this case, the central remnant proto-Earth is nonaxisymmetric, most likely a tumbling bar, that truncates the inner edge of the swarm and probably continues to lose angular momentum to the debris via gravitational torques. The bar instability ejects about 5% to 10% of the mass and 10% to 30% of the angular momentum for material of terrestrial compressibility. The Moon represents only a fraction of the original ejected mass. The rest must be lost, carrying with it most of the system's initial angular momentum. The details of this loss process remain to be worked out.

Most of the elements of this revised fission picture were described by Durisen and Scott (1984), who based their arguments on the numerical simulations in Durisen *et al.* (1985). In this paper, we have presented new simulations that confirm some of the speculations in Durisen and Scott (1984) and that present a consistent picture for fluid bodies with a range of compressibilities.

The two most serious dynamical objections to this modified fission theory are: (1) the difficulty of imparting enough angular momentum to the proto-Earth to reach dynamic instability and (2) the suppression of dynamic instability if the proto-Earth is a solid (or an extremely viscous fluid) rather than a low viscosity fluid. By our current understanding of planetesimal accretion, as discussed elsewhere in this volume, it seems that one or a few favorable impacts by large planetesimals would be necessary to reach angular momentum values large enough for even the pear-shaped instabilities to occur. The addition of fission to the impact origin theory, while displeasing by an economy of hypotheses principle, does have some advantages in establishing the geocentric swarm and preventing its rapid reaccretion and in effectively mixing mantle material into the ejecta. In this case, fission loses its standing as an independent theory but remains a process that must be further considered. The suppression of dynamic fission instabilities in a viscoelastic solid, however, cannot be overcome. The proto-Earth probably had to be largely molten for fission to occur on a dynamic time scale. Whether this was true remains to be determined from detailed models of accretion and heat transport in the early Earth.

Acknowledgments. The authors would like to thank A. Binder, D. Black, P. Bodenheimer, A. Boss, P. Cassen, W. Hartmann, R. Reynolds, E. Scott, J. Tohline, S. Weidenschilling, and J. Wood for useful comments, support, and criticism. L. David assisted in the computation of equilibrium models. Part of this work was supported by NSF grant AST 81-20367 and NASA-Ames/Indiana University Consortium Agreement NCA2-1R335-401. One of us (R.H.D.) was a National Research Council Associate at NASA-Ames Research Center during the preparation of this manuscript.

5. References

Binder A. B. (1974) On the origin of the Moon by rotational fission. *The Moon, 11*, 53–76.

Bodenheimer P. and Ostriker J. P. (1973) Rapidly rotating stars—VIII. Zero viscosity polytropic sequences. *Astrophys. J., 180*, 159–169.

Boss A. P. (1980) Protostellar formation in rotating interstellar clouds—I. Numerical methods and tests. *Astrophys. J., 236*, 619–627.

Boss A. P. and Mizuno H. (1985) Dynamic fission instability of viscous protoplanets. *Icarus*, in press.

Chandrasekhar S. (1969) *Ellipsoidal Figures of Equilibrium.* Yale University, New Haven. 252 pp.

Darwin G. H. (1879) On the precession of a viscous spheroid and on the remote history of the Earth. *Phil. Trans. Roy. Soc. Part II, 170*, 447–530.

Durisen R. H. and Scott E. H. (1984) Implications of recent numerical calculations for the fission theory of the origin of the Moon. *Icarus, 58*, 153–158.

Durisen R. H. and Tohline J. E. (1980) A numerical study of the fission hypothesis for rotating polytropes. *Space Sci. Rev., 27*, 267–273.

Durisen R. H. and Tohline J. E. (1985) Fission of rapidly rotating fluid systems. In *Protostars and Planets II* (D.C. Black, ed.). Univ. Arizona, Tucson, in press.

Durisen R. H., Gingold R. A., Tohline J. E., and Boss A. P. (1985) The binary fission hypothesis— A comparison of results from finite difference and smoothed particle hydrodynamics codes. *Astrophys. J.*, in press.

Gingold R. A. and Monaghan J. J. (1977) Smoothed particle hydrodynamics—Theory and application to nonspherical stars. *Mon. Not. Roy. Astron. Soc., 181*, 375–389.

Gingold R. A. and Monaghan J. J. (1978) Binary fission in damped rotating polytropes. *Mon. Not. Roy. Astron. Soc., 184*, 481–499.

Gingold R. A. and Monaghan J. J. (1979) Binary fission in damped rotating polytropes II. *Mon. Not. Roy. Astron. Soc., 188*, 39–44.

Hartmann W. B. (1986) Moon origin: The impact-trigger hypothesis, this volume.

Lucy L. B. (1977) A numerical approach to the testing of the fission hypothesis. *Astron. J., 82*, 1013–1024.

Lucy L. B. (1981) The formation of binary stars. In *Fundamental Problems in the Theory of Stellar Evolution* (D. Sugimoto, D.Q. Lamb, and D.N. Schramm, eds.), pp. 75–83. Reidel, Dordrecht.

McKinnon W. B. and Mueller S. W. (1984) A reappraisal of Darwin's fission hypothesis and a possible limit to the primordial angular momentum of the Earth (abstract). In *Papers Presented to the Conference on the Origin of the Moon*, p. 34. Lunar and Planetary Institute, Houston.

O'Keefe J. A. (1966) The origin of the Moon and the core of the Earth. In *The Earth-Moon System*, pp. 224–233. Plenum, New York.

O'Keefe J. A. and Sullivan E. C. (1978) Fission origin of the Moon—Cause and timing. *Icarus, 35*, 272–283.

Ringwood A. E. (1960) Some aspects of the thermal evolution of the Earth. *Geochim. Cosmochim. Acta, 20*, 241–259.

Tassoul J.-L. (1978) *Theory of Rotating Stars*. Princeton University, Princeton. 506 pp.

Tohline J. E. (1980) Fragmentation of rotating protostellar clouds. *Astrophys. J., 235*, 886–881.

Weidenschilling S. J., Greenberg R., Chapman C. R., Herbert F., Davis D. R., Drake M. J., Jones J., and Hartmann W. K. (1986) Origin of the moon from a circumterrestrial disk, this volume.

Wise D. U. (1966) Origin of the Moon by fission. In *The Earth-Moon System*, pp. 213–223. Plenum, New York.

The Binary Fission Origin of the Moon

ALAN B. BINDER

Code SN4, NASA/Johnson Space Center, Houston, TX 77058

The concept that the Moon formed from the Earth by classical binary fission was put forth over a century ago by Darwin and is supported by the basic fact that, while the Moon lacks a large core like the Earth, the Moon's bulk, major oxide composition is very similar to that of the terrestrial mantle. Like all lunar origin theories, there are a number of unresolved problems concerning the fission model. These include: how the proto-Earth acquired sufficient angular velocity to fission; how the fission process actually proceeded; and how the Earth-Moon system lost its excess angular momentum after fission. Despite these uncertainties, the compositional similarities between the Earth's mantle and the bulk Moon make the fission model a candidate for the origin of the Moon. The proposed sequence of events in the formation of the Moon by binary fission is as follows: (1) Toward the end of both accretion and the formation of the core, the proto-Earth was driven beyond its rotational stability limit (2.65-hour-period) by the accretion of more mass and/or the core formation event. (2) At that point the proto-Earth changed its figure from that of a stable Jacobi ellipsoid to an unstable, bowling-pin-shaped Poincaré figure. (3) As the proto-Moon, whose initial mass was 10 to 20 times its current value, began to form and broke free, it was thermally unstable and began to explosively change to a vapor state. The latter effect was due to the proto-Moon's high temperature (~3000° to 4000°C) and the drop of internal pressure as the protolunar material detached from the Earth. These two effects caused the Moon to lose the majority of its initial mass within a few hours after it formed. Most of this mass was reaccreted by the Earth. (4) Some of this mass escaped not only from the Moon, but also from the Earth-Moon system via orbital perturbations. The mass that escaped from the Earth-Moon system carried a considerable amount of angular momentum with it, thereby partially accounting for the current deficiency of angular momentum of the Earth-Moon system. (5) After this "catastrophic" phase of mass and angular momentum loss, the Moon was still very hot (2000° to 2500°C) and had an extended atmosphere of volatile-rich, evaporated silicates. Because of the near proximity of the nascent Moon to the Earth, this atmosphere filled the lunar Roche lobe and escaped through the L_1 point into the Earth's Roche lobe, carrying with it most of the Moon's volatile components. (6) Most of this material fell onto the Earth, but some of it was lost from the Earth-Moon system, carrying away additional angular momentum. The loss of a total of only 1.4 lunar masses of material by this process and that discussed under (4) above accounts for the entire angular momentum deficiency of the Earth-Moon system. (7) During the L_1 phase of mass and volatile element loss, any

Fe that was present in the proto-Moon as it fissioned, Fe that was accreted by the proto-Moon after fission, and Fe that formed by the reduction of FeO as the Moon devolatilized formed a small core and caused the siderophile elements to undergo a second stage of depletion (the first occurred as the Earth's core formed before fission), thereby giving them their nonterrestrial depletion pattern. (8) As a result of the above processes, the Moon reached its current mass, was depleted in volatile elements, had a nonterrestrial siderophile element depletion pattern, and began its initial petrological differentiation into a lower mantle, mare basalt source region, and crust within a few hundred to thousand years after fission.

1. Introduction

The theory that the Moon formed by the binary fission of mantle material from a rapidly rotating proto-Earth is the oldest (proposed by Darwin in 1878) and currently among the least accepted of the lunar origin models. Before Apollo, this theory was based on the facts that the Moon's mean density is equal to that of uncompressed terrestrial mantle material, and any lunar core is at least an order of magnitude smaller than the terrestrial core. As a result of Apollo, it appears that the Moon's nonvolatile, lithophile element composition, as well as other compositional and physical characteristics, are consistent with a lunar origin by fission. However, there are a number of arguments—many of them dynamical—against this model.

In the following I will review the major arguments for and against the binary fission model and outline questions that need to be answered in order to determine if the Moon could or could not have formed in this way. I will limit my discussions to the classical form of binary fission in which the Moon formed from a mass of mantle material that detached from the rapidly spinning proto-Earth. However, recent work suggests that the fission process may have resulted in the formation of a ring of circumterrestrial material rather than in a single proto-Moon (e.g., Durisen and Tohline, 1985; Durisen and Gingold, 1986). As these authors suggest, the Moon could have accreted from this ring of fissioned material and therefore this model has much in common (both positive and negative aspects) with the classical form of binary fission. However, these studies have not yet covered the entire range of fission modes and the results of other dynamicists do support the classical model to some extent (e.g., Lucy, 1977; Gingold and Monaghan, 1979). Thus classical binary fission cannot now be excluded as a possible model of lunar origin and therefore needs to be considered further, especially in view of the compositional similarities between the Moon and the terrestrial mantle.

2. Dynamics

2.1. Initial rotational period of the proto-Earth

In order for fission to have occurred, the proto-Earth must have been rotating with a period of about 2.65 hours (e.g., Wise, 1963). According to accretion models, the initial rotation period of the Earth was only 10 to 15 hours (Giuli, 1968; Harris, 1977), values that agree with that of 10 hours or more derived from a simple interpretation of an angular momentum density plot of solar system bodies (Hartmann and Larson, 1967). If these conclusions hold, the Earth's rotation was far too slow for fission to have occurred. The question that needs to be resolved is, Do the accretion models and the simple interpretation of the angular momentum density plot hold?

First, consider the accretion models. If the Moon was captured (e.g., MacDonald, 1964; Kaula and Harris, 1973), formed as the result of a giant impact (Hartmann and Davis, 1975; Cameron and Ward, 1976), or formed by fission (O'Keefe, 1968, 1969, 1970, 1972; Wise, 1963, 1969; Binder, 1974, 1978, 1980), the Moon's initial orbit was only a few Earth radii (R_e) from the Earth. In several of these models, the initial lunar orbit was essentially at the Roche limit, i.e., at 3 R_e. Under this initial condition and given the current angular momentum of the Earth-Moon system, the Earth's initial rotation period was only 5 hours—a factor of 2 to 3 shorter than is possible according to the models that are based on pure accretional processes.

The binary accretion model is the only one that, at first appearance, does not contradict the >10-hour initial rotation period of the accretional model. However, if the binary accretion model is correct and the initial rotation period of the Earth was 10 to 15 hours, conservation of angular momentum requires that the Moon accreted at a distance of 30 to 50 R_e from the Earth. These distances are inconsistent with empirical data that indicate that the Earth-Moon distance was only about 20 to 25 R_e as late as 3×10^9 years ago (Lambeck and Pullan, 1980; Binder, 1982) and therefore that (1) the Moon was at $\ll 20$ R_e at 4.6×10^9 years ago and (2) the Earth's initial period was $\ll 8$ hours.

Thus, no matter how the Moon formed, the Earth's rotation period after lunar formation was considerably shorter than is allowed by pure accretional modeling. Since accretional models are based on the assumption that the Earth formed by the direct accretion of solid material from heliocentric orbits, the evidence suggests that these assumptions are wrong. Somehow the Earth acquired at least a factor of 2 more angular momentum than is allowable by pure accretional modeling.

In the case of the impact model, this difficulty is surmounted by assuming that the impacting body was as large or larger than Mars and that the impacting body brought the missing angular momentum into the system (Cameron, 1984). However, according to Cameron, the Moon formed mainly from material from the impacting body and not from terrestrial mantle material. If this conclusion is verified, it becomes necessary to make the ad hoc assumption that the composition of the mantle of the impacting body was similar to that of the terrestrial mantle after core formation. Whether these (and other) questionable assumptions are justifiable or not remains to be determined as this currently popular model is further investigated.

Given that all other lunar origin models definitely require that the Earth's initial rotation period was higher than that allowed by pure accretion modeling and that, despite its current popularity, the giant impact model may very well be wrong, dynamicists need to reexamine the accretion models to determine if and how the proto-Earth could have obtained an initial period of between 5 and 2.65 hours. For example, the possibility that the Earth formed by the collapse of a large accretion disk and therefore acquired more angular momentum than by direct accretion needs to be investigated. Until this is done and it is shown with irrefutable negative results that the proto-Earth could not have a 2.65-hour rotation period, fission cannot be excluded as a possible mode of lunar formation.

Second, the meaning of the angular momentum distribution in the solar system is somewhat obscure. The current angular momentum density of the Earth-Moon system falls on a line drawn through the points for Jupiter, Saturn, and the asteroids in a plot of angular momentum density vs. mass (e.g., MacDonald, 1963; Hartmann and Larson, 1967; see also Fig. 1). These and other authors have argued that this line represents the maximum angular momentum density a solar system body could acquire during formation. If this is the case, then it is argued that this plot shows that the initial Earth-Moon system was deficient in angular momentum by a factor of about 3 compared to that needed for the fission of the proto-Earth. However, as Durisen pointed out in his review comments, the angular momentum density plot ". . . lumps together objects which almost certainly achieved their present rotation states by completely different but poorly understood mechanisms." Furthermore, Fig. 1 is an updated version of the Hartmann and Larson plot, with upgraded values for Uranus and Neptune and new data for the Pluto-Charon system. If we accept for the moment that the line drawn through Jupiter, Saturn, the Earth-Moon system, and the asteroid points may represent the *mean* angular momentum density of solar system bodies, rather than the maximum value, we note the following. First, Neptune lies as far below this line as the Earth-Moon system would lie above it if the Moon formed by fission. Second, the Pluto-Charon system (which is the second candidate

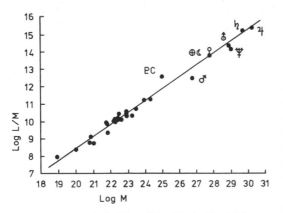

Fig. 1. Angular momentum density (L/M) vs. mass (M) of solar system bodies. The filled circles with log L/M values below about 11.5 are the data for the asteroids. The data for the individual planets, the Earth-Moon system, and the Pluto-Charon system are given by filled circles and are identified by their astronomical symbols. The L/M value the Earth-Moon system would have had if the Moon formed by fission is given by an open circle connected to the current Earth-Moon data point.

for a fission pair of bodies in our solar system; Lin, 1981) lies as far above the mean line as Mars lies below. These observations make it seem quite possible that the line does define the *average* initial angular momentum density of solar system bodies (rather than the maximum value) and that individual bodies deviate from this mean line by up to a factor of 4. If so, the proto-Earth and proto-Pluto could have been rotating fast enough to have formed their high mass satellites by fission. In any case, the exact meaning of a log angular momentum density vs. log mass plot like Fig. 1 is sufficiently uncertain so that one cannot use it to determine if high initial rotation rates for the proto-Earth and other solar system bodies are possible or not. This is the case until the mechanisms that caused the planetary rotational rates are much better understood.

2.2. Classical fission dynamics

Classical dynamical model. As pointed out in the introduction, the most recent numerical studies of certain modes of fission suggest that, rather than a single secondary body being detached, a ring of fissioned material forms around a terrestrial type primary body [e.g., see the review paper by Durisen and Tohline (1985) and the paper by Durisen and Gingold (1986) in this volume]. This ring of fissioned material could then have accreted to form the Moon. However, somewhat earlier numerical studies suggest that something more like the classical version of fission may also occur (e.g., Lucy, 1977; Gingold and Monaghan, 1979), assuming different conditions. Though these results indicate that the fission process may be different from the classical view and more work is needed to understand the full range of possible fission modes, I am limiting my review in the following to the concepts proposed for the classical version of the binary fission model (see also Fig. 2).

As reviewed and discussed by Wise (1963, 1969) and O'Keefe (1968, 1969, 1970, 1972), classical rotational fission is thought to have occurred as the proto-Earth reached its rotational stability limit (2.65-hour period) toward the end of accretion and/or due to the decrease in the moment of inertia of the proto-Earth as the core formed (Wise, 1963). Prior to having reached the stability limit, the proto-Earth passed from a mildly oblate Maclaurin spheroid (when the rotation period was very much longer than 2.65 hours) to a Maclaurin spheroid whose oblateness was such that the polar to equatorial axial ratio was 7:12 and the period was near 2.65 hours. At that time, any increase in the rotational angular momentum would have caused the proto-Earth to become a triaxial Jacobi ellipsoid whose moment of inertia was larger than that of the limiting Maclaurin spheroid and therefore whose rotational period became longer than the 2.65-hour limit. Continued addition of angular momentum (via continued accretion) to the Jacobian proto-Earth caused it to again spin up to a 2.65-hour period. At that point the axial ratios of the ellipsoid were 8:10:23 and the Jacobian changed over to a bowling-pin-shaped Poincaré figure with the nascent Moon forming at the small, bulbous end of the figure. As

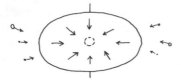

Maclaurin Oblate Sphere – 7:12 , Period – 2.6 hr

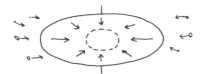

Jacobian Triaxial Ellipsoid – 8:10:23

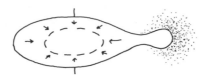

Poincaré Figure

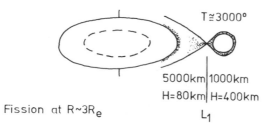

$T \cong 3000°$

5000km|1000km
H=80km|H=400km

Fission at $R \sim 3R_e$

L_1

Fig. 2. Stages of the fission of the proto-Moon from the proto-Earth as discussed in detail in the text. Prior to fission (as shown in the top two illustrations), the proto-Earth increased its rate of rotation as a result of continued accretion of matter and as the iron core formed. As fission began to occur (the third illustration), the massive proto-Moon, which was forming at the small end of the Poincaré figure, began to lose matter (depicted by dots) during a few-hour-long, catastrophic mass loss phase. This loss of mass was caused mainly by the proto-Moon's thermal instabilities. After fission (the bottom illustration), the proto-Moon continued to lose mass (depicted by dots) through the L_1 point for several hundreds of years. The distance from the surfaces of the Earth and proto-Moon to the L_1 point, and the scale heights of the evaporated silicate atmospheres (depicted by dots) of the Earth and proto-Moon, are given for the case where their temperatures were 3000° C.

the neck of the "bowling pin" broke, the main part—the Earth—contracted back to a Jacobi ellipsoid or a Maclaurin spheroid (depending on the size of the new Moon; Binder, 1980). At that point the proto-Moon, whose mass was up to 20% of that of the Earth (O'Keefe, 1969), was orbiting the Earth in a nearly circular, equatorial orbit just outside the Roche limit at about 3 R_e. In this model, the angular momentum of the proto-Earth at the point of fission was about four times that of the current Earth-Moon system, assuming the proto-Earth was a homogeneous body (Wise, 1963). However, allowing for the reduction of the moment of inertia caused by the core and compressional effects, the angular momentum of the proto-Earth was about three times that of the current system.

Empirical evidence. It is reasonable to approach the question of whether or not fission can occur from the empirical side by asking the question, Is there any empirical evidence suggesting that there are additional bodies in the solar system that may have formed by binary fission?

Pluto and its satellite Charon are an excellent candidate for a pair of bodies formed by fission (Lin, 1981). From the dynamical standpoint, this pair is like the Moon and Earth; the Charon/Pluto mass ratio is very large (~1/10) compared to all other planet-satellite pairs in the solar system. The distance between Pluto and Charon is relatively small (17,000 km or 11 to 13 Pluto radii) and they both have locked rotation periods equal to their mutual period of revolution (6.39 days). The latter shows that most' of the initially higher rotational velocity of both bodies has been turned into orbital velocity via tidal action and that the two were much closer together earlier in their history. Given the currently somewhat uncertain data on the mass ratio of the pair and Pluto's diameter, and the relatively good data on their separation and periods, the angular momentum of this system is such that if Pluto and Charon were combined into one body, it would have a rotation period of 7 to 11 hours. The critical rotation period of this body would be 5 to 8 hours, as calculated using Darwin's criterion, i.e.,

$$\omega^2/2\pi\rho G = 0.187 \qquad (1)$$

where ω is the angular velocity, ρ is the density, and G is the gravitational constant. Thus the proto-Pluto was rotating initially at or close to the period needed for fission. Also, as noted from Section 2.1, the angular momentum density of the Pluto-Charon system is a factor of 4 over the mean line through the solar system bodies in Fig. 1. These observations all make Charon a likely candidate for a fission-born satellite.

The possibility that Charon was formed by fission does not demonstrate that binary fission is dynamically possible, nor that the Moon was formed by fission. However, the existence of two pairs of large secondary/primary mass ratio terrestrial-type bodies out of five in the solar system suggests that whatever process formed them was statistically likely (perhaps common) and strengthens the case for continued studies of the dynamics of fission as such a process. The former observation also argues against models that are stochastic in nature, e.g., the giant impact model of the origin of the Moon.

2.3. Excess angular momentum problem

There are two sides to the angular momentum problem of the Earth-Moon system. The first, discussed in Section 2.1, is concerned with the possibility that the proto-Earth had enough angular momentum for fission to have occurred. The second, discussed here, is that if the proto-Earth did have this amount of angular momentum, where did it go? The Earth-Moon system now has only one-third of that needed if the Moon formed by fission.

In their discussions of fission, both Wise (1969) and O'Keefe (1969, 1970, 1972) pointed out that the newly fissioned Moon had a temperature in the range of a few thousand degrees and therefore that the proto-Moon had a hot atmosphere of evaporated silicates. O'Keefe (1969) also pointed out that the fission process most likely produced a proto-Moon with an initial mass of 10 to 20 times the current lunar mass. Wise (1969) suggested that the temperature was high enough so that a large fraction of the original lunar mass was lost from the system via thermal escape, thereby carrying away the excess angular momentum of the system. O'Keefe (1969) made the more plausible suggestion that this atmosphere was able to expand beyond the equipotential surface in which the L_2 point is found and could escape the system, carrying away the excess angular momentum of the system.

I have suggested that the loss of mass and angular momentum via the hot, silicate atmosphere was dynamically somewhat different than these investigators proposed on the basis of the following two observations (Binder, 1978, 1980). First, in semi-detached binary star systems mass is transferred from the primary to the secondary through the L_1 point (e.g., Kopal, 1978). Second, as a consequence of that transfer, some matter is lost from the system, carrying considerable angular momentum with it (e.g., Amnuel and Guseinov, 1982; Chau and Nelson, 1983). These observations led me to suggest that there were two phases that led to the loss of matter and angular momentum from the early Earth-Moon system.

The first phase is proposed to have occurred mainly during the fission process and continued a few hours (?) after fission was complete. First-order calculations suggest that this phase, which I have termed the catastrophic phase, came about because the forming proto-Moon underwent pressure release or retrograde boiling (Binder, 1980), and the fission process may have produced a swarm of small bodies as well as a main body (Wise, 1963, 1969; Wise's speculation finds some support in a fission model of Gingold and Monaghan, 1979. The result of this model calculation was a binary with a mass ratio of 0.28 containing 91% of the mass of the system, with the remaining 9% of the mass in a ring of debris). The forming proto-Moon should have been thermally unstable since at 2000° to 4000°C [the proposed range of initial temperatures of the proto-Earth (Ringwood, 1970) and therefore of the proto-Moon (Wise, 1969; O'Keefe, 1969, 1970, 1972; Binder, 1978, 1980)] most silicates are in a vapor state at low pressure but are liquids at high pressures. The pressure gradient in the outer layers of the large, hot, Jacobian proto-Earth was high. As the Jacobi ellipsoid began to change over to a Poincaré figure, the pressure gradient in the forming, small bulbous end of the figure had to decrease. Thus a large fraction of the material that was beginning to fission off to form the nascent Moon should have changed into vapor to form an extensive silicate atmosphere. It seems most likely that this process would have proceeded so fast that the forming atmosphere could not have been contained within the Roche lobe of the forming proto-Moon and therefore could not have been slowly bled off to the Earth through the L_1 point. Rather, it seems likely that this atmosphere expanded beyond the forming proto-Moon's Roche lobe and probably well beyond the L_2 equipotential surface, thereby escaping the proto-Moon.

I have proposed that, while a large fraction of the material lost from the forming proto-Moon via the proposed thermal effect (and any in a fission debris ring) was mainly recollected by the Earth, some of this material went into orbit around the Earth-Moon system and eventually escaped from the system, carrying away the excess angular momentum with it (Binder, 1980). The proposed mechanism is as follows.

In a normalized system where the unit of mass is 1 lunar mass (m), the unit of distance is 1 R_e, the unit of angular velocity (ω) is 1/day, and noting that the momentum of inertia factor of the Earth is close to 1/3, the angular momentum of the Earth-Moon system is

$$1/3 \ (81 \ m) \ R_e^2\omega + m \ (60 \ R_e)^2\omega w/27.3 = 159 \ m \ R_e^2\omega \qquad (2)$$

The first term in equation (2) is the angular momentum due to the Earth's rotation (mass $= 81$ m) and the second is due to the Moon's orbital motion (radius $= 60 \ R_e$ and period $= 27.3$ days). As discussed in Section 2.2, the angular momentum of the proto-Earth at the point of fission was about three times that of the current Earth-Moon system, or 480 m $R_e^2\omega$. So about 320 $R_e^2\omega$ of momentum are missing from the system. As suggested above, some of the material that escaped the forming proto-Moon probably went into orbit around the Earth-Moon system. Perturbations caused this material to gain angular momentum from the Earth and Moon (as discussed shortly in regard to material lost from the Moon during the second phase of lunar mass loss) until it reached the point where solar perturbations would have removed it from the Earth-Moon system, i.e., the radius of action at 192 R_e (Kuiper, 1961). The orbital angular velocity at 192 R_e is $\omega/159$. So the angular momentum lost from the system per lunar mass would have been 236 m $R_e^2\omega$ at 192R_e, and the entire 320 m $R_e^2\omega$ of missing angular momentum could be accounted for if only 1.4 lunar masses of material were lost by this mechanism during the proposed catastrophic mass loss phase (Binder, 1980). Given that some 10 to 20 lunar masses of material were initially fissioned from the proto-Earth, a loss of 1.4 lunar masses from the system does not seem unreasonable.

After the short, catastrophic mass loss phase, the proto-Moon had a mass of a few lunar masses and had cooled enough (to 2000° or 2500°C?) so that the rate of evaporation of silicates into the hot atmosphere equaled the rate at which they could be bled off to the Earth through the L_1 point; i.e., the evaporated atmosphere just filled the lunar Roche lobe. I call this the L_1 phase and it probably lasted several 100 years, i.e., the time needed for the proto-Moon to cool to the liquidus (Binder, 1978, 1980). As I pointed out earlier, the vaporized silicates would have streamed through the L_1 point into the Earth's Roche lobe, but the reverse could not occur (Binder, 1978, 1980). This is the case since (1) the distance of L_1 from the lunar surface was only 1000 km or 2 to 3 lunar atmosphere scale heights and (2) the L_1 distance from the Earth's surface was at a minimum of 5000 km or >25 terrestrial atmosphere scale heights. Thus the lunar atmosphere easily reached L_1, but the terrestrial silicate atmosphere did not. So material could only pass from the lunar Roche lobe into the terrestrial lobe (Fig. 2) in exactly the same way as occurs in semi-detached binary stars.

Furthermore, it appears likely that, although most of the material passing through L_1 struck the Earth, a portion of the matter misses the Earth and went into orbit around both bodies, as is observed in semi-detached binary star systems (Kopal, 1978). According to calculations by Kopal, such material is perturbed into larger orbits and spirals outward from the pair. In the absence of other effects, this material reaches a maximum distance from the pair and then spirals back. However, in the case of stars, the stellar winds cause this material to be lost from the system once it spirals out to large distances. In the case of the Earth-Moon system, solar perturbations would have caused this material to be lost from the system if it reached the radius of action at 192 R_e (Kuiper, 1961). As discussed above, any material lost via this effect at 192 R_e carried 236 m $R_e^2\omega$ of angular momentum per lunar mass from the Earth-Moon system. Thus, only 1.4 lunar masses of material need have been lost by this L_1 phase mechanism alone to account for the 320 m $R_e^2\omega$ of missing momentum from the Earth-Moon system.

In summary, on the basis of first-order calculations, I have suggested that the loss of a total of 1.4 lunar masses of material (out of the 10 to 20 lunar masses of initial fissioned protolunar material) from the proto-Moon and the Earth-Moon system during (1) an initial few-hour-long catastrophic mass loss phase and (2) a several-hundred-year-long L_1 mass loss phase could account for the current angular momentum deficiency of the Earth-Moon system if the Moon formed by fission. The proposed models need to be tested by more detailed calculation to determine if the mechanisms can account for the loss of mass from the proto-Moon and the loss of some 320 m $R_e^2\omega$ of angular momentum from the Earth-Moon system. The lower two illustrations in Fig. 2 show schematically the loss of mass and angular momentum from the Earth-Moon system as during the two proposed phases.

2.4. Orbital inclination

It has been argued (e.g., Gerstenkorn, 1969) that since the Moon's orbit is not presently in the plane of the Earth's equator, the Moon could never have been in an equatorial orbit and therefore could not have formed by fission. However, this argument (1) applies equally well to all models of lunar formation except pure capture and (2) is most certainly incorrect. First, Rubincam (1975) has shown that the current inclination of the lunar orbit may be explained dynamically, even though the postfission orbit was equatorial. Second, a nonzero inclination of the Moon's orbit may have been caused by early, large impacts as discussed by W. K. Hartmann at the Conference on the Origin of the Moon, from which this volume resulted. Third, as discussed by C. Yoder at the same conference, calculations of the inclination and other characteristics of the lunar orbit are not valid using current theory at Earth-Moon distances of less than about 10 R_e. Thus, the nonequatorial orbit of the Moon cannot be used to argue against the fission origin of the Moon.

3.   Compositional Considerations

3.1.   Iron core and siderophile elements

The absence of a significant (>5% by weight) lunar core, while the terrestrial core contains 31% of the Earth's mass, has been one of the major facts behind the fission model. This difference in the core sizes would be a direct result if fission occurred near or at the end of the formation of the terrestrial core. In fact, Wise (1963, 1969) proposed that the core formation process (which reduced the moment of inertia of the Earth and therefore caused it to rotate faster) initiated the fission process. Since fission must have occurred after the core had largely formed, the bulk of the Moon's core-forming iron and siderophile elements are in the terrestrial core. However, the fission model does not exclude the possibility that the Moon has a small core, since there are three possible sources of metallic iron that would have contributed to form a small lunar core of a fissioned Moon (Binder, 1974).

First, some residual metallic iron may have been in the upper part of the proto-Earth when fission occurred. This was probably the case since (1) the transfer of metallic iron from the terrestrial mantle to the core was not necessarily 100% complete when fission occurred and (2) accretion certainly tailed off slowly so that metallic iron may have been continually added to the terrestrial mantle for some time after the core began to form. Second, since accretion tailed off slowly, some iron may have been added to the Moon by accretion after fission. Third, the proposed devolatilization of the proto-Moon (discussed in the following section) would have resulted in the reduction of FeO.

As a result of a combination of these three processes, it seems likely that there should have been sufficient iron in the nascent Moon for a small iron core to have formed (Binder, 1974). The formation of this small lunar core would have resulted in the residual lunar siderophile elements undergoing a second stage of depletion and having the nonterrestrial depletion pattern (Binder, 1974). This second depletion of the lunar siderophiles has been more recently discussed by Rammensee and Wänke (1977), Ringwood (1978), Newsom (1984), and others. Newsom (1984) concludes that if the lunar core is smaller than about 2%, the material that formed the Moon must have come from the Earth's mantle (unless metal was lost from the precursor material); if the core is larger than about 2%, an origin from the Earth's mantle is unlikely unless it had higher concentrations of siderophiles than at present. Thus the question as to the size of the lunar core is important in determining if fission is a viable model. In any case, one can currently state that the existence of an iron core containing <2–3% of the lunar mass and the nonterrestrial, siderophile element depletion pattern are consistent with the fission model.

3.2. Volatile element depletion

The observed depletion of volatile elements was a predictable consequence of the fission of the Moon from a hot proto-Earth (Wise, 1963; O'Keefe, 1969, 1970, 1972), although the loss mechanism is somewhat different than these authors proposed. In the general case the hot, molten proto-Moon may have lost its volatiles via thermal escape, but it probably recollected them in the same way that has been suggested for the H_2 in Titan's atmosphere (Hunten, 1977). The reason is that the volatiles would have simply gone into geocentric orbits that were essentially identical to that of the proto-Moon. A torus of volatiles would have formed along the proto-Moon's orbit in the same way that a torus of H_2 exists along Titan's orbit (Broadfoot et al., 1981). Unless this torus was swept away by the solar wind (which was not possible if the proto-Moon was within the geomagnetic field), or unless the material was ionized by solar UV radiation (which would not have been possible before the solar system was swept clean of gas and dust), the volatiles would have been quickly recollected by the proto-Moon as it moved through the torus.

However, as reviewed in Section 2.3, if the hot proto-Moon was within a few R_e of the Earth, the lunar atmosphere of evaporated, *volatile-enriched* silicates would have (1) filled the lunar Roche lobe, (2) passed through the L_1 point into the Earth's Roche lobe, and (3) been captured by the Earth (Binder, 1978, 1980). This model has been criticized by Taylor (1984), who points out that elemental Sm, Eu, Tm, and Yb are relatively volatile in comparison to the rest of the REE (Fig. 3). Thus, if the Moon lost its volatiles by the proposed mechanism at temperatures $\geqslant 2000°C$, the Moon's REE would be depleted in Sm, Eu, Tm, and Yb, and this is not the case. However, the proto-Moon is expected to be at very high temperatures ($\geqslant 2000°$ to $4000°C$) only during the short (up to a few hours) catastrophic mass loss phase, during which the mass loss rate was so high (a few lunar masses/hour or more) that fractional evaporation effects would have been negligible. Later, during the much more quiescent, long-time-scale (a few hundred years) L_1 phase, fractional vaporation effects would have become important. However, at the expected pressures and temperatures of the evaporated lunar atmosphere during this phase ($10 < P < 10^4$ bar and $1500 < T < 2500°C$; Binder, 1978, 1980), the REE are present as oxides and not as metals as in the solar nebula (which had pressures of only 10^{-3} to 10^{-4} bar). The melting and boiling points of *all* the REE oxides are above $2200°C$ and $2500°$ to $3000°C$, respectively (Fig. 3), and thus at temperatures below $2500°C$

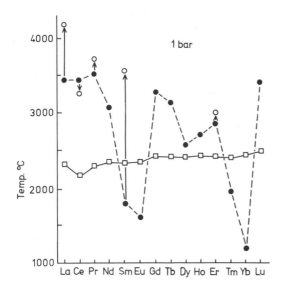

Fig. 3. The melting and boiling points of the REE at 1 bar pressure. The filled circles (connected by a dashed line) are the boiling points for the REE in metallic form. The open squares (connected by a continuous line) and the open circles are the melting points and the boiling points, respectively, of the REE in oxide form. The arrows show the differences in the boiling points between the REE in metallic and oxide forms. The data are from Samsonov (1973).

the REE would not show any significant depletion of Sm, Eu, Tm, and Yb with respect to the other REE.

Similarly, Kreutzberger *et al.* (1984) have argued on the basis of experimental data that the depletion pattern of Na, K, Rb, and Cs is not consistent with the fission origin depletion model. However, the Kreutzberger *et al.* experiments were carried out at lower temperatures and pressures and with different elemental concentrations than are expected for the evaporated lunar atmosphere, so their results do not apply to the proposed model.

In summary, given the observation that vaporized material is transferred from the Roche lobe of one astronomical body to a second body, it is reasonable to propose that a hot, evaporated silicate atmosphere of a fissioned proto-Moon would be transferred to—and largely captured by—the Earth during the relatively quiescent, L_1 phase of mass loss from the proto-Moon (Section 2.3). It is also reasonable to propose that the evaporated silicates would be enriched in the more volatile components and that this process would have led to a depletion of lunar volatiles. However, the proposed model needs verification by both more complete theoretical modeling and laboratory experimentation on the relative volatilities of the REE and the volatile elements under the correct pressure and temperature conditions ($10 < P < 10^4$ bar and $1500 < T < 2500°C$).

3.3. Bulk lunar composition

The second major compositional consequence of fission of the Moon from the Earth would be that the bulk composition (excluding volatiles, Fe, and siderophiles) of the Moon would be very similar to that of the parental terrestrial mantle. As a result of (1) the presence of a massive (~10% of the lunar mass) feldspathic crust and Ti-rich mare basalts, (2) a misinterpretation of the initial heat flow data that suggested that the Moon contained twice as much U and Th as does the Earth (Langseth *et al.*, 1973), and (3) the concept that only the outer few hundreds of kilometers of the Moon were initially molten (e.g., Wood *et al.*, 1970), most lunar scientists came to the initial conclusion that the Moon's bulk composition was enriched in refractory elements and therefore was very different from that of the terrestrial mantle. However, as shown as early as 1974, the Al, Ca, and Ti content of terrestrial mantle material is quite adequate to account for the thick, feldspathic lunar crust and the ilmenite-rich mare basalts if the Moon was initially totally molten and underwent fractional crystallization (Binder, 1974). The argument that the Moon was initially totally molten is supported by recent data as reviewed in an accompanying paper (Binder, 1986). Furthermore, the heat flow data were reinterpreted, yielding values that are compatible with a bulk lunar U and Th composition equaling that of the terrestrial mantle (Conel and Morton, 1975; Binder, 1975; Langseth *et al.*, 1976; Rasmussen and Warren, 1985).

The most recent estimates of the major oxide composition of the bulk Moon are all very similar to that estimated for the terrestrial mantle (e.g., Ringwood, 1979; Binder, 1983; Delano, 1984; Warren, 1984). The only major oxide that seems to deviate is FeO, which seems to be enriched in the Moon by up to a factor of 2 relative to the Earth. Assuming that this is correct [Warren (1984) has argued against the FeO enrichment], it might speak against a fission origin of the Moon. However, the estimates of the terrestrial mantle composition are based largely on the composition of the uppermost mantle. Given the uncertainties in the composition of the lower mantle, it is quite possible that the terrestrial mantle is more iron-rich than we now believe. Also, the current estimates may not be valid for the earliest part of geological history. Jagoutz and Wänke (1982) have argued that the early terrestrial mantle was richer in FeO than today, that FeO has been reduced, and that Fe was incorporated in the core as the Earth evolved. This concept has been challenged by Newsom (1985). Thus, although it is not now clear if the possible differences in the FeO contents of the Earth's mantle and the Moon are significant in terms of the fission model, this question needs to be answered. In any case, it has become relatively clear that the bulk major oxide composition of the Moon is very close to that of the terrestrial mantle, a finding that is certainly compatible with the fission origin of the Moon.

Similarly, the bulk nonvolatile, nonsiderophile trace element content of the Moon is essentially equal to that of the Earth's mantle. This has already been mentioned in the case of U and Th, as derived from the heat flow data, and is also the case for at least the REE, Ba, and Sr as discussed in an accompanying paper (Binder, 1986).

Finally, as has been known since the Apollo 11 mission, the stable isotopic ratios for O, C, Si, and S are equal to or within the terrestrial range of values (Levinson and Taylor, 1971). This clearly indicates that there is a close genetic relationship between the Earth and the Moon.

In summary, the bulk composition of the Moon is very nearly equal to that of devolatilized, terrestrial mantle material. The differences between the Moon's bulk composition and that of the Earth's mantle may be explained by the evolution of both bodies after fission, i.e., the loss of volatiles via transfer to the Earth through the L_1 point while the Moon was still very hot (e.g., Binder, 1978, 1980), the second depletion of the lunar siderophiles as the Moon's small Fe core formed (e.g., Binder, 1974; Newsom, 1984), and the possible differences in the FeO content of the terrestrial mantle as a function of depth and/or time (Jagoutz and Wänke, 1982). If these compositional similarities between the Earth's mantle and the Moon do not favor the fission model, they at least are certainly consistent with it.

4. Summary of the Classical Fission Model

According to the classical fission model, the Moon formed and evolved as follows:

1. Toward the end of the accretion, the superheated (~3000° to 4000°C) proto-Earth was rotating near the stability limit, i.e., with a period slightly longer than 2.65 hours. The proto-Earth had the form of a triaxial Jacobi ellipsoid. As accreting material added additional angular momentum to the proto-Earth and/or as the moment of inertia of the proto-Earth decreased as the Fe core was forming, the proto-Earth finally reached the 2.65-hour rotation limit and became rotationally unstable.

2. The proto-Earth took on the form of a bowling-pin-shaped Poincaré figure, with the nascent Moon forming at its small end. The initial mass of the proto-Moon was 10% to 20% of that of the proto-Earth. As the proto-Moon developed and broke free from the small end of the Poincaré figure, the majority of the initially large protolunar mass was lost mainly as a result of the proto-Moon's high temperature and the large drop of internal pressure as the proto-Moon detached from the proto-Earth. The majority of the material shed from the forming Moon fell back onto the Earth. A fraction of this matter escaped the Earth-Moon system, carrying away considerable angular momentum. At the end of the short (few-hour?) fission process, the nascent Moon had a mass a few times its current mass, was revolving around the Earth at about 3 R_e in a nearly circular, zero inclination orbit, had a temperature in the 2000° to 2500°C range, and had a bulk composition identical to that of the Earth minus the Fe and siderophiles lost to the Earth's core.

3. Due to the high temperature of the proto-Moon, it had an extensive atmosphere of vaporized silicates enriched in the more volatile components. Also, due to the close proximity of the proto-Moon to the Earth, this hot lunar atmosphere filled the lunar Roche lobe and poured through the L_1 point into the Earth's Roche

lobe. Most of this volatile-rich material fell onto the Earth. However, some fraction of this material missed the Earth, went into orbit around the Earth and Moon, and was lost from the system via solar perturbations. In doing so, additional angular momentum was lost from the Earth-Moon system. The loss of only 1.4 lunar masses of material (out of the initial 10 to 20 lunar masses of fissioned material) from the Earth-Moon system during the initial, catastrophic mass loss phase and this more quiescent L_1 mass loss phase would explain the current deficiency of a factor of 3 in the angular moment of the Earth-Moon system. Also, the loss of a few lunar masses of volatile enriched material (mainly) to the Earth during the L_1 phase may explain the observed depletion of lunar volatiles.

4. As the proto-Moon lost mass and devolatilized during the L_1 phase, FeO in the Moon was reduced. This newly formed Fe, any residual Fe that was still in the terrestrial mantle at the time of fission, and any Fe accreted by the Moon after fission sank to form a small lunar core. The lunar siderophiles therefore underwent a second core-forming event that resulted in their nonterrestrial depletion pattern.

As a result of these processes, the Moon had obtained its present mass and bulk compositional characteristics within a few hundred to thousand years after it formed.

Acknowledgments. I thank Drs. R. Durisen and G. J. Taylor for their constructive reviews of this paper. This work was done while the author was a senior NRC Fellow at the NASA/Johnson Space Center.

5. References

Amnuel P. R. and Guseinov O. H. (1982) On the evolution of low mass binary systems. *Astrophys. Space Sci., 86*, 91–106.

Binder A. B. (1974) On the origin of the moon by rotational fission. *The Moon, 11*, 53–76.

Binder A. B. (1975) On the heat flow of a gravitationally differentiated moon of fission origin. *The Moon, 14*, 53–76.

Binder A. B. (1978) On fission and the devolatilization of a moon of fission origin. *Earth Planet. Sci. Lett., 41*, 381–385.

Binder A. B. (1980) The first few hundred years of evolution of a moon of fission origin. *Proc. Lunar Planet. Sci. Conf. 11th*, pp. 1931–1939.

Binder A. B. (1982) The moon: Its figure and orbital evolution. *Geophys. Res. Lett., 9*, 33–36.

Binder A. B. (1983) An estimate of the bulk, major oxide composition of the moon (abstract). In *Workshop on Pristine Highland Rocks and the Early History of the Moon* (J. Longhi and G. Ryder, eds.), pp. 17–19. LPI Tech. Rpt. 83-02, Lunar and Planetary Institute, Houston.

Binder A. B. (1986) The initial thermal state of the moon, this volume.

Broadfoot A. L., Sandel B. R., Shemansky D. E., Holberg J. B., Smith R., Strobel D. F., McConnell J. C., Kumar S., Hunten D. M., Atreya S. K., Donahue T. M., Moos H. W., Bretaux J. L., Blamont J. E., Pomphrey R. B., and Linick S. (1981) Extreme ultraviolet observations from Voyager 1 encounter with Saturn. *Science, 212*, 206–211.

Cameron A. G. W. (1984) Formation of the prelunar accretion disk (abstract). In *Papers Presented to the Conference on the Origin of the Moon*, p. 58. Lunar and Planetary Institute, Houston.

Cameron A. G. W. and Ward W. R. (1976) The origin of the moon (abstract). In *Lunar Science VII*, pp. 120–122. The Lunar Science Institute, Houston.

Chau W. Y. and Nelson L. A. (1983) Non-conservative evolution of low-mass, close binaries with gravitational radiation and systemic mass losses. *Astrophys. Space Sci., 90*, 245–260.

Conel J. E. and Morton J. B. (1975) Interpretation of lunar heat flow data. *The Moon, 14*, 263–290.

Delano J. W. (1984) Abundances of Ni, Cr, Co, and major elements in the silicate portion of the moon: Constraints from primary magmas (abstract). In *Papers Presented to the Conference on the Origin of the Moon*, p. 15. Lunar and Planetary Institute, Houston.

Durisen R. H. and Gingold R. A. (1986) Numerical simulations of fission, this volume.

Durisen R. H. and Tohline J. E. (1985) Fission of rapidly rotating fluid systems. In *Protostars and Planets* (D. C. Black, ed.). Univ. of Arizona Press, Tucson, in press.

Gerstenkorn H. (1969) The earliest past of the earth-moon system. *Icarus, 11*, 189–207.

Gingold R. A. and Monaghan J. J. (1979) Binary fission in damped rotating polytropes II. *Mon. Not. Roy. Astron. Soc., 188*, 39–44.

Giuli R. T. (1968) On the rotation of the earth produced by gravitational accretion of particles. *Icarus, 8*, 301–323.

Harris A. W. (1977) An analytical theory of planetary rotation rates. *Icarus, 31*, 168–174.

Hartmann W. K. and Davis D. (1975) Satellite-sized planetesimals and lunar origin. *Icarus, 24*, 504–515.

Hartmann W. K. and Larson S. M. (1967) Angular momenta of planetary bodies. *Icarus, 7*, 257–260.

Hunten D. M. (1977) Titan's atmosphere and surface. In *Planetary Satellites* (A. J. Burns, ed.), pp. 420–437. Univ. of Arizona Press, Tucson.

Jagoutz E. and Wänke H. (1982) Has the earth's core grown over geological times? (abstract). In *Lunar and Planetary Science XIII*, pp. 358–359. Lunar and Planetary Institute, Houston.

Kaula W. M. and Harris A. W. (1973) Dynamically plausible hypotheses of lunar origin. *Nature, 245*, 367–369.

Kopal U. (1978) *Dynamics of Close Binary Systems*. Reidel, Boston. 501 pp.

Kreutzberger M. E., Drake M. J., and Jones J. H. (1984) Origin of the moon: Constraints from volatile elements (abstract). In *Papers Presented to the Conference on the Origin of the Moon*, p. 22. Lunar and Planetary Institute, Houston.

Kuiper G. P. (1961) Limits of completeness. In *Planets and Satellites* (G. P. Kuiper and B. M. Middlehurst, eds.), pp. 575–591. Univ. of Chicago Press, Chicago.

Lambeck K. and Pullan S. (1980) The fossil bulge hypothesis revisited. *Phys. Earth Planet. Inter., 22*, 29–35.

Langseth M. G., Keihm S. J., and Chute J. L., Jr. (1973) Heat flow experiment. In *Apollo 17 Preliminary Science Report*, pp. 9–1 to 9–24. NASA, Washington, D.C.

Langseth M. G., Keihm S. J., and Peters K. (1976) Revised lunar heat flow. *Proc. Lunar Sci. Conf. 7th*, pp. 3143–3171.

Levinson A. A. and Taylor S. R. (1971) *Moon Rocks and Minerals*. Pergamon, New York. 222 pp.

Lin D. C. N. (1981) On the origin of the Pluto-Charon system. *Mon. Not. Roy. Astron. Soc., 197*, 1081–1085.

Lucy L. B. (1977) A numerical approach to the testing of the fission hypothesis. *Astron. J., 82*, 1013–1024.

MacDonald G. J. R. (1963) The internal constitutions of the inner planets and the moon. *Space Sci. Rev., 2*, 473–557.

MacDonald G. J. F. (1964) Tidal friction. *Rev. Geophys., 2*, 467–541.

Newsom H. E. (1984) Constraints on the origin of the moon from molybdenum and other siderophile elements (abstract). In *Papers Presented to the Conference on the Origin of the Moon*, p. 13. Lunar and Planetary Institute, Houston.

Newsom H. E. (1985) Did the earth's core grow through geological time? (abstract). In *Lunar and Planetary Science XVI*, pp. 616–617. Lunar and Planetary Institute, Houston.

O'Keefe J. A. (1968) Fission hypothesis for the origin of the moon. *Astron. J. Suppl., 73*, 10.

O'Keefe J. A. (1969) Origin of the moon. *J. Geophys. Res., 74*, 2758–2767.

O'Keefe J. A. (1970) The origin of the moon. *J. Geophys. Res., 75*, 6565–6574.

O'Keefe J. A. (1972) The origin of the moon: Theories involving formation with the earth. *Astrophys. Space Sci., 16*, 201–211.

Rammensee W. and Wänke H. (1977) On the partition coefficient of tungsten between metal and silicate and its bearing on the origin of the moon. *Proc. Lunar Sci. Conf. 8th*, pp. 399–409.

Rasmussen K. L. and Warren P. H. (1985) Megaregolith thickness, heat flow, and the bulk composition of the moon. *Nature, 313*, 121–124.

Ringwood A. E. (1970) Origin of the moon: Precipitation hypothesis. *Earth Planet. Sci. Lett., 8*, 131–140.

Ringwood A. E. (1978) Origin of the moon (abstract). In *Lunar and Planetary Science IX*, pp. 961–963. Lunar and Planetary Institute, Houston.

Ringwood A. E. (1979) *Origin of the Earth and Moon*. Springer-Verlag, New York. 295 pp.

Rubincam D. P. (1975) Tidal friction and the early history of the moon's orbit. *J. Geophys. Res., 80*, 1537–1548.

Samsonov G. V. (1973) *The Oxide Handbook*. IFI/Plenum, New York. 524 pp.

Taylor S. R. (1984) Tests of the lunar fission hypothesis (abstract). In *Papers Presented to the Conference on the Origin of the Moon*, p. 25. Lunar and Planetary Institute, Houston.

Warren P. H. (1984) The bulk-moon MgO/FeO ratio: A highland perspective (abstract). In *Papers Presented to the Conference on the Origin of the Moon*, p. 19. Lunar and Planetary Institute, Houston.

Wise D. U. (1963) An origin of the moon by rotational fission during core formation of the earth's core. *J. Geophys. Res., 80*, 1547–1554.

Wise D. U. (1969) Origin of the moon from the earth: Some new mechanisms and comparisons. *J. Geophys. Res., 74*, 6034–6045.

Wood J. A., Dickey J. S., Jr., Marvin U. B., and Powell B. N. (1970) Lunar anorthosites and a geophysical model of the moon. *Proc. Apollo 11 Lunar Sci. Conf.*, pp. 965–988.

VI. Theories and Processes of Origin 2:

Considerations Involving Large Bodies in the Environment of Primordial Earth, and Chances for Close Approaches or Impacts

Accumulation of the Terrestrial Planets and Implications Concerning Lunar Origin

G. W. WETHERILL

Department of Terrestrial Magnetism, Carnegie Institution of Washington, 5241 Broad Branch Road, N.W., Washington, DC 20015

In order to help provide a context for understanding lunar formation, 28 new three-dimensional simulations of terrestrial planet formation from a gas-free planetesimal swarm have been carried out. The natural orbital and collisional evolution of 500 initial planetesimals ranging in mass from 5.7×10^{24}g to 1.1×10^{26}g is followed until only final planets in noncrossing orbits remain. The results are in general agreement with the number, size, and orbits of the observed terrestrial planets, but also show considerable variation of stochastic origin. These results are combined with 11 simulations using 500 bodies of equal initial mass presented earlier, as well as with some other numerical studies, to conclude that for a wide range of initial conditions, terrestrial planet accumulation was characterized by giant impacts, ranging in mass up to 3 times the mass of Mars, at typical impact velocities of ~9 km/sec. These large planetesimals and the impacts they produce are sufficient to explain the unexpectedly large angular momentum of the Earth-Moon system. Such an accumulation history should have had profound effects on the early thermal, physical, and chemical history of the terrestrial planets and their atmospheres.

1. Introduction

Because of the need for a context in which to evaluate the relative importance and probability of alternative lunar formation processes, a meaningful discussion of the way the Moon was formed requires understanding of the formation of the terrestrial planets as well.

In this paper terrestrial planet formation from a heliocentric swarm of planetesimals under conditions in which nebular gas drag was of minor importance will be investigated. Particular emphasis will be placed on the question of the existence of planetesimals large enough to provide the angular momentum of the Earth-Moon system in a single event. A more complete discussion of this subject would also require exploration of the consequences of alternative modes of formation of the planets [i.e., accumulation in the presence of a gaseous nebula (Hayashi *et al.*, 1985) or as a result of massive gravitational instabilities ("giant gaseous protoplanets") in the solar nebula (Cameron, 1978)].

The requirement that the formation of the Moon must be considered in the context of terrestrial planet formation could be criticized as being a useless truism, in that analogous statements could be made about the relationships between the formation of the terrestrial planets, the solar system, the sun, the galaxy, and the universe. This criticism has some validity, and the burden of relating all of these processes to one another is not light.

On the other hand, these relationships are not all identical to one another. For example, if as considered in this paper, the terrestrial planets formed by accumulation of smaller planetesimals, the characteristic time scale for the formation of those bodies in heliocentric orbit is 10^7–10^8 years (Safronov, 1969; Hayashi *et al.*, 1985), longer than the time scale for the formation of the sun of 10^5–10^6 years (e.g., Larson, 1978). So if the Earth formed by accumulation, at least in some approximate way the formation of the sun and terrestrial planets can be considered as consecutive separate events. In contrast, the characteristic time for accumulation of material in geocentric orbit is 10–100 years, many orders of magnitude shorter than the time scale for accumulation of the Earth. Therefore, the time-limiting factor in the formation of the Moon would be that of the supply of material from an external source, whether that source be a heliocentric population of large or small bodies, or the Earth itself. The formation of the Earth and Moon would thus be more concurrent than consecutive, and it is not possible to discuss them as separate, independent events.

In the context of accumulation of the Earth from planetesimals, the question of the formation of the Moon reduces to that of understanding in the broadest way the supply of material from the heliocentric planetesimal swarm to gravitationally bound geocentric orbit and the natural evolution of the resulting geocentric swarm of material, i.e., the extent to which it will accumulate to form a relatively large object with characteristics resembling the observed Moon or will ultimately be swept up by the Earth or lost back into heliocentric orbit. There are a number of possible sources of external material, and to some degree all of these must contribute to the orbiting geocentric population. These include:

1. Capture of intact planetesimals of all sizes from heliocentric orbit into geocentric orbit (Gerstenkorn, 1969; Nakazawa *et al.*, 1983);

2. Capture of heliocentric bodies following their fragmentation by collisions with other heliocentric planetesimals or with preexisting geocentric bodies (Ruskol, 1975; Harris and Kaula, 1975);

3. Capture of heliocentric bodies following disruption by close encounters with the Earth, either tidally (Öpik, 1972; Mitler, 1975) or by shallow grazing impacts;

4. Retention in geocentric orbit of material derived from the Earth, as a consequence of rotational instability (Darwin, 1880; Ringwood, 1972) or by "splashing" of ejecta

from giant impacts (Hartmann and Davis, 1975; Cameron and Ward, 1976; Thompson and Stevenson, 1983).

In most of the references cited, the particular process for populating geocentric orbit was considered in a limited way, i.e., as being the sole or at least the dominant source of lunar material. A major outstanding task, not attempted in this paper, is that of consistently and quantitatively evaluating the relative contribution of every one of these various sources over the entire range of planetesimal masses, in the context of observational evidence and theoretical models such as the one presented in this paper.

An observational fact of central importance is the large angular momentum ($\sim$3.4 $\times$ 10^{41} c.g.s.) of the present Earth-Moon system. During the accumulation of the Earth from planetesimals with relative velocities equal to one-third or one-half the escape velocity of the largest bodies, as expected from analytical theory (Safronov, 1969; Stewart and Kaula, 1980) and numerical calculations (Wetherill, 1980a), impacts imparting angular momenta of either sign are very nearly equally probable, and the average angular momentum of the resulting planet will be essentially zero. In contrast, the present angular momentum of the Earth-Moon system is about 10% of the value expected if all the impacts had provided angular momenta of the same sign, requiring a remarkable lack of symmetry in the impact history of the Earth during its formation. Harris (1977), following the earlier work of Giuli (1968), showed that only for relative velocities of planetesimals strongly concentrated toward an implausibly and unstably low value ($\sim$0.5 km/sec) can an angular momentum as large as one-half that of the Earth-Moon system be explained as the average result of accumulation.

It therefore seems very likely that the angular momentum of the geocentric swarm and the resultant Earth-Moon system was not acquired simply as a result of averaging over a large number of small bodies, but was caused by stochastic fluctuations in the angular momentum input of an accumulation history involving very large objects, only partially smoothed out by contributions from smaller bodies by means of the various mechanisms listed above.

For this reason, in relating lunar origin to Earth formation it is particularly important to understand the masses and velocities of the largest accumulating bodies that failed to become terrestrial planets in their own right. This question was addressed by Safronov (1969), who concluded, on the basis of the inclination of the Earth's rotation axis, that the largest body to impact the Earth was $\sim$6 $\times$ 10^{24}g, insufficient to provide angular momentum fluctuations of the required magnitude. Safronov proposed as a theoretical basis for this conclusion his approximate result that the ratio of the largest mass (m_1) (the planet itself) to the second largest mass (m_2) asymptotically approached

$$\left(\frac{m_1}{m_2}\right) \rightarrow (2\Theta)^3 \tag{1}$$

where Θ is the Safronov number

$$\Theta = \frac{1}{2} \frac{V_e^2}{V^2} \tag{2}$$

V_e is the escape velocity of the largest body of the swarm and V is the mean relative velocity of the swarm relative to a body in a circular Keplerian orbit. For $\Theta \sim 5$, Safronov then calculates a mass ratio of ~ 1000, in agreement with his conclusion based on the inclination of the rotation axis. Subsequently (Wetherill, 1976) it was shown that much larger second largest masses were consistent with the Safronov theory because: (1) The Safronov number was likely to be ~ 2 during the final stages of accumulation; (2) The asymptotic limit (equation (1)) is approached only as the mass of the growing planet approaches infinity, whereas the actual growth of the mass ratio ceases when the planet grows to its quite finite final mass. When this is taken into consideration, it is found that within the framework of the Safronov analytical theory second largest masses in the range of 10^{26} to 10^{27} g are permitted.

All of this work considered only the growth of a single planet, rather than the simultaneous accumulation of the observed number of terrestrial planets. In the framework of theories limited in this way, it is not possible to address even such a simple question of whether the second largest body should be considered as an impactor or a separate terrestrial planet. Answers to questions of this kind require multiple planet accumulation calculations that can provide information regarding the radial distribution of mass and velocity of the planetesimal swarm up to the termination of terrestrial planetary growth.

Numerical "theories" of the late stages of multiple planet accumulation at ~ 1 AU have been presented, both in two dimensions (Cox and Lewis, 1980) and in three dimensions (Wetherill, 1980a,b). The present paper represents an extension of this earlier work. The general approach is fundamentally the same as that taken in another recent contribution (Wetherill, 1985), but the calculations reported here are all new, and are based on an initial distribution in which the masses of the bodies are distributed over a range of values, rather than assuming equal initial masses, as was done earlier.

2. The Initial State

The initial conditions for these calculations should be those that may plausibly be expected to arise from an earlier stage of accumulation of large planetesimals from smaller bodies. Because it was not practical to use more than 500 bodies in the present calculations, these initial conditions were chosen to match as well as possible those found by earlier workers at the stage of growth where the largest bodies of the swarm were in the range 10^{25}–10^{26} g.

These earlier stages have been studied by Safronov (1962a, 1969) and Nakagawa *et al.* (1983). Safronov obtained a solution to the transport equation describing the gas-free coagulation of an initial population of equal mass bodies:

$$N(m_1\tau) = \frac{N_o(1-\tau)}{2\pi^{\frac{1}{2}}} \, m_o^{0.5} m^{-1.5} e^{1-\tau^{\frac{1}{2}}(m/m_o)} \tag{3}$$

N is the number of bodies of mass m in an interval dm, where N_o and m_o are the initial number and mass of the bodies, and $\tau = 1 - N/N_o$. As the growth of a planet goes to completion, τ varies from 0 to 1. Equation (1) represents a power law distribution, truncated by a steep exponential term. Zvyagina *et al.* (1973) extended this result to include the effects of fragmentation, with the result that the power law index in equation (3) increases from 1.5 to about 1.8.

Nakagawa *et al.* (1983) treated the growth of a swarm of initially equal mass bodies in the presence of solar nebula gas by use of a Fokker-Planck transport equation, thereby permitting the effects of radial diffusion to be included.

The differential mass distributions at the stage of growth where the maximum mass is $\sim 10^{26}$ has been calculated from the results of these workers (Fig. 1). Nakagawa *et al.* also find the slope of the distribution to be steep in the vicinity of the maximum mass.

The Safronov equation (1) was obtained under the assumption that the gravitational cross-section of a growing body was proportional to the third power of its radius (R), representing an "average" between the geometric value of R^2 and the low-velocity limiting value of R^4 given by the conventional two-body gravitational cross-section expression

$$\sigma_g = \pi R^2 \left(1 + \frac{V_e^2}{\bar{V}^2}\right) = \pi R^2 (1 + 2\theta) \tag{4}$$

where V_e is the escape velocity (proportional to R) and $\bar{V}$ is the mean relative velocity of the accumulating bodies. The result of Nakagawa *et al.* did not involve this assumption, but simply used equation (4).

Despite the differences in the underlying assumptions, both of these investigations lead to growth characterized by a "marching" steep front in the vicinity of the largest body, followed by a mass distribution at lower masses with a slope corresponding to concentration of most of the mass toward the high mass end of the distribution.

A numerical study by Greenberg *et al.* (1978) led to the conclusion that the steep slope at the upper end of the mass distribution found by Safronov would be greatly modified as a consequence of "early runaway accretion." It was found that after the first large bodies of mass 10^{18}–10^{21}g formed, the mean velocity of

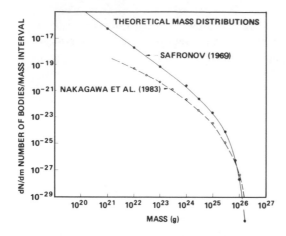

Fig. 1. Calculated differential mass distributions of a gas-free (Safronov, 1962a, 1969) and a gas-rich (Nakagawa et al., 1983) planetary swarm at the stage of growth when the largest mass is $\sim 10^{26}$g.

the swarm did not increase, as a consequence of most of the mass of the swarm still residing in the smallest bodies. By equation (4), the increase in Ve at constant relative velocity $\bar{V}$ then led to a runaway growth of bodies near the high mass end of the distribution. Greenberg *et al.* terminated their calculations at a largest mass of $\sim 10^{23}$g, because at about this mass the gravitational cross-section, equation (4), became larger than the range of interaction of the bodies, as determined by their eccentricities.

The work of Greenberg *et al.* (1978) thus did not extend to masses of $\sim 10^{25}$g, required for the initial distribution of the present work. These workers speculate, however, that continuation of this runaway might lead to a situation where the first few $\sim 10^{23}$–10^{24}g bodies might become the "embryos" of the final planets, and all the remaining mass could remain in the form of relatively small bodies during the final stages of accumulation.

The conclusions of Greenberg *et al.* (1978), as well as those of Greenberg (1980), should be reexamined in the light of subsequent work (Wetherill and Cox, 1984, 1985) on the validity of the two-body approximation in calculations of planetesimal encounters at low values of $\bar{V}/V_e$. It is found that when $\bar{V}/V_e \lesssim 0.1$, strong perturbations and collisions occur between bodies moving in trajectories that, in the absence of their mutual gravity, would not intersect one another. As a consequence, the low value of eccentricity found does not in itself preclude continuation of growth because of isolation of the bodies. These same calculations show, however, that below $\bar{V}/V_e \approx 0.03$, the gravitational cross-section increases less rapidly than R^3 (the value assumed in Safronov, 1969), and therefore much less rapidly than R^4, as expected from the two-body gravitational cross-section (equation (3)) that provided the basis for the runaway accretion.

The work of Safronov (1962b, 1969), as well as subsequent investigations (Wetherill, 1980a,b; Stewart and Kaula, 1980), clearly show that for a swarm of bodies of uniform mass $\bar{V}/V_e \lesssim 1$. If, as found by Greenberg *et al.*, the eccentricity of the swarm does not increase, $\bar{V}/V_e$ will already fall below 0.03 when the radius of the largest bodies is $\gtrsim 33$ times the initial radius, i.e., its mass is $\gtrsim 4 \times 10^{20}$g. Therefore there is no reason to expect further runaway between the larger bodies to continue even as far as 10^{23}g, let alone until planet-size embryos have formed.

In selecting the initial state for the present calculations, there is an additional conceptual matter that should be addressed. The work of Safronov and Nakagawa *et al.* assume a continuous density distribution, whereas the actual distribution must terminate at a particular largest value of the mass. When the effects of low values of $\bar{V}/V_e$ are included, during the initial stages of growth these largest bodies will interact over only a limited radial range. For example, when the largest body is as large as 10^{25}g, the results of Wetherill and Cox (1984, 1985) show that at the low values of $\bar{V}/V_e$ proposed by Greenberg *et al.* it will interact with bodies over a distance of $\sim 10^{-2}$ AU, and at the higher values of $\bar{V}/V_e$ expected from the Safronov theory, ($\Theta = 3$) interactions will occur over a somewhat larger range of 0.03 AU. In either case, up to this stage of growth, the theoretical mass distributions refer only to a local portion of the swarm. The steep upper cutoff calculated by Safronov and Nakagawa *et al.* may be expected to occur in every one of these concentric zones. For this reason, it is believed that the most plausible global initial mass distribution to use, starting at the stage at which the largest mass is in the 10^{25}–10^{26}g range, is one in which there are a fairly large number of bodies of similar mass at the upper end of the mass distribution, with a power law distribution extending as far down in mass as permitted by the limitation to 500 initial bodies.

In the principal calculations reported here, in accordance with the foregoing discussion, an initial state was chosen, consisting of 500 bodies ranging in mass from a minimum value of 5.7×10^{24}g ($3 \times$ mass of Ceres) up to a maximum value of 1.1×10^{26}g ($1.5\times$ mass of the Moon), and distributed according to a power law

$$dN \propto M^{-1.83}dM \qquad (5)$$

The total mass was taken to be equal to 1.26×10^{28}g and the material density to be 3.80 g/cm^3. This distribution is illustrated in Fig. 2 (curve A). The limitation to 500 bodies was imposed by computer time and memory. It represents a 5-fold increase over earlier 100-body calculations (Wetherill, 1980a,b). The initial eccentricity was assigned a random value between 0 and 0.05, the initial inclination a random value between 0 and 0.025, and the initial semi-major axis was randomly sampled between 0.7 and 1.1 AU, weighted so that the number of bodies per unit surface area was constant. It has been shown earlier that an initial distribution of semi-major axes similar to this is necessary to match the present energy and angular momentum of the terrestrial planets (Wetherill, 1978). Because it is found that only a few percent of the mass, energy, and angular momentum is lost from the system

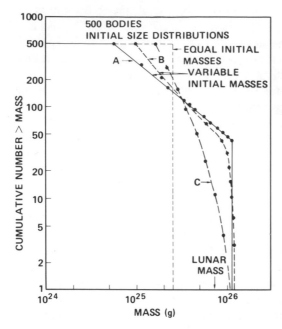

Fig. 2. Cumulative initial mass distribution. The points on the curves represent the actual values of the initial masses used in the calculations, and the number of bodies with masses equal to or greater than these values. Curve A is the power-law used for 11 new accumulation calculations reported in this paper. Curve B is a variant variable mass initial state used for one additional case. Curve C assumes the distribution found by Nakagawa et al. (1983) applies to the entire swarm, rather than only locally. The dashed line labelled "Equal initial masses" represents the mass used in an earlier study (Wetherill, 1985) in which all the initial masses were equal.

during its growth (for a wide range of initial states), this matching of the initial and final values of these quantities is essential.

As mentioned above, an initial state containing a number (taken to be 42) of bodies of largest mass is assumed to be the most plausible one consistent with earlier work on this problem. In order to test the sensitivity of the results to the validity of this assumption, a number of alternative initial distributions were also calculated.

These include five cases in which the distribution, calculated from that of Nakagawa et al. (1983), is assumed to be valid globally, rather than only locally (curve C of Fig. 2). One case in which a somewhat smoother version of distribution A was chosen was also calculated (curve B). The following initial distributions were also studied: (1) those in which the largest bodies had evolved to 10^{26}g at the inner edge of the swarm, but only to smaller values, ranging down to $\sim10^{25}$g at the outer edge of the swarm; (2) those with initial eccentricities averaging 0.0005 instead of the standard value of 0.025; (3) those with an order of magnitude gap between the mass of the largest bodies and the remainder of the swarm.

It was found that all of these initial states led to final states similar in number and position to one another, and all involved "giant" impacts of bodies comparable or greater than the mass of Mars upon the largest of the final planets found. In an earlier study (Wetherill, 1985) more directed to the phenomenon of giant impacts, the initial state was chosen so as to maximize the growth required to produce giant impactors; i.e., all of the 500 bodies were assigned an initial mass as small as possible, i.e., 2.5×10^{25}g, about one third the mass of the Moon. As detailed in Section 5, this initial state again resulted in similar size distribution and frequency of giant impacts.

It is possible to choose distributions at the stage where the largest bodies have masses $\sim 10^{26}$g that lead to final states that differ significantly from those found for the distributions described above. For example, an initial state in which the semi-major axes were so spread out that the energy and angular momentum of the present terrestrial planets could not be matched (0.4 to 1.4 AU) not surprisingly led to different final states (too many planets) but did not affect the frequency of impacts from very large "second largest bodies" in any noticeable way. The only initial state that has to date been identified that leads to a significantly smaller number of giant impacts is one in which only two bodies of largest mass $\sim 10^{27}$g each are assumed to be present initially, the remainder of the 500 bodies all being of the order of 10^{25}g in mass. Even in this case, several "Earth" impacts ranging from 2×10^{26}g to 5×10^{26}g were found for each case calculated. If the terrestrial planets indeed evolved through such an intermediate state, then the importance of giant impacts may be expected to be reduced, but not eliminated. When evaluated in the light of recent work (Wetherill and Cox, 1984, 1985), previous work on this problem does not at present provide a basis for expecting an intermediate state of this kind, but attention should be given to this possibility.

The principal limitations imposed by the restriction to 500 bodies is the need to neglect the much larger number of smaller bodies that must have been present in the swarm. As a result of experience gained by studying the details of the accumulation of the smallest bodies of the distributions studied, it is concluded that it is essentially impossible that any of these even smaller bodies would nucleate the growth of bodies that would then outstrip the growth of the largest bodies, those in the $\sim 10^{26}$g mass range. The fate of these even smaller bodies must be to be captured by tbe larger bodies, or be lost to the swarm by extensive fragmentation. They could conceivably have the effect, neglected in the present calculations, of causing the swarm as a whole to be sufficiently dissipative to damp eccentricities enough to cause the number of final planets to be greater than the number observed. This matter requires further study. Preliminary calculations, in which the ratio of gravitational acceleration to collisional damping is artificially increased, strongly suggests that in a more dissipative swarm, the emergence of excess final planets precedes the disappearance of giant impacts.

Starting with the initial states shown on curves A, B, and C of Fig. 2, the natural evolution of the system was calculated by a Monte Carlo technique ultimately based on that of Arnold (1965), generalized to include a large number of mutually interacting bodies rather than simply a single small body interacting with the present system of planets. The evolution of the system is assumed to result from close two-body encounters between planetesimals in crossing orbits. (Orbits are said to be crossing when intersection of the two orbits is possible for some value of the arguments of perihelion and longitudes of the nodes of the two bodies.) When the encounter distance between two planetesimals becomes less than the sum of their physical radii, the bodies are assumed to merge to form a larger body with mass equal to the sum of the masses. Otherwise, an encounter between two bodies results in gravitational perturbation to new orbits. As the calculation progresses the number of bodies becomes smaller (unless the effects of tidal fragmentation are included as discussed later). Eventually, only bodies in noncrossing orbits will remain, the calculation is then terminated, and the surviving bodies are considered to be the final planets resulting from that particular accumulation calculation.

3. Possible Final States

The possible outcomes of the calculation are various numbers of final planets in noncrossing orbits, the total mass of which is equal to or less than the initial mass of the swarm. Both this and previous work (Wetherill, 1980a,b, 1979) show that the most relevant factor determining the actual outcome is the balance between the strength of the gravitational perturbations, which tend to increase the eccentricity and the relative velocity, and collisional damping, which tends to decrease these quantities. If gravitational perturbations are relatively strong, orbital crossing will be maintained until only a few large bodies remain. If gravitational perturbations are weak, relative to the effects of damping, the "planetary embryos" become isolated from one another prematurely and an excessive number of undersized final planets result. This effect has been illustrated by calculation of an extreme case in which the amplitudes of all gravitational perturbations were set to zero. The 500 initial bodies with eccentricities ranging from 0 to 0.05 evolved into 11 final planets of mass $>10^{26}$, of which the 6 largest all had masses between 1×10^{27}g and 2×10^{27}g. Six bodies of mass $<10^{26}$g also survived. As a result of collisions the initial eccentricities were reduced to an average value of 0.0053. Using the method of calculation of the present work the number of final planets could become arbitrarily large for very low values of initial eccentricity. This would not be the case for a real planetesimal swarm, because when the eccentricity becomes very low, the two-body approximation used here is no longer valid, and encounters can occur even for noncrossing orbits (Giuli, 1968; Nishida, 1983; Wetherill and Cox, 1984, 1985). These studies show that for nearly concentric orbits encounters can occur when the bodies are within about four Lagrangian sphere radii (R_L). For bodies of equal mass M, this is given by:

$$R_L = \left(\frac{M}{3M_\odot}\right)^{1/3} a \tag{6}$$

where $M_\odot$ is the solar mass and a is the semi-major axis. The number (n) of final planets of equal mass that can be accommodated within an interval of semi-major axes extending from a_1 to a_2 will be:

$$n = \left(\frac{3M_\odot}{M_T} \right)^{1/2} \left(\frac{1}{4} \right)^{3/2} \left[ln \ \frac{a_2}{a_1} \right]^{3/2} \tag{7}$$

where M_T is the total mass of the terrestrial planets (1.2×10^{28}g). A swarm that cannot accelerate itself to higher eccentricities cannot spread radially either, so it is appropriate to use the initial boundaries of the swarm for a_2 and a_1, with the result that n = 27. Thus, if for some reason damping completely prevailed over perturbations, initial eccentricities were zero, and spacings were optimal, a final outcome could be as many as 27 planets with masses of 4×10^{26}g (Mercury-size) instead of the observed 4 bodies. At the other extreme, as few as two final planets could simultaneously satisfy the angular momentum, energy, and mass conservation constraints.

The primary question addressed here reduces to that of learning just where in this range of ~27 to 2 final planets the evolution of the swarm will end, using plausible assumptions regarding the amplitude of the gravitational perturbations and the degree of collisional damping. As discussed earlier, another important question concerns the size of the largest bodies that fail to become planets, but whose identity is instead terminated by impacting the final planets.

Whether or not one should expect the outcome of a "simulation" of this kind to resemble in much detail the observed planets is not clear. As considered here, planetary accumulation becomes a highly chaotic stochastic process. Imperceptible differences in prior conditions can make the difference between two large bodies becoming isolated final planets or colliding in a giant impact to form one planet. It seems likely that actual planetary systems also are formed in this chaotic manner, and that we are observing only one out of many final configurations of similar probability that approximately resemble one another only in the total number of final planets, heliocentric distance, and impact history. The range of outcomes found in the present calculations can be viewed either as exhibiting defects in the modelling (which undoubtedly are present) or as affording a preview of terrestrial planet systems that are yet to be discovered.

4. Method of Calculation

Starting with the initial mass and orbital distribution described, the spontaneous orbital and collisional history of the system was followed using a Monte Carlo approach based on that of Arnold (1965) for evolution of small planet-crossing bodies in the present solar system. Those bodies in crossing orbits were identified, and their probability of encountering one another within a given number (usually 5, sometimes 7) of two-body gravitational collision was calculated using the Öpik (1951) collision formula. An encounter was considered to have actually occurred between each pair of bodies during a first (10^4 year) time step if the probability of an encounter during this interval was greater than a number between 0 and 1 chosen at random. The minimum unperturbed separation between the bodies, in units of the gravitational collision radius, was then chosen at random, appropriately geometrically weighted so that the probability of an encounter in the interval $\Delta\rho$ was proportional to ρ, whereas ρ is the minimum separation distance in these units. If the separation distance was less than the gravitational radius, an impact was scored. The mass of the larger body was augmented by that of the smaller body (identified arbitrarily if the two bodies were of equal mass). The smaller body was then removed from the swarm. A new orbit was then calculated for the larger body by use of angular momentum conservation in the reference frame of the center of mass of the two bodies.

If an impact did not occur, the mutual two-body scattering of the encountering bodies was calculated using the expressions used by Öpik (1951) and Arnold (1965), assuming impact at the selected distance and at a random azimuth on a target circle fixed on one of the bodies and oriented perpendicular to their relative velocity vector. The changes in the three relative velocity components were calculated in a reference frame in which the center of mass moves on a circular Keplerian orbit at the heliocentric distance of the encounter. These relative velocities were then converted into new heliocentric velocities, and new perturbed orbital elements were calculated for both of the bodies. The validity of calculating perturbations in this way has been investigated by numerical integration of the three-body equations of motion (Wetherill and Cox, 1984). As expected from Safronov's analytical theory, almost all of the encounters are in the range of relative velocity/escape velocity >0.35, for which the two-body algorithm is a good approximation.

Practical requirements of computational time constrain the encounter distance for which perturbations can be calculated to values less than 10 (= "K") gravitational radii, and when all tradeoffs are considered (e.g., use of a larger number of bodies), lead to optimal limits of K = 5 to 7 gravitational radii. On the other hand, because they occur often, the effect of encounters out as far as 4 Lagrangian radii (equation (6)), which is typically hundreds of gravitational radii, have been shown to be important (Wetherill and Cox, 1984). Arnold (1965) presented a way of approximately including the effect of the very large number of these very small distant perturbations without significantly increasing the time required for computation. This is accomplished by weighting the probability of encounters between K/2 and K gravitational radii relative to closer encounters in such a way that each of these moderately distant encounters included the root-mean-square average effect of the many weaker encounters out to the edge of the sphere of influence. Comparison with numerical integration (Wetherill

---

and Cox, 1984) showed that this general procedure is valid, but that it was better to extend this "Arnold extrapolation" to 10 sphere of influence radii and to gradually augment the amplitude of the perturbations for values of (escape velocity/relative velocity) less than 0.5, in such a way that they are doubled when $V/Ve = 0.1$. These modifications were made in the present work.

Other minor modifications, directed toward more appropriate treatment of the rare values of $V/Ve < 0.1$, were to increase the probability of impact by a factor of 4 when V/Ve is between 0.1 and 0.03 and to limit the maximum two-body gravitational cross-section to 3000 (corresponding to $V/Ve = 0.018$) (Wetherill and Cox, 1985).

After all the encounters found to occur during the first time step were calculated, new encounter probabilities with all the other bodies were calculated for those bodies whose orbital elements had changed. The above procedure was then repeated for a second time step. The duration of the time steps were automatically adjusted so that for the pair of bodies with the highest encounter probability during the time step, the probability of two encounters remained about 20%. For almost all the bodies, the probability of two encounters was very much less than this.

The procedure was repeated until all the remaining bodies were in noncrossing orbits. Typically ~2000 time steps were calculated, requiring about 15 hours of VAX 11/780 time using a floating point accelerator and with 4 Mbyte of storage available. In addition to collisions, bodies were assumed to be lost from the terrestrial planet system when their aphelion crossed the present orbit of Jupiter. Typically 5% of the mass was lost from the system in this way, providing *a posteriori* justification for use of a swarm with initial mass 5% greater than that of the present terrestrial planets.

An important variation of the above procedure consisted of allowing a limited degree of "tidal disruption" to take place for 5 of the cases calculated. The "limited degree" consists of the requirement that in order to fragment, the bodies must be nearly completely melted, and the center of the smaller body must pass within 2.5 radii of the larger body, somewhat within the Roche limit of ~3 planetary radii. The requirement of melting is based on the work of Mizuno and Boss (1985), who showed that for viscoelastic bodies, even at low viscosities ($<10^{13}$ poise) disruption is not found to occur during the short time available of the flyby encounter. Melting was assumed to occur if the total thermal input during the accumulation exceeds 3×10^{10} ergs/g. Whenever an impact occurred it was assumed that 25% of both the kinetic energy of impact calculated in the center of mass frame and the previously accumulated heat of the smaller body is permanently retained within the larger body.

When the two criteria for tidal disruption are met, it was assumed that the smaller body disrupts into the smallest bodies included in the initial swarm (5.68×10^{24}g). All of the gravitational energy required for this disruption is subtracted from the kinetic energy of relative motion in the center of mass frame, and the remaining kinetic energy, if any, is used in calculating the heliocentric orbits of the fragments. If the remaining kinetic energy is negative, the bodies are assumed to combine, as in a collision.

This way of determining if the smaller body is melted is only a crude approximation to an obviously complex physical problem. At the present stage of investigation of planetary accumulation, however, this is believed to be adequate. It will be seen in the next section that when tidal disruption is included in the way described above, the effect on the final outcome of the accumulation is minor. Only if tidal disruption occurs much more easily than is required by these criteria will it be consequential. This possibility was explored by removing the requirement of melting entirely. As will be discussed later, the effect of this was noticeable, but not dramatic. Only if tidal disruption could occur even more readily, i.e., for solid bodies $<5 \times 10^{24}$g in mass, will this phenomenon limit the growth of "second largest" impacting bodies to the size proposed by Safronov (1969).

In addition to tidal disruption, two other fragmentation processes may occur, simple collisional mass loss and rotational instability. Collisional mass loss will occur when the kinetic energy of the fragmented bodies exceed their gravitational binding energy. These hypervelocity collisions are highly dissipative, and resemble explosions more than colliding brittle bodies, such as marbles. Only collisions at several times the escape velocity result in net mass loss (O'Keefe and Ahrens, 1977). Such collisions do occur, but with the exception noted below, preliminary calculations suggest they are not frequent enough to affect the outcome of these calculations. These modes of disruption were not included in the calculations detailed in this paper. Doing so would introduce both conceptual and practical difficulties that would complicate the present discussion without compensating enlightenment. These effects will be discussed in a subsequent publication.

The exception is the case of "Mercury." In many of the calculations a small innermost final planet is found in the vicinity of 0.4 to 0.5 AU. During the final stage of accumulation relative velocities in this region become fairly high, e.g., ~15 km/sec. With the material densities assumed, the escape velocity of a body with the mass of Mercury will be ~4 km/sec. Impacts of bodies of comparable mass on these bodies at this relative velocity are sufficiently energetic to disrupt and disperse them. In these calculations, some impacts of this kind were found. The initial swarm already contained some bodies massive enough to disrupt "Mercury" at the velocity of the final stages of accumulation. It is difficult to say whether or not these catastrophic impacts on a small final planet are an artifact of the calculation, resulting from the "graininess" of the size distribution and failure to include a mechanism that will disrupt bodies smaller than Mercury as the velocity of the swarm increases. This matter requires further attention. It is possible that, considering the small size of the planet, somewhat smaller but still relatively massive impacts are responsible for the depletion of Mercury in silicates.

More often, sufficient angular momentum is imparted during the collision to cause the resulting combined body to be rotationally unstable with respect to equatorial mass loss. Only rarely, however, is the impact sufficient to cause the projectile mass to be gravitationally unbound. Calculations have been carried out for greatly exaggerated mass losses of 50% following rotational instability, and the effect is not obvious. Even less plausible 100% mass loss causes the accumulation process to be occasionally Sisyphean, without affecting the final outcome markedly.

5. Results of Calculations

The calculated evolution of an accumulating planetesimal swarm, the final outcome of which resembles the present terrestrial planets, is depicted in Fig. 3. Initially all of the bodies are confined to a narrow zone of semi-major axes and eccentricities (as well as inclinations), as shown in Fig. 3a. After only 0.5 m.y. (Fig. 3b), nine bodies in the small "planet" mass range (2×10^{26}g-10^{27}g) have been formed, the mutual perturbations of the planetesimals have "pumped up" the mean eccentricity to about three times its initial value, and a number of bodies have begun to diffuse beyond the original boundaries of the swarm.

After 2.2 m.y. (Fig. 3c), the bodies that will turn out to be the analogs of the present planets Earth and Venus are the largest bodies in the swarm, and their eccentricities are well below the mean eccentricity of the smaller bodies. The eccentricities of the smaller bodies consist more or less of a Gaussian distribution centered in values corresponding to a fairly low Safronov number (equation (2)) of 1.6. Fourteen small "planets" now exist. On the average, their eccentricities are lower than those of the very smallest bodies. On the other hand, many of them have aphelion-perihelion distances comparable to the heliocentric spacings of the bodies that will evolve into the final planets. Even at this early time, the concept of unique feeding zones for each "planetary embryo" has little usefulness.

At 10.9 m.y. (Fig. 3d) "Earth" and "Venus" have reached 84% and 42% of their final masses. Note the 1.2×10^{27}g object near "Venus." Five million years later, it will strike "Venus" with an impact velocity of 13 km/sec. The kinetic energy of the impact in the center of mass frame is ~65% of the gravitational potential energy of the combined body. An impact this energetic is somewhat unusual, but not extremely so. Another object of interest is the body that will ultimately become the analog of Mercury. At this stage of accumulation, it has reached 37% of its final mass of 3.9×10^{26}g (~72% of its final radius). It is the final planet with the largest semi-major axis.

At 31 m.y. (Fig. 3e) the "clean-up" epoch of accumulation has been entered. 81% of the initial mass of the system has been incorporated into the final planets. The eccentricities (and inclinations) of the remaining bodies are quite large, and some of them are evolving into "asteroidal" orbits with semi-major axes of ~2.0.

Other than the final planets, the largest body remaining has a mass of 3.0×10^{26}g (~ present mass of Mercury). Seventeen million years later, it will strike the Earth, and represent the last impact of a body with mass $>10^{27}$ on that planet. "Mercury" is now beyond the present semi-major axis of Mars. During the next 3 million years a series of close encounters with "Mars," "Earth," and "Venus" will perturb it to a semi-major axis of 0.58 AU, inside the orbit of "Venus." Its survival as a final planet was simply a matter of luck. It could just as well have struck a planet during this journey through the inner solar system. By 64 m.y., the clean-up is in its final stages (Fig. 3f). "Mercury" is near its final heliocentric distance.

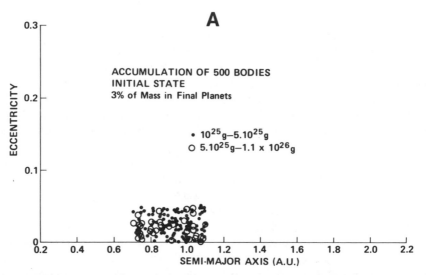

Fig. 3. "*Snapshots" of the evolution in mass semi-major axis, and eccentricity during the course of evolution for a case that led to a final state resembling the present terrestrial planets. See text for description and discussion of these results. (a) Initial state. In order to avoid crowding, only one-half of the bodies with mass $>5 \times 10^{25}$g were plotted, and none of the 210 bodies with mass $<5 \times 10^{25}$g were plotted. Solid circles, 10^{25}g–5×10^{25}g; open circles, 5×20^{25}g–1.1×10^{26}g. (b) Time = 0.5 m.y. Small solid circles, 10^{25}g–5×10^{25}g; large solid circles, 5×20^{25}g–2×10^{26}g (approximately lunar-size bodies); open circles, 2×10^{26}g–10^{27}g (small "planets"). The position of the "embryos" of the final planets are indicated, and named by analogy between the calculated final distribution and the observed terrestrial planets. One hundred thirty-seven bodies with mass $<10^{25}$ are not plotted. (c) Time = 2.2 m.y. All bodies plotted. Small solid circles, 5×10^{24}g–5×10^{25}g; large solid circles, 5×10^{25}g–2×10^{26}g; open circles, 2×10^{26}g–10^{27}g; solid squares, $>10^{27}$g (large terrestrial planets). (d) Time = 10.9 m.y. Points have same definition as in Fig. 2, curve C. (e) Time = 31 m.y. Final sweep-up of residual planetesimals has begun. (f) Time = 64 m.y. All final planets are in their final positions. A population of bodies remain, mostly with high velocities and large semi-major axes. (g) Time = 239 m.y. Planets have reached their final masses and positions. Three small remaining bodies will be ejected from the solar system by "Earth" and "Mars" perturbations.*

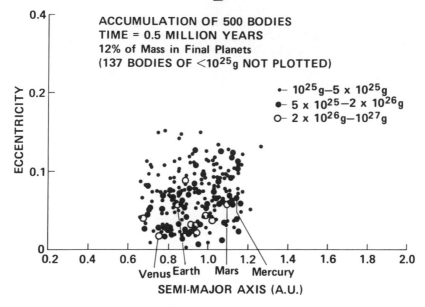

B

ACCUMULATION OF 500 BODIES
TIME = 0.5 MILLION YEARS
12% of Mass in Final Planets
(137 BODIES OF $<10^{25}$g NOT PLOTTED)

$\bullet$– 10^{25}g–5×10^{25}g
$\bullet$ 5×10^{25}–2×10^{26}g
$\circ$– 2×10^{26}g–10^{27}g

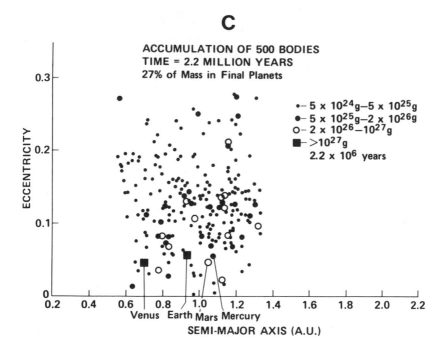

C

ACCUMULATION OF 500 BODIES
TIME = 2.2 MILLION YEARS
27% of Mass in Final Planets

$\bullet$– 5×10^{24}g–5×10^{25}g
$\bullet$– 5×10^{25}g–2×10^{26}g
$\circ$– 2×10^{26}–10^{27}g
$\blacksquare$– $>10^{27}$g
2.2×10^{6} years

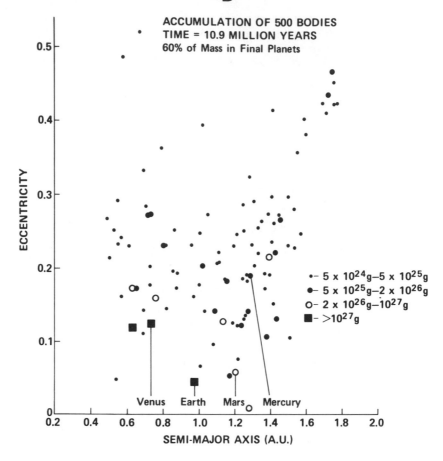

D

ACCUMULATION OF 500 BODIES
TIME = 10.9 MILLION YEARS
60% of Mass in Final Planets

Legend:
- • – 5×10^{24}g–5×10^{25}g
- ● – 5×10^{25}g–2×10^{26}g
- ○ – 2×10^{26}g–10^{27}g
- ■ – $>10^{27}$g

Venus Earth Mars Mercury

ECCENTRICITY (y-axis)

SEMI-MAJOR AXIS (A.U.) (x-axis)

The last impact is that of a 5×10^{24}g body on "Mars" at 202 m.y. The three small remaining bodies will be ejected from the solar system following encounters with "Earth" and "Mars" (Fig. 3g).

The final states of 12 accumulation calculations, 7 without tidal disruption and 5 in which tidal disruption is allowed if the projectile is melted, are shown in Figs. 4 and 5. In these cases, the final distribution of bodies does not resemble the terrestrial planets as well as the case described by Fig. 3 (corresponding to no. 4 of Fig. 4). All of these results have the common characteristic, however, that the final number of planetary bodies with masses greater than 10^{26}g (1.4 lunar masses) is small, almost always four or less. (In two cases a fifth body of mass 1 to 2×10^{26}g was found.) In half of the cases the number of "large" planets ($>2 \times 10^{27}$g) is three instead

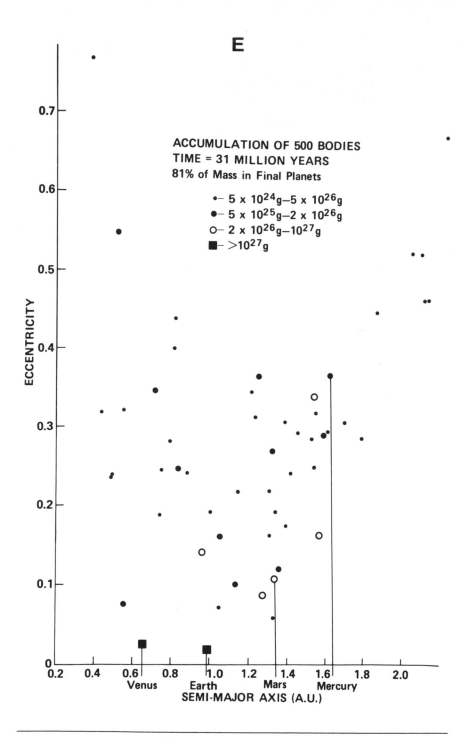

F

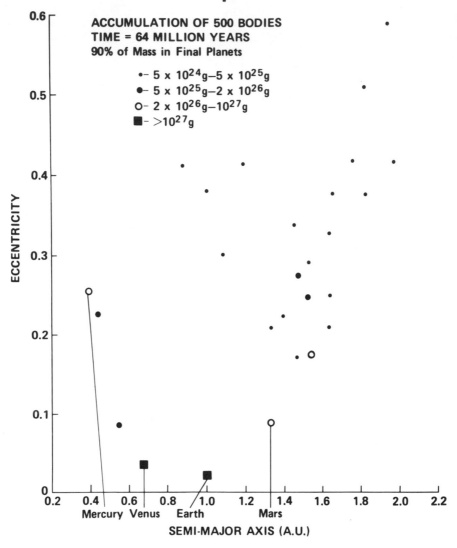

ACCUMULATION OF 500 BODIES
TIME = 64 MILLION YEARS
90% of Mass in Final Planets

• – $5 \times 10^{24}g$ – $5 \times 10^{25}g$
● – $5 \times 10^{25}g$ – $2 \times 10^{26}g$
○ – $2 \times 10^{26}g$ – $10^{27}g$
■ – $> 10^{27}g$

ECCENTRICITY

SEMI-MAJOR AXIS (A.U.)

Mercury Venus Earth Mars

of the observed two. As discussed in Section 2, it is not known if this variable outcome is the result of the approximate nature of the modelling or represents physically realistic stochastic variability in the accumulation process. If planets did form this way, however, it appears difficult to escape a highly stochastic evolution.

One feature that may well be an artifact of the calculation is the result that the outermost planet-size body ("Mars") is systematically two to four times more

G

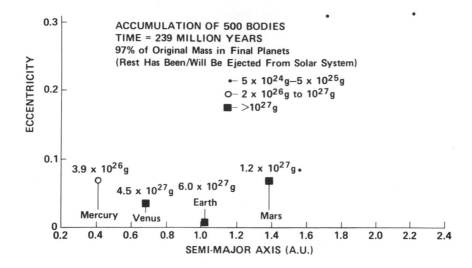

massive than the actual planet Mars. The occasional lingering presence of subplanetary-size bodies with semi-major axes beyond "Mars" may be a related phenomenon. In the present solar system, the orbital evolution of bodies in orbits of this kind is strongly controlled by the combined effects of Mars perturbations and the ν_6 secular resonance (Wetherill, 1979). The latter has not been included in this calculation. If, however, as seems likely, the major planets existed during the formation of the terrestrial planets, at least similar secular resonance regions should have been present. In analogy to the present solar system, these would reduce the stability of moderately and highly eccentric orbits beyond that of the Earth, reduce the number of planetesimals available to "Mars," and remove the more distant residual planetesimals.

The principal effect of including tidal disruption is a larger number of small residual planetesimals, mostly with semi-major axes beyond 1.5 AU. If the effects of the ν_6 resonance had been included, it is unlikely that these orbits would be stable on a >200 m.y. time scale. During the final stages of accumulation it is also likely that bodies this small would be vulnerable to collisional destruction.

The final states of 10 additional calculations (Wetherill, 1985) in which the 500 initial masses were all equal are shown in Figs. 6 and 7, with and without tidal disruption. Differences between these results and those shown in Figs. 3 and 4 are not manifest.

Impacts of mass $>2 \times 10^{26}$g on "Earth" for the 12 accumulations shown in Figs. 4 and 5 are shown in Fig. 8. (In a few cases, identification of "Earth" in the final outcome was ambiguous and a reasonable but arbitrary choice was made.

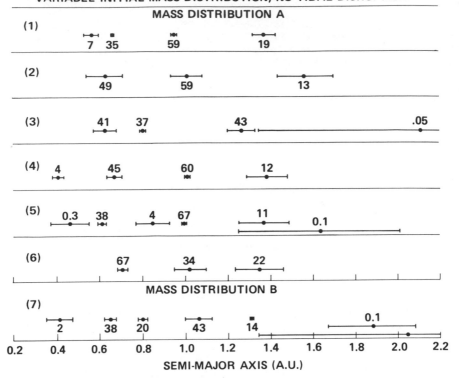

Fig. 4. *Outcome of six accumulation calculations using the variable mass initial state "A" and one using initial state "B" (Fig. 1). The final positions of the planets are shown as points; the attached bars indicate their aphelia and perihelia. The numbers beside each point indicate the final mass in units of $10^{26}g$.*

Because giant impacts occur on all large planets, a different choice would not affect the conclusions.)

Typically, one or two impacts of bodies more massive than Mars occur for each accumulation, and about three more massive than Mercury. The same result has been found and presented elsewhere (Wetherill, 1985) for the 10 calculations shown in Figs. 6 and 7. These giant impacts occur most frequently after the accumulation has proceeded for 1–15 m.y. In this time interval from 15% to 70% of the mass of the Earth had already formed. The distribution of large impacts for an initial state in which the distribution laws of Safronov (1962a) and Nakagawa *et al.* (1983) are assumed to apply globally (Fig. 2, curve C) are shown in Fig. 9.

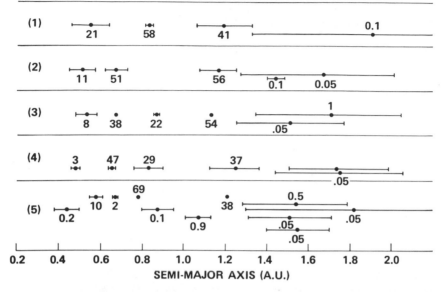

OUTCOME OF ACCUMULATION CALCULATIONS
POWER LAW INITIAL STATE
TIDAL DISRUPTION IF PROJECTILE IS LIQUID

Fig. 5. Outcome of five accumulation calculations using initial state "A". Tidal disruption occurs if a previously melted body passes within 2.5 planetary radii of a larger body.

More detailed step-by-step examination of the calculated growth of the planetesimals, as well as numerical experiments with variant models, suggests that the phenomenon of giant impacts is not confined to initial states similar to those calculated but is a more general phenomenon, including planetary systems having quite different values of mass, energy, and angular momentum than our terrestrial planets.

Although not supported by actual calculations, it should be valuable to try and understand better what is actually going on here, rather than simply accepting "what the computer tells me." The fundamental circumstances that lead to these impacts seems to be a primordial initial state containing many small bodies of similar (10^{16}–10^{18}g) mass, together with the local nature of the accumulation process. No little planetesimal is labelled "Earth" or "Venus." As Victor Safronov often says: "All planetesimals are created equal."

At any stage of growth, planetesimals are effective in perturbing and colliding only with bodies in crossing orbits, or in the case of very low eccentricity orbits, bodies within less than about five times the distance of the nearest Lagrangian points. For the terrestrial planets, eccentricities large enough to permit widespread exchange of material throughout the entire swarm do not occur until a number of $>10^{27}$g

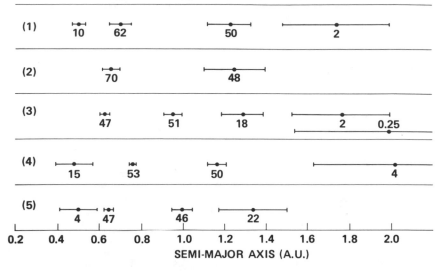

OUTCOME OF ACCUMULATION CALCULATIONS
EQUAL INITIAL MASSES (500 BODIES 2.5 x 10²⁵g)
NO TIDAL DISRUPTION

*Fig. 6. Outcome of five accumulation calculations using an initial state of equal mass (2.5 ×
10²⁵ g) bodies. Giant impacts for these calculations (and those of Fig. 6) have been discussed
elsewhere (Wetherill, 1985), but these detailed final state distributions have not been presented
before.*

bodies have formed. Prior to this final stage, growth of bodies at one heliocentric
distance do nothing to prevent growth of bodies of similar mass in other regions
of the heliocentric swarm. In the earliest stage of accumulation studied by Greenberg
et al. (1978), at first the growing planetesimals accumulate material only from a
region ≈10⁻³ AU in width, and hundreds of bodies of mass ≥10²³ can be expected
to form. As these bodies gradually perturb one another into orbits of higher eccentricity
they will continue to accumulate preferentially from their neighbors, leading to a
smaller number of larger bodies in adjacent zones that are still of comparable mass
to one another. As the process of growth continues this repeated collision of bodies
of comparable mass is responsible for the giant impact phenomenon.

Multiplanet accumulation hasn't been studied in a gas-rich nebula, but from the
work of Hayashi and his colleagues (e.g., Hayashi *et al.*, 1985), it appears that
this mode of accumulation also starts from many small bodies and proceeds locally.
If so, collisions between bodies of comparable mass should also be expected during
the accumulation of the planets, ultimately leading to giant impacts.

Rather than considering giant impacts as a somewhat radical suggestion, if one
is skeptical about the reality of the phenomenon, a good starting point would be

to consider it a normal phenomenon that one should, at least naively, expect during planetary formation. Then if it does not occur, it should be possible to find some mechanism that will thwart this natural tendency by breaking the initial symmetry imposed by the similarity of the original masses. This mechanism must be one that will permit the growth of a few large bodies and at the same time preclude the growth of somewhat smaller bodies.

A favorite mechanism of this kind has been "runaway accretion," whereby compounding of the enhanced gravitational cross-section of a larger accumulating body permits its growth to greatly outstrip the growth of its neighbors. Runaway accretion is a local phenomenon *par excellence*, however. It will function only for bodies in low velocity (i.e., low eccentricity) orbits that thereby fail to interact with bodies in distant regions of the heliocentric swarm. As discussed in Section 2, it may be expected to be important in facilitating the rapid growth of ~10^{21}g bodies within an initial concentric zone. This mechanism can also work quite well in excessively formal particle-in-a-box calculations because the effect of the bodies really being at different heliocentric distances is simply ignored in this approximation. In

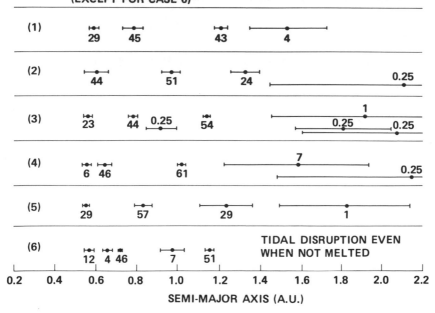

Fig. 7.   Outcome of six accumulation calculations with equal initial masses. Tidal disruption when melted: (1)–(5). Melting not required for case (6).

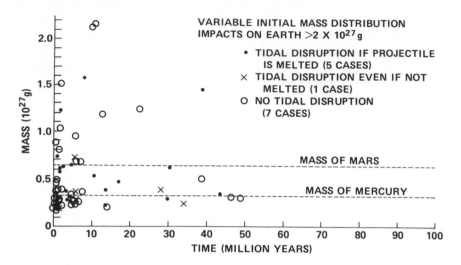

Fig. 8.   Giant (>2 × 10²⁶g) impacts on Earth for 13 calculations. Open circles, variable initial mass distribution with no tidal disruption (seven cases); solid points, variable initial mass disruption, tidal disruption if projectile is melted (five cases); cross, equal initial mass distribution, melting not required for tidal disruption (one case).

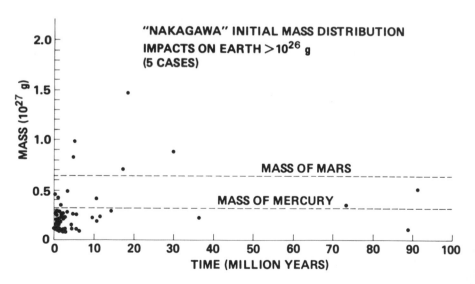

Fig. 9.   Giant (>10²⁶g) impacts for five calculations assuming the mass distribution of Nakagawa et al. (1983) to apply to the whole swarm (curve C, Fig. 2).

this way erroneously short growth times can be formally calculated. These have no meaning because in order for more distant bodies to merge, eccentricities must increase so much that the low velocities that permit runaway accretion disappear. Therefore these short times for final growth of the terrestrial planets should be simply considered as artifacts of the otherwise often useful particle-in-a-box approach.

What is actually required is a mechanism that will permit a slightly larger body to frequently destroy its neighbors of comparable mass. In this way the population could be reduced to a few large bodies and many very small ones. Tidal disruption during flyby within the Roche limit has been proposed as such a mechanism (Wetherill, 1976). The work of Mizuno and Boss (1985) implies that tidal disruption will be ineffectual in this regard. Other possibilities—collisional mass loss, rotational instability, and grazing encounters—have been considered. Although they do occur, preliminary calculations suggest that they also appear to be insufficient. If there is no potent mechanism for removing bodies of comparable mass before they have a good chance to impact, the presence of large planetesimals that fail to become planets and the resulting giant impacts should be regarded as normal characteristics of planetary accumulation.

The occurrence of these giant impacts will cause large stochastic fluctuations in the growth rate of the terrestrial planets. Nevertheless, by averaging the results of several calculations, the underlying characteristic time for accumulation can be discerned, as shown in Figs. 10 and 11.

It should be noted that the time scale for growth is not significantly dependent on heliocentric distance, in contrast with the prediction of the analytic theory of the growth of a single planet from its own private "feeding zone" (Safronov, 1969). The final accumulation of "Mars" and "Mercury" do appear to be somewhat delayed. As mentioned earlier, in the case of Mars, this result may be an artifact caused by neglect of secular resonances in this region of the solar system.

In accordance with conventional opinion, the time scale required for nearly complete growth of the terrestrial planets by gas-free accumulation is $\sim 10^8$ years. It should be emphasized, however, that during the first ten million years the growth is much more rapid. Bodies larger than 10^{27}g in mass are formed within ~ 1 m.y., and 50% growth in mass (and 79% in radius) occurs within 7 m.y. For this reason, there is at present no theoretical basis for believing that the time scale for gas-free accumulation of the terrestrial planets is any longer than that expected for accumulation in a gaseous nebula. It is possible that future studies of multiplanet accumulation in a gaseous nebula will lead to even shorter time scales, but this is not the case for the single-planet studies that have been published.

The contrary conclusion given by Hayashi *et al.* (1985) appears to be based on several misunderstandings. In the review by Hayashi *et al.*, the results of Nakagawa *et al.* (1983) are discussed. In this work it was found that a planet at the Earth's heliocentric distance growing in a gas-rich swarm would reach a mass of 10^{27}g

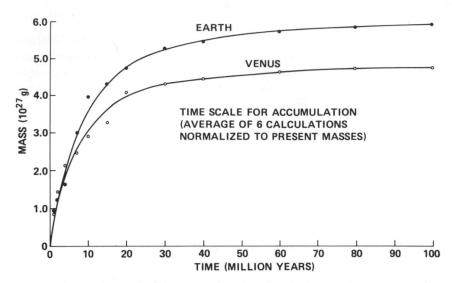

Fig. 10.   Time scale for growth of "Earth" and "Venus" based on averages of six calculations for which the final state resembled the present distribution of these planets well enough for the analogy to be meaningful. The percentage of the final mass was calculated for each time, normalized to the present masses of Earth and Venus, and then averaged.

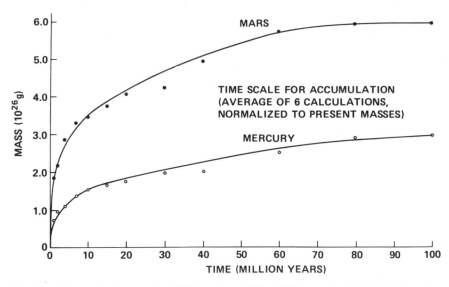

Fig. 11.   Time scale for growth of "Mercury" and "Mars". Same method of calculation as in Fig. 8.

after 6 m.y. In the present investigation, as well as earlier gas-free investigations involving 100 initial bodies (Wetherill, 1980a,b), this mass was reached in an even shorter time of 1 to 2 m.y. The calculation of Nakagawa *et al.* was terminated at a mass of 10^{27}g, and a total growth time for the Earth of 10^7 years was found by simply scaling the time of growth from 10^{27}g to the final mass of the Earth as $m^{1/3}$. In both the gas-rich and the gas-free cases the rate of accumulation can be expected to decrease with time because orbits with low collision lifetimes become preferentially populated. In addition, the rate of growth will decrease for the trivial reason that the amount of material remaining in the swarm will become small. Therefore gas-rich accumulation should also have a long-lived "tail" similar to that for the gas-free case (Figs. 8 and 9).

Hayashi *et al.* interpret this supposed difference in time scale as the consequence of the present author ignoring tidal disruption and radial migration of planetesimals. Mizuno and Boss have shown that tidal disruption is very unlikely to occur to the extent assumed by Nakagawa *et al.* (1983). Radial migration is intrinsic to the Monte Carlo calculations reported both here and earlier (Wetherill, 1985, 1980a,b, 1978) and was explicitly studied, discussed, and specially illustrated in three of these references.

6. Discussion, Conclusions, and Conjectures

With the framework of this general approach to the formation of the terrestrial planets from planetesimals in the absence of a dynamically significant gas phase, it is found that midway in the growth of an accumulating population of heliocentric planetesimals the swarm should have contained a number of large bodies, i.e., ~100 of approximate lunar mass, ~10 of mass greater than Mercury, and several greater than the mass of Mars, in addition to those that survived as final planets. No effective mechanism for removal of these bodies from the swarm is known other than impacts usually onto a final planet. About a third of these large bodies will strike the Earth. The orbital and relative velocity distribution of these impacting bodies is also provided by these calculations.

The presence of these planetesimals in the swarm provides opportunities for large stochastic fluctuations in the supply of material and angular momentum to geocentric protolunar orbit. For example, a number of lunar-size bodies are available for intact capture, capture following a fragmentation event while within Earth's sphere of influence, or tidal disruption (if the body is liquid). If it can be shown quantitatively that this number of bodies in the orbits found is sufficient to cause these events to be probable during accumulation of the Earth, then this event can be considered to be a way of forming the Moon that is consistent with accumulation of the terrestrial planets from planetesimals. Rough calculations of these probabilities suggests to the author that this is not the case. Nevertheless, at least for smaller planetesimals, these processes require careful study, as they may well play a significant role in furnishing material to the geocentric swarm, impacting the Moon, and influencing its geocentric orbital evolution. Even though single events of this kind may be incapable of forming the Moon, these processes should not be dismissed on the grounds that "there is no need for them." To some presently uncertain degree, they should have happened without any regard as to whether or not we need them.

It is particularly interesting that these large planetesimals provide in a natural way the giant impacts proposed by Hartmann and Davis (1975) and Cameron and Ward (1976) as a way of forming the Moon. In spite of earlier demonstration that impacts of this magnitude were entirely consistent with Safronov's theory (Wetherill, 1976; Greenberg, 1979), there has been a tendency, at least until recently, to regard these suggestions as too extreme to be taken seriously or even as "Velikovskian." Although it would be presumptuous to conclude that these large planetesimals and impacts were inevitable consequences of planet formation, their probable occurrence imposes the obligation of explicitly considering their consequences in any discussion of the early history of the Earth and the Moon. These include more than the question of formation of the Moon by ejecta following giant impacts discussed by Cameron and Ward (1976, 1978), Thompson and Stevenson (1983), and by these and other authors in this volume. These impacts should also have had a major effect on the thermal and chemical evolution of all the terrestrial planets, as well as that of their atmospheres (Cameron, 1983) and their rotation.

Although no multiplanet accumulation calculations of the kind presented here are available for the case of gas-rich accumulation, there are good reasons for believing that giant impacts should accompany this mode of planet formation as well. Further investigation of this question is needed, particularly inasmuch as it appears probable that gas was actually present in the terrestrial planet region for 10^5 to 10^6 years, during which time significant planetary growth occurred.

Assuming that most of the nebular gas was removed by ~5×10^5 years it still seems plausible that more modest quantities of nebular gas (probably mass and chemically-fractionated) remained for longer times, if only as a consequence of evaporation of volatile residual planetesimals scattered in from the outer solar system (Whipple, 1976). Because of its low concentration, such residual nebular gas could not produce the massive -10^{26}g atmospheres calculated by Hayashi and his coworkers for gas-rich accumulation, but nevertheless may have contributed significantly to the primordial atmospheres of the planets. As proposed by Cameron (1983), the inert gas content of the atmospheres of Earth and Venus may have been similar prior to the removal of the terrestrial atmosphere by the giant impact that formed the Moon. This would imply that the inert gases such as CO_2 and H_2O were not in the Earth's atmosphere at the time of the impact, which, if true, is an important fact. One might speculate further along the lines that it may well be possible to remove terrestrial planet atmospheres by giant impacts smaller than those required to form the Moon. The variability of the inert gas contents of terrestrial atmospheres could then simply reflect the timing of the last atmosphere-removing impact, possibly more or less dated by the $\sim 10^8$ year formation age of the Earth (Wetherill, 1975; Pepin and Phinney, 1976).

Acknowledgments. This renewed numerical effort was inspired by the discovery of Mizuno and Boss that it was not necessary to calculate and store the orbits of the myriad fragments that would be produced by tidal disruption, inasmuch as this process appears unlikely to occur. I wish to thank these colleagues for this, as well as for valuable discussions. This work would also not have been possible but for the collegial effort provided by Kent Ford, Alan Linde, Alan Boss, and Larry Finger in sustaining and improving the DTM-Geophysical Laboratory computing facilities. The assistance of Mary Coder in preparing the manuscript is appreciated. This work was partially supported by NASA grant NSG7397, and is part of a larger program at DTM supported by NASA grant NAGW 398.

7. References

Arnold J. R. (1965) The origin of meteorites as small bodies. *Astrophys. J., 141*, 1536–1556.

Cameron A. G. W. (1978) Physics of the primitive solar nebula and the giant gaseous protoplanets. In *Protostars and Planets* (T. Gehrels, ed.), pp. 453–487. Univ. of Arizona, Tucson.

Cameron A. G. W. (1983) Origin of the atmospheres of the terrestrial planets. *Icarus, 56*, 195–201.

Cameron A. G. W. and Ward W. R. (1976) The origin of the moon (abstract). In *Lunar Science VII*, pp. 120–122. The Lunar Science Institute, Houston.

Cox L. P. and Lewis J. S. (1980) Numerical simulation of the final stages of terrestrial planet formation. *Icarus, 44*, 706–721.

Darwin G. H. (1880) On the secular changes in the elements of the orbit of a satellite revolving about a tidally distorted planet. *Phil. Trans. Roy. Soc., 171*, 713–891.

Gerstenkorn H. (1969) The earliest past of the Earth-Moon system. *Icarus, 11*, 189–207.

Giuli R. T. (1968) On the rotation of the Earth produced by gravitational accretion of particles. *Icarus, 8*, 301–323.

Greenberg R. (1979) Growth of large, latestage planetesimals. *Icarus, 39*, 140–151.

Greenberg R. (1980) Numerical simulation of planet growth (abstract). In *Lunar and Planetary Science XI*, pp. 365–367. Lunar and Planetary Institute, Houston.

Greenberg R., Wacker J., Chapman C. R., and Hartmann W. K. (1978) Planetesimals to planets: a simulation of collisional evolution. *Icarus, 35*, 1–26.

Harris A. W. (1977) An analytical theory for the origin of planetary rotation. *Icarus, 31*, 168–174.

Harris A. W. and Kaula W. M. (1975) A co-accretional model of satellite formation. *Icarus, 24*, 516–524.

Hartmann W. K. and Davis D. R. (1975) Satellite-sized planetesimals and lunar origin. *Icarus, 24*, 504–515.

Hayashi C., Nakazawa K., and Nakagawa Y. (1985) Formation of the solar system. In *Protostars and Planets II*. Univ. of Arizona, Tucson, in press.

Larson R. B. (1978) The stellar state: formation of solar-type stars. In *Protostars and Planets* (T. Gehrels, ed.), pp. 43–57. Univ. of Arizona, Tucson.

Mitler H. E. (1975) Formation of an iron-poor moon by partial capture, or: yet another exotic theory of lunar origin. *Icarus, 24*, 256–268.

Mizuno H. and Boss A. P. (1985) Tidal disruption of viscous planetesimals. *Icarus*, in press.

Nakagawa Y., Hayashi C., and Nakazawa K. (1983) Accumulation of planetesimals in the solar nebula. *Icarus, 54*, 361–376.

Nakazawa K., Komuro T., and Hayashi C. (1983) Origin of the Moon—Capture by gas drag of the Earth's primordial atmosphere. *Moon and Planets, 28*, 311–327.

Nishida S. (1983) Collisional processes of planetesimals with a protoplanet under the gravity of proto-sun. *Progr. Theor. Phys., 70,* 93–105.

O'Keefe J. D. and Ahrens T. J. (1977) Meteorite impact ejecta: Dependence of mass and energy lost on planetary escape velocity. *Science, 198,* 1249–1251.

Öpik E. J. (1951) Collision probabilities with the planets and the distribution of interplanetary matter. *Proc. Roy. Irish Acad., 54A,* 165–199.

Öpik E. J. (1972) Comments on lunar origin. *Irish Astron. J., 10,* 190–238.

Pepin R. O. and Phinney D. (1976) The formation interval of the Earth (abstract). In *Lunar Science VII,* pp. 682–684. The Lunar Science Institute, Houston.

Ringwood A. E. (1972) Some comparative aspects of lunar origin. *Phys. Earth Planet. Inter., 6,* 366–376.

Ruskol E. L. (1975) On the origin of the moon. In *Proceedings of the Soviet-American Conference on Cosmochemistry of the Moon and Planets,* pp. 638–644. Nauka, Moscow. (Translation in NASA SP-370, 1977, pp. 815–822.)

Safronov V. S. (1962a) A particular solution of the coagulation equation. *Dokl. Akad. Nauk USSR, 147,* 64.

Safronov V. S. (1962b) Velocity dispersion in rotating systems of gravitating bodies with inelastic collisions. *Vopr. Kosmog., 8,* 168.

Safronov V. S. (1969) *Evolution of the Protoplanetary Cloud and Formation of the Earth and Planets.* Nauka, Moscow. Translated for NASA and NSF by Israel Program for Scientific Translations, 1972, NASA TT F-677.

Stewart G. R. and Kaula W. M. (1980) Gravitational kinetic theory for planetesimals. *Icarus, 44,* 154–171.

Thompson A. C. and Stevenson D. J. (1983) Two phase gravitational instabilities in thin disks with application to the origin of the moon (abstract). In *Lunar and Planetary Science XIV,* pp. 787–788. Lunar and Planetary Institute, Houston.

Wetherill G. W. (1975) Radiometric chronology of the early solar system. *Ann. Rev. Nucl. Sci., 25,* 283–328.

Wetherill G. W. (1976) The role of large bodies in the formation of the earth and moon. *Proc. Lunar Sci. Conf. 7th,* pp. 3245–3257.

Wetherill G. W. (1978) Accumulation of the terrestrial planets. In *Protostars and Planets* (T. Gehrels, ed.), pp. 565–598. Univ. of Arizona, Tucson.

Wetherill G. W. (1979) Steady-state population of Apollo-Amor objects. *Icarus, 37,* 96–112.

Wetherill G. W. (1980a) Numerical calculations relevant to the accumulation of the terrestrial planets. In *The Continental Crust and its Mineralogical Deposits* (D. W. Strangway, ed.), pp. 3–24. Geol. Soc. Canada Spec. Paper 20.

Wetherill G. W. (1980b) Formation of the terrestrial planets. *Ann. Rev. Astron. Astrophys., 18,* 77–113.

Wetherill G. W. (1985) Giant impacts during the growth of the terrestrial planets. *Science, 228,* 877–879.

Wetherill G. W. and Cox L. P. (1984) The range of validity of the two-body approximation in models of terrestrial planet accumulation. I. Gravitational perturbations. *Icarus, 60,* 40–55.

Wetherill G. W. and Cox L. P. (1985) The range of validity of the two-body approximation in models of terrestrial planet accumulation. II. Gravitational cross-sections. *Icarus,* in press.

Whipple F. L. (1976) A speculation concerning comets and the earth. *Mem. Soc. Roy. Liege, Ser. 6, IX,* 101–111.

Zvyagina Y. V., Pechernikova G. V., and Safronov V. S. (1973) A qualitative solution of the coagulation equation taking into account the fragmentation of bodies. *Astron. Zh., 50,* 1261–1273.

Giant Impactors: Plausible Sizes and Populations

WILLIAM K. HARTMANN AND S. M. VAIL

Planetary Science Institute, 2030 E. Speedway, Suite 201, Tucson, AZ 85719

The concept of large impactors striking the planets late in their formation has been widely developed and supports the impact-trigger hypothesis of lunar origin. One line of evidence comes from the distribution of planetary spin and orbital properties, especially obliquities. These properties are too varied to have resulted if all accreting mass were in very small planetesimals. We have constructed solar system models with different sources of large impactors. We study stochastic variation in obliquities and rotation periods resulting from each source. The models suggest that the largest impactor masses often lay in the range of 0.3–20% of target planetary masses, explaining spin properties. Present asteroid mass ratios, relative to Ceres, are consistent with these figures. Stochastic variations of impact outcomes are considerable; large impactors could thus trigger formation of a relatively large satellite for at least one, but not all, planets. A promising solar system model has each terrestrial planet affected primarily by its own local planetesimals, ranging up to several percent of each planet's mass; however, in the outer solar system, Jupiter planetesimals with as much as $0.02 M_{Jupiter}$ were widely scattered gravitationally by Jupiter, so that Uranus and Neptune interacted with objects of unusually high relative mass. Such a model predicts the low obliquity of Jupiter and high obliquity of Uranus, and allows the stochastic possibility of a Moon-spawning impact on Earth. A "Mars-size" Earth-impactor ($\sim 10\% \, M_\oplus$), posited by several impact-trigger models of lunar origin, is thus plausible.

1. Introduction

Orbital and rotational characteristics of the solar system may be thought of as regularized to first order by cumulative statistical effects of the accretion of innumerable small planetesimals, but with irregularities produced by stochastic impacts of a few larger bodies. This view has been widely developed, with variations on the theme (Shmidt, 1958; Safronov, 1966, 1969; Giuli, 1968; Hartmann and Davis, 1975; Harris and Ward, 1982; Hartmann, 1984, 1986; Wetherill, 1985). Among the observed regularities are prograde rotations (Giuli, 1968; Harris, 1977; Weidenschilling, 1984), modest obliquities, and nearly circular orbits. Among the irregularities are Uranus' high obliquity (attributed to large impacts by Safronov, 1966) and Earth's relatively large angular momentum (attributed to impact by Cameron and Ward, 1976), as well as Earth's unusual Moon.

The purpose of this paper is to investigate the largest sizes of planetesimals required to explain spin properties of planets, in the context of the impact-trigger hypothesis of lunar origin; in particular, we investigate the effects of stochastic variations. This is done by assuming the existence of various families of large impactors, each family with selected source region in the solar system, and hence approach velocity (at large planetocentric distance); we then evaluate the range of obliquities and rotation periods each can produce. These results are then compared with relevant planets. (Data on Pluto are poor; Mercury and Venus are of little help since they are tidally affected, though Venus perhaps lost much of its initial angular momentum due to an early large impact.)

2. The Impact Model

A computer model was developed to simulate a gravitational encounter in a two-body system, planet and impactor. The planet rotates in a specified initial period. A run of 500 impacts was used to determine the most likely value and range of resulting angular momentum of the planet, following collision with a planetesimal of m_1/m, considerably less than unity. This follows Safronov's (1972) terminology, where

$$m_1/m = mass_{largest\ impactor}/mass_{target\ planet} \qquad (1)$$

The total angular momentum is added to the resulting body of $m + m_1$, with angular momentum coefficient $k = (angular\ momentum)/mR^2\omega$ equal to that observed in the present-day planet. This choice of k assumes core formation before impact; R = planet radius; ω = angular velocity. The model allows planetesimals to approach from random directions at specified speeds, and computes the final angular momentum after collision, giving the obliquity and period. In each run of 500 impacts, the obliquities were divided into 180 1°-wide bins, the periods into 200 bins, covering whatever range of periods was generated in that run. "Most frequent value" is defined as the value of obliquity or period in the most populated bin. Our focus here is only partly on the most frequent value itself (which varies somewhat by randomness from one run to another), but also on the scatter around the most frequent value, within one 500-impact run, due to random variations in impact parameter.

The program allows us to specify parameters of the target planet, such as mass and radius. The program also allows us to specify the planetesimal properties, including approach velocity and mass ratio relative to the target planet. Speed must be sufficiently great to satisfy the two-body approximation (i.e., solar perturbations are assumed insignificant).

The planet's initial axis is assumed to be perpendicular to the fundamental plane of the solar system. The planet is visualized as surrounded by an imaginary sphere of large radius through which the impactor passes at a randomly selected point, p. Assuming that the planetesimals have been stirred by near encounters, we allowed random distribution of p's on the celestial sphere and and allowed random orientations of the velocity vector of the planetesimal, specifying only the magnitude of the velocity. A planetesimal crossing a target planet's sphere at a random point and with a randomly oriented velocity vector will impact the planet, provided it lies within a circular capture cross section perpendicular to the line of motion, having radius

$$S = R\sqrt{1 + 8\pi GR^2\rho/3v^2} \qquad (2)$$

where R = planet radius, ρ = planet density, and v = approach velocity at infinity.

For each impacting planetesimal that passes through circle S, its angular momentum is added to the planet's angular momentum vector, thus determining the new obliquity and rotation period of the target planet. (For simplicity we have assumed that ejecta mass and angular momentum lost to the system is negligible. Future studies of giant impact mechanics may clarify this.)

3. Two Solar System End-Member Models

Safronov (1969) and some others have considered each planet, one at a time, and calculated the most likely size of impactor to have created the observed obliquity, assuming a specified approach velocity at large distance. We take the inverse of the same problem: We consider a fixed population of planetesimals and consider the range of results on the ensemble of planets. We use this approach to focus on the overall planetesimal population, looking for dynamically plausible populations that could produce observed planetary properties.

Two types of calculations were made, corresponding to two end-members in a range of possible models of solar system evolution. Model A, the local planetesimal model, assumes that each planet was accompanied by a population of planetesimals formed *near its own orbit;* the largest of these grew to certain characteristic values of m_1/m. Each planet would have been accompanied by a second largest body, third largest body, etc. Sometimes the largest bodies were ejected from the local zone (hitting a different planet), so that the largest impactor was not necessarily the largest planetesimal of the local zone. This produced some stochastic fluctuations in m_1/m. The impactors in model A were derived from nearby circular orbits, but are assumed to have been scattered by near-encounters before impact; thus they are assumed to approach with moderate velocities. To represent model A, therefore, we calculated distributions of obliquities and periods resulting from impacts at a specified approach velocity and plotted these as a function of m_1/m to see what m_1/m best corresponds to the observed solar system.

Model B, the single-source model, assumes that the largest available planetesimals, hence the more effective ones, were those that grew among the giant planets and were scattered out of the outer solar system (henceforth OSS) by near-encounters with the giants. In both models, appropriate approach velocities were drawn from Hartmann (1977a). For example, in a model B example (to be discussed below; Fig. 4), approach velocities for the OSS-scattered bodies approaching the various planets, from Mercury through Neptune, were assumed to be 60, 40, 30, 25, 17, 12, 8, and 7 km/s. In the case of Earth, approach at 30 km/s was deemed to represent model B. A comment on the minimum approach velocities used here: We have not applied our model to approaches at less than 5 km/s, because solar perturbations lead to breakdown of our two-body assumption. However, as shown by Safronov (1969) and others, a typical Earth-impactor has had several near-misses, so that typical impactor approach velocities are pumped up to roughly one-half Earth's escape velocity, even for locally originating planetesimals. Objects scattered from other planet's zones would have considerably faster approach velocities. Actual approaches to Earth must have been at no less than ~5 km/s for model A, and higher for model B, where impactors are scattered from further away by other planets.

Figure 1 shows a step in a representative calculation: histograms of the 500 obliquities and periods resulting when a body of $m_1/m = 0.025$ approaches a planet at 20 km/s (during 500 test cases). The most frequent value is identified, along with dashed lines giving a measure of scatter: the range that includes two-thirds of the values. The assumed initial period and obliquity are 10 hours and 0°.

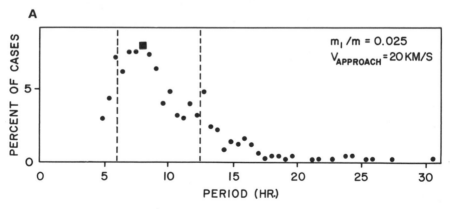

Fig. 1. Histograms showing results of 500 impacts (one "run") onto Earth by a planetesimal with 0.025 M_{Earth} approaching at 20 km/s. (a) Obliquities; (b) periods. The most frequent value is boxed. Dashed lines show a range of scatter defined to include two-thirds of all values (i.e., two-thirds of the values less than the most frequent value, and two-thirds of those greater).

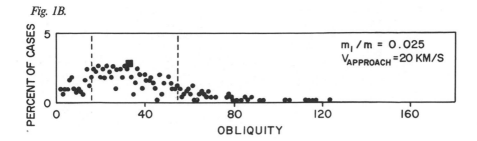

In Fig. 2, we show four sets of obliquity and period data, giving the most frequent values for impacts at 5, 10, 20, and 30 km/s. The first two values are relevant to model A; the second pair are more relevant to model B. These are plotted as a function of m_1/m. Target parameters were those of an Earth accreted from much smaller bodies, with 0° obliquity and a 10-hour initial period. These numbers are consistent with proposals in the literature. (Preimpact periods appropriate to accretion processes are discussed in more detail in Section 5.) The figure shows that the most likely obliquity increases with impactor mass, and the most likely period decreases.

We now ask what impactors would be most likely to change the assumed preimpact obliquity and period, about 0° and about 10 hours, to the postimpact values characteristic of the primordial Earth-Moon system.

The postimpact obliquity of Earth at the moment when the lunar material was close to, or separating from, Earth was less than the present 23.5°. Burns (1982), quoting Goldreich (1966), shows that when the Moon was close to Earth, Earth's

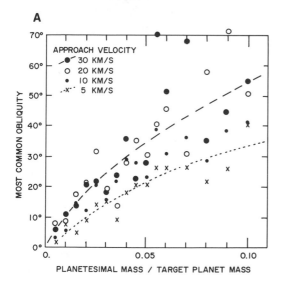

Fig. 2.   Dependence of impact-induced obliquity and period on mass of impactor. Target planet has physical parameters of Earth, with initial obliquity 0° and initial rotation period 10 hours. Each plotted point is the most frequent value in a run of 500 impacts at specified impactor mass. Results apply approximately to any planet, but assume Earth's specific density and coefficient of angular momentum. Dotted line shows smoothed fit for approaches at 5 km/s; dashed line, for approaches at 30 km/s.

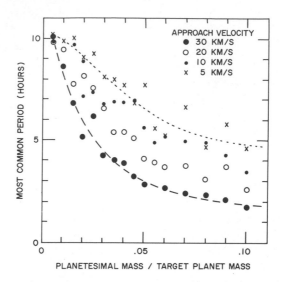

Fig. 2B.

PLANETESIMAL MASS / TARGET PLANET MASS

obliquity value approached the range of 10°–15°. Similarly, the postimpact period of primordial Earth, corresponding to the angular momentum of the Earth/Moon system, was 4.1 hours.

What objects produce these results? Figure 2 shows that in the velocity range of 5–30 km/s, impactors of around 0.01–0.04 $M_\oplus$ would be the most likely to produce this obliquity. Somewhat larger impactors, around 0.03–0.12 $M_\oplus$, are the size that would be most likely to produce a period of about 4.1 hours, depending on approach speed and initial period. This important result, that planetesimals of about 0.01–0.12 $M_\oplus$ can well explain Earth's spin properties, is consistent with the size of impactor (~0.1 $M_\oplus \simeq M_{Mars}$) suggested in several versions of the impact-trigger theory of lunar origin.

For our purposes, it is important to go beyond Figs. 1 and 2 to assess what planetesimal size range has a significant probability of producing the proposed values of obliquity and period. In particular, the above definitions make it easy to consider which planetesimals have a two-thirds probability of producing the proposed result. Figure 3 presents information similar to that in Fig. 2, but includes information on the range of stochastic variation in a model-A-like scenario with approach velocities of 5 km/s. The filled circles show the most frequent values; the central solid line is a smoothed fit to them. The shaded band contains two-thirds of the results. The outer envelope (extent of vertical bars) shows the maximum and minimum values achieved in each 500-impact run (see caption).

"Observed" obliquities and periods are shown at the right margins. These include the values for primordial Earth mentioned earlier. We see that the desired 10°–15° primordial obliquity would fall in the two-thirds probability range for impactors

A

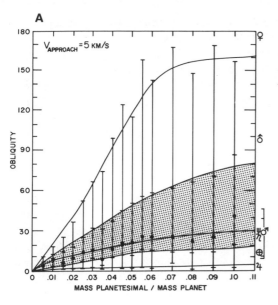

Fig. 3.   Indications of stochastic variation in obliquity and period occurring when impactors approach Earth at 5 km/s. Earth has assumed initial obliquity of 0° and initial rotation period of 10 hours. Black dots show the most frequent values occurring during individual 500-impact runs. Shaded zones include two-thirds of the values, centered on the most frequent value. Vertical bars show the range from minimum to maximum (99.8% of the values; outer envelope curves represent this probability level after smoothing stochastic variations). Results apply approximately to other planets. (See text for discussion.)

of ~0.01–0.04 $M_⊕$ approaching Earth at 5 km/s. Harris and Ward (1982) review the fact that Mars' obliquity oscillates from 11°–38° presently, and probably from about 9°–46° in the past, prior to Tharsis volcanism. They note that a primordial Martian obliquity (i.e., following the large impact) anywhere in this range could produce the Mars we see. These ranges are plotted for Mars. For the lower end

B

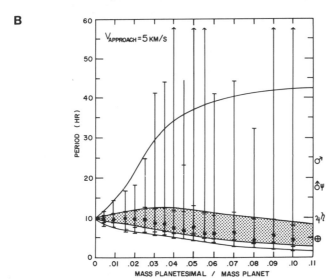

of this range, a projectile with a few percent of a Martian mass would do; for the larger obliquities, mass of 0.03 to >0.12 could have been likely.

In the case of giant planets, approach velocities in model A would be higher than the 5 km/s adopted in Fig. 3; however, in our similar calculation for approach velocity at 20 km/s, the most probable obliquity for $m_1/m = 0.1$ has risen to only 40°, compared with the 30° in Fig. 3. Therefore, we can now go beyond Fig. 2 and suggest that families of model A planetesimals, with largest bodies on the order of a few percent to as much as a few tenths of the masses of the target planets in their zones, would be the most plausible to produce the "observed" obliquities and periods of Earth, Mars, Jupiter, Saturn, Uranus, and Neptune. In evaluating such a statement, it is important to remember that a given planetesimal can produce a wide range of results, depending on its impact parameters. A 20-km/s impactor with mass ~0.12 M_{planet}, for example, has about a one-sixth chance of creating an obliquity as large or larger than Uranus' value of 98°; the same impactor would be most likely to produce obliquities around 42°. A family of such planetesimals might thus have produced one Uranus out of six planets, plausibly consistent with observation.

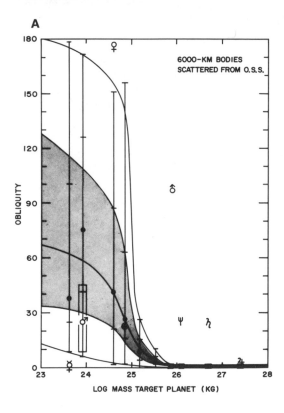

Fig. 4. Results of a solar system model that attempts to explain obliquities and rotation periods by impacts of 6000-km-diameter (m = 1.75 × 10²³ kg; ρ = 1547 kg/ m³) planetesimals scattered out of the outer solar system by the giant planets. Symbols are as in Fig. 3. (See text for discussion.)

Fig. 4B.

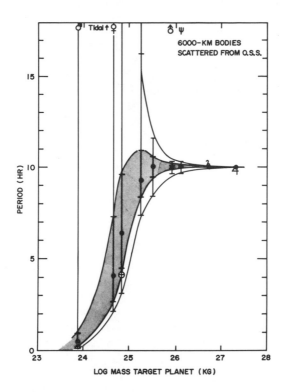

Model B calls for planetesimals drawn from a single population in the OSS, mixed by giant planet encounters, and then scattered into the ISS. The impactors on any one planet are likely to be as large as those on another, independent of planet size, but with stochastic variation due to the actual planetesimals' size distribution. To study this situation, at least crudely, we assumed a fixed planetesimal mass and relatively high (cometary) approach velocities in the ISS.

Figure 4 shows an example of results. A set of rock-and-ice planetesimals with diameters of 6000 km (diameter $= 0.88$ D_{Mars}; $m = 1.75 \times 10^{23}$ kg ≈ 0.03 M_{Earth}) was scattered into the ISS. This size was chosen to be large enough (or perhaps more than large enough) to allow the observed obliquities of Earth, Mars, and Venus, which are plotted in Fig. 4. Figure 4 shows that such bodies are nonetheless too small to account for the OSS obliquities or periods of Saturn, Uranus, and Neptune (denoted by their symbols). Conversely, bodies large enough to account for those cases would wreak havoc during ISS impacts. A pure model B thus fails.

Figure 5 shows a composite model that appears realistic and promising: aspects of model A apply in the ISS, B in the OSS. We assume that in the ISS impactors grew in each planet-forming zone to the size of 0.035 $M_{local\ planet}$ as suggested by

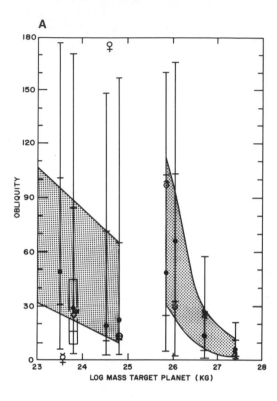

A

model A calculations. In the OSS, the largest impactors were those scattered by Jupiter, with characteristic largest mass of $3(10^{25})$ kg = 0.016 $M_{Jupiter}$ = 0.35 M_{Uranus}. These latter bodies were scattered and mixed in the OSS during creation of the Oort cloud, but were not frequent impactors in the ISS. Therefore, while the ISS is best modeled by local impactors associated with each planet, the OSS is best modeled by a single swarm.

As seen in Fig. 5a, this idealized (and undoubtedly oversimplified) model is remarkably successful in placing the obliquities of Mars, Earth, Uranus, Neptune, Saturn, and Jupiter within the shaded two-thirds probability zone. Only Venus requires a stochastically less probable event to drop its angular momentum to the point where tides take over; a review by Harris and Ward (1982) notes that this primordial obliquity probably was >90°. Mercury's initial state is indeterminate but must be <90°. Figure 5b shows that the model is less successful in dealing with periods, assuming 10–15-hour periods and present masses.

A particularly attractive characteristic of this model is that Uranus has the highest probability of any planet of obtaining an unusual obliquity, because if it draws from a single population of large OSS impactors, it is likely to have the highest m_1/m ratio, since it has the lowest m value of the giants.

Fig. 5B.

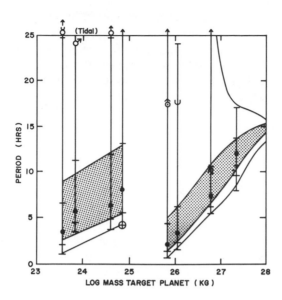

Another attractive feature of this model is that this scenario, constructed in order to match observed obliquities, also follows from models of OSS planetesimal scattering. Weidenschilling (1975) studied the scattering of OSS planetesimals by Jupiter and found that few are scattered into the ISS. The frequency of perihelia decreases rapidly with decreasing solar distance, and is at least two orders of magnitude less for Earth than for Mars. "The flux at Venus and Mercury. . . was negligible" (Weidenschilling, 1975). Since Uranus and Neptune planetesimals generally need to work their way inward past Saturn and Jupiter in order to ever approach the ISS, the ISS is also protected from them. In summary, the ejection of OSS planetesimals (i.e., comets) into hyperbolic or Oort cloud orbits generally mixes the OSS bodies and ejects the vast majority of the OSS planetesimals without their having much probability of hitting a terrestrial planet. None of this is inconsistent with ISS cratering by a small fraction of the OSS debris, as postulated by Kaula and Bigeleisen (1975), Wetherill (1975), and others. Although a small fraction of the total OSS swarm may be perturbed into the ISS, Weidenschilling's model shows that the probability of the largest few individual bodies in the swarm interacting with a terrestrial planet before hitting a giant or being ejected is negligible. Therefore the scenario in Fig. 4, with giants hit by a mixed OSS population, but with the ISS planets dominated by their own planetesimals, is plausible.

A related phenomenon is that Jupiter's scattering cross section is so large compared to its radius (see equation (2)) that it has a high chance of ejecting any specific approacher. For this reason, Jupiter is less likely than other planets to be hit by the largest planetesimal from its own feeding zone. Thus even if every feeding zone produced a largest planetesimal with mass, say 0.035 M_{planet}, just as in the ISS, Jupiter would be more likely than any other giant to have its largest impact involve not its biggest planetesimal, but one of its more abundant, intermediate-size planetesimals (0.016 $M_{Jupiter}$ in Fig. 5). At the same time, the scattering of remaining "intermediate-size" Jupiter planetesimals throws them onto orbits crossing other giants' orbits, so that a smaller giant like Uranus may get a hit by an object bigger than its own zone's largest planetesimal (0.35 M_{Uranus} in Fig. 5). This effect may explain why Jupiter was apparently hit by a body with smaller m_1/m than other planets.

4. Comparison with Other Results

Our result, that planets may have been hit by bodies with a few percent mass ratios, is supported by other studies, as shown in Table 1. Only a few years ago such models appeared extreme, but the trend in the literature has been in this direction. Indeed, the most recent calculations of Wetherill (1985) propose orbital migrations of planet-sized bodies in the ISS and sporadic impacts by bodies with 30% planetary mass or more. In Table 1, column 1 lists the planetesimal mass ratios m_1/m derived

TABLE 1. Mass Estimates: Largest Planetesimals Hitting Each Planet.*

	This Paper	Safronov, 1972		Harris and Ward, 1982		Hartmann and Davis, 1975	Observation
	(1) (Fig. 5 Model solar system)	(2) (1 large body)	(3) (Power Law Mass Distribution)	(4) (Obliquity)	(5) (e's, i's)	(6) Accretion Theory	(7) Asteroids/Ceres
Mercury	–	–	–	–			
Venus	–	–	–	–			
Earth	1–10	1	0.3	0.3	4	0.05–43	
Mars	3–5	1	0.2	0.4	70		
Asteroids							16,15,9,4,4,3
Jupiter	~2	0.5	0.03	0.06	0.04		
Saturn	~5	6	4	9	0.08		
Uranus	~35	8	7	10	0.07		
Neptune	~29	2	0.7	1	0.003		

*m_1/m expressed as percent of target planet mass.
Columns:
(1) Terrestrial planets from Figs. 1–4; giant planets from model in Fig. 4 with $2(10^{25})$ kg impactors.
(2) Entire obliquity attributed to one large impact.
(3) Obliquity results from power law distribution of impactors; cumulative no. $\alpha\ m^{-2/3}$.
(4) Obliquities derived from impacts of growing planetesimals with upper limit mass cutoff; power law mass distribution.
(5) Orbital eccentricities and inclinations attributed to impacts as in (4).
(6) Theoretical model of accretion of 2nd, 3rd, 4th...largest bodies during accretion of planet.
(7) Masses of 2nd, 3rd, 4th. . . largest asteroids, relative to Ceres, assuming same density.

from our "composite" solar system model in Fig. 5. They lie in the range of 1–35%. Column 2 gives a result by Safronov (1969), similar to our model in that it attributes obliquities to a single giant impact on each planet; his solar system model calls for mass ratios of 0.05–8%. Our result for Uranus is higher because of our suggestion of a single, mixed OSS planetesimal swarm, in which Uranus, the smallest giant, receives proportionally the largest hit. Columns 3–5 are models calling for a power law distribution of large planetesimals; the largest impact is not so distinct from the next largest. Here, mass ratios range from 0.03–10%.

Harris and Ward (1982) include an example of a model that attributes orbital eccentricities and inclinations to impacts (column 5). Results call for similar mass ratios, 0.06–4%, but with a whopping 70% for Mars. This is provocative in the context of Weidenschilling's (1975) model, which calls for Mars to lie in the inner fringe of the flux of planetesimals scattered from the OSS. If this were true, Mars would be likely to receive a much larger hit relative to its size than other planets, just as required by these results. Thus Mars' orbit may have been affected by an unusually large impactor that produced moderate obliquity and altered the orbit. Could such an impact have produced the notable hemispheric geologic asymmetry on Mars? Wilhelms and Squyres (1984) have suggested such an event, based on geological structure.

Hartmann (1977b) reported additional unpublished calculations by D. R. Davis and himself, using power law mass distributions and accounting for obliquities with mass ratios of 1–3%. Orbital eccentricities and inclinations were modeled by impactors with ratios typically in the range of 1–20%.

Other kinds of evidence are available. Hartmann and Davis (1975) modeled accretionary growth of second largest, third largest, etc., planetesimals under a variety of assumed density, velocity, and size-distribution regimes. As noted in column 6, four models that produced an Earth-sized planet produced second largest bodies with m_1/m of 0.05%, 2%, 4%, and 43%. Similarly, Wetherill's recent work (1985; personal communication, 1984) follows simultaneous accretion in systems of 500 bodies, finding that bodies of lunar to 3× Mars size (mass ratios 1–32%) would be likely to strike Earth well within its first 100 m.y.

Hartmann (1977b) noted an additional line of evidence: From the direct evidence of impact basins on the planets and satellites, we know that a remnant population of planetesimals with roughly power-law-size distribution extending up to 150 km diameter was still hitting terrestrial planets as long as 500 m.y. after their formation. Since the flux at that time (determined from crater counts) was only some 10^{-6} of the flux required to form the planets (see Hartmann, 1986), we can be confident of a vastly greater population of objects at the end of planet formation; it probably extended the power law distribution to larger sizes.

Finally, if the asteroid belt can be taken to represent a fossil tableau of aborted accretion, we have direct observational evidence of the high mass ratios of secondary bodies sharing a "feeding zone" with a major body. If Ceres could be forced to sweep up the rest of its neighbors, the six largest impactors would have mass ratios of 3–16%. Even if Ceres swept up all the rest of the asteroid belt before being hit by a Vesta or Pallas that swept up no mass in the interim, the Vesta and Pallas mass ratios would be of the order of 10% of that of the new "planet." The asteroids thus not only give evidence of impactors with large mass ratios, but also give evidence quantitatively consistent with the ratios deduced from spin and orbit properties.

5. Comments and Summary

The goal of this paper was to use observed rotational characteristics to examine the plausible sizes of the largest planetesimals available in the primordial solar system. In particular, we wished to see if the sizes of planetesimals required to produce these characteristics are consistent with the impact-trigger hypothesis of lunar origin, in which a relatively large impactor hits Earth to spawn the Moon. Can we claim evidence that such bodies were available in the primordial solar system? We conclude the answer is yes. Our results and those of other authors suggest that the largest impactors' masses often lay in the range ~ 0.3–20% M_{planet}.

Let us comment on the less resolved part of our argument: the poor modeling of rotation periods as a result of large impacts. Perhaps not too much can be made of this; the periods of Uranus and Neptune are uncertain, those of Mercury and Venus are irrelevant. Conceptually, we assumed a preimpact spin rate to represent the rate produced by the accretion of many small planetesimals. Generally, such accretion produces very little angular momentum (Weidenschilling, 1984). Giuli (1968) and Harris (1977) sought special orbital eccentricities to produce prograde spins of about 10–15 hours during accretion of innumerable small bodies. Our adoption of preimpact periods of 15 hours in Fig. 5b produces a fair fit to several observed planets, but the problem needs more work.

First, other preimpact spins can be considered, since we do not adequately understand how accretion produces average prograde spins, and the Giuli-Harris 10–15-hour periods are in the nature of limiting cases. We made preliminary investigations of the periods by starting with much longer periods. These are not promising; Uranus and Neptune, for example, get spun up to higher rotation speeds than Saturn and Jupiter. Because of Jupiter's high mass, it tends to spin up the least and therefore favors starting with initial periods of not much longer than 10 hours.

Second, the giant planets may have attained spins in quite a different manner; the impact models may be less relevant to them. They have rock, metal, and ice cores estimated at 13–19 Earth masses, and perhaps smaller rock-metal cores of only a few Earth masses (Hubbard, 1984, pp. 264–288). The giants' rotations may have been fixed during the cores' capture of massive extended atmospheres from the solar nebula, instead of by accretionary processes (Weidenschilling, personal communication, 1985). In this case, the present impact models are not relevant to giants. Nonetheless, a late impact could have altered obliquity and period after gas accretion. Further work is needed to understand planetary rotation evolution during accretion, especially during gas capture onto giant cores.

Third, in reality, there was not a clear distinction between cumulative small impacts and a giant impact. Rather, a size distribution of planetesimals hit. The final spin was probably dominated by the *largest* impact, but only if it occurred late enough in accretion not to have had its effects erased by subsequent planetesimal or gas accretion. If the largest planetesimal available to a planet was used up by an early impact, that planet's spin may have evolved back to a less distinctive state, while if the planetesimal hit very late (perhaps growing to even larger size in the meantime) it may have produced a distinctive spin state that could not be later erased. Indeed, the great effectiveness of large impactors in spinning targets up to periods of only a few hours (Fig. 2b) may be the solution to the problem that small-particle accretion produces little net spin (Weidenschilling, 1984) and may explain the curious fact (Hartmann, 1983, p. 205) that the planets and asteroids seem to be generally spun up to periods only a factor of ~3–5 times longer than rotational instability.

While rotations remain problematic, the observed obliquities are highly consistent with planets being hit by impactors with several percent of their mass. They are well matched by the composite model in Fig. 5. In this view, the spin states of Venus, Earth, and Uranus were particularly affected, but lie within a plausible range of stochastic variation.

Our results support a variety of earlier studies, all suggesting that impactors of some ~0.003 and 0.2 M_{planet} affected the various planets. Such a hypothesis is extremely attractive in providing not only the right order of rotational and orbital parameters, but also in providing the right degree of stochasticity to explain variations among planets. A hypothesis that insists on planetesimal debris (after planet growth to present masses) involving only much smaller bodies fails not only to provide the right obliquities and periods, but also the observed degree of variation among planets. In the context of this book, our study finds it highly plausible that Earth was struck by a body of between several M_{Moon} and M_{Mars} (~0.03–0.12 $M_\oplus$) with enough energy and angular momentum to dislodge mantle material and form the present Earth-Moon system, consistent with the impact-trigger model of lunar origin.

Acknowledgments. Special thanks to D. R. Davis for early calculations (referenced in Hartmann, 1977) that contributed to the development of this paper, and to S. J. Weidenschilling for helpful discussions during its preparation. Thanks also to Floyd Herbert, R. G. Greenberg, and C. R. Chapman for helpful suggestions. Paula Watson-McBride, Alice Olson, and Alix Ott assisted in preparing the manuscript for publication. This work was supported by the NASA Planetary Geology and Geophysics programs, NASA contract nos. NASW-3718 and NASW-3516. This is PSI Contribution No. 204. PSI is a division of Science Applications International Corporation.

6. References

Burns J. (1982) The past solar system. In *Formation of Planetary Systems* (A. Brahic, ed.), pp. 458–468. Cepadues Editions, Toulouse.

Cameron A. G. W. and Ward W. (1976) The origin of the moon (abstract). In *Lunar Science VII*, pp. 120–122. The Lunar Science Institute, Houston.

Giuli R. (1968) On the rotation of the Earth produced by gravitational acceleration of particles. *Icarus, 8*, 301–323.

Goldreich P. (1966) The history of the lunar orbit. *Rev. Geophys., 2*, 661–685.

Harris A. (1977) An analytical theory of planetary rotation rates. *Icarus, 31*, 168–174.

Harris A. and Ward W. (1982) Dynamical constraints on the formation and evolution of planetary bodies. *Ann. Rev. Earth Planet. Sci., 10*, 61–108.

Hartmann W. K. (1977a) Relative crater production rates on planets. *Icarus, 31*, 260–276.

Hartmann W. K. (1977b) Large planetesimals in the early solar system. In *Comets, Asteroids and Meteorites* (A. Delsemme, ed.), pp. 277–281. Univ. of Toledo, Ohio.

Hartmann W. K. (1983) *Moons and Planets*. Wadsworth, Belmont, CA. 509 pp.

Hartmann W. K. (1984) Stochastic ≠ ad hoc (abstract). In *Papers Presented to the Conference on the Origin of the Moon*, p. 39. Lunar and Planetary Institute, Houston.

Hartmann W. K. (1986) Moon origin: The impact-trigger hypothesis, this volume.

Hartmann W. K. and Davis D. R. (1975) Satellite-sized planetesimals and lunar origin. *Icarus, 24*, 504–515.

Hubbard W. B. (1984) *Planetary Interiors*. Van Nostrand Reinhold, New York.

Kaula W. and Bigeleisen P. (1975) Early scattering by Jupiter and its collision effects in the terrestrial zone. *Icarus, 25*, 18–33.

Safronov V. S. (1966) Sizes of the largest bodies falling onto planets during their formation. *Soviet Astron. AJ, 9*, 987–991.

Safronov V. S. (1969) Evolution of the protoplanetary cloud and formation of the Earth and the planets. *NASA TT F-677*, translated 1972 by Israel Program for Scientific Translation, Jerusalem. 206 pp.

Shmidt O. Yu. (1958) *A Theory of the Origin of the Earth*. Foreign Languages Publishing House, Moscow. 139 pp.

Weidenschilling S. J. (1975) Mass loss from the region of Mars and the asteroid belt. *Icarus, 26*, 361–366.

Weidenschilling S. J. (1984) The lunar angular momentum problem (abstract). In *Papers Presented to the Conference on the Origin of the Moon*, p. 55. Lunar and Planetary Institute, Houston.

Wetherill G. W. (1975) Late heavy bombardment of the moon and terrestrial planets. *Proc. Lunar Sci. Conf. 6th*, pp. 1539–1562.

Wetherill G. W. (1985) Occurrence of giant impacts during the growth of the terrestrial planets. *Science, 228*, 877–879.

Wilhelms D. E. and Squyres S. W. (1984) Martian hemispheric dichotomy may be due to a giant impact. *Nature, 309*, 138–140.

Mechanical Models of Close Approaches and Collisions of Large Protoplanets

W. M. KAULA* AND A. E. BEACHEY

Department of Earth and Space Sciences, University of California, Los Angeles, CA 90024

Close approaches and collisions of protoplanets are modeled by sets of spherical bodies—typically about 90 per protoplanet—that start in closest packing within fluid equilibrium figures beyond the Roche limit, and whose orbits are numerically integrated. Inelastic collisions are modeled as imposing corotation of bodies identified as clustering, conserving linear and angular momentum. Purely elastic collisions are modeled as having reversals of momentum upon clustering twice those for inelastic. Provision is made for intermediate degrees of elasticity. The model proved invalid for the elastic cases: there always develop instabilities that are properties of the set of spherical bodies, not of the continuum protoplanet. Inelastic close approaches in hyperbolic orbits never resulted in tidal disruption for all cases tested. Modelings of inelastic collisions were for two pairs of masses: (1) 0.1 $M_\oplus$ and 0.9 $M_\oplus$; (2) both 0.5 $M_\oplus$, and for approach velocities ranging from 2.5 to 6.6 km/sec. Placing material in orbit around the main mass appears to require a combination of the unequal mass ratio with the approach velocities above 6.0 km/sec, corresponding to perturbation from the asteroid belt or the vicinity of Mercury.

Introduction

The formation of the Moon is necessarily dependent upon the formation of the Earth. Furthermore, the formation of the Moon is probably associated with the later stages of the Earth's growth; i.e., protolunar matter was placed in orbit around the proto-Earth when the latter was probably more than 50% as massive as the final Earth. It is difficult to maintain protolunar matter as satellites around the proto-Earth while two orders-of-magnitude more of matter are falling into the planet:

*Now at the National Geodetic Survey, NOS/NOAA, Rockville, MD 20852

The satellites would be much disrupted by the infalling matter, as well as being gravitationally drawn by the growing proto-Earth (although the preexistence of a disk favors formation by capture of infall; see Harris and Kaula, 1975).

Association of the Moon's formation with the later stages of the Earth's growth means that the solar nebula had already evolved appreciably by the time the Moon formed. If a sizable nucleus for the Earth had already formed, then most of the nonvolatile matter in the inner solar system would also be collected in sizable bodies, all else being equal. The main question then becomes the effective definition of "sizable": "planetesimal" ($\sim 3 \times 10^{20}$ kg), "lunar" ($\sim 10^{22}$ kg), or "protoplanetary" ($\gtrsim 3 \times 10^{23}$ kg)? The simplest assumption is truly "protoplanetary": If an Earth nucleus of more than 3×10^{24} kg has formed, then one would expect most remaining volatiles to be in other bodies more than 3×10^{23} kg in mass. To hypothesize otherwise requires special explanation: There must be a mechanism to suppress growth of bodies other than the proto-Earth. Such a mechanism is most obviously associated with the proto-Earth itself, and two have been proposed: (1) runaway growth of the proto-Earth, because of the advantage of its gravitational cross-section for capture (Safronov, 1972); (2) tidal disruption of sizable bodies, because the cross-section for close approach sufficient for disruption is considerably greater than the cross-section for collision (Öpik, 1972).

Hypothesis (1), runaway growth, has been tested in the computer experiments of Wetherill (1980, 1986), who finds that it is true only for rather narrow zones, resulting in a population of sizable bodies much more numerous than the present planets. Hence some other mechanism, such as scanning secular resonances (Ward, 1981), must be invoked to bring these bodies together.

Hypothesis (2), tidal disruption, has not been adequately tested. Mitler (1975) and Hayashi et al. (1985) treated the passing body as completely inviscid. A more realistic modeling is thus desirable. This modeling should take into account the dissipation that must occur in any asymmetric distortion of a planetary body, as has also been done by Mizuno and Boss (1985).

The hypothesis of creating the Moon by a major impact in the proto-Earth has raised cosmochemical objections that trace element abundances in the Moon are difficult to reconcile with origin from the Earth's mantle (Drake, 1986). If, however, the lunar matter came predominantly from the impacting body, then it could plausibly have compositional differences dependent on differing histories of the two colliding bodies, rather than on circumstances of the collision itself. Such a concentration of impactor matter in the Moon is plausible for purely mechanical reasons: The matter with highest angular momentum density prior to the collision should tend to be thrown into orbit. This conjecture is not fully testable by a model that does not include the thermal aspects of collisions—the vaporization and consequent expanding gas cloud—but may be constrained in some respects by a purely mechanical model.

Model Characteristics

The model chosen is to represent each of the two protoplanets in an encounter by a set of spherical bodies. The continuity of mass condition is approximated by not allowing any pair of bodies to approach within a prescribed distance of each other. The main reason for choosing this representation was economy of computer time. The representation by relatively few bodies implies that the critical wavelength of tidal disruption is rather long, as originally inferred by Roche (Jeans, 1928). But taking thermodynamic factors into account leads to a different conclusion (Mizuno and Boss, 1985): There can be mass loss by shredding an atmosphere.

Table 1 gives the properties of the models calculated. Both planets are assumed to have rotation axes normal to the plane of their motion. Runs are normally started at a small distance beyond the Roche limit, typically 2.9 planet radii. The starting configuration of both planets is assumed to be those appropriate for tidally distorted fluids, on the argument that tidally induced strains in an elastic body

$$\epsilon = hGMR/r^3g \tag{1}$$

exceed the yield strain of 10^{-3}. In (1), h is the displacement Love number (2.5 for a homogenous fluid, but 0.1 to 0.6 for the elastic case applicable here), G is the gravitational constant, M is mass of the disturbing planet, R and g are radius and gravity of the tidally stressed planet, and r is the distance between planet centers.

These starting configurations were determined by calculating which of a set of closest packed spheres would have centers within the tidally distorted ellipsoid. The closest packed spheres are referred to rectilinear axes u_1, along the line between protoplanet centers; u_2, in the planet of motion; and u_3, normal thereto. Planes of bodies were arranged parallel to the 1–2 plane. In the 3-direction the spacing

TABLE 1. Collision Cases Tested and Outcomes

Protoplanet Masses		Approach	Approach	Pericenter	Protoplanet 1		
1-Smaller $M_\oplus$	2-Larger $M_\oplus$	Velocity km/sec	Offset $R_\oplus$	Overlap $R_\oplus$	Merge	Orbit	Escape
0.1	0.9	2.95	4.79	0.00	0	0	91
0.1	0.9	2.95	3.95	0.43	91	0	0
0.1	0.9	6.6	2.22	0.23	62	29	0
0.5	0.5	2.52	5.85	0.00	0	0	88
0.5	0.5	2.52	4.94	0.43	88	0	0

$M_\oplus = 5.97 \times 10^{24}$ kg; $R_\oplus = 6371$ km.
The outcomes are the eventual locations of the spherical body elements.

between these planes is $\sqrt{2/3}$ D, where D is the sphere diameter. In the 2-direction, the spacing between lines in the same plane is $\sqrt{3}$ D/2, and adjacent lines in successive planes are offset by D/2 $\sqrt{3}$. In the 1-direction, the spacing between bodies in the same line is D, and adjacent bodies in successive lines are offset D/2. Thus the center-to-center distances of a body's nearest neighbors in adjacent lines of the same plane is $[(D/2)^2 + (\sqrt{3} D/2)^2]^{1/2} = D$; and in adjacent planes, $[(D/2)^2 + (D/2\sqrt{3})^2 + (\sqrt{2/3} D)^2] = D$. A parallel-piped array of bodies sufficient to encompass the protoplanet is set up, and bodies are retained that fall within the tidal envelope:

$$(u_1/a)^2 + (u_2/b)^2 + (u_3/b)^2 < 1,$$

$$a = (1 + \epsilon) R/E,$$

$$b = (1 - \epsilon/2) R/E,$$

$$E = (1 - 3\epsilon^2/2 + \epsilon^3/4)^{1/3}$$

where ϵ is defined by (1) with h $= 2.5$ (Jeans, 1928; Munk and MacDonald, 1960).

Normally D/R is set at 0.4. For a Mars: Earth-like collision, $M_1 = 0.1$ $M_\oplus$, $M_2 = 0.9$ $M_\oplus$, and starting $r_o = 3.0$ R_2, this procedure leads to the number of bodies per protoplanet being $n_1 = 91$ and $n_2 = 88$.

In addition to M_1, M_2, R_1, R_2, and r_o, important input parameters are v_a, the approach velocity; d_a, the approach offset; and ω_1, ω_2: the rotation rates of the protoplanets prior to approach. The radii R_i throughout are taken to give mean densities the same as the present Earth, 5.52 gm cm^{-3}. The rotation rates ω_1, ω_2 are set as zero, to clarify how much spin-up came from the collisions.

Initially, approach velocities corresponding to perturbations from no further than Mars (2.95 km/sec) or Venus (2.52 km/sec) were considered because of the argument based on the oxygen isotopes that the lunar matter must come from close to the Earth's. However, these cases did not yield orbiting matter, so consideration was extended to higher velocities, equivalent to coming from the inner asteroid belt or almost as far as Mercury: 6.6 km/sec.

Setting the center-to-center distance of closest approach as P $= R_1 + R_2$ and using conservation of energy and angular momentum to obtain to offset distance d_a

$$d_a = P [1 + G(M_1 + M_2)/P/v_a^2]^{1/2}$$

gives as the maximum offset distance d_a for category (1) 4.791 $R_\oplus$ = 30,524 km and for category (2) 5.850 $R_\oplus$ = 37,270 km. The angular momentum H is given by

$$H = M_1 M_2 v_a d_a / (M_1 + M_2)$$

An idea that was explored is the collision number, defined as the ratio of the angular momentum of approach to that required for rotation instability of a homogenous planet of mass equal to the sum of the two protoplanets and the same density. The critical angular momentum corresponding to exchange of instabilities between Maclaurin and Jacobi ellipsoids is (Jeans, 1928, p. 210)

$$H_c = 0.30375 \, (G \, M^3 R)^{1/2}$$

which is 0.92×10^{29} kg km^2/sec for a homogenous body of the Earth's mass M and radius R. Taking $M = M_1 + M_2$ and $R = (R_1^3 + R_2^3)^{1/3}$, we thus have, for example, a maximum collision number of 0.5283 for the 0.1 $M_\oplus$:0.9 $M_\oplus$, 2.95 km/sec (i.e., from Mars) case, compared to 0.6114 for the 6.6 km/sec (i.e., from 2.91 AU) case; the higher v_a is largely offset by the smaller d_a. For 0.5 $M_\oplus$:0.5 $M_\oplus$, 2.52 km/sec (i.e., from Venus) 1.5306 is obtained, but the computer model does not give any breakup and scattering into orbits upon collision. Clearly other factors than angular momentum are involved: in particular, greater concentration of impact effects in a small portion of the material, as occurs at the higher mass ratio.

Computational Procedure

Once set up, the computation proceeds by numerical integration. For body i of protoplanet j, the acceleration is:

$$a_{ij} = G \sum_{l=1}^{a} m_l \sum_{k=1}^{n_l} \nabla \, (1/r_{ijkl}), \quad ij \neq kl.$$

where

$$r_{ijkl}^{\;2} = (x_{ij} - x_{kl}) \cdot (x_{ij} - x_{kl})$$

and m_l is the mass of a body in protoplanet l. The time step Δt for integration is controlled by the minimum r_{ijkl}. If this minimum distance drops below $\alpha(D_j + D_l)/2$, where α is typically 0.6 (operative usually only at the instant of major collision), the time step is shortened; if α gets above 1.0, the step is lengthened.

The main difference from typical N-body integrations is the handling of collisions implict in $r_{ijkl} < (D_j + D_l)/2$. Given the position, velocity, and acceleration of body i at time t as $\mathbf{x}_{oi}$, $\mathbf{v}_{oi}$, and $\mathbf{a}_{oi}$, define estimates at time $t + \Delta t$:

$$\mathbf{x}_{1i} = \mathbf{x}_{oi} + \mathbf{v}_{oi}\Delta t + \mathbf{a}_{oi}(\Delta t)^2/2$$

$$\mathbf{v}_{1i} = \mathbf{v}_{oi} + \mathbf{a}_{oi} \cdot \Delta t.$$

If $|\mathbf{x}_{1i} - \mathbf{x}_{1j}| < r_m$, a prescribed minimum distance, a collision occurs. In reality, there is a rapid change of acceleration $\mathbf{a}$ and velocity $\mathbf{v}$ within the interval Δt. In the computation, it is simplest to calculate adjusted velocity $\mathbf{v}_{Ai}$ and acceleration $\mathbf{a}_{Ai}$ at time t giving conservation of linear and angular momentum of the entire cluster of bodies closer than r_m to their nearest neighbors, while expanding to give a minimum interbody distance r_m (a crude form of the continuity condition). Define

$$\mathbf{v}_{Ei} = \mathbf{v}_{Ai}(t) + \mathbf{a}_{Ai}(t) \, \Delta t \tag{2}$$

$$\mathbf{x}_{Ei} = \mathbf{x}_{oi} + \mathbf{v}_{Ai} \, \Delta t + \mathbf{a}_{Ai} \, (\Delta t)^2/2 \tag{3}$$

where $\mathbf{v}_{Ei}$ and $\mathbf{x}_{Ei}$ satisfy the continuity and momentum conditions for a coherent spinning body. The linear momentum condition gives a velocity $\mathbf{v}_c$ for the centroid of the cluster

$$\mathbf{v}_c = \sum_i m_i \mathbf{v}_{1i} / \sum_i m_i$$

and the centroid location

$$\mathbf{x}_c = \sum_i m_i \mathbf{x}_{1i} / \sum_i m_i$$

To satisfy the continuity condition, the simplest procedure is a linear expansion from the centroid

$$\mathbf{x}_{Ei} = \mathbf{x}_c + r_m \, (\mathbf{x}_{1i} - \mathbf{x}_c) / |\mathbf{x}_{1i} - \mathbf{x}_{1j}|_{min}$$

To obtain the spin of the cluster, the moment-of-inertia tensor is required

$$I_{jk} = \sum_i m_i \left[\delta_{vk} \sum_l (x_{1\,Ei} - x_{1c})^2 - (x_{jei} - x_{jc}) (x_{kei} - x_{kc}) \right]$$

Whence, by conservation of angular momentum

$$\boldsymbol{\omega}_c = \mathbf{I}^{-1} \sum_i m_i \, (\mathbf{x}_{ic} - \mathbf{x}_c) \times (\mathbf{v}_{ic} - \mathbf{v}_c)$$

Thence

$$\mathbf{v}_{Ei} = \mathbf{v}_c + \mathbf{w}_c \times (\mathbf{x}_{Ei} - \mathbf{x}_c)$$

and from equations (2) and (3)

$$\mathbf{v}_{Ai} = 2\,(\mathbf{x}_{Ei} - \mathbf{x}_{oi})/\Delta t - \mathbf{v}_{Ei}$$

$$\mathbf{a}_{Ai} = 2\,(\mathbf{x}_{oi} - \mathbf{x}_{Ei} + \mathbf{v}_{Ei}\Delta t)/(\Delta t)^2$$

An additional calculation carried out was to compute the change in kinetic energy occurring upon collision

$$\Delta K = \tfrac{1}{2}\sum_i m_i \left[v_{Ei}^{\,2} - v_{li}^{\,2} \right] \tag{4}$$

This energy change was distributed among the members of the cluster in proportion to the absolute values of their energy changes, $|m_i(v_{Ei}^2 - v_{li}^2)|$, since a body is shocked as much gaining a given amount of momentum as in losing it. Equation (4) is imprecise in that it neglects the potential energies, but this change should be slight within the one second or so of an integration time step Δt.

The foregoing equations apply to a perfectly inelastic body. For a perfectly elastic body, the velocity change arising from the collision is taken as twice that of the perfectly inelastic

$$\underset{\text{elastic}}{\mathbf{v}_{Ei}} = \mathbf{v}_{li} + 2\,(\underset{\text{inelastic}}{\mathbf{v}_{Ei}} - \mathbf{v}_{li})$$

In the elastic case, the differing velocities require shifting the positions $\mathbf{x}_{Ei}$ of cluster members orthogonal to the axis of angular momentum so as to conserve angular momentum. Considerable ingenuity was expended thereon to no avail for the problem of interest.

The integration was terminated when either (1) a specified number of integration steps had passed; (2) a specified number of bodies had either clustered or escaped beyond the Roche limit; (3) the integration step became too small, due to some instability; or (4) one of the monitored quantities had drifted too much. These monitored quantities were the total energy, angular momentum, and linear momentum.

Results

The inelastic cases were tried and their results are summarized in Table 1. As mentioned, elastic cases developed instabilities that were properties of the artifice of a set of spheres, rather than the protoplanets they are intended to represent.

The inelastic cases where collision did not occur are rather uninteresting. Cohesion of both bodies was always maintained, with no tidal disruption, in confirmation of Mizuno and Boss (1985). We did not obtain any small percent of loss at the equator as they did in some cases, because we did not provide for an equation of state allowing for vaporization into an atmosphere. These results are reassuring in that the representation by a relatively small number of bodies would be expected to err on the side of instability (as demonstrated by the elastic cases).

The variation in outcome of inelastic cases resulting in collision does not relate to any one readily identifiable variable. The collision number, the ratio of angular momentum to that required for instability of a single homogenous protoplanet of the same total mass and density, was examined as such, but it is clear that other parameters, such as the mass ratio of the protoplanets and the approach velocity, are also significant. The mass ratio is apparently important because it concentrates impact effects in a smaller portion of the matter. The importance of doubling the approach velocity from about 3 km/sec to 6 km/sec is harder to interpret, since the energy acquired by impact, $GM_1M_2/(R_1 + R_2)$, is appreciably larger than the original energy, $(M_1 + M_2)v_a^2/2$.

A result of interest in the cases that do place material into elliptic orbits, 0.1 $M_\oplus$:0.9 $M_\oplus$ and 6.6 km/sec, is that this material comes from the outer parts of the smaller protoplanet, as shown in Fig. 1. This result suggests that if a collision occurs between protoplanets that are already differentiated, the material placed in orbit is deficient in iron. Also, this material is appreciably heated, according to the summation of energy dissipated in collisions (equation (4)), so that it would be expected to be depleted in volatiles.

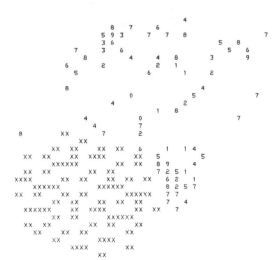

Fig. 1. Collision of protoplanets of masses 0.1 $M_\oplus$ and 0.9 $M_\oplus$ at approach velocity of 6.6 km/sec, leading to a pericenter distance of 1.2 $R_\oplus$ or an overlap of 0.3 $R_\oplus$. Distribution of the 89 spherical elements of the smaller protoplanet is shown by the numbers, which are the original radial loci rounded off to the nearest tenth of the protoplanet radius. Distribution of the 88 spherical elements of the larger protoplanet is shown by the X's.

Discussion

The computations described here support the hypothesis of lunar origin by major collision in that they indicate that tidal disruption does not occur upon passage within the Roche limit if the rheology is dissipative, as is generally believed. As in Mizuno and Boss's calculation, the dissipation is implicit, arising from constraints placed upon the relative motion of elements in contact with each other.

As mentioned above, it is reassuring that cohesion was maintained upon noncollision of inelastic protoplanets by our representation, since it might be expected to develop instabilities. It did, in fact, as symptomized by a slight warming from application of (4) to jostlings between bodies away from actual collision of the protoplanets. But this noise was kept to a minor level by the combination of the short time step, one second, and the imposition of cohesion and corotation upon clustering. The main defect of our representation is probably that maintaining the identities of nearly incompressible spheres squelched the mixing of material that would actually occur as the result of liquefaction and vaporization in high velocity impacts.

It would, of course, be desirable to relate viscosity to the dissipation. We do not see how to do this without adopting a continuum representation, such as Mizuno and Boss's. More fundamental is a better understanding of how inelastic actual material would be upon close approaches, particularly if a body were hot enough to be fluid. In this case, there might be tidal disruption involving larger pieces than found by Mizuno and Boss (1985). Furthermore, if relative velocities between elements were rather small—less than 0.002 km/sec, a rather gentle nudging—collisions could plausibly be somewhat elastic.

Our results partially support the collision model of Cameron (1986) in indicating that it is feasible to place material in orbit around a protoplanet by a major collision. The range of conditions under which it would occur needs to be explored further. One parameter we have not examined at all is preexisting spin of the protoplanets, ω_1 and ω_2. It is yet to be defined whether there are conditions under which it can occur at approach velocities less than 3 km/sec, as desirable to be consistent with geochemical differences between the Moon and the Earth (Drake, 1986). The main question affecting whether the Moon formed from a major collision or a circumterrestrial accretion disk (Weidenschilling et al., 1986) remains the size distribution of the protoplanet population in the terrestrial zone. A population dominated by planetesimals much smaller than the Earth, as required for circumterrestrial accretion, appears to require a suppressor of growth for which a plausible hypothesis is lacking at present.

Acknowledgments. This research was supported by NASA grant 05-007-002.

References

Cameron A. G. W. (1986) The impact theory for origin of the Moon, this volume.

Drake M. J. (1986) Is lunar bulk material similar to Earth's mantle?, this volume.

Harris A. W. and Kaula W. M. (1975) A co-accretional model of satellite formation. *Icarus, 24,* 516–524.

Hayashi C., Nakazawa K., and Nakagawa Y. (1985) Formation of the solar system. In *Protostars and Planets II* (D. C. Black and M. S. Matthews, eds.), pp. 1100–1154. Univ. of Arizona Press, Tucson.

Jeans J. (1928) *Astronomy and Cosmogony.* Cambridge University Press, 428 pp. (Republished by Dover, 1961).

Mitler H. J. (1975) Formation of an iron-poor moon by partial capture. *Icarus, 24,* 256–268.

Mizuno H. and Boss A. P. (1985) Tidal disruption and the origin of the moon. *Icarus, 63,* 109–133.

Öpik E. (1972) Comments on lunar origin. *Irish Astron. J., 10,* 190–238.

Safronov V. S. (1972) Evolution of the Protoplanetary Cloud and Formation of the Earth and Planets. *NASA TT 71-50049,* NASA, Washington, D.C.

Ward W. R. (1981) Solar nebula dispersal and the stability of the planetary system: I. Scanning secular resonance theory. *Icarus, 47,* 234–264.

Weidenschilling S. J., Greenberg R., Chapman C. R., Herbert F., Davis D. R., Drake M. J., Jones J., and Hartmann W. K. (1986) Origin of the Moon from a circumterrestrial disk, this volume.

Wetherill G. W. (1980) Formation of the terrestrial planets. *Ann. Rev. Astron. Astrophys., 18,* 77–113.

Wetherill G. W. (1986) Accumulation of the terrestrial planets and implications concerning lunar origin, this volume.

VII.   Theories and Processes of Origin 3:
Lunar Formation Triggered by Large Impact

Moon Origin: The Impact-Trigger Hypothesis

WILLIAM K. HARTMANN

Planetary Science Institute, 2030 E. Speedway, Suite 201, Tucson, AZ 85719

The impact-trigger hypothesis proposes that an unusually large planetesimal impact on the early Earth initiated lunar formation. Several findings in the mid-1970s combined to produce this hypothesis: shortcomings of earlier theories; Apollo discoveries of similarities in Earth-mantle and lunar material; and theoretical studies showing plausibility of giant impactors (see Section 1). Such an impact would dislodge hot, volatile- and iron-poor material to form the Moon. Section 2 lists lunar properties favoring this hypothesis: lunar iron and volatile deficiency; angular momentum of the Earth-Moon system; similar O isotopes, bulk iron contents, and densities of Earth's mantle and the Moon. A common objection, that the impact is *ad hoc*, is refuted in Section 3: stochastic large impacts are an acceptable, if not essential, part of planet formation. Section 4 shows that the intense early bombardment averaged during Earth's formation was several 10^9 times the present meteoritic mass flux, consistent with a giant impact. Section 5 discusses three different physical effects possibly involved in lofting debris into a circumterrestrial cloud: forward launch of debris as a result of planetesimal shearing or jetting of ejecta in a near-tangential impact, rotational spinup, and ordinary cratering ejection. These effects can combine in different ways to create different impact-trigger submodels. Section 6 estimates an impactor mass of a few M_{Moon} to $\sim M_{Mars}$ from three independent lines of evidence. In Section 7, numerical models of ejecta velocity distributions indicate that a few lunar masses of debris could have been ejected at speeds between circular and escape velocities; second burn gas expansion effects may increase this value and help the material achieve orbit.

1. Background

A major goal rationale of Apollo lunar exploration was to learn the origin of the Moon. This failed, partly because anticipated 4.6-b.y.-old "genesis rocks" were mostly destroyed by the extreme bombardment rate persisting from 4.0–4.6 b.y. ago. A second reason for lack of success, as seen in retrospect, was concentration on three classical theories of lunar origin: (1) Capture—capture of a planetesimal, formed elsewhere in the solar system, into Earth orbit; (2) Fission—spontaneous ejection of upper mantle material into a circumterrestrial swarm due to rotational instability, probably during core formation; (3) Coaccretion—formation of the Moon by accretion in a circumterrestrial nebula, in which differentiation is needed to produce lunar composition. None of these theories of origin appeared satisfactory. The capture event is unlikely, and no source region is known to produce bodies of Moon-like composition. The classical fission process does not explain the energy and angular momentum needed to eject the necessary material.

Coaccretion models, such as those developed by Ruskol (1972) and Harris and Kaula (1975), have been more successful, but have not satisfactorily explained how incoming interplanetary material gets differentiated into the iron- and volatile-poor material observed in the Moon. Even newer models of this differentiation, such a compositional filtering by a circumterrestrial swarm (Weidenschilling *et al.*, 1986), involve a debatable chain of processes including planetesimal melting, differentiation, subsequent fragmentation that pulverizes silicate mantles but preserves iron cores, capture of the silicate fraction into a circumterrestrial swarm, and an angular momentum input adequate to maintain the swarm. Furthermore, most such coaccretion models evolve so smoothly from the postulated initial conditions that there is no explanation of why they did not produce satellites of similar relatively large size around other planets. Nonetheless, the dynamical and chemical evolution of a circumterrestrial swarm is an important aspect of the impact-trigger model, to which all work on coaccretion and compositional filtering is relevant.

The impact-trigger model was first suggested by Hartmann and Davis (1975) and independently by Cameron and Ward (1976). (Historical note: This model was presented by Hartmann at the Conference on Satellites of the Solar System that was held at Cornell University in the summer of 1974; in a response from the floor, Cameron noted that he and Ward were pursuing a similar model, with similar positive results.) Hartmann and Davis (1975) were principally concerned with establishing the plausibility of the required giant impacts, and presented simple accretion models to demonstrate that as the largest planetesimal in a given solar system zone grows to planetary dimensions, the next largest planetesimals can grow to considerable size. We followed the growth of the ten largest bodies in a swarm, simultaneously, and found that

"In the case of Earth-sized planets, the models suggest second-largest bodies of 500–3000 km radius, and tens of bodies larger than 100 km radius. Many of these interact with the planet before suffering any fragmentation events with each other. Collision of a large body with Earth could eject iron-deficient crust and upper mantle material, forming a cloud of refratory, volatile-poor dust that could form the moon." (Hartmann and Davis, 1975)

Wetherill (1985, 1986; personal communication, 1984) recently strengthened this part of the argument, with new calculations modeling the simultaneous growth of numerous large bodies. Nonetheless, the presence of large enough impactors is not yet proven, because existing accretion models only poorly follow the final growth from hundreds of kilometers to planetary dimensions (Greenberg, 1979). One of the most sophisticated of these models (Greenberg et al., 1978) notes that once a growing embryo is large enough ($\sim 10^3$ km), it tends to sweep material out of its own zone faster than the feeding occurs, complicating further modeling of the critical transition stage, leading from Ceres-sized embryos to Mars-sized bodies.

In spite of these theoretical difficulties, empirical evidence such as planetary obliquities supports the contention that most planets were struck by bodies containing at least a few percent of their own mass (Hartmann and Vail, 1986). Hartmann and Davis (1975) concluded that Earth received a larger-than-average impact:

> "This model can thus account for the iron depletion, refractory enrichment, and volatile depletion of the moon, and at the same time account for the moon's uniqueness; the moon may have originated by a process that was likely to happen to one out of nine planets."

Cameron and Ward (1976) did not deal with the plausibility of large planetesimal impacts, but considered two additional attractive aspects of the impact model. First, they pointed out that a large, near-tangential impact could provide the large specific angular momentum of the Earth/Moon system. Assuming that Earth had zero initial spin and that the planetesimal had "collisional velocity. . . close to 11 km/s," they found that an impactor with a mass "comparable to that of Mars" could explain the angular momentum of the Earth/Moon system. Second, they concluded that a massive cloud of hot gas would help loft material from the site. Because gas pressure gradients could produce nonballistic orbits, they proposed that refractory silicates and other materials would condense into particles whose sizes and compositions were position- and time-dependent during expansion. "A substantial portion. . . should go into orbit" on eccentric orbits, and would form a disk in which gravitational instabilities would "produce clumping of particles" that would form the Moon.

Ringwood (1979), in his major synthesis of work on the origin of Earth and the Moon, accepted the impact-trigger model, perhaps with several modest impactors, as the most promising way to generate an orbiting disk of hot, condensing debris. Ward and Cameron (1978) and Thompson and Stevenson (1983) pursued the evolution of the disk of debris. Ward and Cameron considered evolution within the Roche limit and pointed out that angular momentum transfer within the disk, due to the viscosity of the medium, would drive some material outward beyond the Roche limit. In an example, they found that if a 2 lunar mass cloud were expelled into a disk by impact, "a large amount of material would drift out beyond the Roche limit quickly where satellite formation becomes possible." Thompson and Stevenson found that in a two-phase hot disk, due to reduction in sound speed over single-phase systems, gravitational instability is facilitated. They found dramatic gravitational instability in a massive hot disk, such as might be ejected by a large impact. This "could lead to formation of the moon. The formation timescale could have been as short as ca. 100 years and the moon could have formed completely molten."

Little further modeling resulted from the impact-trigger hypothesis of lunar origin until the Conference on the Origin of the Moon, held in Kona, Hawaii in October of 1984.

2. First-Order Properties and Constraints on Lunar Origin

The Kona meeting revealed that a full-fledged paradigm of geochemical constraints on lunar origin has not yet emerged. Significant differences of opinion exist, for example, on the degree and plausible causes of alteration of Earth-mantle material that would be needed to create Moon-like material. Listed here are some "first-order" lunar characteristics to be explained. The impact-trigger hypothesis appears to offer fruitful possibilities in each case.

1. Iron deficiency and gross similarity to Earth's upper mantle: The pre-Apollo observation of the Moon's low density and consequent iron deficiency led Wise (1963, 1969) and others to hypothesize that the Moon formed from mantle material *after* Earth's core formed. Wise's model called for ejection of the material by spontaneous fission following spinup after core formation—a possibility now widely rejected on dynamical grounds. (The mantle iron deficiency might also be explained if inhomogeneous accretion occurred, leading to early formation of iron cores and later accretion of iron-poor silicates; this possibility has not been adequately explored.) After Apollo, lunar rock geochemistry led to the consensus that the lunar material crudely resembles Earth's mantle; second-order differences were viewed with different degrees of concern by different theorists. There was disagreement in the geochemical community over whether the Moon could be made primarily from mantle material, and over what processing would be needed. One view is that of Wänke and coworkers (Wänke and Dreibus, 1984), that the Moon bears an indelible fingerprint of Earth-mantle origin. The gross similarity of lunar material to Earth-mantle material is ideally explained if the Moon originated with a substantial fraction of material from Earth's mantle, but requires special additional assumptions in theories invoking coaccretion or circumterrestrial swarms of nonimpact origin. In view of the shortcomings of the classical fisison model, the impact-trigger model appears to be the only model capable of dislodging Earth-mantle material.

2. Volatile depletion: The volatile depletion pattern of the Moon has always been difficult to explain in detail. However, at a first-order level, it appears consistent with a strong heating of most lunar material, probably in pulverized form to allow volatile escape, perhaps to temperatures of 1400–1800 K, and possibly additional chemical processing. By "first-order" I mean that strong heating appears to be a good starting point for a theoretical explanation; additional effects probably must be invoked to explain the complete pattern. In a giant impact, large masses of material would likely be vaporized, while additional entrained, pulverized dust might be ejected at a wide range of temperatures. Efficiencies of additional processes, e.g., recondensation of dust, gas escape, and thermal differentiation of gas species are uncertain. The hypothesis of an impact ejecting hot, finely disseminated material thus appears to be a step forward in understanding lunar volatiles (e.g., Ringwood, 1979), but the chemistry of impact processing clearly requires further study.

3. *Angular momentum considerations:* As Cameron and Ward (1976) and Cameron (1984, 1985) emphasize, a giant impact provides a plausible mechanism to explain the unusually high value of angular momentum in the Earth/Moon system, relative to other planets. "Unusually high" requires further comment. MacDonald (1963) published a log-log curve of angular momentum/mass (L/m) for planets. The curve passed near the values for Mars and the giants, leaving the Earth/Moon system anomalously high. MacDonald concluded that a model with the primordial Earth or primordial Earth/Moon system having this high value would require a special explanation. Many subsequent authors, such as Cameron and Ward, adopt this view (see Binder, 1986, for additional discussion). Hartmann and Larson (1967) and Hartmann (1983a) pointed out that if asteroids are added to the diagram a new best-fit line is defined, and it passes closer to the Earth/Moon system value, as well as the giants; Mars is low. They pointed out that the new line is inconsistent with equipartition of rotational energy by collision (as might be expected among smaller asteroids as a result of collisional equilibrium). The new line appears consistent with a tendency for primordial planets to spin up to periods averaging around one-third that needed for instability, i.e., around 10 hours, but such periods are not well understood. Could the cumulative effect of accretion have spun up the planets toward this modest prograde spin? The net angular momentum brought into a system by low eccentricity or high-eccentricity-randomized planetesimals is near zero (Weidenschilling, 1984; Herbert and Davis, 1984). Giuli (1968) and others have shown that impacts of selected families of moderate-eccentricity planetesimals could yield a prograde spinup during accretion to values around 10–15 hours (Giuli, 1968; Harris, 1977), but this spinup process is questionable at best because of its selection of certain orbits; it appears inadequate to reach the angular momentum of the Earth/Moon system, corresponding to a period of 4.1 hours for a single, proto-Earth/Moon body with Earth's moment of inertia. Large impacts would have helped to spin up the planets from the low angular momentum states produced by accretion of innumerable small bodies (Hartmann and Vail, 1986). Indeed, a large impact is the ideal mechanism to produce Earth's final spinup to the effective period of

4.1 hours, matching the angular momentum of the present system (Cameron and Ward, 1976; Hartmann and Vail, 1986).

   4. Oxygen isotope ratios: Clayton *et al.* (1973) discovered that materials formed in different parts of the solar system have different oxygen isotope ratios, and developed this evidence for a variety of meteorites and other materials (Clayton, 1981). Relative to terrestrial materials, C3 carbonaceous chondrites are by far the most alien, followed by C2 and C1 carbonaceous chondrites. Common chondritic and iron meteorite types are next, having isotope ratios near but distinctly discriminable from Earth (Clayton, 1981). SNC meteorites, which are probably from Mars, also fall near but distinct from the Earth data, and considerably closer to Earth than L chondrites, for example (Clayton and Mayeda, 1983; Clayton, 1981). Finally, lunar samples and enstatite meteorites fall on the chemical mass fractionation line characteristic of Earth materials and are indistinguishable from Earth (Clayton and Mayeda, 1975; Clayton *et al.*, 1984).

TABLE 1.   Zonal Chemistry of Solar System.

Type Material	Asteroid and Meteorite Evidence	Oxygen Isotope Evidence	Inferred Zone of Origin
Carbonaceous material	C-type asteroids carbonaceous chondrite materials concentrated in outermost belt (and outer solar system)	Far removed from terrestrial materials	Outer belt to Jupiter region
Chondrites, irons, dunite (olivine cumulate), meteorites	Probably similar to asteroids in midbelt regions, such as classes S, M, A	Similar to, but clearly distinct from, terrestrial materials	Midbelt
Mars rocks (SNC meteorites)	Identified from SNC meteorites	Quite similar to, but distinct from, terrestrial materials	Mars zone
Earth, Moon, enstatite chondrites and enstatite achondrites	Higher temperature mineral suites than many meteorites. E-type asteroids may match enstatite chondrites; concentrated in inner belt; thrown there from Earth zone? (Wetherill, 1977)	Indistinguishable from Earth materials	Earth zone

These findings agree beautifully with the emerging evidence of zonal compositional structure in the solar system (Gradie and Tedesco, 1982), as shown in Table 1. Carbonaceous chondrites are almost certainly related to asteroid types from the outer belt and giant planet regions. Many chondrites are probably similar to asteroid materials from the middle and inner belt. Mars (SNC meteorites) is closer to Earth in distance and composition. Enstatite meteorites may have originally formed near Earth's zone. In summary, the *O-isotope data require that the Moon formed from material that originated in the same terrestrial planet "feeding zone" that contributed material to the Earth, and not as far away as the "feeding zone" of Mars.*

5. *Digression on the relevance of enstatite meteorites:* Four independent facts related to enstatite meteorites combine in striking juxtaposition: (1) E asteroids' spectra resemble, or are at least consistent with, spectra of enstatite achondrite meteorites (Zeliner *et al.*, 1977). (2) "...in the innermost region of the belt, the Hungaria region, the population is dominated by types E and R" (Gradie and Tedesco, 1982; the three E's are in the region from 1.9–2.7 AU). (3) Enstatite achondrites and enstatite chondrites are the only meteorite types with O-isotopes identical to Earth's, suggesting they formed in or near Earth's zone. (4) Wetherill (1977) showed that ~0.4% of the original bodies formed in Earth's zone were perturbed by near planetary encounters into long-lived orbits on the *inner edge* of the asteroid belt, and that "some...residual Earth material should be stored in the inner (edge of the) asteroid belt for more than 4.5 b.y." Taken together, these four findings *suggest* that enstatite achondrites and enstatite chondrites, together with material of bulk Earth composition, are good candidates for the suite of materials accreting in Earth's zone of the solar system. In addition to O-isotope ratios, the total elemental iron contents of enstatites are more compatible with Earth's bulk composition than are the iron contents of other chondrites (contrary to the widespread assumption of ordinary chondritic materials as primitive Earth-building material):

whole Earth:	35% iron by weight (Mason, 1958)
enstatite chondrites:	22–35% iron by weight (Dodd, 1981, p. 19; Mason, 1962, pp. 89, 164)
other ordinary and C chondrites:	19–28% iron by weight (Dodd, 1981, p. 19)

These figures show the plausibility of a similarity between the enstatite chondrites and the Earth-building planetesimals, and are consistent with a trend of decreasing iron away from the sun.

6. *Bulk iron content:* It is common to emphasize the different iron oxidation states among the Earth's mantle (partly oxidized), Moon (more oxidized), enstatite chondrites (almost totally reduced), etc., as indicating a lack of generic connection. However, the position taken here is that we have a complete lack of understanding of how the oxidation states were established. As will be discussed later, numerous mechanisms, such as a time-varying mantle composition, or chemical interactions

in a hot, circumterrestrial cloud, may have caused differences between the oxidation states of the Moon and present-day mantle.

The estimated bulk *elemental* iron content of the Earth's mantle and the Moon are:

Earth mantle: 7% iron by weight (Mason, 1958, p. 50)
Moon: 7–9% iron by weight (see below)

The lunar entry is calculated by assuming a metal core (0.908 Fe by weight; density = 7.9 g/cm^3 = iron meteorites) of 200–300 km radius and, in the remaining lunar mantle, 0.069 Fe by weight (Earth mantle number; Mason, 1958) to 0.085 Fe by weight (Taylor, 1982, p. 396, model for primitive lunar mantle). The similarity is predictable if the Moon formed from ejected upper mantle material (especially if some projectile iron were added), but is an odd coincidence in other theories.

7. *Density:* The mean densities of the Moon (3.344 ± 0.002 g/cm^3) and upper mantle (3.3 to 3.4 g/cm^3) are virtually identical (Ringwood, 1979, p. 165). This is directly explained if the Moon formed from ejected upper mantle material, but is an odd coincidence in other theories.

In summary, a number of first-order parameters—iron and volatile deficiency, gross similarity to Earth's upper mantle, angular momentum, oxygen isotope ratios, estimated iron content, and density—give strong circumstantial evidence that the Moon originated from material blown out of the Earth. The first-order parameters are eminently consistent with the impact-trigger hypothesis and less directly predicted by other hypotheses.

3.   Stochastic ≠ Ad Hoc: Plausibility of a Giant Impact

The impact-trigger hypothesis of lunar origin has sometimes been relegated to a hypothesis of last resort, with the assertion that it is *ad hoc*. Many lunar origin theorists have felt constrained by Occam's razor to avoid postulating a role for such stochastic events. "Stochastic" connotes that the outcome has a strong dependence on an element of chance; this is contrary to the outcome of the accretion of the innumerable smaller planetesimals, whose net effect is more predictable due to statistical smoothing of fluctuations.

Some classes of influential events in solar system history are class-predictable but not event-predictable: i.e., we believe the class of events occurred, but we cannot determine times and magnitudes of individual events. These events are stochastic, but not *ad hoc*. Giant impacts are class-predictable, in the sense that there are growing grounds to believe that in addition to innumerable collisions with mass ratio $m_{planetesimal}/m_{planet} \ll 0.01$, there were a few impacts in which this mass ratio was a few percent or more. In some cases it was large enough to alter the nature of the finished planet.

An example of the problem of class-predictable events in planetary science is the probable Cretaceous-ending asteroid impact. Since the 1960s, asteroid statistics have implied such events every few 10^7-10^8 years, but we could not convincingly tie specific geologic effects to specific impacts. In the absence of such evidence, impacts of this size tended to be ignored; as scientists, we should have pursued the geologic and climatic consequences of these class-predictable events instead of waiting for iridium-rich layers to take us by surprise.

Analogously, we shold not ignore class-predictable mega-events in studying lunar origin. As long ago as 1958, O. Yu. Shmidt emphasized the statistical cumulative effects of innumerable small planetesimals, explaining the tendencies toward regularity in the solar system: prograde rotation, modest obliquities, inclinations, and eccentricities. Superimposed were the effects of large-scale stochastic events, emphasized by Safronov as early as 1966. The magnitude and timing of each planet's largest impact may have been crucial to that planet's development (Hartmann, 1977). Theoretical and observational results support the importance of large-scale stochastic events (impacts and close encounters?) in explaining differences among planets: (1) Safronov (1966) attributed obliquities to impacts of masses up to a few percent of planetary mass (5% for Uranus; 0.1% for Earth). (2) D. R. Davis and I (Hartmann, 1977) found similar results for planetary eccentricities and inclinations. (3) Hartmann and Vail (1986) studied effects of impactors on obliquity and period, concluding that impactors up to a few percent of a planetary mass are plausible. Items (1)–(3), together with the work discussed in Section 1, imply planetesimals with diameter $\geq$ 4000 km ($\geq$ 3% M_{Earth}) striking planets during the planet-forming period. Several other lines of evidence prove the existence of catastrophic collisions, albeit of lesser magnitude. (4) Craters and multi-ring basins give direct evidence of impacts nearly big enough to have disrupted the target bodies: the craters reach 34–40% of planetary diameter in the cases of Mimas, Tethys, and Phobos; Voyager analysts (Smith *et al.*, 1981, 1982) concluded that Mimas and other moons were disrupted and reassembled during the early intense bombardment. Ring systems are coming to be widely viewed as debris of such catastrophes. (5) The largest basins, dating back to at least 3.9 b.y. ago, were formed by objects believed to have diameters of ~150 km. Survival of such large bodies ~1–2 b.y. after planet formation implies more numerous and, by the nature of power law mass distribution, even larger interplanetary bodies during planet formation. These observations confirm the importance of catastrophic impact processes and at the same time indicate the stochastic nature of the most destructive event in each system: mass distributions based on observed power law approximations (asteroids, craters, fragmentation events in nature), or theoretical runaway growth models (accretion processes), imply very few of the largest impactors, or even a uniquely "largest" impactor, considerably larger than the rest.

Impacts and close encounters with large objects during planet formation are thus class-predictable. A stochastic, class-predictable event, such as a large impact capable of triggering ejection of Earth-mantle material into a circumterrestrial cloud, should not be rejected as *ad hoc* or too catastrophic, nor should it be set aside as a "theory of last resort." Rather, a way to deal with such an event scientifically is to investigate its consequences; if they can be shown to fit the known constraints on lunar origin well, the proposed event becomes a viable concept among hypotheses of lunar origin.

4. The Intense Early Bombardment

The present meteoroidal mass flux onto Earth is some 10^2–10^3 tons/day (Dodd, 1981, p. 2). Crater counts on dated lunar and terrestrial surfaces prove that this rate averaged much higher in the first half-billion years of Earth/Moon history; the earliest rate that can be directly determined, about 4.0 b.y. ago, is some 10^3 times higher than the present value (Hartmann, 1972; Neukum *et al.*, 1975). Wetherill's (1977) modeling of the sweepup of planetesimals, and hence the meteoroidal flux decline, is in excellent agreement with the observed decline in cratering rate (Hartmann, 1983a, p. 163). As objects on short-lived orbits were swept up, the half-life of impactor populations slowly lengthened, as observed in the cratering data. In accord with this, the bombardment rate during planet formation was enormous. The entire Earth's mass accumulated in about 10^7–10^8 years; isotopic data prove that the interval was <150 m.y. (Pepin and Phinney, 1976). Intervals of 30 m.y. or less are widely discussed. In order to accumulate the Earth in 30 m.y., the mass flux is about 0.5–$5(10^9)$ times the present flux (Hartmann, 1980). I adopt the factor $2(10^9)$.

In the following discussion, the mass distribution of this flux is assumed to resemble that found today: roughly a power law. This distribution, which is known among asteroids, meteorites, and meteoroidal dust, is viewed as a natural consequence of accretion and collisional fragmentation processes (Chapman and Davis, 1975; Hartmann, 1969). A somewhat different mass distribution may have applied early in planetesimal accretion, when the largest bodies approached only some 10^2 km (Greenberg *et al.*, 1978), but I assume here that near-misses with multiple planet-sized bodies pumped up speeds, producing fragmentation, hence extending the power-law-like mass distribution, characteristic of modern asteroids, to relatively large bodies by the close of the planet-forming period.

Figure 1 shows an estimate of the primordial impact flux as a function of mass. This is obtained by taking the present flux curves (obtained by crater counts, asteroid surveys, meteorite fall rates, etc.—see Dohnanyi, 1972) and multiplying by the $2(10^9)$ factor derived above. I extrapolate to larger masses than those that made the largest observed craters and basins, and adjust the shape and maximum mass (i.e., that corresponding to one Earth-impact during the formation interval) to assure accumulation of no more than one Earth mass in the specified time interval. The adjustment involves a slight downward bend in the curve at high mass, consistent with a slight downward bend suggested by the statistics of the largest craters (Hartmann, 1984). In this sense the curve is somewhat conservative: the downward adjustment gives fewer large bodies than would be indicated by a linear extrapolation.

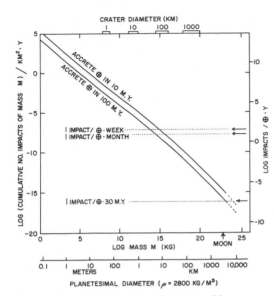

Fig. 1.   Cumulative mass distributions of impactors during the accretion of Earth as reconstructed in text. Horizontal scales give planetesimal mass, diameter, and rough crater size for plausible velocities. Vertical scales give impact rate (~2 × 10⁹ × present rate) in two equivalent units. Formation of Earth in 10–100 m.y. would imply a largest impactor of superlunar size (lower right) and weekly to monthly formation of nearly 100-km-scale craters. The latter craters' debris would create a thin, steady-state, circumterrestrial swarm of fluctuating density, with superimposed density surges from larger impacts. The Moon's origin is viewed here as a consequence primarily of the largest single impact (Weidenschilling et al., 1986, gives a further discussion of the evolution of the swarm).

It is important to note that since the largest known craters in the solar system (~1000 km diameter) involve impactors of only some 10^{18}–10^{19} kg, we have no observational knowledge of the size distribution of larger impactors. However, consistent with Section 3, it is intriguing to note that the curves in Fig. 1 extrapolate smoothly to one impact of a roughly Mars-sized body during Earth's formation interval. My view is that such arguments cannot specify the precise size of the largest impactor ever to hit Earth, but we have shown that it was plausibly of the order of several lunar masses or larger.

5.   Impact Mechanics and the Nature of the Impactors

If it could be shown that such an impact would eject primarily only Earth-mantle material into orbit (for example, if the impactor were vaporized and it ejected both vaporized and solid Earth-mantle material, and if most of the vapor phase or its resultant condensates escaped from the Earth/Moon system), then the origin of the Moon would be largely solved. Heated, pulverized Earth-mantle dust would accrete in orbit. This possibility was an initial attraction of the impact-trigger hypothesis; as seen in Section 2, such material would already make a first-order match to lunar composition in terms of bulk density, iron content, volatile loss, etc.

However, many models of giant impacts (see Melosh and Sonett, 1986; Cameron, 1986; and other works in this volume) suggest that most of the ejecta is vaporized planetesimal material. On the other hand, some studies (O'Keefe and Ahrens, 1977; Ahrens, personal communication, 1985) suggest that the fraction of mass vaporized depends primarily on impactors' velocity, not size, implying that the percentage of vaporized ejecta in a giant impact should not necessarily be 100%. If the ejecta were mostly vapor, and most of the vapor were from the Earth's mantle, we would have to consider recondensing mantle material. However, formation of the Moon from recondensing dust is problematic; volatility would be the primary control, and it seems dubious whether lunar composition would result. If much of the vapor phase came from the planetesimal, the planetesimal's iron content becomes a problem. Once we lose the elegance of starting with mantle material, we face the classic problem of older theories: accounting for the lunar silicate/iron ratio.

Four phenomena may help produce lunar composition, even if the ejected material is primarily gas. First, Melosh's model (Melosh and Sonett, 1986) calls for the highest velocity ejecta to be material jetted from the contact region between Earth's mantle and the impactor mantle. Thus, the material that reaches orbit may be a mixture of mantle-like dust and gas from both bodies. This would solve much of the composition problem. Secondly, the dust component, though initially small, may be selectively left in orbit as the hot gases thermally dissipate. This helps avoid making the Moon entirely of condensates from a gas. Third, the compositional filter process in the resulting circumterrestrial swarm (Weidenschilling et al., 1986) may increase the swarm's silicate/iron ratio. Fourth, Ringwood (1979, p. 244) points out that the impactor may have hit a deep magma ocean on Earth, enhancing the ratio of Earth/planetesimal material over that of an impact into a solid surface. This favors Earth-mantle composition.

I regard as still open the question of how much heated, pulverized, solid Earth-mantle dust could be carried into a circumterrestrial cloud during a giant impact. Gas expansion will accelerate dust by "second burn" drag effects. Models that eject primarily gas and virtually no high-speed dust have yet to demonstrate that they are successful when scaled down to smaller events: they need to account for the thick ejecta blankets stretching $\sim 10^3$ km from lunar basins. Just as significantly, these models need to show that they are consistent with the compelling evidence that lunar and Martian rocks have been ejected into space with negligible thermal alteration.

How might the giant impact's outcome be affected by the geometric impact parameters, as well as impact mechanics? The answer to this question subdivides the impact-trigger hypothesis into three subsets of the original impact-trigger idea. These echo aspects of the three classical theories of fission, capture, and coaccretion. Figure 2 shows the three variants at the corners of a triangle diagram intended to nurture further thought. Each corner represents the dominance of a particular physical effect: in (a) it is shearing (mechanical or tidal) of the planetesimal and forward jetting of ejecta associated with a nearly tangential impact; in (b) it is spinup and consequent fission due to angular momentum input from a nearly tangential

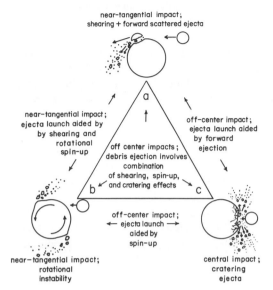

near-tangential impact;
shearing + forward scattered ejecta

near-tangential impact;
ejecta launch aided by
by shearing and
rotational
spin-up

off center impacts;
debris ejection involves
combination
of shearing, spin-up,
and cratering effects

off-center impact;
ejecta launch aided
by forward
ejection

off-center impact;
← ejecta launch →
aided by
spin-up

near-tangential impact;
rotational
instability

central impact;
cratering
ejecta

Fig. 2. A new trilogy of lunar origin models, derived from the impact-trigger hypothesis. In model (a), a grazing impact results in formation of an orbiting swarm by shearing of the planetesimal and forward ejection of the debris. In (b), the nearly tangential impact imparts enough angular momentum to make Earth rotationally unstable, resulting in immediate mass ejection by a fission-like process. In (c), the near-vertical impact launches ejecta into an orbiting swarm, primarily by the cratering process. Any point in the triangle represents a combination of the effects. See text for discussion.

impact; in (c) it is the more symmetric ejection of a massive swarm of debris associated with cratering, without invoking a near-tangential impact. The diagram is a continuum; the real process of lunar origin could be represented as a dot on any edge (combining two effects) or in the interior (combining all three effects). Only corner (c) represents a precisely vertical impact. A point on line (a)–(b) would represent a nearly-tangential impact with forward jetting of some debris or planetesimal fragments into orbit, followed by equatorial fission of additional debris during a subsequent short interval. (The two sets of debris would not necessarily be coplanar.) A point on line (b)–(c) represents an off-center impact where both cratering and rotational spinup play a role in launching debris. A point on line (a)–(c) represents an off-center impact where cratering debris attaining orbit are aided by asymmetric forward ejection. A point inside the triangle would indicate an off-center impact in which crater debris are forward-ejected, but materially aided in reaching orbit by the rotational state of the planet.

Study needs to be directed to all of these effects. Classical cratering models tend to deal primarily with (c), which may not be adequate to get enough ejecta into orbit. Some further remarks on the "end member" effects at the three corners may be useful:

1. Much of the planetesimal might pass by in a brushing encounter, either escaping or going into orbit. Some models have disposed of the planetesimal's iron core in this way, assuming it escapes while planetesimal mantle debris from the grazing impact enters orbit. Other models call for capture of some farside planetesimal material

that shears off and enters orbit but is minimally altered by impact processing. Models triggered in this way echo classical capture models.

2. Earth, finding itself spinning faster than the stability limit, would immediately begin deforming, possibly fissioning according to the Darwinian mode (Binder, 1986) or possibly spinning off a disk of material as pictured by Durisen and Gingold (1986). Excess angular momentum would have to be lost from the system, possibly dissipated in escaping gas. Models triggered in this way echo classical spontaneous fission models.

3. If the impact were highly efficient in ejecting either solid debris or gases from which solids condensed (especially by the jetting of surface materials from each body; see Melosh and Sonett, 1986), it could produce a circumterrestrial swarm in which the Moon could accrete. Since the atmosphere is of negligible thickness compared to the cratering region, some fraction of material in the expanding chaotic cloud, with motions affected by gas pressure and perhaps mutual gravitational interactions, may be effectively launched from above the atmosphere and directed into an orbiting disk-shaped swarm. Further evolution of a swarm produced by any combination of the three effects echoes the classical coaccretion models (see Weidenschilling *et al.*, 1986).

How might the giant impact's outcome be affected by planetesimal composition? Negligibly, if the planetesimal contributes a negligible fraction of the final lunar mass. But many models (see other papers in this volume) propose that much of the ejected swarm comprises planetesimal material. Most models have assumed, explicitly or implicitly, a nearby origin for the impactor. In that case, as seen in Section 2, we might expect a bulk impactor composition resembling that of Earth or enstatite chondrites. This obviates any problem of O-isotope ratios but such an impactor might have an iron core whose final disposition needs consideration. Approach velocities of a local planetesimal would be modest, perhaps 5–20 km/s, as a result of scattering by multiple, preimpact encounters.

Alternatively, the impactor could have been gravitationally scattered from a more distant zone of the solar system. In this case, we would expect different composition: ordinary chondritic or even a carbonaceous icy object from the outer solar system (henceforth OSS). Although the flux of scattered OSS bodies fell off rapidly inside Mars' orbit (Weidenschilling, 1975; see Hartmann and Vail, 1986), many such bodies may have been scattered into our region (Kaula and Bigeleisen, 1975). Phobos and Deimos may be examples. The largest OSS planetesimals were probably much larger than those of the inner solar system, and were likely to approach Earth with much larger velocities (30 km/s or more) than the local planetesimals.

OSS planetesimals are perhaps the second most likely source of impactor and offer radically different possibilities from local planetesimals. They would be a likely source for large bodies and the energy/kg would be much greater, allowing dramatic possibilities for ejecting large amounts of mass. Their compositions would be an opposite extreme from that of local bodies. The silicate materials would be carbonaceous, perhaps as in C- or D-class asteroids, and they would be at least 50% ice by mass (Hartmann, 1980). They might resemble Callisto, Rhea, Phoebe, Chiron, or Pluto.

Collisions with such bodies should be modeled along with those of terrestrial-type impactors. They may offer a better chance of driving dust into orbit because of the "second burn" effects of expansion of the enormous masses of volatiles vaporized. The cloud of (dissociated) water vapor (or other hydrogen-rich compounds) would offer an interesting environment for chemical alteration of the ejected or recondensed dust, especially oxidation/reduction reactions. O-isotope ratios would be a problem, but perhaps it can be shown that much of the vaporized planetesimal ice would escape the system, and that most of the silicates forming the resultant Moon came from terrestrial mantle materials carried aloft by the expanding gas.

6. Impactor Size

Lacking precision in models of giant impacts, we cannot specify exactly the impact parameters and impact mass necessary to generate a circumterrestrial swarm in which the Moon could have grown. However, three guidelines converge on the order of magnitude of necessary mass. First, impactors of at least two lunar masses appear necessary to generate the obliquities observed among planets (Hartmann and Vail, 1986). Second, as shown in Fig. 3, proto-Earths with initial rotation periods longer than 5–8 hours could be spun up to the observed angular momentum of the Earth/

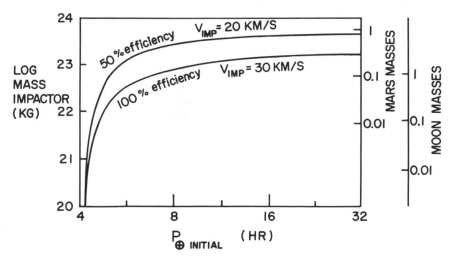

Fig. 3. Dependence of impactor mass, necessary to impart observed angular momentum of the Earth/Moon system, on initial rotation period of proto-Earth. Two different approach velocities and efficiencies of angular momentum transfer are shown.

Moon system by tangential impactors of roughly lunar to Martian mass with plausible velocities. The smaller sizes would apply if impact velocities were high (OSS impactors?) or if proto-Earth spun faster; larger ones if the impact were less tangential or less efficient in transferring angular momentum. This boundary condition is somewhat less stringent, incidentally, than the conclusion by Cameron and Ward (1976) and Cameron (1984) that the impactor's mass was comparable to that of Mars. Third, in the giant impact regime, although little is known about the ratio of ejecta mass/impactor mass, studies show this ratio decreasing from values around 10^3 for lab-scale hypervelocity impacts to values nearer 1 as impactor size increases, given a constant target material. Extrapolations of scaling relationships for rock and fluid target materials by Holsapple and Schmidt (1982) suggest ratios of the order of 1–10 for impactors of lunar to Martian size, even at velocities as high as 30 km/s. Only a fraction of this would reach orbital speed (see Section 7). Because of uncertainties in target material behavior for giant impacts it is hard to apply these data, but an impactor of at least several lunar masses appears necessary to generate and orbit the minimum amount of material necessary to make the Moon.

An alternative exists. A smaller impactor could generate a swarm of sublunar mass; this swarm could then grow by capturing interplanetary dust. This variant of the impact-trigger hypothesis may seem philosophically less attractive than ejection of a full lunar mass: if the swarm has much less than 1 M_{Moon}, an additional mechanism must be posited to bring the captured material to a composition that resembles the Moon. In particular, iron must be lost, and problems of the old capture theory reappear. Nonetheless, as we learn more about giant impact mechanics, models that require less initial mass ejection may become more attractive. Weidenschilling *et al.* (1986) consider such models, in which much of the lunar mass comes from silicate dust selectively captured by the swarm as denser iron cores of fragmented planetesimals pass on through. In the model discussed by Weidenschilling *et al.*, the initial circumplanetary swarm must come from somewhere, and the model becomes more attractive if not every planet has one, since not every planet has an Earth-like moon; a giant impact may be just what is needed to initiate the Weidenschilling-type swarm.

7. Toward a Model of the Velocity Distribution; How Much Material Went Into Orbit?

The question of the total mass and velocity distribution of the material exiting the giant impact site is vexing but critical to the impact-trigger hypothesis. Together with an analysis of acceleration during gas expansion, it determines how much material can end up orbiting in a swarm where the Moon can be formed.

We can treat the ejected material in four steps. In step A we imagine an impactor of mass M_I and radius R_I, striking the early Earth, mass M_E, at velocity V_{impact}. We then know the impact energy, surface gravity g, etc. We could then apply some model of cratering to estimate the total mass M_{ej} of material ejected from the impact site. Needless to say, our understanding of giant impact mechanics is very imperfect; some attempts at giant impact modeling appear in this book. One crude way to estimate the order of magnitude of M_{ej} is to use the gravity-dependent, target-dependent scaling formalisms developed by Holsapple and Schmidt (1982) and Housen et al. (1983). Figure 4 shows the results for the widest available range of materials, extrapolated into the M_I region of interest. For powder-like to water-like targets, the calculated M_{ej}'s for Mars-scale impactors hitting at 12 to 30 km/s range from about 0.04–100 lunar masses. The behavior of Earth's bulk material is uncertain. A general form of the Holsapple-Schmidt laws can be adapted from Schmidt (1980):

$$\log M_{ej} = \log M_I - \alpha \log R_I + 2 \alpha \log V_{impact} - \alpha \log g + (\log A - 0.5079\alpha) \quad (1)$$

For a representative rocky target, limestone, $\alpha = 0.627$ and $A = 0.337$ (Schmidt, 1980, after Bryan, 1980). For a projectile of $M_I = 0.8\ M_{Mars}$ with density 2000 kg/m^3, striking at 22 km/s, this gives a total ejecta mass of $M_{ej} = 5.5\ M_{Moon}$.

In step B, we attempt to calculate the distributions of velocities in the material leaving the immediate impact site. As noted by Cameron and Ward (1976) and others, these velocities may subsequently be increased by the "second burn" effect

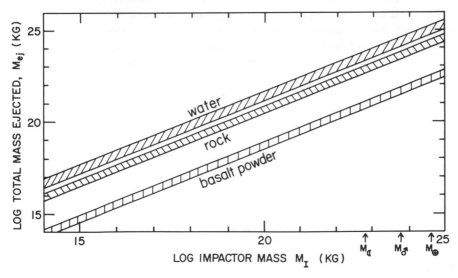

Fig. 4. Calculated total masses of ejecta for three different target materials, based on extrapolation of Holsapple-Schmidt scaling laws (from data cited in Hartmann, 1985). Lower edge of each bar corresponds to $v_{impact} = 12$ km/s; upper edge, $v_{impact} = 30$ km/s. Sand targets overlap lower edge of rock curve, which is based on limestone results by Bryan et al. (1980).

of acceleration due to gas expansion. Step C could deal with this augmentation of the initial ballistic velocities. We could add a step D, which could evaluate the temperature contours within the ejecta as a function of time, examining both the expanding gas and any entrained dust. The initial, high-speed jets, for example, may be the hottest material. This step is needed in evaluating the chemical history of the material. Steps C and D are beyond the scope of this paper.

Steps A and B are partly decoupled in an important sense: Step B simply says that if we have an ejecta mass of M_{ej} from an impact at V_{impact}, it will have a velocity distribution of such-and-such. We might be entirely mistaken in step A about the size of the impactor, M_I, required to produce this material, but given the existence of an ejecta cloud, M_{ej}, we could still calculate the distribution of velocities.

To carry out step B, the following formalism may be pursued (Hartmann, 1985). An approximate order-of-magnitude homology in ejecta pattern, as scaled to crater diameter, extends 6 or 7 orders of magnitude in size, from my gravity-dependent lab craters to larger craters and basins on the Moon and planets. We assume it extends another 1–1½ orders to ejecta from giant impacts. The rough homology suggests that distances achieved by any given fraction of the ejecta (e.g., the fastest 40%) are roughly proportional to the diameter of an "idealized" cavity, corresponding to the crater itself at diameters up to 100 km, but corresponding to a somewhat adjusted diameter for larger features. Cratering studies indicate that the ratio of (idealized cavity diameter)/(projectile diameter) in hypervelocity gravity-scaled impacts decreases from the order 30 to the order 1–2 for basin-scale features.

While the homology may not be exact, the striking similarity of ray and continuous ejecta patterns around, for example, Copernicus and lab-scale craters gives some confidence that we can scale initial ejecta velocity distributions from small- to large-scale impacts, in the following way: We want to distribute the ejecta mass, M_{ej}, from step A (equation (1)), launched from the rim of this cavity, according to the best data on velocity distributions. Let f(v) = fraction of ejecta travelling faster than v; R_{rim} = radius of idealized cavity; and r = distance ejecta is thrown at velocity v on the planet in question, assuming specified launch angle.

Hartmann (1985) found experimentally, within a few crater diameters of rim,

$$f(v) \propto (r/R_{rim})^{-\sigma} \tag{2}$$

Hartmann (1985) quoted values of $\sigma = 1.0 \pm 0.5$ and 1.9 ± 1.0 for two target powders over a range of impact speeds from 5–2000 m/s. Lab data (Hartmann, 1985) together with theoretical work (Melosh, personal communication, 1984) suggest that σ declines (broadcasting ejecta more widely) as v_{impact} increases from subsonic to hypervelocity regimes, so the lower value may be preferred here.

In independent but related work, McGetchin *et al.* (1973), Pike (1974), and Settle *et al.* (1974) described the variation of ejecta thickness, t, with distance from crater center, r, and found

$$t \propto (r/R_{rim})^{-B} \tag{3}$$

These authors all choose $B = 3.0$. Hartmann (1985, Table 4) cites other values that have been discussed. By integrating

$$\begin{aligned} M_r &= \text{mass ejected beyond distance r} \\ &\propto \int t r \, dr \end{aligned} \tag{4}$$

from $M_r = M_{ejecta}$ at $r = R_{rim}$ to $M_r = 0$ at a distance r_{max}, corresponding to maximum ejecta velocity, we have

$$f(r) \equiv \frac{M_r}{M_{ej}} = \frac{r^{2-B} - r_{max}^{2-B}}{R_{rim}^{2-B} - r_{max}^{2-B}} \tag{5}$$

The maximum velocities are expected to be of the order $1/2 \ V_{impact}$, and we can show (Hartmann, 1985) that within a few crater diameters

$$f(r) \cong (r/R_{rim})^{2-B} \tag{6}$$

This gives the identity (from equation (2))

$$\sigma \cong B - 2 \tag{7}$$

The "best value" of $B = 3$ corresponds to $\sigma = 1$, and the range of B values cited in the literature, to $0.5 < \sigma < 2.3$. This agrees remarkably with my experimental results of $\sigma = 1.0$ and 1.9, given that the other authors were dealing with ejecta from basins seven orders of magnitude bigger than my craters.

Assuming launch from near the rim at 41° to the horizontal (Hartmann, 1985), we can combine the equations to give the velocity distribution:

$$f(v) = \frac{(.101v^2 + R_{rim})^{-\sigma} - (.101v_{max}^2 + R_{rim})^{-\sigma}}{R_{rim}^{-\sigma} - (.101v_{max}^2 + R_{rim})^{-\sigma}} \tag{8}$$

The reader may object at this point that these calculations apply to ballistic solid ejecta and are irrelevant because some models of giant impact call for virtually all ejecta to be gaseous. Ejecta would then be entirely controlled by gas motions; the initial crater-exit velocities, irrelevant. Three responses: (1) Even in energetic impacts, as long as the target is not small enough to be disrupted or vaporized, the shock wave dissipates and the rim of the cavity is defined by an energy density where material is barely broken loose, not vaporized; some solid material must be ejected. (2) The present formalism does not deny a gas component, but proposes to decouple its consequences (step C) from an estimate of the initial dust/gas velocities (present step, B). The gas-expansion motions are superimposed on the initial velocity field calculated here, though they may eventually dominate. The cavity is thus viewed as a black box from which, regardless of impact details, a dust/gas mixture is ejected with a certain initial velocity distribution estimated here. Steps B and C both need improved modeling. (3) We know that gravity-scaled impacts up to the size of Imbrium basin ("crater" diameter 700–1300 km) emplace massive ejecta blankets of *solid* debris. The calculation performed here is consistent with their properties. If it is to be claimed that the solid/gas ratio in a giant impact is zero, it must be shown how the transition occurs from basins to the giant impact. Such a claim is inconsistent with the evidence that solid, minimally-altered rocks have been blown off the Moon and Mars. Basins, with their ejecta blankets, give the best link between traditional cratering theory and the results of giant impact.

We now follow the formalism for the example of an impactor of 5.14 (10^{23}) kg = 0.086 M_{Earth} = 0.8 M_{Mars} = 7.0 M_{Moon}, approaching Earth at 20 km/s. The impact velocity (tangential contact) would be 22 km/s. Impactor density is taken as 3600 kg/m^3, consistent with enstatite chondrite material. The total impact energy is 1.24 (10^{32}) joules. About half of this goes into the target, Earth (Hartmann, 1983b). This is some $6(10^4)$ times the energy needed for catastrophically fragmenting a mass of rock comparable to the Earth, assuming an "impact strength" of basalt, some 340 joules/kg (Greenberg and Hartmann, 1977, defined impact strength as the energy/mass needed for catastrophic disruption—disruption when the largest fragment = half the original target mass). We do not know that early Earth, with its plastic interior and perhaps molten/solid core, would have behaved like rock during impact, but this datum suggests that a large mass may have been pulverized or fractured.

On the other hand, even the total energy is only 48% of that needed to overcome the impact strength *and* disperse the total mass of material to infinity. Thus, from simple energy considerations alone, we anticipate substantial disruption of Earth, but not dispersal of much of its mass.

As an estimate of the total mass ejected, equation (1) gives M_{ej} = 4.6 (10^{23}) kg = 6.2 M_{Moon} = 0.9 M_I, if the target behaves like rock. As seen in Fig. 4, fluid-like targets are expected to eject an order of magnitude more material. We thus estimate a total ejecta mass of M_{ej} = 8 (10^{23}) kg = 11 M_{Moon} = 1.6 M_I. The ejection of more than the impactor's mass appears reasonable, especially since much of the impactor may be vaporized.

To be conservative, assume that the idealized cavity has only the radius of the projectile. (A hemisphere of this radius in Earth's crust, plus the projectile, would amount to a mass about 1.9× that calculated for M_{ej}. Our result above is plausible; the projectile may not excavate an entire hemisphere of this diameter, but make a shallower cavity of this size.) We also assume that the maximum velocity attained by any ejecta is $V_{impact}/2$. Finally, we assume $\sigma = 1.1$ as a compromise among the suggested values.

Figure 5, curve A, shows results computed for this impact with these assumed parameters, which are believed to be near the most probable values. Table 2 lists the parameters used in this and a number of other model calculations. Note that

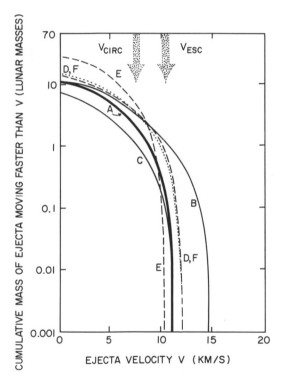

Fig. 5. Velocity distributions for the ejecta from large impacts. The calculations are described in the text and the parameters are given in Table 2.

the top half of the table gives data needed to estimate the total ejected mass. The bottom half is a virtually independent part of the formalism, a calculation of how this mass is distributed in velocity. Even if we should find that our linkage between impactor mass and ejecta mass (top half) is incorrect, the distributions of the listed ejecta masses by velocity (bottom half) may remain valid (and vice versa).

TABLE 2. *Examples of Models.*

Case Code Name	"3A"	"3B"	"2D"	"4"	"5A"	"5B"	"6"	"7"
Mass of Impactor M_i								
(Mars masses)	0.80	same	0.48	0.90	1.6	same	4	3
(Lunar masses)	7.0	same	4.2	7.9	14	same	35	26
Mass of Proto-Earth								
(Present-day Earth masses)	1	same	0.6	0.9	0.9	same	same	0.6
Impact Velocity V_i (km/s)	22	same	21	24	17	same	14	10
Circular Velocity of Proto-Earth (km/s)	7.9	same	6.9	7.6	same	same	same	7.0
Calculated Total Ejecta Mass (lunar masses)								
Rock target	6.2	same	5.0	7.9	8.2	same	11	7
Water target	25	same	30	48	48	same	64	43
Adopted Total Ejecta Mass (lunar masses)	11	same	6.8	7.9	14	same	27	15
Estimated Idealized Cavity Radius R_{rim} (projectile radii)	1.0	same	same	1.1	same	same	same	same
Estimated Maximum Ejecta Velocity Cutoff/V_i	1/2	2/3	1/2	1/2	1/2	2/3	2/3	1
Assumed Velocity Distribution Exponent σ	1.1	same	same	same	same	same	same	same
Calculated Ejecta Mass at $V_{circ} < V_{ej} < V_{esc}$ (lunar masses, neglecting gas expansion second burn effects)	1.7	1.6	0.9	1.3	1.3	2.7	4.8	3.5
Calculated Kinetic Energy in Ejecta/Total Impact Energy	10%	13%	10%	6%	9%	14%	24%	26%
Letter Code in Fig. 5	A	B	C	–	–	D	E	F

Given the impact configuration (Fig. 6), with the mass of the projectile much larger than atmospheric thickness, we assume that much of the material between $v_{circular}$ and v_{escape} has at least the potential of entering the orbital swarm. As the planetesimal hits and decelerates toward zero net relative velocity, and as the shock wave propagates into it, some material may explode off the outer surfaces from above the atmosphere, making orbit entry easier. (Many models, however, call for the impactor effectively to bury itself in Earth; see Melosh and Sonett, 1986). If thermal expansion "second burn" effects occur, material would be shifted toward higher velocities, and more material would be lofted into the circumterrestrial swarm, some of it attaining orbit. Viscous collisional effects will drive the inner part of the swarm inward, while the outer part moves outward, perhaps allowing accretion

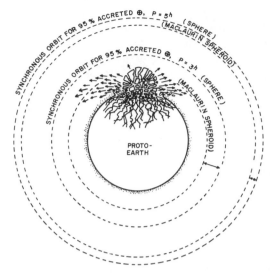

Fig. 6. Circumstances of the giant impact. Initial high-speed jetting comes from the mantle-mantle interface, but if the planetesimal breaks up in tangential impact or before full interpenetration, much additional mass may be ejected from an effective launch point well outside Earth's atmosphere (schematically shown by stippling). Synchronous orbits are shown for Earth-rotation periods of 3 hours and 5 hours, both for a spherical model Earth and adjusted for the effects of Maclaurin spheroid shape appropriate to the period. (Latter curves after Weidenschilling, personal communication, 1984). The latter curves are closer to Earth, facilitating lunar formation outside the synchronous orbit, where it must accrete in order to evolve outward tidally.

to occur beyond the synchronous point. Evolution of the swarm is beyond the scope of this paper, and is treated by Weidenschilling *et al.* and other authors in this volume.

In the impact just considered, no ballistic material leaves faster than v_{escape} because of the assumed velocity cutoff at $v_{impact}/2 = 11$ km/s. Some 1.7 M_{Moon} moves faster than $v_{circular}$ and thus has the potential to enter orbit. The total kinetic energy in all the ejecta is 9.8% of the initial impactor energy, consistent with data on energy partitioning (reviewed by Hartmann, 1983b, 1985). A larger impactor could be invoked to yield more mass between $v_{circular}$ and v_{escape}, if necessary to the model.

These are attractive results from the point of view of the impact-trigger hypothesis. If they are correct, an impactor of 0.8 M_{Mars} or somewhat larger could loft more than a lunar mass into a circumterrestrial swarm. The debris would be mantle material from Earth and impactor, largely pulverized, partially recondensed, and low in volatiles and iron.

Figure 5 and Table 2 show results of some other calculations to clarify the effect of changing some parameters. Curve B shows a result for the same collision, but with the maximum ejecta velocity raised to $2v_{impact}/3$. The total mass moving at orbital speed ($v_{circular} < v < v_{escape}$) is 1.6 M_{Moon}, nearly the same as before. Some 13% of the impact energy appears in the kinetic energy of ejecta.

There is no requirement that the impact occurs after Earth reached its present size, except that the core must have formed. Curve C shows a calculation for an impactor and a proto-Earth with 60% of the previously assumed masses, 0.48 M_{Mars} and 0.6 M_{Earth}, respectively. The approach velocity is 20 km/s, and the impact velocity is 21.4 km/s. The total ejecta mass is estimated at $5(10^{23}$ kg) = 6.8 M_{Moon}, about 1.4× that calculated for a rock target, but only about 23% of that calculated for water. Maximum ejecta velocity is taken as 11 km/s; other parameters are as before, with the idealized cavity equaling the impactor in diameter. In this model, the total mass moving at orbital speed is 0.9 M_{Moon}, and the ejecta contains 10% of the impact energy.

The first four models in Table 2, including curves A, B, and C, deal with impactors smaller than Mars, moving at relatively high speeds. The high speed has two aspects in the model: It provides high specific energy to help maximize ejecta, and, under our assumption that maximum velocities are one-half to two-thirds v_{impact}, it pumps adequate ejecta into the range $v_{circular}-v_{escape}$. However, the reader may note that these impact speeds are relatively high for local planetesimals (with Earth-like O isotopes). A body with aphelion between Earth and Mars would approach at a few km/s and impact at speeds more like 12 km/s, though multiple scattering by Earth could pump the value up. The remaining models, therefore, deal with larger impactors at slower speeds.

Curves D and E illustrate models with impactors of 1.6 and 4 Mars masses, hitting at 17 and 14 km/s, respectively. It is estimated that some 14 and 27 lunar masses are ejected; with the assumption that maximum ejecta velocities are two-thirds impact velocity, these models yield 2.7 and 4.8 lunar masses reaching orbital speed. Some 14% and 24% of the available energy (in the center of mass coordinate system) is consumed in ejecta. In these two cases, the planetesimal is so large, it is a major building block of Earth. In D, about half a lunar mass escapes the system, leaving a "completed" Earth-Moon system within a few percent of the present mass. In E, no mass escapes ballistically, and the remaining system is 30% more massive than the Earth-Moon system; mass loss by gas expansion would have to be invoked in such an impact.

This leads to a promising model, F. Here, the impact velocity is reduced even further: a large planetesimal of 3 Mars masses has accreted locally, approaches a growing proto-Earth at only about 5 km/s, and impacts at about 10 km/s. This planetesimal has 0.32 Earth masses, and the proto-Earth has reached 0.69 Earth masses, giving enough total mass to complete the system. Because of the low impact velocity, it must be assumed that the fastest ejecta reaches a cutoff speed about equal to the impact speed, to loft some material ballistically faster than circular velocity. This is not inconsistent with known cratering mechanics, though a lower cutoff could be chosen as in earlier models, and the remaining velocity increment ascribed to "second burn." In this model, about 3.5 lunar masses is lofted ballistically between $v_{circular}$ and v_{escape}, neglecting any "second burn" effects. Some 26% of the available energy appears in ejecta kinetic energy.

Note that the first four models in Table 2 invoked small projectiles hitting a relatively large target proto-Earth at relatively high speeds, required to eject enough mass; these models resembled ordinary cratering phenomena more than the last four models in Table 2. As we require more local planetesimals, velocities go down and the collision resembles more of a double planet collision than a cratering event. The result is harder to model from classical cratering theory, and involves a chaotic mass of vapor-charged, interacting debris, on the borderline of being gravitationally bound. Interestingly, while we started with a fairly small mass ratio (0.08–0.10) in the first four cases in Table 2, the high mass ratios of the final four models (0.19, 0.19, 0.46, and 0.52) echo an earlier conclusion about asteroids (Hartmann, 1979a,b): namely, that "collisions between comparable-sized bodies are a special class of collisions, rarer than other collisions, but producing interesting products, such as . . . co-orbiting binary pairs or swarms." The unusual co-orbiting planetary pair, Earth/Moon, may involve such an origin.

Because of the uncertainties in modeling giant impacts, I consider all these models as demonstrating the plausibility more than the necessity of the impact-trigger hypothesis. The spectrum of models, from smaller, high-speed impactors (more distant origin?) to large, low-speed, local impactors, remains a valid regime for further work. The models partition plausible fractions of the initial impact energy into kinetic energy of ejecta: 6–26% in the eight cases. For experimental impacts into rock at 5–6 km/s, O'Keefe and Ahrens (1977) cite values of 10–53%. For gravity-scaled impacts into sand at 6 km/s, Breslau (1970) found 30–53%; and for gravity-scaled impacts into powders I (Hartmann, 1985) found the fraction increasing with impact velocity, reaching about 3–30% at about 2 km/s. All eight models are lower limit cases, in the sense that the second burn effect is not considered; it may loft more mass than given by the ballistic calculations in Table 2 and Fig. 5. Models E and F are also conservative in the sense that they assume an ejecta mass less than that of the impactor; the models of Cameron (1986) and of Melosh and Sonett (1986) suggest that roughly half to all of large impactors are vaporized and ejected.

Our proposed step C, treatment of the second burn augmentation effects of the gas phase, remains to be done. Meanwhile, we must remain flexible in visualizing effects of both gas expansion and mutual interactions of fragments in the chaotic ejecta cloud, which would help prevent a fraction of the debris from pursuing trajectories back into Earth. Certainly, a "golf ball analogy," in which ejecta is viewed as following a ballistic path after an impulse delivered at Earth's surface, is too simple. Indeed, the latest calculations of Melosh (personal communication, 1985) show masses of vapor ejected at temperatures of a few thousand degrees and then expanding.

Other uncertainties remain in our formalism: (1) The values of ejected mass, M_{ej} (equation (1)), are very uncertain due to extrapolation of scaling laws. (2) Even if the extrapolation is correct, the behavior of the target material is uncertain. These two uncertainties may require upward or downward adjustment of M_{ej} from any one impactor, probably by as much as an order of magnitude. If a more fluid behavior is favored, the total ejecta could be nearly an order of magnitude greater, and many lunar masses (~17 in the first example) would move between $v_{circular}$ and v_{escape}. Ringwood's (1979) suggestion of an impact into a magma ocean (with a partially melted mantle) needs further study. It helps justify my choice of a compromise between the rock and fluid target parameters in Table 2. Also, one could plausibly compensate a modest downward adjustment in M_{ej} (by as much as an order of magnitude) by upward adjustments in M_I or V_{impact}. (3) We assumed the smallest plausible idealized cavity size, equaling the projectile diameter. This controls the mean velocity of ejecta; raising it would raise velocities and boost more ejecta into orbit. (4) The result is sensitive to the assumed maximum velocity of the ejecta. For example, in our assumed impact at 22 km/s, if v_{max} were reduced from one-half to one-third v_{impact}, then v_{max} would be only 7.3 km/s, slightly below $v_{circular}$, and no material could be orbited without the second burn effect. (Again, we note that solid SNC meteorites require the possibility of solid ejecta accelerated off Mars to at least 5.0 km/s.) If the impactor were from a more distant region of the solar system (raising problems of O isotope ratio if much impactor debris survives), or if it was scattered several times among the inner planets, v_{impact} could be higher, raising v_{max} in the ejecta and lofting more material orbital speeds. (5) The value of σ, the exponent in the velocity distribution, controls the amount of ejecta partitioned toward high velocities and is uncertain. We chose the most probable value, but more or less mass would achieve orbit if this variable were adjusted.

8. Degrees of Freedom in Accounting for Lunar Composition

The present model offers a number of possible mechanisms for arriving at lunar composition after a giant impact on Earth ejects material. (1) Most of the ejected material may be Earth-mantle material already close to lunar composition. (2) The composition of Earth's mantle may be evolving in time; the upper mantle composition at the time of impact may have been different (more lunar?) than at present. (3) An uncertain fraction of the ejecta may be planetesimal material. (4) The planetesimal may be biased toward the planetesimal mantle (also lunar-like) because of jetting from the contact region. (5) If Earth's surface were a magma ocean, the ratio of Earth/planetesimal material ejected may be enhanced over that for a solid Earth (Ringwood, 1979, p. 244). (6) Depending on the composition of hot gas ejected, oxidation-reduction and other chemical processing may occur as the material is ejected. (7) Volatiles are selectively lost due to the high temperature of the ejecta. (8) Enrichment of refractories may occur during condensation of cooling gas in some fraction of the total ejecta. (9) The orbiting swarm may capture an uncertain additional mass fraction of interplanetary material. (10) This capture may involve compositional sorting of the incoming interplanetary material along the lines described by Weidenschilling *et al.* (1986). (11) Differential capture of interplanetary materials from different source regions, with different approach velocities and capture cross sections, leads to different compositional evolutions of Earth and Moon (Hartmann, 1976). This last effect alters the relative amounts of local and distant materials in Earth's and Moon's most recently added surface layers, which we have preferentially sampled on the Moon in the Apollo and Luna programs.

All of these mechanisms deserve further study before we can say whether lunar composition can be explained by the impact-trigger hypothesis.

9. Summary and Remaining Problems

The origin of the Moon is an old mystery. This paper shows that large planetesimals were present at the scene, and had both the opportunity and the capability of triggering the deed by impacting the early Earth.

A host of questions remain, many of which are tractable. Can further modeling of accretion shed light on the expected sizes of the second-largest bodies growing in each planetary zone and the expected size of the largest impactors in each planet's history? What are the detailed mechanics of the corresponding giant impact? How would the impact of a high-velocity icy body from the outer solar system differ from the impact of a low-velocity local body? How do the crater-scaling laws extrapolate into this regime? Does the Earth respond to impact as a "rocky" or "fluid" material? What are the dust/gas ratio and temperature contours in the evolving ejecta? If the initial velocity distributions of crater-exiting material discussed here are relevant, what are the best values of parameters such as idealized cavity radius R_{rim}, velocity cutoff v_{max}, and velocity distribution exponent σ (equation (7))? How much does second burn gas expansion accelerate the ejecta? Is ejecta energy lost to turbulence? In short, how much material from a given impact remains aloft in a circumterrestrial cloud?

Once a cloud is formed, what are its dynamics and chemical evolution? Is expansion beyond the Roche limit and synchronous point achieved so that a Moon can accrete and move outward? What are the roles of gas, solid, and possible liquid phases? What is the timescale for accretion? The pioneering studies by Ward and Cameron (1978) and Thompson and Stevenson (1983) should be extended.

Regarding chemistry, how much chemical alteration is necessary to match lunar composition? Can lunar composition be accounted for, element by element, by considering only processing of terrestrial material, or is extraterrestrial material required? Is the extraterrestrial material attributable to the impacting planetesimal, or is subsequent capture of additional material by the swarm required? Can experimental data be generated on volatile loss patterns for mantle materials as a function of temperature?

Acknowledgments. Thanks are expressed to the attendees of the Kona Conference on the Origin of the Moon and to members of the Tucson Consortium (see Weidenschilling *et al.*, 1986) for their informative and vigorous debates on lunar origin. I appreciate reviews and suggestions by T. Ahrens and P. Nicholson. Thanks also to Paula Watson-McBride, Alix Ott, Karla Hankey, and Renee Dotson for facilitating manuscript preparation and review in the face of tight deadlines. This work was supported by NASA contracts NASW-3516, NASW-3718, and NASW-2909 in the Planetary Geology and Planetary Geophysics and Geochemistry programs. This paper is PSI Contribution No. 206. PSI is a division of Science Applications International Corporation.

References

Binder A. B. (1986) The binary fission origin of the Moon, this volume.

Breslau D. (1970) Partitioning of energy in hypervelocity impact against loose sand targets. *J. Geophys. Res.,* 75, 3987–3999.

Bryan J., Burton D., Lettis L., Morris L., and Johnson W. (1980) Calculations of impact crater size vs. meteorite velocity (abstract). In *Lunar and Planetary Science XI,* pp. 112–114. Lunar and Planetary Institute, Houston.

Cameron A. G. W. (1984) Formation of the prelunar accretion disk (abstract). In *Papers Presented to the Conference on the Origin of the Moon,* p. 58. Lunar and Planetary Institute, Houston.

Cameron A. G. W. (1985) Formation of the prelunar accretion disk. *Icarus, 62,* 319–327.

Cameron A. G. W. (1986) The impact theory for the origin of the Moon, this volume.

Cameron A. G. W. and Ward W. (1976) The origin of the Moon (abstract). In *Lunar Science VII,* pp. 120–122. The Lunar Science Institute, Houston.

Chapman C. R. and Davis D. R. (1975) Asteroid collisional evolution: evidence for a much larger early population. *Science, 190,* 553–556.

Clayton R. N. (1981) Isotopic variations in primitive meteorites. *Phil. Trans. Roy. Soc. London, A303,* 339–349.

Clayton R. N. and Mayeda T. (1975) Genetic relations between meteorites and the Moon. *Proc. Lunar Sci. Conf. 6th,* pp. 1761–1769.

Clayton R. N. and Mayeda T. (1983) Oxygen isotopes in eucrites, shergottites, nakhlites, and chassignites. *Earth Planet. Sci. Lett., 62,* 1–6.

Clayton R. N., Grossman L., and Mayeda T. (1973) A component of primitive nuclear composition in carbonaceous meteorites. *Science, 182,* 485–488.

Clayton R. N., Mayeda T., and Rubin A. E. (1984) Oxygen isotopic compositions of enstatite chondrites and aubrites. *Proc. Lunar Planet. Sci. Conf. 15th,* in *J. Geophys. Res., 90,* C245–C249.

Dodd R. T. (1981) *Meteorites.* Cambridge University Press, Cambridge. 368 pp.

Dohnanyi J. S. (1972) Interplanetary objects in review: Statistics of their masses and dynamics. *Icarus, 17,* 1–48.

Durisen R. H. and Gingold R. A. (1986) Numerical simulations of fission, this volume.

Giuli R. (1968) On the rotation of the Earth produced by gravitational accretion of particles. *Icarus, 8,* 301–323.

Gradie J. and Tedesco E. (1982) Compositional structure of the asteroid belt. *Science, 216,* 1405–1407.

Greenberg R. and Hartmann W. K. (1977) Impact strength: A fundamental parameter of collisional evolution (abstract). In *American Astronautical Society Division of Planetary Science, 8th Annual Meeting,* p. 15. American Astronautical Society, Honolulu.

Greenberg R. (1979) Growth of large, late-stage planetesimals. *Icarus, 39,* 141–150.

Greenberg R., Wacker J., Hartmann W., and Chapman C. (1978) Planetesimals to planets: Numerical simulation of collisional evolution. *Icarus, 35,* 1–26.

Harris A. W. (1977) An analytical theory of planetary rotation rates. *Icarus, 31*, 168–174.

Harris A. W. and Kaula W. (1975) A co-accretional model of satellite formation. *Icarus, 24*, 516–524.

Hartmann W. K. (1969) Terrestrial, lunar, and interplanetary rock fragmentation. *Icarus, 10*, 201.

Hartmann W. K. (1972) Paleocratering of the Moon: Review of post-Apollo data. *Astrophys. Space Sci., 12*, 48–64.

Hartmann W. K. (1976) Planet formation: Compositional mixing and lunar composition anomalies. *Icarus, 27*, 553–560.

Hartmann W. K. (1977) Large planetesimals in the early solar system. In *Comets, Asteroids, Meteorites* (A. Delsemme, ed.), pp. 277–281. Univ. of Toledo Press, Toledo, Ohio.

Hartmann W. K. (1979a) A special class of planetary collisions: Theory and evidence. *Proc. Lunar Planet. Sci. Conf. 10th*, pp. 1897–1916.

Hartmann W. K. (1979b) Diverse puzzling asteroids and a possible unified explanation. In *Asteroids* (T. Gehrels and M. Matthews, eds.), pp. 466–479. Univ. of Arizona Press, Tucson.

Hartmann W. K. (1980) Surface evolution of two-component stone/ice bodies in the Jupiter region. *Icarus, 44*, 441–453.

Hartmann W. K. (1983a) *Moons and Planets.* Wadsworth, Belmont, CA. 509 pp.

Hartmann W. K. (1983b) Energy partitioning in impacts into regolith-like powders (abstract). In *Lunar and Planetary Science XIV*, pp. 281–282. Lunar and Planetary Institute, Houston.

Hartmann W. K. (1984) Does crater "saturation equilibrium" occur in the solar system? *Icarus, 60*, 56–74.

Hartmann W. K. (1985) Impact experiments 1: Ejecta velocity distributions and related bodies from regolith targets. *Icarus, 63*, 69–98.

Hartmann W. K. and Davis D. R. (1975) Satellite-sized planetsimals and lunar origin. *Icarus, 24*, 504–515.

Hartmann W. K. and Larson S. (1967) Angular momenta of planetary bodies. *Icarus, 7*, 257–260.

Hartmann W. K. and Vail S. (1986) Giant impactors: Plausible sizes and populations, this volume.

Herbert F. and Davis D. R. (1984) Models of angular momentum input to a circumterrestrial swarm from encounters with heliocentric planetesimals (abstract). In *Papers Presented to the Conference on the Origin of the Moon*, p. 53. Lunar and Planetary Institute, Houston.

Holsapple K. and Schmidt R. (1982) On the scaling of crater dimensions, 2: Impact processes. *J. Geophys. Res., 87*, 1849–1870.

Housen K., Schmidt R., and Holsapple K. (1983) Crater ejecta scaling laws: Fundamental forms based on dimensional analysis. *J. Geophys. Res., 88*, 2485–2499.

Kaula W. and Bigeleisen P. (1975) Early scattering by Jupiter and its collisional effects in the terrestrial zone. *Icarus, 25*, 18–33.

MacDonald G. (1963) The internal constitutions of the inner planets and the Moon. *Space Sci. Rev., 2*, 473–557.

Mason B. (1958) *Principles of Geochemistry.* Wiley, New York. 329 pp.

Mason B. (1962) *Meteorites.* Wiley, New York. 274 pp.

McGetchin T. R., Settle M., and Head J. (1973) Radial thickness variation in impact crater ejecta: Implications for lunar basin deposits. *Earth Planet. Sci. Lett., 20*, 226–236.

Melosh H. J. and Sonett C. P. (1986) When worlds collide: Jetted vapor plumes and the Moon's origin, this volume.

Neukum G., Konig B., and Fechtig H. (1975) Cratering in the Earth-Moon system: Consequences for age determination by crater counting. *Proc. Lunar Sci. Conf. 6th*, 2597–2620.

O'Keefe J. D. and Ahrens T. (1977) Impact-induced energy partitioning, melting, and vaporization of terrestrial planets. *Proc. Lunar Sci. Conf. 8th*, pp. 3357–3374.

Pepin R. O. and Phinney R. O. (1976) The formation interval of the Earth (abstract). In *Lunar Science VII*, pp. 682–684. The Lunar Science Institute, Houston.

Pike R. J. (1974) Ejecta from large craters on the Moon: Comments on the geometric model of McGetchin et al.. *Earth Planet. Sci. Lett., 23,* 265–271.

Ringwood A. E. (1979) *Origin of the Earth and Moon.* Springer-Verlag, New York. 295 pp.

Ruskol E. (1972) Origin of the Moon III. *Sov. Astron. AJ, 15,* 646.

Safronov V. (1966) Sizes of the largest bodies falling onto planets during their formation. *Sov. Astron. AJ, 9,* 987.

Schmidt R. M. (1980) Meteor crater: Energy of formation—implications of centrifuge scaling. *Proc. Lunar Planet. Sci. Conf. 11th,* pp. 2099–2128.

Settle M., Head J., and McGetchin T. (1974) Ejecta from large craters on the Moon. *Earth Planet. Sci. Lett., 23,* 271–274.

Shmidt O. Yu. (1958) *A Theory of the Origin of the Earth.* Foreign Languages Publishing House, Moscow. 139 pp.

Smith B. A. and Voyager Imaging Team (1981) Encounter with Saturn: Voyager 1 imaging science results. *Science, 212,* 163–191.

Smith B. A. and Voyager Imaging Team (1982) A new look at the Saturnian system. *Science, 215,* 504–537.

Taylor S. R. (1982) *Planetary Science: A Lunar Perspective.* Lunar and Planetary Institute, Houston. 481 pp.

Thompson A. C. and Stevenson D. J. (1983) Two-phase gravitational instabilities in thin disks with application to the origin of the moon (abstract). In *Lunar and Planetary Science XIV,* pp. 787–788. Lunar and Planetary Institute, Houston.

Wänke H. and Dreibus G. (1984) Geochemical evidence for the formation of the moon by impact induced fission of the proto-earth (abstract). In *Papers Presented to the Conference on the Origin of the Moon,* p. 48. Lunar and Planetary Institute, Houston.

Ward W. R. and Cameron A. (1978) Disc evolution within the Roche limit. *Proc. Lunar Planet. Sci. Conf. 9th,* pp. 1205–1207.

Weidenschilling S. J. (1975) Mass loss from the region of Mars and the asteroid belt. *Icarus, 26,* 361–366.

Weidenschilling S. J. (1984) The lunar angular momentum problem (abstract). In *Papers Presented to the Conference on the Origin of the Moon,* p. 55. Lunar and Planetary Institute, Houston.

Weidenschilling S. J., Greenberg R., Chapman C. R., Herbert F., Davis D. R., Drake M. J., Jones J., and Hartmann W. K. (1986) Origin of the Moon from a circumterrestrial disk, this volume.

Wetherill G. W. (1977) Evolution of the Earth's planetesimal swarm subsequent to the formation of the Earth and Moon. *Proc. Lunar Sci. Conf. 8th,* pp. 1–16.

Wetherill G. W. (1985) Occurrence of giant impacts during the growth of the terrestrial planets. *Science, 228,* 877–879.

Wetherill G. W. (1986) Accumulation of the terrestrial planets and implications concerning lunar origin, this volume.

Wise D. U. (1963) An origin of the Moon by rotational fission during formation of the Earth's core. *J. Geophys. Res., 68,* 1547–1554.

Wise D. U. (1969) Origin of the Moon from the Earth: Some new mechanisms and comparisons. *J. Geophys. Res., 74,* 6034–6045.

Zellner B., Leake M., Morrison D., and Williams J. (1977) The E asteroids and the origin of the enstatite chondrites. *Geochim. Cosmochim. Acta, 41,* 1759–1767.

The Impact Theory for Origin of the Moon

A. G. W. CAMERON

Harvard-Smithsonian Center for Astrophysics, 60 Garden Street, Cambridge, MA 02138

A discussion is given of the essential ideas contained in the suggestion of W. R. Ward and the author that a very large impact on the proto-Earth involving a body with about the mass of Mars or greater led to the formation of an accretion disk about the proto-Earth, and that from the dissipation of this disk the Moon was formed.

The Angular Momentum Argument

In 1975 Hartmann and Davis (1975) noted that the spectrum of bodies bombarding the proto-Earth could have contained a number of quite large objects, having masses extending up to 10^{-1} Earth masses. They also noted that the collision of a large object with the proto-Earth would eject material into space, forming "a cloud of hot dust, rapidly depleted in volatiles." They then stated their expectation that the particles would interact and collapse into the equatorial plane where a satellite could form. Their paper suggested that an impactor of two lunar masses might be adequate, but contains an interesting footnote. When Hartmann first presented these ideas at a 1974 meeting, I objected that such a theory did not solve the angular momentum problem, and that this could only be done with a single collision involving a body having at least 10^{-1} Earth masses (i.e., about the mass of Mars). That idea was then under development by W. R. Ward and me and was presented the next year (Cameron and Ward, 1976). The two theories are often thought to be essentially the same, but in fact there is a profound difference in their approaches to the problem.

The question Ward and I started with was how big an object would have to be such that, striking the proto-Earth a glancing blow, it would impart to the proto-Earth and to any material that might be placed in orbit a total angular momentum equal to the angular momentum of the present Earth-Moon system. The answer was that the projectile should be about 0.1 Earth masses, or about the mass of Mars. Since the geometry of the collision was chosen in this question in an optimum way, the real answer to this question is that the *minimum* mass of the projectile

---

should be about the mass of Mars, and that the actual mass could be substantially larger if the collisional impact was more centrally directed.

Note that we did not ask how much angular momentum should be involved in the collision in order to allow one Moon mass to be in orbit just outside the Earth's Roche lobe. This is only about one-sixth of the present angular momentum of the Earth-Moon system. In order that the Moon can have moved out by tidal interaction to its present orbit, the spin of the Earth must have had at least the other five-sixths of the angular momentum of the present Earth-Moon system. Thus the result required of the collision is not only that enough prelunar material must be placed in proto-Earth orbit, but the proto-Earth itself must be induced to spin rapidly. Since the angular momentum involved here is very large compared to typical planetary spin angular momenta (scaled for the mass of the Earth), this spin is much more likely to be induced in a single large collision than in a series of smaller ones in which the impact points are likely to be spread randomly over the face of the target.

This single collision with a major body became the central theme of the ideas developed by Ward and me. It satisfied the angular momentum problem by definition, and the major question that perhaps one should ask is whether it is plausible that such a major body should have been present in the early solar system. We were also very much concerned with the question of how any of the material that is ejected from the proto-Earth could have gotten into orbit, since all material ejected from the Earth on a trajectory leaving the surface must return to below the surface unless escape velocity has been achieved. We realized that at the collisional velocities involved, large quantities of rock vapor would be produced, and hence that, in addition to the usual gravitational forces in the problem, acceleration of material could take place by means of gas pressure gradients.

On the other hand, Hartmann and Davis started with what they regarded as a plausible planetary formation picture and tried to derive a lunar formation scenario from it. The Hartmann and Davis picture is consistent with ours only in the case where their collision involves a body of the same large mass. They also did not address the need for nongravitational forces in the problem of placing material into orbit. These two issues were the starting points of our own approach.

The Question of Plausibility

Is it plausible that an extra planetary body with about the mass of Mars or more should have been wandering around in the inner solar system at an appropriate time to have participated in our postulated collision? Let us recall that there are two main theories of planetary accumulation.

One of these theories, which has a majority of adherents, assumes that the evolution of the solar nebula reached a stage when any turbulence had died away, so that the small solid particles present in the gas of the nebula would have a chance

---

to drift downward toward the midplane under the influence of gravity. When these particles have formed a rather thin disk, according to the theories developed by Safronov (1972) and by Goldreich and Ward (1973), the thin disk would become unstable against gravitational clumping, and would form objects with roughly the mass of asteroidal bodies. There then starts a process of hierarchical accumulation, in which a spectrum of masses develops in any region of the inner solar system, with the growth of some one body outdistancing the others. George Wetherill has been a leading investigator of this line of thought, and he stated at the Conference on the Origin of the Moon (from which this volume resulted) that it seems quite plausible to him that the second largest body could be 0.1 or more of the mass of the largest body. Thus there seems to be no reason to challenge the plausibility of our suggested impact theory of lunar formation within the framework of this general theory of planetary accumulation.

My own view of the planetary accumulation process (which seems to have rather few adherents) is that the history of planetary accumulation is a much richer and more complicated subject. We start with the earliest stages in the accumulation of the solar nebula, in which the infalling gas is cold, turbulent, clumpy, and only weakly gravitationally bound together. Under these conditions gravitational instabilities in the gas are likely to form giant gaseous protoplanets having something like the mass of Jupiter, plus or minus at least a factor of a few (DeCampli and Cameron, 1979). As the mass of the solar nebula increases, the local temperature rises rapidly, and the envelopes of the giant gaseous protoplanets are evaporated (Cameron et al, 1982). However, remnants of planetary mass may remain behind if the condensed particles originally present in the gas of a protoplanet have a chance to precipitate downward to the center of the protoplanet, forming a molten body of very refractory materials. These protoplanetary remnants become the nuclei for further planetary accumulation, and there may be more of them than the present number of inner planets. The solar nebula becomes very hot as the sun is formed from its dissipation, and then it cools and turbulence dies away. As in the "standard" picture, asteroidal-sized bodies can then form, containing elements of medium volatility, and these are primarily collected by the planetary nuclei already formed to complete the planetary formation process. In this picture there are probably no masses within a few orders of magnitude of the planets themselves in the inner solar system, but of course there can be extra planets.

A body in an eccentric Earth-crossing orbit is likely to collide with the Earth after a time on the order of 10^8 years (comparable to typical cosmic ray exposure ages of iron meteorites). That number gives the age of the solar system at which I would expect extra planets to be eliminated by collision. Of course, it is also the time when a major collision would be expected on the "standard" picture. I would expect all of the inner planets, including extra ones, to be formed molten with immediate iron cores, and the accumulation of planetesimals to have provided enough of an atmosphere to have given the planets a relatively cool radiating surface, and thus to have kept them molten over the 10^8 year interval. I believe the proponents of the "standard" picture would believe the planets to have been formed warm but perhaps in general not molten, and maybe that is the major consequence of our different views on the scenario leading to a major collision with the proto-Earth.

Formation of the Accretion Disk

If the center of the impact-induced explosion is at or below the surface of the proto-Earth, then all material emerging from the site of this collision, if on a ballistic trajectory, would either escape from the proto-Earth or fall back onto the surface of the proto-Earth. Neither outcome would place material in orbit and hence no such material could participate in the formation of the Moon. The key to the resolution of this problem lies in the fact that much of the colliding material would be vaporized.

Ahrens and O'Keefe (1972) have discussed the shock vaporization of a variety of materials. The threshold for vaporization in different rocks is predicted by them to lie in the range 7 to 12 km/sec impact velocity, with most values in the range 9 to 11 km/sec. The threshold for iron appears to be a little higher. It is obvious that these thresholds will be significantly reduced if the impact occurs in already molten material, which I believe to be probable. It is expected that the collision of a Mars-sized projectile with the proto-Earth will occur at the escape velocity, 11 km/sec or higher, possibly as much as 14 to 15 km/sec, depending on the eccentricity of the projectile orbit.

Normally, when a small projectile hits a large target, a majority of the impact energy can be dissipated within a large volume in the target. In this case the impact velocity is substantially above the vaporization threshold and the bodies differ only by a factor of about two in radius. Therefore a great deal of vapor should be produced in the collision. The gas in the vapor cloud rising above the surface of the proto-Earth is subject to acceleration by pressure gradients as well as by gravity. This is the crucial difference that makes the formation of an accretion disk around the proto-Earth possible. For example, if the projectile hits the proto-Earth tangentially at 12 km/sec, and if the gas cloud were to receive an equal contribution of mass from the proto-Earth and from the projectile, then the gas would have a mean transverse velocity of 6 km/sec, which is slightly suborbital. The gas on the forward edge of the cloud then need only receive nearly 2 km/sec additional velocity in the direction of the collision due to pressure gradient acceleration to reach orbital velocity. It should also be clear that the expected yield of material in orbit is much less than half of the vaporized mass. It should also be clear that if the gas components derived from the projectile and the proto-Earth do not mix efficiently, then a considerable amount of the gas derived from the projectile may be lost from the system because it exceeds escape velocity, while most of that derived from the proto-Earth will fall back on the surface.

An estimate of the yield of material in proto-Earth orbit as a function of the collision conditions is thus highly desirable. I have recently made estimates of this yield using a particle-in-cell hydrodynamic code; the results were reported at the Conference on the Origin of the Moon and are published (Cameron, 1985).

The particle-in-cell method is a rather crude technique for determining the hydrodynamic behavior of a gas with an irregular three-dimensional geometry. The procedure can at best make semiquantitative predictions about the flow of gas in a system, and it is difficult to represent nonideal behavior in the equation of state of the gas. Nevertheless it is a reasonable method to use in the first investigation of a problem that exhibits relatively little symmetry. The available space is subdivided into cells, and there should be enough particles present so that in the active region of the problem there are several particles in each cell.

In any one time step of a problem, the positions of the particles are advanced with the particles traversing ballistic orbits subject to acceleration due to the gravitational field. At the end of a time step the position and motion of the center of mass in each cell is computed, and the energy associated with random motions with respect to the center of mass is extracted from the cell. The lowest energy state of a cell involves rigid rotation around the center of the mass. The extracted energy from a cell is then fed back into the particles randomly but in a way that preserves the angular momentum. Thus at the end of a time step the mean motion of the fluid in each cell is preserved but the motions of the particles are randomized.

Imagine two adjacent cells sharing a common boundary but with different numbers of particles. The randomization of the motions of the particles will then send more particles across the common boundary from the rich cell to the poor cell than will be sent in the reverse direction. In this way the scheme acts to accelerate mass down pressure gradients.

This scheme was used to predict the behavior of the colliding system for a wide variety of initial conditions. The velocity of impact was varied from 11 to 14 km/sec. The initial state of the shocked vapor was varied from "hot" (odd numbered particles given the velocity of the proto-Earth and even numbered particles given the velocity of the projectile) to "cold" (all particles given the same mean velocity). The source region of the gas was varied from just the region of overlapping hemispheres of the colliding bodies to essentially the entire volume of the projectile, including the overlapping region. The prediction of interest was the amount and angular momentum of the gas that settled into orbit about the proto-Earth, neither escaping from the system nor falling back onto the surface of the proto-Earth. Because the disk that is formed is subject to dissipation, with mass flowing both inward toward the proto-Earth and outward toward the Roche lobe, a "successful" disk formation was deemed to be one in which the mass in orbit was at least twice that of the Moon and the angular momentum was at least twice that of a lunar mass in orbit just beyond the Roche lobe.

The results showed that these conditions can be achieved under a wide variety of conditions. For the best yields the gas should be neither too "hot" nor too "cold," the collisional velocity of the projectile should be a few km/sec greater than the escape velocity minimum, and most of the volume of the projectile should be part of the source region for the vapor that is produced. In all cases the majority of the mass that goes into orbit following the collision is derived from the projectile.

These initial conditions are optimally met for a projectile in orbit about the sun in a proto-Earth-crossing trajectory of significant eccentricity. This is consistent with a time scale for the collision on the order of 10^8 years after the formation of the solar nebula.

The collision should lead to the loss of the original atmosphere of the proto-Earth (Cameron, 1983). Judging from the fact that Venus has an atmospheric content of rare gases that is very large compared to the Earth, one can conclude that the loss of the atmosphere occurred after the bulk of the accumulation of volatile-containing planetesimals had occurred. This is another argument in favor of the collision happening about 10^8 years after formation of the solar nebula.

Dissipation of the Accretion Disk

Once the accretion disk has been formed, it is subject to severe dissipation. The column density of the disk is great enough that self-gravitation is important, and there is instability against local clumping due to gravitational instabilities (Ward and Cameron, 1978). However, inside the Roche limit, the gravity of the proto-Earth is strong enough to shear apart any such clumping that occurs. This dissipates a lot of energy and hence leads to a spreading inward and outward of the mass in the disk. To the extent that the temporary mass fluctuations can raise small tides in the proto-Earth, there can also be a small amount of angular momentum transfer from the spin of the proto-Earth to the orbital motion of the disk material.

Ward and I estimated the dissipation time of the disk to be months to years. However, Thompson and Stevenson (1983) pointed out that such a rapid dissipation would release energy at a sufficient rate to completely volatilize the material in the disk. This would obviously slow the dissipation rate to a value that would allow some of the material to remain in condensed form. Thompson and Stevenson also pointed out that in a mixture of the gaseous and condensed phases of a substance ("froth"), sound speed is greatly reduced relative to the value in either phase alone. This enhances the gravitational instability. They estimated the dissipation time of the disk to be on the order of a century.

So far there has not been a careful examination of what happens at the outer edge of the disk as it recedes past the Roche limit. One possibility is that a gravitational instability occurs that forms a body of not very great mass, and that this continues to accrete matter as it emerges or is perturbed past the Roche limit. Another possibility is that a small body forms, as before, and then recedes from the proto-Earth because of tidal interaction with both the proto-Earth and the accretion disk. Then another body can form and recede in its turn, and so on. If the last body to form in this way were to be the most massive, then it would recede fastest and proceed to sweep up its predecessors. Otherwise this last scenario might be in trouble.

Discussion

The general theory outlined above requires much more detailed examination to expose its strengths and weaknesses and to fill in the gaps.

The iron core of the projectile is probably not significantly vaporized. Assuming, as stated above, that projectile is a differentiated planetary body, then the iron core should have concentrated and extracted the siderophile elements from throughout the body of the projectile. Hence the accretion disk and the Moon should be depleted in these elements, as is observed to be the case.

The deposition of mass into the proto-Earth by the collision should release a great deal of energy, which I have estimated to raise its surface temperature to about 6000 K (Cameron, 1983). Dissipation within the accretion disk probably maintains the temperature near 2000 K in the vicinity of the Roche lobe. Hence when the Moon forms it should do so in a temperature field that will have prevented most of the elements of medium volatility from condensing with any significant abundances. Again, this is observed to be the case.

It should be noted that the accretion disk need not form precisely in the equatorial plane of the proto-Earth initially, nor should one expect the orbit of the Moon to lie in that plane today. The accretion-disk-forming collision adds a very large angular momentum vector to the proto-Earth, which will nevertheless add to the previous angular momentum vector. Thus the initial plane of the disk will differ from that of the equator. There will be a large equatorial bulge, and there will thus be a tendence to align the two angular momentum vectors. The rate at which this happens needs to be investigated. It is likely that a great deal of matter escapes from the proto-Earth system after the collision (so my simulations have generally indicated), but this matter will go into independent Earth-crossing orbits and will before too long probably recollide with the proto-Earth, further disturbing the angular momentum vector of the proto-Earth.

It has generally been assumed that the equality of the oxygen isotope ratios in the Earth and the Moon indicates that the Moon was formed out of material that condensed at the Earth's orbital distance from the sun. Until the origin of the variations in oxygen isotope ratios is properly understood, this principle will remain empirical, but it seems reasonable to accept it provisionally. It seems to me that this requirement is plausible with either of the two models of planet accumulation.

There are a number of other aspects of lunar composition, such as the iron oxide content of its mantle, that were discussed at the Conference on the Origin of the Moon. In particular, Ringwood felt that if the bulk of the Moon could not be made out of terrestrial material, then the projectile should be considerably larger than Mars to meet chemical restraints. I see no problems with that; the calculations that I did were for a Mars-sized body, but this is only a lower limit, and the projectile could perhaps be as large as one-third of the mass of the Earth without getting into trouble with the dynamics.

Acknowledgments. This research was supported in part by the National Aeronautics and Space Administration under grant NCR 22-007-269.

References

Ahrens T. J. and O'Keefe J. D. (1972) Shock melting and vaporization of lunar rocks and minerals. *The Moon, 4,* 214–219.

Cameron A. G. W. (1983) Origin of the atmospheres of the terrestrial planets. *Icarus, 56,* 195–201.

Cameron A. G. W. (1985) Formation of the prelunar accretion disk. *Icarus, 62,* 319–327.

Cameron A. G. W. and Ward W. R. (1976) The origin of the Moon (abstract). In *Lunar Science VII,* pp. 120–122. The Lunar Science Institute, Houston.

Cameron A. G. W., DeCampli W. M., and Bodenheimer P. H. (1982) Evolution of giant gaseous protoplanets embedded in the primitive solar nebula. *Icarus, 49,* 298–312.

DeCampli W. M. and Cameron A. G. W. (1979) Structure and evolution of isolated giant gaseous protoplanets. *Icarus, 38,* 367–391.

Goldreich P. and Ward W. R. (1973) The formation of planetesimals. *Astrophys. J., 183,* 1051–1061.

Hartmann W. K. and Davis D. R. (1975) Satellite-sized planetesimals and lunar origin. *Icarus, 24,* 504–515.

Safronov V. S. (1972) *Evolution of the Protoplanetary Cloud and Formation of the Earth and Planets,* translated from the Russian. Israel Program for Scientific Translations, Tel Aviv.

Thompson A. C. and Stevenson D. J. (1983) Two-phase gravitational instabilities in thin disks with application to the origin of the Moon (abstract). In *Lunar and Planetary Science XIV,* pp. 787–788. Lunar and Planetary Institute, Houston.

Ward W. R. and Cameron A. G. W. (1978) Disc evolution within the Roche limit (abstract). In *Lunar and Planetary Science IX,* pp. 1205–1207. Lunar and Planetary Institute, Houston.

Short Note: Snapshots from a Three-Dimensional Modeling of a Giant Impact

W. BENZ AND W. L. SLATTERY

Los Alamos National Laboratory, Los Alamos, NM 87545

A. G. W. CAMERON

Harvard-Smithsonian Center for Astrophysics, 60 Garden Street, Cambridge, MA 02138

We present results of a series of numerical simulations of an impact between the proto-Earth and an object of the size and mass of Mars (Benz *et al.*, 1985). The simulations were done using the "Smoothed Particle Hydrodynamic" (SPH) method (Gingold and Monaghan, 1982) using a total of 2048 particles.

The model includes self-gravity, shock heating, material under tension, and the possibility for material to be either vapor, liquid, or solid. However, to keep the problem tractable we did not include: shear strength (most of the simulations were started with molten planets), radiative transfer, or energy losses due to radiation (for timescale reasons). For convenience both planets were assumed to be made of granite (new models including iron cores are in progress). Finally, we modeled the thermodynamics using the Tillotson equation of state (Tillotson, 1962).

The simulations leading to the formation of a prelunar accretion disk of about 1 to 3 lunar masses and angular momentum corresponding to the Earth-Moon system are the ones started with a low relative velocity (less than 4 km/sec at infinity) and a large impact parameter. Starting with molten or solid planets does not make any difference. After the impact, part of the impactor forms a clump orbiting around the proto-Earth. This clump, however, is on a very eccentric orbit, bringing it back well inside the proto-Earth's Roche limit; it is therefore destroyed and spread out into a disk. It is worth noting that the material ending in the clump originates completely from the side of the impactor opposite to the proto-Earth at the time of the impact.

Figures 1 and 2 show snapshots of events taking place between the impact and the formation of a clump in orbit. Initial relative velocities at infinity are respectively

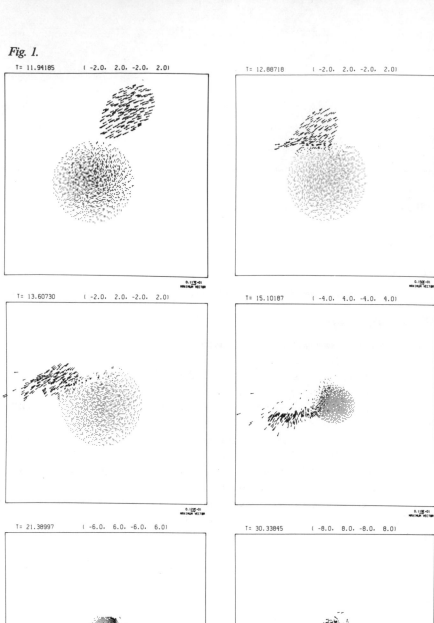

Fig. 1.

T= 11.94185   (-2.0, 2.0, -2.0, 2.0)

T= 12.88718   (-2.0, 2.0, -2.0, 2.0)

T= 13.60730   (-2.0, 2.0, -2.0, 2.0)

T= 15.10187   (-4.0, 4.0, -4.0, 4.0)

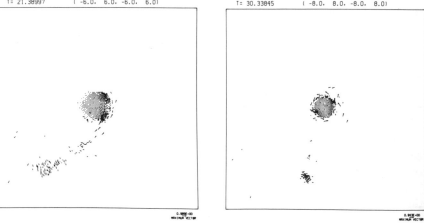

T= 21.38997   (-6.0, 6.0, -6.0, 6.0)

T= 30.33845   (-8.0, 8.0, -8.0, 8.0)

Fig. 2.

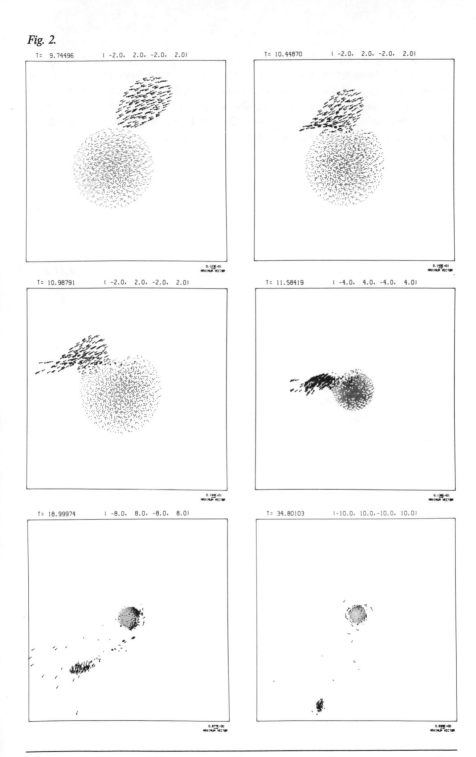

0 km/sec (Fig. 1) and 3 km/sec (Fig. 2); all planets were considered as molten. The impact parameter in Fig. 2 is about 1.2 times larger than in Fig. 1. Plotted are velocity vectors at particle locations projected in the plane defined by the center of the two planets. Time is given in the upper line of each frame (1 time unit = 18.2 minutes). Note also the change of scale between frames. The coordinates of the four corners of the box are also given in the upper line (in units of 8008 km).

References

Benz W., Slattery W. L., and Cameron A. G. W. (1985) The origin of the moon and the single impact hypothesis I. *Icarus*, in press.

Gingold R. A. and Monaghan J. J. (1982) Kernel estimates as a basis for general particle method in hydrodynamics. *J. Comput. Phys., 46*, 429.

Tillotson J. H. (1962) Metallic equation of state for hypervelocity impacts. *General Atomic Report GA3216.*

When Worlds Collide: Jetted Vapor Plumes and the Moon's Origin

H. J. MELOSH AND C. P. SONETT*

Department of Planetary Sciences and Lunar and Planetary Laboratory, University of Arizona, Tucson, AZ 85721

Collision of a Mars-size planetesimal with the proto-Earth could explain both the anomalously large angular momentum of the Earth-Moon system and the similarity between the Moon's bulk composition and the Earth's mantle. A detailed mechanical analysis of such a collision between two comparably-sized spheres as a function of the impact parameter (a measure of the obliquity of the impact) and the radius ratio of the spheres indicates that a total of up to 0.40 projectile mass is ejected as a plume or jet of high-speed, dense, hot vapor when collision velocities exceed 11 km/s. The plume's velocity in the target's frame of reference is comparable to the impact velocity, and its mass is a peaked function of impact parameter. The vapor is concentrated in a fan that diverges only a few tens of degrees in azimuth from the extrapolated flight direction of the projectile and for moderately oblique impacts is elevated circa 20° above the surface of the target planet. The vapor plume is composed of an approximately equal mixture of target and projectile material. All jetted vapor is derived from the outer half of the projectile and a comparable depth in the target. The extrapolation to the Earth-Moon system thus yields a vapor cloud free of either proto-Earth or projectile core material, explaining the Moon's depletion in iron. Volatiles present in the Earth or projectile are lost to space during the condensation of the hot vapor as it expands adiabatically into vacuum. The initial hydrodynamic expansion is expected to inject a fraction of this material into orbit.

The Mega-impact Hypothesis

The angular momentum per unit mass of the Earth-Moon system has long been known to be anomalously large compared to the other terrestrial planets. Conventional planetary accretion models that envision protoplanet growth by the accumulation of large numbers of small (circa 10-km-diameter) planetesimals do not predict

---

*Carnegie Fellow currently on leave at the University of Edinburgh

significant angular momentum densities in any of the terrestrial planets (Harris and Ward, 1982) because the encounter probabilities with planetesimals that impart either prograde or retrograde rotation are roughly equal. Hartmann and Davis (1975) first suggested that the impact of a large planetesimal with the proto-Earth late in the accretion process could both provide the extra angular momentum and eject volatile- and iron-poor material that might form the Moon. The other terrestrial planet's spin states are thought to be determined by late-stage impacts of smaller planetesimals than that responsible for the Earth-Moon system.

Compared to the Earth's mantle, the Moon is strongly depleted in both iron and volatile elements. However, aside from these profound differences, the Moon's bulk composition is rather similar to the Earth's (see Drake, 1986, for a recent review). This similarity prompted Ringwood (1970, 1972, 1984) to suggest that impacts on the proto-Earth somehow boiled off Earth mantle material and injected it into orbit. Ringwood's model has been severely criticized on dynamical grounds (Wood, 1977), mainly because the many small impacts he proposes could not plausibly impart a net orbital angular momentum to the cloud of vaporized mantle material. The hypothesis of a *single* large impact, however, removes the angular momentum difficulty while still allowing a large component of Earth mantle material to be incorporated in the Moon.

The major difficulty with models that attempt to derive the Moon from the Earth by impact processes is that ejected debris that follows Keplerian orbits from the instant of ejection must, according to standard ballistics, either escape the Earth entirely (hyperbolic orbits) or reimpact the Earth after completing only part of an orbit. Elliptical orbits return to their starting points, which, by hypothesis, lie on the Earth's surface. Stevenson (1984) epitomized the situation by calling on a "second burn," in analogy with the burn-coast-burn strategy of injecting spacecraft into high orbits. Cameron and Ward (1976), however, pointed out that if *vapor* (as opposed to solid material) was ejected, the vapor would expand under the influence of both pressure gradients and gravity. Condensates from a hydrodynamically expanding vapor cloud would not decouple from the gas and enter Keplerian orbits until some time after the impact, when they are already high above the Earth. There is thus a possibility that some of the ejected material will achieve stable, closed orbits around the Earth. Stevenson (1984) also suggested that the complex and changing gravitational field induced by a massive impactor could aid in inserting the ejected debris into orbit.

Neither Cameron and Ward (1976) nor Stevenson (1984) examined the mechanical conditions of a very large impact. Their work is mainly concerned with the evolution of the vapor after ejection. There is a difference of opinion about the origin of this vapor: Cameron and Ward (1976) and Cameron (1985) suppose that it is predominantly projectile material, while Ringwood (1984), arguing from geochemical premises, supposes that it is predominantly Earth mantle material.

This paper examines the impact event itself. An approximate calculation provides details on the amount of the highest speed, most highly shocked ejecta and its ejection pattern, speed, angles of launch, etc. as a function of the impact parameter and the ratio of the sizes of the colliding protoplanets. We find that in a certain range of parameters, a great deal of vapor (up to circa 40% of the projectile mass) may be ejected at speeds approaching the impact speed, consisting of an approximately equal mixture of projectile and target mantle material. The results of these computations are in accord with the mega-impact hypothesis.

There are many types of ejecta produced in an impact event. The vapor plume that is the most likely source of the Moon's material is produced by a process known as "jetting," a type of crater ejecta that may be new to many readers. The classic "ejecta curtain" plays no role in the injection of debris into Earth orbit: its velocity is simply too low. Appendix A reviews the current understanding of each class of high-speed ejecta from an impact and evaluates the potential contribution of each to an orbital debris cloud. Readers unfamiliar with recent work on impact cratering may find this review useful in establishing connections between the jetting process described in this paper and the other processes occurring during an impact.

Jetting from the Collision of Unequal Spheres

Jetting occurs during the early contact stage of an impact between irregular objects. The solid surfaces of the projectile and target impinge obliquely upon one another (Fig. 1). When the angle between the two surfaces is sufficiently large, a jet of high velocity material squirts out from the intersection. This material is the fastest moving and most highly shocked of all classes of ejecta.

Jetting was originally described by Birkhoff *et al.* (1948) as a hydrodynamic phenomenon. Later workers (Walsh *et al.*, 1953; Allen *et al.*, 1959; Al'tshuler *et al.*, 1962) recognized the importance of shocks in determining a minimum angle, θ_{cr}, below which jetting does not occur. Most investigations of jetting are concerned

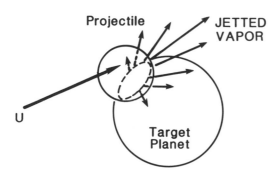

Fig. 1. Schematic illustration of the collision between the proto-Earth and a Mars-size protoplanet. During the early contact phase of the collision, a plume of vapor jets from the interface between the two impacting bodies. Some of this vapor may achieve a stable orbit around the proto-Earth and eventually condense, forming the Moon.

with the symmetric two-dimensional oblique impact of flat plates where a steady state solution is appropriate, although Walsh *et al.* (1953) briefly discuss both asymmetric jets and time dependent flows. A few numerical investigations (e.g., Harlow and Pracht, 1966) have addressed more general aspects of jetting such as the jet width and velocity for realistic equations of state. Kieffer (1975, 1977) derived values of the minimum angle as a function of velocity for geologic materials and pointed out that even low-velocity (1–2 km/s) collisions could produce melt because of the pressure multiplication that occurs in jetting configurations.

Symmetric jets

Most discussions of jetting assume that the jet arises from the oblique convergence of two thin plates of identical compositions, illustrated in Fig. 2. The two plates in Fig. 2a each move with velocity v_p perpendicular to their faces and strike each other at half-angle θ. We call this coordinate frame the "standard" frame since most publications cite the velocity v_p (many authors call this the "laboratory" frame; however, in an impact, this coordinate system moves with respect to the target, so we avoid this usage). The net rate of approach of the plates to one another, perpendicular to the axis of symmetry, is $2\,v_p\cos\theta$. Figure 2b illustrates the other commonly used coordinate system, the "collision" frame. This frame is related to the standard frame by a velocity translation along the symmetry axis. Its principal virtue is that the stagnation point of the flow is stationary in this frame. The two

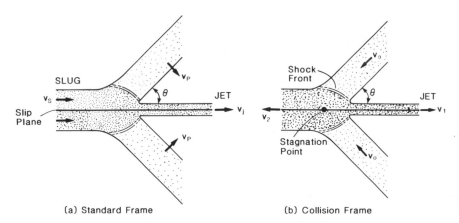

Fig. 2. Jetting in the symmetric collision of two plates of equal thickness. The plates intersect at half-angle θ and velocity v_p in the "standard frame." The "collision frame," in which the stagnation point is at rest, is shown in the right half of the figure. The plane of symmetry of each figure is the "slip plane." The velocity jumps discontinuously across this plane in an asymmetric collision.

incident plates' velocity vectors are here parallel to their faces. Simple vector analysis shows these velocities are related by $v_o = v_p \cot \theta$. Harlow and Pracht (1966) define a constant $f = v_1/v_o$, where v_1 is the jet velocity in the collision frame. The jet velocity in the standard frame, v_j, is thus

$$v_j = v_p \ (\csc \theta + f \cot \theta) \tag{1}$$

The ratio f is nearly 1 in most circumstances, making analysis of jetting velocities particularly simple.

The maximum pressure of the jetted material is computed from the observation that at the stagnation point the total kinetic energy (per unit mass) of the colliding plates $(1/2 \ v_o^2)$ is converted into enthalpy (PV). Table 1 illustrates these maximum pressures for a number of geologic materials and "impact velocities" $2 \ v_p$ following the method of Kieffer (1977). It is clear that velocities on the order of Earth's escape velocity (circa 10 km/s) are sufficient to completely vaporize nearly all geologic materials that enter the jet (where, following Ahrens and O'Keefe, 1977, we assume that "complete vaporization" takes place at pressures above about 500 GPa). Without jetting to enhance the shock pressures, 10-km/s-impacts would not produce significant amounts of vapor (Croft, 1982).

The quantity of jetted material has been estimated only for the steady convergence of thin plates. In this case it is derived from momentum conservation in conjunction with the jet and "slug" velocities (Birkhoff et al., 1948). Harlow and Pracht (1966) noted that the jet width is linearly related to the angle of incidence θ above the critical angle θ_{cr}. They derived an empirical formula for the jet width w in terms of the plate thickness t:

$$w = t \ (\theta - \theta_{cr})/(\pi - \theta_{cr}) \tag{2}$$

TABLE 1. Increase of Maximum Pressure in Jetting.

Material*	Impact Velocity, $2v_p$							
	5 km/sec		10 km/sec		15 km/sec		20 km/sec	
	Normal Pressure GPa	Jet Pressure** GPa	Normal Pressure GPa	Jet Pressure GPa	Normal Pressure GPa	Jet Pressure GPa	Normal Pressure GPa	Jet Pressure GPa
Aluminum	61	107	170	474	323	1118	523	2040
Quartz	40	109	136	454	288	1034	497	1850
Bronzitite	67	129	184	583	349	1387	564	2544
Dunite	68	129	198	559	390	1301	644	2356

*Material data from Kieffer (1977).
**$\theta_{cr} = 17.5°$ is assumed.

Although this equation strictly applies to thin plates only, we will use it to estimate the width of the jet ejected from the collision of unequal spheres. This is the weakest assumption of our jetting calculation and is made only as a matter of necessity: there are no estimates of jet width for nonsteady flows. An obvious extension of our work would be to perform the necessary three-dimentional hydrodynamic computation numerically and thus eliminate this weak point.

We estimate the quantity of material jetted from the collision of two spheres of radii $r_1 > r_2$ by replacing the plate thickness t in (2) by the radius of the smaller planet, r_2. The width w thus computed is converted into a mass flux per unit length of intersecting surfaces by multiplying the width by the jet velocity and *uncompressed* density. Integration of this flux over the total duration of jetting and over all azimuths (see Appendix B for details) yields the total mass ejected by jetting, illustrated in Fig. 4.

Asymmetric jetting

Even when the projectile and target are composed of the same material (which we assume throughout this discussion), it is generally impossible to find a coordinate frame in which both the projectile's and target's surfaces are perpendicular to their velocity vectors, as in the standard coordinate frame (Fig. 2a). There is nearly always a component of velocity parallel to the plane of symmetry. These velocity components may be either in the plane of Fig. 2 or perpendicular to it (parallel to the two plates' line of intersection). This situation can be treated by assuming that the junction of the two plates is an impenetrable *slip surface* across which the tangential velocity jumps discontinuously (Walsh *et al.*, 1953). Although, in reality, turbulent mixing of material in the two plates might be expected, this prescription leads to approximately correct amounts of mass and momentum transferred into the jet. The new "standard" coordinate frame is one in which the two plates have equal, but opposite, velocity components parallel to the symmetry plane in addition to the old vectors v_p. The jet velocity in this frame is still given by (1) and its width by (2). The major difference is that because of the slip surface (which extends into the jet), the velocity of the jetted material contributed by each plate is different and its mass flux is therefore different: projectile and target do not contribute equal quantities of material to the jet. The detailed equations describing this partitioning are written in Appendix B. Table 2 lists this partition for a number of impact parameters.

Geometry

The general geometry of a collision between two unequal spheres is shown in Fig. 3. The obliquity of the impact is described by the impact parameter b, the offset between the sphere's centers projected perpendicular to the line of approach. This parameter ranges between b = 0 (for a head-on collision) to $b = r_1 + r_2$

TABLE 2. Parameters of Jetted Material*, $r_2/r_1 = 0.5$.

Impact Parameter b/r_2	Total Mass Jetted Mass of Projectile	Ejection Velocity** Impact Velocity	Mean Launch Azimuth*, Degrees	Launch Elevation** Angle, Degrees	Fraction of Jetted** mass from projectile
0	0.25	1.12 ± 0.4	90 (symmetric)	18.5 ± 2.2	0.29 ± 0.14
0.2	0.25	1.15 ± 0.4	69	19.3 ± 3.5	0.33 ± 0.13
0.4	0.31	1.12 ± 0.4	48	22.4 ± 5.9	0.39 ± 0.14
0.6	0.41	1.09 ± 0.3	35	26.4 ± 8.2	0.49 ± 0.16
0.8	0.26	1.2 ± 0.3	35	23.7 ± 6.8	0.54 ± 0.19
1.0	0.13	1.29 ± 0.3	34	20.2 ± 5.5	0.56 ± 0.23
1.2	0.04	1.34 ± 0.3	32	16.1 ± 3.8	0.59 ± 0.25
1.4	2×10^{-4}	1.25 ± 0.2	26	10.7 ± 1.9	0.66 ± 0.20

*Critical angle $\theta_{cr} = 17.5°$.
**Mass-weighted quantity (see text). See Appendix for the definition of the variance.

(for a collision so glancing that only one point of each sphere touches). The distance between the sphere's centers along the line of impact is parameterized by d and the velocity of the projectile is U.

The geometry of the intersection between the two spheres changes radically during the early contact phases of the impact. The angle between the sphere's surfaces along the circle of intersection does not depend upon azimuth (defined here as the angle in the plane of the circular intersection of the two spheres, measured clockwise from the projectile's extrapolated direction of motion), but is a strong function of the distance between the sphere's centers. This angle is zero at the first instant of contact, then increases as the spheres interpenetrate. Jetting begins when this angle exceeds θ_{cr}. The velocity of the jet, given by (1) in the standard frame, is evaluated by locally (at each point of the circle of intersection) transforming to the standard frame, remembering that the impact velocity vector's angle with respect to the plane of intersection varies with azimuth. The jet velocity vector is then

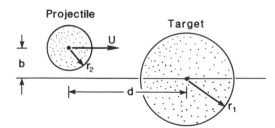

Fig. 3. Geometry of the collision between the proto-Earth, radius r_1, and another protoplanet of radius r_2. The impact parameter is b, the distance between their offset centers is d, and the collision velocity is U.

retransformed from the standard frame into the target planet's rest frame (Fig. 3), where the jet velocity, launch angle (angle between the target planet's surface and the jet), and launch azimuth are computed. The total mass ejected during any stage of the jetting and for each 10° azimuth interval is evaluated using the jet width, equation (2), and velocity.

Since this process yields too many numerical values to report easily, only mass-averaged values of launch angle, azimuth, and jet velocity are reported in this paper. That is, each quantity of interest is computed for a large number of times during the course of jetting and over a range of azimuths. The quantity is then multiplied by the mass ejected in this time and angle interval. This product is accumulated for all times and azimuths and is finally divided by the total jetted mass to obtain the "mass averaged" quantity. The resulting value is thus weighted by the amount of material that is ejected. Variances are also reported with the quantities listed in Table 2.

The detailed geometrical transformations required to describe the collision of two spheres are given in Appendix B. Note that a further constraint must also be applied: jetting only takes place when the intersection of the sphere's surfaces encounters *new*, or previously unshocked, material. Furthermore, "the time-dependent collision can be regarded as a succession of separate (noninteracting) collisions between plane surfaces" (Walsh *et al.*, 1953) only if the angle between the colliding surfaces increases monotonically with time. This condition is invariably satisfied in the collision of spheres. The implementation of these criteria is also described in Appendix B.

Although the critical angle, θ_{cr}, is a monotonically increasing function of impact velocity v_p, Kieffer (1977) shows that it changes slowly in the high velocity (>5 km/s) regime. In view of other approximations that must be made, it is adequate to take $\theta_{cr} = 17.5°$, independent of velocity.

Results

Figure 4 illustrates the quantity of material jetted as a function of impact parameter b and the radius ratio of the colliding spheres, r_2/r_1. It is clear that little material is jetted when a small sphere strikes a large one vertically: this is a close approximation to experimental observation by Shoemaker *et al.* (1963) of a sphere impacting a plane. However, the amount of jetted material can approach 80% of the projectile mass at certain values of the impact parameter, when the projectile and target are of comparable dimensions. The mass of jetted material is a peaked function of impact parameter. The peak occurs at zero impact parameter for equal diameter spheres, but moves to larger impact parameters as the sphere's radius ratio decreases. Its position is given roughly by

$$b_{max} = r_1 - r_2 \tag{3}$$

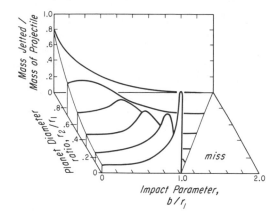

Fig. 4. *Total jetted mass in units of the projectile's mass as a function of the radius ratio r_2/r_1 and normalized impact parameter b/r_1. The jetted mass is a strongly peaked function of the impact parameter.*

The jetted vapor is derived from both the projectile and target in roughly equal proportions. The asymmetry of the jet conditions does produce a slight preponderance of target material in the jet, as shown in Table 2, which lists the mass-averaged values and variances of several parameters in the collision of a sphere with another twice as large.

Since the width of the jet can be at most $r_2/2$, no jetted material is derived from deeper than one-half the projectile's radius. Thus if the projectile has a core that is smaller than this it will not enter the jet, although core material may, of course, be ejected later as part of the lower velocity ejecta curtain. Figure 5 schematically illustrates the provenance of jetted material in both the target and projectile.

Table 2 demonstrates that most of the jetted material leaves the projectile-target interface at speeds comparable to the projectile's impact speed. The earliest ejecta is fastest, but the jet velocity declines rapidly as the angle θ increases. The launch

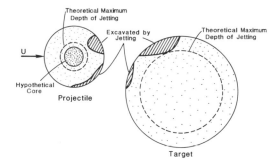

Fig. 5. *Provenance of the jetted vapor. This cross section of the projectile and target planets shows the regions that are ejected in the vapor plume. The maximum depth of excavation is roughly $r_2/2$, so that if the projectile's core is smaller than half its radius, core material does not enter the plume. The cross-hatched areas indicating jetted material are sections through a rather complexly shaped volume. No material is jetted from the point of first contact because the angle between the two surfaces must exceed twice the critical angle, $2\theta_{cr}$, before jetting begins.*

angle (the angle between the target planet's surface and the velocity vector of the jet) is typically low, only a few tens of degrees for impacts at all impact parameters. The azimuthal spread of the jet, however, is a strong function of impact parameter. When the impact parameter is zero the jetted plume is axially symmetric and equal amounts of material are jetted to all azimuths. As the impact parameter increases, the jet becomes a narrow fan in the downrange direction, spreading only a few tens of degrees away from the extrapolated flight direction of the projectile. The jetted plume carries a large net angular momentum in this configuration, which is a necessary precondition for it to condense into an orbiting satellite.

Hydrodynamic Expansion and Condensation of the Jetted Vapor

Jetting produces a high velocity plume of hot, dense vapor that squirts out from the interface between the projectile and the target planet. Initially ejected as a more-or-less narrow fan centered about the projectile's extrapolated direction of motion, the gas quickly expands laterally as rarefactions propagate inward from adjacent free surfaces. This expansion is hydrodynamic: that is, the motion of a small parcel of gas is governed by pressure gradients, not just by gravity, which dominates purely ballistic motion. This pressure gradient acceleration, as Cameron and Ward (1976) first realized, allows some ejected debris to enter orbits that neither leave the Earth nor reimpact it.

When we began the investigation reported in this paper, we had hoped to compute the mass of the jetted vapor that thus achieves orbit about the Earth. Unfortunately, we have found that the gas expands so rapidly that it fills the space between the projectile and target's curved surfaces. Harlow and Pracht (1966) found similar results in their two-dimensional numerical computation of jetting using a perfect gas equation of state. These moving surfaces both confine and accelerate the gas. The flow regime of the gas switches between subsonic in the jet to supersonic as it expands and impacts the surfaces of the adjacent planets, indicating the need for a full three-dimensional numerical computation.

Qualitatively, it can be said that the gas plume expands freely into the vacuum about the Earth under the influence of the large pressure gradient between its core and edge. The maximum speed of expansion, u_{max}, is given by (Zel'dovich and Raizer, 1967, vol. 1, p. 103)

$$U_{max} = \frac{2}{\gamma - 1} C_o \qquad (4)$$

where γ is the ratio of specific heats and c_o is the initial sound speed in the compressed gas ($c_o = \sqrt{\gamma P_o / \rho_o}$, where P_o is the initial pressure in the jet and ρ_o is its initial density). This is the speed of the edge of the gas cloud, however. The *mean* rate of expansion from the cloud's center of mass is lower: about 0.53 u_{max} or, equivalently,

1.58 c_o, for $\gamma = 5/3$. For example, a jet in which the gas was compressed to 700 GPa and that has a density of 6000 kg/m³ and ratio of specific heats of 5/3 exhibits a sound speed of about 14 km/s and a mean expansion rate of 22 km/s—faster than the assumed impact velocity of 12 km/s.

As the gas cloud expands it cools adiabatically. Its temperature T declines as its radius R increases according to the law $T \sim R^{-3(\gamma-1)}$, or $T \sim R^{-2}$ for $\gamma = 5/3$. Condensation of solid silicate particles thus begins after the cloud has expanded to several times its initial diameter (given roughly by the jet width, w; equation (2)). The size of the condensed particles can be estimated from the theories of Raizer (1960) and Lattimer (1982). Since the particle size scales linearly with the initial diameter of the gas cloud, Raizer's (1960) results for the impact of a small iron meteoroid or Melosh's (1982) result for the 10-km-diameter K-T impactor can be used to estimate that the vapor from a Mars-size impactor striking the Earth would directly condense into particles roughly a centimeter in diameter. Condensation begins several Mars radii from the site of impact. As the gas expands and its density decreases these particles decouple from the gas and enter Keplerian orbits around the Earth, some of which close high above its surface and thus remain stable.

The complex series of chemical reactions that may occur in this rapidly condensing silicate vapor are dissimilar to those that are believed to have occurred in the solar nebula because of the rapidity of condensation (in addition to the differences that arise from the very different fugacity of the condensing gas—the solar nebula probably condensed with a large excess of H_2 gas, whereas little or no hydrogen is present in the vapor plume jetted from the proto-Earth). Equilibrium is not reached: condensation is governed by chemical kinetics. The resulting "frozen equilibria" are similar to those observed in rocket exhausts (Altman and Carter, 1960) and supernova explosions (Lattimer, 1982). Nevertheless, it seems clear that volatiles such as water would be largely excluded from the condensing silicates. Even if water ice did condense at first, it would not survive long in contact with hard vacuum.

The complete vaporization attendant upon jetting is thus qualitatively consistent with the Moon's "high temperature condensate" geochemical character. Closer scrutiny in the form of geochemical modelling is required before the jetting model's predictions can be compared to the Moon's actual composition.

The centimeter-size silicate particles (protolunar pebbles?) that find themselves in stable orbits about the Earth initially have a wide range of inclinations. Inclinations may differ by more than several tens of degrees, if the azimuthal spread of the jetted vapor can be used as a guide. Collisions inevitably occur where orbital planes intersect, and many generations of collision, fragmentation, and perhaps vaporization events may cause further geochemical evolution. The rapid loss of the gas phase from this system may inhibit a solar-nebula-like drift of solid particles into the Earth's equatorial plane. The widely postulated lunar "magma ocean" could arise naturally from the heat delivered by the impact of energetic, highly inclined debris onto the growing lunar embryo. However, much more dynamical study of the evolution of a vapor plume in the near-Earth environment must precede firm statements of this kind.

Conclusion

Of all the types of ejecta thrown out by an impact, only the portion that arises from jetting can make a significant contribution to a cloud of debris orbiting about the proto-Earth. Contrary to expectations based on laboratory experiments with planar targets, the calculations reported here indicate that a large quantity of material, up to one-half the mass of the projectile, may be ejected as an early, high-velocity vapor plume from a moderately oblique collision of two comparable size spheres.

The detailed characteristics of such a collision seem to eminently qualify it as the event that created the Earth's Moon. The narrow range of impact parameter over which a great deal of material is jetted makes the formation of a large satellite relatively improbable, perhaps explaining why only the Earth, of the four terrestrial planets, has a large satellite. Another possibility, however, is that the formation of a large satellite requires the ejection of a *vapor* plume: solid debris enters Keplerian orbits and either reimpacts the primary or escapes altogether. The conditions of a 15 km/s impact on the Earth show that, although vapor is readily formed in the jet, it is not produced abundantly by normal impact processes. On a smaller planet, such as the proto-Mars or proto-Mercury, the impact velocity may, according to the arguments of Safronov (1969), have been comparable to their escape velocity and even jetting would not have raised pressures high enough to vaporize much material.

In either case, an impact on Earth would have produced abundant vapor whose subsequent condensation, first into small silicate pebbles in space, then into a larger orbiting body, naturally accounts for the "high temperature condensate" geochemical character of the Moon. The incorporation of large amounts of Earth mantle material in the jet accounts for the otherwise great geochemical similarity between the Earth and the Moon, while some differences may be attributed to the admixture of projectile material in the jet. Since jetting is a near-surface phenomenon, the vapor jet will not include core material, at least not if the projectile's core is smaller than one-half its radius. This may explain the Moon's relative deficiency in iron and low mean density compared to the other terrestrial planets.

The calculations reported here are approximate and depend upon a number of assumptions whose validity is difficult to evaluate. Thus, the assumption of a steady jet, and, more importantly, the use of a flat-plate formula for the jet width, are imperfections that must be removed in future work. Many dynamical questions about the early evolution of the plume must be answered: How much jetted material actually enters closed orbits? Is there a preferential partition of vapor from the target and projectile during the orbital insertion process? Does other ejecta make a significant contribution to debris in the vapor cloud? How much, and what kind, of chemical differentiation takes place in the condensing cloud? These questions, and many others, may await three-dimensional numerical hydrodynamic modeling of the collision event. Nevertheless, the approximate model described here shows that, under the right conditions, it is plausible that jetting ejects a massive plume of vaporized material from the Earth at orbital speeds. It thus lends a higher degree of probability to the hypothesis that the Moon originated from a collision between the proto-Earth and a very large object.

Appendix A

The high speed ejecta from an impact

Impact cratering is an orderly (but rapid) process in which the kinetic energy of the impactor is transformed into kinetic, internal, and gravitational energy of both the impactor and target material. Ejecta moves away from the impact site with a wide range of speeds and internal energies. In order of decreasing ejection velocity, the quantity, speed, and physical state (solid, liquid, or vapor) of the ejecta classes are:

A.1. Jetted ejecta. The highest speed ejecta is thrown out during the earliest stage of impact cratering, when the projectile first contacts the target (Gault *et al.*, 1968). The oblique convergence of portions of the projectile's surface with the target's surface generates a jet of highly shocked, melted (at 6 km/s impact velocity), or vaporized ejecta that travels *faster* than the projectile's initial velocity. The jetting phenomenon is now well understood in the context of armor-penetrating-shaped charges, but has received little attention from the impact cratering community (with the exception of Kieffer, 1975, 1977) because it seemed to involve only a small fraction of the projectile's mass (<10% projectile mass: Shoemaker *et al.*, 1963). The preceding paper shows that this fraction is substantially larger for oblique impact of comparably-sized spheres than for the vertical sphere-into-plane geometry observed by Shoemaker *et al.*.

A.2. Vapor expansion. Impacts at speeds exceeding 11 km/s may vaporize a portion of the projectile in addition to comparable or larger quantities of the target. A vertical impact velocity of 11 km/s (the minimum speed for Earth impactors) is near the threshold for the production of significant quantities of vapor: larger impact velocities yield much larger amounts (see Ahrens and O'Keefe, 1972, or Croft, 1982, for a summary). Although this vapor expands into the surrounding vacuum at tens of kilometers per second (Singer, 1983) depending upon its initial internal energy, it begins its expansion only after the projectile has buried itself in the target. The vapor is thus likely to form a predominantly symmetrical plume that carries little net angular momentum, even for a moderately oblique impact (but it is difficult to separate this component from jetted material in a very oblique impact). More research, including numerical studies of oblique impacts, are necessary to refine the role of this component. However, because any candidate prelunar material must carry a great deal of angular momentum, we provisionally neglect this ejecta class.

A.3. Spallation. The shock waves generated by the impact spread both through the target and back into the projectile. The peak particle velocity behind these shocks is roughly half the impact velocity (assuming that projectile and target have similar shock impedences) at the site of the impact, but falls off as an inverse power of the distance farther away from the impact site. When either shock wave encounters a free surface, whether of projectile or target, it is reflected as a tensile wave. The resulting stresses easily overcome the tensile strength of even the strongest rocks, fragmenting them into small pieces and ejecting them at a speed of nearly twice the peak particle velocity, by the "velocity doubling rule" (Walsh and Christian, 1955). These fragments are nearly unshocked because of the proximity of the free surface, even though the speed of the fastest spalls is close to the impact velocity (Melosh, 1984).

The total quantity of such lightly-shocked, high-speed ejecta, however, is small: Melosh (1985a) shows that for a spherical projectile impacting a half-space target, a portion of the target equal to only about 1% projectile volume is spalled at speeds exceeding 7 km/s. (This is a liberal estimate that includes all material shocked from 0 to 50 GPa pressure. The projectile is assumed to strike vertically at 15 km/s.) More refined calculations that take into account the sphericity of the Earth are unlikely to increase the mass of spalled ejecta by as much as a factor of 10, which still yields a negligible amount of spalled material. This spalled fraction also includes the material ejected from the antipode of the impact site (Schultz and Gault, 1975).

Since spalled material is lightly shocked, it is in the solid state and follows Keplerian trajectories after ejection. It therefore either escapes the Earth entirely or reimpacts the Earth shortly after ejection. Neither case contributes much to the quantity of orbiting debris.

In summary, the small quantity of lightly-shocked, high-speed, spalled material is unlikely to play any role in the Moon's origin, except insofar as some of it may be swept up in the expanding plume of jetted gas. However, since spalls are ejected only *after* jetting ceases, even this possibility is remote.

A.4. The ejecta curtain. Most of the mass ejected from an impact crater, totalling up to one thousand times the projectile's mass, is expelled during the excavation stage. This includes a large amount of melted and highly shocked (a few to circa 60 GPa) rock material. The excavation flow is initiated by the rapidly-moving shock wave: This narrow shock expands hemispherically away from the impact site. Target material engulfed by the shock is first accelerated to a high velocity, then is released to low pressure. The decompressed target material retains a small residual velocity (typically one-fifth of the maximum particle velocity) because of the thermodynamic irreversibility of the shock wave. This residual velocity is the source of a subsonic excavation flow that eventually opens the crater (Melosh, 1985b). Streamlines of this flow field cut across contours of peak shock pressures (Grieve et al., 1981) so that material ejected at a given velocity contains materials that have been shocked to a variety of maximum pressures. On average, however, the fastest ejecta is more heavily shocked than the slower ejecta.

The maximum velocity attained in the excavation flow is typically only 10% of the impact velocity (one-fifth of the peak particle velocity, which itself is about one-half of the impact velocity). This theoretical expectation (see, e.g., Austin et al., 1981; Maxwell, 1977; O'Keefe and Ahrens, 1977) is confirmed by small-scale impact experiments (Shoemaker et al., 1963; Hartmann, 1985). It is most vividly demonstrated by the restricted range of ejection blankets around large craters on the Moon, Mercury, and Mars, where even the secondary impact craters, which extend well beyond the continuous ejecta deposits, are produced by fragments moving at less than about 1 km/s (Vickery and McConnell, 1985).

The material thrown out as part of the classic "ejecta curtain" thus travels too slowly to be of much importance to the impact origin of the Moon, in spite of the postulated projectile's large size: the *maximum* speed of ejection scales with the projectile's impact velocity, independent of size.* Furthermore, material in the ejecta curtain follows ballistic trajectories and thus must either escape the Earth or fall back upon it—it cannot contribute to an orbiting reservoir of proto-Moon material (again, except insofar as some of it may be swept up by the previously ejected vapor jet).

A.5. Ricochet. A large fraction of the projectile may ricochet from the target for highly oblique impacts. The ricocheted material, traveling at speeds approaching the impact speed, follows ballistic trajectories that begin at the impact site on the Earth's surface and therefore either reimpact the Earth or escape entirely. Nevertheless, ricocheted projectile material may be important because it is ejected early, at the same time as the jetted gas. Acceleration by the expanding gas may thus inject some of this material into orbit. Although this possibility is not emphasized in this paper, its potential importance deserves consideration in a more sophisticated treatment.

---

*This is not true, however, of the still lower speed ejecta in the gravity-dominated cratering regime (Housen et al., 1983) because the final crater size and hence the speed of ejecta that *escapes* the crater depends on projectile size and gravity. A larger proportion of the ejecta escaping a crater made by a large projectile moves with high velocity because low velocity material is trapped within the larger crater. Despite this effect, the *upper limit* to the velocity scales with the projectile's velocity, not its size.

Appendix B

Mathematical details

Referring to Fig. B1, simple trigonometry establishes the following auxiliary quantities, R, α_1, α_2, and β:

$$s = \frac{r_2^2 - r_1^2 + d^2 + b^2}{2\sqrt{d^2 + b^2}} \tag{B1}$$

$$R = \sqrt{r_2^2 - s^2} \tag{B2}$$

$$\alpha_1 = \cos^{-1}\left[\left(\sqrt{d^2 + b^2} - s\right)/r_1\right] \tag{B3}$$

$$\alpha_2 = \cos^{-1}\left[s/r_2\right] \tag{B4}$$

$$\beta = (\alpha_2 - \alpha_1)/2 \tag{B5}$$

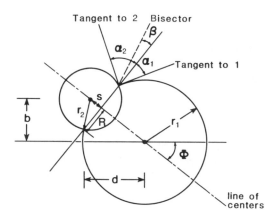

Fig. B1.   *Definitions of distances and angles for the problem of two colliding spheres. The geometry is azimuthally symmetric about the sphere's line of centers, although the impact velocity vector is not. The angle between the sphere's surfaces along the circle of contact is independent of azimuth.*

Using these quantities, the half angle θ included between the surfaces is given by

$$\theta = (\alpha_1 + \alpha_2)/2 \tag{B6}$$

The convergence velocity in the standard frame, v_p, is

$$v_p = (U/2)\sec\theta\,[\cos\Phi\,\cos\beta + \sin\Phi\,\sin\beta\,\cos\phi] \tag{B7}$$

where

$$\Phi = \tan^{-1}(b/d) \tag{B8}$$

The jet velocity in this frame, v_j, equation (1) in the text, is used to define three auxiliary velocities:

$$v_1 = v_j - v_p \sin\theta - (U/2)\,[\cos\Phi\,\sin\beta - \sin\Phi\,\cos\beta\cos\phi] \qquad (B9)$$

$$v_2 = v_p \cos\theta \qquad (B10)$$

$$v_3 = (U/2)\sin\phi\,\sin\Phi \qquad (B11)$$

Equation (B9) is valid even for asymmetric collisions, using the slip plane concept discussed in the text. The coordinate system used to derive these expressions is illustrated in Fig. B2.

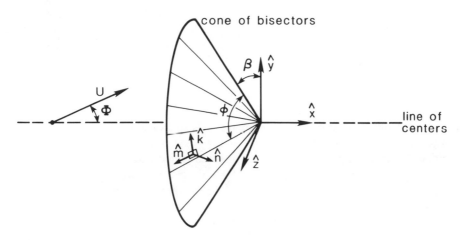

Fig. B2.   Vectors used to determine the projection of the impact velocity vector upon the surface that bisects the intersection between the two spheres. This projection is used to define the velocities in the standard frame at each point on the circle of intersection. The coordinate system is aligned along the line of centers.

The ejection velocity v_e in the target frame is given in terms of v_1, v_2, and v_3 by

$$v_e = \sqrt{v_1^2 + v_2^2 + v_3^2} \qquad (B12)$$

Similarly, the elevation angle ψ and launch azimuth ϕ can be shown (Fig. B3) to be

$$\psi = \sin^{-1}[v_1 \sin\theta - v_2 \cos\theta\,)\,/\,v_e\,] \qquad (B13)$$

and

$$\Phi = \left(\frac{v_1 \cos\beta \sin\phi + v_2 \sin\beta \sin\phi - v_3 \cos\phi}{v_1 \cos\beta \cos\phi + v_2 \sin\beta \cos\phi + v_3 \sin\phi} \right) \tag{B14}$$

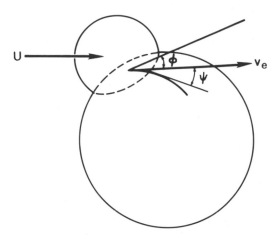

Fig. B3.   Definition of angles used to describe the ejection of the vapor plume. The azimuth, ϕ, is measured clockwise from the projected direction of the projectile's path. The launch elevation, ψ, is measured upward from the target planet's surface at the point of contact. The ejection velocity in the target planet's rest frame is v_e.

Jetting does not occur from all azimuths: the intersection of the two spheres must be advancing into new material if jetting is to occur. A receding intersection, although formally possible, is unphysical because jetting and compression processes have already exhausted the material there. Since the circle of intersection both changes size as the distance between the sphere's centers decreases and changes position with respect to some previous circle of intersection, these two factors must be incorporated into a criterion for jetting. We now define the parameter Δ:

$$\Delta = \left(\frac{sd}{Rb} \right) \frac{\sqrt{d^2 + b^2} - s}{\sqrt{r_1^2 - R^2}} \tag{B15}$$

It can be shown that jetting occurs at all azimuths if $\Delta > 1$, whereas no jetting occurs for $\Delta < -1$. In intermediate cases jetting occurs from azimuth zero (along the extrapolated path of the projectile) to a maximum azimuth ϕ_{max}, where

$$\phi_{max} = 2 \cos^{-1} \sqrt{(1 - \Delta)/2} \tag{B16}$$

The rate at which mass $M(t,\phi)$ is ejected between azimuths ϕ and $\phi + d\phi$ is given by

$$\frac{d^2M}{d\phi dt} = \rho R \, w \, v_a \tag{B17}$$

where the jet width w is given by equation (2) in the text, ρ is the uncompressed density of the target or projectile material (assumed equal) that enters the jet, and v_a is the jet velocity in the collision frame. This velocity is closely related to v_1, i.e.,

$$v_a = v_1 - v_p \cos \theta \cot \theta \qquad (B18)$$

Jetting is absent if $v_a < 0$. This velocity is the *mean* jet velocity. In asymmetric jetting the postimpact velocity of the projectile material is different than the target material, leading to a different quantity of material entering the jet from each body (assuming the width of the jet formed by each material is the same, $w/2$). If the velocity of the jet from the target is v_- and from the projectile is v_+, then it can be shown that

$$v_{\pm} = v_a \pm (U/2) [\cos\Phi \sin\beta - \sin\Phi \cos\beta \cos\phi] \qquad (B19)$$

The fraction of projectile that contributes to the jet, listed in Table 2 for several values of the impact parameter, is thus given at each time by F_p, where

$$F_p = v_- / v_a \qquad (B20)$$

Jetting can potentially begin as soon as contact occurs (depending upon the value of θ_{cr}). For purposes of numerical integration of the ejected mass, it is useful to know the maximum and minimum ranges, d, between the sphere's centers at which jetting can occur. Simple geometry (Fig. B1) shows that first contact occurs at

$$d_{max} = \sqrt{(r_1 + r_2)^2 - b^2} \qquad (B21)$$

The end of jetting is attained when either the top or bottom of the projectile (depending upon the impact parameter) finally buries itself in the target. This happens when

$$d_{min} = \sqrt{r_2^2 - (r_1 - b)^2} \qquad \text{for } b \geq r_1 - r_2 \qquad (B22a)$$

or

$$d_{min} = \sqrt{r_1^2 - (r_2 + b)^2} \qquad \text{for } b \leq r_1 - r_2 \qquad (B22b)$$

Quantities computed during a collision event change rapidly as a function of d, azimuth, time, etc. To reduce the amount of output to manageable proportions, results are reported as *mass-weighted averages*. For a quantity of interest V, its mass-weighted average $\langle V \rangle$ is defined as:

$$\langle V \rangle = \frac{\int V \, dm}{\int dm} \qquad (B23)$$

and its variance, σ, is defined by:

$$\sigma^2 = \frac{\int (V - \langle V \rangle)^2 \, dm}{\int dm} = \frac{\int V^2 \, dm}{\int dm} - \langle V \rangle^2 \qquad \text{(B24)}$$

Acknowledgments. The authors were inspired to begin this investigation in the aftermath of the Conference on the Origin of the Moon, held in Kona, Hawaii on October 13–16, 1984. Our original goal was to discredit the mega-impact hypothesis, and it was with some amazement that we realized that our analysis of such an impact agrees remarkably well with the details of the Moon's origin. H.J.M. acknowledges many enlightening discussions with M. Drake and W. K. Hartmann. He was supported by NASA grant NASW-428. C.P.S. thanks Prof. K. M. Creer and the Department of Geophysics, Edinburgh University for their hospitality.

References

Ahrens T. J. and O'Keefe J. D. (1972) Shock melting and vaporization of lunar rocks and minerals. *The Moon, 4*, 214–249.

Ahrens T. J. and O'Keefe J. D. (1977) Equations of state and impact-induced shock-wave attenuation on the Moon. In *Impact and Explosion Cratering* (D. J. Roddy, R. O. Pepin, and R. B. Merrill, eds.), pp. 639–656. Pergamon, New York.

Allen W. A., Morrison H. L., Ray D. B., and Rogers J. W. (1959) Fluid mechanics of copper. *Phys. Fluids, 2*, 329–333.

Altman D. and Carter J. M. (1960) Expansion processes. In *Liquid Propellant Rockets* (D. Altman, J. M. Carter, S. S. Penner, and M. Summerfield, eds.), pp. 26–63. Princeton Univ. Press, Princeton, NJ.

Al'tshuler L. V., Kormer S. B., Bakanova A. A., Petrunin A., Funtikov A. I., and Gubkin A. A. (1962) Irregular conditions of oblique collision of shock waves in solid bodies. *Sov. Phys. JETP, 14*, 986–994.

Austin M. G., Thomsen J. M., Ruhl S. F., Orphal D. L., Borden W. F., Larson S. A., and Schultz P. H. (1981) Z-model analysis of impact cratering: an overview. In *Multi-Ring Basins* (P. H. Schultz and R. B. Merrill, eds.), pp. 197–205, *Proc. Lunar Planet. Sci. 12A*, Pergamon, New York.

Birkhoff G., Macdougall D. P., Pugh E. M., and Taylor G. (1948) Explosives with lined cavities. *J. Appl. Phys., 19*, 563–582.

Cameron A. G. W. (1984) Formation of the prelunar accretion disk (abstract). In *Papers Presented to the Conference on the Origin of the Moon*, p. 58. Lunar and Planetary Institute, Houston.

Cameron A. G. W. (1985) Formation of the prelunar accretion disk. *Icarus, 62*, 319–327.

Cameron A. G. W. and Ward W. R. (1976) Origin of the Moon (abstract). In *Lunar Science VII*, pp. 120–122. The Lunar Science Institute, Houston.

Croft S. K. (1982) A first-order estimate of shock heating and vaporization in oceanic impacts. *Geol. Soc. Amer. Spec. Paper 190*, pp. 143–152.

Drake M. J. (1986) Is lunar bulk material similar to Earth's mantle?, this volume.

Gault D. E., Quaide W. L., and Oberbeck V. R. (1968) Impact cratering mechanics and structures. In *Shock Metamorphism of Natural Materials* (B. M. French and N. M. Short, eds.), pp. 87–99. Mono, Baltimore.

Grieve R. A. F., Robertson P. B., and Dence M. R. (1981) Constraints on the formation of ring impact structures, based on terrestrial data. In *Multi-Ring Basins* (P. H. Schultz and R. B. Merrill, eds.), pp. 37–57, *Proc. Lunar Planet. Sci. 12A*, Pergamon, New York.

Harlow F. H. and Pracht W. E. (1966) Formation and penetration of high-speed collapse jets. *Phys. Fluids, 9*, 1951–1959.

Harris A. W. and Ward W. R. (1982) Dynamical constraints on the formation and evolution of planetary bodies. *Ann. Rev. Earth Planet. Sci., 10*, 61–108.

Hartmann W. K. (1985) Impact experiments: Ejecta velocity distributions and related results from regolith targets. *Icarus, 63*, 69–98.

Hartmann W. K. and Davis D. R. (1975) Satellite-sized planetesimals and lunar origin. *Icarus, 24*, 504–515.

Housen K. R., Schmidt R. M., and Holsapple K. A. (1983) Crater ejecta scaling laws: fundamental forms based on dimensional analysis. *J. Geophys. Res., 88*, 2485–2499.

Kieffer S. W. (1975) Droplet chondrules. *Science, 189*, 333–340.

Kieffer S. W. (1977) Impact conditions required for formation of melt by jetting in silicates. In *Impact and Explosion Cratering* (D. J. Roddy, R. O. Pepin, R. B. Merrill, eds.), pp. 751–769. Pergamon, New York.

Lattimer J. M. (1982) Condensation of grains. In *Proceedings of the Conference on the Formation of Planetary Systems* (A. Brahic, ed.), pp. 189–282. Cepadues-Editions, Toulouse.

Maxwell D. E. (1977) Simple Z-model of cratering, ejection, and the overturned flap. In *Impact and Explosion Cratering* (D. J. Roddy, R. O. Pepin, and R. B. Merrill, eds.), pp. 1003–1008. Pergamon, New York.

Melosh H. J. (1982) The mechanics of large meteoroid impacts in the Earth's oceans. *Geol. Soc. Amer. Spec. Paper, 190*, 121–127.

Melosh H. J. (1984) Impact ejection, spallation and the origin of meteorites. *Icarus, 59*, 234–260.

Melosh H. J. (1985a) Ejection of rock fragments from planetary bodies. *Geology, 13*, 144–148.

Melosh H. J. (1985b) Impact cratering mechanics: Relationship between the shock wave and excavation flow. *Icarus, 62*, 339–343.

O'Keefe J. D. and Ahrens T. J. (1977) Meteorite impact ejecta: Dependence of mass and energy lost on planetary escape velocity. *Science, 198*, 1249–1251.

Raizer Y. P. (1960) Condensation of a cloud of vaporized matter expanding in a vacuum. *Sov. Phys. JETP, 37*, 1229–1235.

Ringwood A. E. (1970) Origin of the Moon: The precipitation hypothesis. *Earth Planet. Sci. Lett., 8*, 131–140.

Ringwood A. E. (1972) Some comparative aspects of lunar origin. *Phys. Earth Planet. Inter., 6*, 366–376.

Ringwood A. E. (1984) Origin of the Moon (abstract). In *Papers Presented to the Conference on the Origin of the Moon*, p. 46. Lunar and Planetary Institute, Houston.

Safronov V. W. (1969) Evolution of the protoplanetary cloud and formation of the Earth and planets. *NASA TTF-667*, translated in 1972 by the Israel Program for Scientific Translation, Jerusalem. 206 pp.

Schultz P. H. and Gault D. E. (1975) Seismic effects from major basin formation on the Moon and Mercury. *The Moon, 12*, 159–177.

Shoemaker E. M., Moore H. J., and Gault D. E. (1963) Spray ejected from the lunar surface by meteoroid impact. *NASA TN D-1767*, NASA/Ames Research Center, Moffett Field, CA.

Singer A. V. (1983) Effect of an impact-generated gas cloud on the acceleration of solid ejecta (abstract). In *Lunar and Planetary Science XIV*, pp. 704–705. Lunar and Planetary Institute, Houston.

Stevenson D. J. (1984) Lunar origin from impact on the Earth: Is it possible? (abstract). In *Papers Presented to the Conference on the Origin of the Moon*, p. 60. Lunar and Planetary Institute, Houston.

Vickery A. M. and McConnell J. M. (1985) Size-velocity distribution of ejecta fragments on planetary bodies (abstract). In *Lunar and Planetary Science XVI*, pp. 879–880. Lunar and Planetary Institute, Houston.

Walsh J. M. and Christian R. H. (1955) Equation of state of metals from shock wave measurements. *Phys. Rev., 97*, 1544–1556.

Walsh J. M., Shreffler R. G., and Willig F. J. (1953) Limiting conditions for jet formation in high velocity collisions. *J. Appl. Phys., 24*, 349–359.

Wood J. A. (1977) Origin of Earth's Moon. In *Planetary Satellites* (J. Burns, ed.), pp. 513–529. Univ. of Arizona Press, Tucson.

Zel'dovish Ya. B. and Raizer Yu. P. (1967) *The Physics of Shock Waves and High-Temperature Hydrodynamic Phenomena.* Academic, New York. 916 pp.

Short Note: A Preliminary Numerical Study of Colliding Planets

M. E. KIPP

Sandia National Laboratories, Albuquerque, NM 87185

H. J. MELOSH

Lunar and Planetary Lab, University of Arizona, Tucson, AZ 85721

We have performed preliminary numerical computations of the early stages in the impact of a Mars-size protoplanet with the proto-Earth and evaluated its potential for the creation of a Moon-size satellite. We used the code CSQ II on a Cray I computer at Sandia National Labs. The code is two-dimensional, so this computation actually represents the eccentric collision of two cylinders, not spheres. The finer resolution available in a 2D code makes this a reasonable approximation for a first effort. The mesh we used is 120×120 cells, where each cell is 250 km square.

Both planets are assumed to have iron cores with radii equal to one half the planet radius. Since no silicate equation of state is presently available in CSQ II, we approximated the planet's mantles by aluminum. Aluminum is a reasonably good approximation to olivine in both density and Hugoniot relation (compare the data for Twin Sisters and Mooihoek Dunite in McQueen *et al.*, 1967, Fig. 12, with the aluminum data of Al'tshuler *et al.*, 1960). Since the computation did not include gravity, the orbital evolution of the ejected plume could not be followed further than its initial stages. Our primary interest here is the physics of ejection where gravity is not of great importance.

The results are illustrated in Fig. 1a–d. The first frame, Fig. 1a, shows the two planets just before contact. The projectile planet is exactly half the size of the target proto-Earth. Figure 1b documents the beginning of a fast forward-moving jet of hot, highly shocked vapor and a slower, cooler backward-moving jet. These jets evolve further in Fig. 1c. The tip of the fast jet is at a temperature of about 10,000°K, has an average density near 60 kg/m^3, and is traveling in excess of 20 km/sec. This material escapes the Earth entirely and is not of interest for the origin of the

Moon. This vapor cloud continues to translate and expand in Fig. 1d. It is composed predominantly of projectile mantle material.

The "neck" of the hot plume in Fig. 1d, however, is more likely to become trapped in Earth orbit. Its velocity of circa 10 km/sec is less than Earth escape velocity (11 km/sec), but higher than low Earth orbital velocity (7.7 km/sec). Its

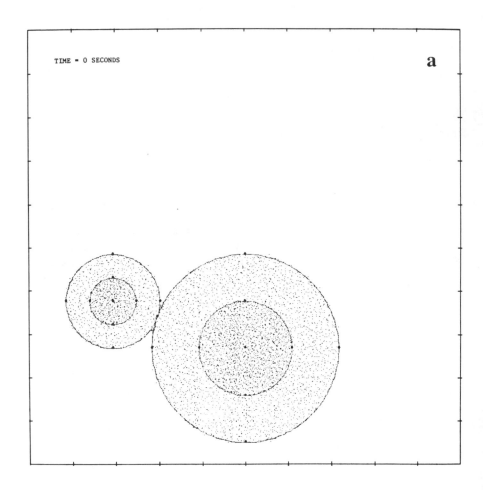

Fig. 1a–d   Four stages in the collision of a Mars-size protoplanet with the proto-Earth at 15 km/sec. Each frame is separated by about 6 minutes from the previous one. The stipple density is proportional to the material density. The initial velocity of the projectile is horizontal (parallel to the x-axis) from left to right in (a), thus making this an oblique impact with an impact parameter equal to one half of the proto-Earth's radius.

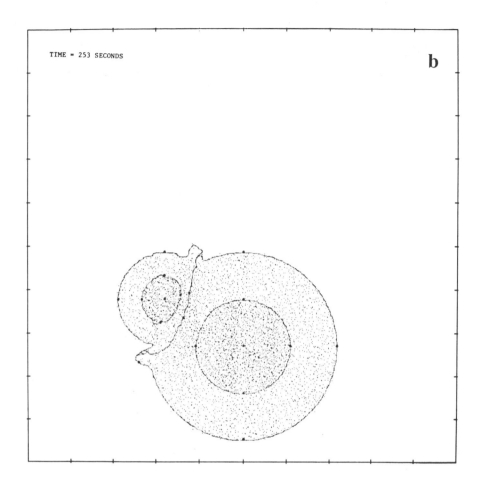

TIME = 253 SECONDS

b

density ranges between 300 and 1000 kg/m^3 and its temperature is in the vicinity of 6000°K. Although it is difficult to extrapolate this to three dimensions, it seems probable that at least one lunar mass is ejected in this "neck." It contains roughly equal amounts of projectile and target mantle material. Neither planet's core is ejected in the jet, although the projectile's core remains at more than 6000°K after the release wave decompresses it. It will thus eventually vaporize and mix with vaporized mantle material from both the projectile and target. This vapor, however, moves too slowly to attain orbit directly.

These preliminary numerical computations are in substantial agreement with the conclusions of Melosh and Sonett (1986). They thus support the possibility that

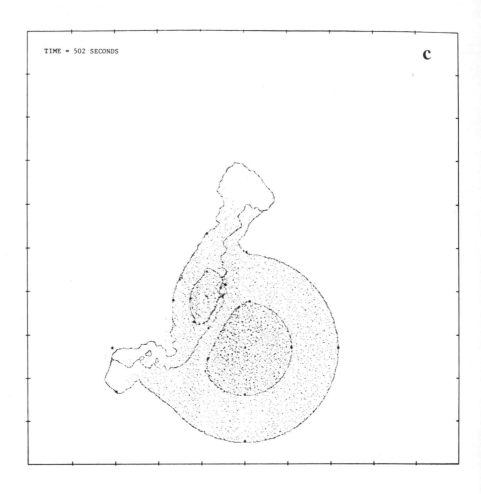

TIME = 502 SECONDS

c

the Moon originated from material ejected in a collision between the Earth and a Mars-size protoplanet early in solar system history.

References

Al'tshuler L. V., Kormer S. B., Bakanova A. A., and Trunin R. F. (1960) Equation of state for aluminum, copper and lead in the high pressure region. *Soviet Physics JETP, 11*, no. 3, 573–579.

McQueen R. G., Marsh S. P., and Fritz J. N. (1967) Hugoniot equation of state of twelve rocks. *J. Geophys. Res., 72,* 4999–5036.

Melosh H. J. and Sonett C. P. (1986) When worlds collide: Jetted vapor plumes and the Moon's origin, this volume.

TIME = 750 SECONDS

d

Geochemical Evidence for the Formation of the Moon by Impact-Induced Fission of the Proto-Earth

H. WÄNKE AND G. DREIBUS

Max-Planck-Institut für Chemie, Abteilung Kosmochemie, Saarstrasse 23, D-6500 Mainz, F.R. Germany

The composition of the Earth's mantle is the result of a complex not yet fully understood and not generally agreed upon accretion history of the Earth. As a consequence, the composition is unique, the only possible exception being the mantle of Venus. There is no doubt that the silicate phase of the Moon differs from the Earth's mantle in respect to its lower content of volatile and moderately volatile elements. This depletion can be very easily accounted for by loss of these elements during accretion of the Moon. We have put together a number of observations that show, on the other hand, that the silicate phase of the Moon and the Earth's mantle are for a number of highly characteristic elements very similar to each other. Both the terrestrial as well as the lunar mantle are depleted in V, Cr, and especially Mn. Relative to C1 abundances the terrestrial and lunar mantles are depleted in all siderophile elements, including the only moderately siderophile element W. The observed depletion of W in the lunar mantle requires equilibration with metal in an amount exceeding that permissible for the Moon. The terrestrial mantle has overabundances of Co and especially Ni, as well as of all highly siderophile elements in respect to the concentrations expected under equilibrium with its metallic core. The lunar mantle shows slightly higher abundances for elements with metal/silicate partition coefficients $<10^2$ and almost identical abundances for Co (D ~ 200), while, at least in the source regions of the lunar mare basalts, the concentrations of highly siderophile elements (D > 200) decrease parallel to the increase of the effective metal/silicate partition coefficient of the elements in question. The increasingly larger depletion of siderophile elements with higher metal/silicate partition coefficients is a clear indication of equilibration with small amounts of metal and its subsequent removal by segregation. A close genetic link between the Earth's mantle and the Moon is an almost unavoidable consequence. The most plausible model for the origin of the Moon in line with geochemical and cosmochemical constraints is an impact-induced "fission" of the proto-Earth. Impact-induced splash off of the Earth's mantle material can also account for the presence of unfractionated Earth's mantle material in lunar highland breccias. Formation of the Moon from material of the Earth's mantle has been advocated for almost two decades, following Ringwood (1966a,b).

1. Introduction

Formation of the Moon from material of the Earth's mantle (after core formation) has been long discussed in order to account for the low density of the Moon (see Ringwood, 1979). In fact, fission of the proto-Earth by rotational instability was the first quantitative model for the origin of the Moon (Darwin, 1880). The two major problems encountered by the fission model due to rotational instability are (1) the required delay of the core formation of the proto-Earth until it has almost reached its final mass and (2) the excessive amount of angular momentum corresponding to an Earth with a rotation period of 2.5 hours as required for fission.

Recently, collision-induced fission became a most plausible physical process for the formation of the Moon (Hartmann and Davis, 1975; Cameron and Ward, 1976; Ringwood, 1979). According to Wetherill (1976), the largest object impacting on the Earth during its accretion was in the order of several lunar masses. Model calculations of Boslough and Ahrens (1983) have shown that 0.2 lunar masses impacting on the Earth can vaporize one lunar mass. A large fraction of the vaporized material will be placed in orbit around the Earth and subsequently form the Moon in a fast time scale. However, as only a small fraction of the evaporated material will remain in orbit, impact of a much larger object is required to form the Moon by such a process (Stevenson, 1985; Melosh, 1985).

Although impact-induced fission of the Earth is physically possible and plausible, it is not the only possibility for formation of the Moon. In fact, formation of the Moon independent of the Earth has been continuously advocated by several authors (Ganapathy and Anders, 1974; Drake, 1983). In this paper we have put together and added new geochemical arguments for the formation of the Moon from material of the Earth's mantle and, in particular, evidences for an impact-induced fission.

2. Oxygen Isotopes

From the work of Clayton and Mayeda (1975), it is well known that lunar samples lie within the analytical uncertainties, on the mass-fractionation line defined by terrestrial material. This fact became even more significant since it is known that eucrites as well as the SNC meteorites (shergottites, nakhlites, and chassignites) fall distinctly off the terrestrial oxygen isotope fractionation line. Like the Earth and Moon, the EPB (eucrite parent body, possibly Vesta; Consolmagno and Drake, 1977) and the SPB (Shergotty parent body, probably Mars; Becker and Pepin, 1984) underwent magmatic fractionation.

3. Refractory Elements

Contrary to ideas generated immediately after the first lunar samples were received, it is now generally accepted that the bulk Moon is not significantly enriched in refractory elements compared to the Earth's mantle (Ringwood and Kesson, 1977a; Wänke *et al.*, 1977; Morgan *et al.*, 1978a; Ringwood, 1981; Wänke and Dreibus, 1982; Delano, 1985b).

4. Major Elements

With the exception of FeO, the major element chemistry of the Moon is nearly identical with that of the Earth's mantle. The lower FeO content of the latter (Moon: FeO = 13.9%; Ringwood, 1977; Earth: FeO = 7.6%; Wänke *et al.*, 1984) may be due to a gradual transfer of some FeO into the core during the first few hundred million years of the Earth's history (Jagoutz and Wänke, 1982). As discussed by Ringwood (1979), disproportionation of FeO would be the responsible process. Other oxides such as MnO, etc., may be similarly affected.

The metal formed in this way in the lower mantle will dissolve large amounts of oxygen. The argument raised by Newsom *et al.* (1985) against a transfer of iron from the mantle into the core after the main core formation event holds only in respect to a pure metal or sulphide phase, but not at all for a Fe-FeO liquid as proposed by Jagoutz and Wänke (1982). The partition coefficients relative to silicates of such a Fe-FeO liquid are highly different to those of a pure metal or sulphide phase. Growth of the Earth's core by such a process would rather quickly cease parallel with the available heat for driving mantle convection after accretion and main core formation event. Hence, larger changes in the FeO content of the Earth's mantle can only be expected for the first few hundred million years of the Earth's history.

Addition of FeO from the projectile also has to be taken into account in the case of impact-induced fission. In this respect it could be that the lower FeO content of the present Earth's mantle indicates incomplete mixing of the mantle at the time of the impact. If the collision partner was a Mars-sized object, it could have supplied a substantial fraction of the oxidized volatile-rich component B in the model of Ringwood (1984) and Wänke (1981). As recently proposed by Delano and Stone (1985), the FeO content of the upper mantle, being identical with that of the Moon shortly after impact, may have been lowered by convective mixing with the (at that time) FeO- and siderophile-element-poor lower mantle.

In this respect, we would like to emphasize that addition of 10% of today's Earth's mass to the proto-Earth in one impact event is not expected to lead to a full equilibration of the metal and sulphide phase of the projectile with the silicates of the Earth's mantle.

5. Mn, Cr, and V

Under normal planetary conditions (ol + opx + cpx being the major Fe-bearing phases), the liquid-solid partition coefficient of FeO is only slightly above 1. Therefore, the higher FeO and MnO contents of the Moon are directly evident from the higher concentrations of these elements in lunar mare basalts (LB) compared to the concentrations in terrestrial basalts (TB) (LB ~ 18% FeO and ~ 0.26% MnO; TB ~ 11% FeO and 0.18% Mn).

However, more important than the differences in the FeO and MnO contents of LB and TB is the large depletion of MnO in both objects relative to the chondritic abundance (Fig. 1). For the Earth's mantle the depletion of Mn and also of Cr

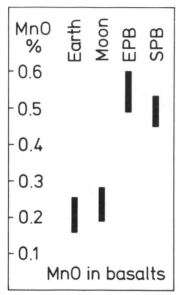

Fig. 1.   MnO in basalts of the Earth, Moon, EPB, and SPB. While MnO is not depleted on EPB and SPB, both the terrestrial and the lunar mantles are severely depleted in MnO.

and V was first noted by Ringwood (1966b). Ringwood and Kesson (1977a) favoured a depletion mechanism based on the higher volatility of Cr and Mn compared to Si as observed in the case of CM, CO, and CV chondrites (Kallemeyn and Wasson, 1981). Dreibus and Wänke (1979) presented evidence that makes the depletion of Cr, Mn, and V by volatility less likely. Basalts from the EPB that formed under lower oxygen fugacity than terrestrial basalts, as well as those from the SPB that formed under higher oxygen fugacity, indicate no depletion of Mn on these planetary objects (Fig. 2). For both the EPB and SPB, depletion of moderately volatile elements, such as Na or K, that is comparable to or higher than that of the Earth is beyond any doubt. In the case of Cr, the volatility argument was never very convincing, and it fails completely in the case of V, which is in fact a refractory element, and its abundance should be compared to that of Al or Ca.

The most likely explanation for the depletion of Mn and that of Cr and V in the Earth's mantle is their removal into the Earth's core, either in reduced form as metals or sulphides or as oxides (Wänke, 1981). All three elements (Mn, Cr,

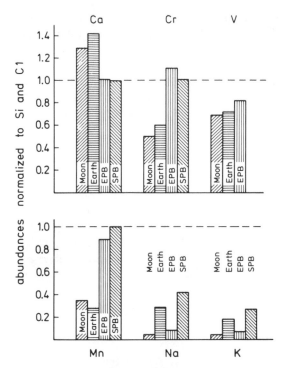

Fig. 2. In addition to Mn, Cr and V are also depleted in the Earth's mantle relative to C1 abundances. Vanadium is a refractory element and its abundance on the Earth and the Moon has to be compared with the abundance of refractory elements such as Ca. The silicate phase of the Moon shows similar depletion. In spite of strong depletions of the moderately volatile elements Na and K on the EPB and SPB, no depletion of Mn and Cr and only a slight depletion of V is observed on these objects. Abundance data for Moon: Wänke et al. (1977); for Earth: Wänke (1981); for the EPB: Dreibus et al. (1977); for the SPB: Dreibus and Wänke (1984).

and V) show increasing siderophile tendencies with decreasing oxygen fugacity. Since metal-silicate partition coefficients are about a factor of 10 lower for Mn than for Cr and V, removal of Mn as sulphide is suggested, as the stability of MnS by far exceeds that of the sulphides of Cr, V, Ni, and Fe (Rammensee et al., 1983). We also note that enstatite chondrites contain most of their Mn and Cr in the form of sulphides (Keil, 1968). In the case of Mn, the most volatile of these three elements, depletion by volatilization may also have played some role. Removal into the core in reduced form was originally proposed by Ringwood (1966b). Removal as oxides could be expected if the Earth's core contains a large amount of dissolved FeO as proposed by Ringwood (1977).

Independent of whatever the cause of the depletion of Mn and that of Cr and V in the Earth's mantle was, this depletion is a very characteristic feature of the Earth's mantle and is strongly coupled to the accretion mode of the Earth. It is striking that the Moon shows very similar depletions of all three elements. Because of the small mass of the Moon and core mass, which is at least ten times smaller, almost none of the depletion mechanisms discussed for the Earth's mantle could operate on the Moon. Removal of Mn, etc. into the core in reduced form requires inhomogeneous accretion and core formation during accretion because Mn, Cr, and V will not be stable in metallic or sulphide form in equilibrium with larger amounts of FeO. Removal of these elements in oxidized form (i.e., take-up of FeO and MnO, etc. by metallic iron), requires high temperatures and pressures (Ohtani and Ringwood, 1984).

6. Siderophile Elements

Especially important and widely discussed are the abundances of siderophile elements in the Earth's mantle and the Moon.

6.1. Phosphorus

Newsom and Drake (1983) have argued that the difference in the P content between the mantles of the Earth and Moon favours an independent origin for both bodies. They based their argument on a correlation diagram of P vs. La. For the Earth's mantle, P correlates reasonably with La. Detailed studies in this laboratory showed that P correlates best with Nd (Weckwerth, 1983; and Weckwerth *et al.*, 1983). However, for the Moon no such correlation exists (Fig. 3). The P/La correlation

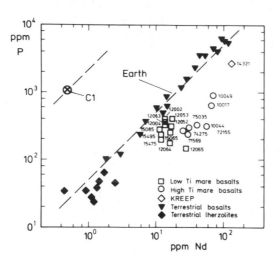

Fig. 3. Phosphorus vs. Nd. As found by Weckwerth (1983), P correlates best with Nd for terrestrial samples. Lunar basalts do not show such a correlation. However, low-Ti mare basalts show only slightly lower P/Nd ratios than terrestrial basalts.

shown in the paper by Newsom and Drake (1983) for lunar samples is in fact a mixing line that is due to the KREEP component mechanically admixed in many highland breccias (Dowty *et al.*, 1974; Warner *et al.*, 1974). In order to treat this problem in a realistic way, we put only one data point (14321) for the KREEP component in the P/Nd diagram of Fig. 3, as sample 14321 is one of the highland breccias with the highest KREEP content. All samples containing KREEP diluted by other components would plot on a mixing line fixed in its position by sample 14321.

As demonstrated in the P-Nd diagram of Fig. 3, the less-fractionated low-Ti mare basalts have variable P-Nd ratios, almost reaching the terrestrial correlation line with the lowest ratios less than a factor of 2 below the Earth's mantle ratio.

6.2. Cobalt

In the past, the opinions about the abundances of Co and other moderately siderophile elements in the source regions of lunar basalts differed drastically (Wänke et al., 1978; Delano and Ringwood, 1978a,b; Anders, 1978). However, the good correlation of Co vs. MgO + FeO in both terrestrial basalts and pristine lunar highlands rocks first quoted by Wänke et al. (1979) should be conclusive (Fig. 4a). The correlation line of pristine lunar highland rocks (Warren and Wasson, 1978) indicates for these rocks a Co abundance of only a factor of 1.7 below

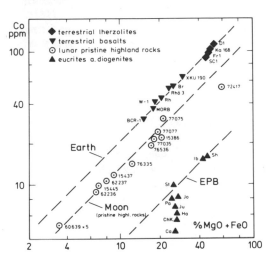

Fig. 4a. An excellent correlation of Co with (MgO + FeO) was observed by Wänke et al. (1983) for terrestrial rocks and pristine lunar highland rocks. The latter are only depleted by a factor of 1.7 relative to terrestrial samples. Eucrites and diogenites are, however, depleted by at least a factor of 6. Data for pristine lunar highland rocks are from Warren and Wasson (1978, 1979); for dunite 72417 from Laul and Schmitt (1975); all other data on lunar and terrestrial rocks are from the Mainz laboratory.

the value of the Earth's mantle (Fig. 4a). Some low-Ti mare basalts also plot along the correlation line of pristine highland rocks; the others have higher Co concentrations almost reaching the terrestrial correlation line (Fig. 4b).

The lunar highland breccias contain varying amounts of meteoritic component. Using Ir to monitor the meteoritic component, the indigenous Co concentrations can be obtained, assuming a C1 Co/Ir ratio for the meteoritic component (Wänke et al., 1978). In Fig. 4b only those highland breccias were included for which the correction for the meteoritic component is less than 50%; in fact, for most of the highland breccias used in Fig. 4b, the meteoritic component amounts to less than 30% of their total Co content. As seen from Fig. 4b, several of the highland breccias plot on the terrestrial correlation line, others plot between the terrestrial correlation line and that of the pristine lunar highland rocks. Many of the highland breccias

---

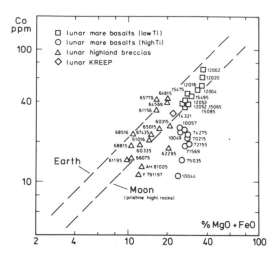

Fig. 4b. As in the case of P (Fig. 3), Co is only slightly depleted in low-Ti mare basalts compared to terrestrial samples. After correction for the meteoritic contribution (assuming a C1 Co/Ir ratio for the meteoritic component), lunar highland breccias plot on or close to the terrestrial correlation line. Data for lunar highland breccias 64569 and 65779 from Wasson et al. (1977); for 68516 from Laul and Schmitt (1973); for lunar meteorite ALHA 81005 from Palme et al. (1983); for Yamato 791197 from Ostertag et al. (1985); all other data from the Mainz laboratory.

especially those returned by Apollo 16 are thought to contain appreciable amounts of "unfractionated primary matter" (Wänke *et al.*, 1974, 1975, 1976). We will return to this point in the last section.

The almost vertical array of the data points of lunar mare basalts in Fig. 4b seems to indicate an equilibration with small amounts of metal, which on its subsequent segregation extracts Co to a varying degree. This supposition is substantiated by the fact that the depletion of Co is almost parallel to the depletion of phosphorus as demonstrated in Fig. 5. Obviously the effective metal-silicate partition coefficients of P and Co are about equal. In this respect the incompatible character of P has to be taken into account (Newsom and Drake, 1983). As shown by Wänke *et al.* (1971), lunar mare basalts do indeed contain small amounts of metal rich in Co. Using a hand-magnet, a total of 4.32 mg metal from rock 12053 was separated, which corresponded to a metal concentration of 0.036%. The Co concentration in this metal of 1.36% yields a contribution of 4.9 ppm to the 39 ppm Co measured in this rock. In the case of Ni the portion in the metal phase is even higher. High-Ti mare basalts have considerably lower Co concentrations as compared to low-Ti basalts. The formation of high-Ti mare basalts involves separation of ilmenite-rich cumulates from evolved Ti-rich magmas (Ringwood and Kesson, 1976). Crystallization of ilmenite may be accompanied by formation of metal during oxidation of Ti^{3+} to Ti^{4+} (Schreiber *et al.*, 1982). However, aside from segregation of metal, olivine segregation would also lead to a slight depletion of Co, and since P enters olivine more readily than Nd, it would also lead to a depletion of P relative to Nd (Weckwerth, 1983).

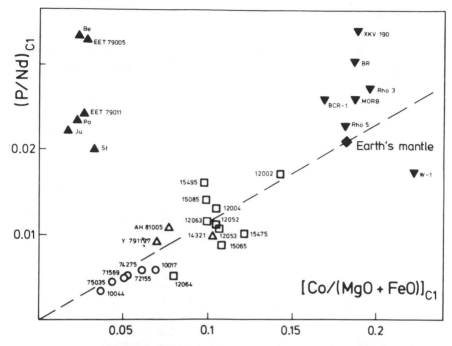

Fig. 5. C1 normalized P/Nd vs. Co/(Mg + FeO). The depletion of Co in lunar mare basalts is probably due to the presence of small amounts of metal since the Co depletion is clearly accompanied by that of P. Note that all lunar samples plot close to the (P/Nd)/[Co/(MgO + FeO)] ratio of the Earth's mantle. Terrestrial samples scatter around the data point of the Earth's mantle, while eucrites plot in a very different area. For analytical data and symbols see Figs. 4a,b.

6.3. Nickel

As shown in Fig. 6, the Ni/Mg ratio varies in terrestrial samples; however, the data points follow quite closely a trend line that is mainly due to fractionation of olivine. The measured Ni concentrations in lunar highland breccias containing meteoritic component were corrected using Ir as monitor element and assuming a C1 Ni/Ir ratio of 22,500 (Palme *et al.*, 1981). Again, we only used samples for which the correction of the meteoritic component was less than 50%. As in the case of Co, we find that many Apollo 16 breccias plot along the line representing the Ni/Mg ratio of the Earth's mantle defined by primitive mantle nodules (Jagoutz *et al.*, 1979). A considerable portion of Co and almost all of the Ni in the highland breccias plotted in Figs. 4b and 6 is contributed by a Mg-rich component that we believe to represent unfractionated primary matter (Wänke *et al.*, 1974, 1975, 1976). We will return to this point in the last chapter.

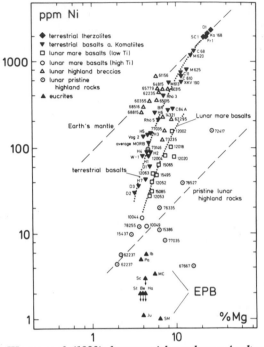

Fig. 6.   Nickel vs. Mg. The dotted lines represent least squares fits of data points of terrestrial and lunar mare basalts, respectively. The lower dashed line is a least squares fitted linear correlation line for pristine lunar highland rocks. The upper dashed line indicates the Ni/Mg ratio of the Earth as obtained from most primitive mantle nodules (Jagoutz et al., 1979). This plot indicates that low-Ti lunar mare basalts have about a factor of 4 lower Ni content than terrestrial basalts with a similar Mg content. After correction for the meteoritic component (assuming a C1 Ni/Ir ratio), lunar highland breccias have Ni/Mg ratios identical to those found for the Earth's mantle. For analytical data see Figs. 4a,b; for pristine lunar highland rock 78527 see Warren et al. (1983); for terrestrial samples see Arndt and Nesbitt (1984), Dietrich et al. (1978), Gladney and Colleen (1983), and Hertogen et al. (1980); for eucrites see Chou et al. (1976), McCarthy et al. (1973), and Morgan et al. (1978b).

The Ni/Mg ratios of lunar mare basalts are about a factor of 4 lower than those of terrestrial samples with identical Mg contents. As in the case of Co, most of this difference is probably due to equilibration with small amounts of metal and its segregation during the genesis of lunar mare basalts. In accordance with our results, Delano and Lindsley (1983) and Delano (1985a,b) have obtained depletion factors of 1.2 and 4.0 respectively for Co and Ni from investigations on lunar volcanic glasses. The relatively high Ni/Mg ratios of lunar mare basalts have been also pointed out by Ringwood and Kesson (1977b).

Pristine lunar highland samples have almost constant Ni/Mg ratios. The exceptions are olivine-rich rocks like dunite 72417 that have noticeably higher Ni/Mg ratios. Relative to the Earth's mantle, pristine lunar highland rocks show a Ni depletion by a factor of 38, while for Co we find from the Co/MgO + FeO diagram a depletion factor of 1.7.

The metal/silicate partition coefficient of Ni depends quite strongly on the temperature (Schmitt, 1984). Because of the strong preference of Ni for olivine compared to silicate melt, the relevant metal/silicate partition coefficient depends on the amount of solid olivine in the melt. To the contrary, the metal/silicate partition coefficient of Co of about 200 (Schmitt, 1984) is independent of temperature and only slightly influenced by the presence of solid olivine. With a metal/silicate partition coefficient for Co of 200 and the Co depletion in pristine lunar highland rocks a factor of 1.7, relative to the Earth's mantle, we find that 0.35% metal would account for the observed depletion. In the case of a totally molten magma ocean, the metal/silicate melt partition coefficient D(Ni) would be about 3000 (Schmitt, 1984), yielding upon segregation of 0.35% metal a depletion of Ni by a factor of 12, compared to the observed Ni depletion factor of 38. However, most of the pristine lunar highland rocks have been formed from more advanced liquids, i.e., after olivine crystallization and will have further reduced Ni contents. The actual metal/silicate partition coefficient was probably somewhat different from those applicable for a pure metal because of the uptake of elements like P and S.

In connection with the discussion in the last chapter, we especially want to point out the very low Ni content of eucrites (Fig. 6). Chemical equilibration between the metal and the silicate phase was obviously achieved in the case of the EPB, but not in the case of the Earth and the Moon.

6.4. Tungsten

Tungsten behaves in oxidized form as a highly incompatible element. From their W-La correlation diagram Wänke et al. (1974) deduced almost identical W concentrations in the mantles of Earth and Moon. Rammensee and Wänke (1977) have measured the metal/silicate partition coefficient of W and found that in the case of a totally molten Moon a metal content of 26% would be required to account for the observed W depletion on the Moon. They concluded that the similarity of the concentration of W in the mantles of Earth and Moon strongly argues for a close genetic relationship between Earth and Moon. Newsom and Drake (1982) have pointed out that the amount of metallic iron required to achieve the observed depletion of W is reduced if one assumes a low degree of partial melting. However, for a reasonable FeO content of 13% in the lunar mantle the required W depletion can only be achieved by a very low degree of partial melting of less than 5% and a low Ni content of the metal phase.

As pointed out by Ringwood and Seifert (1984), the lunar metal is expected to be Ni-rich and, in fact, even the metal phase found in lunar dunites (72415 and 72417; Dymek et al., 1975) contains as much as 30% Ni. As shown by Rammensee and Wänke (1977), the metal/silicate partition coefficient of W decreases substantially with increasing Ni content of the metal phase.

Tungsten is in fact more incompatible than La (Palme and Wänke, 1975) and, as shown by Palme and Rammensee (1981a), correlates better with U than with La. The W/U correlation diagram (Fig. 7) indicates that the W concentration in

---

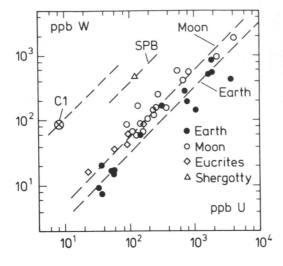

Fig. 7. *Tungsten vs. U. In the case of lunar samples, all data points refer to mare basalts except for the highest and second highest points, which refer to the KREEP-rich breccias 14321 and 62235, respectively. The correlation lines indicate slightly higher W abundances in the silicate phase of the Moon compared to the Earth's mantle. The fact that eucrites plot close to the lunar correlation line seems fortuitous. All data are from the Mainz laboratory.*

the lunar mantle even exceeds that of the Earth by a factor of 1.8. Clearly the W/U ratios of both Earth and Moon are not very well constrained; nevertheless, it seems certain that W is slightly more abundant in the lunar mantle compared to the terrestrial mantle. It is conceivable that siderophile element abundances are higher in the Moon than in the terrestrial mantle, if one takes into account either a transfer of iron together with some portions of other siderophile elements from the Earth's mantle into the core during the first few hundred million years, or the addition of siderophile elements from the projectile in an impact-induced fission process.

6.5. Molybdenum

As shown by Newsom and Palme (1984), Mo behaves in oxidized form as a moderately incompatible element and, like P, correlates best with Nd. The Mo/Nd ratio (Mo/Nd = 0.0016) in lunar samples as measured by Newsom (1984a) is a factor of 27 lower than in terrestrial samples (Mo/Nd = 0.043). At lunar oxygen fugacity conditions, Mo is a siderophile element with a metal/silicate partition coefficient of 1100 (Schmitt, 1984). As in the case of P and W, presence of solid olivine would raise the effective metal/silicate partition coefficient. Hence, it is no surprise that one finds in lunar mare basalts Mo depletions relative to terrestrial basalts that exceed even those of Ni in spite of its lower metal/silicate melt partition coefficient. Contrary to the case of Mo, the effective metal/silicate partition coefficients of Ni decreases if solid olivine is present. It is of special interest to note that the mare basalt sample 12002 that has the highest Co/MgO + FeO and highest Ni/Mg ratios also exceeds in its Mo/Nd ratio all other lunar samples measured by

Newsom (1984a). Loss of siderophile elements by equilibration with small amounts of metal was obviously smallest in the case of 12002, or its precursor material.

We propose the following scenario to account for the siderophile elements in pristine lunar highland rocks. About 0.5% metal probably formed by auto-reduction (see section 7) equilibrated with the almost totally molten magma ocean. This metal on its segregation extracted all siderophile elements according to their metal/liquid silicate partition coefficients. On cooling with olivine as cumulate phase, the Ni concentration in the liquid was further reduced. Dunite 72417 with its noticeably higher Ni/Mg ratio compared to other pristine rocks may be an example of this cumulate phase. Dymek et al. (1975) have shown that rock 72417 contains 0.17% metal with a Ni content of 23.8%. Different samples of this rock show variable Ni concentrations; the sample 72417 used in Fig. 6 is the one with the lowest Ni content (Ni = 160 ppm) analysed by Laul and Schmitt (1975). On further cooling, the other pristine highland rocks with their varying chemistry formed (Palme et al., 1984). In general the siderophile elements in pristine lunar highland rocks can be clearly explained by assuming concentrations of siderophile elements in a lunar magma ocean identical to those in the Earth's mantle.

7. "Primary" Matter in the Lunar Highlands and the Formation of the Moon

The Mainz group (Wänke et al., 1974, 1975, 1976, 1977) has repeatedly advocated the existence of an unfractionated primary component in the lunar highlands. The Mg/Ni correlation (see Fig. 6) observed for a number of highland breccias was in fact the first indication for the primary component that was then substantiated by mixing computations using geochemical as well as cosmochemical (C1 abundance ratios for refractory lithophile elements) constraints. From Figs. 4b and 6 it is evident that after correction for the presence of the meteoritic component, using Ir as a monitor element, both Co as well as Ni are present in the primary component in concentrations identical to that of the Earth's mantle.

In Table 1 we have listed the major element composition of the primary component as calculated by Wänke et al. (1977), assuming a C1 Mg/Si ratio of 0.91 and, as calculated by Wänke and Dreibus (1982), assuming a Mg/Si ratio of the upper mantle of the Earth of 1.04 (Jagoutz et al., 1979). Comparing the composition obtained for the primary matter with the most recent estimates of the lunar bulk composition one finds a perfect agreement.

When the first papers on the observation of unfractionated primary matter in lunar highland breccias were published, it seemed difficult to reconcile this observation with the general models of the formation and evolution of the Moon. How could it be that one finds on top of a thick, highly fractionated feldspathic crust material that obviously has escaped any significant fractionation? Impact-induced fission provides an excellent scenario to account also for the primary component in highland breccias.

TABLE 1. Comparison of the Composition of the Bulk Moon with that of the Unfractionated Primary Component Observed in Lunar Highland Samples (Wänke et al., 1977).

	Moon (Silicate portion)		Unfractionated primary component	
			Mg/Si = C1	Mg/Si = Earth's mantle
	Ringwood (1977)	Delano (1985)	Wänke et al. (1977)	Wänke and Dreibus (1982)
SiO$_2$ (%)	44.6	41.9	45.6	44.2
MgO	33.4	34.3	32.4	35.5
FeO	13.9	16.7	13.0	12.7
Al$_2$O$_3$	3.7	3.58	4.6	3.76
CaO	3.4	2.86	3.8	3.15
MnO	—	0.23	0.18	0.16
Cr$_2$O$_3$	0.4	0.38	0.4	0.37

Collision of the proto-Earth at its late stage of accretion with a large object will vaporize considerable portions of the Earth's mantle as well as of the projectile itself. The highly turbulent cloud surrounding the remaining central body will be loosely coupled to its surface by gas friction. In this way angular momentum could be transferred (Stevenson, 1984). On cooling, condensation will occur, allowing some fractionation of elements according to their volatility. In this way the lower concentrations of all elements more volatile than Na can be explained. However, we do not expect larger fractionation of elements less volatile than Na. In this respect Melosh and Sonett (1986) state: "We note only that the processes that take place are not similar to those that are believed to have occurred in the solar nebula because of the rapidity of condensation. Equilibrium is not reached: condensation is governed by chemical kinetics." In addition to the partial loss of volatiles we expect that auto-reduction will occur as FeO will partially decompose, forming small amounts of metal. In orbit around the Earth, the Moon could accrete on a time scale fast enough to explain the formation of a huge magma ocean. In respect to heat loss, even the most optimistic calculations on the heating of the Moon during accretion by Kaula (1979) failed to create a lunar magma ocean on formation of the Moon from material in heliocentric orbit.

It has been argued by Kreutzberger et al. (1984) and Drake (1986) that the Moon has a higher Cs/Rb ratio than the Earth, which, if correct, would argue against a depletion process of the alkaline elements due to volatility. It is true that MORB and ocean island basalts have lower Cs/Rb ratios as lunar basalts (Hofmann and White, 1983). However, we do not believe that the Cs/Rb ratio of MORB, etc., which are derived from a reservoir of the Earth's mantle that is highly depleted in the most incompatible elements such as Cs, reflect the true Cs/Rb ratio of the whole Earth. As shown in Fig. 8, there are many types of terrestrial rocks with considerably higher Cs/Rb ratios as observed in MORB. Even if we accept the

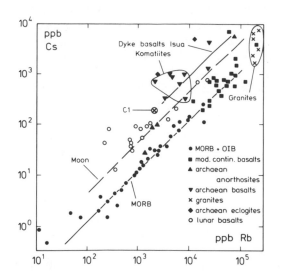

Fig. 8. Cesium vs. Rb in lunar
and terrestrial samples. The Cs/Rb
ratio in lunar samples is about a
factor of 4 higher as in terrestrial,
MORB, and ocean island basalts,
whereas terrestrial Archaean
anorthosites and basalts, as well as
especially Archaean eclogites, have
Cs/Rb ratios higher than lunar
samples. The Cs/Rb ratios of
modern continental basalts and
granites range between the lunar
and the MORB-ratios. Data for
MORB and OIB samples are from
Hertogen et al. (1980) and
Hofmann and White (1983); all
other data are from the Mainz
laboratory.

proposition that these high Cs/Rb ratios are due to crustal contamination, we are faced with the fact that large terrestrial reservoirs—the crust being then one of these reservoirs—with high Cs/Rb ratios must exist. In this respect, it is important to note that a constant ratio is also observed for the pair Nb/U in MORB and ocean island basalts that differs considerably from the C1 ratio. But as these two are refractory lithophile elements, it is very doubtful that the ratio observed in these rocks actually reflects the ratio valid for the Earth's mantle as a whole (Jochum et al., 1985). There are other indications that constant element ratios in modern mantle-derived rocks are under certain circumstances not identical to the ratio in the primitive mantle (White et al., 1985). The use of correlated elements to deduce abundance ratios for the whole planet is only justified in those cases in which identical ratios are observed for all types of igneous rocks. As seen from Fig. 8, this is not so in the case of the Cs/Rb.

It is conceivable that only a fraction of the total mass originally ejected finally ended up in the Moon. The major part probably fell back on Earth, and some fraction may have left the gravity field of the Earth while it was coming into heliocentric orbits. It is also conceivable that in addition to one large moon, one or several smaller moons were formed. Objects in orbits outside the orbit of the proto-Moon would be swallowed by the proto-Moon on its tidal retreat from the Earth. Such a scenario would be in accordance with the model advocated by Runcorn (1983), in which the large lunar basins were formed by impacts of lunar or terrestrial satellites. This would require that the orbits of these satellites are stable for several hundred million years. One could also visualize impacts of material that originally left the gravity field of the Earth and was stored in heliocentric orbits.

Formation of the lunar basins by objects made out of material identical to that of which the proto-Moon was formed would clearly add unfractionated material on top of the solidified crust. Primary matter was mainly observed in Apollo 16 breccias. This landing site is dominated by material ejected by the Imbrium event (Stöffler et al., 1984). It seems that highland breccias from areas far away from the big maria do not contain the primary component found in Apollo 16 breccias. The first lunar meteorite ALHA 81005 (Palme et al., 1983) could only barely be fitted into the mixing diagram formed by Apollo 16 and some Apollo 17 breccias, while the second lunar meteorite Yamato 791197 (Ostertag et al., 1985) cannot at all be fitted into the mixing diagram. Obviously, this sample does not contain appreciable amounts of primary matter. Wänke et al. (1976) noticed a weak correlation of La with Ni indicating that the primary and the KREEP component were added to many highland breccias in fixed proportions and, hence, were probably added in one event (Palme et al., 1984). Interestingly, the two lunar meteorites do not contain any noticeable admixture of KREEP, and they also do not contain significant amounts of Ni of nonmeteoritic origin.

Prior to the detailed trace element studies of lunar samples it was thought that all that is needed to understand our Moon is to explain the Moon's obvious depletion in metallic Fe as well as in total Fe. It has been argued that the Moon could be formed by a kind of disruptive capture of a differentiated object in such a way that only the silicate portion of it would stay in orbit around the Earth (Wasson and Warren, 1979). However, the silicate phase of such an object would have been equilibrated with FeNi during segregation of the metal phase, i.e., core formation. If the Moon had formed in such a way and the differentiated object would have been volatile-poor (low sulphur content) its silicate phase chemistry would be reflected by that of the eucrites. In the case of a volatile-rich differentiated object (high sulphur content), its chemistry would be that of the Shergotty parent body (SPB) (Dreibus and Wänke, 1984; Wänke and Dreibus, 1985). In Fig. 9 we have compared the concentrations of a number of crucial elements in basalts from the Moon and Earth with basalts from the EPB and SPB with similarly low degrees of fractionation. Both eucrites and shergottites differ from terrestrial and lunar basalts in their higher abundances of Mn (no depletion). Eucrites match the lunar basalts in respect to P and almost also to W; however, because of the equilibration of their source material with large amounts of FeNi, they are too low in Co and far too low in Ni. Shergottites, on the other hand, match lunar basalts in respect to Co and Ni because of the high amount of FeS on the SPB (Dreibus and Wänke, 1984; Wänke and Dreibus, 1985). Tungsten and P were extracted by the sulphur-rich FeNi alloy into the core only to a small degree, yielding a high abundance of P and W in the silicate phase. As shown by Jones and Drake (1982), the partition coefficient of W, for example, is strongly influenced by addition of sulphur to the metal phase.

The silicate phase of any differentiated object in the size range of up to 10–20% of the Earth's mass will always have much lower concentrations of siderophile

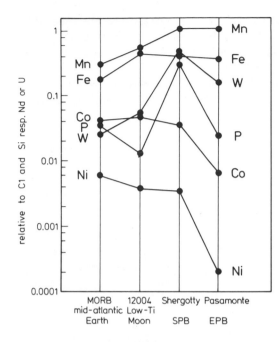

Fig. 9. The abundances of Fe, Mn, W, P, Co, and Ni in basalts from the Earth, Moon, SPB, and EPB underlines the close compositional similarity of the Earth's mantle with the silicate phase of the Moon. Basalts from SPB and EPB have characteristically different compositions. For data from MORB sample, see Hertogen et al. (1980) and Palme and Rammensee (1981b); for Pasamonte see McCarthy et al. (1973); all other data are from the Mainz laboratory.

elements such as Ni and Co than the Moon. This is due to the simple effect of the chemical equilibrium between the metal phase or FeNi-FeS phase and the silicate phase as governed by the respective partition coefficient. Only objects comparable in mass to that of the Earth can have mantle concentrations of siderophile elements in excess of those given by the metal/silicate, sulphide/silicate partition coefficients, respectively. The excess amounts can either be the result of inhomogeneous accretion with core formation parallel to accretion due to high temperatures reached during accretion (Mainz model; Wänke, 1981). In the homogeneous accretion model the excess of siderophiles might be explained by the uptake of oxygen in the metal phase at high temperatures and pressures, if the segregation of metal can be sufficiently delayed (Canberra model; Ringwood, 1977, 1984).

Clearly, the EPB and the SPB may be rather extreme examples and one could well visualize a differentiated object with a sulphur content of its core between that of the EPB and that of the SPB. Basalts from such an object would plot somewhere on the lines connecting the data points of Shergotty and Pasamonte. But as can be seen from Fig. 9, they would not come much closer compositionally to the lunar basalts. We conclude that the Moon cannot be formed by disruptive capture of a differentiated object.

Terrestrial basalts reflect the composition of the lunar basalts in all details, except their lower FeO and MnO contents and their higher contents of volatile and the highly siderophile elements. There is no doubt that small amounts of metal were responsible for the lower abundances of siderophile elements (Wänke *et al.*, 1978, 1979; Palme, 1980; Wolf and Anders, 1980; Palme *et al.*, 1984). In this respect we have to remember that metallic iron is stable under lunar mantle conditions, but not under upper Earth mantle conditions.

Figure 10 illustrates the abundance of siderophile elements in the lunar mantle relative to the Earth's mantle. As emphasized by Wänke *et al.* (1983) and Newsom

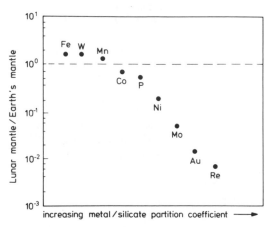

Fig. 10. Relative to the present (upper) Earth's mantle, the lunar mantle shows an overabundance of Fe, W, and Mn and a depletion of Co, P, Ni, Mo, Au, and Re. The sequence in the abundance ratios of these elements is parallel to their effective metal/silicate partition coefficients. Note that the depletions observed are deduced from lunar mare basalts and are therefore upper limits. It seems likely that before the equilibration with small amounts of metal (<0.5%) and its subsequent segregation, the lunar mantle contained siderophile elements in concentrations equal to or even slightly above the concentrations observed for the present upper Earth's mantle. Data for Mo from Newsom (1984a); for Au and Re from Chou et al. (1983) and Wolf et al. (1979).

(1984b), the depletion of siderophile elements in the lunar mantle is parallel to the effective metal/silicate partition coefficients of the respective elements. For Fe, Mn, and probably also W (all elements having metal/silicate partition coefficients < 100), we see even higher abundances on the Moon. Hence, it seems most likely that the Moon actually was originally somewhat richer in siderophile elements compared to the present Earth's mantle. However, as emphasized in Section 4, it seems likely that at the time of its formation the Moon was identical in its siderophile element content with the Earth's mantle or with that portion of the Earth's mantle into which the projectile was mixed.

8. Conclusion

The close similarity of the composition of the lunar mantle with the very characteristic composition of the Earth's mantle makes a close genetic relationship of the Moon with the Earth's mantle an almost unavoidable consequence. Impact-induced splash off of the Earth's mantle material from which the Moon was formed is the model that best fits the geochemical observations. The presence of unfractionated primary matter compositionally identical to the Earth's mantle in the lunar highland seems to indicate that besides the proto-Moon at least one smaller satellite formed that was swallowed by the proto-Moon on its tidal retreat from the Earth on a comparatively small time scale.

Acknowledgments. This work was carried out within the Forschergruppe Mainz, supported by the Deutsche Forschungsgemeinschaft.

9. References

Anders E. (1978) Procrustean science: Indigenous siderophiles in the lunar highlands, according to Delano and Ringwood. *Proc. Lunar Planet. Sci. Conf. 9th*, pp. 161–184.

Arndt N. T. and Nesbitt R. W. (1984) Magma mixing in komatiitic lavas from Munro Township, Ontario. In *Archean Geochemistry* (A. Kröner, G. N. Hanson, and A. M. Goodwin, eds.), pp. 99–114. Springer Verlag, Berlin.

Becker R. H. and Pepin R. O. (1984) The case for a Martian origin of the shergottites: Nitrogen and noble gases in EETA 79001. *Earth Planet. Sci. Lett., 69*, 225–242.

Boslough M. B. and Ahrens T. J. (1983) Shock-melting and vaporization of anorthosite and implications for an impact-origin of the Moon (abstract). In *Lunar and Planetary Science XIV*, pp. 63–64. Lunar and Planetary Institute, Houston.

Cameron A. G. W. and Ward W. R. (1976) The origin of the Moon (abstract). In *Lunar Science VII*, pp. 120–122. The Lunar Science Institute, Houston.

Chou C.-L., Boynton W. V., Bild R. W., Kimberlin J., and Wasson J. T. (1976) Trace element evidence regarding a chondritic component in howardite meteorites. *Proc. Lunar Sci. Conf. 7th*, pp. 3501–3518.

Chou C.-L., Shaw D. M., and Crocket J. H. (1983) Siderophile trace elements in the Earth's oceanic crust and upper mantle. *Proc. Lunar Planet. Sci. Conf. 13th*, in *J. Geophys. Res., 88*, A507–A518.

Clayton R. N. and Mayeda T. K. (1975) Genetic relations between the Moon and meteorites. *Proc. Lunar Sci. Conf. 6th*, pp. 1761–1769.

Consolmagno G. J. and Drake M. J. (1977) Composition and evolution of the eucrite parent body: Evidence from rare earth elements. *Geochim. Cosmochim. Acta, 41*, 1271–1282.

Darwin G. H. (1880) On the secular changes in the orbit of a satellite revolving around a tidally disturbed planet. *Philos. Trans. Roy. Soc. London, 171*, 713–891.

Delano J. W. (1985a) Abundances of cobalt and nickel in the silicate portion of the Moon, this volume.

Delano J. W. (1985b) Mare volcanic glasses, II: Abundances of trace Ni and the composition of the Moon (abstract). In *Lunar and Planetary Science XVI*, pp. 179–180. Lunar and Planetary Institute, Houston.

Delano J. W. and Lindsley D. H. (1983) Mare glasses from Apollo 17: Constraints on the Moon's bulk composition. *Proc. Lunar Planet. Sci. Conf. 15th*, in *J. Geophys. Res., 88*, B3–B16.

Delano J. W. and Ringwood A. E. (1978a) Indigenous abundances of siderophile elements in the lunar highlands: Implications for the origin of the Moon. *Moon and Planets, 18*, 385–525.

Delano J. W. and Ringwood A. E. (1978b) Siderophile elements in the lunar highlands: Nature of the indigenous component and implications for the origin of the Moon. *Proc. Lunar Planet. Sci. Conf. 9th*, pp. 111–159.

Delano J. W. and Stone K. (1985) Siderophile elements in the Earth's upper mantle: Secular variations and possible cause for their overabundance (abstract). In *Lunar and Planetary Science XVI*, pp. 181–182. Lunar and Planetary Institute, Houston.

Dietrich V., Emmermann R., Oberhansli R., and Puchelt H. (1978) Geochemistry of basaltic and gabbroic rocks from the West Mariana Basin and the Mariana Trench. *Earth Planet. Sci. Lett., 39*, 127–144.

Dowty E., Prinz M., and Keil K. (1974) Very high alumina basalt: A mixture and not a magma type. *Science, 183*, 1214–1215.

Drake M. J. (1983) Geochemical constraints on the origin of the Moon (abstract). In *Lunar and Planetary Science XIV*, pp. 162–163. Lunar and Planetary Institute, Houston.

Drake M. J. (1986) Is lunar bulk material similar to Earth's mantle?, this volume.

Dreibus G. and Wänke H. (1979) On the chemical composition of the Moon and the eucrite parent body and a comparison with the composition of the Earth; the case of Mn, Cr, and V (abstract). In *Lunar and Planetary Science X*, pp. 315–317. Lunar and Planetary Institute, Houston.

Dreibus G. and Wänke H. (1984) Accretion of the Earth and the inner planets. *Proc. 27th Intern. Geol. Congr., Vol. 11, Geochemistry and Cosmochemistry*, pp. 1–20, VNU Science Press, Utrecht.

Dreibus G., Kruse H., Spettel B., and Wänke H. (1977) The bulk composition of the Moon and the eucrite parent body. *Proc. Lunar Sci. Conf. 8th*, pp. 211–227.

Dymek R. F., Albee A. L., and Chodos A. A. (1975) Comparative petrology of lunar cumulate rocks of possible primary origin: Dunite 72415, troctolite 76535, norite 78235, and anorthosite 62237. *Proc. Lunar Sci. Conf. 6th*, pp. 301–341.

Ganapathy R. and Anders E. (1974) Bulk composition of the Moon and Earth, estimated from meteorites. *Proc. Lunar Sci. Conf. 5th*, pp. 1181–1206.

Gladney E. S. and Colleen E. B. (1983) Compilation of elemental concentrations in eleven United States Geological Survey rock standards. *Geostandards Newsletter, 7*, 3–226.

Hartmann W. K. and Davis D. R. (1975) Satellite-sized planetesimals and lunar origin. *Icarus, 24*, 504–515.

Hertogen J., Janssens M.-J., and Palme H. (1980) Trace elements in ocean ridge basalt glasses: Implications for fractionations during mantle evolution and petrogenesis. *Geochim. Cosmochim. Acta, 44*, 2125–2143.

Hofmann A. W. and White W. M. (1983) Ba, Rb, and Cs in the Earth's mantle. *Z. Naturforsch., 38a*, 256–266.

Jagoutz E. and Wänke H. (1982) Has the Earth's core grown over geologic times? (abstract). In *Lunar and Planetary Science XIII*, pp. 358–359. Lunar and Planetary Institute, Houston.

Jagoutz E., Palme H., Baddenhausen H., Blum K., Cendales M., Dreibus G., Spettel B., Lorenz V., and Wänke H. (1979) The abundances of major, minor and trace elements in the Earth's mantle as derived from primitive ultramafic nodules. *Proc. Lunar Planet. Sci. Conf. 10th*, pp. 2031–2050.

Jochum K.-P., Hofmann A. W., and Seufert H. M. (1985) Nb, U, Zr, Y in oceanic basalts and Nb/U, Zr/Nb, Zr/Y ratios of the mantle. *Terra Cognita, 5*, 276.

Jones J. H. and Drake M. J. (1982) An experimental geochemical approach to early planetary differentiation (abstract). In *Lunar and Planetary Science XIII*, pp. 369–370. Lunar and Planetary Institute, Houston.

Kallemeyn G. W. and Wasson J. T. (1981) The compositional classification of chondrites—I. The carbonaceous chondrite groups. *Geochim. Cosmochim. Acta, 45*, 1217–1230.

Kaula W. M. (1979) Thermal evolution of Earth and Moon growing by planetesimal impacts. *J. Geophys. Res., 84*, 999–1008.

Keil K. (1968) Mineralogical and chemical relationships among enstatite chondrites. *J. Geophys. Res., 73*, 6945–6976.

Kreutzberger M. E., Drake M. J., and Jones J. H. (1984) Origin of the Moon: Constraints from volatile elements (abstract). In *Papers Presented to the Conference on the Origin of the Moon*, p. 22. Lunar and Planetary Institute, Houston.

Laul J. C. and Schmitt R. A. (1973) Chemical composition of Apollo 15, 16 and 17 samples. *Proc. Lunar Sci. Conf. 4th*, pp. 1349–1367.

Laul J. C. and Schmitt R. A. (1975) Dunite 72417: A chemical study and interpretation. *Proc. Lunar Sci. Conf. 6th*, pp. 1231–1254.

McCarthy T. S., Erlank A. J., and Willis J. P. (1973) On the origin of eucrites and diogenites. *Earth Planet. Sci. Lett., 18*, 433–442.

Melosh H. J. (1985) When worlds collide: Jetted vapor planetesimals and the Moon's origin (abstract). In *Lunar and Planetary Science XVI*, pp. 552–553. Lunar and Planetary Institute, Houston.

Melosh H. J. and Sonett C. P. (1986) When worlds collide: Jetted vapor plumes and the Moon's origin, this volume.

Morgan J. W., Hertogen J., and Anders E. (1978a) The Moon: Composition determined by nebular processes. *Moon and Planets, 18*, 465–478.

Morgan J. W., Higuchi H., Takahashi H., and Hertogen J. (1978b) A "chondritic" eucrite parent body: Inference from trace elements. *Geochim. Cosmochim. Acta, 42*, 27–38.

Newsom H. E. (1984a) The abundance of molybdenum in lunar samples: New evidence for a lunar metal core (abstract). In *Lunar and Planetary Science XV*, pp. 605–606. Lunar and Planetary Institute, Houston.

Newsom H. E. (1984b) The lunar core and the origin of the Moon. *EOS (Trans. Am. Geophys. Union), 65*, 369–370.

Newsom H. E. and Drake M. J. (1982) Constraints on the Moon's origin from the partitioning behavior of tungsten. *Nature, 297*, 210–212.

Newsom H. E. and Drake M. J. (1983) Experimental investigation of the partitioning of phosphorus between metal and silicate phases: Implications for the Earth, Moon and eucrite parent body. *Geochim. Cosmochim. Acta, 47*, 93–100.

Newsom H. E. and Palme H. (1984) The depletion of siderophile elements in the Earth's mantle: New evidence from molybdenum and tungsten. *Earth Planet. Sci. Lett., 69*, 354–364.

Newsom H. E., White W. M., and Jochum K. P. (1985) Did the Earth's core grow through geological time? (abstract). In *Lunar and Planetary Science XVI*, pp. 616–617. Lunar and Planetary Institute, Houston.

Ohtani E. and Ringwood A. E. (1984) Composition of the core, I. Solubility of oxygen in molten iron at high temperatures. *Earth Planet. Sci. Lett., 71*, 85–93.

Ostertag R., Stöffler D., Palme H., Spettel B., Weckwerth G., and Wänke H. (1985) Lunar meteorite Yamato 791197: A weakly shocked regolith breccia from the far side of the Moon (abstract). In *Lunar and Planetary Science XVI*, pp. 635–636. Lunar and Planetary Institute, Houston.

Palme H. (1980) The meteoritic contamination of terrestrial and lunar impact melts and the problem of indigenous siderophiles in the lunar highland. *Proc. Lunar Planet. Sci. Conf. 11th*, 481–506.

Palme H. and Rammensee W. (1981a) The significance of W in planetary differentiation processes: Evidence from new data on eucrites. *Proc. Lunar Planet. Sci. Conf. 9th*, pp. 949–964.

Palme H. and Rammensee W. (1981b) Tungsten and some other siderophile elements in meteoritic and terrestrial basalts (abstract). In *Lunar and Planetary Science XII*, pp. 796–798. Lunar and Planetary Institute, Houston.

Palme H. and Wänke H. (1975) A unified trace-element model for the evolution of the lunar crust and mantle. *Proc. Lunar Sci. Conf. 6th*, pp. 1179–1202.

Palme H., Spettel B., Wänke H., Bischoff A., and Stöffler D. (1984) Early differentiation of the Moon: Evidence from trace elements in plagioclase. *Proc. Lunar Planet. Sci. Conf. 15th*, in *J. Geophys. Res., 89*, C3–C15.

Palme H., Spettel B., Weckwerth G., and Wänke H. (1983) Antarctic meteorite ALHA 81005, a piece from the ancient lunar crust. *Geophys. Res. Lett., 10*, 817–820.

Palme H., Suess H. E., and Zeh H. D. (1981) Abundances of the elements in the solar system. In *Landolt-Börnstein*, Vol. 2, *Astronomy and Astrophysics* (K. Schaifers and H. H. Voigt, eds.), pp. 257–272. Springer Verlag, Berlin.

Rammensee W. and Wänke H. (1977) On the partition coefficient of tungsten between metal and silicate and its bearing on the origin of the moon. *Proc. Lunar Sci. Conf. 8th*, pp. 399–409.

Rammensee W., Palme H., and Wänke H. (1983) Experimental investigation of metal-silicate partitioning of some lithophile elements (Ta, Mn, V, Cr) (abstract). In *Lunar and Planetary Science XIV*, pp. 628–629. Lunar and Planetary Institute, Houston.

Ringwood A. E. (1966a) Chemical evolution of terrestrial planets. *Geochim. Cosmochim. Acta, 30*, 41–104.

Ringwood A. E. (1966b) Mineralogy of the mantle. In *Advances in Earth Science* (P. Hurley, ed.), pp. 357–398. MIT, Boston.

Ringwood A. E. (1977) Composition of the core and implications for origin of the Earth. *Geochem. J., 11*, 111–135.

Ringwood A. E. (1979) *On the Origin of the Earth and Moon*. Springer Verlag, New York. 295 pp.

Ringwood A. E. (1981) Geophysical and cosmochemical constraints on properties of mantles of the terrestrial planets. In *Basaltic Volcanism on the Terrestrial Planets*, pp. 633–699. Pergamon, New York.

Ringwood A. E. (1984) The Earth's core: Its composition, formation and bearing upon the origin of the Earth. *Proc. Roy. Soc. London, A395*, 1–46.

Ringwood A. E. and Kesson S. E. (1976) A dynamic model for mare basalt petrogenesis. *Proc. Lunar Sci. Conf. 7th*, 1697–1722.

Ringwood A. E. and Kesson S. E. (1977a) Composition and origin of the Moon. *Proc. Lunar Planet. Sci. Conf. 8th*, pp. 371–398.

Ringwood A. E. and Kesson S. E. (1977b) Basaltic magmatism and the bulk composition of the Moon. II. Siderophile and volatile elements in Moon, Earth and chondrites. Implications for lunar origin. *The Moon, 16*, 425–464.

Ringwood A. E. and Seifert S. (1984) Nickel-cobalt systematics and their bearing on lunar origin (abstract). In *Papers Presented to the Conference on the Origin of the Moon*, p. 14. Lunar and Planetary Institute, Houston.

Runcorn S. K. (1983) Lunar magnetism, polar displacements and primeval satellites in the Earth-Moon system. *Nature, 304*, 589–596.

Schmitt W. (1984) Experimentelle Bestimmung von Metal/Sulfid/Silikat-Verteilungs-koeffizienten geochemisch relevanter Spurenelemente. Thesis, University of Mainz.

Schreiber H. D., Balazs G. B., Shaffer A. P., and Jamison P. L. (1982) Iron metal production in silicate melts through the direct reduction of Fe(II) by Ti(III), Cr(II), and Eu(II). *Geochim. Cosmochim. Acta, 46*, 1891–1901.

Stevenson D. J. (1984) Lunar origin from impact on the Earth: Is is possible? (abstract). In *Papers Presented to the Conference on the Origin of the Moon*, p. 60. Lunar and Planetary Institute, Houston.

Stevenson D. J. (1985) Implications of very large impacts for Earth accretion and lunar formation (abstract). In *Lunar and Planetary Science XVI*, pp. 819–820. Lunar and Planetary Institute, Houston.

Stöffler D., Bischoff A., Borchardt R., Burghele A., Deutsch A., Jessberger E. K., Ostertag R., Palme H., Spettel B., Reimold W. U., Wacker K., and Wänke H. (1984) Composition and evolution of the lunar crust in the Descartes highlands, Apollo 16. *Proc. Lunar Planet. Sci. Conf. 15th*, in *J. Geophys. Res., 90*, C449–C506.

Wänke H. (1981) Constitution of terrestrial planets. *Philos. Trans. Roy. Soc. London, A303*, 287–302.

Wänke H. and Dreibus G. (1982) Chemical and isotopic evidence for the early history of the Earth-Moon system. In *Tidal Friction and the Earth's Rotation II* (P. Brosche and J. Sündermann, eds.), pp. 322–344. Springer Verlag, Berlin.

Wänke H. and Dreibus G. (1985) The degree of oxidation and the abundance of volatile elements on Mars (abstract). In *Lunar and Planetary Science XVI*, pp. 28–29. Lunar and Planetary Institute, Houston.

Wänke H., Baddenhausen H., Blum K., Cendales M., Dreibus G., Hofmeister H., Kruse H., Jagoutz E., Palme C., Spettel B., Thacker R., and Vilcsek E. (1977) On the chemistry of lunar samples and achondrites. Primary matter in the lunar highlands: A re-evaluation. *Proc. Lunar Sci. Conf. 8th*, pp. 2191-2213.

Wänke H., Dreibus G., and Jagoutz E. (1984) Mantle chemistry and accretion history of the Earth. In *Archaean Geochemistry* (A. Kröner, G. N. Hanson, and A. M. Goodwin, eds.), pp. 1-24. Springer Verlag, Berlin.

Wänke H., Dreibus G., and Palme H. (1978) Primary matter in the lunar highlands: The case of the siderophile elements. *Proc. Lunar Planet. Sci. Conf. 9th*, pp. 83-110.

Wänke H., Dreibus G., and Palme H. (1979) Non-meteoritic siderophile elements in lunar highland rocks: Evidence from pristine rocks. *Proc. Lunar Planet. Sci. Conf. 10th*, pp. 611-626.

Wänke H., Dreibus G., Palme H., Rammensee W., and Weckwerth G. (1983) Geochemical evidence for the formation of the Moon from material of the Earth's mantle (abstract). In *Lunar and Planetary Science XIV*, pp. 818-819. Lunar and Planetary Institute, Houston.

Wänke H., Palme H., Baddenhausen H., Dreibus G., Jagoutz E., Kruse H., Palme C., Spettel B., Teschke F., and Thacker R. (1975) New data on the chemistry of lunar samples: Primary matter in the lunar highlands and the bulk composition of the Moon. *Proc. Lunar Sci. Conf. 6th*, pp. 1313-1340.

Wänke H., Palme H., Baddenhausen H., Dreibus G., Jagoutz E., Kruse H., Spettel B., Teschke F., and Thacker R. (1974) Chemistry of Apollo 16 and 17 samples: Bulk composition, late stage accumulation and early differentiation of the Moon. *Proc. Lunar Sci. Conf. 5th*, pp. 1307-1335.

Wänke H., Palme H., Kruse H., Baddenhausen H., Cendales M., Dreibus G., Hofmeister H., Jagoutz E., Palme C., Spettel B., and Thacker R. (1976) Chemistry of lunar highland rocks: A refined evaluation of the composition of the primary matter. *Proc. Lunar Sci. Conf. 7th*, pp. 3479-3499.

Wänke H., Wlotzka F., Baddenhausen H., Balacescu A., Spettel B., Teschke F., Jagoutz E., Kruse H., Quijano-Rico M., and Rieder R. (1971) Apollo 12 samples: Chemical composition and its relation to sample locations and exposure ages, the two component origin of the various soil samples and studies on lunar metallic particles. *Proc. Lunar Sci. Conf. 2nd*, pp. 1187-1208.

Warner J. L., Simonds C. H., and Phinney W. C. (1974) Impact-induced fractionation in the lunar highlands. *Proc. Lunar Sci. Conf. 5th*, pp. 379-397.

Warren P. H. and Wasson J. T. (1978) Compositional-petrographic investigation of pristine nonmare rocks. *Proc. Lunar Planet. Sci. Conf. 9th*, pp. 185-217.

Warren P. H. and Wasson J. T. (1979) The compositional-petrographic search for pristine nonmare rocks: Third foray. *Proc. Lunar Planet. Sci. Conf. 10th*, pp. 583-610.

Warren P. H., Taylor G. J., Keil K., Kallemeyn W., Rosener P. S., and Wasson J. T. (1983) Sixty foray for pristine nonmare rocks and an assessment of the diversity of lunar anorthosites. *Proc. Lunar Planet. Sci. Conf. 13th*, in *J. Geophys. Res., 88*, A615-A630.

Wasson J. T. and Warren P. H. (1979) Formation of the Moon from differentiated planetesimals of chondritic composition (abstract). In *Lunar and Planetary Science X*, pp. 1310-1312. Lunar and Planetary Institute, Houston.

Wasson J. T., Warren P. H., Kallemeyn G. W., McEwing C. E., Mittlefehldt D. W., and Boynton W. V. (1977) SCCRV, a major component of highland rocks. *Proc. Lunar Sci. Conf. 8th*, pp. 2237-2252.

Weckwerth G. (1983) Anwendung der instrumenellen β-Spektrometrie im Bereich der Kosmochemie, insbesondere zur Messung von Phosphorgehalten. Diplom., Max-Planck-Institut für Chemie, Mainz.

Weckwerth G., Spettel B., and Wänke H. (1983) Phosphorus in the mantle of planetary bodies. *Terra Cognita, 3*, 79-80.

Wetherill G. W. (1976) The role of large bodies in the formation of the Earth and Moon. *Proc. Lunar Sci. Conf. 7th*, 3245-3257.

White W. M., Newsom H., Jochum K.-P., and Hofmann A. W. (1985) Siderophile-chalcophile element abundances, core formation and mantle Pb isotope evolution. *Terra Cognita, 5*, 273.

Wolf R. and Anders E. (1980) Moon and Earth: Compositional differences inferred from siderophiles, volatiles, and alkalis in basalts. *Geochim. Cosmochim. Acta, 44*, 2111–2124.

Wolf R., Woodrow A., and Anders E. (1979) Lunar basalts and pristine highland rocks: Comparison of siderophile and volatile elements. *Proc. Lunar Planet. Sci. Conf. 10th*, 2107–2130.

Composition and Origin of the Moon

A. E. RINGWOOD

Research School of Earth Sciences, Australian National University, Canberra, A.C.T. 2601, Australia

A powerful discriminant between alternative hypotheses of lunar origins is provided by siderophile element abundances. The siderophile signature of the Earth's mantle was established by a combination of several complex processes involved in the separation of a large metallic core in a planetary-sized body and is unique to the Earth. Siderophile signatures of the mantle of the eucrite parent body and the shergottite parent body (Mars?) differ from the Earth in several important respects. Lunar siderophile element geochemistry is reviewed in detail. The average abundances of Cr, V, Mn, Ni, Co, W, P, S, Se, and Te are similar within a factor of two in the bulk Moon and in the Earth's mantle. Copper and Ga can be added to this list if appropriate corrections for their loss as volatiles are applied to the lunar abundances. The mafic component of the lunar highland crust displays a similar siderophile abundance pattern. The similarity between this group of siderophiles in Earth and Moon and its essential dissimilarity with the siderophile signatures of eucrites and shergottites argues strongly in favour of derivation of the Moon from the Earth's mantle. If attention is restricted to elements whose abundance is best established, it is found that Cr, V, Mn, Co, and W are very similar (±30%) in terrestrial and lunar mantles, whereas P, Ni, Mo, Re, and Au are depleted by factors of 1.7, 3, 25, 80, and 80 respectively. This sequence corresponds to the order of increasing siderophile character of these elements. Whereas the nickel abundance in the lunar mantle is depleted threefold compared to the Earth's mantle, parallel studies demonstrate that the average Ni concentration for the *bulk Moon* is also very similar to that of the Earth's mantle. This implies that the nickel deficit in the mantle is compensated by the presence of a small (0.4%) nickel-rich lunar core (~40% Ni). A core of this size readily explains the observed depletions of P, Ni, Mo, Re, and Au in the lunar mantle. The hypothesis of terrestrial origin for the Moon is thus shown to be capable of providing a quantitative explanation of the abundances of the less-volatile siderophile elements in the Moon. Hypotheses that relate the Moon to the mantles of the parent bodies of differentiated meteorites are quite unable to explain the observed signatures. It is concluded that the siderophile evidence definitively requires a terrestrial origin of the Moon. The most promising mechanism for removing terrestrial material from the Earth's mantle arises from the impacts of many large planetesimals (100–1000 km) at a late stage of accretion after core formation. The presence of a primitive terrestrial atmosphere that corotated with the Earth also played a key role in the formation of the Moon. The primordial Earth rotated with a period of four to five hours and the corotating primitive atmosphere extended beyond the corresponding geosynchronous orbit at about 2.5 Earth radii. Impacts of large planetesimals evaporated several times their own masses of the Earth's mantle and shock-melted considerably more. The impact clouds of shock-melted spray and vapour were accelerated to high velocities. Significant proportions of these clouds of liquid spray and condensed vapour were ejected beyond the geosynchronous limit, where the solids accreted to form planetesimals. The corotating primitive atmosphere provided the means of transferring angular momentum from the Earth to the planetesimals via viscous coupling. The planetesimals accordingly spiralled outwards beyond Roche's Limit where they accreted to form the Moon.

1. Introduction

It is well known that the densities of terrestrial planets vary substantially, even when corrected to a common pressure to allow for the effects of differential self-compression in their own gravitational fields. Jeffreys (1937), Urey (1952), and many others explained the density variation by assuming that these planets are composed of varying proportions of silicate ($\rho_0 = 3.3$ g/cm^3) and nickel-iron ($\rho_0 = 7.9$ g/cm^3) phases, each phase being essentially of constant composition. According to their model, Mercury and Mars contain about 65 and 20 wt % of metal phase respectively, while Earth and Venus possess intermediate metal contents. On the basis of this hypothesis, the Moon (density 3.334 g/cm^3) would contain less than 5% of metal phase.

Urey (1952, 1957a,b) proposed that physical processes occurring in the solar nebula prior to accretion of the planets had caused variable degrees of fractionation of iron particles from silicate particles in different regions of the nebula. Subsequent accretion of planets in particular regions reflected these preexisting metal-silicate inhomogeneities. Some kind of metal-silicate fractionation is clearly necessary to explain the high density of Mercury and the low density of the Moon as compared to Mars, Earth, and Venus. Moreover, the major role attributed to metal-silicate fractionation in the solar nebula was supported during the 1960s by measurements that seemed to show that the abundance of iron in the sun was about a factor of five lower than is found in Earth, Mars, Venus, and chondritic meteorites. Urey's interpretation accordingly became widely accepted and has been reflected in the subsequent cosmogonic models of Anders (1968, 1971), Grossman (1972), and many others.

These hypotheses were questioned by Ringwood (1959, 1966a) on the grounds that (in his opinion) the mechanisms invoked as causes of the iron-silicate fractionations were contrived and were physically implausible. Accordingly, he sought to develop a cosmogony that minimized the role of this process. Ringwood proposed that the relative abundances of iron and common lithophile elements (Mg, Si, Ca, Al) were the same in Mars, Earth, and Venus, and were similar to those in chondrites and in the sun. Differences in density between Mars, Venus, and Earth were caused by different oxidation states, a variable that was readily explicable on cosmochemical grounds. Ringwood (1966b) reexamined the evidence on abundances of elements in the sun and concluded that the precision of the existing data base did not justify the earlier conclusion that iron was strongly depleted.

Supporting evidence for this interpretation was soon forthcoming. A more accurate determination of the relative abundance of iron in the sun by Garz *et al.* (1969) showed that it was similar to chondrites. New space probe mesurements provided precise values of the densities of Venus and Mars and of the Martian moment-of-inertia. It was readily demonstrated (as summarized by Ringwood, 1979) that the gross physical (and chemical) properties of Venus, Earth, and Mars were explicable in terms of models based on chondritic abundances of major elements (except for differing mean oxidation states), with Mars being substantially more oxidized than Earth and Venus. More recently, Ringwood (1984) concluded that the metallic cores of the Earth and Venus probably contain considerable amounts of dissolved FeO. This implies that the mean oxidation states of Venus, Earth, and Mars are more similar than had previously been supposed. Indeed, they could be essentially identical. The only planet displaying the effects of major metal/silicate fractionation is Mercury. The high density of Mercury, implying a large iron content, may be due to specialized conditions of accretion owing to its location nearest to the sun (e.g., Ringwood, 1966a; Cameron, 1985).

The considerations discussed above have an important bearing on the origin of the Moon. During the 1960s it seemed reasonable to many scientists to treat the Moon as representing an extreme case of the general process of iron/silicate fractionation that was believed to have occurred between the sun and the planets, and between the planets themselves. We now recognise that the Moon must have accreted in a region of the solar system between Earth, Mars, and Venus, and probably close to the Earth. Currently, there is no evidence requiring iron/silicate fractionation between these planets and the sun. From this perspective, the high depletion of metallic iron in the Moon must be recognized as a truly remarkable phenomenon. It was this recognition that led Ringwood (1960, 1966a) to reject a normal "planetary" origin for the Moon and to revive Darwin's (1880, 1908) hypothesis that the material now in the Moon was derived by a special process from the Earth's mantle after the core had formed.

On the other hand, many lunar scientists have continued to hope that a satisfactory solution to the problem of iron-fractionation will ultimately emerge, and have accordingly preferred other hypotheses of lunar origin that they believe may be more compatible with certain dynamical and chemical constraints. According to the *binary planet hypothesis*, the Moon was formed from the same metal-silicate reservoir as the Earth; however, specialised physical processes led to preferential accretion of metal by the Earth and to accretion of the Moon from silicate material in orbit around the Earth. The *capture hypothesis* requires that the Moon accreted independently of the Earth and was subsequently transferred into orbit via a close approach accompanied by major dissipative processes.

2. Composition of the Moon

It has long been known that the mean density of the Moon is similar to that of the Earth's upper mantle. Indeed, it was this similarity that provided the impetus for Darwin's fission hypothesis, which in turn suggested that the chemical composition of the Moon should resemble that of the Earth's mantle. A wealth of new information on the chemical composition of the Moon has since become available as a result of the Apollo project. This evidence demonstrates the existence of major similarities and major differences in chemical composition between the Moon and Earth's mantle, both of which must be explained by any acceptable theory of lunar origin.

Major elements

About half of the Moon and the Earth's mantle (by weight) is composed of oxygen. Clayton et al. (1976) showed that the oxygen isotopic compositions of lunar and terrestrial basalts were identical. In contrast, oxygen in all classes of meteorites except the enstatite chondrites and achondrites differs significantly from oxygen in the Earth and Moon. It is particularly notable that oxygen from shergottites, which are widely believed to be derived from Mars, is displaced from the terrestrial fractionation line. These differences in ^{16}O, ^{17}O, and ^{18}O abundances are believed to be caused by both an inhomogeneous distribution of ^{16}O within the solar nebula and by chemical fractionations, probably as a result of different temperatures of equilibration with nebula gas (Clayton et al., 1976). The identity of oxygen isotopic compositions in Earth and Moon is strongly suggestive of a genetic relationship between the two bodies (Clayton and Mayeda, 1975).

Early data from the lunar heat flow experiment suggested that the uranium and thorium abundances within the Moon are about four times higher than in the Earth's mantle. This was responsible for a plethora of lunar compositional models in which calcium, aluminum, and related elements that possess similar condensation behaviour in the solar nebula to U and Th were likewise enriched in the Moon by factors of two to five.

Ringwood (1976) and Ringwood and Kesson (1977a) showed experimentally that these models encountered several serious petrologic and geochemical difficulties. For example, they were unable to explain the low mean calcium contents of the pyroxenes in the lunar crust. Other workers who attempted to model the crystallisation behaviour of refractory-rich lunar compositions were driven to consider nonchondritic Ca/Al and rare-earth ratios for the Moon (Longhi, 1981; Nyquist et al., 1977), which are highly implausible on several grounds (e.g., Warren, 1983).

The primary reason for invoking these refractory-rich lunar compositions was removed when an error was discovered in the early heat-flow measurements (Langseth *et al.*, 1976). The significance of lunar heat-flow data was reviewed by Ringwood (1977a, 1979) and Rasmussen and Warren (1985), who showed that the revised values were consistent with terrestrial abundances of uranium and thorium in the lunar interior.

Ringwood (1976, 1977a, 1979) and Ringwood and Kesson (1977a) concluded on the basis of experimental investigations into the petrogenesis of mare basalts and the lunar crust that, with the exception of FeO and volatile elements, the bulk chemical composition of the Moon was similar to that of the Earth's mantle. This conclusion has been supported by other workers, using entirely different data and arguments (e.g., Dreibus *et al.*, 1977; Wänke *et al.*, 1978; Warren and Wasson, 1979; Delano, 1986).

There is, nevertheless, one important difference in chemical composition between the Moon and the Earth's mantle. Seismic P- and S-wave velocities together with petrologic data strongly suggest that the lunar mantle contains about 12–16% of FeO (Ringwood, 1977a, 1979; Wänke *et al.*, 1977; Goins *et al.*, 1979; Buck and Toksöz, 1980; Nakamura *et al.*, 1982; Delano, 1986). This is substantially higher than the terrestrial mantle, which contains around 8% FeO (e.g., Ringwood, 1979) and has been cited by Drake (1983) as a difficulty for hypotheses that seek to form the Moon from material derived from the Earth's mantle.

Recent investigations of the composition of the Earth's core have shown that it probably contains about 40 wt % of dissolved FeO (Ringwood, 1977b, 1979, 1984; McCammon *et al.*, 1983). These authors concluded that the FeO now in the core was originally accreted as a component of the silicate phase, which was correspondingly richer in FeO than the present mantle. During core formation, FeO in the mantle dissolved in the core near its upper boundary, accompanied by strong convection and effective mixing throughout the mantle. The FeO/(FeO + MgO) molar ratio of mantle silicates fell from an initial value of about 0.3 to about 0.12 (the present value) during the core formation process, which may have exceeded 10^8 years. Jagoutz and Wänke (1982) and McCammon *et al.* (1983) pointed out that if the Moon had formed from material removed from the Earth's mantle before the core formation process was complete, it would contain more FeO than is now present in the mantle. Thus, the evidence that the FeO content of the Moon is higher than the present FeO content of the Earth's mantle is not inconsistent with fission-type hypotheses for the origin of the Moon. [An alternative explanation of the high lunar FeO content has been presented by Wänke and Dreibus (1986) and is summarized in Section 5].

Volatile elements

It is firmly established that the Moon is strongly depleted in volatile elements as compared with the abundances of these elements both in chondrites and in the Earth's mantle. The depletions mainly occur in elements that condense from the

---

solar nebula at temperatures below 1200°K (at a total pressure of 10^{-4} atm) based on the calculations of Grossman and Larimer (1974). The depletion patterns are complex and range over four orders of magnitude compared to primordial abundances (Fig. 1). They are highly dissimilar to the depletion patterns of volatile elements occurring in chondritic meteorites (Fig. 2). This does not lend support to cosmogonic hypotheses such as those of Ganapathy and Anders (1974), which interpret the compositions of planets and the Moon in terms of fractionation processes displayed by chondrites. However, with the very important exception of manganese, the depletion patterns of volatile elements in the Moon bear a strong qualitative resemblance to those in differentiated meteorites such as eucrites. This has led some workers to postulate that the Moon was derived from an earlier generation of asteroidal-sized planetesimals that had experienced melting and differentiation processes similar to those displayed by the eucrite-diogenite-pallasite meteoritic association (e.g., Smith, 1974; Wood and Mitler, 1974).

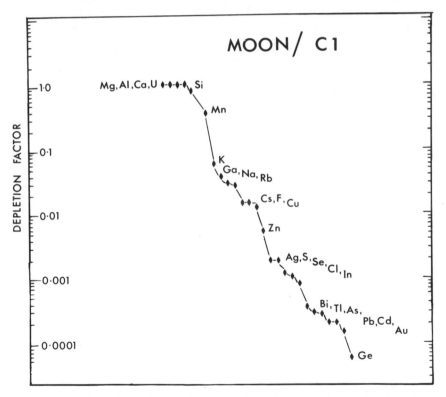

Fig. 1.   Comparison of abundances of (mainly) volatile elements in the source regions of low-Ti mare basalts with corresponding abundances in Cl chondrites (after Ringwood and Kesson, 1977b).

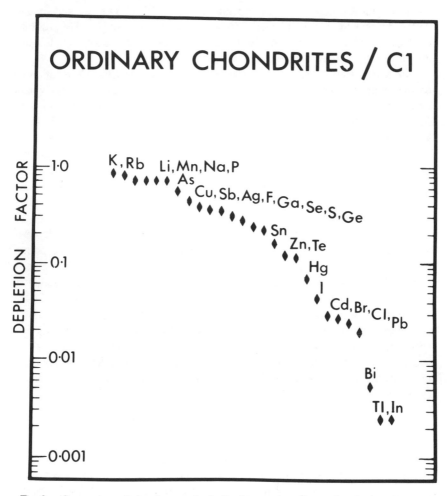

Fig. 2. Comparison of abundances of volatile elements in ordinary chondrites with corresponding abundances in Cl chondrites (after Ringwood and Kesson, 1977b).

Dreibus and Wänke (1979) pointed out that manganese is depleted in the Moon by a factor of about four compared to primordial abundances, whereas the parent body of eucrites did not experience significant Mn depletion. This represents an embarrassment for hypotheses that seek to relate lunar chemistry to that displayed by differentiated meteorites. Moreover, shergottites, which are believed to be derived from Mars, do not display significant depletion of manganese (Dreibus et al., 1982). The depletion of manganese in the Moon thus appears to be a very significant characteristic, as yet unexplained by advocates of binary planet and capture hypotheses. On the other hand, the abundance of manganese in the Moon is very similar to that in the Earth's mantle (Ringwood and Kesson, 1977b; Dreibus and Wänke, 1979) and is thus consistent with fission-derived models of lunar origin. (The possibility that depletion of manganese in the Earth's mantle is caused by its siderophile nature rather than the volatility is discussed in Section 3.)

If the Moon were formed from material derived from the Earth's mantle, it is necessary that the process occurred at sufficiently high temperatures so that volatile elements were selectively removed by evaporation and failed to recondense in the Moon. Current hypotheses of this type (e.g., Ringwood, 1970; Hartmann and Davis, 1975; Cameron and Ward, 1976) appear capable of providing these conditions. Kreutzberger et al. (1984) pointed out that the order of increasing volatility of alkali elements from silicate melts is $Cs > Rb > K > Na$. They argued that if the Moon were formed from the Earth's mantle, accompanied by devolatilization, the residual alkali metals should be depleted in the Moon (compared to Earth) in the above sequence. Since this does not seem to be the case, they concluded that fission-type hypotheses provide an inadequate explanation of the alkali abundances. However, Ringwood (1966a) had already suggested that volatile elements were incorporated into the Moon by a mechanism differing from that envisaged by Kreutzberger et al. According to his model, the Moon was formed by coagulation of a sediment-ring of largely devolatilized planetesimals, ultimately derived from the Earth. It is likely that the sediment-ring would have captured a small proportion ($< 1\%$) of late, oxidized planetesimals containing chondritic abundances of alkali metals from heliocentric material accreting on Earth. Part of the lunar volatile inventory may have been derived from this source. Related suggestions regarding the source of lunar volatiles have been made by Wasson (1971) and Ganapathy and Anders (1974).

Discussion

Most probably, the major element composition of the Moon was very similar to that of the Earth's mantle during the later stages of core formation but before the transport of FeO from mantle to core had been completed. This is consistent with fission-related hypotheses of lunar origin, as is the identity of oxygen isotopic compositions in Earth and Moon. These characteristics could also be explained in principle by the binary planet and capture hypotheses, since it would be reasonable to assume that the Moon accreted in an orbit very similar to that occupied by the Earth, and from silicate phases that also possessed similar compositions. However, these hypotheses rely on complex *ad hoc* mechanisms to explain the depletion of iron and volatiles in the Moon. These latter characteristics appear more explicable in terms of fission-related hypotheses.

The detailed characteristics of volatile depletion patterns impose additional constraints. Versions of the binary planet or capture hypotheses based on interpretations of fractionation patterns in chondritic meteorites do not readily explain the observations.

Alternative versions of these hypotheses based upon analogies with differentiated meteorites are more successful, but fail to explain the depletion of manganese in the Moon. Hypotheses relating the Moon to the terrestrial mantle readily explain the Moon's manganese depletion, but do require additional *ad hoc* assumptions in order to explain some significant characteristics of the lunar volatile pattern.

Protagonists of the principal hypotheses of lunar origin have weighed the previous evidence on lunar bulk composition and volatile abundances differently and with a conspicuous lack of consensus. One is therefore led to enquire whether there is any other source of compositional evidence that can be satisfied unambiguously by only one of the existing hypotheses of origin. The author believes that the answer is provided in the siderophile geochemistry of Earth and Moon.

3. Siderophile Geochemistry of Earth's Mantle and Moon

The terrestrial pattern

The abundance patterns of siderophile elements in the Earth's mantle display some interesting features that have been discussed recently by Ringwood (1984) and Newsom and Palme (1984) [see also Ringwood (1966a), Ringwood and Kesson (1977b), Sun (1982), and Chou *et al.* (1983)]. Some of these features are summarized below. We restrict consideration to siderophiles that are relatively involatile so that their present abundances in the mantle primarily reflect the varying degrees to which they have been sequestered in the core.

1. Nickel, Co, Cu, Fe, Ga, W, Mo, and P are present in the mantle at levels ranging between 4% and 16% percent of their primordial Cl abundances (normalized to Mg). Little relationship exists between the partition coefficients for these elements between silicate and metallic iron phases and their abundances in the mantle. For example, the partition coefficients for W and Mo between $Fe_{90}Ni_{10}$ alloy and a basaltic liquid at 1600°C are about 200 and 10,000, yet their abundances in the mantle differ only by a factor of two (Rammensee and Palme, 1982; Newsom and Palme, 1984). Analogous differences exist in the cases of Ni, Co, and Cu.
2. Nickel, Co, Cu, Mo, and Ge are greatly overabundant in the mantle as compared to expectations based on experimentally-measured partitions between silicate and metallic iron phases.
3. Chromium and vanadium, which display distinct siderophile tendencies at temperatures above 1500°C (Brey and Wänke, 1983; Rammensee *et al.*, 1983), are significantly depleted in the mantle by factors of 2 and 1.5*, respectively.

---

*Using the recent compilation of primordial abundances in Cl chondrites compiled by Anders and Ebihara (1982).

These depletions probably reflect their partial entry into the core (Dreibus and Wänke, 1979).

4. It is possible that the fourfold depletion of manganese in the mantle is caused not by its volatility, as previously discussed, but by its preferential solubility in an Fe-O melt at high pressures during segregation of the core (Dreibus and Wänke, 1979). This possibility will also be considered in subsequent discussions.

5. Platinoid elements, Re and Au, are present in the mantle at a uniform level of 0.7% of the primordial abundance (Chou *et al.*, 1983).

The above patterns have evidently been caused by a combination of several complex processes, as yet incompletely understood, that occurred during core formation within the Earth. They include:

1. Segregation within the Earth of a large iron-rich metallic core amounting to 32% of its mass.

2. Partitions of siderophiles between silicate and metallic phases were affected by the very high pressures and temperatures prevailing within a body of planetary dimensions.

3. These partitions were also probably influenced by the presence in the core of about ten percent of light elements, among which oxygen probably predominates.

4. Physical mixing of an oxidized nebula condensate into the mantle under conditions that did not permit it to equilibrate with metal phase also influenced the abundances of moderately siderophile elements.

5. Introduction of platinoids, Re and Au, seems to have occurred by a mixing process analogous to that in (4) above, but clearly decoupled from it.

The particular combination of processes that generated the siderophile signature of the Earth's mantle is most unlikely to have operated to the same extent in all planets and differentiated planetesimals. In consequence, the resultant siderophile signature in the Earth's mantle is almost certainly unique to the Earth (or to an Earth-sized planet such as Venus).

Support for this conclusion is provided by the siderophile signatures observed in the parent bodies of eucrites and shergottites. The principal classes of nonbrecciated differentiated meteorites [eucrites, diogenites, pallasites (and perhaps the III-AB irons)] were probably derived from one or more asteroidal bodies that experienced extensive melting and differentiation (reviewed by Dodd, 1981). Oxygen isotope measurements show that at least the first three of these classes of meteorites are genetically related. Petrogenetic relationships between the groups have been investigated by Stolper (1977) and geochemical relationships by Dreibus and Wänke (1980). The bulk composition of the parent body may have resembled a devolatilized chondrite (except for oxygen isotopic composition). The amount of metal phase in the parent body(s) that segregated to form Ni-Fe cores and pallasitic regions is not well constrained. In view of the extremely low Ni and Co contents of the silicate phases (see below), a substantial quantity ($\geq 10\%$) of metal must have been present. As an example of differentiation within a small parent body at low pressures and at temperatures below $1300°C$, the siderophile signature of the eucrite parent body presents an interesting contrast to the terrestrial siderophile signature. Principal features are:

1. In contrast to the Earth, siderophile abundances seem to have been established by equilibrium distributions between metal and silicate phases (Stolper, 1977).
2. In consequence, Ni, Co, Cu, and Ga are depleted in the mantle of the eucritic parent body by factors about 30, 6, 3, and 3 respectively, as compared to Earth's mantle (Ringwood and Kesson, 1977b; Jones, 1984).
3. The abundances of phosphorus and tungsten in the eucritic parent body and Earth's mantle are similar (Dreibus and Wänke, 1980). However, molybdenum is relatively depleted in eucrites by more than a factor of ten (Newsom and Palme, 1983).
4. The abundances of chromium, vanadium*, and manganese in the silicate portion of the eucrite parent body are close to the chondritic values, whereas the terrestrial mantle abundances are substantially depleted (as noted earlier).

It is of particular interest to compare the siderophile signature of the parent body of the shergottites (SNC meteorites) with the terrestrial signature in view of the strong evidence for a Martian origin of SNC meteorites (e.g., Wood and Ashwal, 1981). The mass and the central pressure of Mars are about one-tenth those of the Earth. The Martian core amounts to about 20% of the planet's total mass, and its liquid portion is believed to consist of an Fe-O-S alloy (Ringwood and Clark, 1971; Dreibus et al., 1982).

Stolper (1979) showed that the siderophile signature of shergottites and their parent body resembles that of terrestrial basalts and the Earth's mantle much more than do eucrites and their parent body. This is not unexpected in view of the probable complexity of the core formation process in a planetary-sized body. Nevertheless, there are some important differences between the siderophile signatures of the shergottite parent body (SPB) and the terrestrial mantle (Dreibus et al., 1982; Stolper, 1979), e.g.:

1. Tungsten and phosphorus are 5 to 10 times more abundant in the mantle of the SPB as compared to the terrestrial mantle
2. Copper is relatively depleted by a factor of 5–10 in the mantle of the SPB.

---

*Dreibus and Wänke (1982) actually concluded that vanadium was slightly depleted ($\times$ 0.8) in the eucrite parent body. However, when their model is corrected for the vanadium that would enter olivine at the redox state of the eucrite parent body as shown by the partition measurements of Ringwood and Essene (1970), a chondritic V/Mg abundance is obtained.

3. Chromium and manganese are present in primordial abundances in the SPB mantle, whereas they are depleted by factors of 2 and 4 in the terrestrial mantle.

The preceding discussion highlights the unique nature of the siderophile signatures in the mantles of planets and planetesimals that have differentiated to produce substantial metallic cores. It is therefore of considerable interest to study the siderophile signature of the Moon, in which a metallic core probably amounts to less than two percent of its mass (e.g., Ringwood and Seifert, 1986).

Siderophiles in lunar and terrestrial basalts

Ringwood and Kesson (1976, 1977b) carried out a comparative study of available data on siderophile abundances in lunar and terrestrial basalts. In the case of the Moon, data from low-titanium mare basalts were selected because these were believed to have experienced a simpler petrogenetic evolution, thereby carrying a more direct memory of their source region in the lunar mantle than do high-titanium basalts (see accompanying paper by Ringwood and Seifert, 1986). For similar reasons, terrestrial oceanic tholeiites were selected for this comparison, rather than the more highly fractionated terrestrial continental tholeiites.

Ringwood and Kesson concluded that the average abundances of a particular group of siderophile elements (Fe, Co, Ni, W, P, Ir, Os, S, and Se) were similar within a factor of about two between lunar and terrestrial basalts. They pointed to the unique factors discussed above, which were responsible for the siderophile signature of the terrestrial mantle and its derived basalts, and continued: "These factors could not possibly have operated within the Moon, a body in which the core, *if present* amounts to less than 10% of the proportional size of the Earth's core and which, if present, must have formed under very different conditions from the Earth's core, and in a pressure field extending to a maximum of only 47 kilobars, as compared to 3.6 megabars for the Earth. The similarity in siderophile elements between the Moon and Earth's mantle therefore implies that *the Moon was derived from the Earth's mantle after the Earth's core had segregated*." Despite the controversy generated at the time, I believe this interpretation has been fully substantiated by evidence that has subsequently become available.

Ringwood and Kesson (1976, 1977b) also recognized that a second group of siderophiles (Ga, Cu, Ge, As, Ag, Au, Sb, and Re) were depleted in lunar basalts compared to terrestrial basalts by factors ranging from 5 to 500. They suggested that these depletions were unconnected with siderophile properties, but arose because of the loss of these elements as volatile species during formation of the Moon by the same process that was responsible for the strong depletions of volatile lithophile elements. They recognised that loss of rhenium by this mechanism presented a problem.

Dreibus *et al.* (1976) and Rammensee and Wänke (1977) studied the abundances of tungsten and phosphorus in the Moon and Earth, and independently reached similar conclusions to those of Ringwood and Kesson using closely related arguments. O'Keefe and Urey (1977) have also emphasized the bearing of siderophile element abundances upon the relationship between Earth and Moon, although their argument is basically different to those developed by Wänke, Ringwood, and colleagues.

A considerable body of high quality analytical data has been obtained subsequently on lunar and terrestrial rocks, e.g., Wolf and Anders (1980). Tellurium is now added to the list of elements possessing similar abundances in lunar and terrestrial basalts. Nickel is the only case where new data require a minor modification in Ringwood and Kesson's interpretation. Whereas Delano and Ringwood (1978a,b) confirmed that the mean abundances of Ni in oceanic tholeiites and low-Ti mare basalts agreed within a factor of two, a detailed study by Ringwood and Seifert (1986) showed that the source region of the lunar basalts was actually depleted in Ni by a factor of 3.3 compared to the Earth's mantle. In a parallel study, Delano (1986) estimated a Ni depletion factor in the lunar mantle of 4.0 ± 0.5. Note, however, that Ringwood and Seifert (1986) showed that the Ni content of the *bulk* Moon (including its core) was very similar to that of the Earth's mantle.

Although Wolf and Anders' average values for Ir (and, by inference, Os) were nearly identical in lunar and terrestrial basalts, dispersions in the latter were very broad, and in view of alternative explanations that can be advanced to explain these abundances (e.g., Arculus and Delano, 1981; Chou *et al.*, 1983), they are ignored in subsequent discussion.

Gallium and copper are depleted in low-Ti mare basalts compared to oceanic tholeiites by a factor of about six. The alkalis Na, K, and Rb possess comparable volatilities to Ga and Cu and are depleted to a similar degree. The depletions of Ga and Cu in the Moon can probably be attributed primarily to their volatility. This suggests that the original abundances of these elements in lunar material prior to the event that caused devolatilization were also similar to their abundances in the Earth's mantle.

Chromium, vanadium, and manganese

We have already noted that Cr and V are significantly depleted in the Earth's mantle by factors of 2 and 1.5 respectively, probably because of their partially siderophile nature. Delano (1986) has shown from the compositional systematics of primitive lunar volcanic glasses that the Mg/Cr ratio of their source region is very similar to that of the Earth's mantle. Delano (1979) also noted that Cr and V were depleted in green glass by factors of ~3 and ~1.5 compared to other involatile lithophile elements possessing similar ionic radii and/or charges. He suggested that these depletions may be intrinsic to the Moon, or alternatively, may reflect the occurrence of prior spinel fractionation in the source region of these magmas. Experimental investigations by Ringwood and Seifert (1986) carried out at oxygen fugacities corresponding to those for green glass petrogenesis showed that spinel fractionation could not have been responsible for the Cr and V depletions. Accordingly, these seem to be indigenous.

Wänke *et al.* (1977) derived a lunar compositional model based on highland rocks that yielded a similar Mg/Cr ratio to that of the Earth's mantle. Their model also showed that vanadium is depleted in the Moon to a similar extent as in the Earth. Thus we can add chromium and vanadium to the list of siderophiles possessing similar abundances in Earth and Moon. The genetic significance of this observation was first pointed out by Dreibus and Wänke (1979). The case of manganese has already been discussed. It possesses similar abundances in the Earth's mantle and Moon but is depleted by a factor of four relative to the primordial abundance. The depletion may have been caused by its volatility, or, alternatively, by its siderophile behaviour during core formation in the Earth.

Molybdenum, rhenium, and gold

Molybdenum is more siderophile than nickel but less so than rhenium, gold, and the platinoids (Rammensee and Palme, 1982; Kimura *et al.*, 1974). In sharp contrast to the involatile siderophiles previously discussed, Newsom and Palme (1984) showed that molybdenum is depleted in low-Ti mare basalts by a factor of about 25 as compared to terrestrial basalts. This depletion cannot be explained by volatility and shows that the simple picture developed by Ringwood and Kesson is in need of modification. The recognition of a genuine depletion of molybdenum redirects attention to rhenium and gold, which are relatively depleted by factors of about 80 in lunar basalts (Ringwood and Kesson, 1977; Wolf and Anders, 1980). Whereas Ringwood and Kesson (1977b) and Ringwood *et al.* (1981) suggested that these depletions might be explained by volatility under a restrictive set of conditions, it now seems more likely that they are connected with the highly siderophile nature of Re and Au (Newsom, 1984).

Indigenous siderophiles in the lunar highlands

Interpretation of siderophile abundance patterns in the lunar highlands is complicated by the widespread occurrence of contamination from the intense meteoritic bombardment experienced by these regions early in lunar history. For some years, the prevailing view (e.g., Anders *et al.*, 1973) was that the vast bulk of siderophiles were of meteoritic origin and that any indigenous component was very, very small. However, extensive studies (Wänke *et al.*, 1977, 1978) of the correlations of individual siderophiles with lithophile elements possessing similar crystal-chemical properties (e.g., W-La, P-La, Co-Mg, and Ni-Mg) have demonstrated that most of the tungsten, phosphorus, and cobalt in the lunar highlands is indigenous, and that a substantial proportion of the nickel is also indigenous (see also Ringwood and Seifert, 1986).

Delano and Ringwood (1978a,b) estimated the indigenous abundances of siderophiles in Apollo 16 highland breccias using a very simple procedure. They assumed that the projectiles striking the Moon possessed compositions similar to ordinary chondrites. Meteoritic contamination was subtracted from each sample, making the conservative assumption that *all* of the iridium present had been derived from the meteorite. Residuals so obtained for each element were averaged and corrected to a common Al_2O_3 content (to allow for plagioclase dilution). It was found that the residual abundances of W, Ni, Co, P, Cu, Ga, S, and Se were similar (within a factor of two) to the abundances of these elements in low-Ti mare basalts, which are undoubtedly indigenous. It is striking that such a simple model has proven capable of extracting an ordered and meaningful result from the apparently chaotic data base represented by available chemical analyses of Apollo 16 highland breccias. The agreement between the highland siderophile residuals and the mare basalt abundances strongly suggests that the former represent genuinely indigenous siderophiles. As noted above, this has been demonstrated independently using entirely different techniques in the cases of W, P, Co, and Ni.

4. Significance of the Lunar Siderophile Pattern

The Earth-Moon connection

It seems that the principal characteristics of the siderophile abundance patterns in the Moon and Earth are now well established. Considering mainly the less volatile elements*, we have seen that Cr, V, (Mn), Ni, Co, Cu, Ga, W, Mo, P, Re, Au, S, Se, and Te have been depleted to varying degrees in the Earth's mantle by complex processes associated with the differentiation of a large metallic core. The resultant siderophile signature of the terrestrial mantle is almost certainly unique to the Earth.

The abundances of Cr, V, (Mn), Co, W, P, S, Se, and Te are very similar in the terrestrial and lunar mantles. Copper and Ga could be added to this group if appropriate corrections were made for their partial loss by volatilization during formation of the Moon. A similar abundance pattern is found for the indigenous siderophiles in the lunar highlands. It seems inconceivable that the segregation of a core within the Moon amounting to less than two percent of the lunar mass and under profoundly different physical and chemical conditions to those that governed

---

*S, Se, and Te actually behave as highly volatile elements in the solar nebula and this factor strongly influences their abundance in the bulk Earth. However, under conditions of low hydrogen fugacity they are relatively involatile (Ringwood, 1977). Since it is believed that these conditions were relevant to the formation of the Moon (Ringwood and Kesson, 1977b), these elements are retained in the group under discussion.

the formation of the massive terrestrial core would yield similar mantle siderophile signatures for the two bodies. Moreover, in Section 3 it was noted that the lunar siderophile signature is quite different from the corresponding siderophile signatures in the parent bodies of eucrites and shergottites (Mars?), both of which differentiated to form substantial metallic cores.

Ringwood and Kesson (1976, 1977b), Delano and Ringwood (1978a,b), Dreibus et al. (1976, 1977), and Wänke et al. (1977) concluded that the above evidence could be satisfied only if it was assumed that the material now constituting the Moon had been derived in some manner from the Earth's mantle subsequent to segregation of the Earth's core.

Recent developments have further strengthened this interpretation. If attention is restricted to siderophile elements whose abundances are best known, Cr, V, (Mn), Co, and W are found to possess very similar abundances in the terrestrial and lunar mantles, whereas P, Ni, Mo, Re, and Au are depleted by factors of 1.7, 3, 30, 80, 80 respectively (Wänke et al., 1983; Ringwood and Seifert, 1986; Newsom and Palme, 1984; Wolf and Anders, 1980). This sequence corresponds to the order of increasing siderophile character among these elements. Wänke et al. (1978) pointed out that the agreement between terrestrial and lunar siderophile signatures could be further improved if it was assumed that a very small amount of metal phase had segregated within the Moon. This would not significantly affect the abundances of moderately siderophile elements like Cr, V, (Mn), Co, and W, but might cause significant depletions of the more highly siderophile elements.

The study by Newsom and Palme (1984) of molybdenum depletion in the Moon provided support for this hypothesis. Ringwood and Seifert (1986) then showed that, despite the threefold depletion of nickel in the lunar mantle, the bulk abundance of nickel in the Moon is very similar to that in the Earth's mantle. The nickel budget of the Moon can be readily satisfied if a small, nickel-rich (~40% Ni) core amounting to 0.4% of the lunar mass is present. Newsom (1984) and others pointed out that the observed depletions of Mo, Re, and Au in the Moon could be explained by segregation of a core of this size. Thus the correlation between degree of depletion and siderophile nature is simply explained. This model appears capable of providing a quantitative explanation of the abundances of the less volatile siderophile elements within the Moon.

"Independent planet" hypothesis of lunar origin

Hypotheses in this category, including various binary planet and capture models, encounter the first-order problem of explaining the drastic depletion of metal phase in the Moon, as compared to Earth, Venus, and Mars (see Section 1). In the author's opinion, previous attempts to explain this problem, which rely on some kind of mechanical fractionation of silicates from metal phase, have been contrived. Even if the physical difficulties encountered by this hypothesis could be solved, it encounters serious geochemical problems (Ringwood and Seifert, 1986).

Several workers have previously proposed that accretion of terrestrial planets was preceded by the formation of an earlier generation of asteroidal-sized planetesimals that melted and differentiated to form metallic cores overlain by silicate mantles. The differentiation is believed to be analogous to that experienced by the parent bodies of eucrites, pallasites, and iron meteorites. Chapman and Greenberg (1984) speculate that collisional processes occurring in a circumterrestrial swarm of planetesimals permitted the Earth to selectively capture the metallic cores of these bodies whereas the disintegrated silicate mantles were captured into the circumterrestrial swarm that subsequently coagulated to form the Moon. Related models have been proposed by Smith (1974), Wood and Mitler (1974), and Wasson and Warren (1984).

Models of this type predict that the chemistry of the Moon should be similar to that of the eucritic parent body and also imply that the silicates in the Moon had equilibrated with iron-rich metal at low pressures and at temperatures less than 1300°C before being filtered out by the circumterrestrial ring. The observed composition of the Moon, and in particular the high Co and Ni contents of the lunar mantle and the depletions of Cr, V, and Mn, contradict these hypotheses. Additional geochemical problems are discussed by Ringwood and Seifert (1986). Attempts to explain the P and W contents of the Moon on the basis of the independent-planet hypothesis were made by Newsom and Drake (1983) and Newsom (1984). The resultant models were found to be contrived and quantitatively inadequate (Ringwood and Seifert, 1986). Moreover, they did not provide satisfactory explanations of the abundances of Ni, Co, Cr, V, and Mn in the Moon.

5. Origin of the Moon

Geochemical evidence discussed in this and in the accompanying paper (Ringwood and Seifert, 1986) implies that the Moon was probably formed from material derived from the Earth's mantle after the core had segregated. Provision of the energy and momentum required to remove more than one percent of the Earth's mass and place it into orbit represents a rather formidable challenge for proposed mechanisms of lunar origin.

Early attempts to accomplish this objective were based on rotational fission of the Earth (e.g., Darwin, 1880, 1908; Ringwood, 1960; Wise, 1963; Cameron, 1963; O'Keefe, 1966). However, it seems now to be widely agreed that the angular momentum density required to produce fission of the primitive Earth is difficult to achieve (e.g., Kaula, 1971). Moreover, this mechanism implies that the Moon represents only about 10% of the material that was originally removed from the Earth. This material was presumably vaporized and then recondensed (O'Keefe, 1966). It would be rather difficult to prevent substantial chemical and isotopic fractionation from occurring in this environment and hence the similarity between lunar and terrestrial mantle compositions is not so readily explained (e.g., Mayeda and Clayton, 1980).

Ringwood (1966a, 1970, 1972, 1975) proposed that during the later stages of accretion of the Earth, high temperatures were produced by a combination of rapid accretion, core formation, and thermal insulation by a thick primitive atmosphere. Under these conditions, material from the mantle was evaporated into the primitive atmosphere. The latter was spun out in the Earth's equatorial plane and removed by a combination of processes: (1) high initial rotation rate of the Earth, (2) coupling of the atmosphere to the Earth's rotation via hydromagnetic torques and/or turbulent viscosity, (3) intense solar radiation during the T-Tauri phase of the sun, and (4) turbulent mixing of hydrogen from the solar nebula into the atmosphere, thereby lowering its mean molecular weight.

As the primitive terrestrial atmosphere was removed and cooled, the silicate components were precipitated to form an assemblage of Earth-orbiting planetesimals. Further fractionation according to volatility occurred during the precipitation process, since the more volatile components were precipitated at relatively low temperatures, forming micron-sized smoke particles. These remained viscously coupled to the escaping gases and were hence removed from the system. The Moon then accreted from the sediment-ring of devolatilized Earth-orbiting planetesimals.

Impact models

Tektites are believed to have formed by the effects of meteoritic impacts on the Earth's surface. Ablation studies showed that one particular class, the australites, were accelerated by the impact to velocities of about 10 km/sec (Chapman and Larson, 1963). Thus it is empirically established that impact of meteorites or planetesimals after core-formation has the capacity to remove material from the Earth's mantle and place it in near-geocentric orbit. Theoretical and experimental studies of high-velocity impact processes by Boslough and Ahrens (1983) have also demonstrated the capacity of high-velocity planetesimal collisions to evaporate several times their own mass of material from the Earth's surfce and accelerate it to high velocities.

Hartmann and Davis (1975) pointed out that a planetesimal of 0.5 lunar mass impacting the Earth's mantle at 13 km/sec would impart sufficient energy to eject two lunar masses of the mantle at near-escape speeds. A large proportion of this material could well have been placed in geocentric orbit. Hartmann and Davis hypothesized that, provided one or more of these collisions occurred subsequent to core formation, it would have been possible to produce a geocentric ring of material mainly derived from the Earth's mantle, from which the Moon might have formed.

A related model was proposed by Cameron and Ward (1976), who suggested that, subsequent to core formation, the Earth was hit by a differentiated Mars-sized planetesimal at about 11 km/sec, causing vapourization of extensive regions of the mantles of both bodies. They showed that a substantial proportion of the vapour would have expanded at a sufficient velocity to attain Earth-orbit, and suggested that this was followed by selective recondensation of vapour, thereby forming a ring of devolatilized Earth-orbiting planetesimals that coagulated to form the Moon. The physical basis of impact models for lunar origin has been further investigated by Stevenson (1984). He concluded that orbital emplacement of terrestrial material by this mechanism is possible and that the mechanism is much more efficient for large (≥1000 km radius) impacting planetesimals than for smaller bodies (R < 100 km).

Further speculations on the mechanism of origin

The attractive feature of these impact models as a means of placing material from the Earth into geocentric orbit is that they rely on processes that must *inevitably occur* during the accretion of planets and can supply the enormous amounts of energy required. Nevertheless, although the new models are very promising, they will require further development in order to explain the Moon's chemical composition, as distinct from the dynamics of its formation.

In the author's opinion, a successful model for the terrestrial origin of the Moon should preferably satisfy the following conditions:

1. The proportion of vapourized material that was permanently removed from the Earth-Moon system during formation of the Moon should represent a limited fraction (e.g., <30%) of the mass of the Moon. In models where the proportional loss of volatilized material is much greater than this, resultant chemical and isotopic fractionations between the Moon and Earth could be excessive (as noted earlier).
2. The total proportion of impacting planetesimal(s) incorporated into the Moon should be much smaller than the proportion of terrestrial-derived material; otherwise the observed similarity in siderophile signatures between Earth and Moon would be degraded*.
3. Removal of material from the Earth's mantle occurred at a very late stage of accretion, after all of the metallic iron had segregated from the mantle into the core but at an intermediate stage of transfer of FeO from mantle to core.

---

*Wänke and Dreibus (1986) suggested that the relatively high FeO content of the Moon as compared to the Earth was supplied by the impacting planetesimal(s). About 20% of this component would be required, providing that these late-accreting planetesimals were highly oxidised. In addition, it must be assumed that a small amount of auto-reduction occurred during impact, forming a nickel-rich metal phase that sequestered the highly siderophile elements and subsequently entered the lunar core.

These conditions may be satisfied more readily by a scenario intermediate between the models advanced by Ringwood (1966a, 1970, 1972) and those of Hartmann and Davis (1975), Cameron and Ward (1976), and Stevenson (1984), which are based upon impacts by only one or by a few giant-sized (R ≥ 1000 km) planetesimals.

For many years I have suggested that an important role in lunar origin was played by the primitive terrestrial atmosphere that is produced from degassing of planetesimals accreting on the Earth (Ringwood, 1960, 1966a, 1970, 1972). As further discussed by Ringwood (1975, 1977b, 1979, 1984), the primitive atmosphere (mean molecular weight < 4) continually blew off into the solar nebula during accretion. These models also envisaged the Earth accreting before the gases of the solar nebula had been dispersed. Under these conditions the Earth would also capture a primitive atmosphere of solar composition from the nebula that would mix with the terrestrial atmosphere produced by degassing of planetesimals.

Hayashi and colleagues (e.g., Hayashi, 1981; Hayashi et al., 1985; Sekiya et al., 1981; Nakazawa and Nakagawa, 1981) have carried out extensive investigations of the development of the primitive atmosphere formed when the Earth accreted in the solar nebula and its subsequent dispersal by solar radiation (see also Walker, 1982). The lower convective region of the primitive terrestrial atmosphere would be coupled to the Earth by turbulent viscosity and would corotate with the Earth. Estimates of the extent of the convective region [measured in Earth radii (ER) from the Earth's centre] range from about 9 ER (Hayashi, 1981) to somewhat less than 2 ER (Walker, 1982).

Dissipative processes in the primitive atmosphere would have played an important role in capturing incoming planetesimals and retaining impact ejecta. The atmosphere would extend the effective radius of capture of incoming planetesimals, with a corresponding increase in the net specific angular momentum of accreted planetesimals as compared to the gas-free accretion scenario of Safronov (1972). This factor might have contributed to a relatively short rotation period for the primitive Earth. If the Moon is assumed to be derived from the Earth's mantle, an initial rotation period of four to five hours would be required.

Hayashi's (1981) discussion shows that the corotating region of the primitive atmosphere might have extended beyond the geosynchronous orbit of ~2.5 ER (for a period of four to five hours). It is also possible that the effective radius of the corotating region was increased by hydromagnetic coupling with the Earth. At a late stage of accretion, strong convection within the newly segregated terrestrial core may have caused the generation of a stronger magnetic field than currently exists. Also at this stage, particle and far-UV radiation from the young sun began to remove gases from the primitive solar nebula. Appreciable ionization was caused in the gaseous envelope surrounding the Earth, both by solar radiation and by infall of small planetesimals and micrometeoroids. Hydromagnetic coupling between the Earth and outer regions of the primitive atmosphere might then have caused the latter to corotate with the Earth. We will accordingly assume that an extensive region of the primitive atmosphere, extending outwards well beyond the geosynchronous orbit, corotated with the solid Earth. This region may have developed a disc-like configuration in the Earth's equatorial plane.

At this late stage of accretion, impacts of many large (R > 100 km) but not necessarily giant (R > 1000 km) planetesimals at velocities exceeding 12 km/sec would have caused extensive shock melting and evaporation of material from the Earth's surface. Boslough and Ahrens (1983) showed that such planetesimals (~15 km/sec) would vapourize about 5 times their own mass of target material and would shock-melt 100 times their own mass. The shock-melted material would probably form a spray of droplets that would be largely devolatilized at the high temperatures prevailing. Rapid expansion of the impact cloud would cause acceleration to high velocities (Hartmann and Davis, 1975; Stevenson, 1984). As the cloud expanded and cooled, selective condensation of the less volatile components would have produced liquid droplets that subsequently solidified. Highly volatile elements condensed only at relatively low temperatures, forming smoke particles.

The corotating primitive atmosphere is believed to have played an essential role in transferring angular momentum from the solid Earth to impact-evaporated gases and liquid spray so that a substantial proportion of the ejected material was placed in circular equatorial orbits. The devolatilized droplets (1–10 mm) are believed to have accreted to produce an assemblage of planetesimals. The much smaller proportion (by mass) of smoke particles composed of the volatile components remained viscously coupled to the gas phase and hence was prevented from accreting into planetesimals. The majority of planetesimals probably accreted in orbits smaller than the geosynchronous limit. They would have lost energy via dissipation in the corotating atmosphere and hence spiralled back to Earth. However, it is suggested that a significant proportion of planetesimals were formed beyond the geosynchronous limit. They would have been accelerated by gas-drag, causing them to spiral outwards, beyond Roche's Limit, where they accreted to form a ring of Earth-orbiting moonlets. The Moon, in turn, was formed by coagulation of this sediment ring.

The possibility should not be ignored that the impacting projectiles responsible for removing protolunar material from the Earth's mantle were derived from the outer solar system, rather than from the terrestrial feeding zone. Kaula and Bigeleisen (1975) have shown that if accretion of the Earth had been far advanced prior to the major stage of growth of Jupiter via gaseous instability, the Jovian nucleus could have perturbed a substantial proportion of planetesimals from its feeding zone so that they crossed the Earth's orbit. Impacts by these planetesimals (composed mainly of condensed ices) at velocities between 20 and 30 km/sec, should have been very effective in placing terrestrial material in orbit.

Acknowledgments. The author is grateful to Dr. S. E. Kesson for helpful comments on the manuscript.

References

Anders E. (1968) Chemical processes in the early solar system, as inferred from meteorites. *Accounts Chem. Res., 1,* 289–298.

Anders E. (1971) Meteorites and the early solar system. *Ann. Rev. Astron. Astrophys., 9,* 1–34.

Anders E. and Ebihara M. (1982) Solar-system abundances of the elements. *Geochim. Cosmochim. Acta, 46,* 2363–2380.

Anders E., Ganapathy R., Krähenbühl U., and Morgan J. W. (1973) Meteoritic material on the moon. *The Moon, 8,* 3–24.

Arculus R. and Delano J. (1981) Siderophile element abundances in the upper mantle: evidence for a sulphide signature and equilibrium with the core. *Geochim. Cosmochim. Acta, 45,* 1331–1343.

Boslough M. B. and Ahrens T. J. (1983) Shock melting and vapourization of anorthosite and implications for an impact-origin of the Moon (abstract). In *Lunar and Planetary Science XIV,* pp. 63–64. Lunar and Planetary Institute, Houston.

Brey G. and Wänke H. (1983) Partitioning of Cr, Mn and V between Fe melt, magnesiowüstite and olivine at high pressures and temperatures (abstract). In *Lunar and Planetary Science XIV,* pp. 71–72. Lunar and Planetary Institute, Houston.

Buck W. and Toksöz M. N. (1980) The bulk composition of the Moon based on geophysical constraints. *Proc. Lunar Planet. Sci. Conf. 11th,* pp. 2043–2058.

Cameron A. G. W. (1963) The origin of the atmospheres of Venus and Earth. *Icarus, 2,* 249–257.

Cameron A. G. W. (1985) The partial volatilization of Mercury. *Icarus,* in press.

Cameron A. G. W. and Ward W. R. (1976) The origin of the Moon (abstract). In *Lunar Science VII,* pp. 120–122. The Lunar Science Institute, Houston.

Chapman C. and Greenberg R. (1984) A circumterrestrial compositional filter (abstract). In *Papers Presented to the Conference on the Origin of the Moon,* p. 56. Lunar and Planetary Institute, Houston.

Chapman D. R. and Larson K. H. (1963) On the lunar origin of tektites. *J. Geophys. Res., 68,* 4305–4358.

Chou C., Shaw D. M., and Crocket J. H. (1983) Siderophile trace elements in the Earth's oceanic crust and upper mantle. *Proc. Lunar Planet. Sci. Conf. 13th,* in *J. Geophys. Res., 88,* A507–A518.

Clayton R. N. and Mayeda T. (1975) Genetic relations between the Moon and meteorites. *Proc. Lunar Sci. Conf. 6th,* pp. 1761–1769.

Clayton R. N., Onuma N., and Mayeda T. K. (1976) A classification of meteorites based on oxygen isotopes. *Earth Planet. Sci. Lett., 30,* 10–18.

Darwin G. H. (1880) On the secular changes in the orbit of a satellite revolving around a tidally disturbed planet. *Phil. Trans. Roy. Soc. London, 171,* 713–891.

Darwin G. H. (1908) *Tidal Friction and Cosmogony.* Scientific Papers *2,* Cambridge University Press, London.

Delano J. W. (1979) Apollo 15 green glass: Chemistry and possible origin. *Proc. Lunar Planet. Sci. Conf. 10th,* pp. 275–300.

Delano J. W. (1986) Abundances of Ni, Cr, Co and major elements in the silicate portion of the Moon: Constraints from primary lunar magmas, this volume.

Delano J. W. and Ringwood A. E. (1978a) Indigenous abundances of siderophile elements in the lunar highlands: Implications for the origin of the Moon. *Moon and Planets, 18,* 385–425.

Delano J. W. and Ringwood A. E. (1978b) Siderophile elements in the lunar highlands: Nature of the indigenous component and implications for origin of the Moon. *Proc. Lunar Planet. Sci. Conf. 9th,* pp. 111–159.

Dodd R. T. (1981) *Meteorites.* Cambridge Univ. Press, Cambridge. 368 pp.

Drake M. (1983) Geochemical constraints on the origin of the Moon. *Geochim. Cosmochim. Acta, 47,* 1759–1767.

Dreibus G. and Wänke H. (1979) On the chemical composition of the Moon and the eucrite parent body and a comparison with the composition of the Earth: the case of Mn, Cr, and V (abstract). In *Lunar and Planetary Science X,* pp. 315–317. Lunar and Planetary Institute, Houston.

Dreibus G. and Wänke H. (1980) The bulk composition of the eucrite parent asteroid and its bearing on planetary evolution. *Z. Naturforsch., 35a,* 204–216.

Dreibus G., Spettel B., and Wänke H. (1976) Lithium as a correlated element, its condensation behavior, and its use to estimate the bulk composition of the Moon and the eucrite parent body. *Proc. Lunar Sci. Conf. 7th,* pp. 3383–3396.

Dreibus G., Kruse H., Spettel B., and Wänke H. (1977) The bulk composition of the Moon and the eucrite parent body. *Proc. Lunar Sci. Conf. 8th,* pp. 211–227.

Dreibus G., Palme H., Rammensee W., Spettel B., Weckwerth G., and Wänke H. (1982) Composition of Shergotty parent body: Further evidence of a two component model of planet formation (abstract). In *Lunar and Planetary Science XIII,* pp. 186–187. Lunar and Planetary Institute, Houston.

Ganapathy R. and Anders E. (1974) Bulk compositions of the Moon and Earth, estimated from meteorites. *Proc. Lunar Sci. Conf. 5th,* pp. 1181–1206.

Garz T., Kock M., Richter J., Baschwek B., Holweger H., and Unsold A. (1969) Abundances of iron and some other elements in the sun and in meteorites. *Nature, 223,* 1254–1256.

Goins N., Toksöz M. N., and Dainty A. (1979) The lunar interior, a summary report. *Proc. Lunar Planet. Sci. Conf. 10th,* pp. 2421–2439.

Grossman L. (1972) Condensation in the primitive solar nebula. *Geochim. Cosmochim. Acta, 36,* 597–619.

Grossman L. and Larimer J. (1974) Early chemical history of the solar system. *Rev. Geophys. Space Phys., 12,* 71–101.

Hartmann W. K. and Davis D. (1975) Satellite-sized planetesimals and lunar origin. *Icarus, 24,* 504–515.

Hayashi C. (1981) Formation of the planets. In *Fundamental Problems in the Theory of Stellar Evolution* (D. Lamb and D. Schramm, eds.), pp. 113–128. International Astronomical Union, Cambridge.

Hayashi C., Nakazawa K., and Nakagawa Y. (1985) Formation of the solar system. In *Protostars and Planets II,* Univ. Arizona Press, Tucson, in press.

Jagoutz E. and Wänke H. (1982) Has the Earth's core grown over geologic time (abstract)? In *Lunar and Planetary Science XIII,* pp. 358–359. Lunar and Planetary Institute, Houston.

Jeffreys H. (1937) The density distributions in the inner planets. *Mon. Not. Roy. Astron. Soc., 4,* 498–533.

Jones J. (1984) The composition of the mantle of the eucrite parent body and the origin of eucrites. *Geochim. Cosmochim. Acta, 48,* 641–648.

Kaula W. M. (1971) Dynamical aspects of lunar origin. *Rev. Geophys. Space Phys., 9,* 217–238.

Kaula W. M. and Bigeleisen P. E. (1975) Early scattering by Jupiter and its collision effects in the terrestrial zone. *Icarus, 25,* 18–33.

Kimura K., Lewis R., and Anders E. (1974) Distribution of gold and rhenium between nickel-iron and silicate melts: implications for the abundance of siderophile elements on the Earth and Moon. *Geochim. Cosmochim. Acta, 38,* 683–701.

Kreutzberger M., Drake M., and Jones J. (1984) Origin of the Moon: constraints from volatile elements (abstract). In *Papers Presented to the Conference on the Origin of the Moon,* p. 22. Lunar and Planetary Institute, Houston.

Langseth M. G., Keihm S. J., and Peters K. (1976) Revised lunar heat-flow values. *Proc. Lunar Sci. Conf. 7th,* pp. 3143–3171.

Longhi J. (1981) Preliminary modelling of high pressure partial melting: implications for early lunar differentiation. *Proc. Lunar Planet. Sci. 12B*, pp. 1011–1018.

McCammon C., Ringwood A. E., and Jackson I. (1983) A model for the formation of the Earth's core. *Proc. Lunar Planet. Sci. Conf. 13th*, in *J. Geophys. Res., 88*, pp. A501–A506.

Mayeda T. and Clayton R. (1980) Oxygen isotopic compositions of aubrites and some unique meteorites. *Proc. Lunar Planet. Sci. Conf. 11th*, pp. 1145–1151.

Nakamura Y., Latham G., and Dorman H. J. (1982) Apollo lunar seismic experiment—final summary. *Proc. Lunar Planet. Sci. Conf. 13th*, in *J. Geophys. Res, 87*, A117–A123.

Nakazawa K. and Nakagawa Y. (1981) Origin of the solar system. *Prog. Theoret. Phys., 70*, 11–34.

Newsom H. (1984) The lunar core and the origin of the Moon. *EOS (Trans. Am. Geophys. Union), 65*, 369–370.

Newsom H. and Drake M. (1983) Experimental investigation of the partitioning of phosphorus between metal and silicate phases: Implications for the Earth, Moon and eucrite parent body. *Geochim. Cosmochim. Acta, 47*, 93–100.

Newsom H. and Palme H. (1983) The depletion of molybdenum in the Earth, Moon and in the eucrite parent body (abstract). In *Lunar and Planetary Science XIV*, pp. 556–557. Lunar and Planetary Institute, Houston.

Newsom H. E. and Palme H. (1984) The depletion of siderophile elements in the Earth's mantle: new evidence from molybdenum and tungsten. *Earth Planet. Sci. Lett., 69*, 354–364.

Nyquist L. E., Bansal B., Wooden H., and Wiesmann H. (1977) Sr-isotopic constraints on the petrogenesis of Apollo 12 basalts. *Proc. Lunar Sci. Conf. 8th*, pp. 1383–1415.

O'Keefe J. A. (1966) The origin of the Moon and the core of the Earth. In *The Earth-Moon System* (B. G. Marsden and A. G. W. Cameron, eds.), pp. 224–233. Plenum, N.Y.

O'Keefe J. A. and Urey H. (1977) The deficiency of siderophile elements in the Moon. *Phil. Trans. Roy. Soc. London, A285*, 569–578.

Rammensee W. and Palme H. (1982) Metal-silicate extraction technique for the analysis of geological and meteoritic samples. *J. Radioanal. Chem., 71*, 401–418.

Rammensee W. and Wänke H. (1977) On the partition coefficient of tungsten between metal and silicate and its bearing on the origin of the Moon. *Proc. Lunar Sci. Conf. 8th*, pp. 399–409.

Rammensee W., Palme H., and Wänke H. (1983) Experimental investigation of metal-silicate partitioning of some lithophile elements (Ta, Mn, Cr, V) (abstract). In *Lunar and Planetary Science XIV*, pp. 628–629. Lunar and Planetary Institute, Houston.

Rasmussen K. and Warren P. (1985) Megaregolith thickness, heat flow and the bulk composition of the Moon. *Nature*, in press.

Ringwood A. E. (1959) On the chemical evolution and densities of the planets. *Geochim. Cosmochim. Acta, 15*, 257–283.

Ringwood A. E. (1960) Some aspects of the thermal evolution of the Earth. *Geochim. Cosmochim. Acta, 20*, 241–259.

Ringwood A. E. (1966a) Chemical evolution of the terrestrial planets. *Geochim. Cosmochim. Acta, 30*, 41–104.

Ringwood A. E. (1966b) Genesis of chondritic meteorites. *Rev. Geophys., 4*, 113–174.

Ringwood A. E. (1970) Origin of the Moon: The precipitation hypothesis. *Earth Planet. Sci. Lett., 8*, 131–140.

Ringwood A. E. (1972) Some comparative aspects of lunar origin. *Phys. Earth Planet. Inter., 6*, 366–376.

Ringwood A. E. (1975) *Composition and Petrology of the Earth's Mantle*. McGraw-Hill, New York. 618 pp.

Ringwood A. E. (1976) Limits on the bulk composition of the Moon. *Icarus, 28*, 325–349.

Ringwood A. E. (1977a) Basaltic magmatism and the composition of the Moon 1: Major and heat producing elements. *The Moon, 16*, 389–423.

Ringwood A. E. (1977b) Composition of the core and implications for origin of the Earth. *Geochim. J., 11*, 111–135.

Ringwood A. E. (1979) *Origin of the Earth and Moon.* Springer-Verlag, N.Y. 295 pp.

Ringwood A. E. (1984) The Earth's core: its composition, formation and bearing upon the origin of the Earth. *Proc. Roy. Soc. London, A395*, 1–46.

Ringwood A. E. and Clark S. P. (1971) Internal constitution of Mars. *Nature, 234*, 89–92.

Ringwood A. E. and Essene E. (1970) Petrogenesis of Apollo 11 basalts, internal constitution and origin of the Moon. *Proc. Apollo 11 Lunar Sci. Conf.*, pp. 769–799.

Ringwood A. E. and Kesson S. E. (1976) *Basaltic Magmatism and the Bulk Composition of the Moon, Part II: Siderophile and Volatile Elements in Moon, Earth and Chondrites. Implications for Lunar Origin.* Publication 1221, Research School of Earth Sciences, Australian National University, Canberra. 58 pp.

Ringwood A. E. and Kesson S. E. (1977a) Further limits on the bulk composition of the Moon. *Proc. Lunar Sci. Conf. 8th*, pp. 411–431.

Ringwood A. E. and Kesson S. E. (1977b) Basaltic magmatism and the bulk composition of the Moon, II. Siderophile and volatile elements in Moon, Earth and chondrites: Implications for lunar origin. *The Moon, 16*, 425–464.

Ringwood A. E. and Seifert S. (1986) Nickel-cobalt abundance systematics and their bearing on lunar origin, this volume.

Ringwood A. E., Kesson S., and Hibberson W. (1981) Rhenium depletion in mare basalts and redox state of the lunar interior (abstract). In *Lunar and Planetary Science XII*, pp.s 891–893. Lunar and Planetary Institute, Houston.

Safronov V. S. (1972) *Evolution of the Protoplanetary Cloud and Formation of the Earth and Planets.* English translation, Israel Program for Scientific Translation, Tel Aviv.

Sekiya M., Hayashi C., and Nakazawa K. (1981) Dissipation of the primordial terrestrial atmosphere due to irradiation of the solar far-UV during T-Tauri stage. *Proc. Theoret. Phys., 66*, 1301–1316.

Smith J. V. (1974) Origin of the Moon by disintegrative capture with chemical differentiation followed by sequential accretion (abstract). In *Lunar Science V*, pp. 718–720. The Lunar Science Institute, Houston.

Stevenson D. J. (1984) Lunar origin from impact on the Earth: is it possible (abstract)? In *Papers Presented to the Conference on the Origin of the Moon*, p. 60. Lunar and Planetary Institute, Houston.

Stolper E. (1977) Experimental petrology of eucritic meteorites. *Geochim. Cosmochim. Acta, 41*, 587–611.

Stolper E. (1979) Trace elements in shergottite meteorites: Implications for the origin of planets. *Earth Planet. Sci. Lett, 42*, 239–242.

Sun S. (1982) Chemical composition and origin of the Earth's primitive mantle. *Geochim. Cosmochim. Acta, 46*, 179–192.

Urey H. C. (1952) *The Planets.* Yale Univ. Press, New Haven.

Urey H. C. (1957a) Boundary conditions for the origin of the solar system. In *Physics and Chemistry of the Earth, Vol. 2* (L. H. Ahrens, F. Press, K. Rankama, and S. Runcorn, eds.), pp. 46–76. Pergamon, London.

Urey H. C. (1957b) Meteorites and the Origin of the Solar System. *41st Guthrie Lecture, Yearbook of the Physical Society*, pp. 14–29. Physical Society, London.

Walker J. C. (1982) The earliest atmosphere of the Earth. *J. Precamb. Res. 17*, 147–171.

Wänke H. and Dreibus G. (1986) Geochemical evidence for the formation of the Moon by impact induced fission of the proto-Earth, this volume.

Wänke H., Baddenhausen H., Blum K., Cendales M., Dreibus G., Hofmeister H., Kruse H., Jagoutz E., Palme C., Spettel B., Thacker R., and Vilcsek E. (1977) On the chemistry of the lunar samples and achondrites. Primary matter in the lunar highlands: A re-evaluation. *Proc. Lunar Sci. Conf. 8th*, pp. 2191–2213.

Wänke H., Dreibus G., and Palme H. (1978) Primary matter in the lunar highlands: The case of the siderophile elements. *Proc. Lunar Planet. Sci. Conf. 9th,* pp. 83–110.

Wänke H., Dreibus G., Palme H., Rammensee W., and Weckwerth G. (1983) Geochemical evidence for the formation of the Moon from material of the Earth's mantle (abstract). In *Lunar and Planetary Science XIV,* pp. 818–819. Lunar and Planetary Institute, Houston.

Warren P. (1983) Petrogenesis in a Moon with a chondritic refractory lithophile patterns (abstract). In *Workshop on Pristine Highlands Rocks and the Early History of the Moon* (J. Longhi and G. Ryder, eds.), pp. 80–84. Lunar and Planetary Institute, Houston.

Warren P. H. and Wasson J. T. (1979) Effects of pressure on the crystallization of a "chondritic" magma ocean and implications for the bulk composition of the Moon. *Proc. Lunar Planet. Sci. Conf. 10th,* pp. 2051–2083.

Wasson J. T. (1971) Volatile elements on the Earth and Moon. *Earth Planet. Sci. Lett., 11,* 219–225.

Wasson J. T. and Warren P. H. (1984) The origin of the Moon (abstract). In *Papers Presented to the Conference on the Origin of the Moon.,* p. 57. Lunar and Planetary Institute, Houston.

Wise D. J. (1963) An origin of the Moon by rotational fission during formation of the Earth's core. *J. Geophys. Res., 68,* 1547–1554.

Wolf R. and Anders E. (1980) Moon and Earth: compositional differences inferred from siderophiles, volatiles and alkalis in basalts. *Geochim. Cosmochim. Acta, 44.,* 2111–2124.

Wood C. and Ashwal L. (1981) SNC meteorites: igneous rocks from Mars? *Proc. Lunar Planet Sci. 12B,* pp. 1359–1375.

Wood J. A. and Mitler N. E. (1974) Origin of the Moon by a modified capture mechanism, or: Half a loaf is better than a whole one (abstract). In *Lunar Science V,* pp. 851–853. The Lunar Science Institute, Houston.

VIII.   Theories and Processes of Origin 4:

Models Emphasizing Coaccretion or Evolution of a Circumterrestrial Swarm, of Whatever Origin

Formation and Evolution of a Circumterrestrial Disk: Constraints on the Origin of the Moon in Geocentric Orbit

FLOYD HERBERT

Lunar and Planetary Laboratory, University of Arizona, Tucson, AZ 85721

DONALD R. DAVIS AND STUART J. WEIDENSCHILLING

Planetary Science Institute, 2030 E. Speedway, Suite 201, Tucson, AZ 85719

The angular momentum problem of forming the Moon out of material captured into a circumterrestrial disk from heliocentric orbits is studied using a data base of ~25,000 numerically integrated trajectories of Earth-encountering planetesimals. The aim is to test whether such a disk can form, absorb a significant amount of angular momentum and mass, and, in so doing, enrich itself in lithophilic elements by selective capture ("collisional filtering"). We combine plausible assumed mass-orbital element distributions of incoming planetesimals with this data base in order to calculate as a function of distance from Earth the net geocentric specific angular momentum of disk-encountering material on heliocentric orbits. In all cases considered, the mean specific angular momentum of the incoming material is less than that of the material in the disk at each distance from Earth, although the shortfall is more serious under some assumptions than under others. Thus a permanent disk population does not seem possible. A flow of material through a steady-state disk onto the proto-Earth and/or proto-Moon may be possible if there is some other mechanism (e.g., viscous spreading) capable of redistributing material and angular momentum within the disk. Collisions within Earth's sphere of influence involving bodies on heliocentric orbits is also a plausible mechanism for repopulating the outer disk and thus maintaining its capture cross section. In such a case, enough specific angular momentum may be brought in by the captured material to form a Moon at *small* geocentric distances, provided that most of the mass is initially captured at much *larger* distances. The angular momentum lost from the disk via its material falling onto the Earth may be an important source of the Earth's spin. If the heliocentric planetesimal population consists of both silicate and iron bodies resulting from the collisional disruption of differentiated Earth-zone planetesimals, we find that collisional filtering occurs during either type of capture process. Particular assumptions examined as examples lead to a silicate/iron mass capture ratio of 60.

1. *Introduction*

Coaccretion models of lunar origin are based on the hypothesis that the Moon formed in a geocentric orbit from a portion of the same material that was being accreted by Earth (Shmidt, 1957). The general coaccretion model has two basic variations: (1) material orbiting the proto-Earth remained as a swarm of small particles that selectively captured silicate-rich material while allowing iron-rich matter to pass through (circumterrestrial filter model) (Ruskol, 1960; Safronov and Ruskol, 1977; Wasson and Warren, 1979, 1984; Weidenschilling *et al.*, 1986), and (2) the geocentric swarm coalesced at an early stage into a single body or a few large bodies that subsequently accreted heliocentric planetesimals in Earth orbit (parallel coaccretion model) (Harris and Kaula, 1975; Harris, 1978). Both variations assume that planetesimals on initial heliocentric orbits are captured into geocentric orbit by collisions—either collisions among themselves but within Earth's sphere of influence, or collisions with bodies already in geocentric orbit. Consideration of the dynamics of incoming planetesimal trajectories leads to an immediate problem: about half of the incoming particles are moving on prograde orbits (relative to Earth) while the remainder are on retrograde orbits. Hence to a first approximation it seems that the net angular momentum contribution by incoming planetesimals is nearly zero; thus the material would collapse onto the Earth and drag down any Earth-orbiting bodies with it.

Harris and Kaula (1975) avoided this angular momentum dilemma by arguing that the swarm would quickly accrete into a moonlet that would raise tides on Earth. These tides would transfer angular momentum back into the orbit, thus negating the angular momentum drain on the proto-Moon as it grew in orbit about Earth. They show that good agreement with the present lunar orbit occurs if an initially small Moon ($\sim 10^{-4}$ M_E, M_E being the present mass of Earth) is formed near the proto-Earth when the latter has about one-tenth of its present mass. Refined calculations by Harris (1978), which include mass added to the forming Moon by planetesimals colliding within the sphere of influence (SOI) (whose radius is $\sim 150 R_E$, where R_E is the present Earth radius) and subsequently falling onto the satellite, show that perhaps one-third to one-half of the lunar mass may have been added by this mechanism, and that formation of the Moon began at the time the proto-Earth was about half its current mass.

The principal disadvantage of the Harris-Kaula parallel coaccretion model is that it does not readily explain why the Moon is deficient in iron relative to Earth. Once the geocentric swarm collapses into a large body (or bodies) that captures all of the planetesimal mass, there is no mechanism by which iron in the planetesimals can be excluded from the Moon. Only if the planetesimals themselves are iron-poor will the Moon be iron-poor relative to Earth. The compositional filter model, first proposed by Shmidt (1957) and Ruskol (1960) and expanded upon by Ruskol (1972), Safronov and Ruskol (1977), Wasson and Warren (1979, 1984) and

Weidenschilling *et al.* (1986) provides a mechanism for mechanically segregating iron from silicate material (Fig. 1a). Circumterrestrial material must remain as a disk of small bodies in order to accomplish the iron-silicate separation; a ring of matter does not raise tides on Earth that would pump angular momentum back into the disk. Weidenschilling *et al.* (1986) suggest that Earth's gravity harmonics could cause density waves in the disk; this mechanism has not been examined in detail. Ruskol (1972) and Safronov and Ruskol (1977) avoided the angular momentum difficulty by *assuming* that captured material would have a net specific angular momentum (angular momentum per unit mass) equal to that of particles orbiting Earth with angular velocity equal to Earth's *orbital* motion at the distance of capture. With this assumption, material captured at distances smaller than 80 R_E has insufficient angular momentum to stay in orbit, and thus collapses onto Earth. Only material captured further out contributes to formation of the Moon.

In order to better understand the angular momentum distribution contributed by particles captured from heliocentric orbits at different geocentric distances and to test the assumption of Ruskol described above, we carried out numerical studies

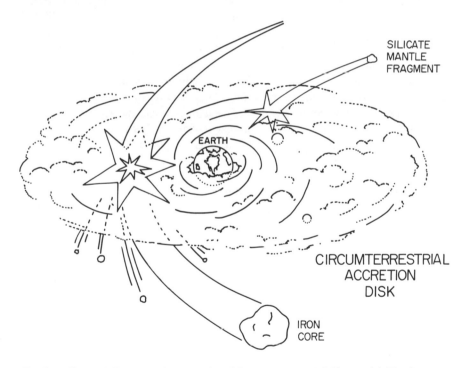

SILICATE
MANTLE
FRAGMENT

EARTH

CIRCUMTERRESTRIAL
ACCRETION
DISK

IRON
CORE

Fig. 1a. Pictorial illustration (not to scale) of the circumterrestrial filter model. The figure shows a massive iron core passing through the circumterrestrial disk while smaller silicate particles are trapped.

of the dynamics of planetesimals approaching Earth from heliocentric trajectories. The principal focus of the investigation is the comparison between the mean specific angular momentum brought in by the heliocentric planetesimals and that possessed by the disk material, assumed to be evolving toward circular orbits. Hence the chief criterion used in this study will be values of $<h(r)>/h_c(r)$, the ratio of mean specific angular momentum (equivalent to tangential velocity) of the heliocentric planetesimal distribution at a geocentric distance r compared to the specific circular keplerian angular momentum. Dimensionless comparisons of this sort are used in order to make the discussion as general as possible and reduce the need for specific assumptions such as values for the total mass in the circumterrestrial disk and heliocentric Earth feeding zone.

Using a set of procedures pioneered by Giuli (1968) we generated a systematic inventory of all Earth-approaching trajectories having eccentricities ≤0.12 using the formalism of the planar, restricted three-body problem (cf. Fig. 1b). Combining several plausible mass vs. orbit-element distributions of the planetesimals with this data base, we are able to compute the net geocentric specific angular momentum brought in to a putative circumterrestrial disk as a function of distance from Earth. For purposes of comparison we repeat this calculation for two disk sizes: 40 and 120 Earth radii (the latter value near to the size of Earth's sphere of influence). In Section 2 we describe our numerical approach to generating the data base while Section 3 presents our assumed mass-element distributions and summarizes our results about the net angular momentum brought in by the infalling planetesimal population. In Section 4 we discuss assumed models for the radial distribution of captured mass together with the total specific angular momentum of each. We apply these results to show that any circumterrestrial disk that is close to Earth will collapse onto Earth if it accretes significant mass from heliocentric planetesimals. Only material captured at large geocentric distances has sufficient net specific angular momentum to form a body in low Earth orbit. We calculate the size of orbit that a body would have if formed totally out of captured matter, under a range of plausible assumptions. We argue in Section 5 that collisions within the SOI between bodies

MATHEMATICAL ANALOG

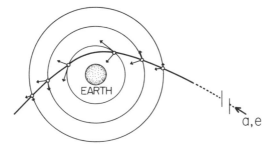

Fig. 1b. Schematic view of calculation of geocentric velocity components at different geocentric distances in our approximation of the scenario of Fig. 1a.

on heliocentric orbits could be an alternate mechanism for capturing material at large geocentric distances and that such collisions would produce iron/silicate fractionation of about the right degree to produce an iron-poor Moon. The final section summarizes our findings and points out areas where further work is needed in order to better understand how the Moon formed.

This paper explores possible mechanisms for material to be captured into geocentric orbit with enough specific angular momentum to remain in orbit. The results of our present work are in contrast with the conditionally negative results given at the Kona meeting (Herbert and Davis, 1984) and the even more pessimistic general interpretation they inspired (cf. Boss and Peale, 1986). The more positive current result comes about principally by extending our earlier calculations to larger geocentric distances. While there remain many unresolved issues concerning the compositional filter model yet to be explored in detail, we no longer feel that an angular momentum deficiency is necessarily a fatal flaw to this scenario.

2. Numerical Approach

In order to study in detail the angular momentum problem of forming the Moon in geocentric orbit, we wish to calculate the mean angular momentum per unit mass imported at different geocentric distances into a hypothetical circumterrestrial disk (CTD, with radius R_D) by impacts of particles on passing heliocentric orbits. In order to make this task feasible we introduce several simplifications and approximations.

We are studying a phase in the accretion of the Earth, hence the mass of the proto-Earth is increasing with time. However, we perform our calculations using a fixed mass for the proto-Earth and assume that our calculations can be scaled to other values of the mass of the proto-Earth. Giuli (1968) found that if one kept Earth's density constant, the distribution of orbits at Earth's surface was approximately invariant as the size of the proto-Earth was varied. Thus in the present calculations we use the present Earth mass and radius but we assume that our conclusions hold—with the appropriate redefinition of Earth radius and mass—for most of the Earth's growth stages.

Another major simplification adopted in the calculation was the assumption that ensemble averages over the three-dimensional orbits intersecting the circumterrestrial disk can be replaced by integrals along two-dimensional orbits lying in the plane of the disk. Each such integral is meant to incorporate the result of the ensemble of inclined orbits with the same a and e as the zero-inclination orbit actually computed. This assumption requires that the plane of the disk be approximately aligned with the ecliptic, and that the bulk of the orbits have small inclination i. The assumption also requires that for each heliocentric a,e and phase of Earth with respect to the planetesimal's perihelion, averaging over all possible lines of nodes, arguments of perihelion and times of perihelion passage are equivalent to, and may be replaced by, averaging over position in the disk-plane orbit ($i = 0$), at least to within a constant factor and for the small values of $|i|$ appropriate to our choices of e.

Given these simplifications, the next task is to determine the values of heliocentric a,e that feed into the three-body orbits intersecting the circumterrestrial disk. For particular values of a and e, the three-body orbit will cross the CTD for a number of ranges of Earth-orbital phase (in effect, the heliocentric longitude relative to the planetesimal's two-body orbit's perihelion). These intervals of phase for which collisions are possible have been termed "bands" by Giuli (1968; however, in his work they referred to Earth-impacting orbits only), whose methodology for locating them has been adapted to the present work.

The significance of the bands is that they select the values of heliocentric orbit parameters for which collisions with the CTD occur. Each band is a projection onto constant a and e of the ensemble of CTD-impacting heliocentric orbits, and is spanned by one degree of freedom whose parameterization is defined below. They are defined by finding their edges, boundaries in parameter space that separate colliding orbits from noncolliding ones. The basic procedure is to define one edge by integration (using a time-reversed outbound orbit) and then to search for the other edge by varying one of the orbit-defining parameters until the resulting orbit becomes once again tangent to the CTD. At tangency, further orbit variation leads to missing the CTD, hence the desired boundary has been found. In the following paragraphs, therefore, we describe our procedure (by comparison with Giuli's) for finding band edges and then add a short description of how we integrate over the band.

The initial, or constructed, edge of each band is found using the same technique that was used by Giuli (1968), illustrated in Fig. 2. A time-reversed orbit is computed that launches off tangent to the edge of the disk at a geocentric azimuthal angle α with velocity (in the nonrotating frame) v_i (see Fig. 2—v_i is analogous to Giuli's v_o but is defined in the nonrotating frame moving in unison with Earth). The angle β between the velocity vector and the clockwise tangent may take on either the

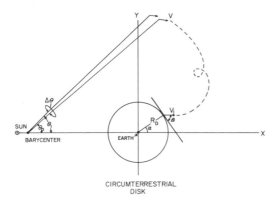

Fig. 2. Geometry of the planar restricted three-body problem and the procedure for determining the width of the bands. Earth and the sun move on circular orbits about their center of mass, and we calculate the motion of a particle of negligible mass moving under the gravitational forces of Earth and the sun.

values of $0°$ or $180°$. The orbit is followed until a two-body orbit criterion is satisfied ($\left| \frac{da}{dt} \right|, \left| \frac{de}{dt} \right| < 10^{-4}$ in the dimensionless units defined by Giuli) at the same time that the heliocentric phase angle θ between the orbiting test particle and Earth becomes larger than a prescribed value ($|\tan \theta| > 0.2$). The first criterion is used to ensure that the a and e of the corresponding heliocentric source orbit are well characterized; all the other orbits of the same band will be constrained to have the same a and e values. The second criterion is for the purpose of making unambiguous the sign of θ_o, the endpoint's value of θ. This is important in a later step in which the sign of variations of θ are compared to the sign of θ itself.

An outbound orbit computed by the procedure just described becomes a grazing impact orbit by reversing time back to the forward direction, i.e., by changing the appropriate signs in the equation of motion and in the change-of-frame relations and restarting integration from the outward orbit's endpoint. When integrated back into the appropriate perigee by our code, the computed orbit typically reproduces the original launch orbit's perigee distance with relative errors of the order of 10^{-10} to 10^{-6}, indicating that the orbit computations are accurate enough to unambiguously characterize the bands. A schematic illustration of an orbit computed using such a procedure is shown in Fig. 2. Note the epicycle resulting from the use of a rotating, Earth-centered coordinate system. Orbits that contain such loops approaching the CTD more closely than some limit (here arbitrarily taken to be 2 R_E) are strongly rescattered by Earth. Such orbits are acutely sensitive to the a and e of their two-body heliocentric source orbits and so are eliminated since they represent minute regions of heliocentric phase space. Inclusion of these orbits would necessitate much more accurate calculations and very elaborate systematization of near-Earth behavior, and yet, as Giuli pointed out, would not change the net mass or angular momentum influx appreciably. Once the initial band edge is found, the next step is to find the other edge by varying θ, while keeping a and e constant, until the heliocentric orbit has swept through the CTD and back out to tangency.

The values of a and e may be kept constant while changing θ by a procedure outlined by Giuli. For a new value of θ, the endpoint of the orbit and its inertial space velocity vector are rotated by an appropriate increment (as indicated in Fig. 2), and the new orbit is then numerically integrated toward the CTD. The value of θ is varied in the direction for which the resulting orbits cross the CTD. The opposite edge of the band, at a new value of $\theta_1 = \theta_o + \Delta\theta$ is defined by the incoming orbit that is once again tangent to the CTD. Thus the internal parameterization of the band is accomplished by varying θ over the interval [θ_o, θ_1].

Following Giuli, we leave out bands for which θ and $\Delta\theta$ have opposite signs. Since each band obviously has two boundaries, we can proceed with deletion knowing that the neglected bands will be picked up by the opposite bounding orbit. This omission of cases accomplishes two desirable ends: it protects us from errors due to a degradation of the criterion for two-body characterization of orbits as $|\theta|$ shrinks, and it reduces by a factor of two the amount of computer time and storage required (via the elimination of duplication).

Having a procedure for finding and systematizing the regions of two-body (a,e,θ) space that feed into three-body orbits colliding with the CTD, it is now possible to calculate moments and, with suitable normalization, average values for dynamical variables. A particular quantity of interest is the mean geocentric angular momentum per unit mass:

$$\langle h(r) \rangle = \frac{\displaystyle\sum_k \iint n(a,e) \int h(r,a,e,\theta) \, d\theta \, da \, de}{\displaystyle\sum_k \iint n(a,e) \, \Delta\theta_k \, (a,e) \, da \, de} \tag{1}$$

Here $\Delta\theta_k(a,e) = \theta_1 - \theta_0$ is the variation of θ over the k-th band at a and e, and $n(a,e)$ is the assumed density of potential impacting planetesimals in two-body heliocentric phase space (assumed independent of θ). The geocentric angular momentum per unit mass at a distance r from Earth of an orbit parameterized by θ and heliocentric a,e is denoted by $h(r,a,e,\theta)$. Other expectation values (kinetic energy, radial velocity, etc.) may be defined analogously, but we shall concentrate on this one.

It should be mentioned that a better approximation of the ensemble averaging of inclined orbits would involve an aberration correction to the angle of incidence on the disk that would favor retrograde impacts over prograde to some variable extent. In addition, the relative frequency of arrival of planetesimals of different a values has also been neglected, producing an overestimate of the contribution of planetesimals with a ≈ 1. To some extent, the neglect of these two effects mutually cancels; the former overestimates and the latter underestimates the specific angular momentum input.

The evaluation of integrals in (1) is performed in two stages. The innermost integral is performed first. Evaluation of $\int h(r,a,e,\theta) \, d\theta$ is relatively easy for $r = R_D$. All the orbits of the band cross the outer edge of the CTD, so orbits may be calculated over a grid of θ values and their values of h summed. But for $r \ll R_D$ only a small fraction of the orbits cross this radius, and so serious sampling errors arise.

The solution adopted is to spline-fit the perigee distance $r_p(\theta)$ and $h(r_p,a,e,\theta)$ as functions of θ. Then at each value of r, the range of θ for which $r_p(\theta) \leq r$ (i.e., the range of θ for which orbits actually cross the radius r) may be found, and the spline approximation for $h(r_p,a,e,\theta)$ is integrated over just this range. This interval also defines $\Delta\theta(a,e)$ for each value of r to insert into the denominator of (1) as well. In this way, the sampling problems of an overly coarse mesh covering the range of θ are avoided.

The outer integration over a and e is complicated by many difficulties. One is that the calculation is performed at points of a mesh of values of the orbit parameters α, β, and v_i. The proposed integration variables a and e are functions of the values chosen for α, β, and v_i, and depend on them in a complicated way. Another is that this dependence is quite nonmonotonic, so that the inverse mapping (a,e,k) onto (α,β,v_i) is multivalued. A third difficulty is that there are enough variables in the problem that a fine mesh in each one would expand the computation time and storage requirements well beyond that which is available. Yet another dilemma is the fact that not every (α,β,v_i) results in a valid escaping orbit or an acceptable band, so that there are many large holes in the multivariable mesh, a fact that makes the previously described spline-function resampling technique difficult to apply to the a,e integration.

The method adopted to deal with these problems is to recast the integral of (1) in terms of the mapping (α,β,v_i) onto (a,e,k). For an arbitrary function f(a,e)

$$\sum_k \iint f(a,e) \, dade = \sum_\beta \iint f(a(\alpha,\beta,v_i), e(\alpha,\beta,v_i)) \left| \frac{\partial(a,e)}{\partial(\alpha,v_i)} \right| d\alpha \, dv_i \qquad (2)$$

where $\left| \dfrac{\partial(a,e)}{\partial(\alpha,v_i)} \right| = \begin{vmatrix} \dfrac{\partial a}{\partial \alpha} & \dfrac{\partial e}{\partial \alpha} \\ \dfrac{\partial a}{\partial v_i} & \dfrac{\partial e}{\partial v_i} \end{vmatrix}$ is the Jacobian determinant of of the derivatives of a

and e with respect to α and v_i for the mapping. The variable β is summed over its two discrete values, 0° and 180°. The derivatives in the Jacobian are estimated numerically as conservatively as possible. That is, whenever possible, three-point interpolation is used to estimate derivatives at the central point; otherwise, the two-point approximation is used. Applying the technique of (2) to integration of (1) for which $f(a,e) = n(a,e) \int h(r,a,e,\theta)d\theta$ and $n(a,e)\Delta\theta(a,e)$ in the numerator and the denominator respectively, is accomplished by a summation over the non-null mesh points in a sampling grid of α, β, v_i values.

The grid of values for the $R_D = 120$ R_E calculations was, with v_e = the two-body geocentric escape velocity at R_D,

$$\left\{ \alpha = 0,5°, 10°, \ldots, 355° \right\} \otimes \left\{ \beta = 0,180° \right\} \otimes \left\{ \frac{v_i}{v_e} = \begin{array}{l} 0.75, 0.8, 0.85, 0.9, 0.95, \\ 1.0, 1.1, 1.2, 1.5, 2.0, 2.5, 3.2 \end{array} \right\}$$

The v_i grid for the $R_D = 40$ R_E case was $\left\{ \frac{v_i}{v_e} = 0.88, 0.90, 0.925, 0.95, 0.975, \right.$

1.0, 1.05, 1.1, 1.2, 1.3, 1.5, 2.0}. Values of α were also in this case evenly spaced 5° apart over [0,355], and β took on the values 0° and 180°. Because of the coarseness of this mesh, this part of the calculation is probably the greatest source of approximation error. The size of the error is best judged by noting the scatter in Fig. 8. At certain ranges of α the orbits become very sensitive to the exact value of α, as was noted by Giuli. For Giuli's calculation (which didn't involve an integration over e) selective finer spacing in a was possible. In the present calculation, a uniform grid was required.

Initial launch velocities (v_i) in the range 0.75 to 3.2 v_e (as above) were used that allow complete sampling of heliocentric eccentricities up to 0.12 and partial sampling to values in excess of 0.2. Launch velocities less than v_e depart from the vicinity of Earth into heliocentric space due to third-body solar perturbations at large geocentric distances. Higher velocity orbits, reflecting larger eccentricities and semimajor axes significantly different from 1 AU, also can come close to Earth, but the net contribution by bodies on such orbits decreases as the encounter velocity increases because capture is difficult at higher energies (though disk erosion will also result).

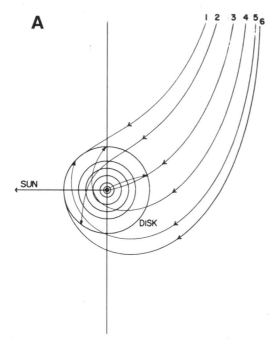

Fig. 3. Selected integrated trajectories as viewed in the rotating coordinate system of the four representative types of bands. Each figure shows the bounding orbits of the associated band together with orbits generated by evenly spaced values of θ (however, the spacing between the bounding orbits and their nearest neighbors in the interior is only half the spacing between the interior orbits themselves). In the first three cases (a–c) the even progression of orbits across the band is apparent. In the complicated case of (d), the progression of orbits folds back on itself. Individual orbits are consecutively numbered for clarity. Trajectory no. 1 is the initial bounding orbit (β = 0° or 180°), while the highest numbered trajectory is the other bounding orbit. Individual cases are: (a) symmetric and retrograde, (b) symmetric and prograde, (c) asymmetric and prograde, and (d) asymmetric and retrograde.

Bands were classified as being "symmetric" if the two bounding trajectories graze the outermost disk with opposite rotational senses, and "asymmetric" if the limiting trajectories impact in the same rotational sense. The two types of bands behave dissimilarly; for example, Giuli found that bands delivering negative $<h>$ are all of the symmetric type. Our calculations yield examples of both symmetric and asymmetric bands delivering both prograde and retrograde $<h>$ (Fig. 3). However, the symmetric, prograde bands and asymmetric, retrograde ones are relatively rare.

The Earth's gravity acts to "focus" orbits composing a band, as is illustrated in Fig. 4. In this figure, the bounding orbits and four intermediate ones are seen approaching Earth for the same case shown in Fig. 3d. The first group of perihelia are nearly unperturbed by Earth, but the second perihelia are perturbed to varying degrees with orbit number 1 being the most perturbed, while number 6 is least affected. On the following perihelion passage, all trajectories pass within R_D of Earth due to the action of Earth's gravity.

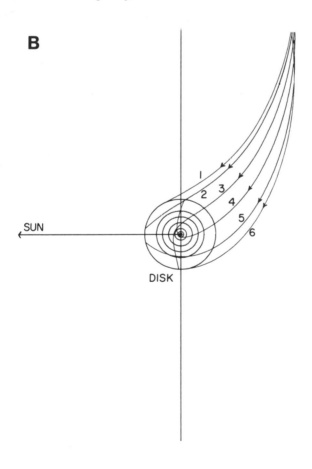

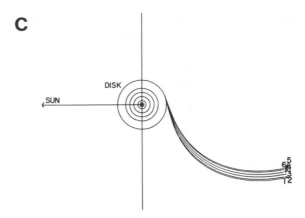

Selected results of the numerical integration of $\int_{\theta_0}^{\theta_1} h(r,a,e,\theta)\,d\theta$ are tabulated in the Appendix, which gives the specific angular momentum and orbit elements for some impacting bands having launch speeds v_i of 0.75 and 2.0 v_e. Figure 5 graphically illustrates the data base in terms of (a,e) space; this figure is a generalization of

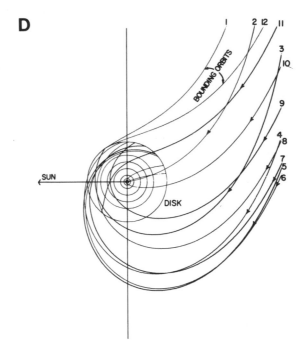

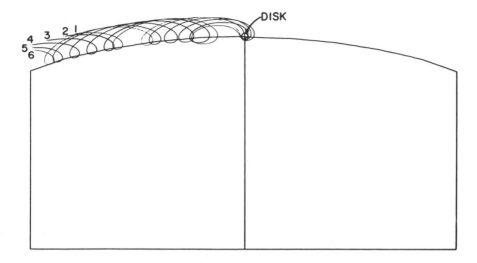

Fig. 4. Representative orbits from the band having a = 1.0152, e = 0.0171 are shown in a coordinate system rotating with Earth, to illustrate the evolution of the orbits prior to encounter with Earth. The numbering scheme is the same as that of Fig. 3. The initial wide spacing of periapse longitudes for band orbits is reduced and focused by the gravitational effects of Earth.

Fig. 13 of Giuli (1968). The arcs shown are loci of constant v_i for varying α and consist in places of two or more retracings of the same curve.

3. Orbit Element Distributions and Their Angular Momentum Contribution

The data base described above gives the net angular momentum per unit mass brought into the CTD by planetesimals moving on heliocentric orbits. Combining this data base with the mass-orbit element distribution n(a,e), we can calculate the resulting specific angular momentum delivered to the CTD as a function of distance from Earth for the entire ensemble of heliocentric orbits. However, the mass-orbit element distribution function of planetesimals moving close to the Earth at the time it was nearly formed has not been well studied to date and is still poorly understood. Giuli, who studied angular momentum transport just to Earth itself, assumed that the total planetesimal mass was uniformly distributed in (a,e) space (up to some limiting value of e). Harris (1977) argued, based in part on Giuli's numerical results, that only planetesimals whose heliocentric two-body orbits were nearly tangential to Earth's orbit (i.e., aphelion or perihelion distance close to 1.00 AU) contribute significant angular momentum to the accreting Earth. Therefore he assumed that the planetesimal mass distribution was uniform in that part of (a,e) space where

A

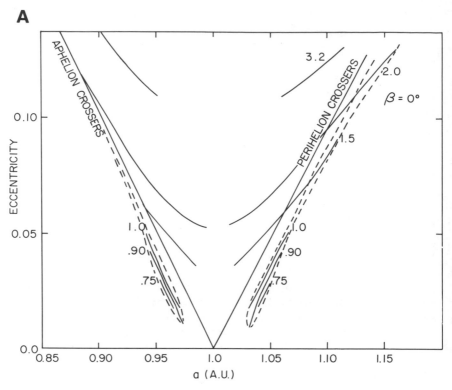

Fig. 5.   Heliocentric orbit elements (a,e) that produce encounters at 120 R_E with values of v_i from 0.75 to 3.2 (units of local escape speed). (a) is for $\beta = 0°$ while in (b) $\beta = 180°$. The dashed line on (a) approximately delineates the region within which orbits bring in a large prograde specific angular momentum. Other orbits deliver a considerably smaller, but generally retrograde, specific angular momentum.

planetesimals bring in significant h and ignored the rest of (a,e) space under the assumption that the net angular momentum was negligible. However, some other parts of phase space can bring in mass with little net angular momentum, reducing the estimate for <h>, hence the Harris calculation may overestimate the net specific angular momentum that would otherwise result from his other assumptions.

The mass-orbit element distribution function is one of the critical quantities required for the calculation of the angular momentum delivered by a CTD. Meaningful studies to investigate n(a,e) are far beyond the scope of this work, so we assumed a variety of forms for n(a,e) in the calculation of the net angular momentum delivered to a CTD. Here we present results for the two extreme distributions in terms of the net angular momentum: (1) a uniform distribution over (a,e) space (up to a limiting value of v_i) that brings in the least angular momentum and (2) a distribution we

B

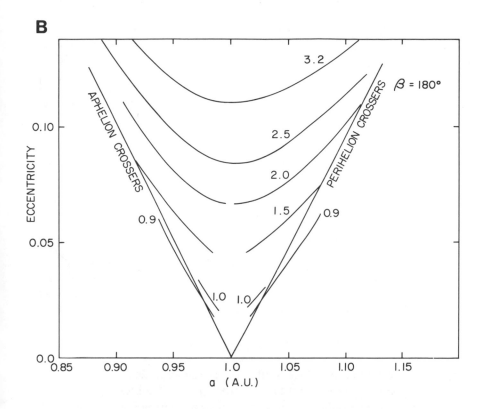

refer to as the "depleted deep-crosser distribution" that contributes the largest specific angular momentum of the distributions studied. A graphical representation of these functions is given in Fig. 6. The uniform distribution is the simplest one to model and one that has been assumed by other authors (Giuli, 1968; Harris, 1977). We now develop the rationale for the second, or "depleted deep-crosser" population.

An important concern in selecting possible choices of n(a,e) is consideration of how planetesimals first encounter the late stage Earth and how their orbits are affected by the presence of a circumterrestrial disk. Presumably proto-Earth swept up material in its immediate vicinity and continued to grow by accreting more distant material. Protoplanetary feeding zones, i.e., regions of orbit element space leading to impacts onto the accreting protoplanet, must either be resupplied with mass as the existing planetesimals are accreted, or else new feeding zones must be formed in order to continue the accretionary process. The resupply process is likely to have been important because the size of the feeding zone accessible to direct impact on proto-Earth (even with a full mass Earth) is small compared with the size of the orbit element space likely to have been populated by terrestrial zone planetesimals (see Fig. 5). We consider it plausible that most material initially approaches Earth on orbits that

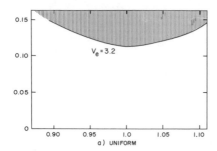

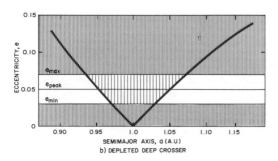

Fig. 6. Graphical representation of the source orbit element distribution functions, n(a,e), for the uniform and depleted deep-crosser planetesimal populations discussed in the text. The heavy hatching shows areas in (a,e) space contributing no planetesimals, while the region indicated by the vertical hatching in the depleted deep-crosser population contains initial orbits whose contribution is depleted by a factor of 10 relative to orbits in unhatched areas. The number of particles in (b) is also varied with the eccentricity of the orbit by using a triangular weighting function with its greatest value at e_{peak} and falling linearly to zero at e_{min} and e_{max}

are nearly tangent to that of Earth's orbit, i.e., planetesimals first interact with Earth near the aphelia or perihelia of their own orbits (Fig. 7). Planetesimal orbits slowly diffuse in (a,e) space until they have an aphelion or perihelion that approaches Earth's orbit. Once the planetesimal can approach Earth closely, then large jumps in orbit elements can occur. Whenever these orbits cross Earth's orbit deeply, subsequent interactions can then occur at any point in the planetesimal's orbit, not just near the apses. However, the geocentric swarm will dampen the scattering process: Weidenschilling *et al.* (1986) estimate that silicate planetesimals will be captured within a few passages through the swarm for assumed swarm densities. Hence capture occurs before their orbits diffuse close enough to Earth for strong scattering to occur, and therefore most material will be added to the swarm from near aphelion or perihelion of their heliocentric orbits. We model this by assuming that in the second version of n(a,e), the mass density in (a,e) space is reduced by an (arbitrarily chosen) factor of 10 for deep Earth-crossing orbits relative to shallow Earth-crossers (i.e., near-tangential encounters); this is the "depleted deep-crosser" distribution for n(a,e) (see Fig. 6).

The mean angular momentum per unit mass brought in by material captured at various geocentric distances is calculated using the angular momentum data base in conjunction with the mass-orbit element distributions via (1) and (2). Figure 8 summarizes these calculations by showing the mean angular momentum per unit

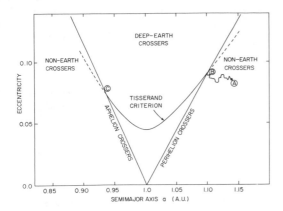

Fig. 7. An initial non-Earth-crossing planetesimal (A) evolves into an Earth-crossing body (B) due to gravitational perturbations by a growing Earth (or other large planetesimals) or due to collision with other planetesimals. Once it can make close encounters with Earth it can jump to any point between B and C along the line labeled Tisserand criterion. However, the CTD leads to capture before the planetesimal has had time to evolve very far from B.

mass brought in by captured material as a function of geocentric distance of collisional capture for the two different mass-element distributions. The other trial distributions n(a,e) gave results that fell between these extremes. Each case is computed for two disk sizes, 40 R_E and 120 R_E. In all cases, there is an increase in the net prograde specific angular momentum with geocentric distance, hence capture at large distances from Earth can permit formation of the Moon in geocentric orbit. The uniform n(a,e) case brings in a relatively small net angular momentum. This is because there is a significant amount of mass brought in on deep Earth-crossing orbits that typically contributes a small retrograde $\langle h \rangle$ that partially offsets the usually larger prograde h contributed by mass from nearly tangential orbits. The depleted deep-crosser n(a,e)

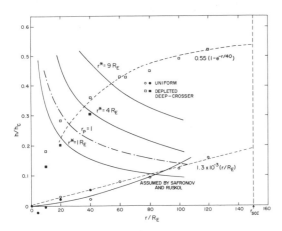

Fig. 8. Mean angular momentum per unit mass, normalized to circular orbit angular momentum, delivered as a function of geocentric distance for the uniform distribution and the depleted deep-crosser distribution. Open symbols are calculations with $R_D = 120$ R_E, while the solid symbols have $R_D = 40$ R_E. Also shown is the angular momentum necessary to produce circular orbits at distances (r^*) of 1, 4, and 9 R_E, e.g., capture with $h/h_c = .115$ at 80 R_E has enough angular momentum

for a 1 R_E circular orbit. The dashed lines represent the numerical approximation to the data used in the text. The dot-dash line is the locus for orbits whose perigee is at 1 R_E; orbits below this line impact Earth in their first orbit.

case, however, has a much larger fraction of mass accreted from nearly tangential orbits, which produces the significantly higher prograde specific angular momentum shown at all geocentric distances. If the synodic frequency of encounter had been incorporated into (1), however, the two distributions would likely have produced more similar results.

The results shown in Fig. 8 reflect only the angular momentum brought in from heliocentric orbit and do not include the angular momentum of the material already in the circumterrestrial disk. If most of the mass of the Moon is captured by collisions between bodies on hyperbolic orbits and bodies already captured moving on elliptic geocentric orbits (free-bound collisions), then the result shown in Fig. 8 can be applied to calculate the orbital angular momentum of the Moon at the time accretion ceased. However, if a significant quantity of material is placed in the swarm by other mechanisms, e.g., the impact-triggered splash model, then the angular momentum distribution of that material would have to be averaged with the results shown in Fig. 8 to find the final angular momentum distribution. We also show in Fig. 8 the distribution of angular momentum for captured material assumed by Safronov and Ruskol (1977) and graphs of the specific angular momentum of orbits that would circularize to indicated orbit sizes. The low net angular momentum primarily results from their assumption that material orbits Earth with Earth's orbital angular velocity. Note that the $r^* = 1$ R_E line (showing specific angular momentum equal to that of a circular orbit at 1 R_E as a function of distance) crosses their angular momentum distribution at ~80 R_E. Consequently all material captured at distances less than 80 R_E would fall onto Earth, as they note in their paper. However, their adopted angular momentum distribution is very close to the one that we calculate using a uniform distribution for n(a,e).

Our calculations indicate that the net angular momentum of captured material depends strongly on the mass-orbit element distribution of the infalling material. This result was first noticed (based on a limited set of integrations) by Giuli (1968) who found that even the sign of the angular momentum was changed (from retrograde to prograde) when the heliocentric source orbits went from circular to having a small eccentricity. Harris (1977) extended this concept by noting from Giuli's work that only planetesimals impacting Earth near perihelion or aphelion of their orbits contributed large (and prograde) angular momentum.

Another approach to calculating the net angular momentum of Earth-approaching bodies is the forward integration of trajectories started well away from the planet. As pointed out above, this method is much less efficient for finding disk-impacting orbits than the backward integration of grazing trajectories. However, it is a useful method for investigating "free-free" collisions between planetesimals on heliocentric orbits. One of us (S. J. Weidenschilling) has carried out simultaneous integrations of trajectories of up to 300 bodies in Earth-approaching orbits. Some fraction of collisions within the Earth's sphere of influence (assumed to result in coagulation) yielded bound geocentric orbits. As described by Weidenschilling et al. (1986), two cases with uniform distributions of a did not yield any excess of prograde orbits; eccentricities ≤ 0.05 yield equal numbers of prograde and retrograde orbits, while zero eccentricity yielded an excess of retrograde captures. A third case corresponded to the "depleted deep-crosser" distribution defined above, i.e., with e's peaked about 0.05 and orbits crossing Earth's (without Earth perturbations) depleted by a factor of 10. Free-free collisions in this case yielded prograde geocentric orbits, with mean angular momentum corresponding to a circular orbit at 30 $R_\oplus$. Due to computer limitations, the number of capture events was low, and the results are subject to large statistical uncertainty. Nonetheless, they are consistent with our results for impacts onto a circumterrestrial disk, which give a net prograde angular momentum for such an orbit distribution.

We note in passing that high-velocity collisions at large geocentric distance can cause mass loss from a circumterrestrial cloud; i.e., high velocity for incoming particles means that matter will be collisionally stripped from the disk onto new heliocentric orbits. The material approaching with high velocity tends to come in on orbits that collectively have little mean specific angular momentum, so removing them from the distribution that contributes mass to the disk will increase the net prograde angular momentum of the remaining captured material. The net efficiency of the capture process then depends on the distribution n(a,e), i.e., the relative mass fraction on high and low collision velocity trajectories. The effect of collisional stripping was tested by assuming that if the collisional energy was greater than three times the orbital energy (an arbitrary cut-off), then all collisional fragments escaped. With this assumption, the net angular momentum at large geocentric distances (>80 R_E) increased significantly, up to 100% of the local circular angular momentum in some cases. We note this to call attention to this effect, but, in order to be conservative, we do *not* include it in the calculations of the next section.

4. Estimates of the Size of the Initial Lunar Orbit

The results of the previous sections show that significant prograde angular momentum can be added to a geocentric accretion disk by material captured from heliocentric orbit, provided that most of the mass is captured at large geocentric distances from orbits that are nearly tangent to Earth's orbit. We will address how material can be captured at these large distances later, but we now consider what would be the size of the final circular orbit if all the captured material coalesced into a single body and only captured mass contributed to the body. Because the angular momentum shortfall is in the incoming population, this approximation is a "worst case" assumption. Captured material will move initially on eccentric orbits around Earth since $\langle h(r) \rangle$ is smaller than $h_c(r)$ for all r (Fig. 8) and, through collisions with either other material already in geocentric orbit or bodies newly arriving from heliocentric orbit, will dissipate orbital energy but conserve angular momentum. Hence the captured material will collisionally damp down into circular orbits whose size is determined by the average net angular momentum of the captured mass.

Although a more realistic model is desirable, we present here a highly simplified analysis of the angular momentum systematics of the evolution of the disk by collisional damping. Our approach to exploring one possible end state of the evolution of some fraction of the disk is to find the radius at which a circular orbit's specific angular momentum equals the mean over the disk fraction. This radius, r^*, can be calculated by combining the distribution of arriving angular momentum (Fig. 8) with a model for the amount of mass captured at different geocentric distances from Earth. We calculate r^* by equating the total captured angular momentum divided by the total captured mass ("net specific angular momentum") to the angular momentum of a unit mass in a circular orbit on the grounds that the angular momentum and mass would be conserved in such a collapse but the energy would not. Let $h(r)$ be net specific angular momentum captured at geocentric distance r and let $dm(r)$ be the mass element captured between r and $r + dr$. Then the net angular momentum, H, brought in by all the material captured between a minimum distance, r_m, and a maximum distance, r_{max}, is:

$$H = \int_{r_m}^{r_{max}} h(r)\, dm(r) \tag{3}$$

and the captured mass is

$$M = \int_{r_m}^{r_{max}} dm(r) \tag{4}$$

We assume that capture can occur anywhere within the sphere of influence, hence $r_{max} = r_s$, the radius of the SOI. Thus H is a function only of the lower limit of the capture distance r_m, i.e.,

$$H(r_m) = \int_{r_m}^{r_s} h(r)\, dm(r) \tag{5}$$

While we have argued that the "depleted deep-crosser" population is a plausible one, it is by no means certain that it is the best representation of the angular momentum distribution of arriving planetesimals. Since the depleted deep-crosser population and the uniform population (Fig. 6) are the extremes in terms of the amount of angular momentum delivered for all the populations considered, we examine both of these in order to find limits on r^*.

Now, from Fig. 8, the angular momentum distribution $h(r)$ may be approximated by

$$h(r) \approx h_c(r) c_1 (1 - e^{-r/\rho}) \tag{6a}$$

for the depleted deep-crosser population and

$$h(r) \approx h_c(r) \, c_2 \, (r/R_E) \tag{6b}$$

for the uniform population where, $h_c(r)$ = circular orbit angular momentum = $\sqrt{GM_Er}$, $\rho = 40 \, R_E$, $c_1 = 0.55$ and $c_2 = 1.3 \times 10^{-3}$.

The amount of mass captured at different distances depends not only on the probability of a collision but also on dynamical quantities such as the relative mass of the colliding bodies, the collision speed, and geometry and the geocentric distance at which the collision occurs (Ruskol, 1972; Weidenschilling *et al.*, 1986). We do not know mass-capture probability as a function of geocentric distance, but we assume two extreme cases for illustration: first, that the mass capture probability per unit area of the disk is constant (constant mass capture distribution case) and the second, in which the capture probability per unit area falls off rapidly with geocentric distance, $\propto r^{-3.5}$ ("peaked" mass capture distribution). The latter is similar to that chosen by Safronov and Ruskol (1977). We outline the procedure for calculating r^* for the case of the "depleted deep-crosser" angular momentum distribution and the uniform mass capture probability (Case I).

$$\bar{h}(r_m) = \frac{H(r_m)}{M(r_m)} = G \, M_E \frac{\displaystyle\int_{r_m}^{r_s} \sqrt{r} \; c_1 \, (1-e^{-r/\rho})r \, dr}{\displaystyle\int_{r_m}^{r_s} r \, dr} \tag{7}$$

Equating $\bar{h}(r_m)$ to the circular orbit angular momentum at r^* gives:

$$\sqrt{r^*/r_s} \; = \frac{2c_1}{r_s^2 - r_m^2} \int_{r_m}^{r_s} r^{3/2} \, (1-e^{-r/\rho}) dr \tag{8}$$

The integral on the right side can be evaluated using the incomplete gamma function, $\gamma(\lambda,\mu) = \int_0^{\mu} t^{\lambda-1} \, e^{-t} \, dt$, whereupon after rearranging terms (8) becomes:

$$\sqrt{r^*/r_s} = \frac{2c_1}{1-(r_m/r_s)^2} \left[2/5 \left(1-\left(\frac{r_m}{r_s}\right)^{5/2}\right) - \frac{(\gamma(5/2, r_s/\rho) - \gamma(5/2, r_m/\rho))}{(r_s/\rho)^{5/2}} \right] \tag{9}$$

Expressions for r^* can be found in an analogous manner to above but using the "peaked" mass capture distribution (Case II), the "uniform" angular momentum capture distribution and the "constant" mass capture function (Case III), and finally the "uniform" angular momentum distribution with the "peaked" mass capture distribution (Case IV).

Figure 9 gives r^* as a function of r_m for all cases. The angular momentum distribution is clearly the dominant factor in determining whether a disk of the type described here can lead to a Moon orbiting Earth. Even the "peaked" mass capture distribution leads to an orbiting Moon, provided the lower bound to the captured mass is ≥ 15 R_E. Such a result is not surprising since from Fig. 8 we see that, on average, material captured at ≤ 20 R_E for the "depleted deep-crosser" case has a perigee distance less than 1 R_E and will thus impact Earth on the first orbit. Such material is removed from the disk very quickly and does not contribute very much mass or angular momentum to the disk. We have been conservative in assuming that such low angular momentum material is averaged into the rest of the disk. Removal

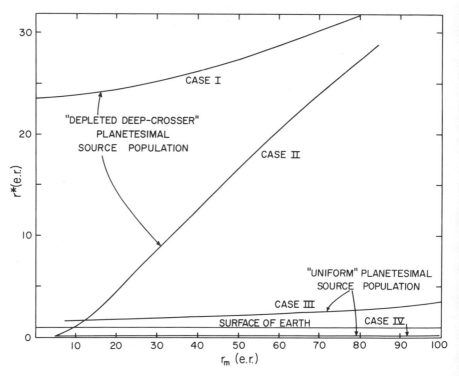

Fig. 9. Size of the circular orbit, r^*, that a single body would have if it were formed out of matter captured between r_m and the edge of the sphere of influence. The specific angular momentum distribution for the uniform and depleted deep-Earth-crosser populations in Fig. 8 are assumed and the results of each are shown for the two models of the mass capture probability distribution discussed in the text.

on the first orbit would remove mass from the disk without any "accretion drag" on the remainder.

However, the "uniform" angular momentum distribution case leads to collapse onto Earth regardless of how mass is captured (except for the extreme case of having all mass captured at large geocentric distances, ≥ 80 R_E). Undoubtedly some of the accreted material will fall onto Earth or be ejected into heliocentric space, and some material splashed off Earth by impacts will be incorporated into the circumterrestrial disk, so the above values of r* are only approximations to the values resulting from a more accurate calculation. Moreover, as noted before, the mass (and angular momentum) of the initial CTD population has not been taken into account. Therefore the values obtained for r* should only be viewed as a guide to the intuition. The important thing to note is that the specific angular momentum input at the edge of the disk for some reasonable guesses of n(a,e) in these scenarios is not negligibly small. In the most favorable cases the mean specific angular momentum can approximate that of circular orbit at tens of Earth radii.

In the foregoing analysis, a number of key issues have been ignored. The most important of these is the source of material at the outer edge of the disk. Two possibilities are evident. One is diffusion of material outward due to either viscous spreading or the statistical spread of impact velocities of collisions in the inner disk regions. This would represent a small fraction of the processed material but might serve to form a disk population that dropped off fairly steeply with radius, but still not entirely to zero.

Another possibility for populating the outer edge of the disk is that of free-free collisions of heliocentric particles within Earth's sphere of influence. As the n-body results show, these can leave preponderantly prograde orbiting bodies in outer regions of the CTD in cases such as the depleted deep-crosser distribution where n(a,e) favors disk development. Our results, too, suggest (by the large values of <h> there) that such capture should be prograde (though our statistical methods are not designed for or accurate for free-free collisions). This material, then, represents material diffusing *inward* from the edge of the CTD, and thus should contribute to making the decline of $\frac{dm}{dr}$ with r less steep.

5. *Implications for Compositional Filtering*

The previous discussion concerned the specific angular momentum that a single accumulated satellite would possess based on the angular momentum brought in by captured heliocentric planetesimals. We now consider the implications of our calculations for the circumterrestrial filter model presented by Weidenschilling *et al.* (1986). That model assumes that a disk of material initially containing ~10% of a lunar mass is distributed within 30 R_E of a late stage Earth (~90% formed).

If capture of the remaining ~90% of the lunar mass occurs by collisions of heliocentric planetesimals with bodies already in geocentric orbit, we may estimate r* for the fully formed Moon using either of the mass capture or angular momentum distributions but with the upper bound to the capture distance being 30 R_E rather than the size of the SOI. (Our assumption of $M = M_E$ is sufficiently close to their assumption of $M = 0.9\ M_E$ that the comparison is valid.) Either assumed capture probability distribution leads to r* $\leq$ 3 R_E; hence the circumterrestrial swarm would collapse to within the Roche limit unless additional angular momentum is added to the disk. Weidenschilling et al. recognized this problem and suggested viscous transport as one possible mechanism for redistributing mass and angular momentum in order to preserve the CTD. Since the angular momentum brought in by the heliocentric planetesimals is always less than the value for a local circular geocentric orbit, "accretion drag" is unavoidable, regardless of the size of the disk. Viability of the model requires viscous spreading or some other mechanism to counteract the accretion drag enough so that the outer edge of the CTD remains at least thinly populated even as most of the material is swept up by the Earth or a low-orbit proto-Moon.

We now consider another possibility: that material is captured at larger geocentric distances than the 30 R_E upper bound assumed by Weidenschilling et al. (1986). Collisions between planetesimals moving on heliocentric precollision orbits (free-free collisions) would be the principal capture mechanism, as suggested by Ruskol (1960). Only a fraction of such collisions lead to capture into Earth orbit, but the captured material, moving on elliptical orbits about Earth, would collide with other such bodies (bound-bound collisions) leading to circularization of orbits at smaller geocentric distances. The Earth-orbiting material would also be hit by heliocentric planetesimals (free-bound collisions) as described by Weidenschilling et al. Understanding the evolution of a CTD demands consideration of all of the above processes, a task beyond the scope of this work.

However, we can estimate the ratio of silicate to iron for the free-free capture process based upon the model for the composition and sizes of planetesimals outlined by Weidenschilling et al. (1986). They assume that the arriving planetesimals are fragments of geochemically differentiated parent bodies that have had their silicate mantles stripped away by collisions, thus exposing the iron core. The iron core is the largest fragment from the disrupted parent planetesimal based on an assumed composition having an iron mass fraction of about 1/3. The silicate material will have a size distribution resulting both from the initial shattering collisions that stripped away the silicate mantles and from subsequent collisions involving the silicate fragments. The largest silicate fragments are assumed to be several times smaller (~5×) than the iron core.

Free-free collisions at large geocentric distances lead to capture into Earth orbit only if the mass of the incoming body is about equal to that of the disk mass that it hits, given the typical Earth-approach speed under discussion here (Weidenschilling et al., 1986). Hence an iron core can be captured only if it collides with another iron core—silicate particles just do not have enough mass to do the job. Silicates will be captured predominantly by collisions with other silicate bodies of about the same size. The mass ratio of the silicate to iron captured by collisions with equal sized particles can be estimated using a simple particle-in-a-box model as:

$$\frac{\Delta m_{Si}}{\Delta m_{Fe}} = \frac{\rho_{Si}}{\rho_{Fe}} \cdot \left(\frac{N_{Si}}{N_{Fe}}\right)^2 \cdot \left(\frac{D_{Si}}{D_{Fe}}\right)^5$$

where ρ, N, and D are the density, number of bodies per unit volume, and diameter of the typical silicate and iron planetesimals that collide within the SOI. As described above, the ratio D_{Si}/D_{Fe} is ~1/5 while N_{Si}/N_{Fe} is ~750 based on the assumed composition of differentiated Earth zone planetesimals. The mass ratio of silicate to iron resulting from collision capture is ~60, using $\rho_{Si}/\rho_{Fe} = 1/3$. This simple model can be parameterized in terms of a single variable R, the size ratio of the typical iron-to-silicate particle. The ratio of silicate to iron captured then becomes

$$\frac{\Delta m_{Si}}{\Delta m_{Fe}} = 12R$$

While the exact ratio depends upon poorly known details of the planetesimal sizes and collision geometry, the general results do imply that iron-silicate fractionation can occur by the process of *collisional filtering*. The basic requirements for this mechanism to be effective are that there be both iron and silicate projectiles, and that the iron bodies be substantially more massive but much less numerous than the silicate ones.

The circumterrestrial filter model proposed here is closer to the original version proposed by Ruskol and coworkers than the version put forth by Weidenschilling *et al.* (1986) in that free-free collisions are an important mechanism for capturing material into geocentric orbit. Harris (1978) also included mass gain by a growing Moon from free-free collisional captures, but neglected the angular momentum brought in by this captured mass. His justification for this assumption was that the net specific angular momentum contributed by infalling mass is only a small fraction (≤ 0.2) of the circular orbit angular momentum at the surface of the Earth, and he argued that this ratio will decrease with increasing geocentric distance. However, our numerical integrations (Fig. 7) show that the ratio of net angular momentum to that of a local circular orbit *increases* with increasing geocentric distance. The reason for this increase is that the larger moment arm at greater geocentric distances more than offsets the decreased velocities and produces an angular momentum contribution that increases faster with distance than does the circular orbit angular momentum.

Third-body effects are important in calculating the total prograde angular momentum imparted by infalling projectiles, but many of the basic ideas can be understood using two-body concepts as pointed out by Harris (1977). He showed that nearly tangential orbits (which deliver the largest angular momentum) do so in the prograde sense by considering exterior orbits (a>1) with perihelia just outside 1 AU and interior orbits (a<1) with aphelia just inside 1 AU. However, two other cases exist for nearly tangential orbits with aphelia just *inside* 1 AU and interior orbits with aphelia just *outside* 1 AU. For these orbits, the net angular momentum, even for two-body orbits, can be prograde or retrograde depending upon details of the collision geometry and therefore bring in mass without much angular momentum, thus decreasing CTD specific angular momentum. We find though, that the latter class of orbits occur only occasionally in our data base, much less frequently than in the data base studied by Harris. Either they are eliminated from impacting due to three-body effects or they are strongly perturbed and make numerous "epicyclic loops" that are then rejected from our data base (Section 2). We note also that even high-velocity, nearly tangential impacting orbits can impart a large prograde angular momentum (Fig. 5). Therefore, the ultimate rotational sense and rate for a body formed by accretion depends principally on the mass-orbit element distribution of the infalling planetesimals.

6. Conclusions

The results of the previous sections indicate that the angular momentum systematics of an accreting circumterrestrial disk continue to severely constrain the possibilities for maintenance of a CTD. In this work, we have systematically quantified the magnitude of the angular momentum problem and described its relationship to the orbit element distribution of the infalling heliocentric planetesimals. The fact that all of the cases that we have computed so far show a net specific angular momentum input at each point of the CTD that is less than the local circular orbit value indicates that no CTD is stable with a permanent resident population whenever there is significant heliocentric mass input. The "accretion drag" of incoming heliocentric planetesimals will, to a first approximation, continually drag disk material to lower geocentric distance. Thus the results of our calculations imply that all component lunesimals (planetesimals that form the Moon) of the CTD must eventually diffuse through the disk and be lost from the inner edge. If a CTD is to persist at all it must do so in a quasi-steady-state form, which processes heliocentrically derived material flowing through it. This possibility, in order to be viable, depends on there being a source of material for repopulation of the outer part of the disk. This requirement may possibly be met by free-free collisions or the outward diffusion of part of the CTD material due to either viscous spreading or the statistical spread of impact angular momentum, as previously discussed.

The calculations of Section 4 indicate that at each accretionary collision both incident and impacted material, on the average, move to considerably lower geocentric distances. Thus material accreting into the disk flows through it and out the inner edge (or falls onto the Earth), experiencing only a few collisions with infalling heliocentric bodies. For this reason, the CTD coaccretion scenario is constrained to a fairly narrow set of possibilities. Either a proto-Moon or several moonlets must exist in the inner region of the disk during some portion of its lifetime in order to capture the mass processed by the disk or else the CTD must be massive in its quasi-steady state in order to "freeze out" to a lunar mass at the end. Otherwise, the disk will play very little role in determining, by compositional filtering, the composition of the Moon. A coaccretionary model of the former type has been proposed by Harris (1978), who postulated a proto-Moon kept at the inner edge of a CTD by tidal torque from Earth, while material diffusing inward through the CTD was swept up by the proto-Moon. Although this is not the only possibility, it seems to satisfy the constraints well.

Judging by the calculations made in Section 4, the following additional characteristics of the CTD scenario are necessary. The CTD must extend out to a large fraction of the radius of Earth's sphere of influence. This is because the accretional drag is least there, and free-free collisions that can populate the disk are the most frequent there. Moreover, the density fall-off with geocentric distance must not be too steep, or the disk will acquire little net specific angular momentum. Of course, since the inner disk material is mostly fed down from the outer regions, the steepness of the fall-off may adjust itself to satisfy this requirement, but the input of material to the outer edge of the CTD must be sufficient to keep the CTD reasonably dense. A self-consistent calculation of CTD population dynamics would be desirable and could be performed given the angular momentum statistics generated in the present work.

The most important constraint on the minimum CTD density is that its total cross section be larger than that of any proto-Moon or moonlets in order for the compositional filter scenario to work. In order for the angular momentum systematics to be favorable, as indicated in Section 3, the density must also be large enough that heliocentric planetesimals passing through the CTD have a high capture probability in order for their distribution to be depleted in the "deep-crosser" region of (a,e) space.

If all these requirements can be met, the compositional filter scenario proposed in the companion paper (Weidenschilling et al., 1986) can work. The demonstration of the possibility of satisfaction of these requirements requires further modeling, however.

Appendix

Selected results from the numerical integrations. We tabulated selected quantities from a subset of cases run on the numerical integration program in order to give the reader interested in the numerical details a sample of our results. These values could serve to validate or provide inspiration for analytical theories (as Giuli's work apparently did for Harris) or to serve as test points for others wishing to carry out further calculations of this type. Tabulated for two impact speeds are bandwidth ($\Delta\theta$), total angular momentum $\left[\int_{\theta_0}^{\theta_1} h(r,a,e,\theta)d\theta\right]$, semi-major axis and eccentricity for the heliocentric orbits contributing bands passing within 120 R_E of Earth. For $v_i/v_e = 0.75$ only $\beta = 0$ orbits exist, while for $v_i/v_e = 2.0$ orbits exists for both $\beta = 0$ and 180, but only the $\beta = 180$ cases are tabulated in order to save space. Additional details are available from the authors.

v_i/v_e	β(deg)	α(deg)	$\Delta\theta$(deg)	Total Angular Momentum ($10^{16} \frac{cm^2\text{-deg}}{sec}$) for:		a(A.U.)	e
				r=120R_E	r=10R_E		
0.75	0	70	1.116	0.673	0.154	1.0387	0.0182
		75	1.285	0.698	0.152	1.0412	0.0218
		80	1.282	0.754	0.070	1.0433	0.0245
		85	1.246	0.789	0.070	1.0449	0.0264
		90	1.211	0.807	0.068	1.0460	0.0276
		95	1.186	0.811	0.067	1.0466	0.0283
		100	1.173	0.811	0.071	1.0468	0.0285
		105	1.174	0.801	0.072	1.0465	0.0281
		110	1.188	0.781	0.075	1.0458	0.0272
		115	1.209	0.750	0.078	1.0443	0.0255
		120	1.200	0.692	0.132	1.0421	0.0226
		125	0.921	0.665	0.000	1.0390	0.0182
		240	−0.042	0.034	0.001	0.9622	0.0196
		245	−0.385	0.436	0.000	0.9651	0.0146
		250	−1.210	0.661	0.184	0.9631	0.0184
		255	−1.312	0.710	0.122	0.9609	0.0219
		260	−1.290	0.766	0.066	0.9591	0.0245
		265	−1.248	0.803	0.067	0.9578	0.0262
		270	−1.211	0.818	0.066	0.9570	0.0274
		275	−1.185	0.820	0.065	0.9565	0.0280
		280	−1.173	0.816	0.065	0.9563	0.0282
		285	−1.175	0.805	0.066	0.9565	0.0279
		290	−1.192	0.786	0.067	0.9572	0.0269
		295	−1.218	0.760	0.760	0.9584	0.0253
		300	−1.226	0.699	0.128	0.9602	0.0225
		305	−1.022	0.662	0.066	0.9629	0.0182
2.0	180	0	−1.453	−0.996	0.008	0.9050	0.1112

v_i/v_e	β(deg)	α(deg)	$\Delta\theta$(deg)	Total Angular Momentum $(10^{16} \frac{cm^2-deg}{sec})$ for:		a(A.U.)	e
				r=120R$_E$	r=10R$_E$		
		5	−1.323	−0.708	−0.002	0.9063	0.1102
		10	−1.210	−0.503	−0.006	0.9082	0.1086
		15	−1.113	−0.350	0.021	0.9108	0.1065
		20	−1.029	−0.247	−0.002	0.9141	0.1039
		25	−0.957	−0.174	−0.022	0.9181	0.1009
		30	−0.896	−0.118	0.001	0.9226	0.0975
		35	−0.845	−0.080	0.004	0.9278	0.0938
		40	−0.801	−0.051	0.005	0.9336	0.0898
		45	−0.764	−0.024	−0.024	0.9400	0.0857
		50	−0.734	−0.008	0.004	0.9470	0.0816
		55	−0.700	0.004	−0.015	0.9544	0.0776
		60	−0.690	0.013	−0.011	0.9624	0.0739
		65	−0.675	0.020	−0.009	0.9707	0.0707
		70	−0.664	0.025	−0.008	0.9795	0.0681
		75	−0.657	0.028	−0.006	0.9886	0.0663
		80	−0.654	0.028	0.100	0.9980	0.0654
		90	0.222	−0.048	−0.001	1.0122	0.0660
		175	1.556	−1.319	0.002	1.1270	0.1175
		180	1.416	−0.976	0.020	1.1261	0.1170
		185	1.294	−0.699	−0.008	1.1241	0.1158
		190	1.187	−0.503	−0.007	1.1212	0.1141
		195	1.094	−0.359	0.014	1.1172	0.1118
		200	1.014	−0.252	−0.004	1.1123	0.1090
		205	0.946	−0.123	0.003	1.1066	0.1057
		210	0.887	−0.084	0.000	1.1000	0.1019
		215	0.837	−0.084	0.003	1.0927	0.0979
		220	0.795	−0.056	0.005	1.0848	0.0936
		225	0.760	−0.028	−0.025	1.0763	0.0891
		230	0.731	0.011	0.003	1.0674	0.0846
		235	0.707	0.002	−0.015	1.0581	0.0802
		240	0.688	0.011	−0.012	1.0485	0.0760
		245	0.673	0.018	−0.010	1.0387	0.0724
		250	0.663	0.023	−0.008	1.0289	0.0693
		255	0.656	0.027	−0.007	1.0190	0.0670
		260	0.654	0.027	0.010	1.0092	0.0657
		265	0.655	0.030	−0.006	0.9996	0.0654
		270	−0.128	−0.011	0.000	0.9952	0.0656
		355	−1.600	−1.389	−0.003	0.9044	0.1117

Acknowledgments. The generous allotment by C. P. Sonett of computer time for this project is a greatly appreciated *sine qua non.* The other members of the TLOG (Tucson Lunar Origin Group) consortium are thanked for their help in the orientation and writing of this paper. We are also grateful for the assistance of Dominique Spaute in the calculations leading to Fig. 9. This work was funded by NASA contracts NASW-3516, NASW-3718, and NAGW-680. The Planetary Science Institute is a division of Science Applications International Corporation. This is PSI Contribution No. 211.

7. References

Boss A. P. and Peale S. J. (1986) Dynamical constraints on the origin of the Moon, this volume.

Giuli R. T. (1968) On the rotation of the Earth produced by gravitational accretion of particles. *Icarus, 8,* 301–323.

Harris A. W. and Kaula W. M. (1975) A co-accretional model of satellite formation. *Icarus, 24,* 516–524.

Harris A. W. (1977) An analytical theory of planetary rotation rates. *Icarus, 31,* 168–174.

Harris A. W. (1978) Satellite formation, II. *Icarus, 34,* 128–145.

Herbert F. and Davis D. R. (1984) Models of angular momentum input to a circumterrestrial swarm from encounters with heliocentric planetesimals (abstract). In *Papers Presented to the Conference on the Origin of the Moon,* p. 53. Lunar and Planetary Institute, Houston.

Ruskol E. L. (1960) The origin of the Moon: Formation of a swarm of bodies around the Earth. *Sov. Astron., 4,* 657–668.

Ruskol E. L. (1972) The origin of the Moon: Some aspects of the dynamics of the circumterrestrial swarm. *Sov. Astron., 15,* 646–654.

Safronov V. S. and Ruskol E. L. (1977) The accumulation of satellites. In *Planetary Satellites* (J. Burns, ed.), pp. 501–512. Univ. Arizona Press, Tucson.

Shmidt O. Yu. (1957) *Four Lectures on the Theory of the Origin of the Earth.* Academy of Science Press, Moscow. 139 pp.

Wasson J. T. and Warren P. H. (1979) Formation of the Moon from differentiated planetesimals of chondritic composition (abstract). In *Lunar and Planetary Science X,* pp. 1310–1312. Lunar and Planetary Institute, Houston.

Wasson J. T. and Warren P. H. (1984) The origin of the Moon (abstract). In *Papers Presented to the Conference on the Origin of the Moon,* p. 57. Lunar and Planetary Institute, Houston.

Weidenschilling S. J., Greenberg R., Chapman C. R., Herbert F., Davis D. R., Drake M. J., Jones J., Hartmann W. K. (1986) Origin of the Moon from a circumterrestrial disk, this volume.

Origin of the Moon from a Circumterrestrial Disk

S. J. WEIDENSCHILLING[1], R. GREENBERG[1], C. R. CHAPMAN[1],
F. HERBERT[2], D. R. DAVIS[1], M. J. DRAKE[2], J. JONES[2],
AND W. K. HARTMANN[1]

[1]*Planetary Science Institute, 2030 East Speedway, Suite 201, Tucson, AZ 85719*
[2]*Lunar and Planetary Laboratory, University of Arizona, Tucson, AZ 85721*

A scenario involving a circumterrestrial disk of small bodies remains a plausible context for considering the origin of the Moon. We present a model that emphasizes processing of material by collisions within a long-lived circumterrestrial disk, motivated in part by the compositional constraint that the Moon is depleted in metal. We address some of the earlier dynamical objections to origin from a disk, such as maintenance of sufficient angular momentum for material of the appropriate composition to accumulate and/or for the Moon to accrete. We find that the relevant physics is not nearly so well-defined as had been thought; indeed, there are several potential solutions. The disk may have originated from collisions of heliocentric planetesimals in near-Earth space, or perhaps from ejection of Earth mantle material by a huge impact onto the Earth. In our model, small, late-stage, Earth-zone heliocentric planetesimals, which are geochemically differentiated bodies with iron cores, undergo collisional and dynamical evolution. Eventually both their silicate and metallic fragments interact with the circumterrestrial disk. The disk captures the silicates, but not the iron, which is eventually depleted by other processes. For some distributions of heliocentric orbits, planetesimals impacting in the outer portions of the disk produce only a modest drag on the disk. Given a gradual mass input, such a disk may be able to maintain a quasisteady state through viscous spreading or other interdisk transport processes despite considerable losses of material and angular momentum to the Earth at its inner edge. Accretion of large bodies within the disk is inhibited by tidal torques and continuing planetesimal bombardment. Eventually the iron-enriched heliocentric population declines, and the diminished bombardment rate allows the Moon to accrete. Our model depends on the late-stage planetesimal population being dominated by small bodies, in contrast to cases dominated by large bodies, which favor impact-triggered lunar origin models. Independent of the validity of this whole scenario, these aspects of disk dynamics are applicable to a wide variety of lunar origin models and to other planetary processes.

1. Introduction: The Circumterrestrial Disk Model

The intensive exploration of the Moon during the Apollo era has thus far failed to yield a consensus on how the Moon originated. Some new perspectives on lunar origin have evolved that differ from the three theories that were being debated fifteen years ago: fission, capture, and coaccretion. The fission model (Darwin, 1880), in which the Moon breaks away from an unstable, rapidly spinning early Earth, can explain the major anomaly of lunar composition—low bulk content of metallic iron—because most material would be derived from the Earth's mantle. It also seems consistent with the observation that the Moon is still retreating from the Earth, but it has severe difficulties with the dynamics of the instability of the primordial Earth and the present orbital inclination. The idea of capturing the fully-formed Moon from heliocentric orbit (Gerstenkorn, 1955) sidesteps the question of how a body of that composition formed in the first place and appears to be an extremely improbable dynamical event. Although the coaccretion model (Shmidt, 1958; Ruskol, 1960, 1963, 1972a, b; Kaula and Harris, 1975) was found to have serious dynamical problems in the 1970s, variants and elaborations of that model have been developed and will be explored in this paper.

More recent thinking about lunar origin, motivated by Apollo data on lunar chemistry and new understanding of primordial processes of accretion and later-stage bombardment in the solar system, has evolved some hybrid models involving elements of three classical models. The lack of a sizable metallic core in the Moon motivated the tidal break-up or disintegrative capture model (Öpik, 1972; Mitler, 1975), in which a differentiated body passing the Earth is broken up by tidal forces, and the mantle material is preferentially captured into Earth orbit. This model may be thought of as involving elements of capture (the passing body is derived from heliocentric orbit), coaccretion (the disintegrated material must reaccrete in Earth orbit), and even fission (in the sense that the Moon is derived from iron-depleted mantle material "fissioned" from a differentiated body, although that body is not the Earth). Recent work (Mizuno and Boss, 1985) suggests that tidal disruption is unlikely during the short duration of a single pass.

Another hybrid model that has recently gained prominent attention (cf. Kerr, 1984) is the impact-trigger model (Hartmann and Davis, 1975; Cameron and Ward, 1976; Stevenson, 1984; Cameron, 1985), in which the Moon forms from the ejecta of an enormous impact with the Earth by a possibly Mars-sized body (see other papers in this volume). The model may be thought of as involving the fission concept, to the degree that the protolunar material is derived chiefly from the Earth's mantle, or the capture model, to the degree that ejecta that goes into Earth orbit is derived from the projectile. The process by which the ejecta subsequently accretes into the Moon has elements in common with the coaccretion model.

In late 1983, we formed an interdisciplinary consortium to address the question of lunar origin from the perspective of meeting both dynamical and cosmochemical constraints. Our model involving origin of the Moon from a circumterrestrial disk (CTD) represents an elaboration of early work on the coaccretion model and includes elements of other models. The chief cosmochemical constraint to be met is the relative lack of iron in the Moon, but we also consider other, more controversial compositional constraints (Section 2). We resort to a physical process to segregate metal from silicate in the protolunar material in the general scenario illustrated in Fig. 1. As the Earth grew by planetesimal bombardment, a circumterrestrial cloud of particles was created from impact-ejected Earth mantle material and/or planetesimal material captured into Earth orbit by mutual collisions within the Earth's sphere of influence. Regardless of its mode of origin, such a cloud would become flattened into a disk-shaped swarm by collisions among the particles. If such a CTD could be maintained, it would continue to capture heliocentric planetesimals (and might be augmented by additional Earth ejecta) until the bombarding population thinned, the Earth stopped growing, and the Moon finally accreted in orbit. If Earth mantle material dominated the swarm when accretion occurred, the geochemical traits of the Moon might approach those expected from simple fission. If the material were dominated by the debris from a single planetesimal (including projectile-enriched

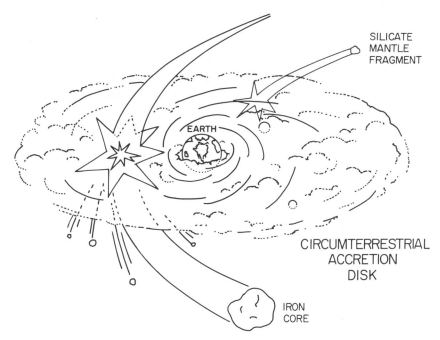

Fig. 1.   Schematic view of the circumterrestrial swarm. Compositional filtering occurs when the swarm is bombarded by heliocentric impacting planetesimals. Silicates are preferentially trapped, while the stronger iron cores pass through.

ejecta from Earth impact), we would have essentially a capture model. If small planetesimals dominated, our model would look more like a coaccretion model.

We emphasize processes involving small planetesimals in this paper for two reasons. First, there is as yet no conclusive understanding as to whether late-stage planetesimal populations are dominated by large, moon- or planet-sized bodies (cf. Wetherill, 1986) or, alternatively, by small (e.g., 1–10-km scale) bodies (cf. Greenberg *et al.*, 1985). The former idea, which is attrative for impact-trigger models for lunar origin, is under active study (see other papers in this volume); our studies offer an alternate perspective. Second, the evolution of small planetesimals yields a natural way to segregate iron from silicates within the protolunar swarm. If the heliocentric population of planetesimals has undergone a history of accretion, differentiation, and collisional breakup analogous to that experienced by the asteroids, then the CTD can act like a filter, preferentially capturing silicate fragments while passing the larger, stronger iron cores until planetary impacts and dynamical perturbations remove them from the inner solar system. We investigate whether the bombarding population's interactions with the disk, or tidal "shepherding" of lunesimals (planetesimals that formed the Moon), can inhibit accretion of the Moon for a duration sufficient to separate the silicates from the iron and for the remaining (iron) planetesimals to be depleted.

Our ideas concerning physical segregation of iron from silicate in a CTD are an outgrowth of earlier work (Ruskol, 1972a, b, 1977; elaborated on by Harris and Kaula, 1975). Wasson and Warren (1979, 1984) noted that instead of considering segregation of small grains of iron and silicates, it is preferable to consider much larger metallic and silicate fragments derived by planetesimals' collisional fragmentation, analogous to that of differentiated asteroids (cf. Greenberg and Chapman, 1983). Since many inner-belt asteroids exhibit differentiated mineralogy, it is plausible that many, possibly all, Earth-zone planetesimals were heated and differentiated by the same early processes. We consider how such a population of planetesimals would collisionally interact with a CTD and whether it is possible for such processes to produce our Moon.

We will show (Section 7) that the heliocentric, differentiated planetesimals collide with each other, fragmenting their silicate mantles but leaving their metallic cores more or less intact. Their orbits diffuse toward the Earth and interact with a preexisting CTD. We address two possible ways of emplacing the initial disk: ejection by a large impact into the Earth (Section 3), and mutual impacts among planetesimals in near-Earth space (Section 4). We show that such a disk can capture heliocentric silicate material but pass most of the iron cores, a process we term "compositional filtering." The bombardment of the CTD by the heliocentric bodies may suppress rapid accretion of the Moon to allow impacts on the Earth and other planets to deplete the metal (we also consider another potential solution, involving tidal and viscous transport processes, to the usual expectation that accretion would be too rapid in such a disk).

A major stumbling block for coaccretion models has been the angular momentum issue. Earth-crossing planetesimals contribute very little net specific angular momentum, and accretion of such bodies by the disk would be expected to lead to its collapse onto the Earth. However, we find that the population of Earth-feeding-zone planetesimals may have a more favorable balance of angular momentum, depending on the distribution of their heliocentric orbital elements. This may permit the disk to process several times its own initial mass without collapsing (Section 5). In addition, viscous and tidal transport processes within the swarm may help to maintain it in orbit (Section 6). We do not envision a stable, static disk, but rather a dynamically evolving one that nevertheless maintains a quasiequilibrium form.

Our work has not yet yielded a comprehensive, definitive model for the origin of the Moon. We have taken some physically plausible starting conditions, have considered what seem to be the most significant processes, and have made progress in addressing the chief objections that have been raised to this general scenario. We are by no means confident that our starting conditions are necessarily correct, and we still have doubts about whether the disk can be adequately maintained. Nevertheless, we have demonstrated that a CTD scenario remains viable. Processes involving a CTD are highly relevant to other models that involve swarms, including both the disintegrative capture and the impact-trigger models.

2. Chemical Constraints

Our discussion of compositional constraints on the origin of the Moon is fairly brief and simple, but not because we consider chemistry unimportant. Many of the current models could be very effectively supported or rebutted if the chemical composition of the Moon were well known. However, lunar bulk composition is extremely difficult to determine since the samples that come from the lunar interior are not undifferentiated (i.e., unprocessed) but, in most cases, are the products of materials that are themselves differentiated (i.e., the cumulates of a lunar magma ocean). Even those samples that appear to be derived directly from the mantle (e.g., lunar volcanic glasses) are not totally unambiguous. For our purposes here, we do not make detailed chemical arguments, but simply state what we believe to be the firmest geochemical constraints and show that filtering in a circumterrestrial swarm is consistent with them. For a more detailed discussion of geochemical constraints, see Drake (1986).

The circumterrestrial filter model involves the processing of currently unconstrained mass fractions from these principal sources through a swarm of debris maintained in Earth orbit. These sources are (1) the Earth's mantle following formation of most of the core, (2) differentiated planetesimals (similar to differentiated meteorite parent bodies or asteroids) of characteristic dimension 1–10 km, and perhaps (3) primitive undifferentiated material.

The chemical composition of the Earth and Moon are known within some limits; the average composition of the various primitive materials that ultimately fed these bodies is unconstrained but *may* be similar to the chondritic meteorites available in our collections. The composition of any contributing differentiated planetesimals is constrained only by the unknown contributions to the present mass of the Moon from the other two reservoirs according to the equation:

$$X_{em}C^i_{em} + X_{pm}C^i_{pm} + (1-X_{em} - X_{pm})C^i_{dm} = C^i_{\mathbb{C}} \tag{1}$$

where X = mass fraction, C^i = concentration of chemical element i, em = differentiated Earth mantle material, pm = primitive material, dm = differentiated planetesimal material, and $\mathbb{C}$ = the Moon.

At first inspection, the problem appears to be undetermined. However, additional constraints such as the constancy of elemental ratios (e.g., K/U, Rb/Cs, CI ratios of refractory elements) can, in principle, permit a solution to be obtained. In practice, there are a relatively small number of elements and elemental ratios with known values, and some of these are subject to considerable uncertainty, as we shall see below.

A brief summary of compositional similarities and differences between the Earth and Moon is given in Table 1. No attempt is made to arrive at a detailed bulk composition for each planetary object. Rather, we discuss those parameters that may be decided with reasonable precision.

*TABLE 1. Compositional Similarities and Differences Between the Upper Mantle of the Earth and the Mantle of the Moon.**

		Earth	Moon
Oxygen isotopes		same	same
Uncompressed density, bulk planet (gcm^{-3})		4.45	3.34
Siderophile element depletions relative to CI			
	W	22 ± 10	22 ± 7
	Ga	4–7	20–40
	P	43 ± 10	115 ± 25
	Mo	44 ± 15	1200 ± 750
	Re	420	10^5
Refractory element abundances	U(ppb)	20	35†
Volatile/Refractory element ratios	K/U	10^4	2500
Volatile element ratio relative to CI	Cs/Rb	1/7	1/2
Mg/(Mg + Fe) ratio		0.89	0.80‡

*After Drake (1986).
†Cannot rigorously rule out a bulk concentration in the Moon as low as 20 ppb.
‡A few estimates based on the most magnesian terrae rocks approach 0.87.

Oxygen isotopes

The Earth and the Moon fall on an identical oxygen isotopic mass fractionation line. All other sampled solar system objects (except the enstatite chondrites and aubrites) fall off this line (Mayeda and Clayton, 1980). The association of many undifferentiated meteorites with the asteroid belt (Greenberg and Chapman, 1983) and the increasingly strong evidence for a Martian origin of the SNC meteorites (e.g., Bogard and Johnson, 1983; Becker and Pepin, 1984) suggest that the Earth and Moon are made predominantly of material originating in the vicinity of 1 AU or closer to the sun. This strong constraint must be satisfied by any model of lunar origin. In particular, the impact-trigger model must either derive most of the lunar material from Earth's mantle ("fission" rather than "capture"), or else the impacting body must have the same oxygen isotopic composition as the Earth. The plausibility of the latter condition depends on the isotopic homogeneity of the terrestrial planet region, which is not known.

Uncompressed density

The low mean uncompressed density of the Moon (3.34 g/cm^3) compared with the Earth (4.45 g/cm^3) has long indicated that the Moon is impoverished in metal, and this has been the principal impetus for the fission model and its variants. If, as argued above, the Earth and Moon were formed from chemically similar materials, a viable model for the Moon's origin must be capable of fractionating metal and silicate.

Siderophile elements

Siderophile elements, which are depleted in the lunar mantle by various factors, place a stronger constraint on the maximum amount of metallic Fe-Ni in the Moon. If the Moon originally had chondritic abundances of refractory siderophile elements (which may be decoupled from metallic Fe-Ni; Drake, 1983), the maximum amount of metal permitted in the Moon is 5.5 wt %, with a more probable value around 2 wt % (Newsom, 1984). If the abundances of refractory siderophiles were subchondritic, the amount of metal is correspondingly reduced.

Refractory elements

It is convenient to use uranium as a typical refractory element since its abundance may be estimated from both geophysical and geochemical considerations. Best estimates are 20 ppb U for the Earth and 35 ppb U for the Moon. However, identical abundances of U in the Earth and Moon cannot be rigorously excluded, and we will not emphasize any possible differences.

Volatile elements

It has been known since the Apollo program that the Moon is impoverished in volatile elements compared with the Earth. For example, there is no evidence for lunar water and the K/U ratio in the Earth is significantly higher than in the Moon (e.g., Taylor, 1982). This volatile depletion cannot be due to simple volatility in high temperature processes. Volatility considerations alone would suggest similar or slightly greater depletions of Cs relative to Rb in the Moon compared with the Earth, if they originally were made of the same material. However, Kreutzberger *et al.* (1986) have shown that the Cs/Rb ratio in the Earth is approximately one-seventh of the CI ratio, while the Cs/Rb ratio in the Moon is approximately one-half of the CI ratio. This observation would appear to rule out simple fission models. It is consistent with the impact-trigger model, provided that ejecta from the projectile constitutes 25–50% of the Moon, and that the projectile's oxygen isotopes (above) are such that the ejecta mixture has oxygen isotopes indistinguishable from Earth's.

Mg/(Mg + Fe) atomic ratio

This important ratio appears to be different in the Earth and the Moon, with the lunar mantle containing almost twice as much FeO as the upper mantle of the Earth. It is possible that the Earth has lost FeO to the core over geologic time (Jagoutz and Wänke, 1982), but there is no evidence for this (Newsom and Palme, 1984). Again, the difficulty in comparing the FeO contents of the Moon and the Earth is that, while we known the Mg/(Mg + Fe) ratio of the Earth's mantle from measuring nodules of mantle material entrained in basalts, lunar basalts come from materials that have been processed and whose Mg/(Mg + Fe) ratio may be different from the Moon as a whole.

In our circumterrestrial filter model, we envision that the Moon formed from materials that accreted into planetesimals within a few tenths of an AU from the Earth. It is reasonable to assume that such planetesimals were similar in bulk composition to the Earth, had a significant amount of metallic iron, and had a very similar oxygen isotopic signature. Of course, the planetesimals that predominantly formed the Moon and the Earth were located in a part of the solar system somewhat distinct from the region beyond 2 AU from which we evidently derive most meteorites today; hence it would not be surprising if there were not samples among our meteorite collections of the precise Earth-zone materials. On the other hand, cosmochemical expectations are that Earth-zone material would be roughly chondritic in bulk chemistry, and there have been suggestions that certain materials, e.g., enstatite meteorites (Wasson and Warren, 1984), were originally formed near 1 AU. There are some intriguing similarities between some compositional traits of the Moon and those of some differentiated meteorites.

As discussed above, our model requires that the planetesimals be differentiated into iron and silicate fractions, hence they must have been heated to at least the Fe-FeS eutectic ($\approx 1000°$ C). These bodies would have had iron cores, silicate mantles, and perhaps basaltic crusts. We do not explore early solar system heat sources in detail here, but note that either electromagnetic induction (Herbert and Sonett, 1980) or ^{26}Al (Urey, 1955) may be capable of raising small planetesimals to high temperatures. Once differentiated and cooled, our planetesimals deliver material to the CTD, and the probability of delivery of a given type of material is very roughly inversely proportional to its depth in the planetesimal. Crustal materials should be quickly and readily stripped while core materials should be protected—both by silicate mantles and by the innately greater strength of iron metal. To the degree that this is true, a CTD that is dominated by captured debris will be depleted in metal and siderophile elements, and may possibly be enriched in those elements that are concentrated in basalts—FeO, KAl_2O_3, U, Th, REE.

Thus in a general way we can meet the constraints of lunar chemistry with a conceptually simple model. A Moon formed from such a CTD (1) should have $^{18}O/^{17}O/^{16}O$ ratios that are very similar to the Earth's; (2) should be depleted in metal and siderophiles; (3) may possibly be enriched in basaltic melt components such as FeO and U; and (4) the primordial heating of small, low-gravity planetesimals could have driven away volatiles. Therefore, the circumterrestrial compositional filter model can meet the strongest chemical constraints without difficulty.

3.   Establishing the CTD: Earth Impacts

In order to model the compositional filter effect of a swarm of material surrounding the proto-Earth, we need to understand how such a swarm could have originated. The depletion of inert gases in the Earth probably excludes the possibility that the Earth formed by gravitational collapse of a portion of the solar nebula; therefore we cannot assume the swarm was a disk-like analog of the solar nebula, such as postulated for the primordial giant planets and their satellite systems. Instead, the disk must have been composed of solid particles, and could have been added at a late stage in the Earth's formation.

Impacts of one or more large planetesimals into the Earth could have played a major role in establishing a circumterrestrial swarm. Although a very large impact might have yielded the Moon more or less directly (the impact-trigger model) or conceivably even caused the Earth to spin up to rotational instability and shed mass into orbit, we emphasize here the role of somewhat smaller impacts. Even though we assume that much of the mass of Earth-zone planetesimals is in kilometer-scale bodies, we might also expect a significant population of bodies intermediate in size between those planetesimals and the nearly grown Earth (cf. Greenberg *et al*, 1978). Such a body, when it impacts the Earth, could perhaps create the required

CTD (of ~0.1 lunar mass, for example) in a manner equivalent to that envisioned for creation of the whole Moon by an even larger projectile in the impact-trigger model. Another potential role of intermediate-sized planetesimals, if there were enough of them of the appropriate sizes, would be the continual repopulation of a CTD by Earth mantle ejecta at a sufficient rate to balance dissipative losses of disk material. A major problem with making an initial disk of less than one lunar mass by a small "splash" is the likelihood that any such disk would extend outward only a few Earth radii. This small size makes it difficult to add much more mass without having the CTD collapse onto the Earth again (cf. Section 5).

If we consider, in the context of our model, that the minimum mass of material to be injected into Earth orbit is ~0.1 lunar mass, then we need a smaller impactor than is required to produce a whole Moon, implying a smaller number of large bodies in the planetesimal population. How large a projectile is required and how large a crater it might produce in the Earth's mantle depends on highly uncertain scaling factors, which must be extrapolated many orders of magnitude from scales for which we have little even indirect evidence. It also depends on the virtually unknown efficiency with which ejected debris may be placed in Earth orbit (cf. Hartmann, 1986; Melosh and Sonnett, 1986). At least one impact sufficient to produce our disk needs to occur during approximately the interval in which the Earth accreted the last few percent of its mass. If there were more than a few impacts of this scale during this interval, then for plausible size distributions it would be probable than an even larger impact would occur, sufficient to emplace a more massive disk, perhaps large enough to form a whole Moon. While this may have occurred if the size distribution were populated by enough large bodies, this scenario blends into the impact-trigger model, which we are not addressing in this paper. If disk accretion processes operate as we discuss later, the apparent lack of disk-produced moons around other planets such as Venus or Mars may imply that impacts sufficient to produce large disks were uncommon. Thus, the idea of emplacing an initial CTD by Earth impact faces similar questions to those facing the impact-trigger model itself. The latter model allows, but does not require, interactions of the disk with additional material added later. The major difference between scenarios involving a small initial swarm and those forming the Moon in a single event is in the assumed size distribution of planetesimals. Another potentially important difference is in the angular momentum contributed by such impacts. The impact-trigger model assumes (or at least allows) that essentially all the angular momentum of the Earth-Moon system is the stochastic contribution of the giant impact (Cameron and Ward, 1976; cf. Hartmann and Vail, 1986). Formation of a less massive disk by a smaller impact or impacts requires that the Earth had acquired this angular momentum earlier, perhaps due to a systematic process during accretion (Section 5, below), and was spinning rapidly at the time of the impact (or impact). This spin would aid the emplacement of "prograde" ejecta into Earth orbit, and would tend to transfer angular momentum to the disk (cf. Dobrovolskis and Burns, 1984). Thus such a disk would tend to be prograde, although the magnitude of this effect is uncertain.

One of us (WKH) has considered the possibility of building up a disk by cumulative effects of innumerable smaller impacts, each ejecting a small swarm of debris into Earth orbit. In principle, such impacts could maintain a continuing flow of material through the disk, even if much of the material fell back onto the Earth on a short timescale. Indeed, under a sufficiently high bombardment rate, the steady-state (although fluctuating) population of debris on high ballistic trajectories could be thought of as an initial swarm. Hartmann (1985) has studied velocity distributions of ejecta from laboratory impact craters in powders. To the degree that these can be scaled toward much larger-scale cratering events, we can estimate the fraction of ejecta from large impacts that is ejected between circular velocity and escape velocity and hence is potentially available for the swarm. Preliminary calculations suggest that for any plausible size distribution and flux history of Earth-zone planetesimals, the maximum mass of such a "steady-state swarm" would be orders of magnitude less than the ~ 0.1 lunar mass we have adopted. These preliminary calculations depend upon uncertain assumptions of the size distribution, efficiency of orbital insertion, and residence time in the swarm, and do not include effects of Earth's rotation.

We tentatively conclude that any Earth-impact origin of a CTD would be due to the largest collision(s), and that successive augmentations by smaller subsequent impacts would be a minor effect. Both the initial size distribution and the detailed impact and ejection processes are highly uncertain. The dimensions of a CTD derived from the Earth may be too small to be compatible with other elements of our scenario described in later sections.

4. Establishing the CTD: Collisions Between Planetesimals

Analytic considerations

If the Earth formed by the accumulation of mass from impacting planetesimals, then doubtless there were also mutual ("free-free") collisions among heliocentrically orbiting planetesimals (those not bound to the Earth), as well. Ruskol (1960, 1972b, 1977) pointed out that such collisions occurring within Earth's sphere of influence could result in capture of matter into a geocentric swarm. If two planetesimals in heliocentric orbits collide, they dissipate some fraction of their kinetic energy. After the collision, some or all of the mass involved may have less than the local escape velocity (more precisely, the Jacobi parameter may be too large for escape from the Hill sphere) and so remain bound in the Earth's vicinity. This matter may stay in orbit or fall onto the planet, depending on the circumstances of the collision. This mechanism was implicitly accepted by Harris and Kaula (1975) as the starting point of their coaccretional model for lunar origin. In this section, we discuss in more detail the formation of such a CTD and its evolution before its accretion into one or a few proto-Moons.

The rate of direct impacts onto Earth is proportional to the number N of planetesimals entering the Hill sphere per unit time, while the rate of mutual collisions is proportional to N^2. Thus, collisional capture into a CTD was presumably more effective in the earlier stages of Earth's accretion, when planetesimals were more numerous. The mass involved in a collision of two planetesimals of diameter d is proportional to d^3, while the probability of their collision scales as their geometric cross section, $\propto d^2$. Suppose that the planetesimals have a power law size distribution, such that the cumulative number of bodies larger than size d is proportional to d^{-q}. Then bodies in each logarithmic increment of size contribute equal amounts of mass to the swarm if q = 3. If q is larger, as we assume in our scenario, the dominant contribution is from the smaller bodies. We will first review earlier work on mutual planetesimal collisions involving a smaller q, in which case the largest bodies contribute most of the mass.

Theoretical size distributions computed by Safronov and his coworkers (Zvyagina and Safronov, 1972; Pechernikova *et al.*, 1976) correspond to values of q $\simeq$ 2.5, in which case the largest planetesimals would contribute most of the mass. For this case, the mass input is not continuous but stochastic, and is dominated by the few largest collisions within Earth's sphere of influence. Such a scenario resembles capture theories in its dependence on infrequent events putting bodies in orbit around the Earth. The angular momentum of a disk formed in this way is also the result of the circumstances of collision, which is essentially a chance event. In this regard, this process is similar to the impact-trigger hypothesis for lunar origin; however, it does not require as much mass to be involved because it delivers both mass and angular momentum more efficiently to the Earth-orbiting swarm. Ruskol (1972b) argued that after a CTD begins to form, "free-bound" collisions (those between heliocentric and geocentric bodies) produce a steeper size distribution in the CTD than that of the original population, which favors capture of smaller bodies. She proposed that this effect could explain the Moon's iron-poor composition, if differences in strength caused silicate bodies to be systematically smaller than iron bodies. A similar model was proposed by Wasson and Warren (1979), and we invoke a variant of the same effect below. However, Ruskol's (1977) own estimate of the steepened size distribution in the CTD is only q $\simeq$ 3 (i.e., an equal mass contribution from each interval of planetesimal size), so an upper size cutoff in the CTD is required that is much smaller than in the heliocentric swarm in order to explain the Moon's composition. That cutoff must be at a small enough size so that the most massive (and iron-rich) planetesimals pass through the disk without being stopped. That scenario resembles our own compositional filter model. However, the Safronov size distribution, dominated by large bodies, leaves very little mass in small bodies available to form the Moon in the late stage of Earth's accretion. Our scenario involves a very different size distribution, dominated by relatively small bodies, in both the heliocentric and geocentric swarms.

We now proceed to a more detailed investigation of the formation of a CTD by planetesimal collisions. It is important to know the efficiency of this process, i.e., what fraction of collisions within the sphere of influence result in capture into geocentric orbits, and whether those orbits have the appropriate properties (e.g., are prograde or retrograde). The first question can be addressed analytically for certain simplifying assumptions. Ruskol (1972b) considered only collisions between bodies of equal mass, and showed that for a given approach velocity v_∞, capture is more probable for collisions deeper within the sphere of influence, i.e., closer to Earth's surface. Here we consider bodies of unequal mass, but still assume that all collisions act as though precisely head-on and that they result in coagulation. An approximate condition for capture is that the center of mass of the two bodies moves at less than the escape velocity at the collision distance (this condition must be modified in the outer part of the sphere of influence, where solar perturbations are important). This guarantees that some mass is captured, even if shattering rather than coagulation is the outcome of the collision. It can be shown that this condition requires that the ratio of masses of the two bodies (m_2/m_1) exceeds a critical value:

$$\frac{m_2}{m_1} > \frac{1 - K}{1 + K} \tag{2}$$

where $K = [(R_\oplus/R_c)/(R_\oplus/R_c + v_\infty^2/v_e^2)]^{\frac{1}{2}}$. Here $R_\oplus$ is the Earth's radius, R_c the distance from the planet's center at which the collision occurs, v_e is the escape velocity from the planet's surface, and $m_2 \leq m_1$. Results are plotted in Fig. 2 for several values of R_c. We see that the commonly quoted "rule-of-thumb," that capture requires a body to collide with another equal to its own mass, is approximately correct, but under favorable circumstances, especially for low v_∞ and R_c, capture is possible for rather small mass ratios (~0.1). The critical mass ratio for capture can be several times higher if m_2 is in geocentric orbit before the collision (Fig. 3).

This criterion does not distinguish whether the captured mass falls onto the planet or attains geocentric orbit, or whether such orbits are prograde or retrograde. These questions cannot be answered analytically. Ruskol (1972b; see also Safronov and Ruskol, 1977) assumed that the geocentric swarm would have specific angular momentum equal to that of "a conglomeration of particles fixed rigidly to the Hill sphere" (i.e., in uniform rotation with angular velocity equal to Earth's orbital motion). This assumption yields prograde orbits for captured material at a mean distance of about 20 $R_\oplus$. However, our numerical studies, described below, show that the actual situation is much more complex. The magnitude, and even the sign, of the mean geocentric angular momentum of the CTD depend on the distribution of heliocentric orbits of the Earth-approaching planetesimals.

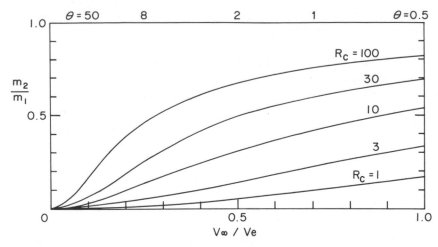

Fig. 2. Mass ratio yielding capture for head-on collision of two unbound planetesimals at various distances within sphere of influence (R_c in Earth radii) as a function of approach velocity V_∞. For low V_∞ (~1 km/sec) capture occurs for mass ratios ~0.1. (θ is the "Safronov number," equal to $1/2\, V_e^2/V_\infty^2$).

Numerical simulations

Planetesimals approaching Earth are subject to perturbations by the sun's gravity, and must be studied in the context of the three-body problem. When approach velocities are low, their trajectories are complex, with loops or cusps; the magnitude and direction of geocentric velocity within the sphere of influence cannot be predicted analytically (cf. Herbert *et al.*, 1986). We performed simultaneous integrations of many trajectories started well outside the sphere of influence. Initial orbits were

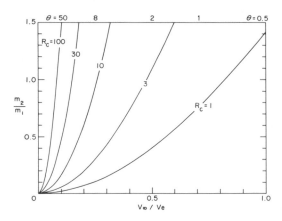

Fig. 3. Mass ratio yielding capture for a heliocentric planetesimal (m_1) overtaking a geocentric swarm particle (m_2) in circular orbit. The hyperbolic orbit of m_1 is assumed tangent to the circular orbit of m_2 at distance R_c. For $V_\infty = 1$ km/sec ($V_\infty \sim 0.1\, V_e$), capture results for m_2/m_1 greater than a few tenths.

both inside and outside that of Earth, so two opposing streams of particles, overtaken and overtaking, passed through the sphere of influence. The program can integrate up to 300 trajectories simultaneously. In order to provide a reasonable collision rate, the calculations were two-dimensional, with all orbits in the ecliptic plane. Also, the effective collision cross section was enhanced, with all approaches closer than 4×10^4 km scored as "collisions." In this numerical model, such collisions resulted in coagulation. This allowed us to determine whether capture into Earth orbit occurred (in reality, such a collision would produce a swarm of debris). The merged particle was assigned the position and velocity of the center of mass of the two particles at the time of collision. Collisions within the sphere of influence were scored as free-free, free-bound, or bound-bound, depending on the precollision status of the colliding particles ("free" means heliocentric and "bound" means circumterrestrial, determined by the value of the Jacobi parameter). For simplicity, all particles had the same initial mass.

Three cases were considered; the number of collisions for each case was limited by available computer time. The first two cases each had a uniform surface density of planetesimals in heliocentric orbits (semi-major axes were chosen randomly from a uniform probability distribution), the first case with approach orbits of zero eccentricity, and the second case with a uniform distribution of eccentricities in the range from zero to 0.05. The zero-eccentricity case yielded 27 captures from 78 free-free collisions (35%), g prograde and 18 retrograde. The second case, with uniform distribution of eccentricity, yielded 14 captures out of 68 collisions (21%), evenly divided between prograde and retrograde orbits. Thus, these two simple orbit distributions showed no tendency to form a prograde geocentric swarm by mutual collisions. The third case corresponded to the "depleted deep crosser" distribution studied by Herbert et al. (1986; see Section 5, below) for its effect on an already-formed CTD. This case features a distribution of eccentricities peaked at 0.05, with a nonuniform distribution of semi-major axes such that Earth-crossing orbits are depleted (due to the planet's growth) by a factor of 10 relative to non-Earth-crossers. This distribution emphasizes the nearly tangential orbits that contribute net angular momentum (Harris, 1977). Of 15 free-free collisions, 4 resulted in capture (27%). All captures were prograde, with mean geocentric angular momentum corresponding to a circular orbit at 30 Earth radii. The small number of events make this figure subject to a large statistical uncertainty, but it is consistent with the results of Herbert et al. for "free-bound" impacts on a CTD, which show a significant net prograde angular momentum (though still less than local circular values) for this distribution. More detailed modeling of collisional capture is clearly needed.

Implications of collisional capture

Our concept of the geocentric disk differs from Ruskol's in several ways. If the Safronov-type power law size distribution prevailed, the disk would be dominated not by the smallest, but rather by the largest bodies colliding within Earth's sphere of influence. Such a disk would not be a steady-state feature during Earth's accretion, but a transient event. The high relative velocities (eccentricities) of planetesimals in Safronov's cosmogony implies that any additional material added to the CTD would have brought in no systematic angular momentum. "Accretion drag" due to this influx of low angular momentum material would cause such a disk to collapse rapidly toward the Earth. Many such disks might have formed sequentially, only to collapse into the planet by accretion drag. Only a disk that formed rather late in Earth's accretion could have survived to form the Moon in that scenario. It is not clear how such a process could have produced an iron-poor composition, since the largest planetesimals would not be depleted in iron.

In our model, the mass input to the CTD is dominated by small bodies, which can achieve collisional fractionation of iron and silicates. If the initial emplacement of the disk was due to many collisional capture events, then the incoming material must have brought with it some amount of systematic prograde angular momentum in order to achieve geocentric orbit. Without such a systematic component, the initial disk could have been emplaced by a chance event, such as an Earth impact or the improbable collision of two large planetesimals. However, we will show in the next section that significant silicate enrichment of such a disk to match the lunar composition probably requires the later-added mass to bring in some systematic prograde angular momentum, as well. That condition in turn constrains the allowed orbits of the incoming planetesimals.

5. The Angular Momentum Budget

How much accretion of heliocentric material can the CTD sustain before collapsing onto the Earth due to the effective drag of this collisional accretion? The problem of the stability of bodies in geocentric orbit subject to a drag force arising from impacting heliocentric projectiles has been recognized by earlier workers, but not studied in detail. Ruskol (1963) presented only a qualitative investigation of how to maintain a geocentric disk. Harris and Kaula (1975) presented a simplified analytic model for the rate of orbital decay of geocentric bodies; they concluded that collapse into the Earth would be averted because a CTD would coalesce into one or a small number of bodies when the satellite/planet mass ratio reached 10^{-4}, and tidal interactions would supply angular momentum from the Earth, which they assumed to be rapidly rotating. On the other hand, our compositional filter requires that the disk avoid coalescing (as described in Section 7) if iron-silicate fractionation occurred in geocentric orbit. Wasson and Warren (1979, 1984) primarily addressed the geochemical aspect of the compositional filter; they did not explore the dynamical difficulties of maintaining the filter in geocentric orbit.

We present here a summary of a numerical study (elaborated on by Herbert *et al.*, 1986) of the orbital evolution of a CTD bombarded by a population of projectiles moving on heliocentric orbits. Herbert *et al.* used the formalism of the restricted three-body (sun-Earth-planetesimal) to calculate the angular momentum and kinetic energy that could be delivered to bodies orbiting the Earth at different distances for various heliocentric orbital element distributions. They then estimated the orbital evolution of the geocentric population in response to this flux, and also compared those numerically-based results with the analytic model of Harris (1977).

The orbit calculations followed the formalism of Giuli (1968), with the exception that orbits intersecting the circumterrestrial disk, rather than just the Earth, were of interest. The objective was to map potential delivery of geocentric angular momentum and energy over a swarm that might range from 1–150 $R_\oplus$. The study was confined to a two-dimensional problem that was assumed to be a reasonable representation of the average three-dimensional case. An inventory of all trajectories that approach within 120 $R_\oplus$ was generated as a function of v_0, a velocity parameter at the largest geocentric ring distance.

Using this database, which can be thought of as a mapping of heliocentric orbits into near-Earth angular momentum and energy, they calculated the angular momentum delivered to a CTD for several distributions of initial heliocentric orbital elements. Figure 4 shows the average specific angular momentum, h(r), brought in by planetesimals as a function of distance from Earth for the two extreme distributions: (1) a distribution uniform in both a and e that contributes the smallest net angular momentum and (2) a distribution uniform in a and with a symmetric "triangular" distribution between 0.03 and 0.08 in e but with Earth crossers depleted by a factor of 10. In all cases, h(r) is less than the angular momentum locally required to maintain circular orbits; for orbital element distribution (1) h is less than 10% of the circular orbit momentum. However, for case (2), the average angular momentum delivered by impacting planetesimals is much greater because heliocentric orbits with perihelia or aphelia that are nearly tangent to Earth's orbit deliver significantly more angular momentum than those that impact from deeply Earth-crossing orbits. The first case delivers an average angular momentum similar to that used by Safronov and Ruskol (1977), who assumed that the average angular velocity of captured material was the same as the Earth's angular velocity about the sun. Their assumption led them to require that all capture occurred outside 80 $R_\oplus$, in accord with Fig. 4. Note, however, that case (2) delivers much more angular momentum and allows much less stringent constraints on the capture distance for the CTD.

Since in each case the angular momentum delivered by planetesimals from heliocentric orbits is less than that of particles moving on circular geocentric orbits, the CTD will collapse toward the Earth as it accretes matter from heliocentric orbit. At any geocentric distance, and for any initial value of angular momentum (h < h_c), there is a minimum orbital radius r* (corresponding to a circular orbit) to which a particle could evolve. Curves of constant r* are shown in Fig. 4. This minimum final orbit is calculated based only on the angular momentum brought in by the planetesimal and does not include orbit shrinkage due to the increasing

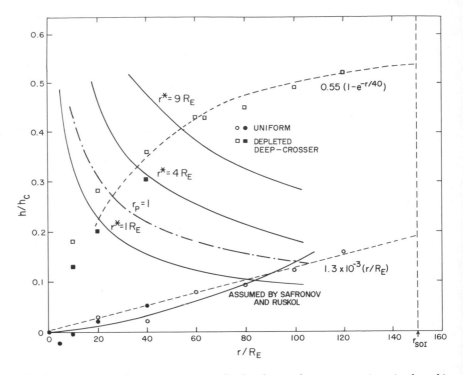

Fig. 4. Average angular momentum normalized to the angular momentum in a circular orbit brought in by particles initially far from Earth as a function of geocentric distance. Two distributions of heliocentric orbit elements for the incoming planetesimals are used, as well as the incoming angular momentum distribution assumed by Safronov and Ruskol (1977). Dashed lines are analytic approximations to the data. Also shown is the size of the circular orbit, r*, having the same angular momentum as larger but noncircular orbits (i.e., h/hc < 1). See Herbert et al. (1986) for details of derivation.

mass of the Earth or the angular momentum contribution of the initial disk. In that sense, the orbits (r*) shown in Fig. 4 are only rough estimates of the final orbits in the CTD.

Several inferences can be made from Fig. 4. First, the mean angular momentum varies significantly depending on the orbital element distribution of the impacting planetesimals. Case (2), which delivers the largest angular momentum to the CTD, also seems on qualitative grounds to be a plausible representation for planetesimal orbits late in the growth of the Earth when Earth-crossing material was largely depleted. Quantitative studies of such distributions have not yet been carried out, but this relatively quiescent scenario, with low eccentricities, is consistent with our assumed size distribution, which is dominated by small bodies. A second inference is that even for favorable orbital element distributions the material likely to remain in orbit about Earth for the longest time is that captured at large geocentric distance, >30 $R_\oplus$. Finally, a CTD will be more nearly stable once the Earth is close to its final size. Herbert et al. (1986) show that the flow of material down to the Earth through such a CTD permits it to process several times its own mass before it requires replenishment from the outside.

The compositional filter model requires that the CTD builds up at least a lunar mass of filtered material, which may require several cyclings of material through the disk. (Some of the segregation of silicates from iron could be achieved directly from the effects of free-free collisions, if the disk is repopulated by material derived from such collisions.) Figure 4 provides information about the possibility of deriving a Moon from a CTD of the form studied by Herbert et al. We require an optimal choice of parameters such that the Moon forms beyond the Roche limit. Consider an initial swarm created when the Earth had already grown to >90% of its final mass, so shrinkage of r* due to growth of Earth's mass would be negligible. Assume the disk had 10% of its final mass and had an effective radius of 30 $R_\oplus$ (i.e., 30 $R_\oplus$ represents some appropriate weighting of the impact probability vs. radius; the actual size of the disk must be somewhat larger). If the subsequently captured material (90% of the final mass) impacted directly on this disk, it would have $\approx$25% of the angular momentum needed to remain near 30 $R_\oplus$. Averaged with the original disk, which had $h/h_c = 1$, the total disk would have had $h/h_c \simeq 0.33$. Thus, the total disk would collapse to 30 $R_\oplus \times (0.33)^2 \simeq 3 R_\oplus$, which is just beyond the Roche limit (if the impacting material brought no net angular momentum, the disk would collapse within the Roche limit if it accreted more than twice its initial mass). More realistically, we might expect the inner part of the disk to collapse onto the Earth, while the outer part remains at a greater distance. The situation may be more favorable than this example, in which we have considered only impacts directly onto the disk; there should be an additional component of material with higher angular momentum due to free-free planetesimal collisions at larger geocentric distances. Thus while drag due to accretion of heliocentric material onto the CTD is an impediment to forming the Moon, the angular momentum problem is not so severe as has been thought in the past (the major obstacle might be the initial emplacement of a CTD several tens of $R_\oplus$ in size). In the next section, we show how viscous redistribution of angular momentum could have stabilized the CTD still further.

6. Viscous and Tidal Evolution of the CTD

The CTD cannot be maintained in a steady state orbit by a purely local balance of angular momentum in the event of continuing heliocentric bombardment, although it could have sustained accretion several times its mass. That accretion could build up enough material to form the Moon, and also satisfy lunar compositional constraints.

However, addition of too much mass would cause the entire disk to collapse onto the Earth. The situation can be improved by transport of both mass and angular momentum within the disk due to its viscosity. This is an effect that has been studied recently in the context of planetary rings, but that was not considered in earlier studies of coaccretion models for lunar origin. A general characteristic of a differentially rotating Keplerian disk is an inward flow of mass combined with an outward flow of angular momentum (i.e., the outermost part of the disk tends to expand). In principle, a steady state is possible if mass is preferentially added to the outer part of the disk. A minimum necessary condition for a steady state is that the specific angular momentum of the added mass equals that of the mass lost from the inner edge of the disk. The lost mass has specific angular momentum $\sqrt{GM_\oplus R_\oplus}$, (where $M_\oplus$ = Earth's mass), characteristic of a surface-grazing orbit. Inspection of Fig. 4 shows that the planetesimal orbital distribution in case 2 has roughly this amount of systematic prograde angular momentum for impacts on the disk at $\simeq 20$ $R_\oplus$. A more sophisticated integration over the disk yields a required size of $\simeq 30$ $R_\oplus$. Thus a disk of that size or larger (most of the "target" area is in the outer part of the disk) can exist in a steady state if viscous transport is effective.

What viscosity would be required, and what physical properties of the disk are implied? The viscous evolution timescale for a disk radius R is $\sim R^2/\nu$, where ν is the kinematic viscosity. Let us suppose that the required disk lifetime is $t_d \sim 10^6$–10^7 yr, and R $\sim 10^{10}$ cm. Than $\nu \sim 10^6$ cm^2/sec ($\pm 1/2$ order of magnitude) gives a viscous evolution time of the same order as the disk lifetime. If the total mass processed through the disk exceeds the instantaneous disk mass, or the disk lifetime is shorter, then ν needs to be higher by the same factor to prevent collapse of the disk. Goldreich and Tremaine (1978) give a formula for viscosity of a particulate disk:

$$\nu \simeq \left(\frac{c^2}{\Omega} \right) \frac{\tau}{(1+\tau^2)} \tag{3}$$

where c is the mean random velocity of the particles, Ω the orbital frequency, and τ the optical thickness of the disk (note that ν is maximized at $\tau \sim 1$ for a given c and Ω). For $\Omega \sim 10^{-5}$/sec and $\tau \sim 1$, a viscosity of $\sim 10^6$cm^2/sec requires that c be a few cm/sec.

The dominant source of random velocities is mutual gravitational scattering of particles in the disk. The infall of material onto the disk does not produce significant stirring of relative velocities in the disk (although it may break up growing agglomerations, and thus suppress accretion temporarily, as described in Section 7). Let us assume that the impacting material strikes the disk at velocity v_i, and mutual collisions of disk particles are inelastic. The energy per unit mass input to the disk is of order v_i^2. The collision rate in the disk is of order $\tau\Omega$, each dissipating energy/mass $\sim c^2$ and continuing for time t_d. Thus $v_i^2 \sim c^2\tau\Omega t_d$, and the stirring velocity due to infall is $c \sim v_i(\tau\Omega t_d)^{-1/2}$. (Note that as used here, and as in most planetary ring literature, τ is really the "dynamical optical thickness," i.e., the optical thickness due to those bodies containing the bulk of the mass. Actual optical thickness might be governed by small, dynamically insignificant particles.) For $v_i = 10$ km/sec and $t_d \sim 10^6$ yr, $c \lesssim 10^{-3}$ cm/sec. Mutual perturbations among the particles cause relative velocities of the order of their escape velocity, or $c \sim 10^{-3}S$/sec, where S is the particle radius, which dominates over infall stirring for $S > 1$ cm. Thus the required viscosity is provided if the characteristic size of the disk particles is a few tens of meters or larger. If the disk were dominated by kilometer-sized particles, as would be the case if the disk mass were one lunar mass and $\tau\sim 1$, the viscous evolution time would be shorter, perhaps only $\sim 10^4$–10^5 yr. In that case, the problem would become an excess of viscous transport, with excessive spreading of the disk, rather than insufficient viscosity to distribute the angular momentum and prevent collapse of the swarm. Thus it is possible that much more than one lunar mass was processed through the disk; the Moon could have been formed from the outermost part of the original swarm.

It may be possible to find reasonable combinations of infall rate, disk size, and particle velocity (or size) that maintain the CTD in an equilibrium state against accretion drag. Whether such an equilibrium would have been stable is problematical. Presumably, the disk parameters would fluctuate considerably in response to variations in the infall rate. More detailed study is needed to determine the disk's probable evolution.

Another mechanism that might have helped prevent collapse of the CTD during compositional filtering and trapping of silicates is angular momentum exchange between the disk and proto-Earth. Disk material falling onto Earth via viscous spreading and/or accretion drag brings specific angular momentum characteristic of a surface-grazing orbit. It is easily shown that the entire present angular momentum of the Earth-Moon system could be accounted for if about 20% of Earth's mass was accreted from a CTD; this is probably an upper limit to the total mass processed through such a disk. However, there is also the possibility of transfer of angular momentum from a rapidly spinning Earth back to the disk via density waves. If the Earth's gravity field is not axially symmetric, density waves will be excited at distances where the orbital period is a multiple of the planetary rotation period. The strongest such resonance is at the synchronous altitude. From Franklin *et al.* (1982), we infer that the disk could be truncated, forming an inner edge at the synchronous altitude, if its viscosity is less than a value

$$\nu_c \sim \Omega R^2 \left(\frac{B-A}{C} \right)^2 \tag{4}$$

Here Ω and R refer to the orbital frequency and radius at the synchronous point, A and B are the two equatorial principal moments of inertia, and C the polar moment of inertia.

For an Earth-sized planet, plausible values of (B–A)/C are in the range 10^{-5}–10^{-6}. Using values of Ω and R appropriate for a 6-hour rotation period, we find that the disk could be truncated at the synchronous point for $\nu \lesssim 10^3$–10^5 cm^2sec. This value is 1–3 orders of magnitude less than the value of ν required to keep the disk from collapsing by redistributing angular momentum within the disk, as discussed earlier in this section. However, somewhat larger values of (B–A)/C are possible; the value of 10^{-5}–10^{-6} refers to asymmetries in the mass distribution that are stable over long times (10^8–10^9 years), while impacts onto the accreting Earth could have caused larger, short-lived asymmetries. There will also be transfer of angular momentum via density waves at other resonances besides the synchronous point. Thus under some conditions density waves could provide a significant source of angular momentum to the CTD, which would help support it during the process of compositional filtering and accretion.

7. Evolution of the CTD and Compositional Filtering of Planetesimal Fragments

In this section we attempt to integrate the processes discussed above in order to examine how they might have led to formation of the Moon. We explore how a circumterrestrial disk might form and grow in mass, how it would interact with the population of heliocentric planetesimals, and what processes might influence the accretion of the Moon from the evolved CTD. The numerical examples that follow are not intended to be a definitive model. Rather, they are simply meant to illustrate that the ranges of parameters that allow compositional filtering are physically plausible. The actual values of such quantities, e.g., sizes and velocities of planetesimals and lunesimals, must be determined by more elaborate modeling.

Heliocentric planetesimals

We assume that the CTD that produced the Moon formed after most of the Earth had accreted (earlier ones might have formed and then collapsed due to accretion drag and/or Earth's mass gain). Of the heliocentric planetesimals remaining at that time, some would have eventually been accreted by the Earth; others could have been scattered into other planetary feeding zones. At least a lunar mass of heliocentric planetesimals must interact with the CTD to form the Moon. Thus we adopt ~0.1 $M_\oplus$ for the total mass of planetesimals remaining in Earth's feeding zone, within a few tenths AU of its orbit. We assume this mass is mostly in bodies ≤ 10 km in size, involving only modest growth from the initial scale of gravitational instabilities in a dust layer in the solar nebula (Goldreich and Ward, 1973). Numerical simulations of gravitational accretion by Greenberg *et al.* (1978) show that by the time the largest bodies reach ~1000 km in size, most of the mass of the system remains in small bodies. It has not been determined whether such a size distribution is preserved until the late stage of accretion, or if it instead evolves into a Safronov-type distribution dominated by large bodies. Wetherill (1986) has constructed self-consistent models of late-stage accretion of planetesimal populations dominated by large bodies. His results have been interpreted (cf. Boss and Peale, 1986) as showing that accretion actually involved such a size distribution. We emphasize that there has been no detailed study linking early and late stages of planetary accretion, and believe that there is insufficient basis for choosing between the case adopted by Wetherill as the initial condition for late-stage growth and the case we explore here.

The range of relative velocities among these late-stage Earth-zone planetesimals is uncertain. Since our scenario has most of the planetesimal mass in relatively small bodies, we do not expect the high orbital eccentricities and relative velocities characteristic of gravitational stirring by numerous massive bodies. Gravitational interactions with the relatively small number of ~100 km-scale bodies in the population would yield velocities of tens of meters/sec (Greenberg *et al.*, 1978). We expect the relative velocities to be dominated by secular perturbations by the Earth and the other nearly-grown planets. Those planetesimals that can make close encounters with Earth will have enhanced velocities, but passages closer to the planet than the radius of the CTD are likely to result in capture or comminution by collision with the disk, rather than gravitational scattering. For our example, we assume relative velocities ~1 km/sec, corresponding to eccentricities of the heliocentric orbits of a few times 10^{-2}.

Planetesimals with these values of size and relative velocity will undergo comminution, rather than accretion, in collisions. They are assumed to be differentiated by early heat sources, with iron cores and silicate mantles (cf. Wasson and Warren, 1984). Mutual collisions would fragment the silicate mantles and strip them from the stronger cores, which would tend to remain intact due to their higher strength (cf. Davis *et al.*, 1985). Rocky materials (impact strength ~10^7 erg/cm^3) would shatter at impacts $\geq$30 m/sec for equal-sized bodies, while iron (~10^9 erg/cm^3) requires a collision with a similar body at $\geq$300 m/sec. An exposed core could survive high-velocity collisions with silicate fragments. It is not clear whether a core would be destroyed by collision with an intact core-mantle planetesimal; most of the impact energy might be absorbed by comminution of the weaker silicate material. Our assumed mass of planetesimals in Earth's zone gives a surface density ~1 g/cm^2; the thickness of the heliocentric swarm for velocities ~1 km/sec gives a space density ~10^{-12} g/cm^3, or number density N of a few times 10^{-29} bodies/cm^3 for planetesimals of initial diameter, d, approximately a few km. The collision rate for a typical planetesimal is ~ d^2 × (relative velocity,v) × (number/volume), ~10^{-5}/year. Since the thickness of the heliocentric swarm is proportional to v, (number/volume) is proportional to $1/v$, so the collision rate is independent of our assumed value of v. Exposed cores are smaller than the original planetesimals; even if all cores are exposed, the rate of core-core collisions is several times lower, $\sim 10^{-6}/$ year. The rate of collisions between silicate fragments varies inversely with their mean size, so the silicate component of the small planetesimals experiences "runaway comminution" in this velocity regime. If comminution continues for much more than $\sim 10^6$ years without significant depletion of the populations of heliocentric cores by other means, e.g., accretion by Earth, then "core-core" collisions would yield smaller iron fragments, reducing the effectiveness of compositional filtering by the CTD. We address this potential problem with the timescale below. More detailed modeling of collisional evolution (cf. Davis *et al.*, 1985) is needed, but for purposes of calculation, we assume that the heliocentric population of fragments consists mainly of intact iron cores ~ 1 km in radius, and comminuted silicate fragments of order 10 m in size.

Initial CTD and its evolution

We now consider further the emplacement of an initial geocentric swarm by free-free collisions (cf. Section 4). The rate of mass buildup is proportional to the rate of such collisions within Earth'ssphere of influence. That rate is of order $N^2 d^2 v$, times the volume of the sphere of influence. For $d \sim 10$ m and the space density $\sim 10^{12}$ g/cm^3 assumed above, the mass input is $\sim 10^{12}$ g/sec, and scales inversely with the assumed value of d. Actually the space density will be somewhat depleted near Earth's orbit, but there is some gravitational focusing even outside the sphere of influence (Herbert *et al.*, 1986). We assume, based on the discussion in Section 4, and the results of Herbert *et al.*, that these free-free collisions result in capture of matter into prograde geocentric orbit at a mean distance of a few tens of Earth radii. An input rate of 10^{12} g/sec corresponds to one lunar mass in $\sim 2 \times 10^6$y due to free-free collisions alone.

Free-bound collisions with the CTD will provide a significant additional mass input. The rate of free-bound collisions is proportional to the area of the CTD. For a disk of radius ~ 30 R$_\oplus$, and allowing for gravitational focusing, our nominal parameters give an impact rate on the disk of a few times $10^{14}\tau$ g/sec, where τ is the normal optical thickness of the disk. Free-bound collisions add mass, but dilute the angular momentum of the CTD, eventually causing some of the mass to be lost by infall to the Earth. Since a particle in the CTD can, on average, accrete a few times its own mass before falling onto the Earth, we can expect an approximate balance between mass gain (due to free-free and free-bound collisions) and loss (due to free-bound collisions) at $\tau \sim 0.01$. Here τ refers to those bodies in the CTD that contribute enough mass to the size distribution to trap a typical heliocentric body that impacts the disk. We do not know the size distribution of bodies in the CTD; this depends on the competition between gravitational accretion and collisional breakup (see below). If the dominant size in the CTD were ~ 1

km, then $\tau \sim 0.01$ corresponds to $\simeq 0.05$ lunar mass in the disk. For the disk to gain mass beyond this point, the "lunesimals" must grow (to keep τ small) or angular momentum transport processes, e.g., ballistic redistribution of mass via free-bound collisions, viscous shear, or density waves induced by Earth's gravitational harmonics, must be effective. Otherwise, the CTD must shrink and collapse toward Earth as it gains mass, although the example in Section 5 showed that such a disk could accumulate a significant fraction of a lunar mass before collapsing, even in the absence of angular momentum transport.

From the discussion of viscosity in Section 5, kilometer-sized bodies in the CTD (with random velocities $\sim 10^2$ cm/sec due to gravitational stirring) give a viscosity $\sim 10^9 \tau$ cm^2/sec. The viscous evolution timescale is $\sim 10^6$y, while the time to double the mass by free-bound collisions is a few times 10^4y (the collapse time will be several times larger). The rate of viscous transport increases with the surface density of the disk (or with τ) in this range, so it seems probably that the CTD can continue to gain mass in its outer part, while losing material to Earth at its inner edge. The mass input to the disk from heliocentric bodies via free-bound impacts amounts to one lunar mass in about ($\sim 10^4/\tau$) years; thus considerably more than the Moon's mass can be processed. During this time, the mass input to the CTD by free-free collisions of heliocentric bodies is a small, but not negligible, fraction (~ 1–10%) of that due to impacts on the disk itself, and continues to supply a significant amount of angular momentum.

Bombardment and accretion in the CTD

In order for compositional filtering to take place, the CTD must be relatively "opaque" to the heliocentric flux of silicate fragments, but more "transparent" to the iron cores. It is possible to meet this condition due to the size difference between the silicate and iron components. From Fig. 3, we see that a heliocentric impacting particle must intercept a significant fraction of its own mass to be trapped by the CTD. We expect the size disk to contain a range of particle sizes due to the competition between accretion and comminution. Saturn's rings may provide an analog of that process (Weidenschilling *et al.*, 1984). In that example, most of the visible cross section is due to small particles, while most of the mass is in the largest bodies. A similar distribution in the CTD would allow the small silicates to be trapped when they impact the disk, while the iron cores would rarely encounter a kilometer-sized or larger lunesimal required to trap it. Thus the cores may survive many passages through the disk, until they impact the Earth.

It is necessary to maintain a "screen" of small bodies in the CTD; i.e., they cannot all accrete into kilometer-sized lunesimals too rapidly. The bombarding flux of heliocentric fragments can accomplish this by disrupting the accreting bodies. An accreting lunesimal would be a gravitationally bound "rubble pile" of small fragments, with very low impact strength. If we assume a strength of $\sim 10^4$

erg/cm^3, characteristic of weakly bonded regolith (Greenberg et al., 1978), then a kilometer-scale body could be disrupted by the impact of a heliocentric impactor of only 5 m radius. Our adopted flux rate would yield an impact of this magnitude on any kilometer-scale lunesimal every few days!

This bombardment would tend to inhibit the accretion of larger moonlets within the CTD, but might not prevent it entirely. Indeed, some growth may be necessary in order to provide enough gravitational stirring for viscous transport of angular momentum. Accretion of larger lunesimals would allow the disk to gain mass without presenting too much cross sectional area in bodies large enough to trap the iron cores. However, it is necessary that the CTD not undergo runaway accretion into a single Moon until the disk has attained about one lunar mass, and the heliocentric population of iron cores has been significantly depleted. Should an occasional body, by chance, manage to avoid collisional disruption, its rate of growth would be limited by depletion of smaller lunesimals in its neighborhood. It would become isolated by a process of "shepherding" familiar in planetary ring dynamics. Random velocities in the CTD are only of the order of the dominant (1 km) size bodies' escape velocity ($\sim 10^2$ cm/sec), even if a single lunesimal grows considerably larger. The local orbital velocity at 10 $R_\oplus$ is $\simeq 3 \times 10^5$ cm/sec, so e $\sim 3 \times 10^{-4}$. The accretion zone width is thus $\sim$ea $\sim 2 \times 10^6$ cm. The mass inside an annulus of that width, assuming a uniform surface density for the CTD, would have been $\sim 10^{21}$ gm, which provides enough mass for only a 100-km-diameter body. Since this size is comparable to the width of the feeding zone, this body could gain a modest amount of additional mass, but then there would be no more mass available in the feeding zone for further growth. Of course, there is the possibility that material would diffuse into the feeding zone. However, such diffusion may be inhibited by tidal torque exerted by the moonlet on the surrounding disk, in the same manner that a shepherd satellite exerts a torque on a planetary ring. A rough criterion for gap formation (Lin and Papaloizou, 1979) is

$$\frac{m}{M} > \frac{\nu}{\Omega a^2} \sim e^2 \tag{5}$$

so an embryo of 100 km diameter would be large enough to exert the torque. The value of ν implied by this value of e is $\sim 10^8$cm^2/s, significantly above that required for viscous evolution of the disk as a whole. Thus local gap clearing is compatible with the need to prevent collapse of the disk. Any large lunesimals located in such gaps would remain coupled to the disk by tidal torques, and their orbits would evolve with the rest of the disk. While these calculations require further refinement, they show that occasional rapid growth of a moonlet would not necessarily lead to runaway destruction of the CTD by coalescence into one or a few bodies.

The end-game

The heliocentric population of silicate bodies declines as it becomes captured by the CTD. Even the iron-core bodies that have, for the most part, escaped capture by the CTD are eventually removed due to perturbations by, or impacts with, the Earth, Venus, and other terrestrial planets. The timescale for this depletion is not clear. The often-quoted value of $\geq 10^7$ years for sweep-up of planetesimals (Safronov, 1972) refers to deep planet-crossing orbits with high eccentricities. Much of the iron complement of the Moon-forming silicates first encounters the Earth in nearly tangential low-velocity orbits with shorter lifetimes, leading to more rapid sweeping up of the bulk of this mass. It may be necessary that this material be significantly depleted, probably by impact onto the Earth, before the Moon accretes, or the Moon might acquire too much residual iron. Whatever iron cores and larger planetesimals remain in heliocentric orbit after the Moon has accreted have as much chance of being accreted onto the Moon as onto any other body, weighted by the cross sectional area of the target body in question. The scenario we have described cannot therefore be expected to yield a Moon totally devoid of iron; it would be expected that a small core would form within such a Moon. It is uncertain what fraction of the accreted Moon would consist of the late-captured heliocentric material. A long-lived "tail" of late-accreted material, including some large bodies, is consistent with our scenario. We note that Wetherill's (1975) estimate of the total mass flux impacting all of the terrestrial planets, accounting for visible craters produced during the "late heavy bombardment," is only $\sim 10^{-3}$ lunar masses, much too small to affect the Moon's bulk composition. Further work is required to assess the final compositional state of the Moon, and the thresholds required to commence the final accretion phase. At some point, the bombardment of the lunesimal swarm—necessary to prevent accretion—declines below a critical threshold, and the Moon rapidly accretes. While Safronov's (1972) formalism predicts an accretion timescale of only a few years, it would actually take much longer to form the Moon. An extended CTD such as we have pictured could not immediately coalesce into a single body. Instead, it would first accrete into an array of moonlets in quasistable orbits, which would eventually collide as their orbits evolved due to tides (cf. Ruskol, 1977). This final assembly process might have yielded such phenomena as the lunar magma ocean and crustal asymmetry.

8. Conclusions

We have demonstrated the possibility that the Moon could have formed from the long-term evolution of a CTD emplaced during the final stages of the formation of the Earth. We have emphasized silicate enrichment (or iron depletion) of lunar material within the disk, due to preferential capture of silicates. Several conditions must be met for this CTD hypothesis of lunar origin to be viable: the initial planetesimals must have been heated and differentiated, their size distribution must have been dominated by small bodies, and their heliocentric orbits must have had rather low ($\lesssim 0.05$) eccentricities. We have argued that it is plausible that much of the mass of the planetesimal population is in small bodies, although this has not been demonstrated directly. The low relative velocities that we have assumed for the planetesimals are consistent with this assumed size distribution. The nature of the Earth-zone planetesimal population during the late stages of the Earth's formulation remains a critical issue; our model will succeed best if there is a balance between a massive population of kilometer-scale planetesimals plus a significant number of larger ones. The latter may be required in order to agitate the heliocentric population of planetesimals so that silicates are smashed and cores exposed, or possibly to produce the initial CTD, if it is due to an Earth impact or chance collision within the sphere of influence. The existence of a moderate number of larger (≥ 100 km) heliocentric planetesimals does not have any direct effect on the CTD, since they are neither captured nor disrupted by passage through the CTD. Like the smaller iron cores, they are eventually accreted by Earth or removed by being perturbed into Jupiter-crossing orbits.

There is still uncertainty concerning the survival of the CTD in a quasisteady state for a sufficient interval to permit the accumulation and compositional filtering of a lunar mass of planetesimal fragments. More than a decade ago, doubts were expressed about the lifetime of a CTD due to the drag effect of collisions with incoming heliocentric planetesimals. Previous arguments for or against such a disk were based on simplistic models of the angular momentum carried by the planetesimals, and neglected the role of viscosity in the CTD. We have shown that the problem is more complicated than it originally appeared. The late-stage orbital characteristics of Earth-zone planetesimals could have been biased in favor of maintaining the disk. Viscous transport of angular momentum also would have acted to prevent its collapse onto the Earth, and may have prevented premature accretion of the Moon. We do not yet understand these processes fully, even with the present-day example of Saturn's rings.

The CTD filter process can explain the relative lack of metallic iron in the Moon. We have no definitive conclusions about how other aspects of lunar composition may be understood by this model, but some general observations are in order. While the main thrust of this model is to derive lunar silicates from differentiated planetesimals, there is room for a significant fraction of Earth mantle material to be incorporated as well (it shares this trait with those versions of the impact-trigger model that emphasize the role of projectile material but that also incorporate some Earth mantle material); this flexibility in the model may help address some specific problems of lunar chemistry. Other compositional differences may be expected between the compositional filtering model and the impact-trigger model because we derive most lunar silicates from silicates differentiated within *very small bodies*. Differentiation within kilometer-scale bodies may result in greater loss of volatiles than from larger bodies. In addition, collisional processes among kilometer-scale bodies subsequent to differentiation may enrich crustal materials over mantle materials in the silicate fraction captured by the swarm (all crustal rocks will be fragmented or eroded from the planetesimals, but some of the bodies may be depleted from the heliocentric population before their deep mantle layers have been thoroughly stripped from the cores). As described in Section 2, the ultimate source for lunar silicates must be Earth-zone planetesimals. Thus, the refractory and volatile abundances in the Moon may be understood in the context of differentiation within small bodies and subsequent collisional evolution.

Resolving the question of the origin of the Moon has implications that extend beyond the Earth-Moon system. Our CTD model involves a relatively quiescent heliocentric swarm with low orbital eccentricities. This condition would tend to limit the degree of mixing between planetary zones, and to preserve isotopic and compositional differences between the various terrestrial planets. In contrast, the planetesimal accretion model of Wetherill (1986) involves the scattering of large bodies throughout the inner solar system; in that case, the differences in composition between, for example, Mercury and Mars must be ascribed to stochastic effects of large-body accretion, rather than to smooth variations that are a function of the planets' present heliocentric distances. It is not clear whether the difference in noble gas contents of Earth and Venus can be reconciled with the degree of mixing between their zones predicted for that scenario (Wetherill, 1981); more detailed accretion models are needed. We have mentioned (Section 2) the constraints on the giant impact model implied by the identical oxygen isotope compositions of the Earth and Moon. That condition can be met if the inner solar system was isotopically homogeneous before planetary accretion, but the SNC meteorites, if they are derived from Mars, suggest otherwise. Eventually, samples from the surfaces of Mercury, Venus, and Mars may be required to clarify the origin of the Moon.

Our preliminary work on dynamical problems involving circumterrestrial disks suggests the need for more detailed analysis. An understanding of these processes is relevant not only to the CTD filter model but also to various other models of lunar origin and the origin of other planetary satellite systems. The fact that some planets, like Mars and Venus, lack massive satellites suggests that circumplanetary disks—if they are commonly created—may not always yield massive moons.

Acknowledgments. This work was performed under the auspices of the Tucson Lunar Origin Consortium. It was supported by grant NASW-3516 to PSI from NASA's Planetary Geology and Geophysics Program, NAG-9-39 to M. J. Drake from NASA's Planetary Materials and Geochemistry Program, and NAGW-680 to the Consortium from NASA's Innovative Research Program. We thank Alix Ott for her efforts in preparation of the manuscript. This is PSI Contribution No. 205. PSI is a division of Science Applications International Corporation.

9. References

Becker R. H. and Pepin R. O. (1984) The case for a Martian origin of the shergottites: Nitrogen and noble gases in EETA 79001. *Earth Planet. Sci. Lett., 69*, 225–242.

Bogard D. D. and Johnson P. (1983) Martian gases in an Antarctic meteorite? *Science, 221*, 651–654.

Boss A. P. and Peale S. J. (1986) Dynamical constraints on the origin of the Moon, this volume.

Cameron A. G. W. (1985) Formation of the prelunar accretion disk. *Icarus, 62*, 319–327.

Cameron A. G. W. and Ward W. R. (1976) The origin of the Moon (abstract). In *Lunar Science VII*, pp. 120–122. The Lunar Science Institute, Houston.

Darwin G. H. (1880) On the secular changes in the elements of the orbit of a satellite revolving about a tidally distorted planet. *Phil. Trans. Roy. Soc. London, 171*, 713–891.

Davis D. R., Chapman C. R., Weidenschilling S. J., and Greenberg R. G. (1985) Collisional history of asteroids: evidence from Vesta and the Hirayama families. *Icarus, 62*, 30–53.

Dobrovolskis A. R. and Burns J. A. (1984) Angular momentum drain: A mechanism for despinning asteroids. *Icarus, 57*, 464–476.

Drake M. J. (1983) Geochemical constraints on the origin of the Moon. *Geochim. Cosmochim. Acta, 47*, 1759–1767.

Drake M. J. (1986) Is lunar bulk material similar to Earth's mantle?, this volume.

Franklin F. A., Columbo G., and Cook A. F. (1982) A possible link between the rotation of Saturn and its ring structure. *Nature, 295*, 128–130.

Gerstenkorn H. (1955) Uber Gezeitenreibung beim Zweikorperproblem. *Z. Astrophys., 36*, 245–274.

Goldreich P. and Tremaine S. (1978) The velocity dispersion in Saturn's rings. *Icarus, 34*, 227–239.

Goldreich P. and Ward W. (1973) The formation of planetesimals. *Astrophys. J., 183*, 1051–1061.

Greenberg R. and Chapman C. R. (1983) Asteroids and meteorites: Parent bodies and delivered samples. *Icarus, 55*, 455–481.

Greenberg R., Wacker J. F., Hartmann W. K., and Chapman C. R. (1978) Planetesimals to planets: Numerical simulation of collisional evolution. *Icarus, 35*, 1–26.

Greenberg R., Davis D. R., Chapman C. R., and Weidenschilling S. J. (1985) Late-stage planetesimals: how big? (abstract). In *Terrestrial Planets: Comparative Planetology*, p. 4. Lunar and Planetary Institute, Houston.

Giuli R. T. (1968) On the rotation of the Earth produced by gravitational accretion of particles. *Icarus, 8*, 301–323.

Harris A. W. (1977) An analytical theory of planetary rotation rates. *Icarus, 31*, 168–174.

Harris A. W. and Kaula W. M. (1975) A co-accretional model of satellite formation. *Icarus, 24*, 516–524.

Hartmann W. K. (1985) Impact experiments I. Ejecta velocity distributions and related results from regolith targets. *Icarus, 63*, 69–98.

Hartmann W. K. (1986) Moon origin: The impact-trigger hypothesis, this volume.

Hartmann W. K. and Davis D. R. (1975) Satellite-sized planetesimals and lunar origin. *Icarus, 24*, 504–515.

Hartmann W. K. and Vail S. M. (1986) Giant impactors: Plausible sizes and populations, this volume.

Herbert F. and Sonett C. P. (1980) Electromagnetic inductive heating of the asteroids and Moon as evidence bearing on the primordial solar winds. In *The Ancient Sun* (R. O. Pepin, J. A. Eddy, and R. B. Merrill, eds.), pp. 563–576. Pergamon, New York.

Herbert F., Davis D. R., and Weidenschilling S. J. (1986) On the origin of the Moon in geocentric orbit: Formation and evolution of a circumterrestrial disk, this volume.

Jagoutz E. and Wänke H. (1982) Has the Earth's core grown over geologic time? (abstract). In *Lunar and Planetary Science XIII*, pp. 358–359. Lunar and Planetary Institute, Houston.

Kaula W. M. and Harris A. W. (1975) Dynamics of lunar origin and orbital evolution. *Rev. Geophys. Space Phys., 13*, 363–371.

Kerr R. A. (1984) Making the Moon from a big splash. *Science, 226*, 1060–1061.

Kreutzberger M. E., Drake M. J., and Jones J. H. (1986) Origin of the Earth's Moon: Constraints from alkali volatile trace elements. *Geochim. Cosmochim. Acta, 50*, 91–98.

Lin D. N. C. and Papaloizou J. (1979) Tidal torques on accretion discs in binary systems with extreme mass ratios. *Mon. Not. Roy. Astron. Soc., 186*, 789–812.

Mayeda T. K. and Clayton R. N. (1980) Oxygen isotopic compositions of aubrites and some unique meteorites. *Proc. Lunar Planet. Sci. Conf. 11th*, pp. 1145–1151.

Melosh H. J. and Sonett C. P. (1986) When worlds collide: Jetted vapor plumes and the Moon's origin, this volume.

Mitler H. E. (1975) Formation of an iron-poor Moon by partial capture, or yet another exotic theory of lunar origin. *Icarus, 24*, 256–268.

Mizuno H. and Boss A. P. (1985) Tidal disruption of dissipative planetesimals. *Icarus, 63*, 109–133.

Newsom H. E. (1984) The lunar core and the origin of the Moon. *EOS (Trans. Amer. Geophys. Union), 65*, 369–370.

Newsom H. E. and Palme H. (1984) The depletion of siderophile elements in the Earth's mantle: new evidence from molybdenum and tungsten. *Earth Planet. Sci. Lett., 69*, 354–369.

Öpik E. J. (1972) Comments on lunar origin. *Irish Astron. J., 10*, 190–238.

Pechernikova G. V., Safronov V. W., and Zvyagina E. J. (1976) Mass distribution of protoplanetary bodies II. Numerical solution of generalized coagulation equation. *Sov. Astron., 20*, 346–350.

Ruskol E. L. (1960) The origin of the Moon: Formation of a swarm of bodies around the Earth. *Sov. Astron., 4*, 657–668.

Ruskol E. L. (1963) The origin of the Moon: The growth of the Moon in the circumterrestrial swarm of satellites. *Sov. Astron., 7*, 221–227.

Ruskol E. L. (1972a) The origin of the Moon: Some aspects of the dynamics of the circumterrestrial swarm. *Sov. Astron., 15*, 646–654.

Ruskol E. L. (1972b) Formation of the Moon from a cluster of particles encircling the Earth. *Izv. Earth Phys., 7*, 99–108.

Ruskol E. L. (1977) The origin of the Moon. In *Soviet-American Conference on Cosmochemistry of the Moon and Planets* (J. H. Pomeroy and N. J. Hubbard, eds.), pp. 815–822. NASA SP-370, NASA, Washington, DC.

Safronov V. S. (1969) Evolution of the protoplanetary cloud and formation of the Earth and the Planets, *NASA TTF-677*, Translated 1972 by Israel Program for Scientific Translation, Jerusalem. 206 pp.

Safronov V. S. and Ruskol E. L. (1977) The accumulation of satellites. In *Planetary Satellites* (J. Burns, ed.), pp. 501–512. Univ. Arizona Press, Tucson.

Shmidt O. Yu. (1958) *Four Lectures on the Theory of the Origin of the Earth*, Acad. Sci. U.S.S.R., Moscow. 139 pp.

Stevenson D. (1984) Lunar origin from impact on the Earth: Is it possible? (abstract). In *Papers Presented to the Conference on the Origin of the Moon*, p. 60. Lunar and Planetary Institute, Houston.

Taylor S. R. (1982) *Planetary Science: A Lunar Perspective*. Lunar and Planetary Institute, Houston. 512 pp.

Urey H. C. (1955) The cosmic abundances of potassium, uranium, and the heat balance of the Earth, the Moon, and Mars. *Proc. Natl. Acad. Sci. U.S., 41*, 127–144.

Wasson J. T. and Warren P. H. (1979) Formation of the Moon from differentiated planetesimals of chondritic composition (abstract). In *Lunar and Planetary Science X*, pp. 1310–1312. Lunar and Planetary Institute, Houston.

Wasson J. T. and Warren P. H. (1984) The origin of the Moon (abstract). In *Papers Presented to the Conference on the Origin of the Moon*, p. 57. Lunar and Planetary Institute, Houston.

Weidenschilling S. J., Chapman C. R., Davis D. R., and Greenberg R. (1984) Ring particles: Collisional interactions and physical nature. In *Planetary Rings* (R. Greenberg and A. Brahic, eds.), pp. 367–415. Univ. Arizona Press, Tucson.

Wetherill G. W. (1975) Late heavy bombardment of the Moon and terrestrial planets. *Proc. Lunar Sci. Conf. 6th*, pp. 1539–1561.

Wetherill G. W. (1986) Accumulation of the terrestrial planets and implications concerning lunar origin, this volume.

Zvyagina E. V. and Safronov V. S. (1972) Mass distribution of protoplanetary bodies. *Sov. Astron.*, *15*, 810–817.

Glossary

*M*any of the terms in this glossary were lifted verbatim from *Planetary Science: A Lunar Perspective* by S. R. Taylor (Lunar and Planetary Institute, Houston, 1982). Others were adapted loosely from *Glossary of Geology*, R. L. Bates and J. A. Jackson, editors (American Geological Institute, Falls Church, VA, 1980) and from *Moons and Planets* by W. K. Hartmann (Wadsworth, Belmont, CA, 1983). In preparing it, the editors followed the philosophy that brief is beautiful.

Ab—An abbreviation for albite, a plagioclase feldspar end member.

Accretion—The growth of planets from smaller objects, one impact at a time.

Achondrite—Stony meteorite lacking chondrules; most are igneous rocks.

Albite—$NaAlSi_3O_8$, the sodium-rich end member of the plagioclase feldspar series of minerals.

Alkali element—A general term applied to the univalent metals Li, Na, K, Rb, and Cs.

Alkali feldspar—Feldspars rich in sodium and potassium, such as albite and orthoclase.

Angular momentum—A property of rotating systems; its value depends on mass (and its distribution), angular velocity, and radius (planetary or orbital). The angular momentum of the Earth-Moon system is the sum of the angular momenta of Earth's rotation and the Moon's orbital motion.

Angular momentum density—Angular momentum normalized to mass.

An—Abbreviation for anorthite, a plagioclase feldspar end member.

Anorthite—$CaAl_2Si_2O_8$, the calcium-rich end member of the plagioclase feldspar series of minerals.

Anorthosite—An igneous rock made up almost entirely of plagioclase feldspar.

ANT—An acronym for the suite of lunar highland rocks: anorthosite, norite, and troctolite.

Assimilation—The incorporation by melting and dissolution of foreign material, for example, surrounding rock, into magma.

Augite—A clinopyroxene rich in calcium and magnesium.

BABI—**B**asaltic **A**chondrite **B**est **I**nitial; the best estimate for the initial $^{87}Sr/^{86}Sr$ ratio of basaltic achondrites, regarded as approximating that of the solar nebula, 0.69897.

Bar—The international unit of pressure (one bar $= 10^6$ dynes/cm$^2 = 10^5$ Pascals).

Basalt—A fine-grained, dark-colored igneous rock composed primarily of plagioclase feldspar and pyroxene; usually other minerals such as olivine and ilmenite are present. Most basalts formed as lavas on planetary surfaces, although some form as dikes and sills.

Basaltic achondrite—Stony meteorites formed from basaltic magmas that resemble terrestrial and lunar basalts; examples are eucrites and howardites.

Big impact hypothesis—see impact-trigger hypothesis.

Binary accretion hypothesis—see coaccretion hypothesis.

Binary fission—see Darwinian fission.

Bouguer gravity—The free-air gravity corrected for the effects of the topography, so it is dependent only on the internal density distribution.

Bow shock—A shock wave in front of a body moving through a fluid or gas.

Breccia—A rock consisting of angular, coarse fragments of rocks and minerals embedded in a fine-grained matrix.

C1, C2 chondrites—see carbonaceous chondrites.

Capture hypothesis—The hypothesis that the Moon formed by gravitational capture of either a fully-formed Moon (intact capture) or by the disruption of one large or numerous smaller objects as they passed within the Roche limit (disintegrative capture).

Carbonaceous chondrites—The most primitive stony meteorites, in which the abundances of all but the most volatile elements (such as H and He) are thought to approximate those of the primordial solar nebula.

Chalcophile element—An element that enters sulfide phases in preference to other types of phases.

Chondrites—The most abundant class of stony meteorites, characterized by the presence of chondrules, which are millimeter-sized, rounded objects composed mostly of olivine and/or low-Ca pyroxene.

CI chondrites—see carbonaceous chondrites.

Classical fission—see Darwinian fission.

Clast—A discrete particle or fragment of rock or mineral; commonly included in a larger rock.

Clinopyroxene—Minerals of the pyroxene group, such as augite and pigeonite, that crystallize in the monoclinic system.

Coaccretion hypothesis—The hypothesis that the accreting Earth accumulated a swarm of orbiting solid objects from which the Moon formed by accretion.

Collisional ejection hypothesis—The idea that the Moon formed as a result of the impact of a large planetesimal; during the impact large amounts of impactor and/or Earth were lifted into orbit. See also impact-trigger hypothesis.

Compatible element—A minor or trace element that partitions readily into crystalline rather than melt phases, i.e., an element that is readily soluble in a solid phase.

Core—Dense, metal- or sulfide-rich central region of a planet. Earth's core is composed principally of metallic nickel-iron; the Moon might have a small core.

Crust—Outer, highly differentiated region of a planet.

Cumulate—An igneous rock composed chiefly of crystals accumulated by sinking or floating from a magma.

Curie temperature—The temperature in a ferromagnetic material above which the material becomes substantially nonmagnetic.

Darwinian fission—A term used to describe a particular mode of formation of the Moon by rotational fission. In this case, the rapidly rotating primitive Earth becomes elongated and a blob detaches to form the Moon.

Differentiation—The process by which planetary bodies develop concentric zones that differ in chemical and mineralogical composition.

Diopside—$CaMgSi_2O_6$, an end member of the pyroxene series of minerals.

Disintegrative capture hypothesis—A version of the capture hypothesis in which one large or numerous smaller planetesimals pass within Earth's Roche limit, are partly disrupted, and portions are captured into Earth orbit, providing the raw material for the Moon.

Distribution coefficient—The ratio of the concentration of a trace element in one phase to its concentration (by weight) in a second phase with which it is in equilibrium; the phases can be solid or liquid, metals or silicates.

Doppler gravity—The primary means of measuring the gravity field on the Moon and other planets by analyzing Doppler frequency shifts of spacecraft communication signals. This yields the line-of-sight component of gravity, i.e., in the direction from the spacecraft to Earth.

Double planet hypothesis—see coaccretion hypothesis.

Dunite—A rock that consists almost entirely of olivine.

Eccentricity—A measure of the amount by which an orbit deviates from circularity.

Ecliptic—The plane of Earth's orbit around the sun.

Eclogite—A dense rock consisting of garnet and pyroxene, similar in chemical composition to basalt.

Ejecta—Materials ejected from a crater by a meteorite impact or volcanic explosion.

En—Abbrevation for enstatite, a pyroxene end member.

Enstatite—The Ca-free, magnesian end member of the pyroxene group, $MgSiO_3$.

Epicenter—The point on a planetary surface directly above the focus of an earthquake or moonquake.

Eucrite—An achondrite meteorite consisting mostly of feldspar and pyroxene; also called basaltic achondrite.

Exsolution—The separation in the solid state of some mineral-pair solutions during slow cooling.

Extinct isotope—A radioactive isotope that existed when the solar system formed, but whose half-life was too short to allow detectable amounts to remain now; examples are ^{129}I and ^{244}Pu.

Fa—Abbreviation for fayalite.

Fayalite—An end member of the olivine mineral group, Fe_2SiO_4.

Feldspar—A group of aluminous silicate minerals. Plagioclase feldspar is a solid solution series between albite, $NaAlSi_3O_8$, and anorthite, $CaAl_2Si_2O_8$.

Feldspathic—An adjective meaning rich in feldspar.

Ferromagnetic—Possessing magnetic properties similar to those of iron.

Ferrosilite—An end member of the pyroxene mineral group, $FeSiO_3$.

Fission theory—The idea that the Moon formed by the spontaneous ejection of Earth's upper mantle material, either as a single mass of material or as a circumterrestrial swarm; see also Darwinian fission, classical fission, and impact-induced fission.

Fo—Abbreviation for forsterite.

Focus—The initial rupture point of an earthquake or moonquake.

Forsterite—An end member of the olivine group of minerals, Mg_2SiO_4.

Fractional crystallization—Formation and separation of mineral phases of varying composition during crystallization of a silicate melt or magma, resulting in a continuous change in composition of the magma.

Fractionation—The separation of elements from an initially homogeneous state into different phases or systems.

Free-air gravity—As applied to the Moon, it is usually synonymous with the observed gravity anomaly. In a strict sense, on the Earth it is gravity reduced to sea level without accounting for the attraction of the rock mass above sea level.

Fs—Abbreviation for ferrosilite.

Fugacity—A thermodynamic function used instead of partial pressure in describing the behavior of nonideal gases and gases dissolved in silicate melts.

Gabbro—A coarse-grained igneous rock made up mostly of augite and plagioclase; it forms when a basaltic magma crystallized at depth.

Gardening—The process of turning over the lunar soil or regolith by meteorite bombardment; it is accompanied by fragmentation and melting of soil constituents.

Garnet—A group of minerals with the general formula $X_3Y_2(SiO_4)_3$, in which the X-site can be occupied by Ca, Mg, Fe^{2+}, and Mn, and the Y-site by Al, Fe^{3+}, and Cr^{3+}. Extensive atomic substitution occurs in the garnet series. Examples of end member garnets are almandite, $Fe_3Al_3(SiO_4)_3$, and pyrope, $Mg_3Al_2(SiO_4)_3$.

Geomagnetic tail—A portion of the magnetic field of Earth that is pushed back to form a tail by the solar wind plasma.

Giant-impact hypothesis—see impact-trigger hypothesis.

Granite—An igneous rock composed chiefly of quartz and alkali feldspar; though common on Earth, granite is rare on the Moon.

Half-life—The time interval during which a number of atoms of a radioactive nuclide decay to one half of that number.

Heat flow—The rate of heat energy leaving a planet's surface per unit area.

Howardite—A type of basaltic achondrite.

Ilmenite—An oxide mineral with the ideal formula $FeTiO_4$.

Impact-induced fission—The idea that fission of the Moon from the Earth could have been triggered by a large impact.

Impact melting—The process by which target rock is melted by the impact of a meteorite, comet, or (during accretion) planetesimal.

Impact parameter—A term referring to the geometry of a collision between two spheres unequal in size; it is the offset between the spheres' centers projected perpendicular to the line of approach. Thus, for a head-on impact this parameter is zero; for a grazing impact it is equal to the sum of the radii of the impactor and target.

Impact-trigger hypothesis—The idea that the Moon formed as a result of a large impact that either directly ejected material into orbit to form the Moon or caused the Earth to spin more rapidly, thus initiating fission.

Inclination—The angle between the plane of a planet's orbit and the ecliptic (Earth's orbital plane), or a satellite's orbit and its planet's equator.

Incompatible element—Minor or trace element that partitions preferentially into melt rather than crystalline phases, i.e., an element not readily soluble in a solid phase.

Intact capture hypothesis—The idea that the Moon formed elsewhere in the solar system and was subsequently captured into an orbit around Earth.

Inviscid—A term used to describe an ideal fluid with zero viscosity.

Ionic radius—The effective radius of ionized atoms in crystalline solids; ionic radii commonly lie between 0.4 and 1.5 Angstroms.

Isochron—A line on a diagram passing through plots of samples with the same age but differing isotope ratios.

Isostasy—Balancing of topography by underlying density. The distribution is such that at some uniform depth the pressure is everywhere constant and beneath this depth a state of hydrostatic equilibrium exists.

Iron meteorite—A class of meteorite composed chiefly of iron or iron-nickel.

Jacobian triaxial ellipsoid—A term applied to bodies that have elliptical cross sections perpendicular to each coordinate axis; the a, b, and c axes of ellipsoids have different lengths and, hence, are not axisymmetric.

Jetting—The process that occurs during the earliest stages of an impact in which highly shocked ejecta is thrown out at the highest speeds.

KREEP—An acronym for a lunar crustal component rich in potassium (K), the rare earth elements (REE), phosphorus (P), and other incompatible elements.

Large impact hypothesis—see impact-trigger hypothesis.

Layered igneous intrusion—A body of intrusive igneous rock that has formed layers of different minerals during solidification; it is divisible into a succession of extensive sheets lying one above the other.

LIL—An abbreviation for large-ion lithophile elements (e.g., K, Rb, U, Th, rare earth elements), which have ionic radii larger than common lunar rock-forming elements; LIL elements usually behave as trace elements in lunar rocks and in meteorites, and all are incompatible elements.

Liquidus—The line or surface in a phase diagram above which the system is completely liquid.

Lithophile element—An element tending to concentrate in oxygen-containing compounds, particularly silicates.

LUNI—Best estimate for initial $^{87}Sr/^{86}Sr$ ratio of the Moon, determined from lunar anorthosites, 0.69903.

Maclaurin oblate sphere—A term applied to objects having surfaces that are ellipses rotated about the axis of rotation; they are, therefore, axisymmetric.

Magma ocean—A hypothetical layer of magma (molten or mostly molten rock) thought to have surrounded the Moon when it formed and from which the original, feldspar-rich lunar crust and at least some mare basalt source regions formed.

Magmasphere—see magma ocean.

Magmatic differentiation—The production of rocks of differing chemical composition during cooling and crystallization of magma by processes such as removal of early formed mineral phases.

Magmifer—A term used to describe a convective, partially molten portion of the lunar mantle; the magmifer might supply melt to an overlying magma ocean.

Magnesium number—The molar ratio Mg/(Mg + Fe); abbreviated mg* or *mg*.

Mantle—The zone of a planet beneath its crust and above its core.

Mare basalt—Basalts that form the lunar maria, which are the dark-colored areas on the Moon. Chemically similar basalts also occur in breccias from the lunar highlands, indicating that such lava flows occurred there, but are not now discernible.

Mascon—Regions on the Moon of excess mass concentrations per unit area identified by positive gravity anomalies and associated with mare-filled multi-ring basins.

*mg**—see magnesium number.

Model age—The age of a rock sample determined from radioactive decay by assuming an initial isotopic composition of the daughter product.

Moment of inertia—A quantity that gives a measure of the density distribution within a planet, specifically, the tendency for an increase of density with depth. It is derived from gravity and dynamical considerations.

Noble gases—The rare gases helium, neon, argon, krypton, xenon, and radon.

Norite—A type of gabbro in which orthopyroxene is dominant over clinopyroxene.

Normal fault—A fault in a planetary surface caused by extension stress.

NRM—Abbreviation for natural remanent magnetization, which is that portion of the magnetization of a rock that is permanent and usually acquired by the cooling of ferromagnetic minerals through the Curie temperature.

Obliquity—The tilt angle between a planet's axis of rotation and the pole of the orbit, i.e., the axis perpendicular to the orbital plane.

Oersted—The cgs unit of magnetic field intensity.

Olivine—A major rock-forming silicate mineral, representing a solid solution between forsterite (Fo), Mg_2SiO_4, and fayalite (Fa), Fe_2SiO_4.

Orthoclase—$KAlSi_3O_8$, an alkali feldspar.

Orthopyroxene—An orthorhombic member of the pyroxene mineral group; most orthopyroxenes are low in calcium.

Partial melting—The process in which rocks melt over a range of temperatures, producing liquids with compositions different from the original unmelted rock and different from the residual unmelted crystals.

Partition coefficient—see distribution coefficient.

Peridotite—An igneous rock composed almost entirely of olivine and pyroxene.

Perturbations—Gravitational effects on orbits caused by planets or other nearby objects.

Pigeonite—A low-calcium, monoclinic pyroxene; it ranges widely in its proportions of the enstatite ($MgSiO_3$) and ferrosilite ($FeSiO_3$) end members, but accommodates limited amounts of the wollastonite ($CaSiO_3$) end member.

Plagioclase—A subgroup of the feldspar group of minerals.

Planetesimal—Bodies from millimeter-size up to hundreds of kilometers in diameter that formed during the planet-forming process; most accreted to form the planets, and the remainder were ejected from the solar system.

Plutonic—A term referring to igneous rocks formed at great depth.

Poincaré figure—A figure resembling a bowling pin; it is the shape acquired by a rapidly rotating proto-Earth just prior to Darwinian fission.

Poise—Unit of viscosity; in cgs 1 poise = 1 dyne sec/cm^2 = 0.1 Pascal-sec.

ppb—Parts per billion (usually by weight); also written ng/g.

ppm—Parts per million (usually by weight); also written ug/g.

P-wave velocity—Seismic body wave velocity associated with particle motion (alternating compression and expansion) in the direction of wave propagation.

Pyroclastic glass—Glass formed by volcanic eruptions in which small particles of lava are spewed out.

Pyroxene—A closely related group of minerals that includes augite, pigeonite, orthopyroxene, etc.

Q—see seismic Q.

Quartz—A mineral consisting of nearly pure SiO_2.

Radiogenic—A term referring to an isotope having been formed from a radioactive parent; for example, ^{206}Pb formed from the decay of ^{238}U, whereas ^{204}Pb is nonradiogenic.

Rare earth elements—A collective term for elements with atomic numbers from 57 to 71, including La, Ce, etc.

Rare gases—The noble gases helium, neon, argon, krypton, xenon, and radon.

REE—An abbreviation for the rare earth elements.

Refractory element—An element that vaporizes at high temperatures; examples are U, Al, Ca, and the REE.

Regolith—Loose surface material, composed of rock fragments and soil, that overlies consolidated bedrock.

Residual liquid—The material remaining after most of a magma has crystallized; it is sometimes characterized by an abundance of volatile constituents.

Roche limit—The critical distance between two bodies with no tensile strength at which tidal forces are so strong that the smaller body is torn apart; for the Earth and Moon, this distance is about 3 Earth radii. However, because bodies have finite tensile strength, a rocky planetesimal would have to pass well inside the Roche limit and remain within it long enough to actually disrupt.

Safronov number—A parameter relating the mass of the largest planetesimal in a collection of planetesimals accreting to form a planet to the mass of the second largest object.

Seismic Q—A measure of the attenuation of seismic wave energy; the higher the value of Q, the less the energy loss.

Siderophile element—An element that preferentially enters the metallic phase.

Silicate—A phase (mineral or liquid) whose crystal structure is controlled by silicon-oxygen bonds.

Solar nebula—The primitive disk-shaped cloud of dust and gas from which all bodies in the solar system originated.

Solar wind—The stream of charged particles (mainly ionized hydrogen) moving outward from the sun with velocities in the range 300–500 km/sec.

Solidus—The line or surface in a phase diagram below which the system is completely solid.

S-wave velocity—Seismic body wave velocity with shearing motion perpendicular to the direction of wave propagation.

Thermal stress—Stress associated with differential expansion or contraction of a planetary body due to heating or cooling.

Thrust faults—A fault in a planetary surface caused by compressive stresses.

Trace element—An element found in very low (trace) amounts; generally less than 1000 ppm.

Troctolite—A type of gabbro in which olivine is dominant over pyroxene.

T-Tauri—An early pre-main-sequence state of stellar evolution, characterized by extensive mass loss from the young star.

Volatile element—An element that vaporizes at relatively low temperatures; examples are K, Na, and Pb.

Subject Index

*Refers to the first page of an article in which this term is a major topic.
ff   Discussion is continued on following pages.

Cadmium   240, 348
Calcium   676
Capture hypothesis   17*ff*, 59, 117, 125*ff*, 263, 264, 300, 453, 679, 688, 689
  by circumterrestrial swarm   724*ff*
  compositional constraints on place of origin   80
  date   482
  disintegration during   80*ff*
  due to collisions within Earth's sphere of influence   743
  dynamics of   31*ff*
  history of hypothesis   30, 76, 471*ff*
  intact   76*ff*
  of outer planet satellites   9
  time constraints on disruption   47
Carbon   240
Carbonaceous chondrites   21, 148
  volatiles   21
Carriers
  ferromagnetic   393
Catastrophic vs. evolutionary processes   587
Cayley formation   395
Cerium   112
Cesium   113, 150, 662, 663, 680
Cesium/rubidium   128
Chemical composition, *see* Composition
Chemical constraints (*see also* Composition)   735
Chemical fractionation   125*, 139, 155*ff*
Chlorine   239, 240, 349
Chondrites   106, 151, 210, 337, 674
  chemical fractionation   149
  composition   149
Chondritic meteorites   674
Chromium   190, 192, 221, 651*ff*, 681, 683, 685, 688, 689
Circumterrestrial swarm   59, 758
  accretion in   755
  angular momentum   38, 86*ff*
  angular momentum input from density waves   751
  angular momentum input from heliocentric bodies   701*ff*, 746*ff*
  capture of heliocentric material   520
  chemical composition   631
  collapse of inner part onto Earth   703, 726
  compositional filtering   34, 85, 723*ff*, 731*ff*
  creation of   42, 487*, 489*ff*, 613, 739*ff*, 741*ff*, 754*ff*
  dissipation timescale   614
  drag   746*ff*
  early suggestions   8
  evolution   731*ff*
  formation of moonlets in   614
  future work   93, 94
  generated by impacts   589, 594
  gravitational collapse   90*ff*
  initial mass   491, 493*ff*
  inward spiralling   87

  iron segregation in   734
  lifetime   86, 750
  mass   613
  nonequilibrium chemistry   631
  numerical model of capture   48*ff*
  orbit inclinations   631
  orbit plane   615
  orbital stability if produced by fission   493
  produced by tidal disruption   45*ff*
  silicate condensates in   631
  sources of material   735
  spun off by rotational instability   39
  survival   703
  temperature history   43, 91
  tidal evolution   749*ff*
  timescale for evolution   750, 758
  triggered by impact   617, 643
  uniqueness of disk formed by fission   493
  viscous spreading   43, 749*ff*
Coaccretion hypothesis   119, 263, 264, 272*ff*, 302*ff*, 701*ff*, 731*ff*
  history   33, 702
Coaccretion   17*ff*, 33*ff*, 580
Cobalt   190, 192, 210*ff*, 231*, 237*ff*, 249*, 250, 254, 655, 656, 681, 683, 684, 686*ff*
  geochemical behavior   259*ff*
Collisional ejection hypothesis (*see also* Impact-trigger hypothesis)   17*ff*, 436, 453
  mechanical models   567*
Collisional filtering   455, 701
Collisions
  comparable-sized bodies   603
  mechanics of   623*ff*
Composition   231*, 279*, 294, 361*, 511*ff*
  as explained by impact-trigger hypothesis   590
  chondrites   105*
  degrees of freedom in explaining   604
  Earth's mantle   105*, 125*, 145*, 158*ff*, 203*, 249*, 250, 280, 281, 649*, 673*, 677
  eucrites   105*
  of chondrites   145*, 149*ff*
  of eucrite parent body   649*
  upper mantle   370
Compositional filter   85, 580, 590, 723*ff*, 755*ff*
  history of concept   702
Condensation sequence   150*ff*, 479, 631
Conductivity
  electrical   378*ff*
Conferences on lunar origin   582
  history   18
  Moon as a proposed site   48
Copper   240, 681, 683*ff*, 687
Core   23, 167, 361*, 412, 423
  cobalt content of   269
  dynamo   394, 397

Earth's 651, 653, 677
   formation 26, 127, 134, 139, 203*, 217ff, 251, 509, 682
   metallic 388ff
   molten silicate 392
   nickel content of 269
   seismic evidence 267ff
Coupling
   core-mantle 392
Cratering mechanics 589ff, 595ff
Craters
   binary 82
Critical escape parameter 458, 461ff
Critical melt fraction 287
Crust 174, 361*
   composition 366
   formation 173*, 279*
   heterogeneities 173*, 193, 197
   highland 175ff
   isostasy 368
   rock types 283ff
   structure 364
   thickness 28

Darwin, G. H. 3, 6ff, 18, 60, 64, 500
Dehydration
   impact 456
   thermal 457
Delaunay, C. E. 3ff
Density 297, 737
   consistent with impact-trigger origin 586
   mean 382
   of planets 674
Depletion patterns 678
Descartes, Rene 4
Differentiation 155, 363ff
Disintegrative capture hypothesis 17ff, 45ff, 117, 153, 263, 264, 272, 273, 302, 453, 689
Distribution coefficients 293
Dunthorne, Richard 4
Dust layer
   in solar nebula 62
Dynamical constraints 59*ff

Earth (see also Proto-Earth) 674, 675, 688
   accretion 48ff
   atmosphere evolution 481ff
   core formation 39, 71, 480ff, 503, 611
   early rotation 6ff
   formation 567ff
   formation time 75
   initial obliquity 555ff
   mantle evolution 738
   mantle viscosity 74
   melting by despinning 476

rotation 35ff
   rotation history 6ff, 476
   rotation input from circumterrestrial disk 701
   viscosity 74
   volatile depletion 22
Earth mantle
   evolution of FeO content 512
Earth-Moon system
   uniqueness 632
Eccentricities
   of planets 533
Ejecta blankets 635
Ejecta curtain
   from impact 634
Electrical conductivity 378ff
Electromagnetic sounding 378ff
Elemental ratios
   of Earth's mantle 154ff
Enstatite chondrites 106
Enstatite meteorites
   as samples of Earth-zone planetesimals 585
Eucrite parent body 152, 251, 664ff
Eucrites 106, 132, 220, 221, 250, 337, 650, 678, 679, 682, 683, 688
   ages 135
Europium 155, 181, 281, 427

FeNi 665
FeO 129, 167, 651, 665, 677
   possible enrichment in Moon 512
Ferrel, William 3ff
Ferroan anorthosites 173*, 174, 176ff, 184, 185, 258, 283, 284, 296, 427
Ferromagnetic carriers 393
FeS 665
Finite strength 429ff, 441
Fisher, Osmond 3
Fission hypothesis 3ff, 7ff, 17ff, 59, 118, 125*ff, 169, 300ff, 439, 453, 499*ff, 680, 689
   advantages of initiating by giant impact 495
   binary fission 488
   classical model 503
   difficulties with model 489, 497
   dynamic problems 40
   formation of disk vs. single satellite 487*ff, 490ff, 493ff, 500
   future work 493ff, 497
   history of hypothesis 39ff, 69ff
   lunar core size as a test of 510
   mass ejected 491, 493ff
   mass loss during 72
   numerical simulations 490ff
   precipitation model 75ff, 127, 690
   role of rotational instability 39ff
Fluorine 239, 240

Fra Mauro formation   395
Fractionation   252
  chemical   689, 691
  date   22
  iron-silicate   674, 675
  isotopic   689, 691

Gabbronorite   178, 183
Gadolinium   281
Gallium   108, 240, 681, 683*ff*, 687
Gas drag
  in solar nebula   62
Gas pressure gradients
  in giant impact   610
Genesis rocks   579
Germanium   218, 240, 681, 684
Gerstenkorn, H.   10
Giant impact, *see* Impact-trigger hypothesis
Giant planets
  initial rotations   564*ff*
Gilbert, Grove Karl   8
Gold   240, 682, 684, 686*ff*
Granophyre   174, 179
Gravitational collapse   61, 611
  in circumterrestrial disk   90
Gravitational instability
  in circumterrestrial swarm   34
Gravity   367*ff*

Halley, Edmund   4
Heat flow   112, 131, 132, 376*ff*, 427, 676, 677
  errors in early data analysis   512
Heliocentric planetesimals
  supplied to Earth   520
History
  capture hypothesis   30, 76, 471*ff*
  coaccretion hypothesis   33, 702
  compositional filter   702
  disintegrative capture hypothesis   45, 80*ff*
  fission hypothesis   39, 69
  impact-trigger hypothesis   42, 47, 88, 580, 609*, 621*ff*
  lunar origin theories   3*ff*, 17*ff*
History of lunar origin theories   3*ff*

Ilmenite gabbro/norite   179
Impact experiments   635
Impact flux (*see also* Intense early bombardment)
  on primordial Earth   588
  role in creating circumterrestrial swarm   739*ff*
Impact mechanics
  in a giant impact   589
  mathematical modeling   636*ff*
  ricochet   635
Impact parameter   627
Impact strength   598

Impact velocity   455
Impact-trigger hypothesis   59, 118, 125*, 134*ff*, 153,
169, 300*ff*, 313, 579*ff*, 649*, 662, 680, 690
  advantages of combining with fission hypothesis   495
  asymmetry in jetting   626
  chemistry of debris   631
  circumterrestrial swarm produced by   631
  coaccretion as an outcome   592
  condensation in gas debris   631
  constraints   501
  difficulties   622
  dust/gas in ejecta   590, 598*ff*
  effect of timing of impact on observed properties   565,
    587
  effects of off-center impacts   591
  effects of rotation   591
  effects on thermal and chemical evolution   548
  ejecta velocity distribution   594*ff*
  escape of debris   612*ff*
  fission as an outcome   592
  formation of proto-Moon embryos from debris   44
  formation of the accretion disk   612*ff*
  future work   93, 605
  gas pressure   88*ff*, 610, 612*ff*, 622, 630*ff*, 633
  history of   42*ff*, 47, 88*ff*, 580, 609*, 621*ff*
  impactor composition   592
  impactor mass needed   556
  impactor size estimates   593*ff*
  iron in debris   44
  mass of ejected material   598*ff*, 629
  mass of projectile   609**ff*
  mechanics of impact   609*
  numerical model of impact   617*, 621*ff*, 643*
  numerical models   600
  plausibility of giant impact   548, 610
  Pluto-Charon system as a counterargument   505
  probability of creating proto-Moon swarms   93
  projectile's iron core   615
  role of jetting   621*
  role of stochastic processes   88
  role of viscous spreading   43
  "second-burn" effect   600, 612*ff*, 622, 630*ff*
  size of impactors   88*ff*
  source of ejecta   589
  source of orbiting debris   617, 645
  submodels of   591
  temperature history of debris   43
  vapor source   622
  Venus rare-gas content as a test   614
Impacts   195*ff*, 287
  alteration of lunar orbit by   494
  comparable-sized bodies   603
  evidence for large   587
  mechanics of   623*ff*
  numerical model   643

Author Index